Lecture Notes in Networks and Systems

Volume 351

The series "Lecture Notes in Networks and Systems" publishes the latest developments in Networks and Systems—quickly, informally and with high quality. Original research reported in proceedings and post-proceedings represents the core of LNNS.

Volumes published in LNNS embrace all aspects and subfields of, as well as new challenges in, Networks and Systems.

The series contains proceedings and edited volumes in systems and networks, spanning the areas of Cyber-Physical Systems, Autonomous Systems, Sensor Networks, Control Systems, Energy Systems, Automotive Systems, Biological Systems, Vehicular Networking and Connected Vehicles, Aerospace Systems, Automation, Manufacturing, Smart Grids, Nonlinear Systems, Power Systems, Robotics, Social Systems, Economic Systems and other. Of particular value to both the contributors and the readership are the short publication timeframe and the world-wide distribution and exposure which enable both a wide and rapid dissemination of research output.

The series covers the theory, applications, and perspectives on the state of the art and future developments relevant to systems and networks, decision making, control, complex processes and related areas, as embedded in the fields of interdisciplinary and applied sciences, engineering, computer science, physics, economics, social, and life sciences, as well as the paradigms and methodologies behind them.

Indexed by SCOPUS, INSPEC, WTI Frankfurt eG, zbMATH, SCImago.

All books published in the series are submitted for consideration in Web of Science.

For proposals from Asia please contact Aninda Bose (aninda.bose@springer.com).

More information about this series at https://link.springer.com/bookseries/15179

Subarna Shakya · Ke-Lin Du · Wang Haoxiang
Editors

Proceedings of Second International Conference on Sustainable Expert Systems

ICSES 2021

Springer

Editors
Subarna Shakya
Institute of Engineering
Tribhuvan University
Kirtipur, Nepal

Wang Haoxiang
Go Perception Laboratory
Cornell University
Ithaca, NY, USA

Ke-Lin Du
Department of Electrical and Computer Engineering
Concordia University
Montréal, QC, Canada

ISSN 2367-3370 ISSN 2367-3389 (electronic)
Lecture Notes in Networks and Systems
ISBN 978-981-16-7656-7 ISBN 978-981-16-7657-4 (eBook)
https://doi.org/10.1007/978-981-16-7657-4

This Springer imprint is published by the registered company Springer Nature Singapore Pte Ltd.
The registered company address is: 152 Beach Road, #21-01/04 Gateway East, Singapore 189721, Singapore

We are honored to dedicate this book to all the technical program committee members, editors and authors of ICSES 2021.

Preface

This proceeding includes the papers presented in the Second International Conference on Sustainable Expert Systems (ICSES 2021), Organized by Tribhuvan University (TU), Nepal with a primary focus on the research information related to Artificial Intelligence (AI), Sustainability, and Expert Systems applied in almost all the areas of industries, government sectors, and academia worldwide. ICSES 2021 will provide an opportunity for the researchers to present their research results and examine the advanced applications in Artificial Intelligence (AI) and the expert systems field. Nevertheless, this proceeding will promote novel techniques to extend the frontiers of this fascinating research field.

Advancement in both the sustainability and intelligent systems disciplines require an exchange of thought and ideas of audiences from different parts of the world. The 2021 International Conference on Sustainable Expert Systems was potentially designed to encourage the advancement and application of sustainability and artificial intelligence models in the existing computing systems. ICSES 2021 received a total of 274 manuscripts, wherein the submitted papers have been peer-reviewed by at least two reviewers drawn from the technical conference committee, external reviewers, and also from editorial board depending on the research domain of the papers. Through a rigorous peer-review process, the submitted papers are selected based on the research novelty, clarity and significance. Out of these, 72 papers were accepted for publication in ICSES 2021 proceedings. It covers papers from several significant thematic areas namely data science, wireless communication, intelligent systems, social media, and image processing.

To conclude, this proceeding will provide a written research synergy that already exists between the intelligent expert systems and network-enabled communities, and represents a progressive framework from which new research interaction will result in the near future. I look forward on the evolving of this conference theme over time and to learning more on the network sustainability and continue to pave the way for applications of network sustainability in the emerging intelligent expert systems.

Dr. Subarna Shakya
Pulchowk Campus, Institute of Engineering
Tribhuvan University
Kirtipur, Nepal

Dr. Ke-Lin Du
Affiliated Associate Professor
Concordia University
Montréal, QC, Canada

Dr. Wang Haoxiang
Go Perception Laboratory
Cornell University
Ithaca, NY, USA

Acknowledgements

We ICSES 2021 are very honored to have the most prominent keynote speaker "Prof. Valentina Emilia Balas, Aurel Vlaicu University of Arad, Head of the Intelligent Systems Research Centre, Romania", who have significant research expertise in their respective specialty.

Further, we would like to thank the technical conference committee and external reviewers, who have worked very hard in reviewing the papers and making valuable suggestions to enhance the research works submitted by authors. We express our sincere gratitude the authors for contributing their innovative research results to the conference. Special thanks go to Springer publications.

Last but not least, we wish to express our thanks to all the members of the ICSES 2021 committee for all their unprecedented efforts in making ICSES 2021 as a successful conference. Finally, our thanks go to the Tribhuvan University (TU), Nepal. It would have been impossible to organize the conference without their moral support. We really hope that all the participants of ICSES 2021 will benefit tremendously from the conference event. Finally, we would like to wish all success in their future research direction.

About the Conference

It is my great pleasure to welcome you all to the Second International Conference on Sustainable Expert Systems (ICSES 2021) held in Nepal on September 17–18, 2021.

The main thrust of ICSES 2021 is to deal with the design, implementation, development, testing, and management of intelligent and sustainable expert systems and also to provide both theoretical and practical guidelines for the deployment of these systems. Due to the well-established track record of intelligent and networked systems, it mandates the implementation of sustainability aspect to become more accessible for the network users. This makes ICSES 2021 an excellent venue for exploring the sustainable computing foundations for the challenging communication network issues.

The success of ICSES 2021 completely depends on the efforts made by the researchers in the field of artificial intelligence and sustainable computing, who have written a submitted research articles on various research topics.

The program committee members and external reviewers deserve sincere accolades, as they have invested more time in assessing and analyzing multiple papers submitted on different research domains to significantly hold and maintain high quality for this conference event. Additional thanks are given to the Springer publications. We hope that you will take advantage of many research insights on intelligent expert systems.

Prof. Dr. Subarna Shakya
Conference Chair
Professor, Department of ECE, Pulchowk Campus,
Institute of Engineering
Tribhuvan University
Kirtipur, Nepal

Contents

E-Learning-Based Collaborative Learning Experience in the Education Paradigm 1
Anuj Bhasin and Surekha Lanka

Using Clean Architecture Principles to Improve the Design and Implementation of the Mobiles Online Platform 11
T. Rachovski, S. Hadzhikoleva, E. Hadzhikolev, and A. Lengerov

Detection of Arrhythmia Using Convolutional Neural Networks 21
Burla Greeshma, Moturi Sireesha, and S. N. Thirumala Rao

The Hybrid Vision Transformer Approach for Hyperpigmentation Nail Disease Detection 31
Krish Kumar, Chandan Kumar, Rahul Nijhawan, and Ankush Mittal

COVID-19 Fake News Detection Using GloVe and Bi-LSTM 43
Chaitanya Kulkarni, P. Monika, S. Shruthi, M. S. Deepak Bharadwaj, and D. Uday

Combinatorial Test Perspective on Test Design of SDN Controllers 57
Rekha Jayaram and R. Krishnan

Technologies Driving Digital Payments in India: Present and Future 71
T. Ravikumar, N. Murugan, J. Suhashini, and M. Sriram

Performance Study of Data Fusion Using Kalman Filter and Learning Vector Quantization 79
Shobha and N. Nalini

Unveiling the Security Vulnerabilities in Android Operating System 89
G. Renjith and S. Aji

A Novel Efficient Energy and Delay Balance Ensemble Scheduling Algorithm for Wireless Sensor Networks 101
B. Srinivasa Rao

Bitcoin Price Prediction Using Machine Learning's Boosting Algorithms 115
Ch. Likhitha Sree, M. Meghana, R. Manjula, and D. Mohan

Anthropometric Measurement Technology Using 2D Images 127
Chinmay Shah, Jaineel Shah, Maruf Shaikh, Harshit Sandhu, and Prachi Natu

A Novel Approach for Sentiment Classification by Using Convolutional Neural Network 143
M. S. Kalaivani and S. Jayalakshmi

Automatic Speech Recognition of Bengali Using Kaldi 153
Subhadeep Guchhait, Arnold Sachith A Hans, and Jacob Augustine

Prediction of Priority to Individual for COVID Vaccine Distribution Using Soft Computing Techniques 167
S. Subbulakshmi, Adarsh R. Nambiar, A. K. Arun, Fawaz Al Faizi, and Vishnu Narayanan Harish

An Experimental Evaluation of Load Balancing Policies Using Cloud Analyst 185
N. Priya and S. Shanmuga Priya

Classification of Ovarian Cyst Using Regularized Convolution Neural Network with Data Augmentation Techniques 199
N. Priya and S. Jeevitha

Generating Game Indefinite Decision Tree in the Banking Sector Using Different Types of Algorithms 211
S. Sridevi and S. M. Chithra

Detection of Pneumonia Using Convolution Neural Networks 229
Sai Hemanth Kumar Kotturi and S. V. N. Sreenivasu

Blockchain Technology in Health Care Sector: An Experimental Approach 245
R. Jaichandran, H. Abdul Shabeer, Tony K. John, K. P. Karan, and Aravind Krishnan

Data Security Framework Using Block Chain Technology 253
R. Jaichandran, R. Shobana, S. Muthamil Selvan, Medagoni Rohan Sai, and Gunupati Lakshmi Sai Vikas Kumar Reddy

Disease Detection Based on Human Eye Images and Analysis Using Multi-agent Systems with SPADE 261
R. Jaichandran, R. Magesh Sunder, K. L. Shunmuganathan, S. Leelavathy, and Vijay Harshan

Distributed Data Balancing on Multi Base Station Wireless Network to Solve the Routing Complexity Issues 273
R. Shobana, R. Jaichandran, V. Sivakumar, G. Devaraj, and A. Anbarasan

RPBK21 Technique Used to Secure the Data 281
K. Shantha Shalini, R. Shobana, Bhupalam Preetham, M. P. Shashank, and Bathula Pavan Kalyan

Classification of Extreme Reviews from Online Products Using RNN Model 297
Satuluri Naganjaneyulu, G. Tarun, Y. Sriram, B. Rama Devi, and P. Manoj

A Novel Text-to-Speech Synthesis System Using Syllable-Based HMM for Tamil Language 305
J. Samuel Manoharan

SOA-Based Middleware Framework for IoT Applications 315
M. Mallegowda, Pooja Sarashetti, and Anita Kanavalli

The Future Warfare with Multidomain Applications of Artificial Intelligence: Research Perspective 329
M. A. Nikitha, B. S. Sai Swetha, Krishna Harika Mantripragada, and N. Jayapandian

Evaluation of Various Classifier Models to Predict the Cardiovascular Disease 343
K. G. Rani Roopha Devi, R. Murugesan, and R. Mahendra Chozhan

Blockchain Advancement for a Safe and Transparent COVID-19 Vaccination 359
S. Vinushree and S. Thejaswini

Learning Conversational Stance from Multiple Views of Social Graph Using Multitask and Other Approaches 373
Bhaskarjyoti Das, Anvitha Poosarla, V. Mahima, and Chetana Kulkarni

Facial Emotion Recognition Using ML Algorithms 389
V. V. S. S. Ramya, Shaik Afifa Reshma, Afrah Samreen, and U. Chandrasekhar

Green Initiative Low-Cost Electric Vehicle 403
K. Gopala Reddy, H. K. Pooja, Saima Zulfa, M. N. V. R. Shwetha, and J. Nanditha

Vision Transformer Framework Approach for Yellow Nail Syndrome Disease Identification 413
Vikas Kumar Roy, Vasu Thakur, and Rahul Nijhawan

An Efficient Fuzzy Logic Cluster Formation Protocol for Data Aggregation and Data Reporting in Cluster-Based Mobile Crowdsourcing 427
M. Renukadevi, E. A. Mary Anita, and D. Mohana Geetha

Utilization of Adaptive K-Harmonic Means Clustering and Trust Establishment in VANETs 447
K. M. Jini, J. Senthilkumar, Y. Suresh, and V. Mohanraj

Design and Development of Blockchain-Based Security and Privacy-Preserving System 459
Nilima V. Pardakhe and V. M. Deshmukh

PSO- and GA-Based Hybrid Algorithm for Feature Selection 475
S. Sandeep Kumar, N. Raghavendra Sai, and M. Jogendra Kumar

System for Verification Using Blockchain: Demonstration of PAN Cards, and College Transcripts Verification 485
Akshay Kale, Prathviraj Walke, Omkar Wanave, Kaushal Bhandari, and S. A. Deshpande

A Review on Data Hiding Mechanism for Enhancing Embedding Capacity 499
Priyanka V. Deshmukh, A. S. Kapse, and V. M. Thakare

Deep Convolutional Neural Network for Brain Tumor and Skin Cancer Detection Over Traditional Neural Networks 509
T. Ashish Narayan, Ch. Anudeep, Pavan Nageswar Reddy Bodavarapu, and P. V. V. S. Srinivas

Design and Development of the Efficient RF Energy Harvesting System 525
P. Kalpana Devi and T. Yuvaraj

A Combined K-Mean Semantic Approach for the Implicit Document Clustering 535
R. S. Rehna

Diagnosis of COVID-19 from Chest X-rays Using CNN and Determination of Its Severity by Text Analysis 545
Gopalam Pujitha, Potu Siva Parvathi, Lukka Venkata Siva Phaneendra, Nimmagadda Snehita, and Rizwan Patan

Safe Encryption Algorithm for Secured Message Communication Using Dcombo: A New Sorting Technique 559
S. Muthusundari, A. Sonya, C. M. Nalayini, A. R. Sathyabama, and P. V. Rajasuganya

Domain Text Classification Using Machine Learning Models 573
Akula V. S. Siva Rama Rao, D. Ganga Bhavani, J. Gopi Krishna, B. Swapna, and K. Rama Sai Varma

Named Entity Recognition Using Stanford Classes and NLTK 583
Akula V. S. Siva Rama Rao, P. V. V. Vamsi, N. Rashmika, K. Hemanth, and K. Aditya Kumar

Enhancing the Quality of Service in MANET Based Optimization on Routing Protocol 599
Nedumaran Arappali, Daniel Getachew Tadesse, Abdulkerim Seid Endris, and P. Karthika

Movie Recommendation Using User Based, Item Based and LDA Analysis 613
Dharsan Vasant and N. Harini

Privacy Preservation Using Anonymity in Social Networks 623
Miryabelli Avinash and N. Harini

Comparison of Algorithms on Breast Cancer Detection in Developing Countries 633
Zahura Zaman, Md. Shakawat Al Sakib, Subarna Akter Liza, Nabanita Saha Joya, Afsana Taslim Farin, and Raka Moni

An Approach for Building Content Recommendation System for Bilinguals 643
A. Bijay, Jahfar Ali, and C. C. Sobin

Upper Bounds of Zagreb Radio Indices 657
Anna Treesa Raj and Joseph Varghese Kureethara

Interruption Detection for Detection and Quantification of QRS Fragmentation Based on Machine Learning and Deep Learning Technique 669
Afsana Banu and K. G. Manjunath

Negative Domination in Networks 679
Athira P. Ranjith and Joseph Varghese Kureethara

Analysis of IoT and Communication Technologies to Develop Waste Management Service Framework for Smart City 691
Mohd Anjum, M. Sarosh Umar, and Sana Shahab

Procedural Impact of Negotiations and Its Variations on Utilization of Resources in Cloud Environment 709
M. N. V. Kiranbabu, Yugandhar Garapati, Velagapudi Sreenivas, and V. Murali Mohan

To Reduce the Water Wastage by Using Water Monitoring System in Cloud-Survey 721
R. Shobana and M. Nithya

KRB02 Method Used to Secure the Generalized Data 729
K. Karthik, S. Rajaprakash, B. Naga Yugandhar Reddy, V. Varun Kumar Reddy, and C. H. Maheswara Balaji

Sentiment Analysis on Tweets Using Emojis to Help the Distressed 737
S. Maneesha, S. Srihari, and H. R. Mamatha

Study and Comparision of "*MIMO-OFDM*" *Under Acoustic Communication Systems* 755
S. Sankar Ganesh and S. Rajaprakash

Comparative Study of Robust Feature Extraction Techniques for ASR for Limited Resource Hindi Language 763
Hetal Gaudani and Narendra M. Patel

Creating Sustainable Human Resource Management Systems for High-Performance Work Culture 777
Shweta Jha

Algorithms: Supervised Machine Learning Types and Their Application Domains 787
N. Divyashree and K. S. Nandini Prasad

A Review on Database and Transaction Models in Different Cloud Application Architectures 809
N. Priya and E. Punithavathy

Effective Design Approach of Yard Trimmer 823
R. Seetharaman, M. Tharun, S. S. Sreeja Mole, and K. Anandan

A Comprehensive Assessment of Privacy Preserving Data Mining Techniques 833
Kundeti Naga Prasanthi and M. V. P. Chandra Sekhara Rao

Mathematical Model for Handling Jamming Attacks in Wireless Sensor Networks 843
E. Jayabalan and R. Pugazendi

Heart Disease Prediction Using Ensemble Model 857
Bikal Adhikari and Subarna Shakya

IoT Technologies and Machine Learning Algorithms on Diabetes Prediction-Survey 869
G. Thilagavathi, N. K. Karthikeyan, G. Priyadharshini, and G. Lavanya

FIGSI—Facial Image Generation for Suspect Identification 877
Prem Bhajaj, Pratik Panchal, Neil Patel, Rucha Shinde,
and Aruna Gawade

Review on Log-Based Anomaly Detection Techniques 893
Pooja Raut, Akanksha Mishra, Shreya Rao, Saloni Kawoor,
Sushila Shelke, Mahendra Deore, and Vivek Kumar

Author Index 907

Editors and Contributors

About the Editors

Prof. Dr. Subarna Shakya is currently a professor of computer engineering, Department of Electronics and Computer Engineering, Central Campus, Institute of Engineering, Pulchowk Tribhuvan University; Coordinator (IOE); and Leader Project (Links in Europe and Asia for engineering, eDucation, Enterprise and Research exchanges), Erasmus Mundus. He received M.Sc. and Ph.D. degrees in computer engineering from the Lviv Polytechnic National University, Ukraine, 1996 and 2000, respectively. His research area includes e-government system, computer systems and simulation, distributed and cloud computing, software engineering and information system, computer architecture, information security for e-government, and multimedia systems.

Dr. Ke-Lin Du is a research scientist at Center for Signal Processing and Communications, Department of Electrical and Computer Engineering, Concordia University, since 2001, where he became an affiliate associate professor in 2011. His research area includes signal processing, wireless communications, and soft computing.

Wang Haoxiang is currently the director and lead executive faculty member of GoPerception Laboratory, NY, USA. His research interests include multimedia information processing, pattern recognition and machine learning, remote sensing image processing, and data-driven business intelligence. He has co-authored over 60 journal and conference papers in these fields on journals such as *Springer MTAP, Cluster Computing, SIVP, IEEE TII, Communications Magazine, Elsevier Computers & Electrical Engineering, Computers, Environment and Urban Systems, Optik, Sustainable Computing: Informatics and Systems, Journal of Computational Science, Pattern Recognition Letters, Information Sciences, Computers in Industry, Future Generation Computer Systems*; Taylor & Francis *International Journal of Computers and Applications* and conference such as IEEE SMC, ICPR, ICTAI, ICICI, CCIS, and ICACI. He is the guest editor for *IEEE Transactions on Industrial*

Informatics, IEEE Consumer Electronics Magazine, Multimedia Tools and Applications, MDPI Sustainability, International Journal of Information and Computer Security, Journal of Medical Imaging and Health Informatics, Concurrency and Computation: Practice and Experience.

Contributors

Bikal Adhikari Department of Information System Engineering, HIST College, Kathmandu, Nepal;
Department of Computer and Electronics and Communication Engineering, Kathford Int'l College of Engineering and Management, Lalitpur, Nepal

K. Aditya Kumar Department of Computer Science and Engineering, SASI Institute of Technology and Engineering (Affiliated To JNTU, Kakinada), Tadepalligudem, Andhra Pradesh, India

S. Aji Department of Computer Science, University of Kerala, Thiruvananthapuram, Kerala, India

Fawaz Al Faizi Department of Computer Science and Applications, Amrita Vishwa Vidyapeetham, Amritapuri, India

Md. Shakawat Al Sakib Department of Computer Science and Engineering, Daffodil International University, Dhaka, Bangladesh

Jahfar Ali LTRC, IIIT Hyderabad, Hyderabad, Telangana, India

K. Anandan PS Division of E & E, TUV SUD South Asia Pvt Ltd., Chennai, India

A. Anbarasan Department of Computer Science and Engineering, Aarupadai Veedu Institute of Technology, Vinayaka Mission's Research Foundation (Deemed to be University), Paiyanoor, Chengalpattu, Tamil Nadu, India

Mohd Anjum Department of Computer Engineering, Aligarh Muslim University, Aligarh, India

Ch. Anudeep Department of Computer Science and Engineering, Koneru Lakshmaiah Education Foundation, Guntur, India

Nedumaran Arappali School of Electrical and Computer Engineering, Wollo University, Kombolcha Institute of Technology, Kombolcha, Ethiopia

A. K. Arun Department of Computer Science and Applications, Amrita Vishwa Vidyapeetham, Amritapuri, India

T. Ashish Narayan Department of Computer Science and Engineering, Koneru Lakshmaiah Education Foundation, Guntur, India

Jacob Augustine Department of CSE, Presidency University, Bangalore, India

Miryabelli Avinash Department of Computer Science and Engineering, Amrita School of Engineering, Amrita Vishwa Vidyapeetham, Coimbatore, India

Afsana Banu Department of Computer Science and Engineering, Siddaganga Institute of Technology, Tumkur, India

Prem Bhajaj Computer Engineering, Dwarkadas J. Sanghvi College of Engineering, Mumbai, India

Kaushal Bhandari Department of Information Technology PVG's COET, Savitribai Phule Pune University, Pune, India

Anuj Bhasin Stamford English Program, Stamford International University, Bangkok, Thailand

A. Bijay SRM University, Amaravati, Andhra Pradesh, India

Pavan Nageswar Reddy Bodavarapu Department of Computer Science and Engineering, Koneru Lakshmaiah Education Foundation, Guntur, India

M. V. P. Chandra Sekhara Rao Department of Computer Science and Engineering, R. V. R. & J. C. College of Engineering, Chowdavaram, India

U. Chandrasekhar Department of Computer Science and Engineering, BVRIT Hyderabad College of Engineering for Women, Hyderabad, India

S. M. Chithra RMK College of Engineering and Technology, Chennai, Tamil Nadu, India

Bhaskarjyoti Das PES University, Bengaluru, Karnataka, India

M. S. Deepak Bharadwaj Dayananda Sagar College of Engineering (DSCE), Bangalore, India

Mahendra Deore Department of Computer Engineering, MKSSS's Cummins College of Engineering for Women, Pune, India

Priyanka V. Deshmukh Department of CSE, SSGMCE, Shegaon, India

V. M. Deshmukh P.R.M.I.T.&R., Badnera, Amravati, India

S. A. Deshpande Department of Information Technology PVG's COET, Savitribai Phule Pune University, Pune, India

G. Devaraj Department of Computer Science and Engineering, Aarupadai Veedu Institute of Technology, Vinayaka Mission's Research Foundation (Deemed to be University), Paiyanoor, Chengalpattu, Tamil Nadu, India

N. Divyashree Department of Information Science and Engineering, Dr. Ambedkar Institute of Technology, Bengaluru, India

Abdulkerim Seid Endris School of Electrical and Computer Engineering, Wollo University, Kombolcha Institute of Technology, Kombolcha, Ethiopia

Afsana Taslim Farin Department of Computer Science and Engineering, Daffodil International University, Dhaka, Bangladesh

D. Ganga Bhavani Department of Computer Science and Engineering, Sasi Institute of Technology and Engineering, (Affiliated To JNTU, Kakinada), Tadepalligudem, Andhra Pradesh, India

Yugandhar Garapati GITAM Deemed to be University, Hyderabad, India

Hetal Gaudani Gujarat Technological University, Ahmedabad, Gujarat, India

Aruna Gawade Computer Engineering, Dwarkadas J. Sanghvi College of Engineering, Mumbai, India

K. Gopala Reddy Electrical and Electronics Engineering Department, Vidyavardhaka College of Engineering, Mysore, Karnataka, India

J. Gopi Krishna Department of Computer Science and Engineering, Sasi Institute of Technology and Engineering, (Affiliated To JNTU, Kakinada), Tadepalligudem, Andhra Pradesh, India

Burla Greeshma Department of Computer Science and Engineering, Narasaraopeta Engineering College, Narasaraopet, Andhra Pradesh, India

Subhadeep Guchhait Department of CSE, Presidency University, Bangalore, India

E. Hadzhikolev University of Plovdiv "Paisii Hilendarski", Plovdiv, Bulgaria

S. Hadzhikoleva University of Plovdiv "Paisii Hilendarski", Plovdiv, Bulgaria

Arnold Sachith A Hans Department of CSE, Presidency University, Bangalore, India

N. Harini Department of Computer Science and Engineering, Amrita School of Engineering, Amrita Vishwa Vidyapeetham, Coimbatore, India

Vishnu Narayanan Harish Department of Computer Science and Applications, Amrita Vishwa Vidyapeetham, Amritapuri, India

Vijay Harshan Department of Computer Science and Engineering, Aarupadai Veedu Institute of Technology, Vinayaka Mission's Research Foundation (Deemed to be University), Paiyanoor, Chengalpattu District, Tamil Nadu, India

K. Hemanth Department of Computer Science and Engineering, SASI Institute of Technology and Engineering (Affiliated To JNTU, Kakinada), Tadepalligudem, Andhra Pradesh, India

R. Jaichandran Department of Computer Science and Engineering, Aarupadai Veedu Institute of Technology, Vinayaka Mission's Research Foundation (Deemed to be University), Paiyanoor, Chengalpattu District, Tamil Nadu, India

E. Jayabalan Department of Computer Science, Government Arts College (A), Salem, Tamil Nadu, India

S. Jayalakshmi Vels Institute of Science, Technology and Advanced Studies, Chennai, India

N. Jayapandian Department of Computer Science and Engineering, CHRIST (Deemed To Be University), Bangalore, India

Rekha Jayaram Department of Information Science and Engineering, Dayananda Sagar College of Engineering, Bangalore, India

S. Jeevitha Department of Computer Application, Shrimathi Devkunvar Nanalal Bhatt Vaishnav College for Women, University of Madras, Chennai, India

Shweta Jha Apeejay School of Management, New Delhi, India

K. M. Jini Anna University, Chennai, India

M. Jogendra Kumar Department of Computer Science and Engineering, Koneru Lakshmaiah Education Foundation, Vaddeswaram, Andhra Pradesh, India

Tony K. John Department of Computer Science and Engineering, Aarupadai Veedu Institute of Technology, Vinayaka Mission's Research Foundation (Deemed to be University), Paiyanoor, Chengalpattu District, Tamil Nadu, India

Nabanita Saha Joya Department of Computer Science and Engineering, Daffodil International University, Dhaka, Bangladesh

M. S. Kalaivani Tagore College of Arts and Science, Chennai, India

Akshay Kale Department of Information Technology PVG's COET, Savitribai Phule Pune University, Pune, India

P. Kalpana Devi Veltech Rangarajan Dr. Sagunthala R&D Institute of Science and Technology, Chennai, India

Anita Kanavalli Department of Computer Science and Engineering, M. S. Ramaiah Institute of Technology, Bangalore, Affiliated to VTU, Bangalore, India

A. S. Kapse Department of IT, AEC, Chikhali, India

K. P. Karan Department of Computer Science and Engineering, Aarupadai Veedu Institute of Technology, Vinayaka Mission's Research Foundation (Deemed to be University), Paiyanoor, Chengalpattu District, Tamil Nadu, India

K. Karthik Department of CSE, Aarupadai Veedu Institute of Technology, Vinayaka Mission's Research Foundation, Salem, India

P. Karthika Department of Computer Applications, Kalasalingam Academy of Research and Education, Krishnankoil, India

N. K. Karthikeyan Coimbatore Institute of Technology, Coimbatore, India

Saloni Kawoor Department of Computer Engineering, MKSSS's Cummins College of Engineering for Women, Pune, India

M. N. V. Kiranbabu Koneru Lakshmaiah Education Foundation, Vaddeswaram, Guntur, Andhra Pradesh, India

Sai Hemanth Kumar Kotturi Department of CSE, Narasaraopet Engineering College, Narasaraopet, Andhra Pradesh, India

Aravind Krishnan Department of Computer Science and Engineering, Aarupadai Veedu Institute of Technology, Vinayaka Mission's Research Foundation (Deemed to be University), Paiyanoor, Chengalpattu District, Tamil Nadu, India

R. Krishnan Department of Computer Science and Engineering, Dayananda Sagar College of Engineering, Bangalore, India

Chaitanya Kulkarni Dayananda Sagar College of Engineering (DSCE), Bangalore, India

Chetana Kulkarni PES University, Bengaluru, Karnataka, India

Chandan Kumar Roorkee Institute of Technology, Roorkee, Uttarakhand, India

Krish Kumar Roorkee Institute of Technology, Roorkee, Uttarakhand, India

Vivek Kumar Veritas Technologies LLC, Pune, India

Joseph Varghese Kureethara Christ University, Bangalore, India

Surekha Lanka Business and Technology Program, Stamford International University, Bangkok, Thailand

G. Lavanya Sri Ramakrishna Engineering College, Coimbatore, India

S. Leelavathy Department of Computer Science and Engineering, Aarupadai Veedu Institute of Technology, Vinayaka Mission's Research Foundation (Deemed to be University), Paiyanoor, Chengalpattu District, Tamil Nadu, India

A. Lengerov Technical University of Sofia, Plovdiv, Bulgaria

Subarna Akter Liza Department of Computer Science and Engineering, Daffodil International University, Dhaka, Bangladesh

R. Magesh Sunder Department of Computer Science and Engineering, Aarupadai Veedu Institute of Technology, Vinayaka Mission's Research Foundation (Deemed to be University), Paiyanoor, Chengalpattu District, Tamil Nadu, India

R. Mahendra Chozhan Chozhan Dental Clinic, Periyakulam, Kodaikanal, Lakshmipuram, India

C. H. Maheswara Balaji Department of CSE, Aarupadai Veedu Institute of Technology, Vinayaka Mission's Research Foundation, Salem, India

V. Mahima PES University, Bengaluru, Karnataka, India

M. Mallegowda Department of Computer Science and Engineering, M. S. Ramaiah Institute of Technology, Bangalore, Affiliated to VTU, Bangalore, India

H. R. Mamatha PES University, Bengaluru, India

S. Maneesha PES University, Bengaluru, India

R. Manjula Electronics and Computer Engineering Department, Sreenidhi Institute of Science and Technology, Hyderabad, Telangana, India

K. G. Manjunath Department of Computer Science and Engineering, Siddaganga Institute of Technology, Tumkur, India

P. Manoj Lakireddy Balireddy College of Engineering, Mylavaram, Andhra Pradesh, India

Krishna Harika Mantripragada Department of Computer Science and Engineering, CHRIST (Deemed To Be University), Bangalore, India

E. A. Mary Anita Department of CSE, School of Engineering and Technology, CHIRST University, Bengaluru, India

M. Meghana Electronics and Computer Engineering Department, Sreenidhi Institute of Science and Technology, Hyderabad, Telangana, India

Akanksha Mishra Department of Computer Engineering, MKSSS's Cummins College of Engineering for Women, Pune, India

Ankush Mittal University of Petroleum and Energy Studies, Dehradun, Uttarakhand, India

D. Mohan Electronics and Computer Engineering Department, Sreenidhi Institute of Science and Technology, Hyderabad, Telangana, India

D. Mohana Geetha Department of CSE, Sri Krishna College of Engineering and Technology, Coimbatore, India

V. Mohanraj IT, Sona College of Technology, Salem, Tamil Nadu, India

Raka Moni Department of Computer Science and Engineering, Daffodil International University, Dhaka, Bangladesh

P. Monika Dayananda Sagar College of Engineering (DSCE), Bangalore, India

V. Murali Mohan Koneru Lakshmaiah Education Foundation, Vaddeswaram, Guntur, Andhra Pradesh, India

N. Murugan PSNA College of Engineering and Technology, Dindigul, Tamil Nadu, India

R. Murugesan Department of Computer Science, C.P.A College, Bodinayakanur, India

S. Muthamil Selvan Department of Computer Science and Engineering, Aarupadai Veedu Institute of Technology, Vinayaka Mission's Research Foundation (Deemed to be University), Paiyanoor, Chengalpattu District, Tamil Nadu, India

S. Muthusundari Department of CSE, R.M.D. Engineering College, Kavaraipettai, India

B. Naga Yugandhar Reddy Department of CSE, Aarupadai Veedu Institute of Technology, Vinayaka Mission's Research Foundation, Salem, India

Satuluri Naganjaneyulu Department of IT, Lakireddy Balireddy College of Engineering, Mylavaram, Andhra Pradesh, India

C. M. Nalayini Department of Information Technology, Velammal Engineering College, Chennai, India

N. Nalini Department of Computer Science and Engineering, Nitte Meenakshi Institute of Technology, Yelahanka, Bangalore, Karnataka, India

Adarsh R. Nambiar Department of Computer Science and Applications, Amrita Vishwa Vidyapeetham, Amritapuri, India

K. S. Nandini Prasad Department of Information Science and Engineering, Dr. Ambedkar Institute of Technology, Bengaluru, India

J. Nanditha Electrical and Electronics Engineering Department, Vidyavardhaka College of Engineering, Mysore, Karnataka, India

Prachi Natu NMIMS, Mumbai, India

Rahul Nijhawan University of Petroleum and Energy Studies, Dehradun, Uttarakhand, India

M. A. Nikitha Department of Computer Science and Engineering, CHRIST (Deemed To Be University), Bangalore, India

M. Nithya Department of CSE, VMKV Engineering College, Vinayaka Missions Research Foundations, Salem, India

Pratik Panchal Computer Engineering, Dwarkadas J. Sanghvi College of Engineering, Mumbai, India

Nilima V. Pardakhe P.R.M.I.T.&R., Badnera, Amravati, India

Rizwan Patan Department of Computer Science and Engineering, V R Siddhartha Engineering College, Affiliated To JNTUK, Vijayawada, India

Narendra M. Patel Birla Vishvakarma Mahavidyalaya, Gujarat, India

Neil Patel Computer Engineering, Dwarkadas J. Sanghvi College of Engineering, Mumbai, India

Bathula Pavan Kalyan Department of CSE, Aarupadai Veedu Institute of Technology, Vinayaka Mission's Research Foundation, Salem, India

Lukka Venkata Siva Phaneendra Department of Computer Science and Engineering, V R Siddhartha Engineering College, Affiliated To JNTUK, Vijayawada, India

H. K. Pooja Electrical and Electronics Engineering Department, Vidyavardhaka College of Engineering, Mysore, Karnataka, India

Anvitha Poosarla PES University, Bengaluru, Karnataka, India

Kundeti Naga Prasanthi Department of Computer Science and Engineering, Lakireddy BaliReddy College of Engineering, Mylavaram, India

Bhupalam Preetham Department of CSE, Aarupadai Veedu Institute of Technology, Vinayaka Mission's Research Foundation, Salem, India

N. Priya Research Department of Computer Science, PG Department of Computer Science, Shrimathi Devkunvar Nanalal Bhatt Vaishnav College for Women, University of Madras, Chennai, India

S. Shanmuga Priya Research Department of Computer Science, Shrimathi Devkunvar Nanalal Bhatt Vaishnav College for Women, Chennai, India

G. Priyadharshini Coimbatore Institute of Technology, Coimbatore, India

R. Pugazendi Department of Computer Science, Government Arts College (A), Salem, Tamil Nadu, India

Gopalam Pujitha Department of Computer Science and Engineering, V R Siddhartha Engineering College, Affiliated To JNTUK, Vijayawada, India

E. Punithavathy Department of Computer Applications, Madras Christian College, Chennai, India

T. Rachovski University of Plovdiv "Paisii Hilendarski", Plovdiv, Bulgaria

N. Raghavendra Sai Department of Computer Science and Engineering, Koneru Lakshmaiah Education Foundation, Vaddeswaram, Andhra Pradesh, India

Anna Treesa Raj Christ University, Bangalore, India

S. Rajaprakash Department of CSE, Aarupadai Veedu Institute of Technology, Vinayaka Mission's Research Foundation, Salem, Tamil Nadu, India

P. V. Rajasuganya Department of Artificial Intelligence and Data Science, Velammal Engineering College, Chennai, India

B. Rama Devi Lakireddy Balireddy College of Engineering, Mylavaram, Andhra Pradesh, India

K. Rama Sai Varma Department of Computer Science and Engineering, Sasi Institute of Technology and Engineering, (Affiliated To JNTU, Kakinada), Tadepalligudem, Andhra Pradesh, India

V. V. S. S. Ramya Department of Computer Science and Engineering, BVRIT Hyderabad College of Engineering for Women, Hyderabad, India

K. G. Rani Roopha Devi Department of Computer Science, Madurai Kamaraj University, Madurai, India

Athira P. Ranjith Christ University, Bangalore, India

Shreya Rao Department of Computer Engineering, MKSSS's Cummins College of Engineering for Women, Pune, India

N. Rashmika Department of Computer Science and Engineering, SASI Institute of Technology and Engineering (Affiliated To JNTU, Kakinada), Tadepalligudem, Andhra Pradesh, India

Pooja Raut Department of Computer Engineering, MKSSS's Cummins College of Engineering for Women, Pune, India

T. Ravikumar CHRIST (Deemed to be University), Bangalore, India

Gunupati Lakshmi Sai Vikas Kumar Reddy Department of Computer Science and Engineering, Aarupadai Veedu Institute of Technology, Vinayaka Mission's Research Foundation (Deemed to be University), Paiyanoor, Chengalpattu District, Tamil Nadu, India

R. S. Rehna Department of Computer Science and Engineering, LBS Institute of Technology, Poojapura, Thiruvananthapuram, Kerala, India

G. Renjith Department of Computer Science, University of Kerala, Thiruvananthapuram, Kerala, India

M. Renukadevi Department of CSE, S.A. Engineering College, Chennai, India

Shaik Afifa Reshma Department of Computer Science and Engineering, BVRIT Hyderabad College of Engineering for Women, Hyderabad, India

Vikas Kumar Roy Computer Science and Engineering, RIT Roorkee, Roorkee, India

Medagoni Rohan Sai Department of Computer Science and Engineering, Aarupadai Veedu Institute of Technology, Vinayaka Mission's Research Foundation (Deemed to be University), Paiyanoor, Chengalpattu District, Tamil Nadu, India

B. S. Sai Swetha Department of Computer Science and Engineering, CHRIST (Deemed To Be University), Bangalore, India

Afrah Samreen Department of Computer Science and Engineering, BVRIT Hyderabad College of Engineering for Women, Hyderabad, India

J. Samuel Manoharan Department of ECE, Sir Isaac Newton College of Engineering and Technology, Nagapattinam, India

S. Sandeep Kumar Department of Computer Science and Engineering, Koneru Lakshmaiah Education Foundation, Vaddeswaram, Andhra Pradesh, India

Harshit Sandhu NMIMS, Mumbai, India

S. Sankar Ganesh Department of CSE, Aarupadai Veedu Institute of Technology, Vinayaka Mission's Research Foundation, Salem, Tamil Nadu, India

Pooja Sarashetti Department of Computer Science and Engineering, M. S. Ramaiah Institute of Technology, Bangalore, Affiliated to VTU, Bangalore, India

M. Sarosh Umar Department of Computer Engineering, Aligarh Muslim University, Aligarh, India

A. R. Sathyabama Department of Information Technology, Velammal Engineering College, Chennai, India

R. Seetharaman Department of ECE, CEG Campus, Anna University, Chennai, India

J. Senthilkumar IT, Sona College of Technology, Salem, Tamil Nadu, India

H. Abdul Shabeer Department of Computer Science and Engineering, Aarupadai Veedu Institute of Technology, Vinayaka Mission's Research Foundation (Deemed to be University), Paiyanoor, Chengalpattu District, Tamil Nadu, India

Chinmay Shah NMIMS, Mumbai, India

Jaineel Shah NMIMS, Mumbai, India

Sana Shahab College of Business Administration, Princess Nourah Bint Abdulrahman University, Riyadh, Saudi Arabia

Maruf Shaikh NMIMS, Mumbai, India

Subarna Shakya Department of Electronics and Computer Engineering, Pulchowk Campus, Lalitpur, Nepal

K. Shantha Shalini Department of CSE, Aarupadai Veedu Institute of Technology, Vinayaka Mission's Research Foundation, Salem, India

M. P. Shashank Department of CSE, Aarupadai Veedu Institute of Technology, Vinayaka Mission's Research Foundation, Salem, India

Sushila Shelke Department of Computer Engineering, MKSSS's Cummins College of Engineering for Women, Pune, India

Rucha Shinde Computer Engineering, Dwarkadas J. Sanghvi College of Engineering, Mumbai, India

R. Shobana Department of Computer Science and Engineering, Aarupadai Veedu Institute of Technology, Vinayaka Mission's Research Foundation (Deemed to be University), Paiyanoor, Chengalpattu District, Tamil Nadu, India;
Department of CSE, Aarupadai Veedu Institute of Technology, Vinayaka Mission's Research Foundation, Salem, India

Shobha Department of Computer Science and Engineering, Nitte Meenakshi Institute of Technology, Yelahanka, Bangalore, Karnataka, India

S. Shruthi Dayananda Sagar College of Engineering (DSCE), Bangalore, India

K. L. Shunmuganathan Department of Computer Science and Engineering, Aarupadai Veedu Institute of Technology, Vinayaka Mission's Research Foundation (Deemed to be University), Paiyanoor, Chengalpattu District, Tamil Nadu, India

M. N. V. R. Shwetha Electrical and Electronics Engineering Department, Vidyavardhaka College of Engineering, Mysore, Karnataka, India

Moturi Sireesha Department of Computer Science and Engineering, Narasaraopeta Engineering College, Narasaraopet, Andhra Pradesh, India

Potu Siva Parvathi Department of Computer Science and Engineering, V R Siddhartha Engineering College, Affiliated To JNTUK, Vijayawada, India

Akula V. S. Siva Rama Rao Department of Computer Science and Engineering, SASI Institute of Technology and Engineering (Affiliated To JNTU, Kakinada), Tadepalligudem, Andhra Pradesh, India

V. Sivakumar Department of Computer Science and Engineering, Aarupadai Veedu Institute of Technology, Vinayaka Mission's Research Foundation (Deemed to be University), Paiyanoor, Chengalpattu, Tamil Nadu, India

Nimmagadda Snehita Department of Computer Science and Engineering, V R Siddhartha Engineering College, Affiliated To JNTUK, Vijayawada, India

C. C. Sobin SRM University, Amaravati, Andhra Pradesh, India

A. Sonya Department of Information Technology, B. S. Abdur Rahman Crescent Institute of Science and Technology, Chennai, India

Ch. Likhitha Sree Electronics and Computer Engineering Department, Sreenidhi Institute of Science and Technology, Hyderabad, Telangana, India

S. S. Sreeja Mole Department of ECE, Christu Jyothi Institute of Technology & Science, Jangaon, India

Velagapudi Sreenivas Dhanekula Institute of Engineering & Technology, Vijayawada, Andhra Pradesh, India

S. V. N. Sreenivasu Department of CSE, Narasaraopet Engineering College, Narasaraopet, Andhra Pradesh, India

S. Sridevi Queens College of Arts and Science for Women, Punalkulam, Pudukkottai, Tamil Nadu, India

S. Srihari PES University, Bengaluru, India

P. V. V. S. Srinivas Department of Computer Science and Engineering, Koneru Lakshmaiah Education Foundation, Guntur, India

B. Srinivasa Rao Gokaraju Rangaraju Institute of Engineering and Technology, Bachupally, Hyderabad, India

M. Sriram CHRIST (Deemed to be University), Bangalore, India

Y. Sriram Lakireddy Balireddy College of Engineering, Mylavaram, Andhra Pradesh, India

S. Subbulakshmi Department of Computer Science and Applications, Amrita Vishwa Vidyapeetham, Amritapuri, India

J. Suhashini PSNA College of Engineering and Technology, Dindigul, Tamil Nadu, India

Y. Suresh IT, Sona College of Technology, Salem, Tamil Nadu, India

B. Swapna Department of Computer Science and Engineering, Sasi Institute of Technology and Engineering, (Affiliated To JNTU, Kakinada), Tadepalligudem, Andhra Pradesh, India

Daniel Getachew Tadesse School of Electrical and Computer Engineering, Wollo University, Kombolcha Institute of Technology, Kombolcha, Ethiopia

G. Tarun Lakireddy Balireddy College of Engineering, Mylavaram, Andhra Pradesh, India

V. M. Thakare Department of CSE, SGBAU, Amravati, India

Vasu Thakur Computer Science and Engineering, RIT Roorkee, Roorkee, India

M. Tharun Department of ECE, CEG Campus, Anna University, Chennai, India

S. Thejaswini Siddaganga Institute of Technology, Tumkur, India

G. Thilagavathi Coimbatore Institute of Technology, Coimbatore, India

S. N. Thirumala Rao Department of Computer Science and Engineering, Narasaraopeta Engineering College, Narasaraopet, Andhra Pradesh, India

D. Uday Dayananda Sagar College of Engineering (DSCE), Bangalore, India

P. V. V. Vamsi Department of Computer Science and Engineering, SASI Institute of Technology and Engineering (Affiliated To JNTU, Kakinada), Tadepalligudem, Andhra Pradesh, India

V. Varun Kumar Reddy Department of CSE, Aarupadai Veedu Institute of Technology, Vinayaka Mission's Research Foundation, Salem, India

Dharsan Vasant Department of Computer Science and Engineering, Amrita School of Engineering, Amrita Vishwa Vidyapeetham, Coimbatore, India

S. Vinushree Siddaganga Institute of Technology, Tumkur, India

Prathviraj Walke Department of Information Technology PVG's COET, Savitribai Phule Pune University, Pune, India

Omkar Wanave Department of Information Technology PVG's COET, Savitribai Phule Pune University, Pune, India

T. Yuvaraj Prathyusha Engineering College, Thiruvallur, India

Zahura Zaman Department of Computer Science and Engineering, Daffodil International University, Dhaka, Bangladesh

Saima Zulfa Electrical and Electronics Engineering Department, Vidyavardhaka College of Engineering, Mysore, Karnataka, India

E-Learning-Based Collaborative Learning Experience in the Education Paradigm

Anuj Bhasin and Surekha Lanka

Abstract This study examined how SMART Boards encourage student-learning engagement in education. Our world is evolving due to technological advancements, and learning classrooms are no exception. In recent years, research has looked at technical advancements in a variety of areas that quantify change. Educationists have densely examined the use of technology in education. Role of one of the most popular technology "smart boards" in different educational fields is studied, and the advantages and complexities of using a SMART board Technology in education is evaluated. SMART Boards enable students to use technology and aid in exploring new concepts, resulting in a more dynamic learning experience. With the use of SMART Boards in the classroom, it is evident that student's interest is learning inclined. Many educators advocate for the use of digital whiteboards in the classroom. The information was collected through articles published on a combination of qualitative and quantitative approaches, including surveys.

Keywords Smart boards · Dynamic learning experience · E-learning · Interactive technology · Collaborative learning · Personal digital assistant (PDA)

1 Background

SMART Technologies, Inc., produces SMART Boards or interactive whiteboards. SMART Board interactive whiteboards were first introduced in 1991 and are now integrated into classrooms, lecture halls, group meetings, and presentations all over the world [1]. SMART Boards encourage users to use their fingers to control the computer using add-on devices and software and increases interactivity. This Canadian-based firm controls more than half of the interactive whiteboard industry.

A. Bhasin (✉)
Stamford English Program, Stamford International University, Bangkok, Thailand
e-mail: anuj.bhasin@stamford.edu

S. Lanka
Business and Technology Program, Stamford International University, Bangkok, Thailand
e-mail: surekha.lanka@students.stamford.edu

S. Shakya et al. (eds.), *Proceedings of Second International Conference on Sustainable Expert Systems*, Lecture Notes in Networks and Systems 351,
https://doi.org/10.1007/978-981-16-7657-4_1

A computer with Notebook software, a projector, and a whiteboard comprises the interactive SMART Board system. The computer images are projected on the board, and they can be easily edited since the board surface is touch-sensitive. User-friendly features enable control of the program both on the machine and on the board itself. Some of the features include inserting notes, highlighting, typing, saving data for later use, and access to the Internet.

2 Introduction

In an ever-increasing number of schools and colleges, educational technology has become an integral part of teaching and learning. Currently, many schools use computers, Internet tools, and projectors; however, SMART Boards are a better alternative and are being used in various fields of education such as engineering, medical science, and business management. Technology gives people opportunities for modern teaching and learning approaches. Despite the abundance of technology available in today's classrooms, it is evident that many teachers are not completely incorporating and using it efficiently [2]. The use of meaningful technology in the classroom facilitates effective instructions, enhances student learning, and changes teacher's pedagogy [3]. Most teachers in classrooms do not have enough time, access, or resources to become proficient in integrating technology into the learning process. Although there is evidence that technology use increases student engagement and academic achievement, many teachers continue to struggle with the realistic integration of new technologies into the curriculum, which is due to a lack of preparation [2]. Technology integration should be used to encourage active learning and engagement and not only as a method to provide information [4]. SMART Boards are a good alternative to traditional whiteboards or flip charts because it projects the data on to the board from the computer's desktop. SMART Board supports various applications such as educational software, Web sites. Smart boards assist teachers in using a student-centered approach to instructions [5].

3 Role of SMART Board in the Education Sector

In recent years, a significant rise in educational standards and creativity in the education sector using SMART Boards has been noted. We cannot argue that technological advancements have made education more interactive and accessible and encouraged the students to participate. The introduction of SMART Boards into the classrooms has revolutionized the teaching and learning experience. SMART Boards not only make learning a fun process but also are very useful for teachers in giving instructions and explaining without wasting a lot of time in writing on the board.

3.1 Promoting Student Engagement

Incorporating all of the new technologies into a classroom may be difficult as technologies are being developed at a fast pace. However, incorporating technologies into the curriculum to help students become more accustomed to new technologies is most beneficial to students. Technology is fascinating and appealing; when used to enhance learning, it engages students and motivates them to learn more. The majority of children are familiar with devices such as laptops, tablets, and smartphones [6]. Any current educational theories take into account the social learner and emphasize the importance of student participation in information creation. Whole-class teaching, constructivism, and constructive learning are three learning philosophies. Student involvement, which is critical to learning, is a common concern among these three learning theories. Previous research has shown that using a SMART Board improves student participation. According to Ref. [7], "educators can use digital resources while maintaining dynamic interaction with the entire class, provide computer-based learning without isolating students, and encourage a higher level of student interaction in both teacher-directed and group-based exchanges." A SMART Board facilitates contact between an instructor and a front-of-class student. In reality, the combination of SMART Boards and software allows the development of engaging classroom activities for students. As a result, SMART Boards promote engagement and interaction, as well as increase student learning [8].

3.2 Improving Motivation

SMART Boards have the ability to engage teaching resources, increase student engagement and motivation, visual illustrations, enhance lectures with audio-visual tools, provide better instructional materials, and adapt to all learning styles. SMART Boards have shown to increase student engagement in a wide body of study [9]. As per the reference of motivation findings from a study conducted in secondary schools, it is concluded that SMART Boards play an important role in inspiring students, concentrating their attention, and improving whole-class learning for both teachers and students. The impact of multimedia and multi-sensory visualization on motivation and work commitment is linked to SMART Boards. The beneficial effect of a SMART Board in the classroom seems to be largely dependent on how it is used. SMART Boards, when used properly, will improve learning and teaching [8].

3.3 Mediation and Sharing Knowledge

Explicit information exchange has a bigger impact on innovation tempo, but implicit knowledge has a bigger impact on innovation consistency. To improve academic institutions' creativity, many sources of information are needed—not only overt knowledge, but also implicit knowledge, such as employee experience, which has become an important feature of academic knowledge. An academic institution's most important asset is the potential to transform implicit information into clear knowledge and communicate that knowledge with others.

4 Applications of SMART Board

SMART Boards are not limited to the education sector but also used in various sectors such as engineering, medical science, and business management to name a few. According to research [10], it is evident that the use of SMART Boards in teaching seventh-grade high school students improved academic performance and helped in remembering the lesson learned. It is evident that SMART Boards have a positive impact on student's academic success in science classes as well. This is reinforced by evidence from the literature [10]. Classrooms fitted with interactive/SMART Boards had a significant impact on students' language learning methods [11]. Indicated that the level of achievement in eighth-grade math classes has risen. SMART Boards were found to increase students' enthusiasm for academic achievement when they were used to teach mathematics [12]. Discovered that students who learned physics using SMART Boards had substantially different academic outcomes than classmates who learned using conventional syllabus [13]. Effectiveness of SMART Board usage under the domain of science and technology aimed to assess not only academic accomplishment but also long-term learning retention.

The functionality of smart classroom applications

As we discussed in the previous section—Role of SMART Boards, collaborative learning is made easier with the smart classroom application suite. Our smart classroom application suite enables students, teachers, and teaching assistants to collaborate on various collaborative learning activities. The suite also includes various collaboration features for teachers, lecturers, and teaching assistants. The following are the features of smart classroom applications [14].

(i) ***Abstract Made More Concrete***
The math tools plug-in for SMART Notebook applications allows you to easily construct visual representations of a variety of concepts, from elementary school fractions to algebraic functions in AP calculus and HS science classes. By reflecting abstract concepts in a more realistic way, these visuals will assist students in comprehending them. Notebook, too, can view 3D objects, some of which are already in the gallery, and others that can be downloaded for free from the Google 3D image library. When these images are combined with a mixed-reality ready document camera, such as the SMART Document Camera, objects in space can be manipulated from all angles.

(ii) ***Single-User Student***
Based on the current time and place, the application suite reminds students of their homework and class schedule. Before and after class, the application module can synchronize lecture notes between a student's personal computer/tablet and desktop device.

(iii) ***Instructor/Teaching Assistant***
Since the desktop computers have the original lecture notes, the program suite synchronizes the lecture notes between the instructor's or teaching assistant's PDA and the desktop computer before and after class.

(iv) ***Communication Between Students***
Students may use the program suite to exchange and distribute their drawing papers. It also allows students to synchronize drawing documents between their PDAs.

(v) ***Contact with Students by the Teacher/Teaching Assistant***
The application suite distributes instructional materials (lecture notes/survey forms/grade sheet/course schedule) from the teacher to the teaching assistant to all students at the appropriate times, such as at the start of a class when the lights are dimmed and the noise level is minimal. Using the application suite, the teacher will develop tests for students and groups.

(vi) ***Communication Between Students and Instructors/Teaching Assistants***

(a) Students may use the application suite to save their questions or complaints in text format on their PDAs. The questions are immediately moved to the teacher's or teaching assistant's PDA when the instructor or teaching assistant is available (in the classroom). Students are encouraged to participate.

(b) Students use the program suite to apply their progress reports in a similar manner. Their notes are handed to the teacher/teaching assistant at the end of each session. Students may use their PDAs to schedule appointments with the instructor, sending the request to the instructor's PDA and receiving approval via the application suite.

(c) Students use the application suite to write exam answers and submit them to the teacher. Students use the application suite to run and view their homework on their PDAs and project it on the computer [14].

(vii) ***Flip the Classroom***
Several innovative professors are experimenting with flipped classrooms, in which videos replace direct instruction and students are given individual time in class to collaborate with their teacher on key learning activities. The flipped classroom is named after the fact that what was previously done in class (the "lecture" is now done at home through teacher-created videos and what was previously homework assigned problems) is now done in class. The SMART Recorder allows teachers to record direct instruction with a single click.

5 Virtual Smart Classroom Solutions—During a Pandemic

Indonesian schools used Google Classroom, WhatsApp, and Zoom Cloud Meeting as the most commonly used distance-learning applications during the pandemic. In a virtual learning setup, the teachers used the three common learning models, namely cooperative, problem-based learning (PBL), and project-based learning (PJBL). Since it followed the systematic learning model, the teacher said the learning model applied was very useful for their work in developing a virtual classroom, enabling students to enhance their comprehension of the content delivered. Each virtual classroom used a different application of learning models to optimize learning and maximize saturation.

According to [15], technology-enhanced ELA classrooms can promote student-centered, constructivist learning environments, which can help students learn more efficiently in the secondary English classroom. When used productively, technology has the potential to support students' higher-order thinking skills, encouragement and commitment. Many benefits of using technology in the classroom were identified in this report, including meeting a wide range of learning styles, differentiating teaching, providing visual aids, providing new experiences, and inspiring students. The instructors, for example, were able to meet students who struggled to read the text individually by using an audio CD. Newby et al. [16] emphasized the value of technical literacy, which is characterized as the ability to use computers and technology to support learning, communication, efficiency, and overall productivity. Technology was seen as an important part of the English curriculum by the teachers, who saw it as a way to help and enhance teaching while also training and encouraging students to become more technologically savvy.

Some data from smart stories are gathered from educators who use smart applications in their classrooms during the pandemic [17].

Applying the flipped classroom paradigm to a large-scale, remote teacher training program: Florence 1 Schools, USA, as educators, we know that using the flipped classroom model, in which students complete learning tasks at home and apply what they have learned in class the next day, is one way to incorporate blended learning.

"Technology offers students a voice in a way that they do not have to use their physical voice to express their thinking or ideas," says a teacher in a Texas classroom. "It is really incredible." Ja'Shyni was a shy student. She was so reserved that on the first day of kindergarten, she refused to tell her teacher her name. Ja'Shyni began using the smart board in her classroom as a teaching tool.

Abigail Davis said it has a lot of advantages for student participation and teamwork in the classroom [18].

As a nascent elementary school teacher, I have had plenty of opportunities to use these interactive whiteboards in the classroom. It is almost impossible to put into terms how rewarding it is to watch students come together and collaborate on a single assignment. Students are more involved than ever before when it comes to reading, writing, and math classes.

6 Good Tech Capabilities and High Education Outcomes Go Together

One connection stands out and has become stronger over time: Schools with high technology skill growth also have the best teaching and learning outcomes. Educators with the best overall skills were ten times more likely to see excellent results [17] (Chart 1).

Chart 1 Five key pillars

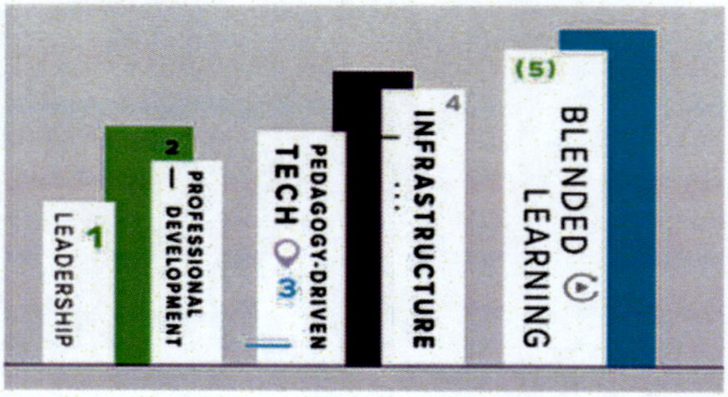

Strong self-reported results are linked to higher percentages of technology used. The majority of educators said their schools had a variety of front-of-classroom hardware: Non-interactive whiteboards and displays, non-interactive panels, and non-interactive projection devices are all examples of digital whiteboards and displays. Lesson design, presentation, collaborative, evaluation, and game-based apps were all mentioned by at least half of those polled as being used in the classroom. There is a close link between the technologies in use, their overall skill growth, and positive outcomes (Chart 2).

Educators who said their schools had good results have used the most of the hardware and software mentioned. The greater the effect of technology on results, the steeper the slope of the purple arrow—led by lesson development, evaluation, and game-based applications.

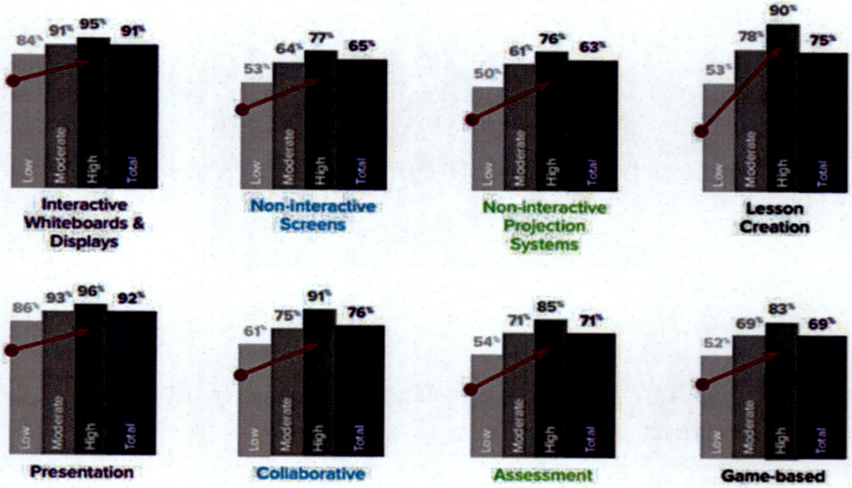

Chart 2 Outcomes of self-reported percentages of technologies

7 Conclusion

SMART Boards have a great deal of potential as an instrument for improving teachers' educational performance and learners' educational accomplishments and thereby improving student achievement. Nevertheless, the argument that this technology will automatically improve a learner's achievement may be incorrect. The study's data gathering was confined to a fixed research sample, and the study's timeframe was also limited. As a consequence, the findings may not be applicable to all teachers. This review paper is intended to serve as a springboard for a more in-depth investigation into the subject, with a focus on cooperative learning in distance education or virtual classroom design, appropriate technology, and online student performance assessment. This review paper expected to serve as a launching pad for a more extensive study on the topic, with a wider scope in cooperative learning in distance education or in virtual classroom design, use of appropriate technology, and assessment of online student performance.

References

1. Springgay J (n.d.) What is the history of the SMART board? Techwalla. https://www.techwalla.com/articles/what-is-the-history-of-the-smart-board
2. Oigara JN, Keengwe J (2011) Pre-service teachers and technology integration with smart boards
3. Culp KM, Honey M, Mandinach E (2003) A retrospective on twenty years of educational technology policy. U.S. Department of Education Office of Educational Technology, Washington, DC
4. McGrail E (2005) Teachers, technology, and change: English teachers perspectives. J Technol Teach Educ 13(1):5–24
5. Momani DM, Alshaikhi TS, Al-Inizi TH (2016) The obstacles of using smart board in teaching English at Tabuk secondary schools
6. Parmeter C (2012) Student learning engagement with SMART boards in reader's workshop. Education masters. Paper 233
7. Al-Saleem BIA (2013) The interactive whiteboard in English as a foreign language (EFL) classroom. Eur Sci J 8(3):126–134
8. Javidi Jelyani S, Janfaza A, Soori A (2014) Integration of smart boards in EFL classrooms. Int J Educ Literacy Stud 2(2):20–23
9. Levy P (2002) Interactive whiteboards in teaching and learning in two Sheffield schools: a developmental study. Sheffield University, Sheffield
10. Aktas S, Aydin A (2016) The effect of the smart board usage in science and technology lessons. Eurasian J Educ Res 64:125–138
11. Schmid EC (2008) Using a voting system in conjunction with interactive whiteboard technology to enhance learning in the English language classroom. Comput Educ 50:338–356
12. Tezer M, Deniz AK (2009) Matematik dersinde interaktif tahta kullanarak yapilan denklem cozumunun ogrenme uzerindeki etkisi [The effect of the equation solution made by using the interactive whiteboard on learning in math class]. In: 9th international educational technology conference, Ankara, pp 500–506
13. Tekin Y (2013) Fizik egitiminde akilli tahta kullaniminin ogrencilerin fizik basarilarina ve fizige karsi tutumlarina etkisinin incelenmesi [Investigation of the effects of using smart board in physics education on students achievements in and attitudes toward physics]. Master thesis, Van Yuzuncu Yil University, Van

14. Stephen S, Gupta S, Karim F, Sheikh I, Wang Y, Wang B (n.d.) Smart classroom: enhancing collaborative learning using pervasive computing technology
15. Flanagan S, Shoffner M (2013) Technology: secondary English teachers and classroom technology use. Contemp Issues Technol Teach Educ 13(3):242–261
16. Newby T, Stepich D, Lehman J, Russell J (2006) Educational technology for teaching and learning. Prentice Hall, Upper Saddle River, NJ
17. https://www.smarttech.com/about-us/customer-stories/customer-stories/cassie. ©2021 SMART Technologies
18. https://medium.com/reviews-r-us/smart-boards-for-education-product-review-7ce422d21835

Using Clean Architecture Principles to Improve the Design and Implementation of the Mobiles Online Platform

T. Rachovski, S. Hadzhikoleva, E. Hadzhikolev, and A. Lengerov

Abstract Mobiles Online is a client–server application that is used to turn "ordinary" web pages into ones suitable for visualization on mobile devices. It is mainly used for legacy web pages, which for pragmatic reasons are expensive and time-consuming to rewrite with adaptive or responsive design. After building the initial prototype of the application, new requirements and functionalities emerged, which proved to be difficult to implement. The problem arose of choosing a more appropriate architecture to minimize code dependencies and facilitate the development of new functionalities. The paper presents the architecture of the Mobiles Online application, based on the concept of the "clean architecture" model. The main goal is to separate and eliminate the dependencies between the modules of an application. An important emphasis is the separation of the application core from other modules and the independence of their functions. This architecture provides flexibility and easy scalability to implement future functions.

Keywords Responsive design · Clean architecture · Mobiles Online

1 Introduction

When using the right architecture, the proper development process and the subsequent maintenance of a software application are easier and more efficient. Often, when building an application, the focus is on the business processes. Features such as

T. Rachovski (✉) · S. Hadzhikoleva · E. Hadzhikolev
University of Plovdiv "Paisii Hilendarski", Plovdiv, Bulgaria

S. Hadzhikoleva
e-mail: stankah@uni-plovdiv.bg

E. Hadzhikolev
e-mail: hadjikolev@uni-plovdiv.bg

A. Lengerov
Technical University of Sofia, Plovdiv, Bulgaria
e-mail: alengerov@tu-plovdiv.bg

S. Shakya et al. (eds.), *Proceedings of Second International Conference on Sustainable Expert Systems*, Lecture Notes in Networks and Systems 351,
https://doi.org/10.1007/978-981-16-7657-4_2

reliability, security, performance, and scalability are also important when choosing the right architecture. In general, applications change over time—there is a need to add new features and modify already existing ones. Choosing the right architecture simplifies the development process so that many software specialists can work on the application—simultaneously or at different stages of its life cycle.

There are Web sites on the Internet that have been developed using old technologies and that do not have an adaptive or responsive design. They are not visualized well on mobile devices with small screens—smartphones, tablets, etc. Building mobile versions or completely rewriting such sites is in some cases quite expensive and time-consuming. This challenged us to develop Mobiles Online—an application for visualization of ordinary web pages that do not have an adaptive or responsive design in mobile devices [1–3]. The difficulties in upgrading the developed software prototype with new functionalities and design presented us with a new challenge—finding a new architectural solution to facilitate this process.

The article presents an approach to building the Mobiles Online application using the concept of clean architecture.

2 Approaches and Methodologies When Creating Software Applications

Building a software application is a complex process consisting of several *main stages*—analysis, design, implementation, testing, and maintenance [4, 5]. Each stage is characterized by certain activities and results.

1. *Software requirements*—at this stage, the requirements to the application are determined, and the possibility for their realization is analyzed. A requirements specification is prepared, which contains all the requirements for the product that must be designed and developed during its life cycle.
2. *Software design*—based on the requirements defined in the analysis, more than one approach to design an application's software architecture is usually proposed and discussed. Appropriate architecture is selected and developed.
3. *Software development*—during this stage, the program code is developed. The goal is to translate the design accurately and consistently into specific programming languages. Well-written code reduces the effort of testing and maintaining the application.
4. *Software testing*—product testing is performed, in which defects are reported, tracked, and corrected. The application is tested repeatedly until it reaches quality standards.
5. *Implementation*—once the application has successfully passed the tests, it is implemented in a real working environment. This usually includes installation, customization, testing, and an additional evaluation period. Employees are trained to work with the software.

6. *Software maintenance*—software maintenance and improvement deal with newly discovered errors. In some cases, setting additional requirements to the application may require a lot of time and financial resources, and this may lead to a change in the design and architecture of the application.

The different ways and sequence of implementation of these stages determine *different methodologies for software development*—agile development, continuous integration, incremental development, rapid application development, spiral development, waterfall development, etc. [6, 7].

Depending on *the techniques and approaches for software application development*, methodologies such as test-driven development (TDD), behavior-driven development (BDD), domain-driven development (DDD), responsibility-driven design (RDD) can be determined.

Test-driven development is a software development process in which the test layer is written before the other layers are built. Once built, the focus of developers is on writing program code to implement features that are sufficient to pass the test. Refactoring is then performed [8, 9].

Domain-driven design is a methodology that centers the development on programming a domain model that focuses on core functionality. The approach is especially suitable for complex domains, where very complex logic must be implemented. The principle is the separation of individual layers, which clearly describes the business requirements and the relationship with the data model [10].

Behavior-driven development is an approach that improves TDD and DDD. It aims to better understand the client and his requirements. As a concept, it is a way of combining business requirements with program code that allows understanding the behavior of the system from the point of view of the business or the end user [11, 12]. In a similar way as in DDD, here also is a separation of individual layers, the idea is to clearly express the functionality of business requirements and interaction with the data model.

Choosing the right methodology for developing a software application is a long and complex process, but in no way replaces choosing the right software architecture.

3 Clean Architecture

The concept of clean architecture has been developed in recent years [13, 14]. It combines the principles of the hexagonal architecture, the onion architecture, and several other variants. They all use dependencies of the DDD and BDD methodologies. The goal is to create a system that allows business logic to be scalable, as well as easier to test.

Clean architecture is a concept introduced by Martin [15]. This approach to software development aims to create an easy-to-read and test code base to build stable, error-free applications. The main priority is to build software that supports the need for frequent refactoring which increases the quality of the software.

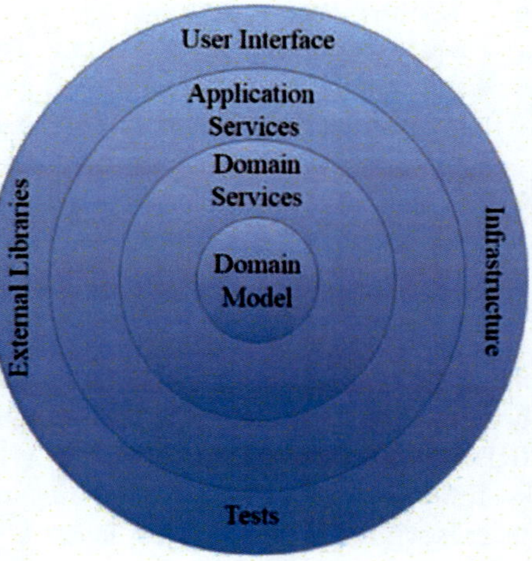

Fig. 1 Clean architecture model

The principles of clean code are becoming increasingly popular among a wide range of software developers.

Figure 1 presents the conceptual model of clean architecture [16]. Unlike other methods, dependencies refer to the outermost circle of the diagram. The application core has no dependencies on the other application layers. It supports data models that are described in the domain model layer.

Domain model and domain services. The basis of the architecture is determined by the domain model. It contains models and objects that clearly describe the business requirements of the application. Linked together, the domain model and domain services form the core of the application.

Application layer contains all of the business logic. It is the most important because it usually performs all the functionality of the application here, including Command Query Responsibility Segregation (CQRS). Each business use case is represented by a single command or request. Commands represent all operations such as creating, updating, and deleting objects that cause the system to enter, modify, and remove data from the system.

The infrastructure layer takes care of external resources such as file systems, databases, web services, SMTP. In case of possible access to external sources, this layer processes the necessary data and transmits it to the other layers.

User interface. The main task of this layer is to provide an interface through which users can use the system. It is the outermost layer of the architecture. Integration to external sources or systems can also be performed here using SOAP API or RESTFUL API.

Test layer. This layer is responsible for testing all functionalities. Each of the other layers, as well as their functionalities, can be easily tested. According to the software testing requirements, test automation pyramid can also be applied here [17, 18], as the tests are divided into three layers:

- UI tests—they relate mainly to the visualization and front-end part of the application;
- Service tests—cover functionality testing;
- Unit tests—testing the functionalities of each module separately.

4 Algorithm of Action of Mobiles Online

Mobiles Online is an application that converts ordinary legacy web pages without responsive or adaptive design into ones that can be well-visualized on mobile devices [1–3]. The application serves as a proxy mediator between the client (desktop or mobile web browser) and the actual server hosting the targeted web page.

The user submits a request for a web page (Fig. 2) by typing its address in Mobiles Online. The request is transmitted to the target web server, which returns the original HTML page. Mobiles Online processes the received data and generates a mobile version of the page, which is provided to the customer.

The business logic of the Mobiles Online application is presented in Fig. 3 and includes the following stages:

- *Stage 1.* The user enters the address of the desired web page, which is sent to the Mobiles Online application. Validation of the address provided by the user is performed.
- *Stage 2.* Mobiles Online downloads and validates the content of the web page. The web page source code is parsed with a standard HTML parser, which automatically creates a DOM tree of its content. Validation is performed on the created DOM object, and if errors occur, the information about them is saved in the database.
- *Stage 3.* Mobiles Online processes the already downloaded content and then forward it in response to the customer.

Detailed information on the methodology of creating mobile versions of web pages with Mobiles Online application, including information about the created prototype and conducted experiments, is published in [19].

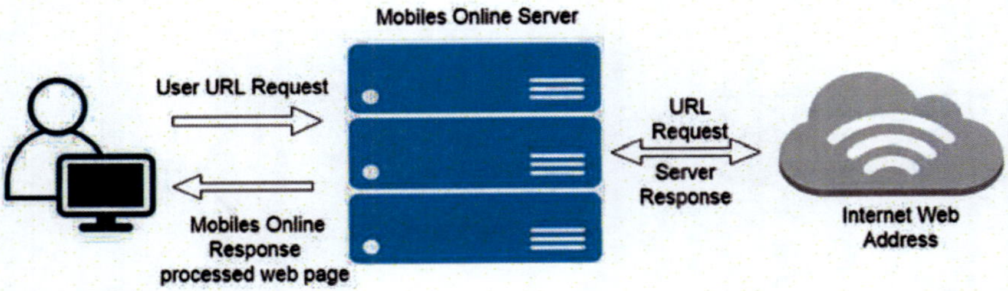

Fig. 2 General model of operation of the Mobiles Online application

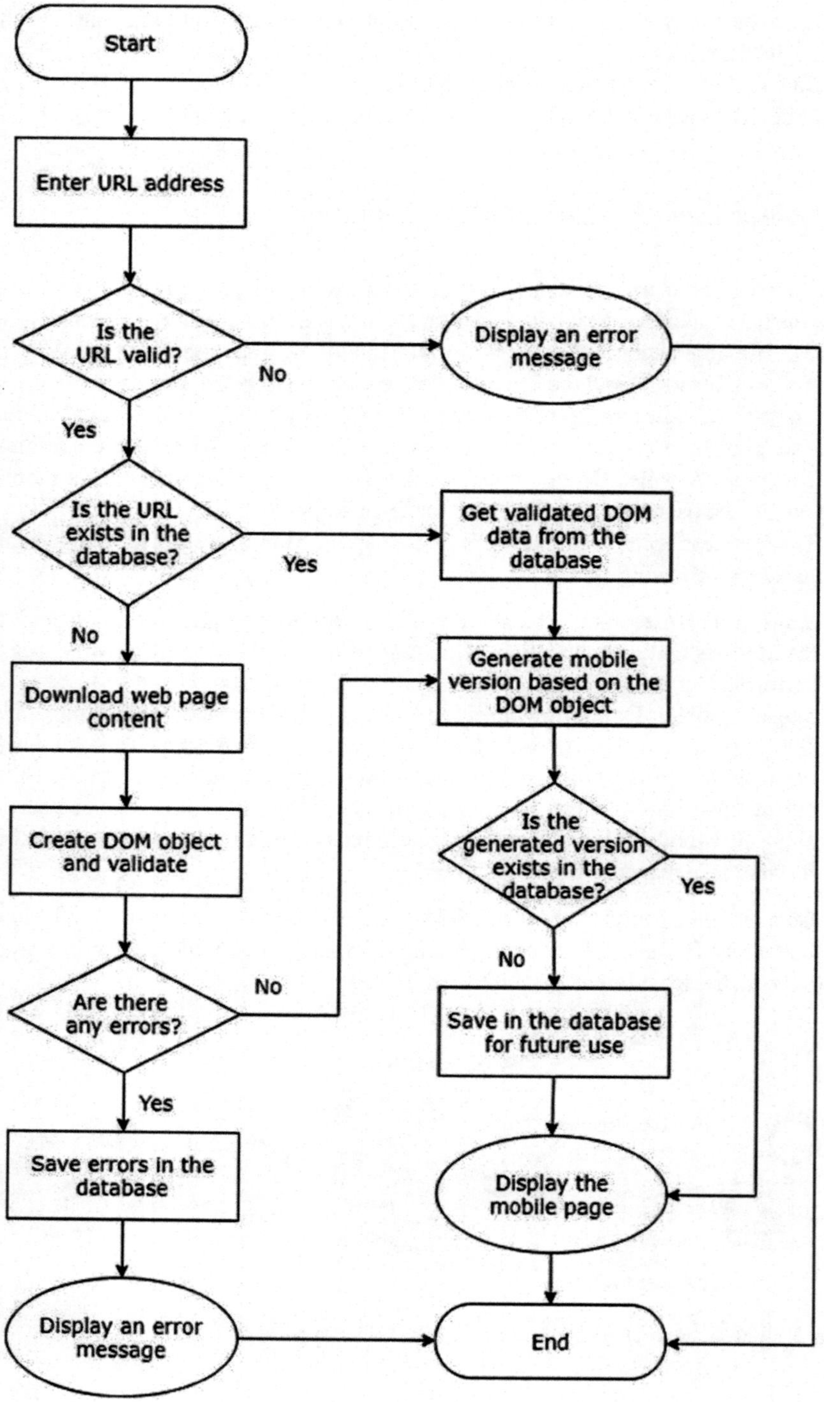

Fig. 3 Business logic of the Mobiles Online application

5 Implementation of Mobiles Online with Clean Architecture

With the expansion of the Mobiles Online project, new requirements and functionalities have appeared. Their implementation turned out to be quite laborious and time-consuming, and the reason for this was the initially chosen architecture of the prototype. After the research and analysis of the different architectures, our choice was focused on clean architecture, which facilitates the process of building and improving software modules.

The implementation of the standard clean architecture framework for the Mobiles Online application is presented in Fig. 4. An important problem that the architecture solves is the speed provided by caching the newly generated pages. This necessitated the separation of two separate components in Mobiles Online, each of which is built on a clean architecture with three main horizontal layers with user functionalities: user interface layer, application layer, infrastructure layer. The testing functionalities are separated in a fourth test layer.

Mobiles Online web interface is the layer of the architecture that takes care of the processing of web requests coming from the client. This part is usually used as a presentation layer. The user makes a request to process a web page. The application checks to see if the page address has already been processed in a previous moment, and depending on that, it moves to the next layer. Provided that a cached version of the generated page exists, the application layer of Component 2 is accessed to retrieve the caches page; otherwise—the application layer of Component 1 is accessed to generate a new page.

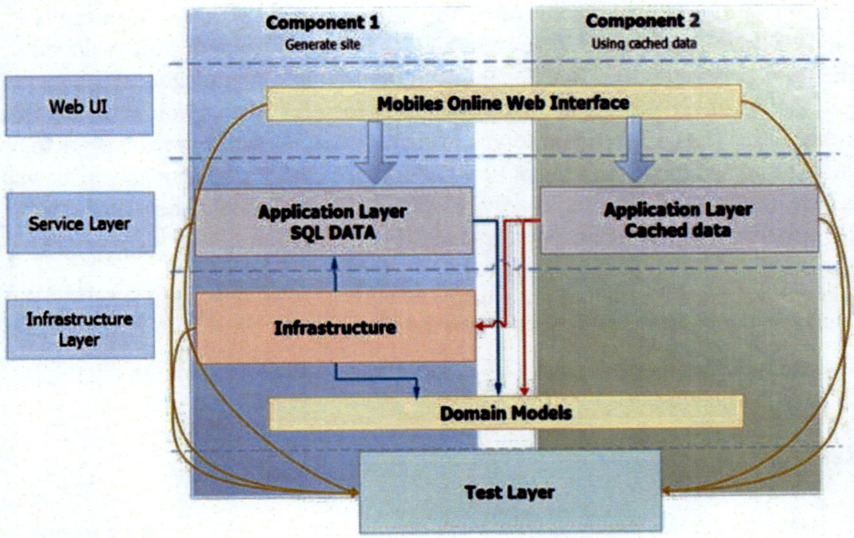

Fig. 4 Mobiles Online model with clean architecture

Component 1 implements the basic business logic of generating a mobile version of a page when requesting a page that has not been accessed through Mobiles Online. The main activities are related to receiving the user's request for a web page, generating a mobile version, and saving it in the database. The functionality is specified in application and infrastructure layers.

In the *application layer* of Component 1, three main activities can be specified:

- Processing off the downloaded HTML content in an appropriate structure;
- Saving the contents in the database;
- Generation of a mobile web page.

Component 1 infrastructure layer—its functionalities are related to the use of an external HTTP client, which downloads the content of the web page requested by the user. The downloaded web page returns to the application layer for processing.

The business logic of Component 2 is related to the management of the pages caches in the database. Component 2's *application layer* saves newly generated pages in the database, as well as searches and retrieves already generated pages when requested by the user. It should be noted here that in order to improve the performance, the use of other approaches is possible [20].

The test layer describes all the tests related to the functionality of the system. It has access to all other layers, which is set as a fundamental idea when building an application with a clean architecture.

6 Conclusion

The choice of a suitable architecture for software application development is an important task that determines the opportunities for its maintenance and development. The need to develop new functionalities in the Mobiles Online application turned out to be a difficult task, which motivated us to look for a solution to use a flexible architecture. The developed prototype, which uses clean architecture, appears to be easier to maintain and implement new and modify existing functionalities, as well as to be integrated with external services. The prototype is still under development. Real-time experiments show good results.

Acknowledgements The work is partly funded by the MU21-FMI-004 project at the Research Fund of the University of Plovdiv "Paisii Hilendarski."

References

1. Hadzhikolev E, Hadzhikoleva S, Rachovski T (2018) Architecture of Mobile Online: an application for creating mobile versions of web pages. In: XXth international symposium on electrical apparatus and technologies SIELA'2018, Bulgaria

2. Rachovski T, Hadzhikoleva S, Hadzhikolev E (2017) Conceptual model of an application for automated generation of webpage mobile versions. TEM J 6(4):877–882
3. Hadzhikoleva S, Rachovski T, Hadzhikolev E (2018) Generalized net model for building responsive design of web pages. In: XXth ınternational symposium on electrical apparatus and technologies SIELA'2018, Bourgas, Bulgaria
4. Bourque P, Fairley R, IEEE Computer Society (2014) Guide to the software engineering body of knowledge (SWEBOK): version 3.0. IEEE Computer Society Press, Washington, DC
5. Ruparelia N (2010) Software development lifecycle models. ACM SIGSOFT Softw Eng Notes 8–13
6. Kuhrmann M, Diebold P, Münch J et al (2017) Hybrid software and system development in practice: waterfall, scrum, and beyond. In: Proceedings of international conference on softwaresystem process, 5–7 July 2017, Preprint
7. Alshamran A, Bahattab A (2015) A comparison between three SDLC models waterfall model, spiral model, and ıncremental/ıterative model. IJCSI Int J Comput Sci Issues 12(1):106–111
8. Beck K (2002) Test driven development: by example. Addison-Wesley Professional
9. Astels D (2003) Test driven development: a practical guide. Prentice Hall
10. Evans E (2003) Domain-driven design: tackling complexity in the heart of software. Addison-Wesley Professional
11. Cisneros L, Maximiano M, Reis C (2018) An experimental evaluation of ITL, TDD and BDD. In: XIIIth ınternational conference on software engineering advances ICSEA 2018, pp 20–24
12. North D (2006) Behavior modification: the evolution of behavior-driven development. Better Softw 8(3)
13. Martin R (2017) Clean architecture: a craftsman's guide to software structure and design. Pearson
14. Kessner A. Explaining clean architecture. https://www.oncehub.com/blog/explaining-clean-architecture
15. Martin R. The clean architecture. The clean code blog. https://blog.cleancoder.com/uncle-bob/2012/08/13/the-clean-architecture.html
16. The clean architecture. https://apprize.best/php/clean/4.html
17. Contan A, Dehelean C, Miclea L (2018) Test automation pyramid from theory to practice. In: International conference on automation, quality and testing, robotics, AQTR 2018, Romania
18. Willett J. The evolution of the testing pyramid. https://www.james-willett.com/the-evolution-of-the-testing-pyramid/
19. Rachovski T, Hadzhikolev E, Hadzhikoleva S (2020) Models and methodologies for automated creating of webpage mobile versions. In: International conference on technics, technologies and education ICTTE 2020, Yambol, Bulgaria, 4–6 Nov 2020
20. Suma V (2020) Resource intensification for mobile devices using the approximate computing entities. J Trends Comput Sci Smart Technol (TCSST) 2(1):26–36

Detection of Arrhythmia Using Convolutional Neural Networks

Burla Greeshma, Moturi Sireesha, and S. N. Thirumala Rao

Abstract Cardiac arrhythmias are irregular heartbeats that are either too fast (tachycardia) or too slow (bradycardia). A minor alteration in the morphology or dynamics of the electrocardiogram (ECG) can induce severe arrhythmia events, which can impair the heart's ability to pump blood and cause shortness of breath, chest pain, exhaustion, and loss of consciousness. A few types of arrhythmias are present infrequently in the ECG signal and, as a result, must be recognised using extensive ECG recordings. Longer ECG records necessitate manual analysis, which takes time and effort. The ANSI/AAMI EC57 guideline identified five categories in heartbeats: non-ectopic or paced beat (N), supraventricular ectopic beat (S), ventricular ectopic beat (V), fusion of a ventricular and normal beat (F), pace beat or fusion of a paced and normal beat (F), and beat which is not categorised (Q). To classify these five types of heartbeats, a technique based on deep neural network has proposed in the study. Physicians benefit greatly from the automatic detection and classification of cardiac arrhythmias. The proposed technique was tested with a patient-specific pattern on the MIT-BIH Arrhythmia Dataset. The overall accuracy for the proposed model achieved was 97.62% with F1-score of 0.88 which is a decent score. The proposed approach performed well in the experiments, indicating that it could be useful in clinical practice.

Keywords Cardiac arrhythmias · Electrocardiogram · Convolutional neural networks

B. Greeshma (✉) · M. Sireesha · S. N. Thirumala Rao
Department of Computer Science and Engineering, Narasaraopeta Engineering College, Narasaraopet, Andhra Pradesh, India

S. N. Thirumala Rao
e-mail: nagatirumalarao@nrtec.email

S. Shakya et al. (eds.), *Proceedings of Second International Conference on Sustainable Expert Systems*, Lecture Notes in Networks and Systems 351,
https://doi.org/10.1007/978-981-16-7657-4_3

1 Introduction

Cardiac arrhythmia is hazardous and, in the majority of cases, fatal [1]. In today's medical practice, skilled cardiologists must carefully examine ECG [2] to diagnose life-threatening cardiac arrhythmias, and ECG also plays a vital role in detecting arrhythmias [3]. Automatic classification of cardiac arrhythmias, on the other hand, can provide objective diagnostic results while also saving cardiologist's time. The algorithm of automatic electrocardiogram (ECG) analysis relies on classification of type of heartbeat type. The ANSI/AAMI EC57 guideline identified five categories of heartbeats: non-ectopic or paced beat (N), supraventricular ectopic beat (S), ventricular ectopic beat (V), fusion of a ventricular and normal beat (F), pace beat or fusion of a paced and normal beat (F), and beat which cannot be categorised (Q) [4]. The benefits improved a lot of interest in classification and diagnosis of ECG signals by computer in hospital and medical community. Arrhythmia identification on ECG uses first-generation algorithmic tools, with cardiologists doing final ocular validation [5]. The ECG signal interpretation is an application of pattern recognition. Pattern recognition goal is to classify a system automatically into one of the several categories. ECG signals were split as heartbeats, and each one of the heartbeats was changed into two-dimensional grayscale pictures that will be sent as input data for the CNN [6]. By simply looking at the printout of ECG waveforms, a skilled cardiologist can diagnose a variety of heart arrhythmias.

The MIT-BIH Arrhythmia Database was the first widely available set of standard test sample for examining arrhythmia detectors [7]. For fully automatic arrhythmia detection or event selection, different methods are proposed for next verification by human experts [8]. Goal of our study is to create diagnose system which is computer assisted which will aid professional cardiologists at offering intelligent, cost-effective, and time-saving ECG arrhythmia diagnostics. By gaining this purpose, the main challenge is that ECG arrhythmia pattern recognition is combined with traditional ECG signal processing techniques. By altering the number of epochs, the suggested system is compared for various activation functions [9]. Right bundle branch blocks (RBBB) are a type of cardiac arrhythmia that can be distinguished, and these are classified from normal and paced beats in their current form [10]. Normal beats are the ECG waveform of a healthy person who is in adult stage; pacemaker produces the artificial heartbeats, and RBBB is one of the arrhythmias which have waveform of ECG with duration of QRS between 0.10, 0.11, and 0.12 s or more, and ischaemic, hypertensive, rheumatic, and pulmonary heart disease, right ventricular hypertrophy, and certain drug intoxication are all linked to a prolonged ventricular activation time or QR interval, as well as right axis deviation. The system can be easily adapted to classify the kinds of similar cardiac arrhythmias.

2 Literature Survey

Arrhythmia is defined as an ECG signal having an irregular rhythm, such as atrial fibrillation, ventricular tachycardia, ventricular fibrillation.

Rajkumar proposed a model with convolutional neural network with only one convolutional layer and ReLU as an activation layer and batch normalisation as regularisation method and classified into seven classes obtained an accuracy of 93.6%.

Rajpurkar and colleagues developed a 34-layer convolutional neural network that uses ECG models to map a sequence of rhythm classes. Using single-lead ECG records, their system surpassed cardiologists in recognising a wide range of heart arrhythmias, and fourteen kinds of arrhythmia were correctly classified with an accuracy of 80%. Ventricular flutter or ventricular fibrillation, an arrhythmia connected to cardiac arrhythmia, was not discovered.

Park and colleagues suggested a system for heartbeat detection and classification. The framework recognises heartbeats based on rhythmic properties and classifies them using KNN algorithm. P wave and QRS complex features, as well as an adaption of the classifier that uses locally weighted regression, a distance metric approach, were used to classify 17 forms of arrhythmia with a 96% accuracy [11].

Kiranyaz and others presented one-dimensional CNN, which is a dedicated CNN model that is trained for a particular patient and is utilised to accurately differentiate long-run ECG data streams. The suggested system is very non-specific due to its basic and fixed parameter nature. It is trained utilising training data that is both common and patient-specific. They were able to accurately classify VEB and SVEB arrhythmias with a 99% accuracy rate [12].

To categorise seven arrhythmias, Tae Joon Jun and others suggested a two-dimensional CNN with 11 layers. Each and every ECG beat is twisted into a two-dimensional grayscale image and fed into a CNN classifier. The evaluation was carried out using tenfold cross-validation method, which uses each ECG recording as test data, giving a 99% [13].

Pourbabaee and others introduced a five-layered CNN as an attribute extractor, and by combining the learned features with other classifiers such as support vector machine, MLP, and KNN, they were able to improve the execution of the network for patient screening when compared to the CNN classifier. With greater RC, CCR, and DOR values, it is shown that the KNN classifier produces greater results [14].

Chen and team have done the research on coronary artery disease to predict using machine learning and built model by taking medical image database and used naïve Bayes and support vector machine algorithm with pooling layers to early predict the coronary artery disease. Though they have gained an accuracy of 90%, neural networks made the machine learning to predict and give accurate results [15].

3 Dataset

The MIT-BIH Arrhythmia Database is a publicly accessible dataset that contains standard investigation material for heart arrhythmia identification. The MIT-BIH Arrhythmia Database is primarily utilised for medical and research purposes in the identification and analysis of various heart arrhythmias. This collection contains 48 recordings, each with half hour samples of ambulatory ECG data recorded on two channels. Out of the 4000 samples that were initially collected, 23 were chosen as normal and 25 were picked as having clinically verified arrhythmias. Out of total number of records from the digitised dataset, 87,554 records are chosen for training and 21,892 for testing the convolutional neural network model.

4 Data Preprocessing

The dataset is unbalanced if we look at the distribution of samples across labels. As a result, resampling samples from the lower-counted class to match the higher-counted class's count. To do so, we will divide the dataset into five sections, each comprising samples from a different class. After that, each dataset is resampled to get a total of 20,000 samples. The five datasets are then combined to create a balanced dataset with a total of 100,000 samples.

The heartbeats in the dataset are divided into five groups, as shown below

- Non-ectopic beats (normal beat)
- Supraventricular ectopic beats
- Ventricular ectopic beats
- Fusion beats
- Unknown beats.

Dataset's class labels are integers ranging (0–4). The class labels are one hot encoded using the keras.utils.to_categorical method because this is a classification task.

5 Proposed System

As the model proposed by Rajkumar and team is able to give accuracy of 93% [9], our proposed model was built by introducing coupled convolutional layers and also using dropout as a regularisation method and ReLU as an activation function to increase the accuracy and able to predict efficiently.

Proposed model consists of coupled convolution of three layers with 32, 64, 128 filters, respectively, and single maxpool layer is used in the model; we have dropout layer with the value 0.5, and the flatten layer next to it which converts the data to 1D.

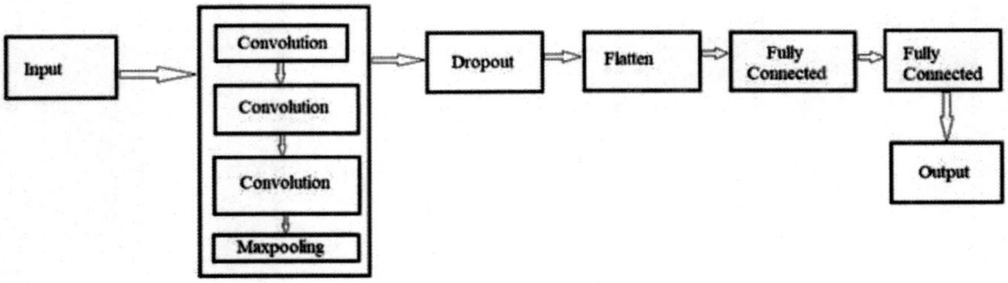

Fig. 1 Proposed model architecture

Also, there are three dense layers; in that, two layers are functions of ReLU activation with units 512 and 1024, respectively, and the third layer contains five nodes with softmax activation functions, corresponding to the five output class labels (Fig. 1).

When the outputs given to train are one hot encoded, the softmax activation method is applied as it converts a vector of n values into a vector with n values that total to 1, representing the probability of each class represented by the n values.

6 Classification Techniques

Classification is one of the supervised learning which can be used on unstructured and structured data. It divides the data to classes and then predicts which class data points will fall into. The data is classified and accurately predicted using CNN, AlexNet, GoogleNet, and VGG-16.

Although there are numerous categorisation algorithms available, it is impossible to say one is superior to the others. It is dependent on the application and the dataset supplied. If the classes are linearly separable, for example, then linear classifier is used. Some of the classification algorithms are:

6.1 CNN

Convolutional neural network (CNN) is one of the deep learning algorithms which takes image as input, assigns importance to different objects in an image, and differentiates between them. Comparing with other algorithms, ConvNet requires less preprocessing. While filters in earlier approaches were hand-engineered, these characteristics are with adequate training by ConvNet. The layers of pooling and nonlinearity do not have any parameters, while layers of convolutional and fully connected will have parameters [16]. Architecture of ConvNet was inspired by organisation of visual cortex and similar to neurons connectivity pattern in human brain. Receptive field, in this, individual neuron responds to stimuli in a small area of the visual field. Many similar fields can be collected on top of each other's to the full visual

Fig. 2 CNN architecture

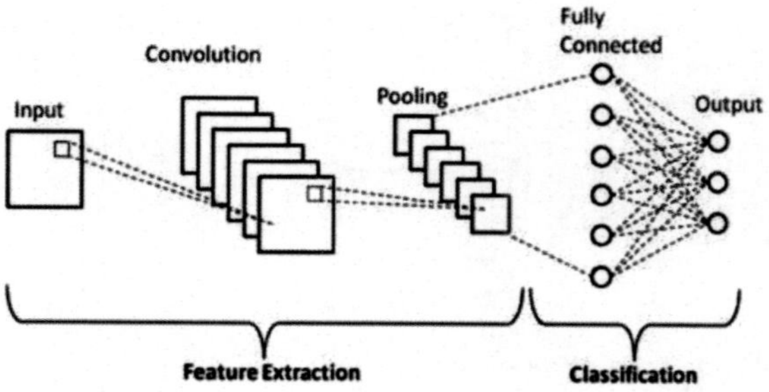

field extent. ConvNet can correctly capture the spatial and temporal connections of a picture using the suitable filters (Fig. 2).

7 Regularisation Techniques

Regularisation is an approach for enhancing the generalisation of a learning system by making minor changes to it. As a result, the performance of model on previously unknown data can be improved. Some of these techniques are:

1. Batch normalisation
2. Dropout.

7.1 Batch Normalisation

It is a method to train the deep neural networks which standardises each mini-batch's input of a layer. It is not only useful for increasing neural network performance, but it is also required for training deep convolutional networks [17]. This stabilises the process of learning and significantly reduces the no. of epochs of training needed to create the deep networks. It is a technique for standardising the inputs to a network that can be used on the activation of a previous layer or the input themselves. It reduces generalisation error by speeding up training by halving or bettering epochs in some circumstances. It also provides considerable regularisation (Fig. 3).

7.2 Dropout

One of the more intriguing sorts of regularisation procedures would be this. It also generates amazing outcomes, and as a result, it is the most often used regularisation technique in deep learning (Fig. 4).

Fig. 3 Batch normalisation

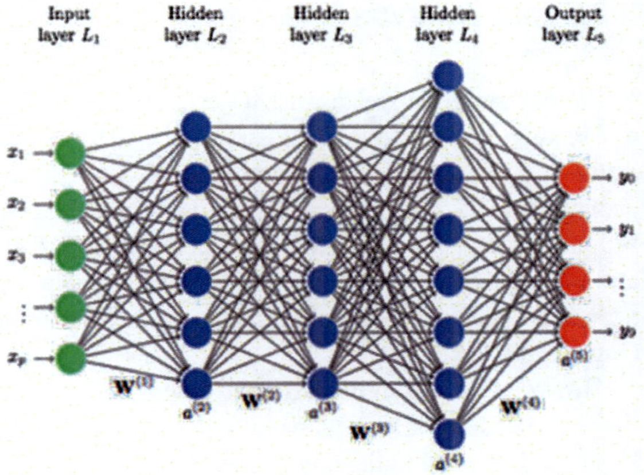

Fig. 4 Dropout

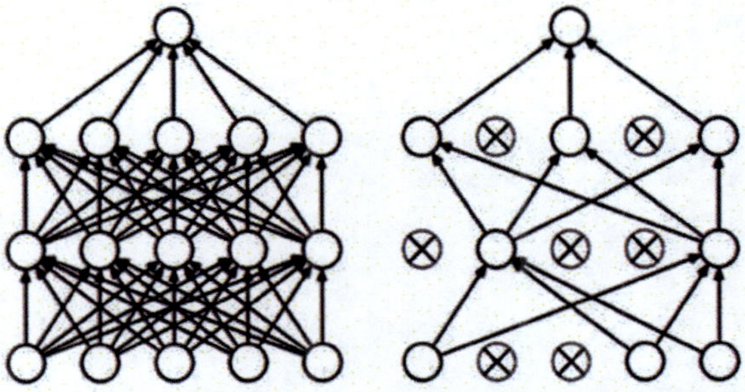

(a) Standard Neural Net (b) After applying dropout.

Dropout is a prominent approach for solving the DNN overfitting problem by discarding units at random throughout the training phase [18]. It selects the certain nodes randomly and eliminates them, together with all of their connections of incoming and outgoing, at each iteration. For result, every iteration will have nodes unique collection, which results in individual outputs. In machine learning, it is called as ensemble approach. As they capture more randomness, ensemble models frequently outperform single models. In the same way, dropout outperforms a traditional neural network model. The chance of deciding how many nodes must be eliminated is the hyperparameter of the dropout function. As we observed from above image, it can be applied on both layers like hidden and input. When we have a structure of big neural network, we can recommend dropout to describe the randomness for these reasons.

8 Result Analysis

In the dataset collected from MIT-BIH, our training data contains 87,554 records and testing data contains 21,892 records, and these are resampled to achieve balance in dataset. The CNN was trained using Google Colaboratory and achieved testing accuracy of 97.62% for our model using dropout mechanism and used coupled convolutional structure in addition to maxpooling. Total 10 epochs were used to build model and obtained training accuracy as 99.41%. We also build other models using batch normalisation and grouping layers having accuracies 96.54% and 94.7%, respectively (Figs. 5 and 6).

However, only training data was balanced, as we remember. This indicates that the test dataset remains uneven, which could cause the high accuracy value. The F1-score is a good indication of the model's usability in the situation of unbalanced datasets.

The model created from training data appears to be performing well in all classes. The performance of each class is shown in the confusion matrix (Fig. 7).

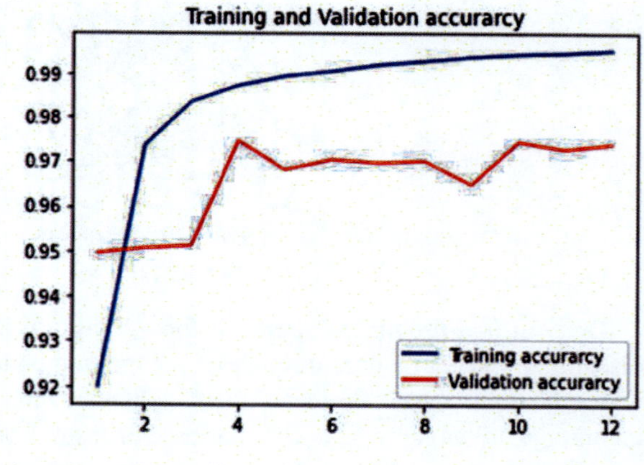

Fig. 5 Accuracy of training and validation for proposed model

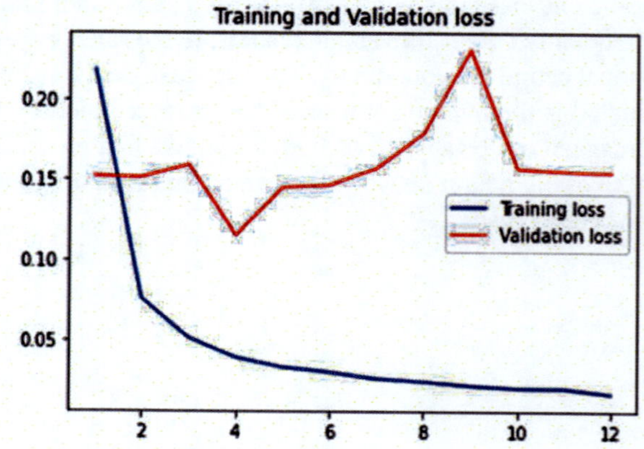

Fig. 6 Loss of training and validation loss for proposed system

Fig. 7 Confusion matrix

<matplotlib.image.AxesImage at 0x7f1f0b17f750>

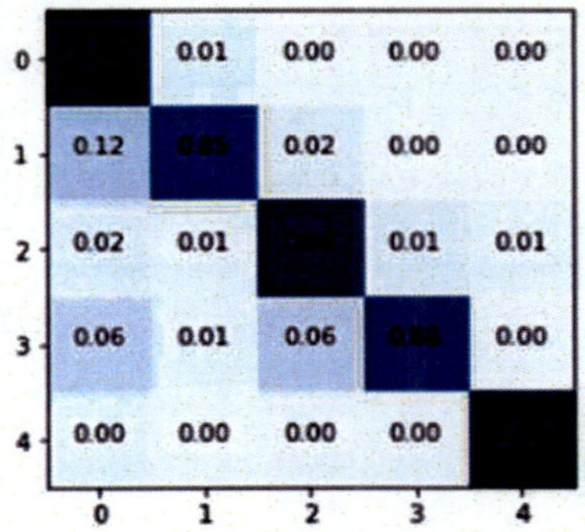

	convolutional layer	Regularization method	Accuracy
Proposed model	Coupled	Dropout	97.62%
Rajkumar [1]	normal	Batch normalization	93.6%

Fig. 8 Model comparison table

As we see test set is also highly imbalanced, but F1-score shows the good performance by scoring 0.88272 (Fig. 8).

9 Conclusion

Finally, by taking the model of Rajkumar [9], an optimised model is built by using coupled convolutional layers and dropout as regularisation method. When compared to previous model, proposed model makes an accuracy of 97.62% which is better than 93.6%. The proposed approach performed well in the experiments, indicating that it could be useful in clinical practice and also in many health-related gadgets like smart bands, watches.

References

1. Khan AH, Hussain M, Malik MK (2021) Arrhythmia classification techniques using deep neural network
2. Isin A, Ozdalili S (2017) Cardiac arrythmia detection using deep learning. Procedia Comput Sci 120

3. Ebrahimi Z, Loni M, Daneshtalab M, Gharehbaghi A (2020) A review on deep learning methods for ECG arrhythmia classification
4. Zhao W, Hu J, Jia D, Wang H, Li Z, Yan C, You T (2019) Deep learning based patient-specific classification of arrhythmia on ECG signal
5. Yildirim O, Talo M, Ciaccio EJ, Tan RS, Rajendra Acharya U (2020) Accurate deep neural network model to detect cardiac arrhythmia on more than 10000 individual subject ECG records
6. Izci E, Ozdemir MA, Degirmenci M, Akan A. Cardiac arrhythmia detection from 2D ECG images by using deep learning technique. https://doi.org/10.1109/TIPTEKNO.2019.8895011
7. Moody GB, Mark RG (2001) The impact of the MIT-BIH arrhythmia database. IEEE Eng Med Biol Mag 20:45–50. https://doi.org/10.1109/51.932724
8. Parvaneh S, Rubin J, Babaeizadeh S, Xu-Wilson M (2019) Cardiac arrhythmia detection using deep learning: a review
9. Rajkumar A, Ganesan M, Lavanya R (2019) Arrhythmia classification on ECG using deep learning
10. Sangaiah AK, Arumugam M, Bian G-B (2020) An intelligent learning approach for improving ECG signal classification and arrhythmia analysis
11. Park J, Lee K, Kang K (2013) Arrhythmia detection from heartbeat using k-nearest neighbor classifier. In: 2013 IEEE international conference on bioinformatics and biomedicine (BIBM). IEEE, pp 15–22
12. Kiranyaz S, Ince T, Gabbouj M (2016) Real-time patient-specific ECG classification by 1-D convolutional neural networks. IEEE Trans Biomed Eng 63(3):664–675
13. Jun TJ, Nguyen HM, Kang D, Kim D, Kim D, Kim Y-H (2018) ECG arrhythmia classification using a 2-D convolutional neural network. arXiv preprint arXiv:1804.06812
14. Pourbabaee B, Roshtkhari MJ, Khorasani K (2017) Deep convolutional neural networks and learning ecg features for screening paroxysmal atrial fibrillation patients. IEEE Trans Syst Man Cybern Syst 99:1–10
15. Chen JIZ, Hengjinda P (2021) Early prediction of coronary artery disease (CAD) by machine learning method—a comparative study. J Artif Intell 3(01):17–33
16. Albawi S, Mohammed TA, Al-Zawi S (2017) Understanding of convolutional neural network
17. Thakkar V, Tewary S, Chakraborty C (2018) Batch normalization in convolutional neural networks
18. Ko BS, Kim H-G, Oh K-J, Choi H-J (2017) Controlled dropout on deep neural networks

The Hybrid Vision Transformer Approach for Hyperpigmentation Nail Disease Detection

Krish Kumar, Chandan Kumar, Rahul Nijhawan, and Ankush Mittal

Abstract Melanonychia is brown or black pigmentation of the nail unit caused by melanin deposition in the nail plate. Melanonychia commonly presents as pigmented band arranged lengthwise along with the nail unit. It typically occurs in dark-skinned people. Identifying nail diseases is still an uninvestigated and challenging struggle in itself. This paper proposes vision transformer (ViT) method which is recently nominated as an alternative of convolutional neural networks (CNNs) and show promise in computer vision task. As far as our knowledge is concerned, this is the first model which has applied a vision transformer for hyperpigmentation nail disease detection. ViT applies the transformer architecture with self-attention to the sequence of image patches. These patches are treated as tokens (words) and then sequenced as linear embeddings as the input without using a convolution layer. This work shows ViT is a feasible tool for the classification of melanonychia nail disease. Due to the absence of a good quality dataset, created a new dataset for examining the performance of the proposed framework. The result indicates that ViT can identify melanonychia nail diseases with an accuracy of 99% on the test images. Also, tested dataset on other state-of-the-art algorithms to find the distinction between proposed approach and accuracy obtained by the current state-of-art algorithm (SVM, random forest, KNN, Naive Bayes, etc.).

Keywords Vision transformer (ViT) · Convolutional neural network (CNN) · Classification · Nail disease · Feature extraction

1 Introduction

A type of protein called keratin helps to form the cells to make the tissue in nails. Keratin plays a significant role in protecting the nails from damage and protects

K. Kumar · C. Kumar
Roorkee Institute of Technology, Roorkee, Uttarakhand, India

R. Nijhawan (✉) · A. Mittal
University of Petroleum and Energy Studies, Dehradun, Uttarakhand, India

S. Shakya et al. (eds.), *Proceedings of Second International Conference on Sustainable Expert Systems*, Lecture Notes in Networks and Systems 351,
https://doi.org/10.1007/978-981-16-7657-4_4

tissues of the toes and fingers. A small change in nail discoloration or defects or any other thing talks about the body's health [1]. Nail diseases have several classifications, and each has its signs and symptoms that may relate to other medical conditions. Some particular types of nail problems like discoloration, hyperpigmentation, malignancies, changes in the nail texture, and other abnormalities hint toward diabetes, psoriatic arthritis, anemia, fungal infection, melanoma, liver disease, malnutrition. For this reason, nail disease detection at early stages is crucial [2].

The nail disease does not depend on age as it may occur at any age. A person may suffer from nail diseases at any age [3]. Mostly, a person whose age lies between 60 and 70 may suffer from nail diseases. Some nail diseases are hazardous, like subungual melanoma (SM), which has an estimated average 5-year survival rate of 16–80%, is not easy to find in their early stages [4]. A nail disease called "Melanonychia" is a brown or black-pigmented nail plate caused by the increased melanin deposition in the nail plate. Longitudinal melanonychia (LM) commonly occurs in more darkly pigmented persons. Recently in China, it was found to be 0.8% in both males and females, while in Poland, it was found to be 19.46%, with the mean age of being 49 years [5]. LM and SM are alike in some aspects, like (1) in hand, and each occurs mainly in thumbs, index finger, or both. (2) Both occur mainly in dark-skinned people. LM is just like an early indication of SM.

Nail disease has become a more focused research topic now; many researchers are now interested in doing deep research to detect nail diseases [6]. Many algorithms distinguish between regular nails and defective nails, like artificial neural networks (ANN) and convolutional neural networks (CNN), which are a part of deep learning. Apart from deep understanding, some other states-of-art models are compared with the deep learning algorithms in terms of quality [7]. The models like support vector machines (SVM), logistic regression (LR), random forest (RF), KNN, and tree [8].

Although promising, CNN-based architecture has some limitations. Firstly, since convolution performs on a fixed window, the model is more concentrated on recognizing local information than long-range spatial relationships between parts of the images and the images as a whole [9]. Other challenges, such as ignoring some local information through max-pooling, inspired researchers to find new methods [10]. Influenced by the success of NLP, recent researchers attempted to combine the self-attention mechanism into CNNs for computer vision tasks [11], some putting back the convolutions entirely. The later models, while hypothetically productive, have not yet been mapped effectively on current hardware accelerators due to the use of specially designed attention patterns [9]. Therefore, in big-scale image recognition, classic ResNet-like architectures are still state-of-the-art [12–14].

The challenges faced by the model are inadequacy of fair quality of nail image dataset. The structure and appearance of nails fluctuate for each individual. To get an efficient result, need a proper and well-organized quality dataset. Withholding all these challenges in mind, come up with an entirely based transformer architecture model called vision transformer (ViT), which has been successfully applied in image classification tasks recently. ViT has recently been proposed as an alternative to CNN's and used to resolve computer vision tasks [9].

ViT is a modified version of the transformer and needs sequence data to work well. ViT first breaks the images into patches, and these image patches are treated as a sequence token and then flatten the patches, generate lower-dimensional linear embedding from the flattened patches, for keeping the information in order and meaningful, positional embedding is done and supply the sequence as an input to the transformer encoder. The whole input sequence is sent into the model, and the attention mechanism will take care of which part needs to focus on.

ViT does not involve convolution layers and converts an image as a sequence of patches to overcome challenges with CNN-based approaches, such as locality and translation invariance. Moreover, the method is dependent on hardware and can be trained on the public ImageNet dataset with much fewer resources than classic CNN ResNet-based architectures [9].

This work demonstrates that ViT is a feasible tool for detecting hyperpigmented nails. The model is pre-trained on the ImageNet-21k dataset and the JFT-300M dataset [9]. To the best of our knowledge now, there are no other works that have explored the use and efficacy of ViT on the classification of medical images, especially for melanonychia (hyperpigmentation) classification.

This paper has been organized in the following order: Introduction (Sect. 1), Literature Review (Sect. 2), Data Collection (Sect. 3), Methodology, Proposed Approach (Sect. 4), Results and Discussion (Sect. 5), and Conclusion (Sect. 6).

2 Literature Review

Recently, many researchers have been presenting attentiveness in identifying various diseases using computer vision techniques. Cicero et al. [15] proposed a deep learning model to classify melanoma disease and uses ResNet as a feature extractor, changes the last three network layers to train them for the classification purpose. Nijhawan et al. [2] have used convolution neural networks (CNN) AlexNet architecture, pre-trained on the ImageNet dataset. Using a combination of three different CNN's to get the various appearance of the human nail, the highest accuracy achieved is 84.58%. Indi and Gunge [16] have presented an image processing (IP) to identify abnormal nails using an image as an input and output the predicted diseases based on their color changes. The average accuracy achieved is 65%. Gandhat et al. [17] have presented a model that extracts a significant feature from the nail portion by using digital image processing technique and examining the extracted nail part to detect the nail disease based on their color. Kenneth et al. [18] proposed a convolution neural network to classify multiple lesion skin disease, non-standardized images were trained on the VGG-16 model. Saranya and Ranichitra [19] have proposed three different segmentation techniques: thresholding, K-means clustering, and watershed segmentation to get the infected region of the nail. The image input has preprocessed by using the union of median and standard filters. Kumuda [20] used segmentation using the watershed method that workout the maxima and minima properties of marker-controlled watershed principles.

Fig. 1 Sample image of nail taken from our dataset

Table 1 Dataset collected

Nail types	Number of images
Hyperpigmentation	100
Normal nails	100

3 Data Collection

The most difficult challenge faced is to gather the dataset. Deep learning requires a robust dataset for training the models. Chances of getting more accuracy improves if the model is trained in a better way. Due to the unavailability of any open-source dataset, we created an image dataset that consists of two classes of images keeping a set of 100 images of each type (Fig. 1).

We have taken a good quality image of two different nails with proper color, shape, and size. Further, divided dataset into two halves: training data consisting of 75% of the dataset and the remaining for testing purposes.

Table 1 shows the organization in the dataset. This table describes the number of images in the dataset concerning different nail classes.

4 Methodology

Used a mechanism similar to the transformer but having a slight deviation from the original transformer, which is pre-trained on the ImageNet-21k dataset and the JFT-300M dataset [9]. First, the input images are divided into patches and then sequenced as linear embeddings as the transformer's input. Now, the image patches are treated as tokens used to train the network in a supervised manner. Used a machine Intel(R) Core (TM) i5-8265U CPU @ 1.60 GHz processor and RAM of 8 GB to implement the proposed framework. Also, implemented other algorithms like SVM, random forest, KNN, and Naive Bayes. in this dataset to compare results with the proposed approach.

4.1 Proposed Vision Transformer Approach

Over the past decade, deep convolutional neural networks (CNN) used to resolve computer vision problems. RNN and CNN dependent on an encoder-decoder scheme stand over the transduction models [21]. Recurrent models do not support parallelization during training due to their linear nature, which causes a problem for learning long-term memory dependencies. Since 2012, CNN-based architecture like VGG, Resnet has solved complex image recognition on ImageNet [22]. However, CNN has some drawbacks, and it requires many layers to capture long-term dependencies in the sequential data. Additionally, max-pooling can lead to a loss of information [23].

When focusing on NLP, attention-based architecture like transformer [24] has been designed to handle language-related work in an efficient way like sentimental analysis and language translation [25]. Because of outstanding achievement, the computer vision community is showing their interest in the transformer to use similar self-attention models for computer vision tasks [26]. The first deployment of ViT for computer vision tasks was launched by Google and applied ViT on an in-house dataset of 300 million images from the JFT dataset [27], making improvements to image recognition prototypes like ImageNet.

The proposed vision transformer (ViT) approach to classify the nail disease. Recently, the vision transformer (ViT) [9] was preferred as the first deep neural network algorithm for large-scale computer vision datasets and applies the transformer architecture with self-attention to sequences of image patches without using convolution layers. It is the first type of model which predicts entirely on attention.

How vision transformer acts as an alternative to CNN

CNN works very well in computer vision tasks. CNN works on ideas where image consists of pixels, one pixel is dependent on neighboring pixels and the next pixel is dependent on the next neighbor pixels. CNN uses filters on the patches of the image to pull out important features [28]. And the model is trained only on the extracted feature but not on the all feature of each pixel of images. However, if the entire image data are trained on the model rather than just training on the important extracted features, the chances of getting model performance are higher. This is happening in the case of a vision transformer. CNN fails to encode relative spatial instruction. This can be the reason that vision transformer performs better than CNN. CNN only looks for the features whether it is present or not in the image pixels. It does not care about the relative position of features.

How vision transformer works on the dataset

Split the hyperpigmented image into equal-sized patches. The patches of images are treated in the same way as tokens (words) in the NLP application. A 2D flattening is done on image patches to reshape them into a vector format that acts as the transformer's input. Then, vectorized image patches are converted to a linear patch embedding. At the same time, a position embedding is inserted into each of these embedding patches and build a compatible network to preserve the positional information of each embedding patch. The transformer encoder [24] comprises multiple encoder blocks, and each block consists of various layers of multiheaded self-attention mechanisms [29].

The ViT architecture is shown in Fig. 2.

Mathematical intuition behind vision transformer

The 1D sequence of input is required in the input layer of the normal transformer. To process the 2 dimensional (2D) image, the image $Y \in S^{I \times X \times D}$ is reshaped into flattened image patches $Y_q \in S^{O \times (Q \times Q \times D)}$, where (I, X) is the original image size; D is the channel count; (Q, Q) is the x and y-coordinates of image patches, and $O = IXQ^{-2}$ is the gives the numbers of patches. The image patches that have flattened are now transformed into E-dimensional vectors through a linear projection and operated in a fixed dimension employing all layers. After projection, positional embedding is done to the patches and combined classification tokens to make it the input sequence of the transformer encoder. The encoder block comprises the multi-layer perceptron (MLP) and multi-head self-attention layers. The self-attention, known as scaled dot-product attention, first of all, calculate a queries set $R = YX_R$, a set of keys $L = YX_L$, and a values set $W = YX_W$, in which $Y \in S^{(O+1) \times E}$ is an input sequence; $X_R \in S^{E \times e}$, $X_S \in S^{E \times e}$, $X_W \in S^{E \times e}$ are the weights which are capable of being trained; and e is the size of the vector. The computation of self-attention is as follows:

$$\text{Attention}(R, L, W) = \text{softmax}\left(\frac{RL^U}{\sqrt{e}}\right)W \quad (1)$$

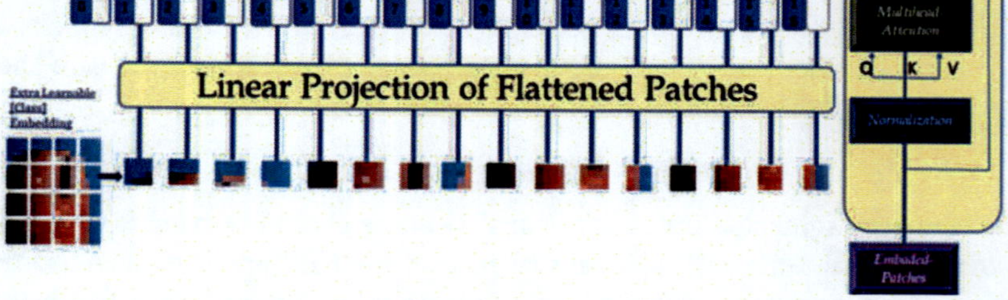

Fig. 2 Vision transformer architecture

Each row of the matrix has passed through the softmax function. In MSA, i heads are appended to self-attention as follows:

$$\text{MSA}(R, L, W) = \text{join}(\text{head}_0, \ldots, \text{head}_{i-1})X \tag{2}$$

$$\text{head}_{i-1} = \text{Attention}(R_{i-1}, L_{i-1}, W_{i-1}) \tag{3}$$

Each head supplies a sequence of size $(O + 1) \times e$. These i sequences are reorganized into an $(O + 1) \times ei$ sequence and MLP reorient into $(O + 1) \times E$. Now, how the transformer encoder treated the image is given as follows:

$$z_0 = \left[y_{\text{class}}; \text{MLP}\left(y_q^1\right); \ldots; \text{MLP}\left(y_q^o\right)\right] + F_{\text{pos}} \tag{4}$$

$$z_m' = \text{MSA}(\text{Norm}(z_{l-1})) + z_{l-1} \tag{5}$$

$$z_m = \text{MLP}\left(\text{Norm}\left(z_m'\right)\right) + z_m' \tag{6}$$

$$g = \text{Norm}\left(z_M^0\right) \tag{7}$$

This was the mathematical background of vision transformer as to how the model processes the image and splits it into patches and makes them flatten and does some vectorization on patches and passes the input sequence tokens to the encoder.

5 Result and Discussion

In this section, the final result has been shown. It is observed that the proposed approach obtained a maximum accuracy as compared to other deep learning models. We have taken four scenarios for accuracy analysis are shown below.

In scenario 1, we used Inception V3 architecture which is the extended version of the inception family that makes various improvements like label smoothing, 7 * 7 factorized convolution. It uses an auxiliary classifier to transmit label information toward the lower network to improve the convergence of a very deep network [30]. Using this architecture, we achieved an accuracy of 94.5% with SVM, 95.9% with neural networks, 89.0% with random forest, 92.4% with Naive Bayes, 82.8% with AdaBoost.

In scenario 2, we used VGG-16 CNN architecture. Instead of having a large number of hyper-parameters, we concentrated on having convolution layers of a 3 × 3 filter with a stride one and using the same padding and max pool layer of a 2 × 2 filter of stride 2. Convolution and max pool layer consistently followed throughout the structure, and at the end, it has two fully connected layers (FC) followed by

SoftMax for the outcome [31]. We achieved an accuracy of 82.8% with SVM, 89.7% with random forest, 80.7% with AdaBoost, 97.2% with Naive Bayes.

In scenario 3, we used VGG19 architecture; we have formulated a 19 layers deep model consists of 16 convolution layers, three fully connected layers, 5 max-pool layers, and 1 SoftMax layer [32]. Fixed-size (224 * 224) RGB images were inserted as input and used kernels of (3 * 3) sizes with a stride of 1 pixel. The dimensional padding used to structure the image's resolution, and max-pooling over a (2 * 2) pixel with a stride of 2 is performed, followed by a rectified linear unit (ReLU) non-linearity to make the model better to classify and improve computational time [33]. Using this deep model technique, get the maximum accuracy of 92% with SVM, 91.7% random forest, 94.5% with Naive Bayes, 87.6% with AdaBoost, 96.6% with neural networks.

In scenario 4, we proposed transformer's new approach, which completely removes recurrence and convolution. It splits the image into patches. These patches are treated as tokens (words) and then sequenced as linear embeddings as the input. Transformers use self-attention, a highly efficient technique that makes it possible to update the embeddings of every word in the input text in parallel. By using this new approach, we get 99% top-5 accuracy on test data.

When comparing the proposed model with deep convolutional neural networks, it has been observed that the result is better in ViT. It is noticed that if the images

Table 2 Accuracy evaluation

Feature extractor	Models	AUC	Accuracy	Precision
VGG-19	SVM	0.975	0.924	0.931
	Random forest	0.981	0.917	0.918
	AdaBoost	0.877	0.876	0.879
	Naive Bayes	0.988	0.945	0.945
	KNN	0.967	0.897	0.915
VGG-16	Random forest	0.958	0.897	0.897
	Neural network	0.985	0.972	0.972
	SVM	0.959	0.828	0.864
	Naive Bayes	0.984	0.972	0.972
	KNN	0.969	0.897	0.903
	AdaBoost	0.808	0.807	0.811
Inception—V3	KNN	0.976	0.952	0.952
	Random forest	0.958	0.890	0.894
	Naive Bayes	0.988	0.924	0.925
	SVM	0.989	0.945	0.945
	AdaBoost	0.828	0.828	0.828
	Neural network	0.988	0.959	0.960
Transformer (PA)	Vision transformer	0.997	0.99	0.986

are tilted and rotated with some degree, then CNN's generally faces complication in classifying the image. In the convolutional layer, all minute detail remembrance is done by high-level neurons [34]. All features are inspected by high-level neurons for its presence/absence. The process of checking whether features are current is executed by striding the image. During this process, ConvNet completely missed out on all the information about the position and combination of the constituent. Information is passed on further to a neuron that might not be able to classify the image. To resolve this problem, we need data augmentation to train the dataset by flipping the image and rotating it by a small amount. We use the same concept in the proposed approach to train the model so that it can give better accuracy in the worst case. Figure 3 shows the results of all four scenarios with the corresponding model.

This figure shows the accuracy of advanced CNN architecture (VGG-19, VGG-16, Inception V3) and the vision transformer (proposed approach) using a bar graph. In advanced CNN architecture, using various algorithms like SVM, KNN, Adaboost, and random forest. The *y*-axis shows the score in terms of the fraction of the corresponding algorithm used in CNN architecture.

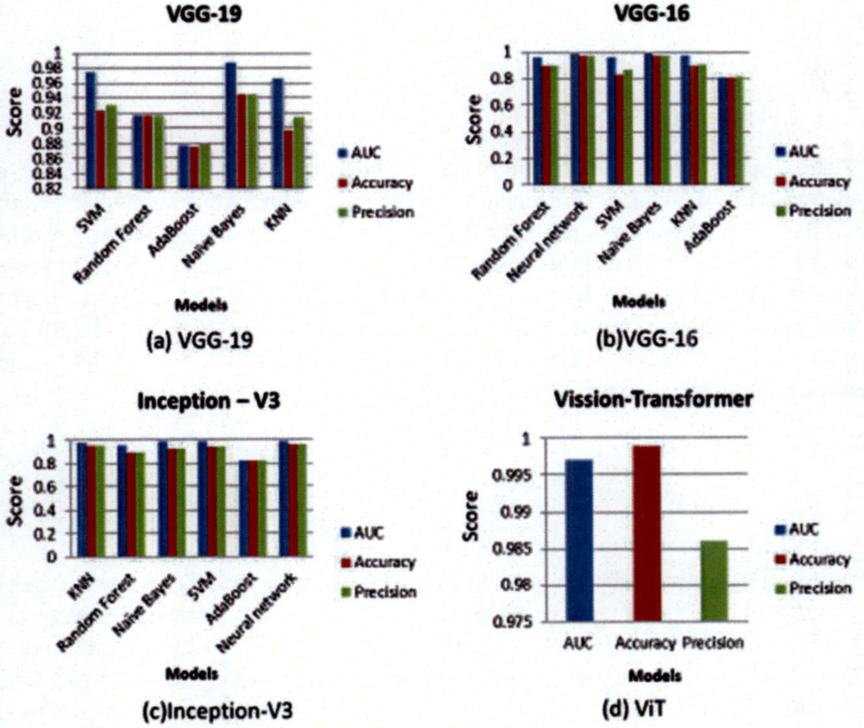

Fig. 3 Binary classification results of melanonychia disease, **a** VGG-19, **b** VGG-16, **c** inception-V3, **d** vision transformer (ViT)

Why vision transformer

ViT uses self-attention, which allows integrating information across the entire image, even in the lowest layer. ViT provides high accuracy with less computation time for training. ViT does not include a convolution layer. The primary idea behind the transformer model is self-attention. It gives the potential to figure out the interdependence between inputs. The variable-sized information is handled by the model with transformer architecture using stacks of self-attention instead of CNNs and RNNs. Applying self-attention between the image pixels in the transformer without the participation of CNN is a big challenge.

ViT has overcome this problem by splitting images into small patches (16 × 16). The transformer encoder module comprises a multi-head self-attention (MSA) layer and a multi-layer perceptron (MLP) layer. The multi-head self-attention layer split inputs into several heads so that each head can learn different levels of self-attention. The outputs of all the heads are then concatenated and passed through the multi-layer perceptron. When using many parameters and when trained with lots of data, these models produce better results than SOTA in image classification or object detection with way simpler models and faster to prepare.

5.1 Limitation of ViT

There is some limitation of ViT, and a natural question arises: can we further improve the capability by making it deeper, just like CNN? Although it seems very clear, the answer may not be minor since ViT works differently from CNN as it works on self-attention and without any involvement of the convolution layer. To settle this question, [35] goes into detail about the scalability of ViTs. As the number of transformer blocks increases, the model performance does not refine accordingly. So, stacking more transformer blocks is inefficient at enhancing ViT model. The feature maps become identical in the top layers of deep ViT models.

6 Conclusion

We have implemented the direct application of vision transformers for the classification of nail disease. The proposed model uses complete self-attention transformer-based architecture that does not use any convolutional layers, which helps us identify the "Melanonychia" nail diseases. This proposed model performs better than advanced CNN architecture (Inception V3, VGG-16, VGG19). In addition, we used data augmentation to enhance classification accuracy. Due to the unavailability of the proper and good quality dataset, this task was challenging to accomplish. While these initial results are enhancing the expectation for future success, many challenges remain. One is to classify other diseases like hair diseases, skin diseases, and teeth diseases, and many other nail diseases using computer vision (ViT).

Moreover, this application can be helpful in the early detection of the nail to take precautions and medication on time. This work can take part in the advancement of the dermatologist department to analyze diseases from patient data to get good treatment to get rid of these types of diseases.

References

1. Goel A, Nijhawan R (2020) A deep learning approach for classification of onychomycosis nail disease. In: Proceedings of ICETIT 2019. Springer, Cham, pp 1112–1118
2. Nijhawan R et al (2017) An integrated deep learning framework approach for nail disease identification. In: 2017 13th international conference on signal-image technology & internet-based systems (SITIS). IEEE
3. Wollina U et al (2016) The diagnosis and treatment of nail disorders. Dtsch Arztebl Int 113
4. Dhaya R (2021) Analysis of adaptive image retrieval by transition Kalman filter approach based on intensity parameter. J Innov Image Process (JIIP) 3(01):7–20
5. Sobjanek M, Michajlowski I, Wlodarkiewicz A, Roszkiewicz J (2014) Longitudinal melanonychia in a northern Polish population. Int J Dermatol 53:e41–e42
6. Muddasani S et al (2021) Nutrition and nail disease. Clin Dermatol
7. Nijhawan R, Sharma H, Sahni H, Batra A (2017) A deep learning hybrid CNN framework approach for vegetation cover mapping using deep features. In: 2017 13th international conference on signal-image technology and internet-based systems (SITIS). IEEE, pp 192–196
8. Nijhawan R, Das J, Balasubramanian R (2018) A hybrid CNN + random forest approach todelineate debris-covered glaciers using deep features. J Indian Soc Remote Sens 46(6):981–989
9. Dosovitskiy A et al (2020) An image is worth 16x16 words: transformers for image recognition at scale. arXiv preprint arXiv:2010.11929
10. Sabour S, Frosst N, Hinton GE (2017) Dynamic routing between capsules. arXiv preprint arXiv:1710.09829
11. Wang X et al (2018) Non-local neural networks. In: Proceedings of the IEEE conference on computer vision and pattern recognition
12. Mahajan D et al (2018) Exploring the limits of weakly supervised pretraining. In: Proceedings of the European conference on computer vision (ECCV)
13. Xie Q et al (2020) Self-training with noisy student improves imagenet classification. In: Proceedings of the IEEE/CVF conference on computer vision and pattern recognition
14. Kolesnikov A, Beyer L, Zhai X, Puigcerver J, Yung J, Gelly S, Houlsby N (2005) Big transfer (BiT): general visual representation learning. In: ECCV, 2020. Enrollment in local colleges
15. Cıcero F et al (2016) Deep learning and convolutional neural networks in the aid of the classification of melanoma. Proc SIBGRAPI
16. Indi TS, Gunge YA (2016) Early-stage disease diagnosis system using human nail image processing. IJ Inf Technol Comput Sci 7:30–35
17. Gandhat S et al (2016) Study and analysis of nail images of patients. Int J Comput Appl 143(13)
18. Thomsen K et al (2020) Deep learning for diagnostic binary classification of multiple-lesion skin diseases. Front Med 7:604
19. Saranya V, Ranichitra A (2017) Image segmentation techniques to detect nail abnormalities. Scholar 2:1
20. Kumuda S (2017) An image pre-processing method for fingernail segmentation. In: 2017 IEEE 2nd international conference on signal and image processing

21. Rawat SS, Bisht A, Nijhawan R (2019) A deep learning based CNN framework approach for plankton classification. In: 2019 fifth international conference on image information processing (ICIIP). IEEE
22. Deng J et al (2009) Imagenet: a large-scale hierarchical image database. In: 2009 IEEE conference on computer vision and pattern recognition. IEEE
23. Chhabra HS, Srivastava AK, Nijhawan R (2020) A hybrid deep learning approach for automatic fish classification. In: Proceedings of ICETIT 2019. Springer, Cham, pp 427–436
24. Vaswani A et al (2017) Attention is all you need. arXiv preprint arXiv:1706.03762
25. Gupta S et al (2019) Classification of lesions in retinal fundus images for diabetic retinopathy using transfer learning. In: 2019 international conference on information technology (ICIT). IEEE
26. Mease PJ et al (2021) Association of nail psoriasis with disease activity measures and impact in psoriatic arthritis: data from the corrona psoriatic arthritis/spondyloarthritis registry. J Rheumatol 48(4):520–526
27. Sun C et al (2017) Revisiting unreasonable effectiveness of data in deep learning era. In: Proceedings of the IEEE international conference on computer vision
28. Zhu C-Y et al (2021) A deep learning based framework for diagnosing multiple skin diseases in a clinical environment. Front Med 8
29. Rezvantalab A, Mitha S, Khademi A (2021) Alzheimer's disease classification using vision transformers
30. Szegedy C et al (2016) Rethinking the inception architecture for computer vision. In: Proceedings of the IEEE conference on computer vision and pattern recognition
31. Varshni D, Thakral K, Agarwal L, Nijhawan R, Mittal A (2019) Pneumonia detection using CNN based feature extraction. In: 2019 IEEE international conference on electrical, computer and communication technologies (ICECCT), pp 1–7. https://doi.org/10.1109/ICECCT.2019.8869364
32. Chandra D, Rawat SS, Nijhawan R (2019) A machine learning based approach for progeria syndrome detection. In: 2019 4th international conference on information systems and computer networks (ISCON). IEEE
33. Nijhawan R et al (2019) A futuristic deep learning framework approach for land use-land cover classification using remote sensing imagery. In: Advanced computing and communication technologies. Springer, Singapore, pp 87–96
34. Nijhawan R, Sharma H, Sahni H, Batra A (2017) A deep learning hybrid CNN framework approach for vegetation cover mapping using deep features. In: 2017 13th international conference on signal-image technology & internet-based systems (SITIS), pp 192–196. https://doi.org/10.1109/SITIS.2017.41
35. Zhou D et al (2021) Deepvit: towards deeper vision transformer. arXiv preprint arXiv:2103.11886
36. Ramachandran P et al (2019) Stand-alone self-attention in vision models. arXiv preprint arXiv:1906.05909

COVID-19 Fake News Detection Using GloVe and Bi-LSTM

Chaitanya Kulkarni, P. Monika, S. Shruthi, M. S. Deepak Bharadwaj, and D. Uday

Abstract Fake news detection is a key use case of Natural Language Processing and Machine Learning. With the advancement of technology, electronic content has become more significant and extensively used than ever before, resulting in a revival in the spurious news as well as tweets that seek quick attention. It is critical to achieve early detection and identification of misinformation in social media. In these difficult times of COVID-19 where the whole world is fighting a global pandemic, it is very important to provide people with genuine news and updates about the prevailing situation. The Twitter is flooded with tweets about COVID-19 and related news, but not all the information is factual and full of veracity. To classify this issue, a Deep Learning model is being proposed to classify misinformation about COVID-19 and related news. Pre-trained state-of-the-art GloVe model was used in the proposed method for converting tokens into vectors. Deep Learning models like RNN, LSTM, and Bi-LSTM were tested to obtain better results after preprocessing the data using basic NLP preprocessing techniques. Bi-LSTM model got a training accuracy as high as 99%. All the results are tabulated and cogently laid out in the latter section.

Keywords COVID-19 · GloVe · LSTM · Bi-LSTM · RNN · NLP

1 Introduction

In the digital era of twenty-first century, social media's popularity is skyrocketing day by day. Digital networks like Facebook, Instagram, and Twitter are rapidly gaining new users at an alarming rate. These platforms provide easy access to lot of individuals to both seek and provide information. Social media has made the lives easy for lots of its consumers by providing easy accessibility to information. But on the other hand, social media provides a fruitful breeding opportunity for misinformation and propaganda. Facebook with more than two billion active users has around 23% of users sharing misinformation either purposefully or unknowingly. According to [1],

C. Kulkarni (✉) · P. Monika · S. Shruthi · M. S. Deepak Bharadwaj · D. Uday
Dayananda Sagar College of Engineering (DSCE), Bangalore 560078, India

S. Shakya et al. (eds.), *Proceedings of Second International Conference on Sustainable Expert Systems*, Lecture Notes in Networks and Systems 351,
https://doi.org/10.1007/978-981-16-7657-4_5

fake news can be defined as "the online publication of intentionally or knowingly false statements of fact." Fake news thrives on the spread of scandals, sensationalism, hoaxes, and false stories as a result of news pieces being shared on social media [2]. Intentional spreading of harm is debated, while social media is being used by political parties and other individuals in spreading propaganda and biased news for their own benefits.

Recent burgeoning of use of social platforms is being used as a vehicle to transport fake news and has significantly increased the risks on the individuals as well as organization on whom news is being shared. Due to the variety and concealment of deceptions, detecting misleading news is difficult if not impossible. Fake news has the potential to have a negative impact as well as inflict harm. By modifying the information streams that are used for news consumption, it alters an individual's decision-making and distorts their perceptions of genuine occurrences. At the organizational level, the impact is more negative because it puts their brand names at danger and may have an impact on how their product or service will be consumed [3]. The problem is exacerbating with the introduction of bots which share news proliferating it without verification. Therefore, it is the need of hour to detect misinformation and fake news from social media and other platforms.

The increased requirement for automated techniques of distinguishing legitimate news from fake news among the enormous volume of data is being addressed by recent breakthroughs in news verification systems. There are numerous techniques to deal with the problem of social media misinformation. Statistical approaches are used to determine the relationship between various aspects of the information, analyze the source of the information, and examine distribution trends. Untrustworthy content is classified using machine learning techniques, and the accounts that share it are analyzed. Various approaches concentrate on establishing strategies for information authentication as well as specific case studies.

COVID-19 [4] was declared as a global pandemic by WHO, there are more than 190 million people affected by the novel SARS-CoV-2 virus [5]. In these hard times, it was seen that people used social media and other online platforms to seek for help and communication as the system failed in many parts of the world. Social media platforms like Twitter were extremely helpful to many individuals seeking help and also individual who helped people for a cause. Twitter was also used to spread awareness and recent news about the corona virus and pandemic [6]. Lot of individuals and organizations intentionally or un-intentionally tried to spread unverified or fake news through these platforms, which is a matter of great risk and concern as Twitter is an active platform to share news and updates which people relied upon. Therefore, it is very important to build a fast and reliable [7] model to classify news.

2 Literature Survey

Bi-LSTM, Support Vector Machine, Naive Bayes, and logistic regression were employed in an ensemble-based approach by Shushkevich and Cardiff in [8]. On the given fake news dataset, their combination of logistic regression and Naive Bayes models gave results that were within 5% of state-of-the-art findings.

Sharif et al. [9] have experimented with several algorithms such as Bi-LSTM, SVM, CNN, and CNN + Bi-LSTM with embedding techniques like TF-IDF and Word2Vec, with the best results coming from SVM with TF-IDF technique. Using five-fold five-model cross validation, Li et al. in [10] suggested an ensemble model comprising of multiple pre-trained models such as BERT, ERNIE, and RoBERTa. In addition, their pseudo label approach substantially improved overall model performance.

Das et al. [11] proposed a heuristic-driven uncertainty-based ensemble framework for fake news detection. They employed an ensemble model consisting of pre-trained models, supplemented by a statistical feature fusion network and a novel heuristic method that incorporated numerous variables available in tweets or news articles as statistical features such as URL domains, username handles, and source. FakeNewsNet and COVID-19 fake news dataset were used to prove the effectiveness of the model. They obtained the best result of F1 score of 0.9073 on FakeNewsNet dataset and 0.989 on the COVID-19 dataset.

3 Dataset Description

The COVID-19 fake news dataset was collected from [12] as a part of challenge provided for a competition held by the organizers. The dataset is collected from various Web applications and social media and they have been manually verified. The dataset labeled as "real" was collected from verified sources which gave factual and right information about COVID-19. The "fake" labeled data were collected from tweets, posts, and articles which were specious information. The dataset is balanced containing 3360 real messages and 3060 messages labeled as "fake." Figure 1 shows the count plot of the data labels. It pictorially represents the counts of data in each category.

Fig. 1 Count plot of COVID-19 fake news dataset

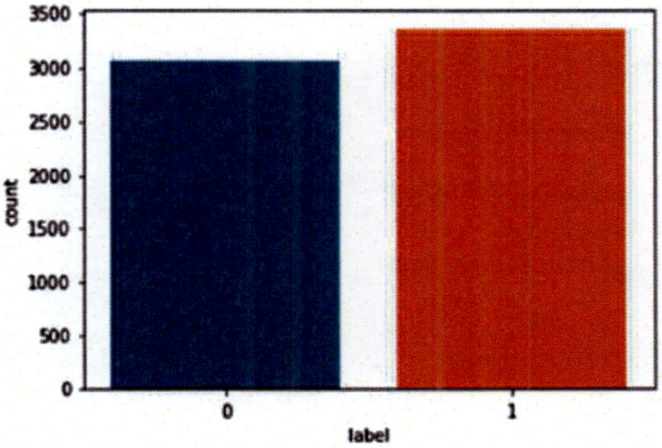

The vocabulary size of the dataset is around 18,100 with more than one lakh total words. The number of words used in fake dataset was 41,977, while real dataset had 64,228 total words in it. The top words used in "fake" labeled messages were {"covid-1375," "coronavirus-1305," "people-295," "trump-227," "pandemic-224"}. The top words used in "real" labeled messages were {"covid-2359," "cases-1565," "new-868," "tests-633," "deaths-522"}.

4 Proposed Method

We propose a three-stage process starting with preprocessing where the raw data collected from the Website is cleaned and prepared for feature extraction, where the cleaned text is converted to vectors or word embeddings, where pre-trained [13] GloVe model is used with 42B tokens and 1.9M vocab. The dimension for each word was chosen to be 300. The 300d vectors were passed through various deep learning models such as conventional recurrent neural network (RNN) [14], long short-term memory (LSTM) [15], and bidirectional long short-term memory (Bi-LSTM) [16]. Sigmoid non-linearity was used in the output layer with one neuron output layer. Adam optimizer was used with learning rate of 0.0001. Detailed description of each stage is given below:

4.1 Preprocessing

Preprocessing is one of the key steps in any NLP task. The raw data collected through various means is passed through a pipeline of NLP preprocessing to get the cleaned data (Fig. 2). This pipeline includes various NLP tasks described below:

Lowering and Punctuation: The raw text collected is first lowered and all punctuations are removed from it to avoid duplications and adding unnecessary punctuations

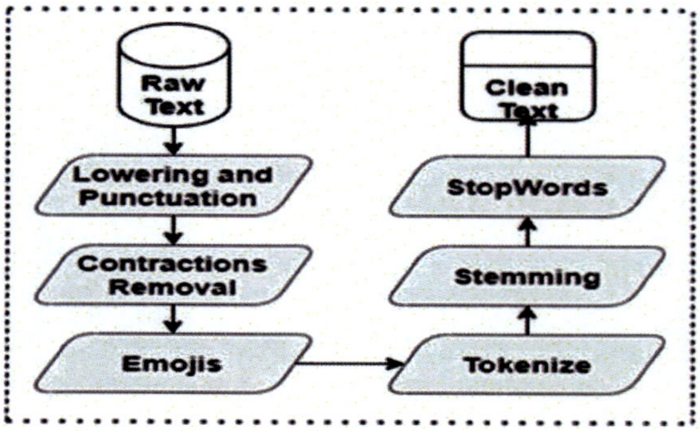

Fig. 2 Pipeline of NLP preprocessing

which does not add any meaning to the text. For instance, consider sentence s1 = {"corona cases rose across country ... India saw one lakh new corona cases"}. This sentence after lowering and punctuation would be s1 = {"corona cases rose across country India saw one lakh new corona cases"}. By doing this duplication of corona is avoided as well storing "..." which do not add any useful information.

Contraction's Removal: Contractions may be defined as words which are shortened by dropping one or two words. For example, "cannot" becomes "can't" and "will not" becomes "won't." The lowered text is passes to a function which removes contractions store the semantic meaning of sentences.

Emojis: The text data used in the proposed model was collected from Twitter. It is seen that in a lot of social media applications user uses emojis to amplify their message, such as "happy 😀" the emojis do not add much of information to the data and our model is not capable of handling emojis, therefore, we drop all the emojis from the data.

Tokenize: In tokenization, the sentences are broken down into smaller tokens usually words. Each token will represent each word in the sentence which is easy to process and feed into the algorithm. For instance, the sentence s1 = {"India got its first home made vaccine"} after tokenization will be {"India," "got," "its," "first," "home," "made," "vaccine"}. These tokenized words can be fed to a word embedding layer to convert it into vectors which is discussed in Sect. 4.2.

Stemming: In stemming, each token is converted to its stem form or base form, this will put all the words with same semantic meaning in the same form therefore reducing the computational cost and time. For instance, consider tokens "choco," "chocolates," "chocolatey" although all three are unique tokens but all three tokens mean the same therefore can end up squandering computational time and capacity and further slow the process and shall be added uniquely to the list of vocabulary.

Stopwords: This is one of the most significant process in data cleaning since it allows us to delete words that do not contribute to the content of the phrase and will not alter the meaning of the sentence if they are removed. Stopwords are pre-defined in libraries like NLTK and Spacy, however, the list can be modified as per the domain by experts.

4.2 Word Embeddings

Word embeddings are the vectorized representation of text, where words with similar meaning have a similar representation. All the tokens in the text have to be converted to word embeddings before feeding it to a machine learning model. The dimension of the vector may be passed in as a hyperparameter while training the model. There are a lot of ways to convert text to vector, commonly used models are BOW, TF-IDF, Word2Vec, and GloVe.

GloVe

GloVe or global vector is a word embedding model presented by researchers at Stanford University in 2014. GloVe uses the similarity score between two tokens as invariant to generate the vector representations. The model makes use of two different models' techniques, including the skip-gram model and the continuous bag of words model (CBOW). GloVe model uses the old theory of word or token cooccurrence, it was presented as an improvement to existing model Word2Vec presented by Mikolvo. GloVe uses both global and local statistical information to generate word vectors. GloVe model is capable of capturing the semantic information accurately. The main issue with skip-gram model was its high computational cost and time (although has good accuracy), whereas CBOW had low accuracies (although low computational time). GloVe tries to combine the best of both and has produced accurate results with low computational time.

GloVe pre-trained model was used in the proposed work. 300-dimension vector was chosen with 42 billion tokens and 1.9 million vocabularies for better results. GloVe was able to vectorize most of the tokens present in the text. However, few tokens which were not converted to vector form was omitted in the training process.

4.3 RNN

Recurrent neural networks are type a feed-forward artificial neural network highly utilized for sequence data like natural languages or time series data (Fig. 3). Recurrent Neural Networks are Neural Networks for which the output from the previous stage is also supplied to the next layer as input, assisting the network in better understanding the relationships. The model analyzes not just the current input, but also what it knows about the previous elements at each element of the sequence. This memory allows the network to learn long-term dependencies in a sequence. Output of a RNN may be represented as

$$h(t) = fh(h(t-1), x(t), \theta h) \quad (1)$$

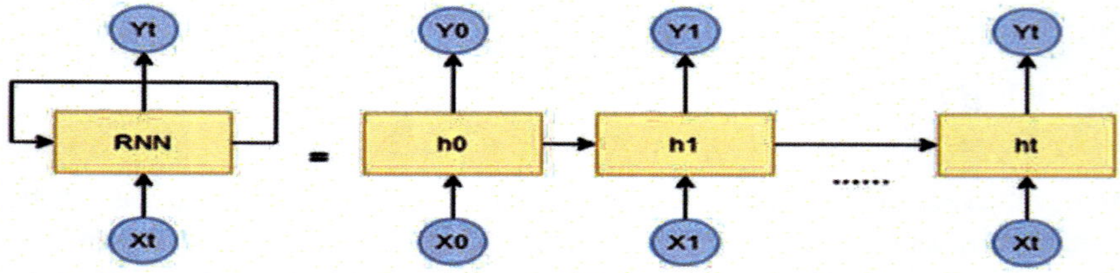

Fig. 3 General architecture of recurrent neural network

where

$h(t)$ is hidden state at time stamp t
$h(t-1)$ is the previous hidden state, i.e., hidden state at time stamp $t-1$
$x(t)$ is input at time stamp t
θh is parameters of function *fh*.

The hidden state of the previous step $h(t-1)$ is fed as input to along with current input $x(t)$ to calculate current hidden state $h(t)$.

The output at any time stamp t is given by

$$Y(t) = fy(ht, \theta y) \tag{2}$$

where θy is parameters of function *fy*.

Recurrent Neural Networks retain state information over time steps, allowing variable length inputs and outputs to be processed. In the scope of a COVID-19 news's overall validity, news comes in a variety of lengths. To determine if a news story is genuine or not, a word will be used as a token and the consequent score of the previous state will be used as input to the current state. RNNs have shown good performance for short articles and news.

The backpropagation algorithm is used to train a Recurrent Neural Network. The minimum of the cost function is established during Recurrent Nural Network training by iteratively taking modest steps toward the direction of negative error derivative with respect to the weights of network. The loss function used for training our RNN is given by

$$H_p(q) = -\frac{1}{N}\sum_{i=1}^{N} y_i \cdot \log(p(y_i)) + (1 - y_i) \cdot \log(1 - p(y_i)) \tag{3}$$

where

y represents the label
$p(y)$ is the predicted probability of the text being 1
N is the total number of samples.

Recurrent Neural Networks suffer with vanishing and exploding gradient descent problems, when the gradients are propagated back to learn long-term dependencies. The gradients corresponding to deeper layers have to undergo matrix multiplication at each time stamp due to each rule and if they have small values very less than 1, then the value may shrink exponentially resulting in very small value making it very difficult to model to learn, this is also termed as vanishing gradient descent problem. On the other hand, if the value is very large, they may get larger and may explode the result in crashing of model, which is also known as exploding gradient descent. Commonly used solutions for both the problem are usage of LSTM or GRU networks which overcome these problems. Other techniques to avoid these problems include using truncated backpropagation, using ReLu activation function and gradient clipping.

4.4 LSTM

Long short-term memory or LSTM networks are widely used as a replacement for RNN networks to avoid problems like vanishing and exploding gradient descent [17]. LSTMs have also proven effective to remember long-term dependencies which RNN could not remember effectively. LSTMs were first introduced by Hochreiter and Schmidhuber in 1997, after which a lot of researchers modified and popularized them in their work. LSTM cell is made up of three gates such as forget gate, input gate, and output gate. LSTM also maintains a cell state which carries information across the cell with little linear modifications (Fig. 4).

Forget Gate. This is the first operation in a LSTM cell, forget gate determines what extent of the previous information should be retained and what extent of information should be discarded. The decision is made by a sigmoid layer. Information from the previous hidden state (h_{t-1}) and current input (x_t) is passed in to the sigmoid layer which outputs a number ranging from 0 to 1, 0 meaning "completely forget the information" and 1 meaning "completely remember the information." This number is then combined with the previous cell state (C_{t-1}) and passed along the LSTM cell. The formula for forget gate is given by

$$f_t = \sigma\left(\left(W_{hf} \times h_{t-1}\right) + \left(W_{xf} \times x_t\right) + b_f\right) \tag{4}$$

where
h_{t-1} is the previous hidden state
x_t is the current input
W_{hf} is the weight associated with the previous hidden state
W_{xf} is the weight associated with current input state
b_f is the bias.

Input Gate. The input gate is in charge of adding information to the cell state. This in turn is a two-step process. Firstly, sigmoid layer is used to regulate what information is to be chosen from input and the previous hidden state which acts like a forget gate. Secondly, there is a tanh layer which creates a vector that possibly could be added

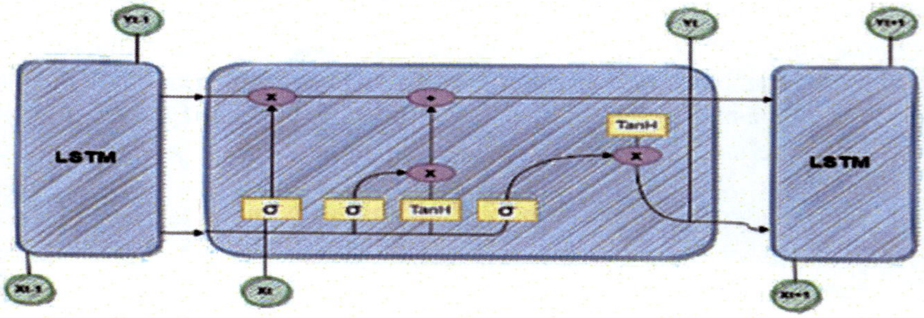

Fig. 4 Architecture of a long short-term memory cell

to the cell state. This output is referred as $\widetilde{C}_t$. The output of a tanh layer varies from -1 to $+1$. The outputs from both the layers (sigmoid and tanh) are multiplied and then added to the cell state. The formula for input gate is given by

$$i_t = \sigma\left(W_i \cdot \left[h_{t-1}, x_t\right] + b_i\right) \tag{5}$$

$$\widetilde{C}_t = \tanh\left(W_C \cdot \left[h_{t-1}, x_t\right] + b_C\right) \tag{6}$$

where
h_{t-1} is the previous hidden state
x_t is the current input
W_i and W_C are the weight
b_i and b_C are biases.

Output Gate. This is the last stage, where we decide what we are going to output from the LSTM cell. The output depends on cell state, previous hidden state, and current input. The cell state is passed through a tanh non-linearity which outputs a vector whose values range from -1 to $+1$. Simultaneously, the current input and the previous hidden state are passed through a sigmoid layer which acts similar to a forget gate. The outputs of these two are multiplied to output only the important and required parts. The resulting output will be sent as the hidden state for next LSTM cell and output (sometimes multiplied by weight and bias may be added). The formula is given by

$$o_t = \sigma\left(W_o\left[h_{t-1}, x_t\right] + b_o\right) \tag{7}$$

$$h_t = o_t * \tanh(C_t) \tag{8}$$

where
h_{t-1} is the previous hidden state
x_t is the current input
W_o and b_o are the weights and bias
O_t is the output
C_t is the cell state.

4.5 Bi-LSTM

Bidirectional long short-term memory or Bi-LSTM is a variant of LSTM network which can learn dependencies in both directions. Bi-LSTM is composed of two LSTM networks, one LSTM runs from in forward direction and learns from past to future, whereas the other LSTM network runs in backward direction and learns from future to past. Bi-LSTMs significantly improve the quantity of information

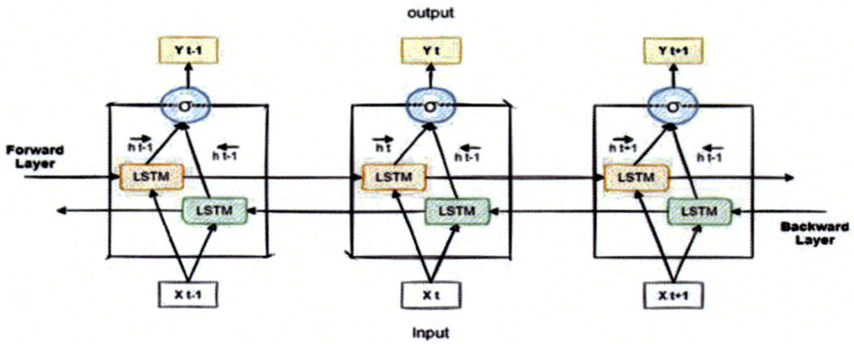

Fig. 5 General structure of a Bi-LSTM network

accessible to the network, increasing the context available to the model. Bi-LSTM is very suited for complex tasks and datasets and is seen to give very good accuracies. The only disadvantage of a Bi-LSTM network and its computational time as it takes a lot of time to train the network. High-end GPUs are required to train huge corpus on a Bi-LSTM model. Figure 5 describes the working of a Bi-LSTM network.

4.6 Dropout

Dropout is one of the key regularization techniques used in all NLP tasks to avoid overfitting. Proposed by Hinton in the year 2012, dropout removes a neuron temporarily from the neural network. There are a lot of techniques to avoid overfitting such as early stopping, data augmentation, and simpler models but dropout has been proven to be effective in RNNs and other sequential models to avoid overfitting. Overfitting is defined as a condition when the model is able to perform good for training data but performs poorly on testing data, in other terms, it has over learnt the training data. The main idea behind dropouts as described by Srivastava in [18] is that "each hidden unit in a dropout-trained neural network must find a way to work with a collection of hidden units selected at random. This should increase the robustness of every hidden unit and encourage it to develop beneficial features on its own, rather than relying on other hidden units to fix its faults."

In the proposed work, a dropout of 0.3 was used in all the networks to avoid overfitting of model and it was seen that without dropout, the difference between training and validation accuracies was large and it was reduced by half after applying dropout technique [19].

Figure 6 shows the flow of proposed method to classify COVID-19 fake news. The raw text from the dataset is cleaned before converting into word embeddings of 300 dimensions using pre-trained GloVe model. Three models were trained to get the best possible results. RNN, LSTM, and Bi-LSTM are popular recurrent neural networks used for sequential data. These models were trained iteratively for 20 epochs with a

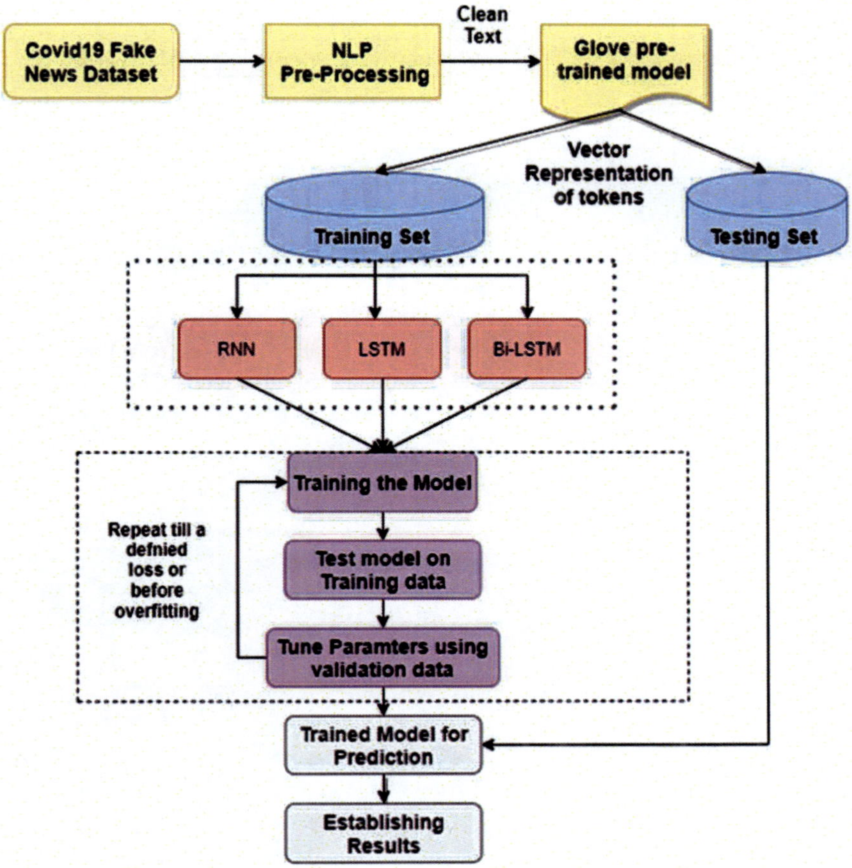

Fig. 6 Proposed methodology to classify COVID-19 fake news using GloVe and Bi-LSTM recurrent neural network

batch size of 32. Parameters of models like dropout ratio, number of cells were tuned for best possible results. The model was tested on the 20% of unseen data separated from the dataset before training and results were established.

5 Results and Discussion

The proposed method was iteratively trained on the COVID-19 fake news dataset to reduce the loss function and increase the accuracy. The accuracies of various models are shown in Table 1. It was observed that Bi-LSTM network performed very well in the training data giving the training accuracy of up to 99%. The time per step and time per epoch are also shown in the table to provide a comparison of computational

Table 1 Training and validation accuracies for various models

Model	Time per epoch (s)	Time per step (ms)	Training accuracy	Validation accuracy
RNN	6	32	90.28	87.54
LSTM	10	64	97.18	90.89
Bi-LSTM	22	137	98.99	91.73

time taken by all the models on Nvidia GeForce GTX GPU of 4 Gb and intel core i7-10750H CPU processor with 16 Gb RAM.

The dataset was split into 80:20, where 80% of the data was used for training, and 20% of the data was used for testing the model. Sklearn's train-test split library was used to accomplish the task. The model was trained for 20 epochs. The plot of loss and accuracy of the model for both testing and training for RNN is shown in Fig. 7, and for Bi-LSTM, it is shown in Fig. 8, respectively.

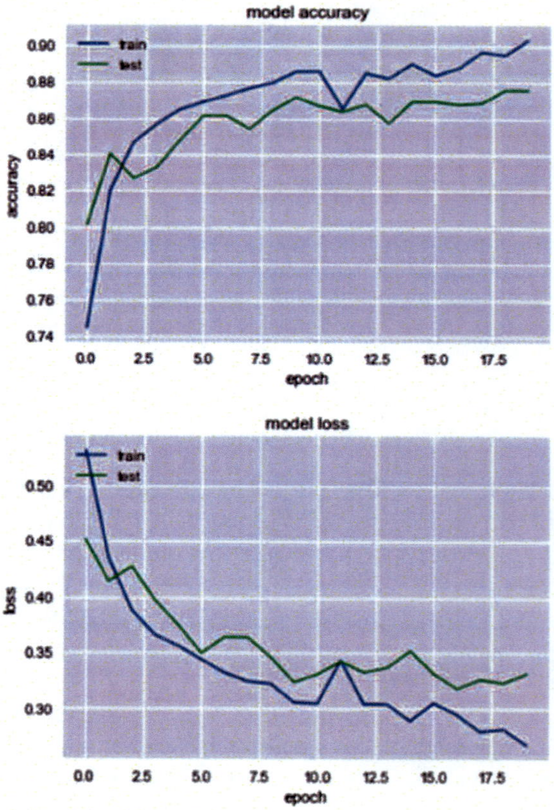

Fig. 7 Plot of accuracy and loss for RNN model

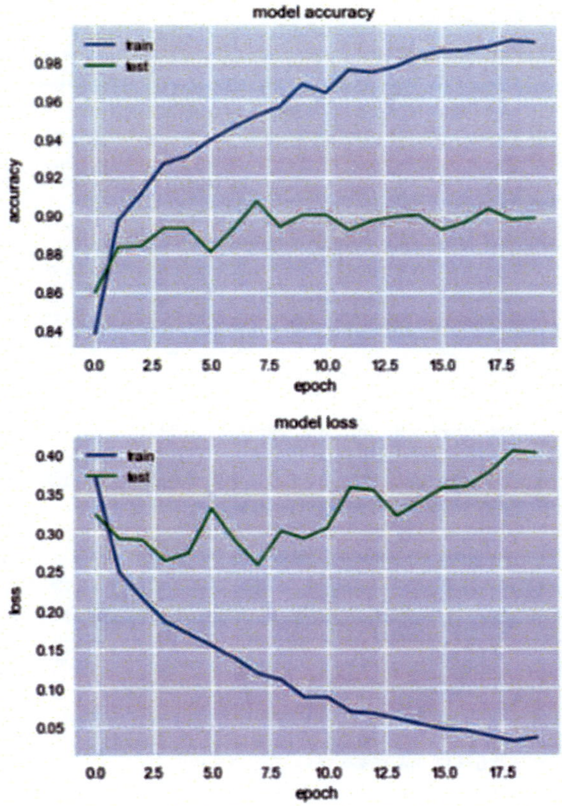

Fig. 8 Plot of accuracy and loss for Bi-LSTM model

6 Conclusion and Future Scope

The aim of this work was to classify fake news from real news related to COVID-19. COVID-19 fake news from Constraints website was used to train and test the model. Various models such as RNN, LSTM, and Bi-LSTM models were trained to compare the results. Bi-LSTM model out-performed other models in training while the results of LSTM and Bi-LSTM model looked similar in testing the model. Although Bi-LSTM model outcasted other models, it took twice the time as LSTM to train the model but huge difference in accuracies was not seen. As Bi-LSTM considers information from both the direction forward and backward and has two LSTM cells, it is expected to take larger training time than conventional LSTM networks. RNN trained the quickest taking half the time as LSTM network and one forth the time of a Bi-LSTM network. Future work may aim at further improving the accuracy of the model. CNN may be utilized along with other sequential models. Attention networks such as transformers, BERT, and GPTs can be fine-tuned to classify the fake news.

References

1. Klein DO, Wueller JR (2018) Fake news: a legal perspective. Australas Policing 10(2)
2. Fisher M (2014) Who cares if it's true? Colum J Rev 52(6)
3. Gross M (2017) The dangers of a post-truth world. R1-R4
4. Sungheetha A (2021) COVID-19 risk minimization decision making strategy using data-driven model. J Inf Technol 3(01):57–66
5. Zivkovic M, Venkatachalam K, Bacanin N, Djordjevic A, Antonijevic M, Strumberger I, Rashid TA (2021) Hybrid genetic algorithm and machine learning method for COVID-19 cases prediction. In: Proceedings of international conference on sustainable expert systems: ICSES 2020, vol 176. Springer Nature, p 169
6. Mohan D, Sheela KA, Sudhakar P (2021) COVID-19 social distancing with speech-enabled online market place for farmers. In: International conference on sustainable expert systems, ICSES 2020, pp 513–525
7. Jony JV, Kumar S (2021) Domain specific fine tuning of pre-trained language model in NLP. In: Proceedings of international conference on sustainable expert systems: ICSES 2020, vol 176. Springer Nature, p 403
8. Shushkevich E, Cardiff J (2021) TUDublin team at Constraint@ AAAI2021—COVID19 fake news detection. arXiv preprint arXiv:2101.05701
9. Sharif O, Hossain E, Hoque MM (2021) Combating hostility: Covid-19 fake news and hostile post detection in social media. arXiv preprint arXiv:2101.03291
10. Li X, Xia Y, Long X, Li Z, Li S (2021) Exploring text-transformers in aaai 2021 shared task: Covid-19 fake news detection in English. arXiv preprint arXiv:2101.02359
11. Das SD, Basak A, Dutta S (2021) A heuristic-driven uncertainty based ensemble framework for fake news detection in tweets and news articles. arXiv preprint arXiv:2104.01791
12. https://constraint-shared-task-2021.github.io/
13. Pennington J, Socher R, Manning CD (2014) Glove: global vectors for word representation. In: Proceedings of the 2014 conference on empirical methods in natural language processing (EMNLP), pp 1532–1543
14. Zaremba W, Sutskever I, Vinyals O (2014) Recurrent neural network regularization. arXiv preprint arXiv:1409.2329
15. Hochreiter S, Schmidhuber J (1997) Long short-term memory. Neural Comput 9(8):1735–1780
16. Vijayakumar T (2020) Posed inverse problem rectification using novel deep convolutional neural network. J Innov Image Process (JIIP) 2(03):121–127
17. Bano S, Pranathi YL, Niharika GL, Sreya GDS (2021) Prognostic of depression levels due to pandemic using LSTM. In: Proceedings of international conference on sustainable expert systems: ICSES 2020, vol 176. Springer Nature, p 11
18. Srivastava N, Hinton G, Krizhevsky A, Sutskever I, Salakhutdinov R (2014) Dropout: a simple way to prevent neural networks from overfitting. J Mach Learn Res 15(1):1929–1958
19. Shakya S, Balas VE, Haoxiang W, Baig Z (2021) Proceedings of international conference on sustainable expert systems: ICSES 2020, vol 176. Springer Nature

Combinatorial Test Perspective on Test Design of SDN Controllers

Rekha Jayaram and R. Krishnan

Abstract OpenDaylight is an open-source Software-Defined Network controller that has made programmable networks possible. It has got together both industry and the OpenDaylight community in delivering programmable and interoperable networks to service providers, enterprises, universities and organizations. It is a continuously evolving project with many different releases, where every new release requires extensive testing, to ensure quality software. Effective regression testing is required to ensure the existing features are not affected. Also a project of this scale will have many test cases, hence an effective selection method is required to ensure a reduced set of test cases for regression testing. The current paper proposes to adopt Combinatorial testing for generating test suite which has reduced number of test cases and better test coverage. Upon successful execution of the combinatorial test suite, the faults identified needs to be localized to help debug them effectively and in less time. This can be achieved through an efficient fault localization technique.

Keywords Combinatorial testing · Fault localization · OpenDaylight · Software Defined Networks

1 Introduction

Software Defined Networking (SDN) has made managing huge amounts of data in data centers possible. SDN has made network management easy and flexible. A fundamental component of SDN is the centralized controller which provides network operators more control over network-management. OpenDaylight (ODL) is an SDN controller which is community-led and is supported by many companies. It is an open-

R. Jayaram (✉)
Department of Information Science and Engineering, Dayananda Sagar College of Engineering, Bangalore, India

R. Krishnan
Department of Computer Science and Engineering, Dayananda Sagar College of Engineering, Bangalore, India

S. Shakya et al. (eds.), *Proceedings of Second International Conference on Sustainable Expert Systems*, Lecture Notes in Networks and Systems 351,
https://doi.org/10.1007/978-981-16-7657-4_6

source framework for enhancing SDN, thus helping service providers, developers, enterprises and a number of organizations. ODL was first released in 2014. There have been many subsequent releases till date. Every release will carry over the project functionalities of the previous releases and a few additions or deletions of its own. Each new version requires extensive integration and regression testing. This paper proposes to apply Combinatorial test strategy to testing of ODL. Combinatorial testing (CT) is a systematic method of testing software. The test cases are developed by considering 2 or more parameters, thus reducing the number of test cases which will be executed. CT samples a large input space into smaller test suites taking care to cover the key parameters and their values combinations. Fault Localization (FL) is the process of identifying parameter combinations that might have contributed to a fault detected. Fault localization methods that analyze the results of It has been proven that combinatorial testing is effective in debugging.

The rest of the paper consists of Sect. 2 that introduces Software Defined Network, its features and an insight into comparing different SDN controllers, stressing on OpenDaylight controller project and the different aspects associated with its testing. Section 3 talks about Combinatorial Testing, test generation strategies, test case execution, Fault localization and the different strategies that help localize faults in combinatorial testing.

2 Software Defined Networks

Increase in network traffic led network operators to look for more control over network-management aspects which would ensure better control over the paths used to deliver traffic. This aspect gave rise to the separating control and data planes, in turn a Software Defined Network (SDN). SDN provides flexibility in communication networks by allowing them to be programmable.

Key features of SDN include separate control and data planes, a centralized controller, open interfaces between control plane and data plane devices and provisions for external applications to program the network [1]. A control plane is responsible for traffic handling and the data plane is responsible for traffic forwarding—depending on the decisions taken by the control plane. The control plane consists of three interfaces or APIs namely:

1. Northbound interfaces which are used by applications to give instructions to the controller
2. Southbound interfaces which is used by the controller to send messages to the data plane and
3. East-west bound interfaces ensures controllers have a common view of the network and coordinate policies.

Openflow is a prominent SDN southbound protocol which provides a flow based solution where every packet forwarding is decided by the controller. It is used to link the controller and switch agents. Openflow applications control the path a packet takes. A good insight into Openflow is given in [2, 3].

Benefits of SDN include better network control, reduced complexity and management costs, centralized control over different vendor devices, improved security, utilization and network performance [4].

2.1 SDN Controllers

A controller is a basic component of the SDN architecture. It consists of pluggable modules that are used to perform various network tasks like collecting information of the availability, capability and statistics of the various devices in the network [5]. It is also referred to as a Network Operating System and plays a key role in deciding if the SDN succeeds or fails. The intelligence of an SDN network is managed by its controller. SDN controller functions as a logically centralized intelligence with an overall view of the network. It also has the ability to reconfigure the network dynamically [4].

A controller is a centralized server that hosts the control plane. A southbound interface is a component of SDN that connects the switches or data plane to the controller. One of the most famous implementations of southbound interfaces is the OpenFlow protocol—an open-source protocol upon which many controllers such as OpenDaylight, Floodlight, POX, NOX, Maestro and others are based. C++, Java and Python are the most common programming languages used mostly to develop controllers. Some of the controllers include:

1. POX—a controller based on python, which is very useful in debugging an SDN network, provide network virtualization and others.
2. Ryu—a controller that is component-based with a group of components that are pre-defined. These components can be modified or expanded for creating customised controller applications.
3. Trema—works on easy-to-write code. It also provides good performance.
4. Floodlight—a module-based controller where each module provides service to other modules.
5. OpenDaylight—an open-source project aimed at creating a robust code that covers many major components of the SDN architecture. It is based on Openflow protocol. It also provides support to many southbound APIs.

Many papers have done extensive comparison of the different controllers. One of the most common parameters used to compare different controllers is its performance. Khondoker et al. [6] have compared POX, Ryu, Trema, Floodlight and OpenDaylight controllers by capturing their properties. The results of the comparison show that OpenDaylight provides high modularity and good documentation. Al-Somaidai and Yahya [7] compares seven different controllers and concludes that OpenDaylight provides good flexibility. Rowshanrad et al. [8] compares OpenDaylight and Floodlight controllers on the basis of QoS parameters of the network and in

Cloud computing and Multimedia fields. The comparison shows that OpenDaylight performs better for networks with low and mid load for any topology. It also gives less latency or delay as compared to Floodlight. Also, OpenDaylight performs better in terms of loss for networks with heavy traffic. Yamei et al. [9] compares OpenDaylight and ONOS controllers, and the results show that OpenDaylight performs better in terms of scalability and reliability due to the presence of flexible plug-ins and also topology discovery and stability.

2.2 *OpenDaylight (ODL) Platform*

OpenDaylight project consortium was hosted by Linux Foundation in 2013. Also called OpenDaylight Platform (since 2015), it is a collective, open-source project which provides a framework for SDN which is supported by the SDN community that includes academicians, researchers, industry and others for enhancing SDN. The project is under Eclipse Public License (EPL), a free and open-source software license. ODL works as a core component of SDN architecture.

Some of the founders of the ODL project include Cisco, Juniper Networks, Ericsson, HP, IBM among others—who committed to providing engineering and financial resources to the project [10]. ODL has seen over 100 deployments with many companies including Comcast, Telefonica, Globe Telecom and others. It is a Java-based, modular and pluggable platform which supports many southbound protocols. ODL platform helps end users and customers in building an SDN controller for their specific needs.

The OpenDaylight architecture as shown in Fig. 1 is multi-layered, with the top layer consisting of applications, and the middle layer gives the framework which shows SDN abstractions. This layer consists of the northbound and southbound APIs, and the bottom layer consists of the physical and virtual devices.

The official web page for ODL is [11]. It gives a complete insight into the OpenDaylight project and all the ODL relevant documentation. The controller project has many different sub-projects under itself which includes AAA, Controller, JSON RPC and others.

2.3 *Testing OpenDaylight*

The first version of Opendaylight was released in 2014. There have been many subsequent releases till date. Every release will carry over the features of the previous releases and a few additions or deletions of their own. Some of these features include authentication, BGP, Faas, Netconf and others. Each of these features is developed using Java as a separate project and then merged together depending upon the version and its features details.

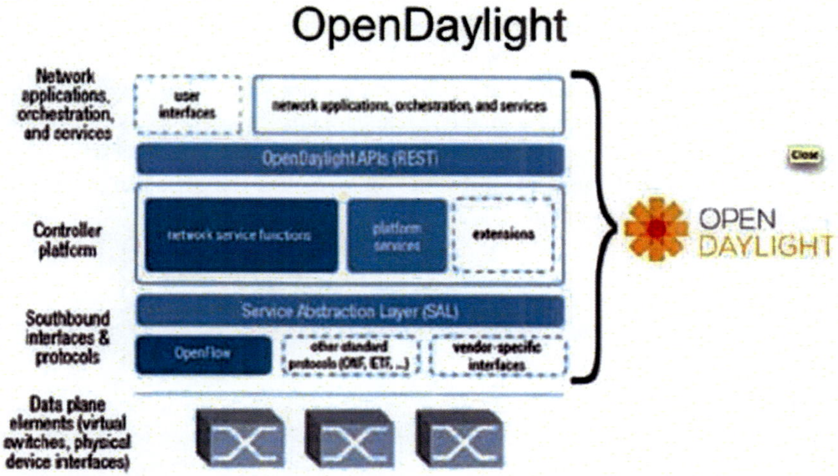

Fig. 1 OpenDaylight architecture [10]

Some of the types of testing done on ODL includes Build Time Testing, Component Testing, System Testing, Integration Testing and Regression Testing.

Build Time Testing uses JUnit framework for writing test cases. These tests are executed during the compilation phase of a project. These tests provide a way for a developer to test the logic which belongs to the given project and verify it is behaving as desired. Build time testing concentrates on the internal implementation classes.

Component Testing also uses JUnit to build test cases. These test cases cover the functionality of a single ODL bundle and are focused on the end-to-end testing of APIs. Component testing does not look into the individual implementation of classes. Here, the focus is on testing the code modules.

System Testing in OpenDaylight is run on all projects that define its essential functionalities. System test jobs run in *continuous integration*. It is performed by installing a version of the ODL controller on a virtual machine and then the testing is done by running the functionality using a test tool. The test cases are built using Robot framework.

Integration Testing in OpenDaylight is performed by integrating the different system test suites. Integration testing of every release of ODL has an integration test job associated with it, available in Jenkins/integration folder. A job also describes the test environment in which the testing needs to be performed. Each integration job has a list of system test suites which will be executed in the order listed. Finally, the clean-up process that needs to be performed after completing executing all the test suites. Upon execution of the integration test job, the status of each job is shown in terms of its last build and an aggregate status (also called weather report) of the recent builds. This status and weather report helps in identifying a stable build.

Regression Testing OpenDaylight projects are developed as an agile project where regression testing plays a key role. After every change or modification, a new build is performed and regression testing will be done. Regression testing in ODL is done by running integration test jobs that describes the different system test suites that needs to be run to test the changes in every build and the order in which they should be executed. These jobs are run periodically. Status, weather reports and code coverage reports are obtained for each build, which helps in identifying a successful build which can be used as the baseline test suite. This baseline test suite can be used to apply combinatorial test strategies.

2.4 Test Environment of ODL

System, integration and regression testing requires *Config Plans* to be developed, which indicate the *Test Environment* that needs to be setup before test execution, variables that need to be enabled to run the test suites and the cleanup activities after the test completes.

Setting up the test environment involves installing a controller distribution in one or more virtual machines upon which functionality testing will be executed using a test tool such as Mininet. The installation procedure for ODL has been well documented in [10].

Finally, *Jobs* are created to indicate the corresponding config plans and the test suites that should be executed to test the particular functionalities. These jobs are created using Yang language and debugged for any errors using Jenkins-sandbox before placing in the corresponding repositories. The jobs have a .yaml extension. A sample job gives an insight into the project name, details of the functionality being tested, the stream it should be tested in and other details.

The different variables such as environment, system, different protocol and others and the common procedures used in testing are defined in a common variable file [12], which will be included in every job as a Resource file. Variables specific to functionalities are mentioned in the corresponding test suites.

2.5 Test Frameworks in ODL

Test cases in ODL are written using JUnit and Robot frameworks. JUnit is used to write test cases used in build time and component testing and robot framework is used to build test cases used in system testing, integration testing and regression testing.

JUnit is an open-source Java library. It provides framework for Java classes where test suite containing test cases for unit testing can be built incrementally. The idea here is to test and code simultaneously—for every few lines of code written, test cases will be written and executed. If the test fails, the bug will be fixed, else more code is written and more test cases are written to test the same. JUnit provides *assertions*, a way to specify the desired outcome and compare it with the actual outcome. Running the JUnit test cases require *setup*, a way of preparation which is run before each test. Upon completion of the testing, cleanup should be done which is known as *teardown*. JUnit installation, building and running test cases, comparison of JUnit with other testing frameworks and other details related to JUnit has been very well documented in [13].

Robot framework is a generic, open-source and industry-supported automation framework that can be integrated with any tool to create flexible automation solutions. The robot test data is written in tabular format. Test cases are written in HTML files, in the form of HTML tables. Lower-level keywords are implemented in test libraries which can be used, while writing test cases. Using these, higher-lever user keywords can be created [14]. Keywords created must be descriptive, in that they should describe WHAT should be done rather than HOW to do it. Descriptive keywords is one of the strongest features of robot framework. Some of the standard keywords include Builtin keywords—for generic testing, Operating System keywords—for operating System tasks, Telnet keywords—used in telnet connection handling and String keywords—used in handling strings. Writing test cases using robot framework follows a natural work flow that takes care of test case preconditions, actions and verification. Keyword description uses real language—which makes the tool simple and easy to use. Reports are generated automatically and also published on web pages. Stresnjak and Hocenski [15] gives a good comparison of automating test cases using robot test cases with manual testing of the same test cases.

2.6 Software Associated with ODL Testing

Some of the software associated with testing of ODL are:

The code coverage tool used in ODL is *JaCoCo*, a free code coverage library for Java. It gives the test coverage in terms of percentage of instructions, branches, complexity, line, method and class covered during testing of every package. The overall coverage report of Jacoco shows that testing coverage of all packages is less than 100% for the parameters [16]. Jacoco is run to check the coverage after completion of every testing. The Jacoco reports are available in SonarCloud.

Bugs tracking is done using *Jira* software. It is an open-source tool which was originated as a bug and issue tracker and has now progressed into a work management tool. Jira can be used for all types of use cases, from requirements and managing test cases to software development using agile methods. Jira software manages the bugs and their statuses using Tickets.

Test jobs are placed in *Jenkins*, an open source, automation server. It provides plug-ins to support development, deploying and automating a project. It is a self-contained, Java-based program. Jobs once placed in jenkins-releng cannot be reconfigured.

The code repository to be used in build time and component testing is available in *GitHub* and *Gerrit*.

The software associated with the different types of testing of ODL are depicted in Fig. 2.

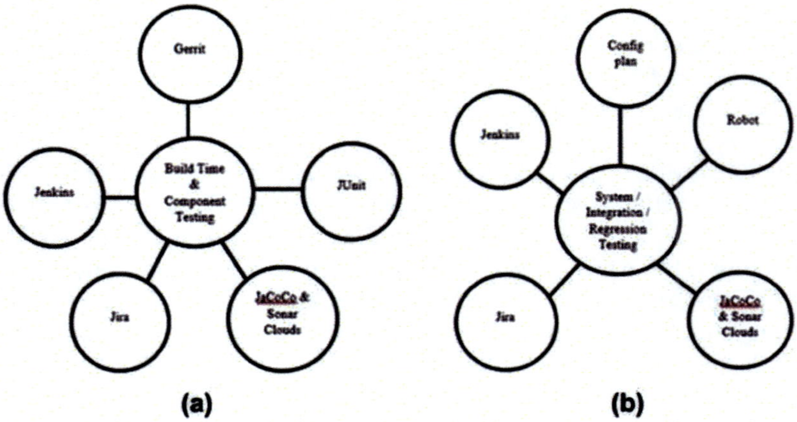

Fig. 2 Software with different types of ODL testing

Upon analysing the testing done on ODL, it has been observed that ODL has not been tested using combinatorial testing methods. Combinatorial testing has been proven as an effective testing technique in different scenarios such as [17–21]. Hence, as future work, we can look into checking the combinatorial test readiness of open-daylight and apply combinatorial test as a testing technique for the same. The next section gives a brief of combinatorial testing.

3 Combinatorial Testing

Combinatorial testing (CT) is a systematic method of testing software using a limited number of test cases, where each test case consists of two or more parameters combined based on different combination strategies. Testing all parameter value combinations is exhaustive in terms of resources. CT samples a large input space into smaller test suites taking care to cover the key parameters and their values combinations. Khalsa and Labiche [22] give a very good insight into CT generating tools, covering arrays generating techniques and coverage strength supported by each tool. The basic approaches to combinatorial testing use configuration parameter value combinations or input parameter value combinations [19]. Configuration parameters are associated with systems and input parameters are associated with applications. CT is a black-box testing technique which requires only a basic knowledge of the system configuration. This helps to identify parameters and their values. Combinatorial Testing involves constructing a Covering Array (CA) that covers all t-way combinations of parameter values atleast once and then execute this CA as a test suite with an aim to trigger faults [20]. A tailored version of the CT process can is given in Fig. 3. Here, IPM refers to Input Parameter Model which is the collection of all the input parameters given to the SUT. Test case generation depends on the strategy used.

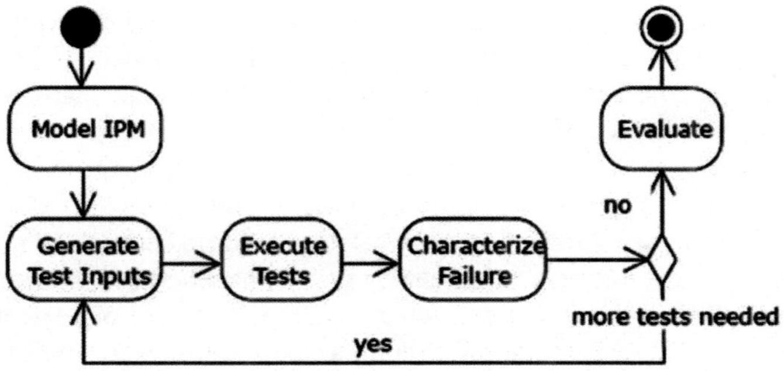

Fig. 3 Combinatorial test process [23]

3.1 *Test Case Generation*

Test case generation strategies include test based strategies that work on developing one test case at a time making sure it covers as many combinations as possible, for example, AETG [24] and Parameter based strategies, that start with "t" parameters and develops a test suite for all the t-wise interactions. It incrementally adds the remaining parameters to the test suite, example, In parameter order [25]. These strategies depend different algorithms used such as:

1. *Greedy algorithm techniques*: In these techniques, each test case comprises of most of the parameter combinations that are not yet covered.
2. *Heuristic search techniques*: These techniques start with an already built test set to which transformations are applied till it covers all the combinations.
3. *Algebraic or Mathematical techniques*: These techniques are based on building large test sets from smaller test sets, known as recursive construction.
4. *Random techniques*: These techniques are ad-hoc methods which select test cases randomly from all the test cases available [20, 26].

Some of the software tools used for test case generation include CATS (Constrained Array Test System), OATS (Orthogonal Array Test System), AETG, IPO, Tconfig, CTS (Combinatorial Test Service), PICT and SST [20].

3.2 *Test Case Execution*

Many aspects contributes to the execution of combinatorial test cases such as:

1. *Modeling the system under test (SUT)*: The efficiency and effectiveness of CT depends on modeling of the SUT, thus making it an very key aspect of CT. Modeling is done by identifying both configuration and user input parameters and the right values for these parameters that may affect the SUT. Parameters

that may have many values require methods such as primary element selection, equivalence partitioning and boundary value analysis. Modeling also requires identifying parameters that will have no interaction or those that have a high interaction with each other and also the constraints that may exist between them.
2. *Test automation*: An important part of test execution is automating the test scripts. A test script gives the sequence of steps that needs to be executed for a given input. Also, for varying inputs, these sequence of steps needs to be executed repeatedly. This can be made possible using automation. A parametrised test script gives different values of the parameters using which the sequence of steps should be executed on the SUT. Test frameworks play a crucial role in automating test cases. Testers can register their test scripts using the pre-defined extension points, E.g., the @Test annotation in JUnit5 [20].
3. *Test oracle*: A test oracle gives the expected result of a test depending on the input given. This oracle helps in deciding if a test has passed or failed depending on whether the actual and the expected outputs are the same or not.
4. *Test case prioritization*: Prioritizing test cases is scheduling the test cases such that the test case with a higher priority gets executed before the others. This helps in executing the important test cases during times of limited resources. Also, prioritized testing may uncover faults early as it guarantees important test cases are executed first.

3.3 Combinatorial Interaction Regression Testing

Combinatorial Interaction Testing (CIT) is an effective and cost efficient test generation technique. CIT is used to systematically sample all t-way combinations of input parameters by testing the parameters together to ensure all combinations are tested atleast once. CIT has been proved to generate smaller test suites, provide better code coverage, with a good ability to detect faults [18]. Two approaches are discussed here that are used for designing Combinatorial regression test suites for software.

First approach is given in [18] where the effectiveness of CIT on regression testing is analyzed across multiple versions of two software subjects. The results obtained suggest that CIT is an effective method of regression testing, in finding faults when compared with an exhaustive test set and also in finding faults earlier. Using this as a benchmark, it can be considered to apply CIT to regression test suites of ODL.

Second approach is given in [17] where ACTS software has been used to design a combinatorial test suite for an IoT based Operating System called Contiki. The proposed approach starts with checking the code coverage for testing done using Combinatorial Coverage Measurement (CCM) tool. Two different scenarios are analyzed where one scenario has a baseline regression test suite and the other does not have a base lined regression test suite. In the first scenario, the existing regression test suite is redesigned using configuration based approach or input parameter based approach and then check the code coverage using a combinatorial code coverage tool. Once a good coverage is obtained, the test suite can be baseline. In the second scenario, where the regression test suite does not exist, a base test suite can be created using Combinatorial approach, say ACTS, which will ensure better code coverage.

Once a software has been tested using CT and faults have been identified, debugging these faults can be done using fault localization techniques. The following section gives a brief of the different fault localization techniques in combinatorial testing to localize faults.

3.4 Fault Localization

The process of Software Fault localization involves identifying the program components such as statements, variables and values that might have contributed to the fault detected. The process can be divided into two major parts: first, to identify suspicious code that may contain bugs and the second, to examine and determine that piece of code actually does contain bugs. If debugging involves only a small percentage of code, then we can say that the fault localization technique used is effective. An effective fault localization technique should identify the line of code where the actual faults resides and also provide clues for fixing the fault [27]. Fault localization techniques can be classified into static or dynamic execution slice-based, program spectrum-based and state-based, statistics-based, machine learning-based and other techniques.

Perez et al. [28] compares different fault localization techniques and shows that Reasoning based approaches which use system models and runtime observations to debug software (such as Model-Based Diagnosis (MBD) technique) provide a very high-quality diagnostic report which will help the developers in doing a root cause analysis of the failure as compared to other techniques. But this technique is not very scalable and also the computational overhead is very high here. In terms of scalability, Delta debugging and coverage-based techniques fare well.

Xuan and Monperrus [29] gives an approach for test case purification which helps in improving the fault localization process. The approach starts with generating a group of test cases for every test case that has failed, it then removes the unrelated statements using test case slicing, does a rank refinement.

3.5 Fault Localization in Combinatorial Testing

Execution of combinatorial test cases may result in failed test cases. In a combinatorial testing context, Fault Localization refers to identifying the *failure-inducing combinations (FIC)* of parameter value combinations. The process of identifying these combinations is called Fault Characterization (FC). A simple method for identifying FIC involves executing all the test cases and then identify the parameter combinations that exist in the failed test cases. However, this approach is very exhaustive. Several approaches have been proposed for identifying these combinations such as:

1. *BEN* is a combinatorial testing-based fault localization tool which takes as input the program source code, the input parameter model and the combinatorial test suite that was executed [30]. Also, the execution status of each test cases is known. The output of BEN ranks the statements of the program code in terms of their likelihood of being faulty. This approach works in two phases. In the first phase, the *failure inducing combination* is identified and in the second phase, the *faulty statement* is identified.
2. *Interleaving* CT (ICT) is a combinatorial test framework proposed to identify the parameter combinations that result in faults [31]. This approach combines test generation and fault localization techniques. It starts by executing a few test cases till a fault is encountered, after which test case execution is stopped and the parameter combinations that caused the failure are identified. Other test cases which were supposed to test the same parameter combinations will not be executed, thus refining the additional test cases generation. Some of the advantages of this approach are that it eliminates redundant test cases, improves the effectiveness of identification of the failure causing schemas and reduces the Masking effect and supports adequate testing.
3. A Fault Localization algorithm known as *Inverse CTD* is proposed in [32]. The concept of the algorithm is to use combinatorics to expand parameters and their values in every direction from a given starting value. These values are then used to generate new test cases. These new tests created will help separate the edges of the bug, and thus identify to the failure-inducing combinations and the parameter values that caused the bug.
4. *LCEC algorithm* is proposed as a fault localization algorithm in [33]. The method used here considers the statement-level entities in program to automatically link the FIC to the suspicious code. The relationship between the input parameters in the FIC and the faulty code is considered to calculate the change-impact analysis of the parameters. Doubtful statements are then used to build cause-effect chains. These chains further help in identifying the faulty code.
5. *Complete Fault Interaction Location (comFIL)* algorithm proposed by [34] helps locate the parameter interactions that cause system failures. Experimental have shown that the algorithm can generate a minor minimal fault interaction set and a high chances of getting safe vales for input parameters.
6. *MIXTGTE* is a fault characterization method proposed by [35] which takes as input an interaction size proposed by the user. This method generates test cases based on the results of previously executed test cases.

Some more fault localization techniques are discussed in [36]. Friedrichs et al. [37] present a comprehensive comparison and evaluation of different fault characterization algorithms and also proposes a comparison infrastructure that allows independent, reproducible and extensible comparison of FCAs.

4 Conclusion

OpenDaylight as an SDN controller offers many possibilities. The different types of testing executed on ODL and the different software and test environment associated with them are discussed here. The features of ODL are being developed in an agile framework requires extensive regression testing to ensure quality software. This paper proposes combinatorial test strategies that can be used to develop effective test suites to test the ODL project. The paper also gives a brief of different fault localization algorithms which can be used to localize faults that may arise from combinatorial testing of ODL.

References

1. Sezer S, Scott-Hayward S, Chouhan PK, Fraser B, Lake D, Finnegan J, Viljoen N, Miller M, Rao N (2013) Are we ready for SDN? Implementation challenges for software-defined networks. IEEE Commun Mag 51(7):36–43
2. Kim H, Schlansker M, Santos JR, Tourrilhes J, Turner Y, Feamster N (2012) Coronet: fault tolerance for software defined networks. In: 20th IEEE international conference on network protocols (ICNP). IEEE, pp 1–2
3. Tammana P, Nagarajan C, Mamillapalli P, Kompella R, Lee M (2018) Fault localization in large-scale network policy deployment. In: 2018 IEEE 38th international conference on distributed computing systems (ICDCS). IEEE, pp 54–64
4. Cox JH, Chung J, Donovan S, Ivey J, Clark RJ, Riley G, Owen HL (2017) Advancing software-defined networks: a survey. IEEE Access 5:25487–25526
5. Raju VS (2018) SDN controllers comparison. In: Proceedings of science globe international conference
6. Khondoker R, Zaalouk A, Marx R, Bayarou K (2014) Feature-based comparison and selection of software defined networking (SDN) controllers. In: World congress on computer applications and information systems (WCCAIS). IEEE, pp 1–7
7. Al-Somaidai MB, Yahya EB (2014) Survey of software components to emulate openflow protocol as an SDN implementation. Am J Softw Eng Appl 3(6):74–82
8. Rowshanrad S, Abdi V, Keshtgari M (2016) Performance evaluation of SDN controllers: floodlight and opendaylight. IIUM Eng J 17(2):47–57
9. Yamei F, Qing L, Qi H (2016) Research and comparative analysis of performance test on SDN controller. In: 2016 first IEEE international conference on computer communication and the internet (ICCCI)
10. Badotra S, Singh J (2017) Open daylight as a controller for software defined networking. Int J Adv Res Comput Sci 8(5)
11. OpenDaylight home page. https://www.opendaylight.org/. Accessed 9 June 2021
12. Test variables. https://github.com/opendaylight/integration-test/blob/63342e0454b82783a6ef7a9b19e1698b9b5cd621/csit/variables/Variables.py. Accessed 9 June 2021
13. Louridas P (2005) Junit: unit testing and coiling in tandem. IEEE Softw 22(4):12–15
14. Pajunen T, Takala T, Katara M (2011) Model-based testing with a general purpose keyword-driven test automation framework. In: 2011 IEEE fourth international conference on software testing, verification and validation workshops. IEEE, pp 242–251
15. Stresnjak S, Hocenski Z (2011) Usage of robot framework in automation of functional test regression. In: Proceedings of 6th international conference on software engineering advances (ICSEA), pp 30–34

16. JaCoCo coverage report. https://jenkins.opendaylight.org/releng/view/Verify/job/controller-maven-verify-2.0.x-mvn35-openjdk11/lastCompletedBuild/jacoco/. Accessed 9 June 2021
17. Patil AH, Goveas N, Rangarajan K et al (2015) Test suite design methodology using combinatorial approach for internet of things operating systems. J Softw Eng Appl 8(07):303
18. Qu X, Cohen MB, Woolf KM (2007) Combinatorial interaction regression testing: a study of test case generation and prioritization. In: 2007 IEEE international conference on software maintenance. IEEE, pp 255–264
19. Kuhn DR, Kacker RN, Lei Y et al (2010) Practical combinatorial testing. NIST Spec Publ 800(142):142
20. Nie C, Leung H (2011) A survey of combinatorial testing. ACM Comput Surv (CSUR) 43(2):1–29
21. Krishnan R, Krishna SM, Nandhan PS (2007) Combinatorial testing: learnings from our experience. ACM SIGSOFT Softw Eng Notes 32(3):1–8
22. Khalsa SK, Labiche Y (2014) An orchestrated survey of available algorithms and tools for combinatorial testing. In: IEEE 25th international symposium on software reliability engineering. IEEE, pp 323–334
23. Bonn J, Fögen K, Lichter H (2019) A framework for automated combinatorial test generation, execution, and fault characterization. In: 2019 IEEE international conference on software testing, verification and validation workshops (ICSTW). IEEE, pp 224–233
24. Cohen DM, Dalal SR, Fredman ML, Patton GC (1997) The AETG system: an approach to testing based on combinatorial design. IEEE Trans Softw Eng 23(7):437–444
25. Lei Y, Tai K-C (1998) In-parameter-order: a test generation strategy for pairwise testing. In: Proceedings of third IEEE international high-assurance systems engineering symposium (Cat. No. 98EX231). IEEE, pp 254–261
26. Kuhn DR, Bryce R, Duan F, Ghandehari LS, Lei Y, Kacker RN (2015) Combinatorial testing: theory and practice. Adv Comput 99:1–66
27. Alipour MA (2012) Automated fault localization techniques: a survey. Oregon State Univ 54(3)
28. Perez A, Abreu R, Wong E (2014) A survey on fault localization techniques
29. Xuan J, Monperrus M (2014) Test case purification for improving fault localization. In: Proceedings of the 22nd ACM SIGSOFT international symposium on foundations of software engineering
30. Ghandehari LS, Chandrasekaran J, Lei Y, Kacker R, Kuhn DR (2015) Ben: a combinatorial testing-based fault localization tool. In: 2015 IEEE eighth international conference on software testing, verification and validation workshops (ICSTW). IEEE, pp 1–4
31. Niu X, Nie C, Leung H, Lei Y, Wang X, Xu J, Wang Y (2018) An interleaving approach to combinatorial testing and failure-inducing interaction identification. IEEE Trans Softw Eng 46(6):584–615
32. Blue D, Hicks A, Rawlins R, Tzoref-Brill R (2019) Practical fault localization with combinatorial test design. In: 2019 IEEE international conference on software testing, verification and validation workshops (ICSTW). IEEE, pp 268–271
33. Ma C, Zhang Y, Liu J (2013) Locating faulty code using failure-causing input combinations in combinatorial testing. In: Fourth world congress on software engineering. IEEE, pp 91–98
34. Zheng W, Wu X, Hu D, Zhu Q (2016) Locating minimal fault interaction in combinatorial testing. Adv Softw Eng 2016
35. Arcaini P, Gargantini A, Radavelli M (2019) Efficient and guaranteed detection of t-way failure-inducing combinations. In: 2019 IEEE international conference on software testing, verification and validation workshops (ICSTW). IEEE, pp 200–209
36. Jayaram R, Krishnan R (2018) Approaches to fault localization in combinatorial testing: a survey. In: Smart computing and informatics. Springer, pp 533–540
37. Friedrichs T, Fögen K, Lichter H (2020) A comparison infrastructure for fault characterization algorithms. In: 2020 IEEE international conference on software testing, verification and validation workshops (ICSTW). IEEE, pp 201–210

Technologies Driving Digital Payments in India: Present and Future

T. Ravikumar, N. Murugan, J. Suhashini, and M. Sriram

Abstract The payments market in India has been witnessing a significant transformation in recent years. The Indian payments market has robustly and consistently been moving towards digitization due to enhanced digital infrastructure, favourable government policies, and initiatives, availability of new technologies, disruptive innovations, and changes in the mindset of the customers. India tops in the world's real-time digital payments with 20.5 billion transactions in the year 2020 despite the adverse effect of the COVID-19 pandemic. This article deals with the growth of the Indian digital payments market and the technologies that drive the digital payments space at present and in the future.

Keywords Digital payments · Technology · Biometric authentication · Soundwave technology · India

1 Introduction

India has rapidly been moving towards a digital economy over a period. The government of India (GoI) has been making numerous efforts to move the Indian economy towards digitization. Digital India program and demonetization are vital efforts towards the digital economy. Enactment of the Payments and Settlement Act in 2007, the establishment of National Payments Corporation of India (NPCI) in 2008, launch of Aadhaar in 2009, introduction of Immediate Payment System (IMPS), Prepaid Payment Instruments (PPIs), Rupay card and Online Payment Gateway Service Providers (OPGSP) in 2010, unveiling of Direct Benefit Transfer (DBT), Aadhaar enabled Payment Services (AePS), and National Automated Clearing House

T. Ravikumar (✉) · M. Sriram
CHRIST (Deemed to be University), Bangalore, India
e-mail: ravikumar.t@christuniversity.in

M. Sriram
e-mail: sriram.m@christuniversity.in

N. Murugan · J. Suhashini
PSNA College of Engineering and Technology, Dindigul, Tamil Nadu, India

S. Shakya et al. (eds.), *Proceedings of Second International Conference on Sustainable Expert Systems*, Lecture Notes in Networks and Systems 351,
https://doi.org/10.1007/978-981-16-7657-4_7

(NACH) in 2011, the introduction of the Merchant Discount Rate (MDR) policy in 2012, unveiling of Unified Payment Interface (UPI), and National Electronic Toll Collection (NETC) in 2016, the launch of Bharat Bill Pay Service (BBPS), Bharat Quick Response (QR) code, and FASTags for toll payments in 2017, and launch of ombudsman scheme for digital transactions and National Common Mobility Card (NCMC) in 2019 are the landmark efforts of GoI towards a cashless economy in India. These efforts have promoted digital financial transactions especially digital payments in India leaps and bounds in recent years.

Digital Payment (DP) is a financial transaction that happens online partially or wholly without the physical exchange of money. Common methods of DPs are contactless payments, debit and credit cards, mobile payments, point of sales and e-wallets. Digital payments have been growing steadily every year as DPs make use of disruptive technologies, and they have gained the confidence of the users. DPs offer convenience, speed, precision, better security, enhanced transparency and cost-effectiveness in their transactions which have attracted the customers towards them. Further, DPs promote financial inclusion, women's emancipation and inclusive growth in a country. Many technologies have contributed to the development of various innovative and suitable DP systems. These technologies are radio-frequency identification (RFID), near-field communication (NFC), application programming interface (API), QR codes, distributed ledger technology (DLT) or blockchain technology and biometric authentication. Furthermore, many technologies are emerging in the digital payment field. This research work discusses the growth of digital payments in India and the technologies that have been used in various digital payments systems at present and in the future.

2 Review of Literature

Digital payment is any type of non-money payment except a paper cheque [1]. The electronic payments system consists of online credit card transactions, online e-commerce payments, e-wallets, e-cash and wireless payment mechanism [1]. An e-wallet is a computer program meant for facilitating online payments or other financial transactions by collecting and validating the credentials of individuals digitally [2]. GPay, Phonepe, Apple pay, Samsung Pay and Amazon Pay are the leading e-wallets [3].

Confidentiality, integrity, authentication, non-repudiation, robustness, efficiency and availability are the features to be emphasized in digital payment systems to overcome security challenges [4]. The key technological factors that influence humans for the usage of mobile payments are perceived security, performance expectancy, facilitating conditions, absorptive capacity, availability and perceived utility [5].

Digital payments were not only driven by positive features of digital payments but also due to negative features of cash [6]: biometric authentication, RFID, NFC, artificial intelligence (AI), QR codes and blockchain in digital payments [7].

3 Methods and Materials

Descriptive is a research design of this study, and the study is based on secondary information available. The secondary information was collected from NPCI, the Reserve Bank of India (RBI), the Ministry of Electronics and Information Technology (MeitY), Digital India, PricewaterhouseCoopers and other relevant sources.

4 Growth of Digital Payments and Technologies Employed

The decade of 2010–20 is called a decade of payments in India. The introduction of an innovative payments system, entry of FinTech companies into the payment space, and a shift in the customer mindset were significant events during this decade [8]. The Indian digital payment space has witnessed robust growth in recent years with a Compound Annual Growth Rate (CAGR) of 23% [9]. This growth is largely contributed by innovative payment products, enhanced smartphone adoption, the pro-digital policy of the government and need for the faster payment methods [9]. More digital payment options are available to the 0 customers in India round the clock and all the days of the week at low or zero cost [8]. India has launched Regulatory Sandbox in 2019 to enhance innovation in the digital payment space [8]. Many innovative payment products are developed and put into operation to facilitate person-to-person, person-to-business, business-to-business, business-to-person, person-to-government and government-to-person digital payments [10].

COVID-19 brought down the volume and value of digital payments during lock-down periods. However, COVID-19 restrictions indirectly pushed digital payments. UPI has been a major contributor to the strong growth of digital payments with an exponential CAGR of 414% [9]. Further, digital payments in India are recognized as a tool to bring financially excluded people into the formal financial system. One-third of Indian households uses any one of the digital payments in India [10]. In India, customers strongly move towards digital payments pre- and post-COVID-19 [10].

India has reached the top of the table in real-time payments in the world in the year 2020 with 20.5 billion transactions and India trumped China and the USA in real-time payments [11]. The growth of digital payments in India since 2010 has been presented in Table 1.

During the last ten years in India, the payment system has undergone a paradigm change. Total digital payments have grown tremendously in India. The total number of digital payments transactions has reached 343,456 lakh transactions in 2019–20 from 9579 lakh transactions in 2010–11. Similarly, the total value of digital payment transactions has grown multifold during the period.

This digital payment growth has been made possible by technologies and pro-policies of the government. Many disruptive technologies have been used in the design of digital payment products. These technology enablers that are used in digital payments in India at present include the Internet, smartphones, NFC, RFID, APIs,

Table 1 Growth of digital payment during 2010 and 2020

Particulars	Volume (lakh)			Value (000' crore)		
	2010–11	2015–16	2019–20	2010–11	2015–16	2019–20
Payment systems						
Large value credit transfers	493	983	1507	48,487	82,457	131,156
Retail payments	9086	58,488	341,949	1308	9589	31,149
Total digital payments	9579	59,472	343,456	49,795	92,046	162,305

Source RBI (2021)

DLT, QR codes, tokenization, Internet of things (IoT), artificial intelligence (AI), and machine learning (ML), cloud, and data analytics (DA). Technology enablers and their benefits are presented in Table 2.

However, many new technologies have emerged in recent times that will take digital payments to the next levels in India. What are the technologies that shape the future of digital payments in India? Biometric authentication, mobile points of sales, cards to codes, enhanced security by AI and soundwave technology are the answer to the above question. Future technologies of digital payments in India and their benefits are presented and discussed (Table 3).

"Digital payments are set to grow continually, and they will be adopted on a larger scale in 2021" [12]. This growth potential of the Indian digital payments space attracted many Tech giants such as Google, Amazon, Apple, Samsung, Flipkart and FinTechs. The growth potential of the Indian digital payments space is $3.6 trillion [13].

Tech giants make use of their customer bases and their behavioural databases collected to provide better digital payment solutions. Tech giants have made strong

Table 2 Present technology enablers and their benefits in digital payments

Technology	Benefits
NFC and RFID	They enable contactless payments
QR codes	They are used for ease of use and speed
Internet of Things	It enables contactless digital payments through smartphones, smartwatches and rings
AI and ML	It enhances security in digital payments and detects frauds
Data analytics	It facilitates in-depth customer behaviour insights and decision making
Cloud	It reduces digital infrastructure costs and improves the quality of service
DLT	It ensures transparency, protects privacy and involves in risk management
API	It enables risk sharing and enhances operational and cost efficiencies (Infosys 2019)
Tokenization	It is the process of covering sensitive data into non-sensitive data to protect privacy and to eliminate the chance of fraud

Source Compiled by the authors

Table 3 Future technology enablers, and their usages and benefits in digital payments in India

Technology	Usage	Benefits
Biometric authentication	It is a verification method that authenticates the transaction after verifying the biological and structural characteristics of the customer such as fingerprint, retina and so on	It reduces fraud in digital payments and avoids identity theft
Mobile points of sales	This technology allows the merchants to take the point of sales beyond their business place to accept the payments	It moves the merchants away from brick-and-mortar businesses. The merchants will be free to accept their business payments at any place
Cards to codes	The EMV (Europay, Mastercard, Visa) technology makes use of codes to do a transaction rather than cards. The codes vary from transaction to transaction	Temporary use of codes enhances the security of a digital payment transaction multifold
AI and ML	AI is ever-improving. Top-rated security powered by AI and ML is applied to detect fraudulent transactions in real time	It reduces fraud
Magnetic secure transmission (MST)	MST creates a magnetic signal to enable quick payments	Quick payment is done without swiping the card
Soundwave technology	Soundwave technology is used in smart speakers and home assistants. Digital payment users can use voice commands to do a transaction through smart speakers and home assistants	It makes digital payments simple and convenient

Source Compiled by the authors

inroads in the Indian digital payments market [12]. These Tech giants have developed new business models and strategies to upscale their business. For instance, digital payments applications provide both bill payments and investments [12] (Fig. 1).

The digital payments market in India will observe 5 key trends in the year 2021. Key trends include the entry of big techs and new business models, wider acceptability of open banking, offline payment innovations, exclusive payment systems for banks and payment companies, and sustained innovations in digital payments [12].

Fig. 1 Key payments trends in 2021. *Source* PwC analysis

4.1 Soundwave Technology: Need and Experiments in India

Digital payments provide many benefits to the users and the providers. The digital payments sector keeps on evolving continuously over time. Many disruptive technologies such as NFC, RFID, API, tokenization, AI, ML, MST, biometric authentication and soundwave technology have contributed to the evolution and growth of digital payments in India. NFC facilitates contactless payments when there is an NFC-compatible device. All modern smartphones have NFC tags [14]. It should be noted that modern smartphones have NFC, but not the previously manufactured smartphones. As a result, approximately 2 billion smartphones are present in the market without NFC technology in India. These smartphone users cannot use contactless payments [14]. So, the digital payments sector needs a disruptive technology that can be accessible by all sections of society. The disruptive technology which is inclusive is "soundwave technology". One of the prominent advantages of sound wave technology is that this technology does not require a smartphone to do a digital transaction. Therefore, this technology can be accessed and used by all mobile phone users. How does soundwave technology work is shown in Fig. 2?

The soundwave technology can be adopted by both vendors and mobile phone users because it is affordable, secure, fast and accurate. More than 668 million users would utilize sound wave technology [15]. Indian banking system already started to experiment with sound wave technology to make a digital financial transaction. HDFC bank, one of the top private sector banks, tied up with Tone Tag, a FinTech company that deals with sound wave technology, to provide a digital payment facility to its customers. Tone Tag has its presence across the globe. 52 million customers use Tone Tag's sound wave technology. Tone Tag has 42 million users and 310 k merchants in India [16]. Ultracash is another FinTech start-up in India that facilitates digital payments between two mobile devices using "very high frequency and inaudible sound waves". The distinct feature of Ultracash is that it does not require any specialized hardware from the vendor's side. The vendors can use their mobile devices to receive digital payments. Chirp, a UK-based FinTech, uses sound wave technology to transact digital payments. Interestingly, Chirp technology does not require Internet, and pairing of the device is also not required. Thus, sound wave

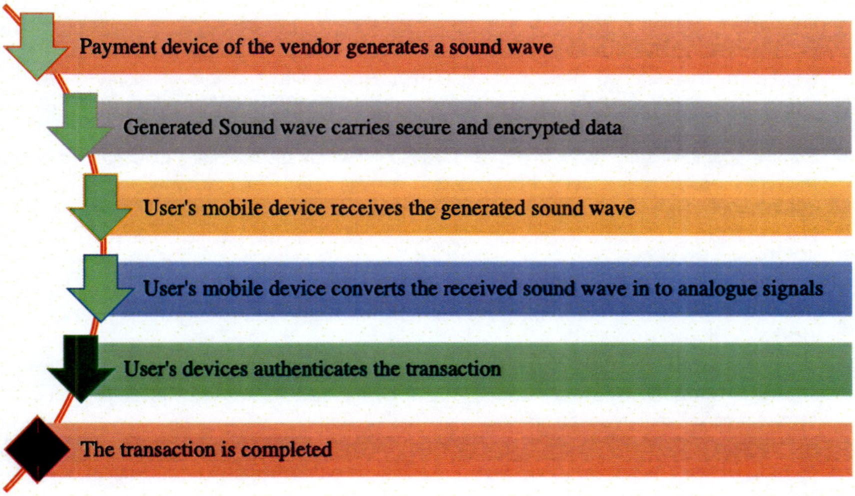

Fig. 2 Operation of sound wave technology in digital payments. *Source* Compiled by the authors

technology has started to penetrate the digital payments space, and it is going to be an inclusive technology in the payments and settlements segment [14].

5 Conclusion

The digital payments market in India has been growing steadily over a period. India is a global leader in the digital payment space. It outperformed two big countries such as China and the USA in digital payments. India is not only leading in numbers of transactions in digital payments, but also surges ahead in terms of technology adoption and innovations in DP products. Further, India has achieved the confidence and trust of the customers through the increased security features, better convenience, and simplicity in DP products. India is leading in the digital payments space even during the COVID-19. Many disruptive technologies and innovations are employed in digital payments in India to reach all sections of society.

References

1. Hord J (2020) How electronic payment works. https://money.howstuffworks.com/personal-finance/online-banking/electronic-payment1.htm
2. Chun SH (2019) E-commerce liability and security breaches in mobile payment for e-business sustainability. Sustainability 11(3). https://doi.org/10.3390/su11030715
3. Alter J (2018) The technology behind mobile payments. Proov. https://proov.io/blog/technology-behind-mobile-payments/

4. Hassan MA, Shukur Z, Hassan MK, Al-Khaleefa AS (2020) A review on electronic payments security. MDPI 1–19
5. Karsen M, Chandra YU, Juwitasary H (2019) Technological factors of mobile payment: a systematic literature review. Procedia Comput Sci 157:489–498. https://doi.org/10.1016/j.procs.2019.09.004
6. Shree S, Pratap B, Saroy R, Dhal S (2021) Digital payments and consumer experience in India: a survey based empirical study. J Bank Financ Technol 5(1):1–20. https://doi.org/10.1007/s42786-020-00024-z
7. Das P (2020) 5 E-wallet technology that will redefine digital payment market. Analytics Insight. https://www.analyticsinsight.net/5-e-wallet-technology-that-will-redefine-digital-payment-market/
8. RBI (2020) Payment and settlement systems and information technology in the reserve bank. CCIL, pp 187–200
9. PwC India (2020) The Indian payments handbook—2020–2025, Dec. [Online]. Available: https://www.pwc.in/assets/pdfs/consulting/financial-services/fintech/payments-transformation/the-indian-payments-handbook-2020-2025.pdf
10. Bijapurkar R, Shukla R, Rai P, Sachdeva V (2020) Digital payments adoption in India, 2020, pp 1–7. [Online]. Available: https://www.linkedin.com/pulse/digital-payments-adoption-india-2020-a-npci-report-ram-rastogi
11. The Economic Times (2020) Transactions 2025
12. PwC India (2021) What lies ahead decoding digital payments trends for 2021
13. CB Insights (2020) From Alibaba to Zynga: 45 of the best VC bets of all time and what we can learn from them. https://www.cbinsights.com/research/best-venture-capital-investments/
14. MEDICI (2020) Sound-based payments as an inclusive technology for the developing world
15. Sharma A (2019) How soundwave technology is helping small businesses with digital payments. Instamojo
16. Pradhan D (2019) ToneTag's sound wave-based digital payments tech is making all the right noises. Inc42

Performance Study of Data Fusion Using Kalman Filter and Learning Vector Quantization

Shobha and N. Nalini

Abstract Data fusion is the process of integrating diverse data sources and provide more accurate, clear, and useful information than any one data source might provide. Data fusion measures are frequently characterized as low, intermediate, or high, depending on the stage of processing at which they are combined. Robust soft LVQ, generalized learning vector quantization, generalized matrix LVQ, and generalized relevance learning vector accuracy with Kalman filter data fusion are implemented in this article.

Keywords Accuracy · Data fusion · Efficiency · Kalman filter · LVQ · Sensor

1 Introduction

Machine learning is one the most exciting techniques. Artificial intelligence uses the data generated by the sensors and maps it to actuators with the help of learning algorithms. The learning algorithms that provide input and output then enable computers or machines to 'learn' without explicitly programming are known as machine learning algorithms. The machine learning algorithms work on a huge volume of data generated in the environment [1–3]. These datasets are often inconsistent, repetitive, and conflict. Therefore, data fusion methods are required to improve the performance and gain [4, 5].

A diagnosis system based on CNN and ensemble learning is discussed in Ilhan et al. [6]. The disadvantage of this method is that deep feature sets created by seven CNN architectures were concatenated for feature level fusion and distributed to a large number of classifiers, limiting efficiency.

Shobha (✉) · N. Nalini
Department of Computer Science and Engineering, Nitte Meenakshi Institute of Technology, Yelahanka, Bangalore, Karnataka, India
e-mail: shobha.p@nmit.ac.in

N. Nalini
e-mail: nalini.n@nmit.ac.in

S. Shakya et al. (eds.), *Proceedings of Second International Conference on Sustainable Expert Systems*, Lecture Notes in Networks and Systems 351,
https://doi.org/10.1007/978-981-16-7657-4_8

Luo et al. [7] describe a multisensor integration and fusion approach. The disadvantage is that the adaptive perspective of fusion is not taken into consideration.

Yan et al. [8] wrote on the significance of trustworthy data fusion and mining are primarily concerned with a holistic data process and analysis that is efficient, accurate, safe, privacy-preserving, and dependable. It has become crucial to the future success of IoT. The processing and merging of IoT data have improved in recent years.

The data fusion was classified into three levels by Steinmetz et al. [9]: low, mid, and high levels. Before a multivariate system, input from several sources is consolidated into a single matrix via a low-level fusion process [10–12]. If the data fusion is not similar in variability and scale, one source of data may have a benefit over the other. The relevant information is merged at a mid-level in the fusion process.

Data fusion is important in data mining and is used in a variety of civilian and military applications, including wildlife habitat monitoring, hazard detection, and surveillance and reconnaissance missions [13].

2 Methodology

Data are gathered from a variety of sources, including text, image, and video. For data fusion, the linear vector quantization and clustering approaches are based on supervised and unsupervised learning, respectively. A hybrid model based on linear vector quantization and Kalman filter is utilized to increase the accuracy and improve time complexity.

In a geospatial application, the multiple sensors capture the data of objects continuously. Due to noise, there would be inconsistency in few records. The data fusion technique is used to refine the records before processing the information. The schematic diagram of the proposed work is given in Fig. 1.

Let us assume two sensors S_1 and S_2 are taking measurement of unknown parameter x. Let m_1 and m_2 be the measure of a parameter with noise N having uniform distribution with standard deviation σ with v as measurement noise.

$$m_1 = p + v_1 \text{ and } m_2 = p + v_2$$

In order to combine the two measurements to produce an optimal estimate of p.

$$\hat{p} = c_1 m_1 + (1 - c_1) m_2$$

Here, the estimate is the linear combination of measurements

$$\hat{p} = \sigma_{22}/(\sigma_{12} + \sigma_{22}) m_1 + \sigma_{12}/(\sigma_{12} + \sigma_{22}) m_2 \tag{1}$$

Let us assume that the measurement m_2 taken at time t_2 was taken and immediately after measurement m_1 which is taken at t_1.

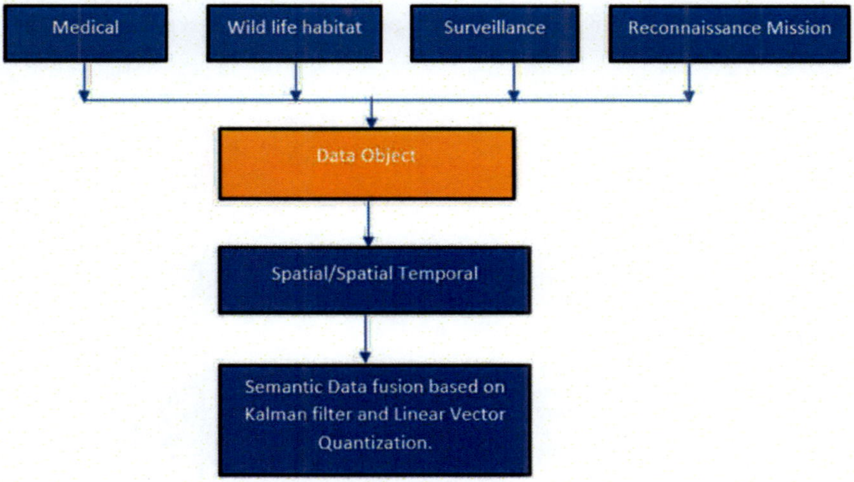

Fig. 1 Data fusion

The estimate at of p is

$$\hat{p}(t_2) = \sigma_{22}/(\sigma_{12} + \sigma_{22})m_1 + \sigma_{12}/(\sigma_{12} + \sigma_{22})m_2 \tag{2}$$

Rewriting the equation

$$\hat{p}(t_2) = m_1 + [\sigma_{12}/(\sigma_{12} + \sigma_{22})][m_2 - m_1] \tag{3}$$

We know that, $\hat{p}(t_1) = m_1$

$$\hat{p}(t_2) = \hat{P}(t_1) + K(t_2)\big[m_2 - \hat{p}(t_1)\big] \tag{4}$$

$$K(t_2) = \sigma_{12}/(\sigma_{12} + \sigma_{22}) \tag{5}$$

Equations (4) and (5) are in Kalman filter form [14].

An estimator is used to the Kalman filter is the most extensively used estimation algorithm. This is needed as sensors with various information rates and correctness are to be consolidated. An optimal estimator combines all the available information. An estimator is optimal when it limits the estimation error in a strictly defined statistical sense based on an optimality criterion.

In estimation, the 'filter' refers to estimating the states or parameters to describe the system at the current time, based on all past measurements.

For the system of linear equation is represented by

$$A \cdot p = y.$$

A general linear system of m linear equations with n unknowns p_i can be written as

$$a_{11}p_1 + a_{12}p_2 + \cdots + a_{1n}p_n = y_1$$

$$a_{21}p_1 + a_{22}p_2 + \cdots + a_{2n}p_n = y_2$$

$$a_{m1}p_1 + a_{12}p_2 + \cdots + a_{mn}p_n = y_m$$

This can be represented by matrix:

$$A = \begin{bmatrix} a11 & a21 & \cdots & a1n \\ \vdots & \vdots & \vdots & \vdots \\ am1 & am2 & \cdots & amn \end{bmatrix} \quad p = \begin{bmatrix} p1 \\ p2 \\ \cdot \\ \cdot \\ pm \end{bmatrix}, \quad y = \begin{bmatrix} y1 \\ y2 \\ \cdot \\ \cdot \\ \cdot \\ ym \end{bmatrix}$$

Let p be the vector of unknowns, m is measurement vector; H is measurement matrix; w is process noise; v measurement noise vector, R measurement covariance matrix, Q is process matrix; $\hat{p}$ is the estimate; P is covariance of $\hat{p}$; k is discrete time point/sample number.

Kalman filter uses measurements m_k from sensors and mathematical model of dynamic system that process w equation and measurement equation

$$pk = Ax_{k-1} + Bu_k + w_{k-1}$$
$$Z_k = Hx_k + V_k,$$

where w and v represent process and measurement noise.

2.1 Kalman Filter Algorithm

This section explains the concepts of Kalman filter algorithm to perform the data fusion. The Kalman filter performs the data fusion using two steps, namely time update and measurement update [15, 16].

Algorithm

Initial estimate: $\hat{p}_{k-1}$ and P_{k-1}

Step 1: Time Update

(i) Next state/value

$$\hat{p}-_k = Ax_{k-1} + Bu_{k-1}$$

(ii) Project error covariance

$$P_k = AP_{k-1}A^{\mathrm{T}} + Q$$

Step 2: Measurement Update

(i) Compute the Kalman gain

$$K_k = P_k H^{\mathrm{T}}\left(HPKH^{\mathrm{T}} + R\right)^{-1}$$

(ii) Update estimate with measurement m_k

$$pk = \hat{x} -_k + K_k\left(m_k - H\hat{p}-_k\right)$$

(iii) Update the error covariance

$$P_k = (1 - K_K H)P^{-k}$$

Repeat Steps 2 and 3 till k = constant/$k = \frac{1}{H}\frac{1}{H}$/$k$ is adjusted by the tuning parameter.

2.2 *Linear Vector Quantization Techniques*

In the proposed work, the various linear vector quantization methods are used for the classification, and performance of the same is analyzed. Learning vector quantization (LVQ) is a statistical pattern classification technique that seeks to learn prototypes for class regions. Hyperplanes between prototypes identify class zones, resulting in Voronoi partitions [17, 18].

Figure 2 shows the architecture of LVQ, it assumes '**n**' number of input units and '**m**' number of output units. The layers 'L_i's are fully interconnected with weights 'W_i' on them.

The parameters that were used are listed in linear vector quantization training methods as well as in the flowchart. Let **X** denotes the training vector (x_1, …, x_i, …, x_n). Let **T** denotes the class for training vector (x_1, …, x_i, …, x_n). The weight vector for ***j*th** output unit is represented by $\mathbf{W_j}$, and class associated with the ***j*th** output unit is denoted by C_j.

To create a learning rule through gradient ascent, Robust Soft LVQ (RSLVQ) uses a statistical objective function. Gaussian power distribution of data in RSLVQ as a result, each class's probability density function is considered to be a Gaussian mixture in this method. The logarithm of the ratio of the valid class's power distribution functions vs the power distribution functions of the invalid classes are employed as a cost function given a data point. This must be exploited to the fullest extent possible. Figure 3 describes the learning vector quantization prediction.

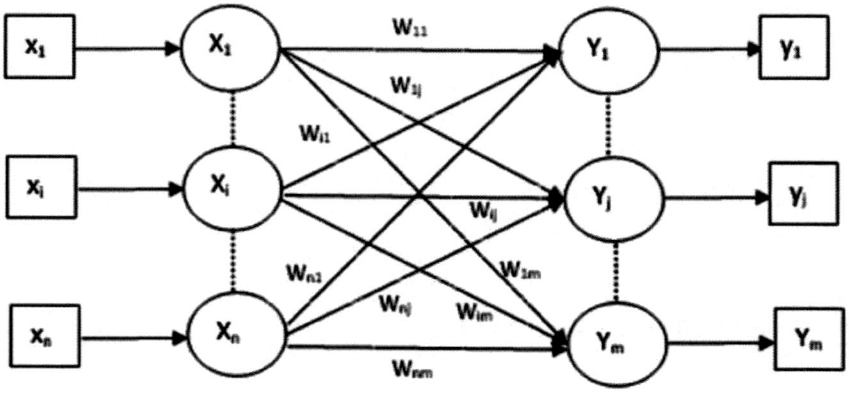

Fig. 2 Learning vector quantization

A cost function is used in generalized learning vector quantization (GLVQ). The cost is defined in this method in such a way that the steepest descent yields a learning rule. To reduce errors, the cost function is applied.

In the generalized matrix linear quantization approach, the generalized distance is used (GMLVQ) Using an adaptive metric, a generalization boundary is calculated for the generalized relevance LVQ (GRLVQ) approach [19].

3 Results

The described fusion architecture has been used on radar datasets. The dataset consists of signals recorded by radars at a fixed area of interest. The dataset has three fields time of signal capture, signal strength, and target ID.

The model is experimented using Python programming. The experimental results are listed in the following diagrams. The graph Fig. 4 compares the accuracy of Robust soft LVQ, generalized learning vector quantization, generalized matrix LVQ, and generalized relevance learning vector with and without Kalman filter. It is observed that the accuracy of the learning vector classifier is improved with the Kalman filter.

The graph Fig. 5 shows the 100% variance covered by the algorithms.

Figure 6 shows the time complexity of algorithms with and without Kalman filtering, and it is observed that generalized learning vector quantization and generalized relevance learning vector perform better with Kalman filtering. Generalized learning vector quantization shows better time complexity as it uses adaptive metrics, and generalized relevance learning vector shows better time complexity because it incorporates an intuitive update rule to measure the relevance of the input data dimensions and permits efficient input pruning.

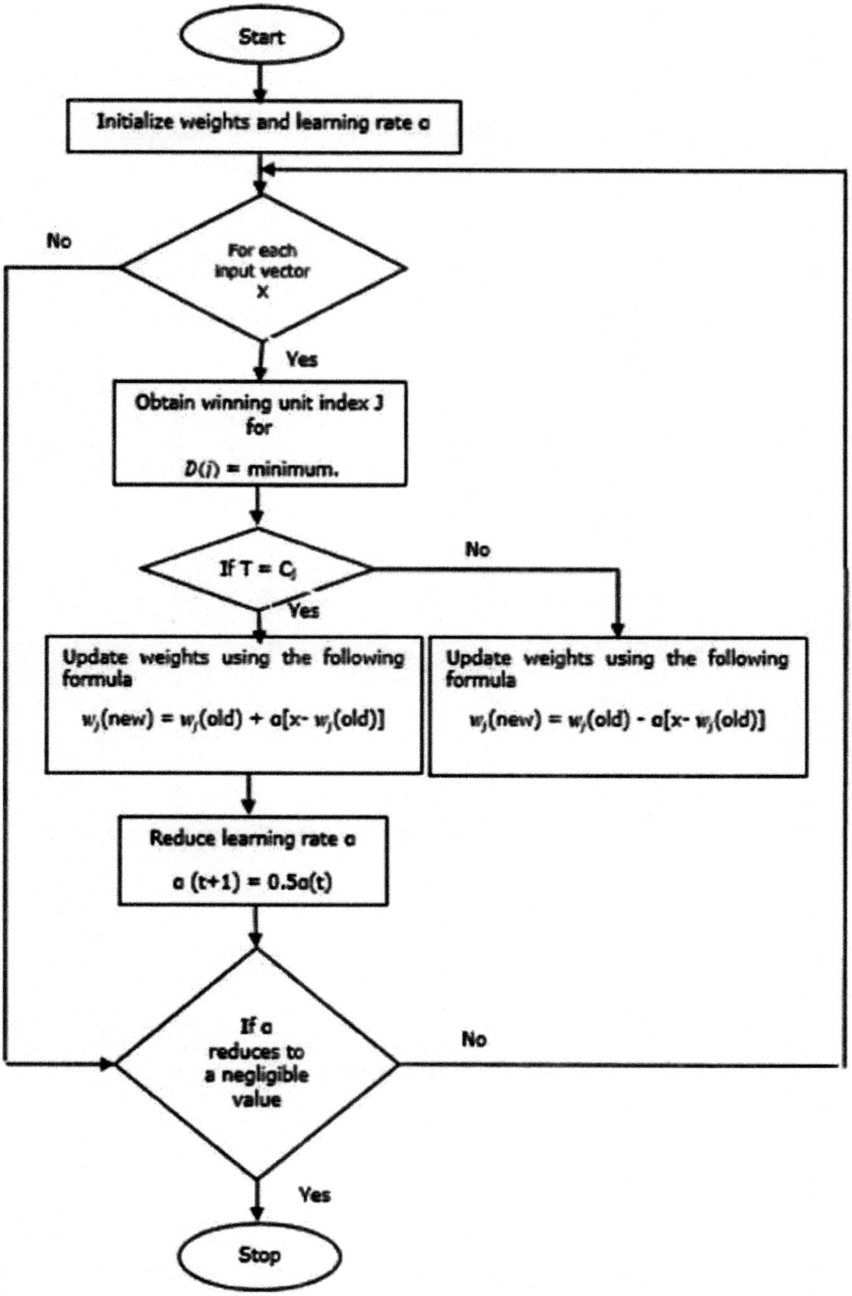

Fig. 3 Flowchart for learning vector quantization prediction

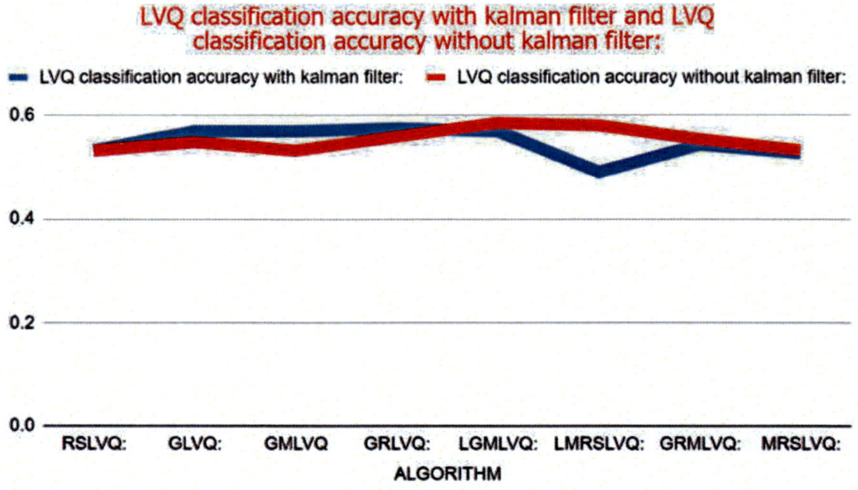

Fig. 4 Accuracy

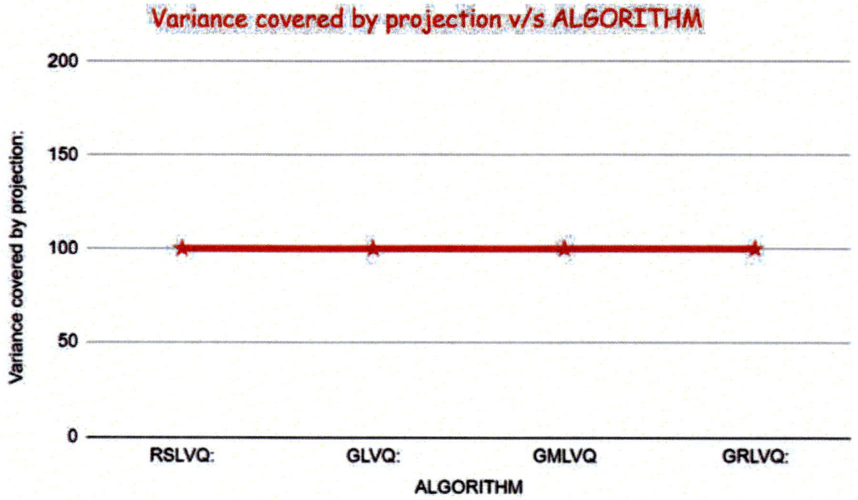

Fig. 5 Variance covered

4 Conclusion

Data fusion is important in machine learning because artificial intelligence deals with large sets of inaccurate data. Using data fusion, the space complexity can be reduced. This results in better classification which in turn increases the accuracy. The data fusion is obtained based on the feature or decision level.

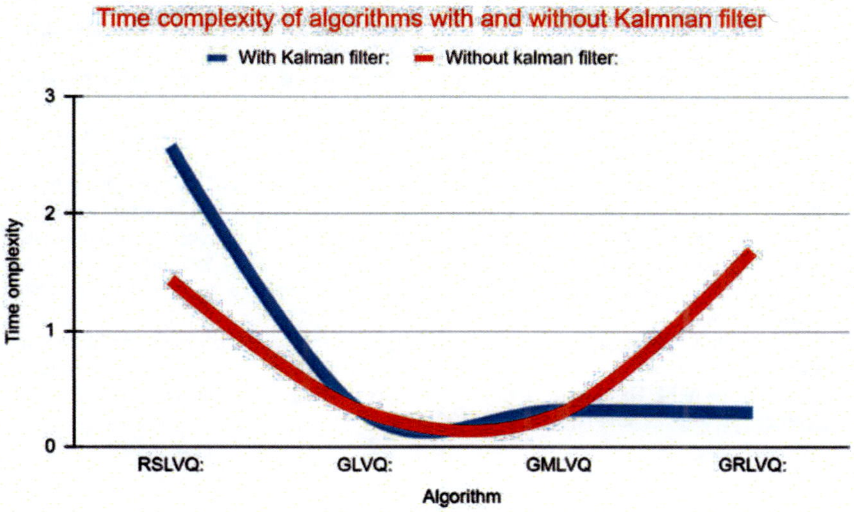

Fig. 6 Time complexity

The proposed work uses feature-based data fusion. This paper compares the accuracy, variance covered by the projection and time complexity for the radar dataset using Kalman filter data fusion and RSLVQ, GLVQ, GMLVQ, and GRLVQ classification techniques. Future work will concentrate on developing models to increase the accuracy.

References

1. Samourkasidis A, Athanasiadis IN (2020) A semantic approach for timeseries data fusion
2. Chen B, Li J, Jin Y (2021) Deep learning for feature-level data fusion: higher resolution reconstruction of historical landsat archive
3. Borras E, Ferre J, Boque R, Mestres M, Acena L, Busto O (2015) Data fusion methodologies for food and beverage authentication and quality assessment—a review
4. Pau LF (1988) Sensor data fusion. J Intell Robot Syst 103–116
5. Palani U, Vasanthi D, Rabiya Begam S (2020) Enhancement of medical image fusion using image processing. J Innov Image Process (JIIP) 2(04):165–174
6. Ilhan HO, Serbes G, Aydin N (2020) Decision and feature level fusion of deep features extracted from public covid-19 data-sets
7. Luo RC, Yih CC, Su KL (2002) Multisensor fusion and integration: approaches, applications, and future research directions. IEEE Sens J 2(2):107–119
8. Yan Z, Liu J, Vasilakos AV, Yang LT (2015) Trustworthy data fusion and mining in internet of things. Future Gener Comput Syst 49:45–46
9. Steinmetz V, SeHvila F, Bellon-Maurel V. A methodology for sensor fusion design: application to fruit quality assessment
10. Wilfried E (2007) A review on system architectures for sensor fusion applications. Springer
11. Wang H, Fang X, Li Y, Zheng Z, Shen J (2021) Research and application of the underground fire detection technology based on multi-dimensional data fusion

12. Chango W, Cerezo R, Romero C (2021) Multi-source and multimodal data fusion for predicting academic performance in blended learning university courses
13. Atluri G, Karpatne A, Kumar V (2018) Spatio-temporal data mining: a survey of problems and methods
14. Musoff H, Zarchan P (2009) Fundamentals of Kalman filtering: a practical approach, 3rd edn. American Institute of Aeronautics & Astronautics
15. Uma Mageswari A, Joseph J, Vinodha R (2012) A comparative study of Kalman filter, extended Kalman filter and unscented Kalman filter for harmonic analysis of the non-stationary signal
16. Pinto AR, Montez C, Araújo G, Vasques F, Portugal P (2014) An approach to implement data fusion techniques in wireless sensor networks using genetic machine learning algorithms
17. Zhou T, Chen M, Yang C, Nie Z (2020) Data fusion using Bayesian theory and reinforcement learning method
18. Deng C, Liu X, Chanussot J, Xu Y, Zhao B (2019) Towards perceptual image fusion: a novel two-layer framework
19. Ghosh A, Biehl M, Hammer B (2006) Performance analysis of LVQ algorithms: a statistical physics approach. Neural Netw 19:817–829

Unveiling the Security Vulnerabilities in Android Operating System

G. Renjith and S. Aji

Abstract Android is one of the biggest multi-featured operating systems in this decade. Android Operating System (AOS) is the preferred choice for most smart use cases. Attacks targeting AOS is increasing every day. The major reason behind these attacks is the vulnerabilities in the AOS source code. AOS vulnerability analysis is very challenging and time-consuming because AOS source code is very heavy, roughly around 93 million lines of code in Android version 11. Hence, the Android security research direction more focuses on malware analysis of Android apps instead of vulnerability analysis of the AOS. Most of the existing Android vulnerability analysis techniques focus on vulnerability analysis of Android apps rather than AOS source code. The rest of the very few pieces of literature available for analysis of AOS source code vulnerabilities are observed to be considered a subset of vulnerabilities reported till 2018 and targets older versions of Android. In this work, we focus on all the officially reported AOS source code vulnerabilities between August 2015 and June 2021, considering the latest Android version in Google Android Security Bulletin. A comprehensive vulnerability analysis is conducted on the source code of all these vulnerabilities. As a consequence, we have identified the latest trends and patterns in AOS vulnerabilities and suggested the recommendations. We have extracted the feature vectors from AOS source code vulnerabilities and presented the cross-section of an active vulnerability from source code to its feature vector format that helps for preparing vulnerability dataset for designing any AI-based vulnerability detection mechanisms. To the best of our knowledge, this is the first work that brings all the aspects of AOS source code vulnerabilities which helps the security researchers to build efficient detection systems.

Keywords Android operating system · Vulnerability analysis · Vulnerability detection

G. Renjith (✉) · S. Aji
Department of Computer Science, University of Kerala, Thiruvananthapuram, Kerala, India
e-mail: renjithg@keralauniversity.ac.in

S. Aji
e-mail: aji@keralauniversity.ac.in

S. Shakya et al. (eds.), *Proceedings of Second International Conference on Sustainable Expert Systems*, Lecture Notes in Networks and Systems 351,
https://doi.org/10.1007/978-981-16-7657-4_9

1 Introduction

There are around 2.5 billion+ Android-activated devices in the market. Apart from running in smartphones and tablets, AOS is also widely deployed in other embedded domains like cameras, wearables, home appliances, IoT, and Automotive. The majority of the AOS source code is open-source software, and hence there is no strict software engineering life cycle or practices followed while designing and writing the software. Design flaws and vulnerabilities are reported due to this reason. AOS includes a mixture of programming languages C, C++, Java, etc., which makes AOS even more complicated. The Android version 11 source code from [1] is downloaded, and lines of code (LOC) is estimated using "cloc" utility. The layer-wise LOC is captured in Table 1. This gives the overall understanding of the size of the AOS source code.

The major problem Android facing is the increasing number of cyberattacks every day. As per the latest statistics [2], an average of 480,000 new malware attacks is detected every month during last year. The primary reasons for these increasing malware attacks are the back doors in the AOS due to the AOS source code vulnerabilities. Even though there are many malware detection and vulnerability mechanisms are available, these are found to be less effective due to the challenges faced in the Android ecosystem. Some of the important challenges are: lack of strict software engineering life cycle, open-source nature of software development, complicated software architecture, heavy code base, major software design and source code changes during every new version migration, diverse hardware supported, third-party App stores etc. The best-recommended approach is to solve the cause of the cyberattacks rather than chasing every new Android apps. Vulnerability analysis and detection are the two important directions in this regard. The limitations in existing Android source code vulnerability research domain are:

1. Lack of availability of automated tools for vulnerability extraction and analysis. Manual extraction and analysis are required for all the reported vulnerabilities in the Android Security Bulletin [3]. This is very time-consuming.

Table 1 Android version 11 layer-wise line of code (LOC)

Layer	LOC in millions
Application	4.764
Framework	10.3
Libraries	50.835
ART	2.551
HAL	6.336
Kernel	18.472
Total	**93.26**

2. Less focus in vulnerability analysis of AOS compared to Android apps. This is because of the difficulty in extracting and studying the vulnerabilities in different layers of AOS such as application, framework, ART, HAL etc.
3. Lack of consideration of vulnerabilities reported after 2018. There are massive changes happening in smartphone and IoT domains in the last three years which are directly impacting Android revisions and source code as well. Android source code is updated every day. There are almost four new Android versions (9, 10, 11, and 12) released after August 2018. The average life span of an Android phone is almost 2–3 years, and the older version of Android smartphones from 2018 is already getting obsolete by now. Hence, it is very important and relevant to analyse the newer vulnerabilities reported for newer versions of Android.
4. Lack of consideration of all the programming languages in AOS. This is because of the difficulty in extracting and analysing the vulnerabilities of different programming languages such as C, C++, and Java.
5. Lack of consideration of differences in the entire source between vulnerable code and its corresponding fixed code. Most of the vulnerability analysis and detection mechanisms focus only on shortcode differences that do not give the vulnerability's actual context or boundary.
6. Lack of consideration efficient deep learning techniques for vulnerability detection. Since the source code can be best represented as a graph feature format, utilizing the latest graph supported algorithms would give more realistic results.
7. Lack of consideration of closed source vulnerabilities. Some of the source code in Android is managed by corresponding chip vendors. The vulnerabilities and their corresponding fixes for those source codes are not publicly available for analysis.
8. Lack of availability of ready to use AOS source code vulnerability graph feature datasets compatible with machine learning or deep learning based detection engines.

In this work, we tried to cover all the limitations noticed in the vulnerability analysis phase. The main contributions are:

1. All the reported Android source code vulnerabilities from all the layers from 2015 till June 2021 are captured.
2. The metadata for all vulnerabilities are manually extracted, analysed, and different trends and patterns are presented.
3. The vulnerable and corresponding fixed source code for all the publicly available vulnerabilities are manually extracted, analysed, and different aspects in Android are discussed.
4. Discussed the design of vulnerability extraction procedure for AOS source and conducted comprehensive experimentation to extract the feature vectors from AOS source code.
5. Selected an active vulnerability in Android source code and presented the cross section of entire feature extraction procedure which helps to prepare the dataset for vulnerable and corresponding fixed code.
6. Recommended a design approach for designing a vulnerability detection mechanism using the extracted graph feature vectors.

The advantages of this vulnerability analysis work are:

1. Helps to understand the latest vulnerability trends and patterns in latest versions of Android till June 2021.
2. Helps to understand how to extract the feature vectors from Android source code vulnerabilities.
3. Helps to understand how to build a vulnerability detection engine for Android.
4. Helps to understand the critical areas or layers that needs immediate research attention.

In Sect. 2 of this work, we discuss the review on related literature in vulnerability analysis and detection in Android. The detailed Android vulnerability trends and patterns are discussed in the vulnerability analysis (Sect. 3). Section 4 presents the cross-section of an active Android vulnerability which helps to understand the feature extraction procedure from Android source code. Section 5 discusses the recommended approaches for vulnerability detection mechanisms in Android. Finally, Sect. 6 concludes our work by giving an insight into the future work.

2 Related Works

In this section, we discuss the related works reported for AOS source code vulnerability analysis and vulnerability detection categories.

Farhang et al. [4] studied the Android-related vulnerabilities that affect the Android operating system. They have explained about vulnerability types, security aspects, affected areas, the time required for fixing the vulnerabilities etc. But, they have only considered a limited number of vulnerabilities reported between August 2015 and November 2016 on the very old version of Android. Siddiquie et al. [5] discussed vulnerabilities that risked the dual persona or multiple user environment in Android 5 and above. They have collected the vulnerabilities from NVD and MITRE CVE databases. They have considered the Android vulnerabilities till 2017. Shewale et al. [6] conducted a study on vulnerabilities that focused on understanding and analysing the vulnerabilities present in Android platform. They have presented the vulnerabilities present in each layer of Android architecture. They have covered the vulnerabilities between old Android version 1.5 and Android 4.4. Jimenez et al. [7] conducted a manual analysis of Android vulnerabilities reported in the National Vulnerability Database. Cause of Android vulnerabilities, vulnerable components, and fix analysis are discussed. But, they have considered the vulnerabilities in the old version of Android between 2008 and 2014.

Wu et al. [8] studied the Android system vulnerabilities and proposed a detection mechanism based on a similarity-based algorithm. Vulnerabilities with different programming languages such as C, C++ and Java from different layers in Android OS are considered for their analysis. But, they have considered the reported vulnerabilities till June 2018. Min et al. [9] proposed a vulnerability detection mechanism for Android framework. A vulnerability mining framework based on dynamic taint analysis is discussed in their work. Vulnerabilities in Java programming language source code is considered. More details about extracted vulnerabilities have not been discussed. Luo et al. [10] proposed a method called CENTAUR that uses symbolic execution of Android framework for vulnerability discovery. Details about the vulnerabilities reported in Android, trends and patterns, performance validation with reported vulnerabilities are not seen.

It is observed that the vulnerability extraction and analysis were done on old versions of Android and between 2008 and 2018. No recent literature are reported with analysis of vulnerabilities identified after 2018 in the newer version of Android. Hence, our vulnerability analysis work bridges all the gaps in vulnerability understanding in Android operating system.

3 Vulnerability Analysis

The Android Security Bulletin releases the security vulnerabilities that are reported in Android operating system and also updates the security patches in order to fix these issues. The vulnerabilities are patched in the AOSP [1] source code repository. The National Vulnerability Database (NVD) [11] is a repository of vulnerabilities and metadata such as checklist references, related security issues, misconfigurations, product names, and metrics for severity and impact. All vulnerabilities in the NVD have been provided with an identifier. CVE is the list of vulnerability entries that contains a unique identification number, a short description, and references. CVE [12] entries are fully synchronized and updated to the NVD database. We manually capture the vulnerabilities from the monthly Android Security Bulletin. Additional metadata about the corresponding CVE ID is collected from CVE and NVD databases. The extracted vulnerabilities are also manually categorized according to various factors and analysed for various vulnerability trends.

The year-wise distribution of Android source code vulnerabilities reported from August 2015 till June 2021 is presented in Fig. 1. It is evident that almost 50% of the vulnerabilities are observed between 2018 till 2021. Hence, it is very important to consider this window for any vulnerability analysis and detection.

The Android version-wise distribution of publicly available Android source code vulnerabilities extracted from Google Security Bulletin is discussed in Fig. 2. Android version 7 has been reported with maximum vulnerabilities. Some of the vulnerabilities listed in the Bulletin did not provide the metadata information about the Android version, which is marked as "Not Specified".

The programming language-wise distribution of publicly available Android source code vulnerabilities is discussed in Fig. 3. It is observed that the highest number of vulnerabilities are reported in components developed in the C programming language in the Android layered software architecture. CPP and Java are second and third positions. As per this info, it is very important to give more focus on vulnerability detection engines that supports C and CPP programming languages.

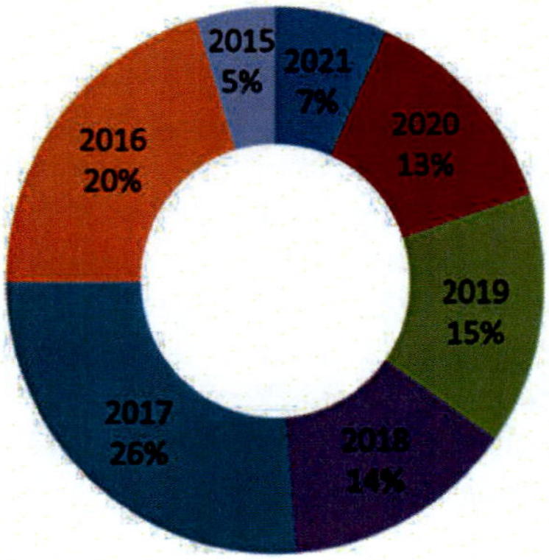

Fig. 1 Year-wise distribution of Android source code vulnerabilities from 2015 to 2021

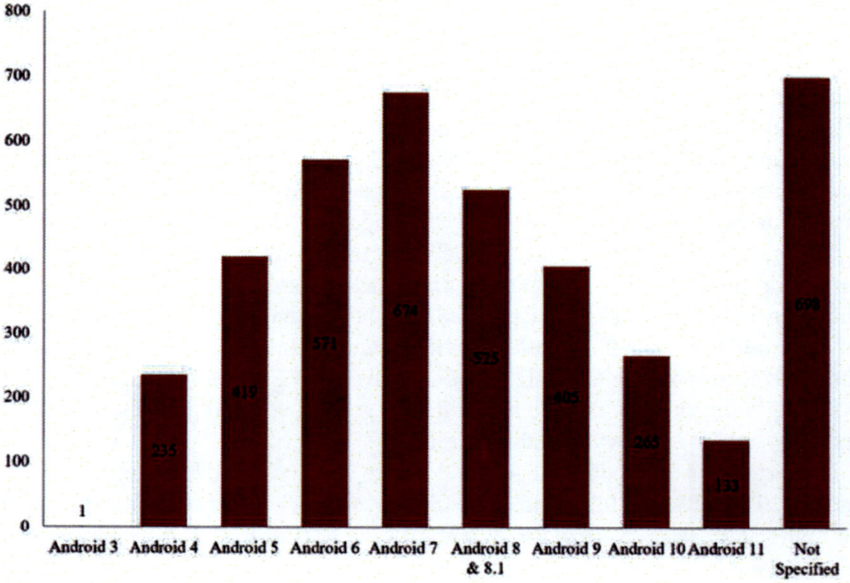

Fig. 2 Android version-wise distribution of source code vulnerabilities

Fig. 3 Programming language-wise distribution of Android source code vulnerabilities

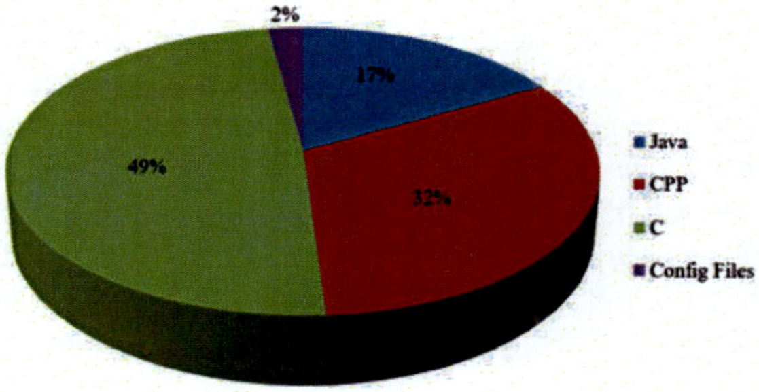

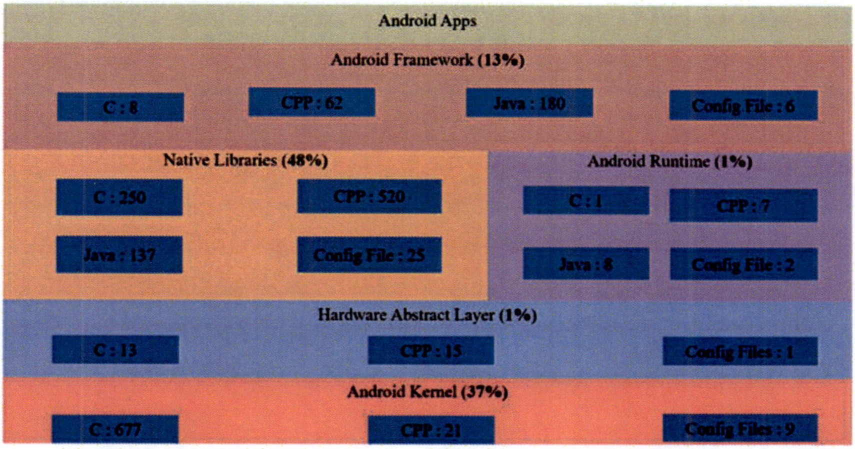

Fig. 4 Android layered software architecture-wise distribution of source code vulnerabilities

Fig. 5 Licensing-wise distribution of Android source code vulnerabilities

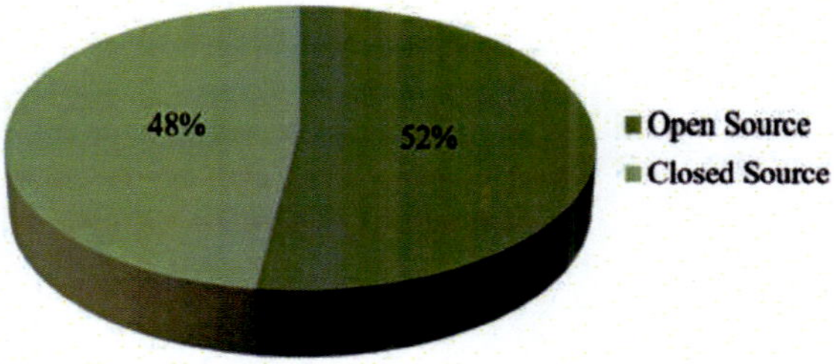

Android layered software architecture-wise distribution of source code vulnerabilities is presented in Fig. 4. Source code of each of the vulnerability IDs (CVE ID) is manually analysed and mapped the layer to which the module/component corresponding to the vulnerable source code belongs. We noticed that Android native libraries and Android kernel are identified with maximum vulnerabilities. Vulnerabilities in Android kernel-space are very dangerous because the existing user-space malware detection mechanisms are not effective for malware that exploits the kernel-space vulnerabilities.

Licensing-wise distribution of source code vulnerabilities is presented in Fig. 5. Even though Android is known to be an Open Source Project (AOSP), but it has a lot of closed source codes. It is identified that closed source codes are mainly coming from chip manufactures for their board support packages, mostly in the kernel-space. Analysis shows that both open source and closed source in Android operating system are equally vulnerable with a slight increase in the open-source codes. But, it is very important to note that closed source vulnerabilities are also potential causes for malware attacks. Hence, it should be also be analysed.

Vulnerability type-wise distribution of publicly accessible Android source code vulnerabilities is discussed in Fig. 6. It is seen that the Escalation of Privilege (EoP) is

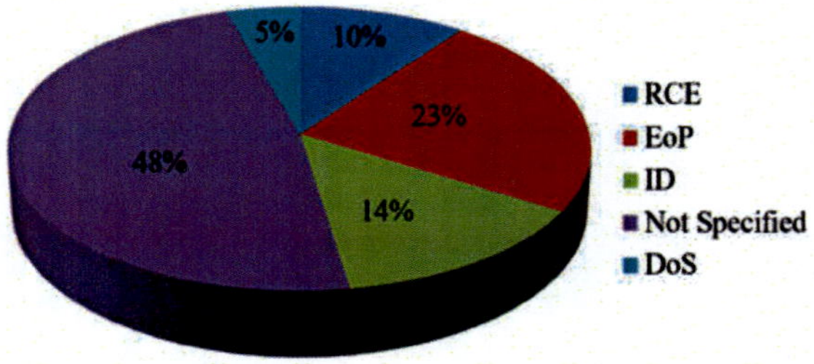

Fig. 6 Vulnerability type-wise distribution of Android source code vulnerabilities

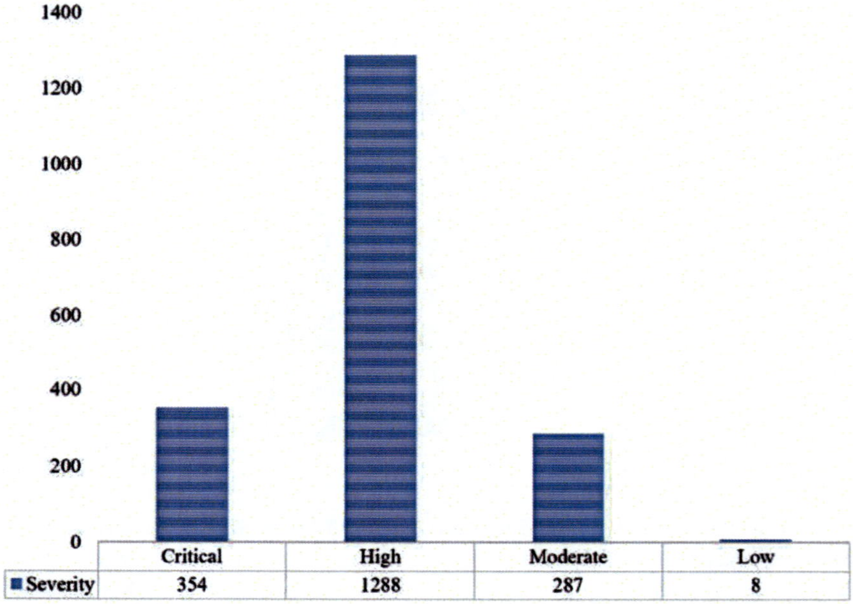

Fig. 7 Vulnerability severity-wise distribution of Android source code vulnerabilities

the most common vulnerability, followed by the remote code execution type (RCE). The least common type of vulnerability is denial of service (DoS), followed by information disclosure (ID). Some of the vulnerabilities reported in closed source components do not have enough metadata information in order to classify their type, and hence they are marked as "Not Specified".

Severity-wise distribution of Android source code vulnerabilities is discussed in Fig. 7. Analysis shows that the highest number of vulnerabilities reported are with "High" Type followed by "Critical", "Moderate", and "Low", respectively.

4 Feature Extraction

For designing any vulnerability detection mechanism, it is important to convert this vulnerable source code to machine learning friendly feature vector formats. The vulnerability extraction for preparing the feature vectors are presented in Fig. 8.

In order to understand more clearly, and officially reported Android source code vulnerability with "CVE-2020-0023" ID [13] and "Critical" severity from Android Security Bulletin is taken and presented the feature extraction in Fig. 9. The vulnerability is in setPhonebookAccessPermission of AdapterService.java, and there is a possible disclosure of user contacts over Bluetooth due to a missing permission check. The impact leads to local information disclosure if a malicious app enables contacts over a Bluetooth connection. The possible fix is to add permission check in setPhonebookAccessPermission of AdapterService.java.

As shown in Fig. 9, the first step is to extract the source code corresponding to the CVE ID. Study the source code and understand the vulnerability, its impact. Define the boundary of the vulnerable code snippet from the overall source code, and it can be a function boundary, class boundary etc. Convert this code snippet to rich intermediate graphs such as CFG, PDG, AST, CPG [14, 19] etc., with the support of conversion tools. Parse the intermediate graphs and extract the graph information and create the output graphs. The information available as strings can be converted to numerical formats using word embedding techniques. Finally, convert these graphs into matrices formats, and this will give the output feature vector for the vulnerable code. Similarly, define the possible fixes for the vulnerable code snippet and continue the same procedure for the fixed code snippet as well. This will give the output feature vector for the fixed code. Repeating the same procedure for entire vulnerabilities will produce the final dataset.

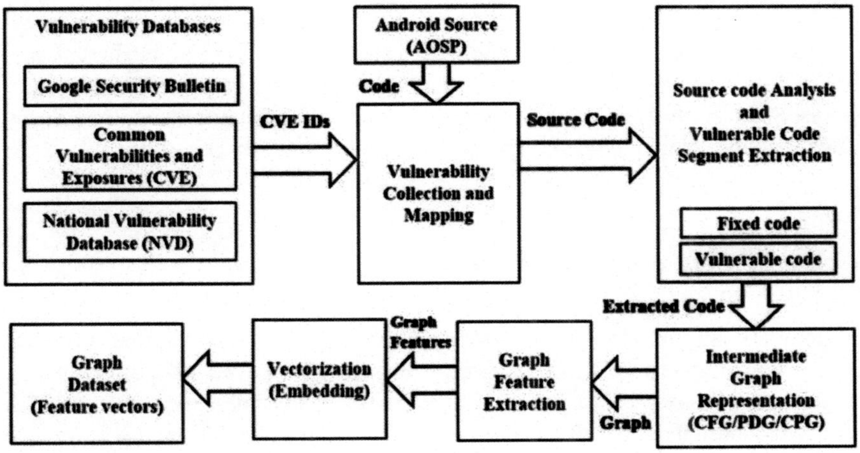

Fig. 8 Vulnerability feature extraction methodology

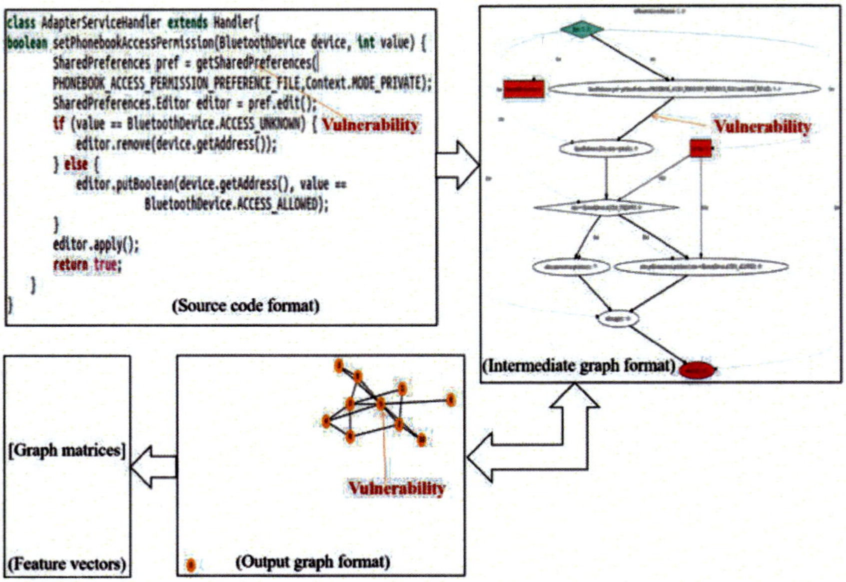

Fig. 9 Cross-sectional view of "CVE-2020-0023" vulnerability ID

5 Vulnerability Detection

Once the dataset is prepared for all the vulnerabilities, the next step is to design a vulnerability detection mechanism. There are multiple options to define the detection mechanism. First recommended mechanism is to use the potential of graph neural network deep learning algorithms such as graph convolutional networks [15] and graph attention networks [16]. Second recommended mechanism is to use the graph embedding mechanisms such as graph2vec [17] algorithm. The extracted graph feature vectors are compatible with graph neural network and graph embedding algorithms. Since Android operating system is mostly used in smartphones, the privacy [18] should be ensured while designing any kind of detection algorithms.

6 Conclusion and Future work

In this work, all the officially reported and publicly accessible Android operating system source code vulnerabilities from August 2015 till June 2021 are manually extracted and analysed. The analysis brings clarity in the following aspects: which version of Android is most affected, which are all layers in Android are more affected, which programming language dominates, which are vulnerabilities are open source and closed source, which type of vulnerability dominates, and which severity is most reported. This information gives a clear direction on where in Android operating system needs immediate attention. Further, we have presented the actual design flow for the feature extraction procedure of the reported vulnerabilities. Also, a cross-sectional view of a real Android vulnerability is identified and demonstrated the flow from source code to the feature vector generation. Finally, we have provided recommendations about the vulnerability detection mechanism. This will help to prepare the vulnerability dataset and then designing a stable and robust machine learning based vulnerability detection mechanism for the Android operating system.

The future work is to publish the Android vulnerable source code feature vectors and dataset for all the reported source code vulnerabilities and then design and develop a machine learning based vulnerability detection mechanism for the latest Android operating system.

References

1. Android open source project (AOSP). https://source.android.com/. Accessed 18 July 2021
2. Statista. Development of Android malware worldwide 2016–2020. https://www.statista.com/statistics/680705/global-android-malware-volume/. Accessed 18 July 2021
3. Google Android security bulletin. https://source.android.com/security/bulletin. Accessed 18 July 2021
4. Farhang S, Kirdan MB, Laszka A, Grossklags J (2020) An empirical study of Android security bulletins in different vendors. In: Proceedings of the web conference. Association for Computing Machinery, New York, NY, pp 3063–3069
5. Siddiquie K, Shafqat N, Masood A, Abbas H, Shahid WB (2020) Profiling vulnerabilities threatening dual persona in Android framework. In: International conference on advances in the emerging computing technologies (AECT), pp 1–6
6. Shewale H, Patil S, Vaibhav V, Deshmukh, Singh P (2014) Analysis of Android vulnerabilities and modern exploitation techniques. ICTACT J Commun Technol 863–867
7. Jimenez M, Papadakis M, Bissyandé TF, Klein J (2016) Profiling Android vulnerabilities. In: IEEE international conference on software quality, reliability and security (QRS), pp 222–229
8. Wu D, Gao D, Cheng EK, Cao Y, Jiang J, Deng RH (2019) Towards understanding Android system vulnerabilities: techniques and insights. In: Proceedings of the 2019 ACM Asia conference on computer and communications security, pp 295–306
9. Min Z, Haimin Y, Ping C, Zhengxing Y (2019) Android software vulnerability mining framework based on dynamic taint analysis technology. In: IEEE 3rd information technology, networking, electronic and automation control conference (ITNEC), pp 2112–2115
10. Luo L et al (2020) Tainting-assisted and context-migrated symbolic execution of Android framework for vulnerability discovery and exploit generation. IEEE Trans Mob Comput 19(12):2946–2964
11. NVD—home. https://nvd.nist.gov/. Accessed 18 July 2021
12. CVE—home. https://cve.mitre.org/index.html. Accessed 18 July 2021
13. Android security bulletin (2020) https://source.android.com/security/-bulletin/2020-02-01. Accessed 18 July 2021
14. Yamaguchi F, Golde N, Arp D, Rieck K (2014) Modeling and discovering vulnerabilities with code property graphs. In: IEEE symposium on security and privacy, pp 590–604
15. Kipf TN, Welling M (2016) Semi-supervised classification with graph convolutional networks. In: CoRR, abs/1609.02907
16. Velickovic P, Cucurull G, Casanova A, Romero A, Lio P, Bengio Y (2018) Graph attention networks. In: Proceedings of the 6th international conference on learning representations (ICLR)

17. Narayanan A, Chandramohan M, Venkatesan R, Chen L, Liu Y, Jaiswal S (2017) graph2vec: learning distributed representations of graphs. In: CoRR, abs/1707.05005
18. Haoxiang W, Smys S (2021) Big data analysis and perturbation using data mining algorithm. J Soft Comput Paradigm (JSCP) 19–28
19. The code property graph. https://plume-oss.github.io/plume-docs/plume-basics/code-property-graph/. Accessed 18 July 2021

A Novel Efficient Energy and Delay Balance Ensemble Scheduling Algorithm for Wireless Sensor Networks

B. Srinivasa Rao

Abstract The main aim of the present research is to propose and implement an efficient sleep scheduling procedure for wireless sensor networks (WSNs) using two popular search techniques such as breadth first search (BFS) and color connected dominated set (CCDS) for reducing energy consumption and delay, when message is broadcasted in WSN. Since, sensor nodes have battery life constraint sleep scheduling that is used to preserve energy and have a prolong life for sensor node for monitoring events. The novelty in the present research problem is ensemble of breadth first search algorithm and color connected dominated set (CCDS) algorithm which was not considered by earlier researchers. Breadth first search (BFS) is implemented to find the minimum distance path from a sensor node and reduce the delay in transmitting the message. Color connected dominated set (CCDS) is used to transmit messages to all nodes without collision and hence minimize the energy consumption. Analysis is made between two algorithms with the same set of nodes.

Keywords WSN · Sleep scheduling · BFS · CCDS · Energy efficiency

1 Introduction

The most important aspect of the WSN is the necessity of long-term involvement and working of sensor node batteries without charging, while monitoring the critical events that effect the efficiency of the WSN. Hence, intelligent techniques are required for effective conservation of the energy of the power sources. Based upon the energy waste in WSN, various types of methods like data reduction, control reduction, energy efficient routing, duty cycling, and topology control have been reported in the literature [1]. Analytically, all these techniques have their advantages and disadvantages at wide range [2]. Obviously, it has been observed that for the enhancement of the lifetime of the power source and in turn of the sensor nodes, it is essential to design energy efficient scheduling algorithms [3]. In that context, sleep

B. Srinivasa Rao (✉)
Gokaraju Rangaraju Institute of Engineering and Technology, Bachupally, Hyderabad 500090, India

S. Shakya et al. (eds.), *Proceedings of Second International Conference on Sustainable Expert Systems*, Lecture Notes in Networks and Systems 351,
https://doi.org/10.1007/978-981-16-7657-4_10

scheduling algorithms played a significant role in reducing the energy consumption and time delay of wireless sensor networks [4]. In general, the sleep scheduling algorithms are used in the form of synchronous, semi-synchronous, and asynchronous mechanisms. In synchronous scheduling mechanism, all sleeping nodes wake up at a time for communication and require additional control traffic. In semi-synchronous mechanism, the nodes form into clusters and the sleeping and wake up takes place within the clusters. But, all these clusters are not in synchronization. In asynchronous sleep scheduling, each node contains its own scheduling as per requirement [5]. Most of the sleep scheduling algorithms fall into any one of these mechanisms. Recently, optimization techniques and heuristic approaches are also become very popular for efficient sleep scheduling mechanisms [6–8]. The main aim of the present research is to propose and implement a novel efficient energy and delay balance ensemble scheduling algorithm for wireless sensor networks using two popular search techniques such as breadth first search (BFS) and color connected dominated set (CCDS) for reducing energy consumption and delay, when message is broadcasted in WSN. The novelty in the present research problem is ensemble of breadth first search algorithm and color connected dominated set (CCDS) algorithm which was not considered by earlier researchers. Breadth first search (BFS) is implemented to find the minimum distance path from a sensor node and reduce the delay in transmitting the message. Color connected dominated set (CCDS) is used to transmit messages to all nodes without collision and hence minimize the energy consumption. In this present paper, the concepts of BFS and CCDS scheduling are illustrated in Sect. 2. The related work has been presented in Sect. 3. The proposed novel sleep scheduling model has been described in Sect. 4. Methodology and experimental parts are given in Sects. 5 and 6, respectively. Section 7 deals with results and discussion parts. Finally, the work is concluded in Sect. 8.

2 BFS and CCDS Scheduling

2.1 BSF Scheduling

The breadth first search scheduling designed from the inspiration of the breadth first search algorithm of the graph theory. The procedure begins with visiting the sensor node (say A) and all of its neighbor sensors. Then, in the subsequent step, the nearest neighboring sensors of A are visited and the same procedure is continued in the next subsequent steps. The algorithm visits all the neighboring sensor nodes of all the sensor nodes in the network and ensures that each sensor is visited exactly once [9]. The proposed BSF scheduling procedure is given in Algorithm 1.

Algorithm 1: BSF scheduling procedure for WSN denoted by V

Start
1. For each node n ε V Do
2. Distance[n] = infinity Pred[n] = −1
3. Color[n] = White
4. Distance[s] = 0 Color[s] = Gray
5. Q = Empty Queue Enqueue(Q,s)
6. While(Qis not empty) Do
7. u = Head(Q)
8. For each neighbor of n of u Do
9. If(Color[n] is White) then
10. Distance[n] = Distance[u] + 1
11. Pred[n] = u
12. Color[n] = Gray
13. Enqueue(Q,n)
14. Dequeue(Q)
15. Color[u] = Black

Stop

By implementing the Algorithm 1, a BSF tree can be constructed that describes the uplink path as follows in the following steps. 1. A sensor node is selected as the central node. 2. Categorization all the sensor nodes into node levels L1, L2, L3, …, etc., as shown in Fig. 1. 3. Each level is depicted by different color. 4. In this process, the neighboring sensors of each sensor are computed as shown in Table 1. 5. From Table 1, BSF tree is constructed for each node as shown in Fig. 2. 6. Then, a routing table is constructed for each node in the WSN as shown in Table 2. Now, each route in the BSF tree routing table defines the uplink paths for that particular sensor node.

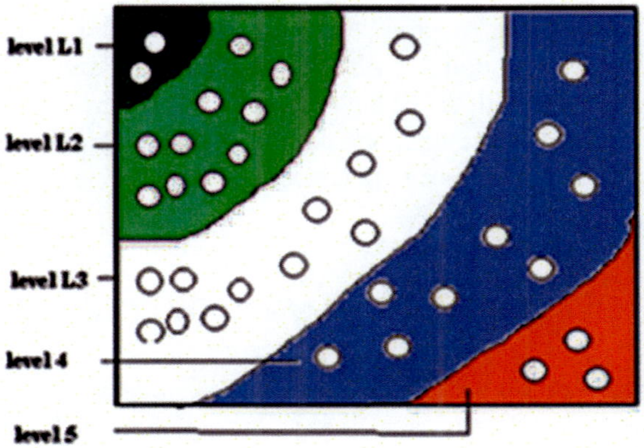

Fig. 1 Levels of sensor nodes

Table 1 Node representation

S. No.	Source node	Neighbor nodes
1	1	2, 3, 4
2	2	1, 6, 5
3	3	1, 4, 5, 8
4	4	3, 1, 7
5	8	3, 7

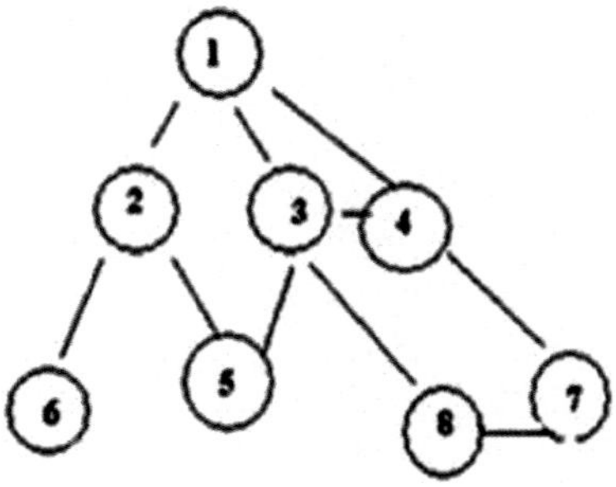

Fig. 2 BFS tree structure

Table 2 Routing table and uplink paths

S. No.	Source node	Uplink path
1	1	$1 \rightarrow 1$
2	2	$2 \rightarrow 1$
3	3	$3 \rightarrow 1$
4	4	$4 \rightarrow 1$
5	5	$5 \rightarrow 2 \rightarrow 1$ or $5 \rightarrow 3 \rightarrow 1$
6	6	$6 \rightarrow 2 \rightarrow 1$
7	7	$7 \rightarrow 4 \rightarrow 1$
8	8	$8 \rightarrow 3 \rightarrow 1$

A typical uplink path in an oblique BSF tree is as shown in Fig. 3. The main advantage with the BFS algorithm is that if there is a path, it will be definitely found. In case of more than one path, it takes the shortest path. It never explores useless paths. Memory is the issue with BSF scheduling.

2.2 *Colored Connected Dominant Set (CCDS) Scheduling*

The CCDS scheduling is used for construction of downlink path in WSN, when a critical event is occurred. The scheduling design is from the inspiration of CCDS concept proposed for WSN that implements the process of self-regulation among the

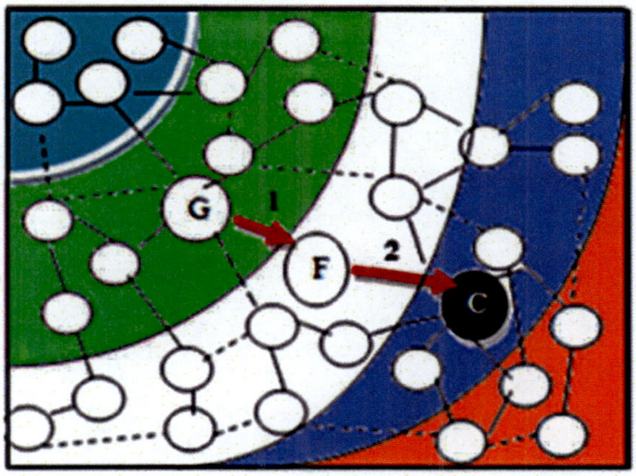

Fig. 3 Uplink traffic path in BFS

nodes [1]. In WSNs, the downlink path construction is different with respect to uplink paths as the communication will be between a long-distance node to the central node and possibility of multiple paths and to select an optimal one. This process requires internal self-regulation among the sensor nodes. In the present, CCDS scheduling different algorithms are combined together that constructs a connected dominating set and serves as a backbone to the WSN. The main objective of the backbone is reduction of communication overhead, enhancement of bandwidth efficiency, minimization of overall energy consumption, and improvement of network effective lifetime in a WSN [1].

2.3 *Construction*

For the construction of CCDS, we have followed the earlier methods proposed by Peng et al. [10]. The construction process involves (i) maximum independent set (MIS) in G, (ii) connected dominated set (CDS), and (iii) internal model controlling (IMC) algorithm [11].

2.4 *Maximum Independent Set (MIS)*

The construction process of the MIS was described elsewhere in detail [10]. Here, the concept is given in Algorithm 2:

Algorithm 2: Construction of maximum independent set

1. Start
2. Divide the nodes of the WSN into levels L1 … Ln
3. Form the binary tree
4. M ← Number HOPs in WSN
5. For HOP = 0 to M

{
Construct MIS
In way that independent nodes present and previous hops are same.
HOP = HOP + 1
}

2.5 *Connected Dominated Set (CDS)*

In any graph G (V, E) having vertices V and edges E, a connector node can be defined using MIS. The sensor nodes that are not in any MIS but have edge with that MIS is denoted as connector nodes. Let U be a subset of V and is an independent set of G and not a superset, then it is denoted by maximal independent set (MIS) of G. "A subset U of V is a dominating set of G's if each node not in U is adjacent to some node in U. If U is a dominating set of G and G[U] is connected, then U is called a connected dominating set (CDS) of G. For color connected dominated set, the color is specified" [1]. Figure 4 shows the downlink traffic path in CCDS. Independent sensor is identified by black color. Connector node is identified by gray color. Rest of the sensor nodes is denoted by white color. The central node C initiates

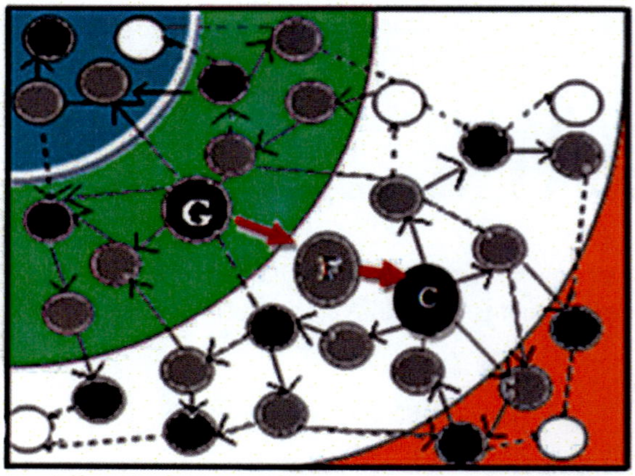

Fig. 4 Downlink traffic path in CCDS

the broadcasting of the alarm packet. The thick lines define the path in CCDS from center to all other sensor nodes. The dashed lines define the path to white node. Here, the role of white nodes is only receiving the message packets. Thus, the white nodes contribute to the energy consumption [1]. The main advantage of the CDS is its capability of serving as virtual backbone to a WSN and minimizes the overheads such as routing and traffic. Vulnerability to the redundancy is an issue to deal with the CCDS.

3 Related Work

Different methods of sleep scheduling for extension of the lifetime of the WSN have been recently reviewed [1–8]. In general, most of the proposed sleep scheduling mechanisms involve the components like target prediction, reduction of awakened sensors, and control active-time of sensor. The energy efficient TDMA sleep scheduling algorithm has the advantages of maximization of the lifetime of WSN, but this mechanism has disadvantages like delay, data overlap, and reduction in channel utilization [12]. The balanced-energy scheduling used WSN sensor redundancy to increase the lifetime and network load balance to improve the efficiency [13], but it has a disadvantage of long-distance communication in the WSN. The DESS algorithm provides reduction in energy utilization and communication delay but having the issues of communication delay and latency [14]. Recently, a new energy efficient reference node selection (EERS) algorithm has been proposed [15] for time synchronization in industrial WSNs. EERS significantly increased the large savings in energy consumption but applicable to large number of nodes at industry level. A multilevel sleep scheduling algorithm was developed adopting clustering concept for wireless sensor networks [16]. Even though the model increases the network lifetime, the asynchronization among the clusters is an issue to consider. An efficient sleep scheduling mechanism was developed for WSN using similarity measure [17]. This model reduces the energy consumption by scheduling the nodes into active or sleeping modes. But this mechanism is not effective in case of sparse distribution of sensor nodes. A heuristic-based delay tolerance and energy saving model were developed for WSN [18] which gives a better performance to other models but it is confined to only mobile base station scenario. A sensor node scheduling algorithm was proposed for heterogeneous WSN [19] that improves network lifetime and regional coverage rate. However, this model is more suitable for only static nodes WSN and may not be effective for mobile nodes. Recently, Mhatre and Khot proposed [20] an energy saving opportunistic routing sleep scheduling algorithm that reduces the energy dissipation effectively but it is confined to only one-dimensional topology networks. Recently, Sinde et al. [21] proposed energy efficient scheduling using deep reinforcement learning that increases the lifetime and reduces the network delay and have shown better performance to the previous models. But the main issue with this model is with the complexity of the deep reinforcement learning. Ant colony optimization algorithm was used for energy optimization of WSN with better energy efficiency [6]. Another scheduling algorithm was proposed [22] using game theory and wake up approach for energy efficiency in WSN. The main disadvantage with this model is many approximations that make the model unrealistic. Recently, a metric routing protocol was proposed for evolution and stability of IoT networks [7]. A heuristic approach-based ant colony optimization multipath routing algorithm was proposed for virtual ad hoc networks to optimize the relay bus and route selection issues [8]. The above two models give an idea for considering some new approaches for efficient scheduling mechanisms for WSN. From the above discussion, we infer that the above-mentioned scheduling algorithms good at some points and have drawbacks in other aspects and stressing for further research in designing new and novel efficient algorithms to enhance the lifetime of the sensors. In all the above algorithms and models, no appropriate attempt has been made to simultaneously balance both energy consumption and delay time. The main advantage of breadth first search in WSN is that during its traversal of the tree in the level by level manner, it classifies tree edges and cross edges, and its time complexity is O(N). This property is very important for efficient routing in wireless sensor networks [9]. Also, breadth first search (BFS) is implemented to find the minimum distance path from a sensor node and reduce the delay in transmitting the message. On the other hand, the connected dominating set (CDS)-based routing is one kind of hierarchical methods has received more attention to reduce routing overhead. Color connected dominated set (CCDS) is used to transmit messages to all nodes without collision and hence minimize the energy consumption [1]. The ensemble of these two technique balances both energy consumption and delay time in wireless sensor network. In the present research work, we propose an efficient sleep scheduling procedure for wireless sensor networks (WSN) using two popular search techniques such as breadth first search (BFS) and color connected dominated set (CCDS) for reducing energy consumption and delay when message is broadcasted in WSN. In the next section, the proposed model has been presented.

4 The Proposed Novel Sleep Scheduling Model

In the present paper, we propose an efficient sleep scheduling procedure for wireless sensor networks (WSN) using two popular search techniques such as breadth first search (BFS) and color connected dominated set (CCDS) for reducing energy consumption and delay when message is broadcasted in WSN. The proposed algorithm is considered in two phases (i) uplink phase, (ii) downlink phase for scheduling. The uplink phase is scheduled by breadth first search scheduling with the inspiration of the breadth first search algorithms of graph theory. The downlink phase is scheduled by color connected dominated set inspired scheduling. Finally, the combination of both BFS scheduling and CCDS scheduling forms the proposed novel efficient sleep scheduling algorithm for reducing energy consumption and delay when message is broadcasted in WSN. The proposed new scheduling is briefly described in the following. A model wireless sensor network that has been deployed for the

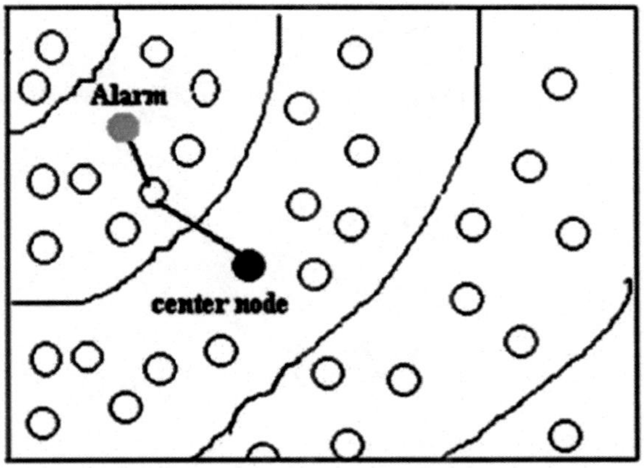

Fig. 5 Construction of uplink path

detection of any critical event is as shown in Fig. 5. It consists of central node (black shaded) also known as center node that has the capability of communication with all the nodes of the network. In the case of detection of a critical or disaster event by any node of the network (denoted by gray shaded node), the gray node sends an alarm to the central node and thus constructs the uplink path. For the construction of the uplink path, BFS scheduling is implemented. In the process of construction of this uplink path, the shortest path from any node to central node is computed by BFS scheduling.

Now, the central node transmits the received alert from the gray node during the uplink phase to all the other sensors in the network. For the construction of the downlink path, the color connected dominated set scheduling is implemented. For the construction of the downlink path, CCDS is constructed using the internal model control (IMC) algorithm [1, 11]. The internal model control (IMC) algorithm is self-regulating processes and characterizes the downlink path while transmitting the alert from the central node to all other sensor nodes of WSN as shown in Fig. 6.

5 Methodology

In this section, the methodology has been discussed for both uplink and downlink phases of the present scheduling algorithm. Initially, deployment of nodes is performed. In order to have a communication in the WSN, route discovery is done using route table entries of all the sensor nodes in WSN. A node initiates route discovery by sending a request to its neighboring first hop nodes to know whether that nodes are located in its path and wait for route reply from the first hop neighbors. Based upon the reply messages from the first hop nodes, the broadcasting node updates its routing table entry destination ID. Sequence number and battery status are updated for latest information about the fresh routes and energy levels of the nodes. After identification of the neighbor nodes, the BSF algorithm is implemented

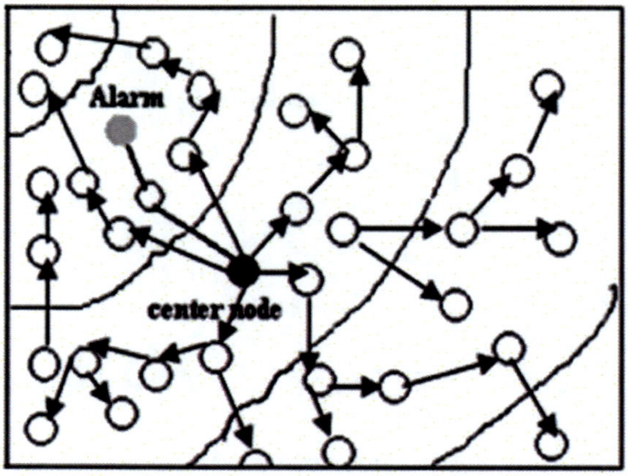

Fig. 6 Construction of downlink path

for the construction of the BSF tree that divides the nodes into different levels. Using these levels, a CCDS is constructed. Also, MIS is constructed for independent nodes of each level. For comparative study, both scheduling algorithms will be inputted with same set of nodes. The methodology diagram for proposed work is as shown in Fig. 7.

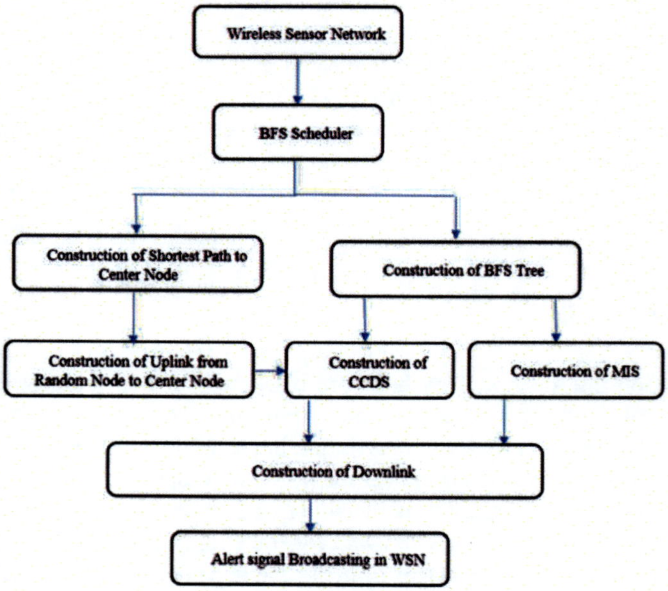

Fig. 7 Methodology diagram for proposed work

6 Experiment

The software requirements are (a) backend: Python 3, (b) frontend: HTML, CSS, and Bootstrap 4. The hardware requirements are RAM—advised to have >32 GB, graphic card, Nvidia GTX 1071, processor—Intel Core i7-8750H, storage—512 GB SSD. Simulator: Network simulators.

7 Results and Discussion

In the present research work, a novel efficient sleep scheduling algorithm for WSN was proposed and experimented on network simulators. The information of the source node and its neighbor has been input and given in Table 3. The implemented results of the BSF scheduling algorithm are presented in Table 4. Similarly, the experimental results of CCDS scheduling algorithm are given in Table 5. Table 6 shows the paths from any sensor node to the central node in WSN. From Table 4, it can be observed that during BSF scheduling some parent nodes have children and others do not have as they are leaf nodes. It is well understood that the no energy releases from leaf nodes. Similarly, from Table 5, it is understood that during CCDS scheduling, the parent nodes not having children would not be involved in message transmission. Comparing Tables 4 and 5, it can be understood that child node and number of the parent nodes of BSF scheduling and CCDS scheduling are not same. Also, childless nodes are more for CCDS. Therefore, finally, it could be understood that the childless parents of BSF do not involve in dissipation of energy as they are leaf nodes and childless nodes of CCBS do not involve in transmission message and in turn conserve energy which is the main objective of the present research work.

Table 6 shows the paths from any sensor node to the central node in WSN. The bold lettered nodes in the table revel the difference in the number of hops. It can be observed that the sensor nodes N12, N14, and N8 have a smaller number of hops in path to the central node N1 for BFS scheduling in comparison with CCDS scheduling.

Table 3 Input neighbors list

S. No.	Nodes	Adjacent list
1	N1	N2, N3, N4, N5, N6, N7
2	N2	N8, N9, N11, N3
3	N6	N7, N18, N19, N17
4	N8	N9, N20
5	N7	N8, N19, N20
6	N4	N15, N14, N13
7	N3	N12, N4, N13
8	N12	N13
9	N13	N14

Table 4 Output of BFS

S. No.	Parent node	Child node
1	N1	N6, N5, N4, N7, N2, N3
2	N6	N19, N17, N18
3	N7	N20, N8
4	N8	–
5	N9	N22, N21, N23
6	N13	–
7	N4	N13, N14
8	N14	–
9	N17	N16
10	N5	N15
11	N15	–
12	N18	–
13	N3	N12
14	N12	–
15	N2	N11, N9

Table 5 Output of CCDS

S. No.	Parent node	Child node
1	N1	N6, N5, N4, N7, N2, N3
2	N6	N17
3	N7	N20, N19
4	N8	–
5	N9	N21, N22, N23
6	N13	N12, N14
7	N4	N13, N15
8	N14	–
9	N17	N16
10	N5	–
11	N15	–
12	N18	–
13	N3	–
14	N12	–
15	N2	N11, N9

At the same time, the nodes N3, N4, N5, N6, and N7 are at one hop distance from N1. Other nodes such as C, D, E, F, and G are one hop away from center node. It is obvious that with the hop count varies with the levels of the nodes in the WSN.

Table 6 Paths representation

Node to center node	BFS path to center node	CCDS path to center node
N15-N1	N15-N5-N1	N15-N4-N1
N13-N1	N13-N4-N1	N13-N4-N1
N12-N1	**N12-N3-N1**	**N12-N13-N4-N1**
N14-N1	**N14-N4-N1**	**N14-N13-N4-N1**
N8-N1	**N8-N7-N1**	**N8-N20-N7-N1**
N9-N1	N9-N2-N1	N9-N2-N1
N17-N1	N17-N6-N1	N17-N6-N1

8 Conclusion

In the present paper, the proposed efficient sleep scheduling algorithm has been successfully designed and implemented. From the experimental results, it can be concluded that to have balance and better energy saving and fastest transmission of message in WSN, the present algorithm has successfully balanced both the BSF and CCDS. Further, research is required for improving the model. In future, we extend this model for both homogeneous and heterogeneous networks of sufficiently large size and compared with other models numerically and analytically.

Acknowledgements The author is thankful to the management of GRIET for their encouragement.

References

1. Liu Z, Wang B, Guo L (2010) A survey on connected dominating set construction algorithm for wireless sensor networks. Inf Technol J 9(6):1081–1092
2. Soua R, Minet P (2011) A survey on energy efficient techniques in wireless sensor networks. In: 2011 4th joint IFIP wireless and mobile networking conference (WMNC 2011), pp 1–9. https://doi.org/10.1109/WMNC.2011.6097244
3. Pagar AR, Mehetre DC (2015) A survey of energy efficient sleep scheduling in WSN. Semanticscholar.org, Corpus ID 212548604
4. Karthihadevi M, Pavalarajan S (2017) Sleep scheduling strategies in wireless sensor network. Adv Nat Appl Sci 11(7):635–641
5. Zhang Z, Shu L, Zhu C, Mukherjee M (2018) A short review on sleep scheduling mechanism in wireless sensor networks. In: Wang L et al (eds) QShine2017. LNICST, vol 234, p 66
6. Chen JIZ, Lai K (2020) Machine learning based energy management at internet of things network nodes. J Trends Comput Sci Smart Technol 2(3):127–133
7. Smys S, Vijesh Joe C (2021) Metric routing protocol for detecting untrustworthy nodes for packet transmission. J Inf Technol 3(2):67
8. Dhaya R, Kanthavel R (2021) Bus-based VANET using ACO multipath routing algorithm. J Trends Comput Sci Smart Technol (TCSST) 3(1):40
9. Akram VK, Dagdeviren O (2013) Breadth-first search-based single-phase algorithms for bridge detection in wireless sensor networks. Sensors (Basel, Switzerland) 13(7):8786–8813. https://doi.org/10.3390/s130708786

10. Peng G, Tao J, Qian Z, Kui Z (2012) Sleep scheduling for critical event monitoring in wireless sensor networks. IEEE Trans Parallel Distrib Syst 23(2)
11. Rivera D, Morari M, Skogestad S (1986) Internal model control—PID controller design. Ind Eng Chem Process Des Dev 25:252–265
12. Laxman P, Rajeev P (2014) Comparative analysis of TDMA scheduling algorithms in wireless sensor networks. https://www.semanticscholar.org/. Corpus ID: 61778529
13. Feng J, Zhao H (2018) Energy balanced multisensory sensory scheduling for target tracking in WSN. Sensors (Basel) 18(10):3585
14. Soumyadip S, Swagatam D, Nasir M, Athanasios VV, Witold P (2012) An evolutionary multi-objective sleep-scheduling scheme for differentiated coverage in wireless sensor networks. IEEE Trans Syst Man Cybern Part C (Appl Rev) 42(6)
15. Mahmoud E, Abd El-Gawad MA, Haneul K, Sangheon P (2020) EERS: energy-efficient reference node selection algorithm for synchronization in industrial wireless sensor networks. Sensors 20:4095. https://doi.org/10.3390/s20154095
16. Hassan S, Nisar MS, Jiang H (2016) Multilevel sleep scheduling for heterogeneous wireless sensor networks. Comput Sci Technol Appl 227
17. Wan R, Xiong N, Loc NT (2018) An energy efficient sleep scheduling mechanism with similarity measure for wireless sensor networks. Hum Cent Comput Inf Sci 8:18
18. Jerew O, Bassan N (2019) Delay tolerance and energy saving in WSN in mobile base station. Wireless Commun Mob Comput 2019. Article ID 3929876
19. Wang Z, Chen Y, Liu B, Yang H, Su Z, Zhu Y (2019) A sensor node scheduling algorithm for heterogeneous wireless sensor networks. Int J Distrib Sens Netw 15:1
20. Mhatre KP, Khot UP (2020) Energy efficient opportunistic routing with sleep scheduling in wireless sensor networks. Wireless Pers Commun 112:1243
21. Sinde R, Begum F, Njau K, Kaijage S (2020) Refining network life time of wireless sensor networks using energy efficient clustering and DRL based sleep scheduling. Sensors 20(5):1540
22. Manikandan KB (2021) Game theory and wake up approach scheduling in WSN for energy efficiency. Turk J Comput Math Educ 12(10):2922

Bitcoin Price Prediction Using Machine Learning's Boosting Algorithms

Ch. Likhitha Sree, M. Meghana, R. Manjula, and D. Mohan

Abstract Cryptocurrency is now gaining a lot of popularity. Bitcoin is the first cryptocurrency that was ever started. Trading of these cryptocurrencies has become legal in most of the countries, and it is expected that most of the countries soon might accept cryptocurrency as tender along with conventional currency. The value of cryptocurrency is variable and predicting the price of it will help people in investments. The cryptocurrency market is open 24/7 and unlike the stock market change of value in cryptocurrency does not vary with the founder or any other factors. The data of bitcoin prices recorded is a time series data, i.e., it changes with time. Its value changes with demand and supply factors. This paper "Bitcoin price prediction using machine learning's boosting algorithms" predicts the future price of the bitcoin for 30 days by analyzing the past trend of bitcoin price and estimates the future demand and supply of bitcoin and predicts the price with the help of machine learning. Our main goal of our project is to predict bitcoin price prediction with higher accuracy than all of the currently existing systems. For this, we are using the newly introduced CatBoost algorithm which came up in 2017.

Keywords Bitcoin · Machine learning · Boosting algorithms · CatBoost · Cryptocurrency · Gradient boosting · CatBoost algorithm

1 Introduction

At present, the demand for various cryptocurrencies is quickly going mainstream. Bitcoin is one such currency that has attracted many investors, researchers, and the media. Being decentralized it is not owned by anyone, and transactions made through bitcoin are not related to any country or a banking system. Bitcoin is an open source and nobody can control bitcoin but everyone can take a part. This type of cryptocurrency came into existence in the year 2009 by a pseudonymous creator Satoshi Nakamoto. Bitcoin can be bought in exchange for services, products, and

Ch. L. Sree (✉) · M. Meghana · R. Manjula · D. Mohan
Electronics and Computer Engineering Department, Sreenidhi Institute of Science and Technology, Hyderabad, Telangana 501301, India

S. Shakya et al. (eds.), *Proceedings of Second International Conference on Sustainable Expert Systems*, Lecture Notes in Networks and Systems 351,
https://doi.org/10.1007/978-981-16-7657-4_11

also other currencies. Furthermore, these coins can also be obtained through the mining process which is known as bitcoin mining. All of these bitcoins can be stored in digital wallets which are similar to having a virtual bank account. All the records are related to the transactions; the timestamp data is stored in a ledger-like account known as blockchain. By the usage of certain machine learning techniques and algorithms, the price prediction of bitcoin can be done accurately which helps in making some better investment decisions. Cryptocurrencies are the new form of digital currencies. The term cryptocurrency is combined from two terms, one being cryptography and the other is currency. These cryptocurrencies are encrypted and transferred from the sender to the receiver just like how the data is transmitted. All the cryptocurrency wallets have a public and a private key. The private key is held by the sender to sign the transactions to send currency to the receiver. The public key is held by the receiver in order to receive the transactions. Transactions with these cryptocurrencies are more secure as they do not need any intermediary such as a bank or any financial institutions. This will exclude the extra transaction processing fee in the case of cryptocurrencies.

1.1 Prediction

The prices of cryptocurrencies vary very rapidly depending on various factors. To make sure a correct investment is being made, predicting the future price of the bitcoin is necessary. The price of bitcoin does not depend on any government changes or business events unlike the stock market. So, to predict the value of bitcoin at least to some possible extent the usage of machine learning and its algorithms are required for an accurate prediction. For this, certain boosting algorithms are being used for price prediction of bitcoin.

1.2 Bitcoin

Bitcoin is the first cryptocurrency that was introduced and is being considered as the most popular one till date. It was introduced for the first time in a paper titled Bitcoin: A peer-to-peer electronic cash system which came out in 2008. The author of this paper and known founder of bitcoin is Satoshi Nakamoto whose identity still remains as a mystery. Its first open-source bitcoin software client was released on Jan 9, 2009. Initially the adoption of bitcoin was very negligible. It was useful for settling international payments as it was quick and easier with the least transactional fee. Later it was used to settle small amount transactions as well but with limited throughput on the ledger the cost of transaction outweighed the transaction amount and the bitcoin price. It was then found that the value of bitcoin varied with a combination of technological breakthroughs, its supply and demand values and powerful networks. With this regard the investment thesis of bitcoin was completely changed and has seen as an alternative for gold, popularly known as digital gold. It became a form of "pristine collateral" for the global economy. On the other hand, bitcoin supported economic freedom and posed a strong protection against censorship, money confiscation and evolution through uncapped inflation.

1.3 Objective

The price of any cryptocurrency is always volatile and the prediction of it for better investments and returns is a very challenging task. With the use of certain machine learning boosting algorithms, the price of the bitcoin can be predicted at least to some extent. The main algorithm which is being used in this work along with other algorithms is *CatBoost Algorithm* which came into practice in the recent years and found to be very effective with the time series data. In this the dataset is being collected from two different sources one from Kaggle and Yahoo Finance (which provides the latest price of bitcoin for every minute) and preprocessed. The main objective of this is that the bitcoin price for the next 30 days is being predicted.

2 Literature Review

Gadey et al. (2020) Price prediction of bitcoin using machine learning [1]:

The authors used the LSTM algorithm to predict the price of the bitcoin, and they analyzed how the usage of this algorithm would affect the bitcoin price by providing better accuracy. In this the price prediction is handled as regression instead of classification and the utilization of LSTM algorithm is being shown.

Reddy and Sriramya (2020) A research on bitcoin price prediction using machine learning algorithms. International Journal of Scientific & Technology Research, 9, 1600–1604 [2]:

In this paper, the authors have compared the accuracy of the model with certain different algorithms and concluded that the linear regression algorithm is found to be more effective than the remaining algorithms. With the usage of linear regression, the authors implemented LASSO and stated that the time complexity while using LASSO is reduced and concluded that LASSO is best among all.

Muniye (2020) Bitcoin price prediction and analysis using deep learning models [3]:

In this particular paper, the authors used certain deep learning models to predict the volatile price of the bitcoin making it easy for the investors to make an investment into the bitcoin. The two deep learning techniques which were implemented were LSTM and GRU; these are used as the prediction models. And the authors concluded that the GRU model is found better for time series data.

Raju and Tarif (2020) Real-time prediction of bitcoin price using machine learning techniques and public sentiment analysis [4]:

In this the bitcoin price prediction is made using tweets and reddit post with the usage of sentimental analysis and machine learning. Here, the authors collected the datasets from two different sources and predicted the price of the bitcoin. And also, the sentiments from the current market have also been taken into consideration. Here, the authors used LSTM and ARIMA and concluded that the LSTM model was more capable of recognizing long-term prices. Sentimental analysis is being performed only from Twitter and Reddit posts, and the price is being predicted.

Jiang (2020) Bitcoin price prediction based on deep learning methods. Journal of Mathematical Finance [5]:

The price of the bitcoin is being gathered per minute and is then being converted into hours and in comparison with other models the authors claimed that multi-layer perceptron (MPL) is least suited for bitcoin price prediction and concluded that the LSTM provides the best prediction when the earlier memory and GRU is included in the model.

Vaidehi et al. (2021) Bitcoin price prediction using machine learning. International Journal of Engineering Technologies and Management Research [6]:

The deep neural networks with the usage of LSTM architecture is being implemented to predict the price of the bitcoin and the working architecture are working model of LSTM for bitcoin price prediction is being implemented.

3 Preprocessing and Implementation

Dataset collection: For the price prediction of bitcoin, the data is being collected from two sources, namely Kaggle and Yahoo Finance. The Yahoo Finance dataset provided minute by minute pre of the bitcoin. And at the final stage only the required attributes are being taken into consideration for the price prediction of bitcoin.

Data cleaning: After gathering the require dataset, the next step is to clean the data from unwanted and NaN values. So, in this step the data is being cleaned from the NaN values.

The outline of the dataset is being given in Fig. 1.

	Timestamp	Open	High	Low	Close	Volume_(BTC)	Volume_(Currency)	Weighted_Price
0	2011-12-31 07:52:00	4.39	4.39	4.39	4.39	0.455581	2.0	4.39
1	2011-12-31 07:53:00	NaN	NaN	NaN	NaN	NaN	NaN	NaN
2	2011-12-31 07:54:00	NaN	NaN	NaN	NaN	NaN	NaN	NaN
3	2011-12-31 07:55:00	NaN	NaN	NaN	NaN	NaN	NaN	NaN
4	2011-12-31 07:56:00	NaN	NaN	NaN	NaN	NaN	NaN	NaN

Fig. 1 Dataset description

In the implementation part, the required libraries are being collected, the cleaned data is being trained and tested in the training part the boosting algorithms are being provided and the model evaluation is being done.

4 Boosting Algorithms

Boosting algorithms are mostly used to improve the prediction accuracy of the model; boosting algorithms combine all the weak learners and combine them to form a final output, i.e., strong learners. Basically, boosting technique uses a collection of predictors in which the learners learn sequentially with simple fitting models for analyzing the data errors. Random trees are fit at every model of the tree to improve the accuracy of the previous trees. The boosting algorithms are being discussed below:

4.1 Gradient Boosting (GBM)

In this boosting technique, all the weak learners are considered as decision trees and this GBM combines the predictions from multiple trees to generate the final predictions. Here, in this gradient boosting technique, every successive decision tree is built on the errors of the previous trees.

4.1.1 Working of Gradient Boosting

The gradient boosting algorithm works on three parameters that include loss function, weak learner for prediction purpose and an additive model to add weak learners to minimize the loss functions.

Loss function: The usage of loss function depends on the problem which is being used and must be differentiable.

Weak learner: In this, the decision trees are considered to be the weak learners. These trees are constructed in a greedy manner and helps in choosing the best split based on the scores.

Additive model: In this, a set of parameters are used such as coefficients or weights. The weights need to be updated after calculating the error or loss to minimize that error. In this we have weak learner subparameters instead of parameters. The basic syntax for gradient boost regression and classification problems is being given in Figs. 2 and 3.

```
# gradient boosting ensemble for making predictions for regression
from sklearn.datasets import make_regression
from sklearn.ensemble import GradientBoostingRegressor
# define dataset
X, y = make_regression(n_samples=1000, n_features=20, n_informative=15, noise=0.1,
# define the model
model = GradientBoostingRegressor()
```

Fig. 2 Syntax for GBM regression

```
# make predictions using gradient boosting for classification
from sklearn.datasets import make_classification
from sklearn.ensemble import GradientBoostingClassifier
# define dataset
X, y = make_classification(n_samples=1000, n_features=20, n_informative=15, n_redu
# define the model
model = GradientBoostingClassifier()
# fit the model on the whole dataset
```

Fig. 3 Syntax for GBM classification

4.2 Extreme Gradient Boosting (XGBM)

This boosting technique is similar to gradient boosting, but an advance version of it. Here, in XGBM, the trees are built sequentially by connecting to the errors of the previous trees. XGBM implements parallel processing which makes it advanced to GBM and also certain regularization techniques which improve the performance and reduce overfitting are being used in XGBM.

4.2.1 Working of Extreme Gradient Boosting

Usually in gradient boosting for regression, the weak learners are regression trees and each of this regression trees maps to a leaf which contains continuous score. XGBoost consists of certain regularized parameters, and this minimizes a regularized objective function which combines a convex loss function. In this the training proceeds iteratively by adding new trees that predict the residuals or errors of prior trees that are then combined with previous trees to make the final prediction; this uses a gradient descent algorithm to minimize the loss when adding new models. The basic syntax for extreme gradient boosting for both regression and classification is given in Figs. 4 and 5.

```
# gradient xgboost for making predictions for regression
from numpy import asarray
from sklearn.datasets import make_regression
from xgboost import XGBRegressor
```

Fig. 4 Syntax for XGB regressor

```
# make predictions using xgboost for classification
from numpy import asarray
from sklearn.datasets import make_classification
from xgboost import XGBClassifier
```

Fig. 5 Syntax for XGB classifier

4.3 LightGBM

The LightGBM boosting algorithm is well known for its speed and efficiency. This boosting algorithm can handle huge amount of data with less time complexity. A leaf-wise grow is being implemented in this algorithm rather than a level-wise growth. The next split of the node is done only on the leaf that has higher delta loss [7].

4.3.1 Working of LightGBM

This algorithm is one of the frameworks of gradient boosting which is based on decision trees which is known for its efficiency and reduces the usage of memory. The LightGBM boosting algorithm splits the tree leaf-wise whereas the other algorithms which grow tree-level wise. A leaf node with maximum delta loss is chosen. But leaf-wise growth may increase the complexity of the model and may lead to overfitting of the model which consists of smaller datasets. The basic syntax for light gradient boosting algorithm is being given in Fig. 6.

4.4 CatBoost

CatBoost is a machine learning open-source algorithm which was implemented in the year 2017. The term CatBoost refers to categorical boosting which works with multiple categories of data such as audio, text, and image. In this algorithm, there is requirement of preprocessing of converging categorical data into numerical data. The CatBoost algorithm itself converts the categorical data into a numerical. The

```
# Creating an object for model and fitting it on training data set
model = LGBMClassifier(model = LGBMClassifier()
model.fit(x_train, y_train)

# Predicting the Target variable
pred = model.fit(x_test)
print(pred)
accuracy = model.score(x_test, y_test)
print(accuracy)
```

Fig. 6 Syntax for LightGBM algorithm

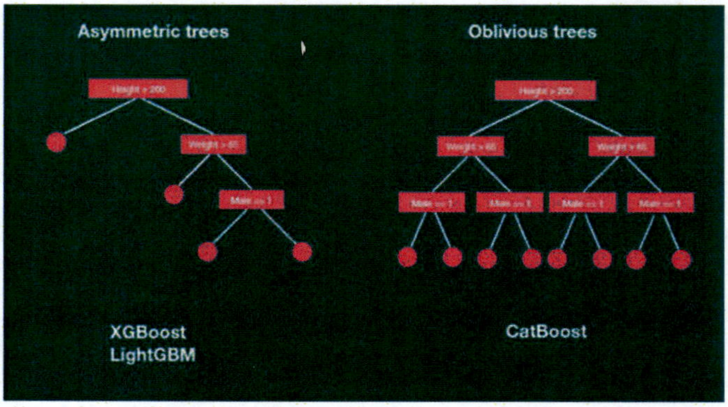

Fig. 7 CatBoost algorithm comparison

CatBoost algorithm is based on gradient boosted decision trees. To prevent over fitting, an overfitting detector is being used [8].

This CatBoost algorithm not only works for classification but also works with the regression problems by importing the required libraries for the CatBoost regression (Fig. 7).

4.4.1 Working of CatBoost Algorithm

CatBoost algorithm directly converts the categorical data into a numerical format with any commands. The CatBoost does it inherently. CatBoost uses one-hot encoding for handling the categorical features. In this, when the models take the dataset and start training a set of decision trees are built consecutively with reduced loss function. Furthermore, in this model the number of trees is controlled by the defined parameters. An overfitting detector is used in this algorithm to prevent overfitting. The basic syntax for CatBoost classification and regression algorithm is being given in Figs. 8 and 9.

Fig. 8 Syntax for CatBoost regressor

```
model_CBR = ctb.CatBoostRegressor()
model_CBR.fit(X_train, y_train)
print(model_CBR)
```

Fig. 9 Syntax for CatBoost classifier

```
model_CBC = ctb.CatBoostClassifier()
model_CBC.fit(X_train, y_train)
print(model_CBC)
```

5 Results

Various algorithms were used for the bitcoin price prediction but the main goal of this paper is to predict bitcoin price with highest accuracy possible, and when compared with the other models, the CatBoost algorithm has been more efficient in predicting the future values and has got highest accuracy when compared with other algorithms. The price of bitcoin is being predicted for the next 30 days given the input data. Here, the input data is given from the year 2013 till may 2021 and the price of next month is being predicted (Table 1).

The predicted bitcoin price for the next 30 days is being visualized in the graph given in Fig. 10.

Table 1 Comparison between the accuracies of each model before and after attribute selection

Algorithms/Model	Accuracy
XGB	0.976
Gradient boosting	0.964
LGBM	0.963
Gradient boosting	0.894
CatBoost	0.983

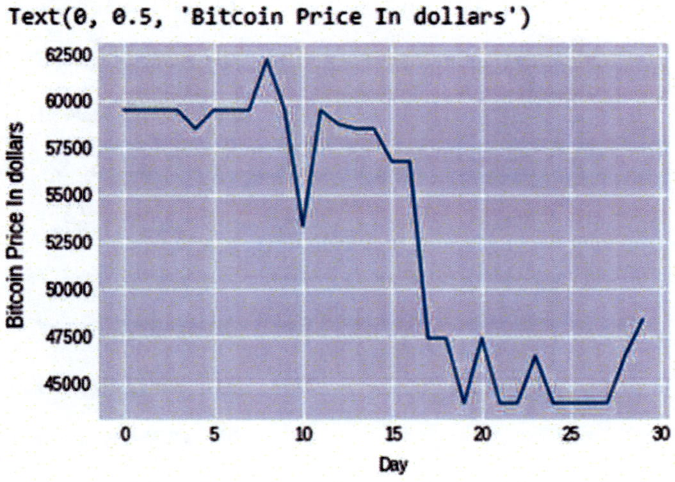

Fig. 10 Bitcoin price trend graph for upcoming 30 days (data visualization)

6 Conclusion

The bitcoin price for the next 30 days is being predicted and interpreted, while the price of the bitcoin depends on many factors; this is one of the methods for the prediction and this solely cannot be taken as reference for investment but can be used to get a gist of the bitcoin price. Predicting the future is always important, and with the help of machine learning algorithms, it can be achieved to some extent. Here, in this project, we have attempted to predict the price of a future price of bitcoin using a machine learning algorithm—CatBoost with an accuracy of nearly 98% (Fig. 11).

	Day	Price
0	1	59473.713784
1	2	59473.713784
2	3	59473.713784
3	4	59473.713784
4	5	58490.800632
5	6	59473.713784
6	7	59473.713784
7	8	59473.713784
8	9	62219.347827
9	10	59473.713784
10	11	53341.687399
11	12	59473.713784
12	13	58725.298585
13	14	58490.800632
14	15	58490.800632
15	16	56743.110536
16	17	56743.110536
17	18	47383.466336
18	19	47383.466336
19	20	43932.456427
20	21	47383.466336
21	22	43932.456427
22	23	43932.456427
23	24	46432.829481
24	25	43932.456427
25	26	43932.456427
26	27	43932.456427
27	28	43932.456427
28	29	46432.829481
29	30	48388.534930

Fig. 11 Bitcoin price for the next 30 day with day and price

7 Future Scope

Initially during the years 2012–2016 the price of the bitcoin was stable, the price of bitcoin in 2017 was raised by 2000%. In coming days with the increase in popularity of cryptocurrency and bitcoin, it would invite a lot of investors on to the cryptocurrency market. Trading of cryptocurrencies has already become popular and is also accepted as legal in most of the countries including India.

In the future, the cryptocurrency may overtake the present exciting currency methodologies and it would always be a wiser decision to invest in the cryptocurrencies based on their market demand. Hence, it is always useful to use certain prediction strategies to obtain the future price of bitcoin of good accuracy so that better investment ideas can be taken into consideration.

References

1. Gadey RS, Thakur N, Charan N, Reddy R (2020) Price prediction of bitcoin using machine learning
2. Reddy LS, Sriramya D (2020) A research on bitcoin price prediction using machine learning algorithms. Int J Sci Technol Res 9:1600–1604
3. Muniye T (2020) Bitcoin price prediction and analysis using deep learning models
4. Raju SM, Tarif A (2020) Real-time prediction of Bitcoin price using machine learning techniques and public sentiment analysis
5. Jiang X (2020) Bitcoin price prediction based on deep learning methods. J Math Finan
6. Vaidehi M, Pandit A, Jindal B, Kumari M, Singh R (2021) Bitcoin price prediction using machine learning. Int J Eng Technol Manage Res
7. Singh A (2020) 4 boosting algorithms you should know
8. CatBoost (2021) (catboost.ai n.d.)
9. Browniee J, Machine learning mastery. https://machinelearningmastery.com/
10. Adam EEB (2021) Survey on medical imaging of electrical impedance tomography (EIT) by variable current pattern methods. J ISMAC 3(02):82–95
11. Chen JIZ, Yeh L-T (2020) Analysis of the impact of mechanical deformation on strawberries harvested from the farm. J ISMAC 2020(3):166–172
12. Haoxiang W, Smys S (2021) Big data analysis and perturbation using data mining algorithm. J Soft Comput Paradigm (JSCP) 3(01):19–28
13. McNally S, Roche J, Caton S (2018) Predicting the price of bitcoin using machine learning. IEEE, Ireland, Dublin
14. Guo T, Bifet A, Fantulin NA (2018) Bitcoin volatility forecasting with a glimpse into buy and sell orders. IEEE
15. S. Yogeshwaran, Maheshwari P, Kaur MJ (2019) Project based learning: predicting bitcoin prices using deep learning. IEEE, Amity University Dubai Dubai, UAE
16. Dwyer GP (2015) The economics of Bitcoin and similar private digital currencies. J Financ Stab 17:81–91
17. Iwamura M, Kitamura Y, Matsumoto T (2014) Is bitcoin the only cryptocurrency in the town? Economics of cryptocurrency and Friedrich A. Hayek

Anthropometric Measurement Technology Using 2D Images

Chinmay Shah, Jaineel Shah, Maruf Shaikh, Harshit Sandhu, and Prachi Natu

Abstract With the change in digital consumerism, customer expectations have also evolved. Present-day user is wary of various confusing size charts, numerous orders of different sizes, the endless trouble of returns, or quite simply about poor fitting garments. In order to take this endless hassle out of the consumer shopping experience, a system for calculating measurements of the human body for perfectly fitting clothes is the need of the hour. While the existing 3D measurement approaches are a possible solution, the high cost of infrastructure make them inconvenient. A possible alternative to traditional size calculation methods is provided by image-based 2D anthropometric measurement systems. This is a convenient and cheaper technique that replace both expensive 3D measurement approaches as well as conventional methods of manual measurement. This paper proposes an efficient and accurate 2D anthropometric measurement system of the body measurement determination process through image segmentation, feature extraction, and dimension calculation.

Keywords Anthropometric · Measurements · Image segmentation · Feature extraction · Body size

1 Introduction

The e-commerce ecosystem has changed the clothing industry considerably. Consumers are becoming increasingly aware and their need is also adapting to the digital market. They do not want to undergo the hassle of ill-fitted clothes and confusing size charts. This, in turn, has led to an increase in mass personalization and requirement of well-fitting clothes. The accuracy of the cloth fitting depends upon the accuracy of body measurements as mentioned in [1]. There are traditional methods of measuring a person using measuring tapes which involves physically

C. Shah (✉) · J. Shah · M. Shaikh · H. Sandhu · P. Natu
NMIMS, Mumbai, India

P. Natu
e-mail: prachi.natu@nmims.edu

S. Shakya et al. (eds.), *Proceedings of Second International Conference on Sustainable Expert Systems*, Lecture Notes in Networks and Systems 351,
https://doi.org/10.1007/978-981-16-7657-4_12

touching the subject using a contact method. These methods, however, are not the most sustainable approaches for the present-day e-commerce ecosystem. Apart from that, the conventional methods have several shortcomings:

- Not efficient in terms of time consumption
- Not adaptable to the changing e-commerce ecosystem
- Leads to inaccurate measurements or fallacies.

These issues can be solved by utilizing techniques that employ three-dimensional scanning. However, the use of such an expensive three-dimensional system is not feasible for the masses as mentioned in [2]. Another alternative is the use of several Kinect sensing technologies for scanning the human body. These systems are slightly more affordable but also have disadvantages like low quality of models reconstructed due to depth perception of poor quality. Owing to these factors, there is a need to develop a two-dimensional automatic anthropometric measurement system that is reliable, affordable, and accurate. In 2019, we explored different methods that can be used to achieve this primary goal. In this paper, we extend the research by proposing a new novel method to achieve the goal.

2 Related Work

2.1 Image Segmentation

In [3], Yoo et al. had proposed an algorithm that makes use of the HLS color space of an image. This algorithm makes an assumption that the background is completely static, i.e., there is only one uniform color in the background. The authors had used the HLS space as it becomes more convenient to compare only one value (hue) rather than comparing all the three components of the RGB space. Mode of the hue values is calculated and since the background is completely static, the mode will always come out to be the background hue value. After getting the mode, the algorithm simply traverses the image pixel by pixel and colors all the pixels black having Huhe value equal to the overall mode that is calculated earlier. Thus, we are left with a segmented human body with a black background. Although this model provides very good accuracy in terms of IoU metrics, the fact that the segmentation occurs only on images with a static background can be obtained at a studio with a green screen behind the subject, thus making it impractical for our use case.

In [4], the authors had proposed an algorithm for the segmentation of human bodies from an image clicked in any natural environment. One assumption was that the person is standing in an almost upright position and in-front of a variable environment. The methodology makes use of the face, skin color, and the approximate dimensions of the body concerning the face. The face guides the construction of the upper body which in turn guides the construction of the lower body. The algorithm makes use of superpixels which are n number of segments in an image. The segments belonging to the human body are determined based on the joint estimation of the foreground and background, which espouses the requirement of shape matching. This methodology is constructed with the help of five different phases.

The first phase is the detection of the face of the person from the image. This helps in determining and reducing the search space for the human body from the image. It also espouses the calculation of the other human parts from the image along with the determination of the skin color. Calculation of the face regions was done with the help of the Viola-Jones face detection algorithm proposed by Viola and Jones in [5]. The implementation of this algorithm can be easily found in the OpenCV library. It helps in achieving both speed and accuracy for this first phase. The pixels that correspond to the face region are calculated by the Viola-Jones algorithm followed by the formation of elliptical regions around the detected face. According to the probabilities of these pixels, the skin color threshold is also calculated.

After the face detection, the second phase involves the multiple segmentation of the image which yields various superpixels as the output. The implementation uses two-level segmentation. This helped them to determine finer segments along with the exposed skin color of the hands and other body parts. The authors evaluated the results with the different numbers of superpixels, i.e., the number of segments. The probabilistic similarity of each pixel in the segments is determined with the help of the segments obtained at each level.

The next step is the detection of the skin of the person. The step can be implemented simultaneously with the previous step as they are independent of each other. The methodology uses an algorithm proposed in [6] which combines two different skin detection models, the adaptive model, and the global skin detection. For combining the results, they use the weighted averaging with the global method having 0.25 weight and the adaptive model having 0.75 weight. The output of this step is an image with a black background and only the visible skin's pixels are restored.

The fourth phase is the determination of the upper body of the human. Using the size of the detected face, estimation of the body using the anthropometric constraints. The most visible part of the human body in the picture is expected to be the torso, as the human is standing in an almost upright position. With the help of the anthropometric constraints, one can get to know the approximate location of the upper body with respect to the face. So, a rectangular mask is created with the help of these anthropometric constraints. After getting this rectangle, we calculate the probability of the segment is a part of the torso or not. This is done by calculating the number of pixels of a segment that are inside the rectangular region. If more pixels are inside the rectangle, then the segment is considered to be the part of the torso.

After the calculation of the upper body, the final step is the calculation of the lower body. The torso helps in initiating the leg sear process. With the help of the anthropometric constraints and the segmented upper body, a similar rectangle for the lower body is calculated. After this, the further process is very similar to that of the upper body segmentation which is already explained in the above paragraph. The result was a fully segmented human body with the color of the background is black.

2.2 *Feature Extraction Using Chain Code*

An automated system for extracting human body feature points from the front and side images is proposed in [7]. The proposed system includes the identification of a feature point by determining the difference between each code in the chain code of the silhouette curves.

Front and side images of subjects were captured and the algorithm was then applied to these images. The authors used a novel approach wherein they used the concept of Freemans' 8 directional chain code in order to derive features points. Their algorithm would traverse the pixel silhouette and find the difference in the direction between the pixels. Hence, pixels with a significant difference between them would be taken as feature points. Initially, they used a 90° directional change and then to locate more points they took into consideration a 45° directional change.

3 Proposed System

The various methodologies addressed in the previous sections had many shortcomings. In Sect. 2.1, there are multiple ways suggested on image segmentation but they have a basic assumption that colors in image are distinguishable, while in Sect. 2.2, the assumption that algorithm is not able to identify relevant feature points. Thus, we are proposing our own state-of-the-art system having the same sequence of computations. For image segmentation, we propose using a semantic segmentation model through which we can easily identify various parts of the human body which will help us anchor the feature extractions. Feature extraction is done with the help of an algorithm proposed by us. With the help of this algorithm, we can extract all the necessary features that are required for the size computations. Size calculation is done by calculating the distances between the pixels and then converting them into inches which are explained in detail in the further Sects. 3.2 and 3.4 (Fig. 1).

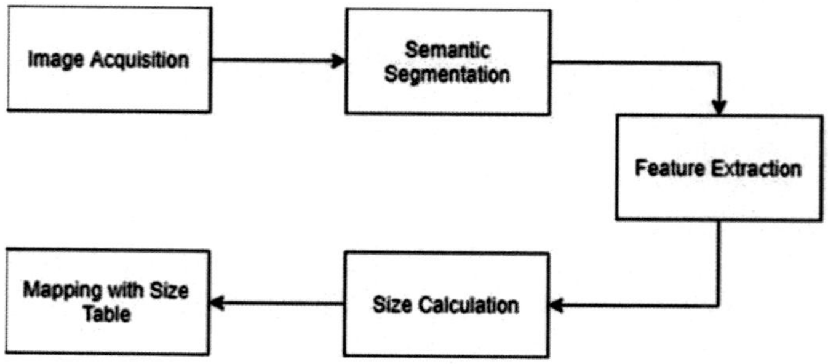

Fig. 1 Proposed system

3.1 Semantic Segmentation

In [8], Long et al. introduced a state-of-the-art system for performing image segmentation using fully convolutional neural network resulting in a significant step in semantic segmentation.

The image taken using a camera is usually taken in a natural setting or inside a house that does not have a constant static background, rather, a highly variable background. To process these images, we propose using a CNN that can segment image, separate the person from the background.

While these images obtained using the image segmentation allows us to separate human body from the background, i.e., binary segmentation and anthropometric measurements require identification of feature points which are facilitated by identification of the parts of the body, which help us find the points of intersection of different physical parts (of the body) helping us to find these feature points, which are essentially semantic segmentation for each body part.

To obtain this goal, we use a pre-trained model by Google's TensorFlow: BodyPix. A CNN, trained on a combination of COCO images and rendered images, allowing us to identify 24 regions of the human body (Figs. 2 and 3).

Fig. 2 Output of semantic segmentation front view

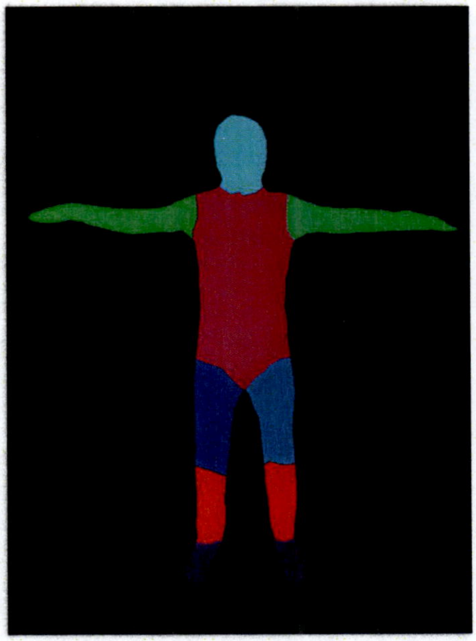

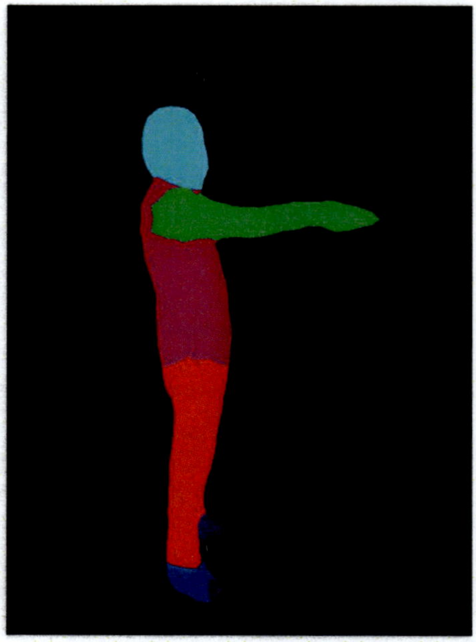

Fig. 3 Output of semantic segmentation side view

3.2 Feature Extraction

After getting the output from the semantic segmentation model, our next step is to obtain the various feature points that will act as anchor points to calculate various sizes. For this, we exploited the various body parts that were segmented and were given different colors in the previous step.

Feature Points from Front View Image: For the extraction of front view feature points, there are two assumptions that we are considering. The first one is that the person is standing in the middle. Although, slight deviation to the left or right can still work. The second assumption is that there is no other person visible in the image. The BodyPix algorithm will segment all the people present in the image and this will lead to anomalies in our feature extraction algorithm.

Top of the head: For calculating the height of the users, first, the topmost blue color pixel is obtained which denotes the top of the users head.

Neck: For calculating the neck feature points, we divide the image into two halves, the first half beginning from the first pixel in that row to the width/2 pixel value and the second half from the width/2 pixel value till the last pixel in that row. For the left neck point, while traversing from top to bottom in the first half of the image, the first pink pixel encountered which denotes the start of the neck. Similarly, for the right neck point, we reiterate the same process (Fig. 4).

Fig. 4 Neck and shoulders feature points

Shoulders and Chest: These points can be calculated in a single go as they both are located at the endpoints of the intersection between the green colored hand and the pink torso. So, for the first half, we check for the range of pixels that have green pixels on the left and pink pixels on the right. Similarly, we followed the same procedure for the second half but with the green pixel being on the right this time. Out of these two ranges of values, we pick the first and the last pixels from each of the two giving us the shoulder points and the chest points on each side.

Inseam Top Point: For calculating the inseam top point, we are checking for the first pixel that contains dark blue pixels (left upper leg) on the top and the black pixel denoting the background region just beneath the groin and between the legs. We can follow the same procedure for the right side as well which will yield almost identical pixels.

Inseam Bottom Point: For calculating the bottom inseam point, we are checking for the last red pixel in the first half of the divided image which is approximately obtained at the center of the ankle. As for inseam, we require only one bottom point, we are not finding the other point in the second half of the image (Fig. 5).

Bottom Point: For the bottom point, we are ignoring the feet and taking the ankle point instead. Since the feet is facing toward the camera, their endpoints will be at a different distance than the other points. So, the ankle point will be much closer to the bottom point than the visible lowest point in the segmented output. In order to obtain this bottom point, we find out the lowest red pixel in the left leg region. Since we are considering the left leg for the inseam calculations, we are taking the left leg for obtaining the bottom point as it would help us in inseam calculations as well.

Waist Points: Another advantage of the BodyPix algorithm is that we can also retrieve the pose of the person. From this, we were able to extract the hip points, and since

Fig. 5 Inseam feature points

the waist is a bit above the hips, we found the row 35 pixel above the hips which was experimentally observed. Then, we calculated the first and the last occurrence of pink pixels and marked them as the endpoints of the waist (Fig. 6).

Feature Extraction from Side View Image: The expected pose for the side view is the person standing in a direction perpendicular to the angle of the camera, and the

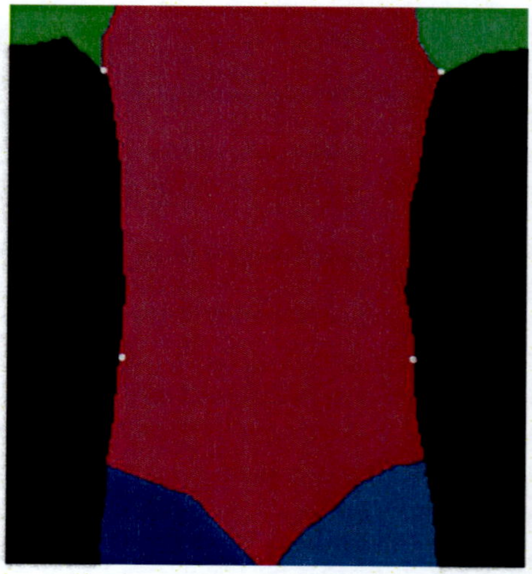

Fig. 6 Waist feature points

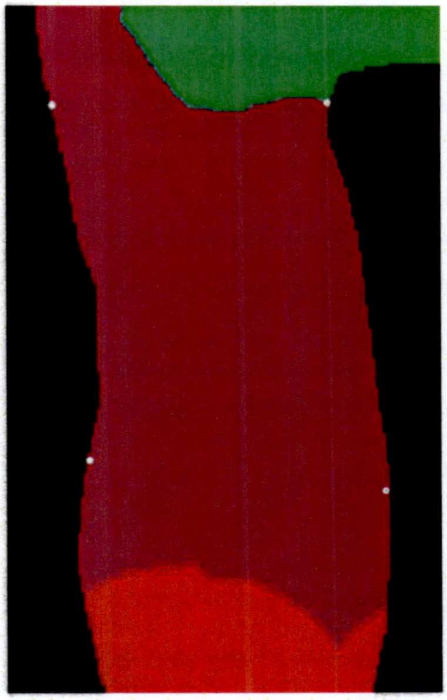

Fig. 7 Chest and waist feature points from side image

hands are held up parallel to the ground. From the side view image, we require only the waist and chest feature points for the calculation of girth measurements.

Waist Points: As mentioned earlier, using the BodyPix algorithm, we can also retrieve the pose of the person. From this, we were able to extract the hip points, and since the waist lies above the waist, we experimentally found the waist two be about 35 pixels above the hip (Fig. 7).

Chest Points: Unlike the front view measurements, where the image was divided into two parts to obtain the two sets of feature points, the side view feature points can be directly obtained. First, to find the rightmost chest point, we check for the range of pixels that have black pixels on the right, pink pixels on the left, and green pixels on the top. Once we obtain the rightmost chest point, we traverse that particular row to find the first pink pixel. Thus, the point obtained would be the left chest point.

3.3 JSON Values Mapping

The output generated by the BodyPix algorithm is a JSON file that contains a dictionary. The three keys of this dictionary are the height, width, and values. The height and width will contain only one value each representing the number of pixels in each row and column of the image, respectively. The values key will have another dictionary with values ranging from 0 to the total number of pixels. The total number of pixels can be calculated by multiplying the height and width of the image. These individual keys will be assigned values ranging from −1 to 24, where −1 denotes the background and the other values signify a particular body part segmented by human body segmentation. Thus, we have a 1D array of such values that need to be interpreted into the original image to get our desired output.

Our first task was to convert the above-mentioned 1D array into a 2D array with height and width as its dimensions. Then, we read the original image using pixel by pixel and check for its corresponding mapping value given in the JSON file. If the corresponding pixel value is −1 (which denotes the background), we color that particular pixel black. Accordingly, we gave color to different body parts that were segmented in the previous implementation. Note that, we did not color all the possible parts of the body but only those that were significant to our use case. The end result was an image with a perfectly segmented human body with a black background in Figs. 2 and 3.

3.4 Size Calculation

After obtaining the feature points by applying the aforementioned algorithm, we calculate the distance, in pixels, between the feature points. Here, we want to calculate five measurements, which are chest, waist, inseam, and shoulder.

In order to calculate the pixel distance between the feature points, we use the formula for the Euclidean distance which is the shortest distance or displacement between two points.

$$d = \sqrt{(y_2 - y_1)^2 + (x_2 - x_1)^2} \quad (1)$$

From the front image, we calculate the "topPoint" and the "bottomPoint" and find the height of the subject in pixels. To find the length of each pixel, we divide the actual height of the user (which is entered by the user itself) by the total number of pixels between the top and bottom point.

$$\text{Each pixel length} = \frac{\text{Actual height in inches}}{\text{Total pixels between top and bottom}} \quad (2)$$

For calculating the vertical distance (i.e., inseam), we are simply multiplying the Euclidean distance by the "each pixel length" value. As we are asking the users to capture their images in 4:3 aspect ratio, for every 4 pixels in length, there will be three pixels in width. Hence, for calculating horizontal measurements, we multiply the Euclidean distance with the "each pixel length" value and the aspect ratio.

$$\text{Vertical distance} = \text{Euclidean distance} \times \text{Each pixel length} \quad (3)$$

$$\text{Horizontal distance} = \text{Euclidean distance} \times \text{Each pixel length} \times \text{Aspect ratio} \quad (4)$$

3.5 Error and Performance Considerations

Two-dimensional anthropometric measurements are a good alternative to traditional methods of body measurement. They overcome the shortcoming of conventional physical systems but have certain errors and performance issues. In these types of systems [9], errors can originate from camera distortion, resolution issues, faulty calibration of the camera, landmarking error, and modeling error.

The error can originate from any of the following steps:

- front and side image preprocessing,
- camera calibration error,
- image segmentation,
- feature extraction,
- body measurement calculation.

There are several performance considerations to be made for accurate body measurements during image acquisition. The test subject should be sufficiently away from the camera such that his entire body frame can be obtained. The subject should also be wearing tight fitted clothes and should stand with his arms and legs extended to avoid inaccuracies during feature extraction. The fulfillment of the above-mentioned criteria will ensure accurate results.

4 Implementation and Results

4.1 Dataset

We have collected 50 samples with each sample having front and side view of the person. Each picture is taken at a distance of 4–5 feet with an aspect ratio of camera being 4:3 (Fig. 8).

4.2 Implementation Details

The model used is ResNet50 with an output stride of 16 using TensorFlow's BodyPix model. ResNet50 offers output stride of 16 and 32, with higher stride essentially reducing the accuracy of the output, leading us to choose the model with output stride of 16, more about it later.

Fig. 8 Sample images from our dataset

4.3 Results

We calculated metrics for 16 different models, with varying parameters for MobileNetV1 and ResNet50 model by changing output stride length and quantization bytes (bytes used for weight quantization).

For ResNet50, we tried output stride of 16 and 32, while for MobileNetV1 with output stride of 8 and 16. For all the models, we tried all quantization bytes of 1, 2, and 4 bytes.

To measure the performance of each model, we use intersection over union (IoU), an accepted metric for semantic segmentation given by

$$\frac{\text{Area of Overlap}}{\text{Area of Union}}$$

In our case, we had labeled each of the images collected individually for semantic segmentation which acts as ground truth while model output from BodyPix acts as predicted output for which we run IoU metric.

In Figs. 9, 10, and 11, we do it for three different classes, Fig. 9 for person segmentation, where we calculate IoU on how well the person's body is segmented from the background. For Figs. 10 and 11, we talk about two different classes, head and the head for which we calculate IoU. In these figures, it is quite evident that MobileNet does not perform as well as ResNet50 with an average of 7% drop in IoU. The models with lower stride length perform way better than one with a higher stride length with almost 2% difference in performance.

Among the metric quantization bytes, the performance again is a variation of 2% in between 1 byte and 4 byte models. In the performance of the resnet50 model with a stride length of 16, the performance variation is minimum in between quantization bytes. Among all these models, ResNet50 with stride length 16 has the best prediction. In between quantization bytes of the said model, due to very minimal changes

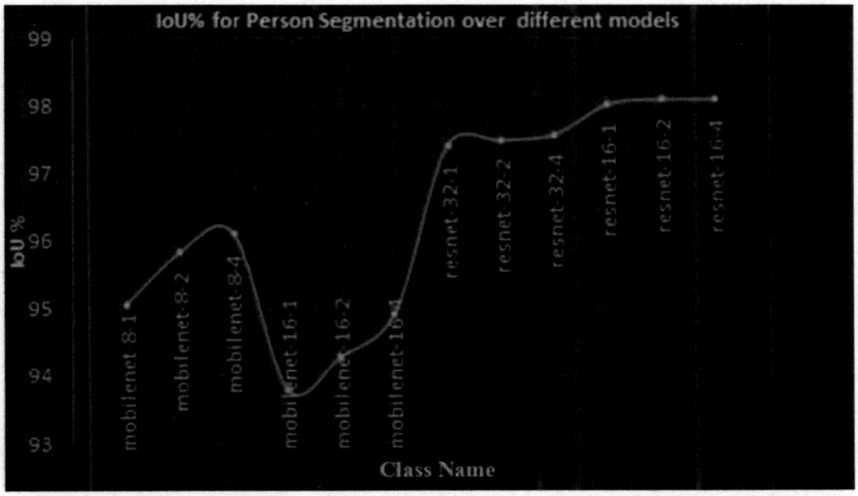

Fig. 9 IoU% for person segmentation over different models

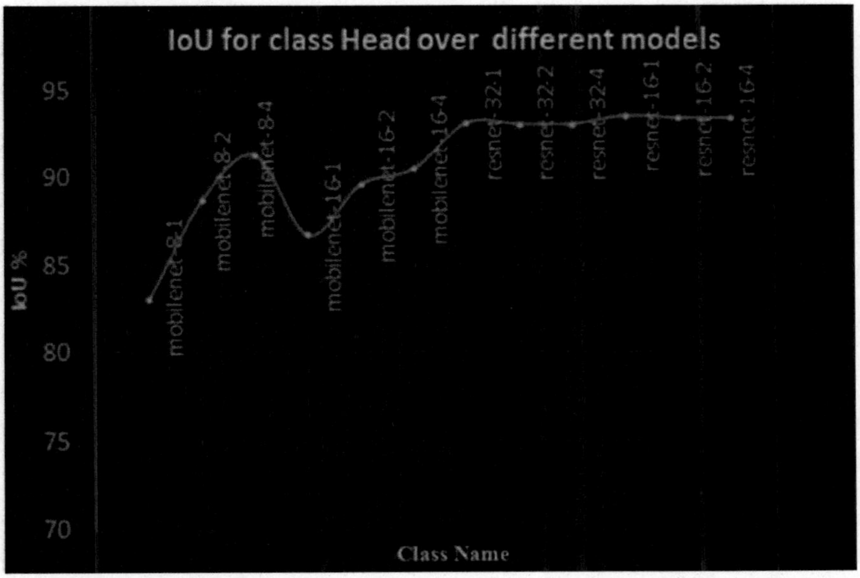

Fig. 10 IoU% for class head

in performance, at the same time, a shift from 1 to 2 bytes results in doubling of model size from 22 to 45 MB while using 4 bytes model takes up 90 MB of just model weights. This results in a clear decision to use ResNet50 with stride length 16 and quantization byte of 1 byte.

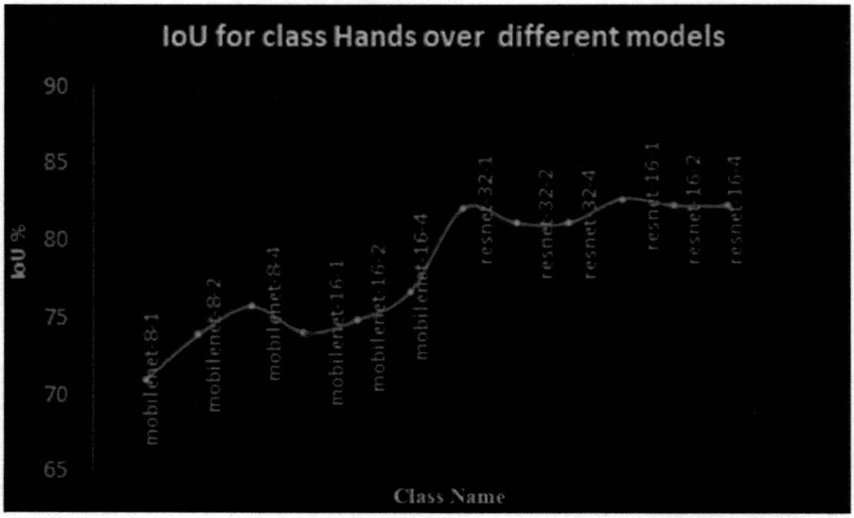

Fig. 11 IoU% for class hands

In Fig. 9, an average performance among MobileNet models is 95% while that of ResNet50 is 98%, which proves that if you want to perform person segmentation, higher accuracy is obtained. In Fig. 10, where we find IoU for head class, while MobileNet has an average of 87% accuracy, ResNet50 has a constant performance of about 93%. Similarly, for hands class in Fig. 11, MobileNet has an average accuracy of 75%, ResNet50 has a constant average of 82%. From the above data, it is quite evident that for an individual class, the model performance is not that constant, even though remains above 70%.

Another conclusion that we are drawing is for person segmentation, ResNet50 with stride length 16 gives us better output even though the difference is just of 0.5%, for all other classes, both ResNet50 with stride length 16 and 32 gives us output within a certain range, thus there is no benefit of using a high-quality model for individual class segmentation, even though there is a clear advantage of using ResNet50 over MobileNet with 7% advantage as mentioned earlier (Table 1).

In order to calculate the performance accuracy of the calculated measurements, we use mean absolute error as the performance metric. It is the average of the absolute difference between the actual size and the predicted size. The shoulders show a mean average error of 0.8 in. which was because of the slight roundness associated with them when we measure. The chest values show an error of 2.5 in. whereas the waist has 1.2 in. mean absolute error. The subjects wearing loose fitted clothes like shirts will affect these calculations as the feature points will always be on the edge of the clothes and not the edge of the body. Thus, wearing tight fitted clothes are advisable for better results. For inseam, we get a mean average error of 1.8 in. This is mostly governed by the fact that the top inseam point of the subject will be below the actual point if they wear loose pants or trousers with drop crotch.

Table 1 Comparison of actual and predicted body dimensions for subjects shown in Fig. 8

Subject	Shoulders (in in.)		Chest (in in.)		Waist (in in.)		Inseam (in in.)	
	Actual	Predicted	Actual	Predicted	Actual	Predicted	Actual	Predicted
1	21	20	42	44	36	37	32.5	30
2	17	17	36	38	32	34	28.5	30
3	18	18	40.5	42	36	36	29	28
4	18	16	38	38	33	34	30	28
5	17	18	34.5	36	32	32	29	27
6	21	20	41	42	36	36.5	32	31

5 Conclusion

Today's competitive e-commerce ecosystem warrants the need for reviewing and developing new reliable, affordable, and accurate technologies for improved customer satisfaction. This leads to the requirement of highly accurate two-dimensional image-based anthropometric systems. This paper consisted of one such system for accurate human body measurements from two-dimensional images. Our system employed numerous efficient algorithms for the three primary stages of segmentation, extraction of feature points, and calculation of anthropometric measurements. It overcomes the hardware requirements and the other shortcomings such as static and background which were discussed in earlier.

The development of such body measurement systems is expensive and time-consuming. A lot of algorithms and methods are proprietary, and information on the same is not available openly to the public as mentioned by Xiaohui et al. [10]. Giving due attention to the possible error and performance issues as well as the deployment issues, further research and work are required for improving the accuracy of this system, before, it is fit for mass deployment.

References

1. Sehgal R, Gupta R, Anand N (2018) Automatic extraction of 3d body measurements from 2d images of a female form. IOSR J Polym Text Eng 5(3):07–17
2. Tong J, Zhou J, Liu L, Pan Z, Yan H (2012) Scanning 3D full human bodies using kinects. IEEE Trans Vis Comput Graph 18(4):643–650
3. Yoo JH, Doosung H, Moon KY (2009) Human body segmentation based on background estimation in modified HLS color space. In: WSEAS International conference. Proceedings. Recent advances in computer engineering, vol 9. WSEAS
4. Tsitsoulis A, Bourbakis NG (2015) A methodology for extracting standing human bodies from single images. IEEE Trans Human-Mach Syst 45(3):327–338
5. Viola P, Jones M (2001) Rapid object detection using a boosted cascade of simple features. In: 2001 IEEE Computer society conference on computer vision and pattern recognition. CVPR 2001, Kauai, HI, USA, pp I–I

6. Conaire CO, O'Connor NE, Smeaton AF (2007) Detector adaptation by maximising agreement between independent data sources. In: IEEE Conference on computer vision and pattern recognition, Minneapolis, MN, pp 1–6
7. Lin YL, Wang MJ (2011) Automated body feature extraction from 2D images. Expert Syst Appl 38(3):2585–2591
8. Long J, Shelhamer E, Darrell T (2015) Fully convolutional networks for semantic segmentation. In: Proceedings of the IEEE conference on computer vision and pattern recognition, pp 3431–3440
9. Meunier P, Shi Y (2000) Performance of a 2D image-based anthropometric measurement and clothing sizing system. Appl Ergon 31(5):445–451
10. Xiaohui T, Xiaoyu P, Liwen L, Qing X (2018) Automatic human body feature extraction and personal size measurement. J Visual Lang Comput 47:9–18

A Novel Approach for Sentiment Classification by Using Convolutional Neural Network

M. S. Kalaivani and S. Jayalakshmi

Abstract Social media platforms facilitate communication and data exchange. There is a substantial number of opinionated information available in digital form. It is essential to validate unstructured Web data in order to extract knowledge from it. Sentiment analysis offers a wide variety of applications across all domains. The primary objective of sentiment analysis is to asses if the input text is positive or negative. When a buyer purchases a product, they submit feedback of the product. These reviews are essential for getting a general sense of how people feel about the product or service. Customer reviews on the Internet help to make purchases. Sentiment analysis results assist businesses, understand customer expectations, and enhance service and product quality. Several deep learning algorithms have been used in this sector with promising results. This paper suggests a deep learning approach for sentiment analysis by using convolutional neural networks.

Keywords Sentiment analysis · Deep learning · Word embedding · Convolutional neural network · Max pooling layer · Fully connected layer

1 Introduction

Sentiment analysis is a linguistic technique from natural language processing, used to filter user reviews, evaluate Web content, and extract information from unstructured data [1]. In many industries, sentiment analysis is used in business to understand consumer feedback and to facilitate better decisions. People nowadays use Instagram, YouTube, blogs, and many other social media outlets to converse their thoughts and ideas. The critiques or comments are driven by personal experiences with a certain product or service [2, 3]. Digital advertising generates a massive quantity of data.

M. S. Kalaivani (✉)
Tagore College of Arts and Science, Chennai, India

S. Jayalakshmi
Vels Institute of Science, Technology and Advanced Studies, Chennai, India
e-mail: jai.scs@velsuniv.ac.in

S. Shakya et al. (eds.), *Proceedings of Second International Conference on Sustainable Expert Systems*, Lecture Notes in Networks and Systems 351,
https://doi.org/10.1007/978-981-16-7657-4_13

Sentiment analysis is critical for interpreting this data, and the results are widely used in a variety of disciplines to foresee useful decisions [4]. Machine learning techniques have been used in sentiment analysis since the 1990s [5]. Deep learning approaches have been implemented in sentiment analysis in recent years.

Input processing with machine learning techniques necessitates sophisticated feature engineering [6]. Each deep learning algorithm [7, 8] has a distinct feature for analyzing data and addressing problems. These models contain several levels in which information may be handled. CNN and RNN are commonly used to categorize sentiments. The application of deep learning models with natural language processing is the prominent topic in recent researches [9]. In [10], the author adopted the CNN model to extract features from the input text as well as a weight-sharing strategy to alleviate computation and training parameters complexity. RNN is designed for data retention over time and is a balance with CNN. To obtain reliable findings, text representation and a relevant deep learning model are required [11, 12]. Input text is rendered as vector values using word embedding techniques. Word embedding methods such as Word2Vec and GloVe are extensively utilized in text classification. Sentiment analysis is categorized into three parts: document-level, sentence-level, and aspect-level. In document-level, the sentiment score for the entire document will be assigned. Each sentence is evaluated at the sentence-level based on the sentiment conveyed. The term "aspect-level" refers an analysis based on the characteristics or aspects presented in the review (Fig. 1).

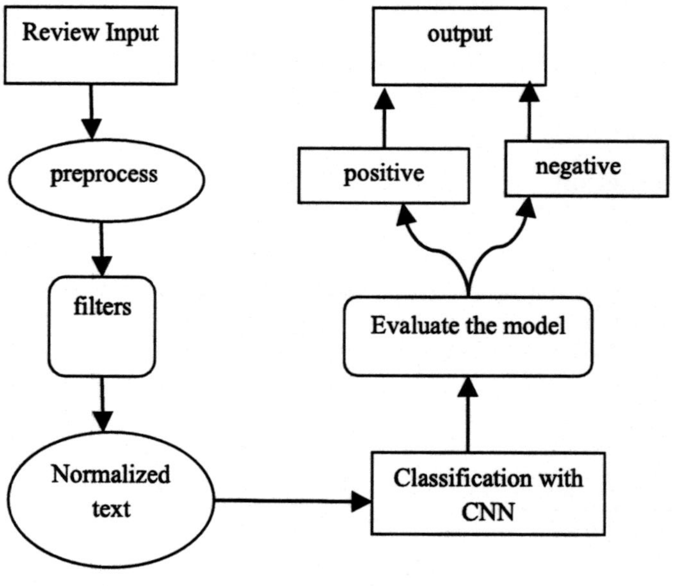

Fig. 1 Sentiment analysis process

The outline of sentiment analysis is depicted in the diagram. In sentiment analysis, there are four key modules. Data collection, preprocessing, feature extraction, and model evaluation. The input dataset for the data collection module is prepared in accordance with the requirements. Preprocessing entails eliminating stop-words, numbers, expanding acronyms, and replacing negation. Attributes or features involved in the training and testing stages are separated in the feature extraction module. In evaluation phase, algorithms are used for analyzing model's accuracy.

2 Background on Sentiment Classification

Sentiment analysis is used to evaluate Internet data and categorize it as positive or negative. With diverse datasets, CNN and LSTM methods yielded superior results. To categorize sentiments, CNN and BiLSTM networks were used in [13]. They applied two trained word vectors to create feature vectors, which serve as LSTM input values. Kapočiūtė-Dzikienė et al. [14] use sentiment analysis in Lithuanian languages, which have a large vocabulary and a variety of dialects. Enhanced results were achieved by merging SVM with the semantic orientation approach. BiLSTM was applied in the study [15] to gather long-term dependence data as well as the location of certain words in a phrase. In this method, CNN and *n*-gram features are combined. A hybrid model with extra convolutional layer BiLSTM was proposed by the author. CNN is used for sentence-level categorization in [16]. CNN relies mostly on local feature extraction and ignores the context of input text. Long short-term memory (LSTM) is an RNN model [17] that is used to create a massive network that effectively leverages memory to prevent gradient issues.

A hybrid model with multiple convolutional layers was proposed by the author [18]. To handle the sequence of characters, an RNN layer is incorporated and CNN performed well in character-level text representation. The paper suggests BiLSTM for text classification in an experimental study [19], by using Chinese product review as input dataset. Other models are outperformed by combining CNN and LSTM. Sun et al. [20] presented a multilevel CNN with LSTM model. To optimize the semantic expression of input phrase, a tree structure with LSTM network was suggested in [21]. By integrating CNN and LSTM with feature learning, the author [22] presented an AC-BiLSTM model for text categorization. The divide-and-conquer technique, suggested by the author [23], employs a neural network-based sequence model to interpret phrases.

Short text sentiment classification is done in [24] with the aid of a key-word vocabulary and LSTM model. A two-layer LSTM network is used to classify text using the output of the feature extraction component. In [22], the preceding and following context of a sequence may be retrieved in the network via backward and forward hidden in BiLSTM. The author of [25] used a combination of pictures and text from the Web to train a hybrid classifier that yielded better outcomes. Bairavel and Krishnamurthy [26] used video data as an input source and extracted features from it. In [27], a hybrid CNNs and BiLSTM were employed to generate special feature vectors from two trained word vectors. The results outperformed baseline models. The author used the LSTM method to describe sentiment analysis on COVID-19, for input dataset from Twitter [28]. The author explored the use of CNN and BiLSTM for Punjabi text sentence analysis [29]. In four classes, the average accuracy was 93.85, 88.53, 83.3, and 95.45%. In [30], the study conducts a comparison research on several deep learning techniques such as DNN, CNN, and RNN and assessed the results. The outcome demonstrates that CNN surpassed other models. This paper includes telecommunication-based sentiment analysis combining CNN with LSTM methods in [31]. As per the results, the majority of complaints are due to technical problems. The author [32] suggested a hybrid technique for sentiment analysis using CNN-LSTM. Restaurant reviews were used as input, and 92% accuracy was achieved.

3 CNN and Word Embedding Techniques

CNN is a multilayer feed-forward network that can be used to reduce the computation time and complexity of an algorithm. CNN is adopted in sentiment analysis to extract local features from input text and understand the features that are involved in sentiment classification. Convolution layer, pooling layer, and fully connected layer are the three significant layers. Sentences or words are transformed to numerical values and fed into convolutional layers as input. To produce input vector values for the text, the Word2Vec and GloVe models are implemented (Fig. 2).

The convolutional layer is used to extract the features from input text. Pooling layer selects and filters the features from the convolutional layer, while the fully connected layer is used to link to the output layer to get the final result. The convolutional filter has the same width as the input matrix. The length of the input sentence n is represented as follow

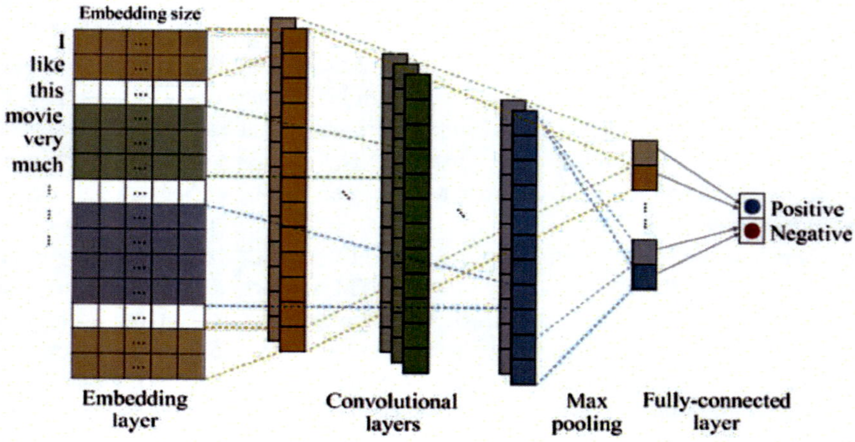

Fig. 2 CNN architecture with single layer

$$x_1 : n = x_1 + x_2 + \cdots + x_n \tag{1}$$

where + represents concatenate operation, x_i represents ith word.

If the input text is smaller than defined length, zero pads are added and feature c_1 is generated as

$$c_1 = f(w.x_1 : h + b) \tag{2}$$

b—bias term, w and $f(.)$—weight of the convolutional filter and nonlinear function. The set c_i denotes feature map, $c = [c_1, c_2, \ldots, ci]$.

Max pooling is applied to get feature map value and extract keywords for convolutional filter. Kim [33] used one convolutional layer with different filter sizes, i.e., 3, 4, and 5 with 100 feature maps, a 300-dimensional vector was generated and connected with positive and negative nodes. Dropout was introduced to the model to make it more consistent. Zhang et al. [34] applied a CNN model for sentiment classification with six convolution layers and three fully connected layers. Temporal convolution is the important module in this architecture. The convolutional layer helps to extract local features efficiently, but it cannot learn the sequence of correlations [35] (Fig. 3).

Assume a discrete input function $g(x) \in [1, l]$, kernel function $f(x) \in [1, k]$, convolution $h(y)$ between $f(x)$ and $g(x)$ is defined as

$$h(y) = \sum_{x=1}^{k} f(x) \cdot g(y \cdot d - x + c) \tag{3}$$

k—kernel size, d—stride, $c = k—d + 1$, $f(x)$—weight function, $g(x)$ and $h(y)$—input and output features. Rectified linear unit (ReLU) used as an activation function to perform max pooling between convolutional layers.

Fig. 3 CNN architecture with word level input

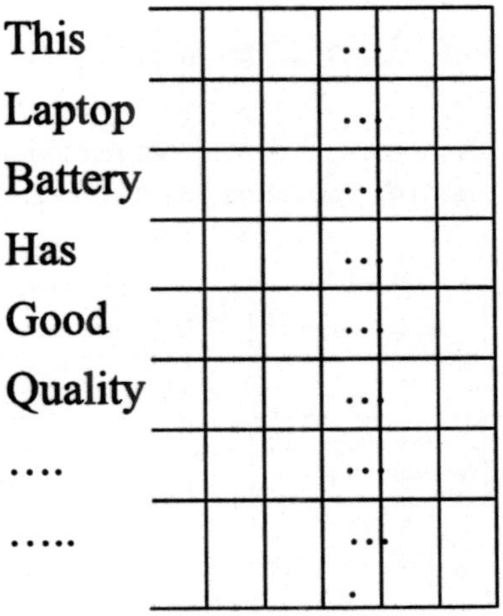

4 Experiment and Analysis

Deep learning models have been employed in a variety of disciplines in recent years. For sentiment analysis, the IMDB dataset with 25,000 reviews was used. In our strategy, we apply TensorFlow with Keras which enables GPU environments. The most essential aspect of natural language processing is data cleaning; the dataset should be exempt of symbols, punctuation, and special characters. Upon preprocessing, the dataset is ready for feature extraction (Fig. 4).

4.1 Dataset

In this study, we used IMDB movie reviews as an input dataset for text sentiment categorization. The dataset was split into two parts: 80% for training and 20% for testing. The ratio of reviews labeled positive and negative in both datasets are equal.

4.2 Preprocessing

The preprocessing phase includes URL and stopwords removal, tokenization, whitespace and punctuation removal. Stopwords are frequently used words in a language. The dataset is organized into structured form and used for tokenization. Tokens are small units of a sentence that are standardized for processing at the next level. During the preprocessing step, a total of 69,545 tokens were extracted. GloVe embedding is used to transform tokens into vector values.

4.3 Word Embedding

The input text is rendered numerically by word embedding techniques. GloVe is a matrix factorization-based word embedding technique. It is a way to acquire vector

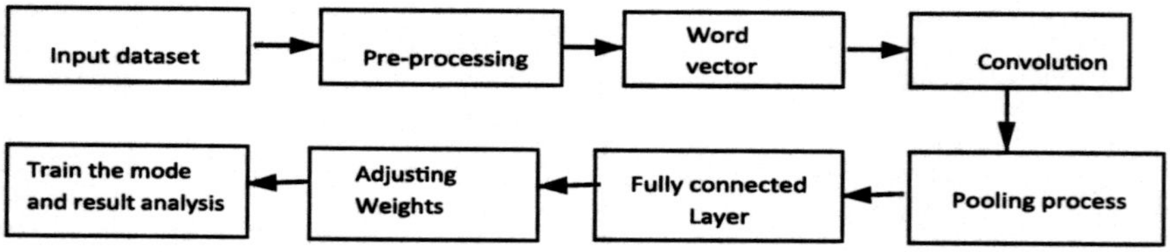

Fig. 4 Sentiment analysis with CNN model

Table 1 Parameter setting in CNN

Parameter	Learning rate	Embedding dimension	Hidden size	Number of layers	Mini-batch size
Value	Le-3	128	128	1	128

values for words using an unsupervised learning algorithm. For sentiment classification, we deployed a CNN-based method in our research. The parameter settings used in implementation are listed in Table 1.

4.4 Embedding Layer

We utilized the Keras module's tokenizer class to build a word-to-index dictionary, where each word is allotted a key and an index value. The list's maximum length is set to 100 characters. If the list length exceeds 100, it will be trimmed to the specified length, or if the value is less than 100, zeros will be appended to the end of the list until it attains the fixed length; this process is known as padding. In our experiment, the embedding matrix acknowledged 69,545 distinct words.

4.5 Result Analysis

We constructed a sequential model with an embedding layer length of 100 and vocabulary size of 69,545 words. For compilation, the Adam optimizer is used and binary cross entropy acts as the loss function. The model is summarized in the table below.

Table 1Model: "sequential_1"

Layer (type)	Output shape	Param #
embedding_1 (embedding)	(None, 100, 100)	6,954,500
conv1d_1 (Conv1D)	(None, 96, 128)	64,128
global_max_pooling1d_1	(Glob (none, 128)	0
dense_1 (dense)	(None, 1)	129

Total params: 7,018,757
Trainable params: 64,257
Non-trainable params: 6,954,500

It is a sigmoid-activated single-dimensional convolutional layer with a kernel size of 5 and 128 features. To limit the feature size, a global max pooling layer is implemented. The model was tested using CNN and GloVe based on the study with efficient parameters (Fig. 5).

The accuracy of 92% in training phase and 84% in testing phase was achieved. The results were preferable to the baseline models. The graph depicts the difference in accuracy and loss between the training and testing datasets.

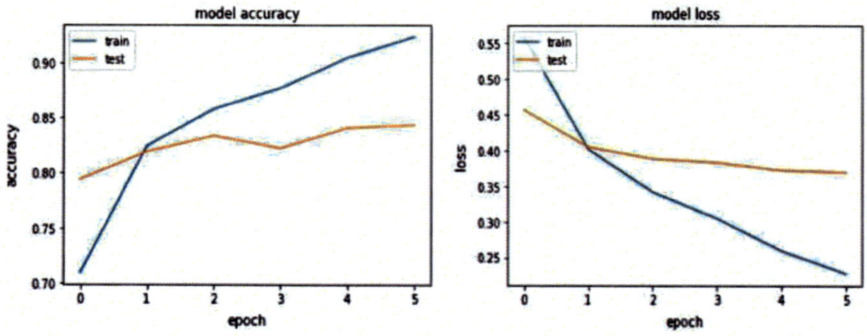

Fig. 5 Model accuracy and model loss in training and testing phases

5 Conclusion and Future Work

The objective of this research is to present a sentiment classification method based on convolutional neural networks. CNN is a unique deep learning network that has been broadly implemented to tackle text-based sentiment categorization issues. This method categorizes sentiments by collecting local dependencies in the context of sentences. In our approach, GloVe technique is employed in word embedding and there are 69,545 tokens identified. CNN effectively recovers text features from the IMDB input dataset and achieved 92% of accuracy. The hyper parameters must be fine-tuned in order to optimize deep learning models. Finally, the proposed model classifies the results effectively for the input dataset. CNN could be coupled with another neural network for further study to capture global dependencies. The NLP technique of POS tagging can be used to strengthen model accuracy. Word embedding technique Word2Vec may be used in conjunction with GloVe, and the results will be examined in future work.

References

1. Liu SA (2015) Mining opinions, sentiments, and emotions. Cambridge University Press, Cambridge, UK
2. Kang P, Seo S, Kim C, Kim H (2020) Comparative study of deep learning-based sentiment classification. IEEE Access 8
3. Salur MU, Aydin U (2020) Novel hybrid deep learning model for sentiment classification. IEEE Access 8
4. Zhou D, Wang R, Jiang M, Si J, Yang Y (2019) Survey on opinion mining: from stance to product aspect 7
5. Pang B, Lee L, Vaithyanathan S (2002) Thumbs up? Sentiment classification using machine learning techniques. In: Proceedings of the ACL-conference on empirical methods in natural language processing. Association for Computational Linguistics, pp 79–86, July 2002
6. Cheng Y, Xiang G, Yao L (2020) Text sentiment orientation analysis based on multi-channel CNN and bidirectional GRU with attention mechanism. IEEE Access 8

7. Smys S, Raj JS (2021) Analysis of deep learning techniques for early detection of depression on social media network—a comparative study. J Trends Comput Sci Smart Technol (TCSST) 3(01)
8. Thilaka B, Sivasankaran J, Udayabaskaran S (2020) Optimal time for withdrawal of voluntary retirement scheme with a probability of acceptance of retirement request. J Inf Technol 2(04)
9. Cheng Y, Sun H, Chen H (2021) Sentiment analysis using multi-head attention capsules with multi-channel CNN and B-GRU. IEEE Access 9
10. Lecun Y, Bottou L, Bengio Y, Haffner P (1998) Gradient-based learning applied to document recognition. Proc IEEE 86(11):2278–2324
11. Sungheetha A (2021) COVID-19 risk minimization decision making strategy using data-driven model. J Inf Technol 3(01)
12. Sivaganesan D (2021) Novel ınfluence maximization algorithm for social network behavior management. J ISMAC 3(01)
13. Wint ZZ, Manabe Y, Aritsugi M (2018) Deep learning based sentiment classification in social network services datasets. In: Proceedings of the IEEE international conference on big data, cloud computing, data science and engineering (BCD), July 2018
14. Kapočiūtė-Dzikienė J, Damaševičius R, Woźniak M (2019) Sentiment analysis of lithuanian texts using traditional and deep learning approaches. Computers 8(1):4
15. Guo Y, Li W, Jin C, Duan Y, Wu S (2018) An integrated neural model for sentence classification. In: Proceedings of the Chinese control and decision conference (CCDC), June 2018
16. Kim K, Chung B-S, Choi YR, Lee S, Jung J-Y, Park J (2014) Language independent semantic kernels for short-text classification. Expert Syst Appl 41(2):735–743
17. Jin C, Li W (2018) Chinese word segmentation based on bidirectional LSTM neural network model. Chin J Inf 32(2):29–37
18. Xiao Y, Cho K (2016) Efficient character-level document classification by combining convolution and recurrent layers. arXiv:1602.00367. http://arxiv.org/abs/1602.00367
19. Zhou K, Long F (2018) Sentiment analysis of text based on CNN and bi-directional LSTM model. In: Proceedings of the 24th international conference on automation and computing (ICAC), Sept 2018
20. Sun B, Tian F, Liang L (2018) Tibetan micro-blog sentiment analysis based on mixed deep learning. In: Proceedings of the international conference on audio, language and ımage processing (ICALIP), July 2018
21. Tai KS, Socher R, Manning CD (2015) Improved semantic representations from tree-structured long short-term memory network. Comput Sci 5(1):36
22. Liu G, Guo J (2019) Bidirectional LSTM with attention mechanism and convolutional layer for text classification. Neurocomputing 337:325–338
23. Chen T, Xu R, He Y, Wang X (2017) Improving sentiment analysis via sentence type classification using BiLSTM-CRF and CNN. Expert Syst Appl 72:221–230
24. Hu F, Li L, Zhang Z-L, Wang J-Y, Xu X-F (2017) Emphasizing essential words for sentiment classification based on recurrent neural networks. J Comput Sci Technol 32(4):785–795
25. Kumar KS, Cheng W-H, Zomaya AY (2020) Hybrid context enriched deep learning model for fine-grained sentiment analysis in textual and visual semiotic modality social data. Inf Process Manage 57(1)
26. Bairavel S, Krishnamurthy M (2020) Novel OGBEE-based feature selection and feature-level fusion with MLP neural network for social media multimodal sentiment analysis. Soft Comput 24(24)
27. Wint ZZ, Manabe Y, Aritsugi M (2018) Deep learning based sentiment classification in social network services datasets. In: Proceedings of the IEEE ınternational conference on big data, cloud computing, data science and engineering (BCD), July 2018
28. Imran AS, Daudpota SM, Kastrati Z, Batra R (2020) Cross-cultural polarity and emotion detection using sentiment analysis and deep learning on COVID-19 related tweets. IEEE Access
29. Singh J, Singh G, Singh R, Singh P (2021) Morphological evaluation and sentiment analysis of Punjabi text using deep learning classification. Univ Comput Inf Sci 33

30. Dang NC, Moreno-García MN, De La Prieta F (2020) Sentiment analysis based on deep learning: a comparative study. Electronics
31. Vieira ST, Rosa RL, Rodríguez DZ, Ramírez MA, Saadi M, Wuttisittikulkij L (2021) Q-meter: quality monitoring system for telecommunication services based on sentiment analysis using deep learning. Sensors
32. Hossain N, Bhuiyan MR, Tumpa ZN, Hossain SA (2020) Sentiment analysis of restaurant reviews using combined CNN-LSTM. In: 2020 11th International conference on computing, communication and networking technologies (ICCCNT)
33. Kim Y (2014) Convolutional neural networks for sentence classification. arXiv:1408.5882
34. Zhang D-G, Tang Y-M, Cui Y-Y, Gao J-X, Liu X-H, Zhang T (2018) Novel reliable routing method for engineering of Internet of vehicles based on graph theory. Eng Comput 36(1):226–247
35. Sakirintam RB, Tanriover O (2021) A ConvBiLSTM deep learning model-based approach for twitter sentiment classification, vol 9, Mar 2021. IEEE

Automatic Speech Recognition of Bengali Using Kaldi

Subhadeep Guchhait, Arnold Sachith A Hans, and Jacob Augustine

Abstract Bengali is a prominent language of the Indian subcontinent. This paper presents a comparison between different Bengali speech recognition models built with the Kaldi and Pytorch toolkits. Deep learning has been employed to improve speech recognition performance, and we explore the performance of the techniques for Bengali datasets. Seven different deep neural techniques have been employed in this paper for comparison. Grapheme to phoneme (G2P) is an important module for Indian languages which helps to decode phones from words in Unicode format. We develop a G2P model for Bangla using RNN, and we have shown that it performs well for the purpose. This research also demonstrated that using Kaldi-based feature extraction with DNN-HMM acoustic models yielded the best WER of 4.16 when combined with the Li-GRU neural network. The aim is to demonstrate the performance of the Bengali language using the current state-of-the-art (Kaldi) method.

Keywords Kaldi · Mel-frequency cepstral coefficients (MFCC) · Deep neural net with hidden Markov model (DNN-HMM) · Gaussian mixture model with hidden Markov model (GMM-HMM) · Long short-term memory (LSTM) · Light gated recurrent units (Li-GRU) · Gated recurrent unit (GRU) · Recurrent neural networks (RNN)

1 Introduction

Deep neural networks have been successfully employed to build a state-of-the-art automatic speech recognition system (ASR). However, there has not been much work done in this area for the Bangla language and for Indian languages in general. Bengali, also known as Bangla, is the world's seventh most widely spoken native

S. Guchhait (✉) · A. S. A. Hans · J. Augustine
Department of CSE, Presidency University, Bangalore, India

J. Augustine
e-mail: jacob@presidencyuniversity.in

S. Shakya et al. (eds.), *Proceedings of Second International Conference on Sustainable Expert Systems*, Lecture Notes in Networks and Systems 351,
https://doi.org/10.1007/978-981-16-7657-4_14

language, India's second most widely spoken language, and Bangladesh's official language. Bengali language is the native language of about 300 million people. Bengali language has 29 consonants, seven vowels, and seven nasalized vowels, and it shares phonology space with Assamese, Odiya, and other Magadhan languages.

In order to build a good ASR, it is essential to have a large and rich speech corpus for the language. A speech recognition system's accuracy is a function of factors such as session variability, inter-speaker variability, and microphone variability. For the ASR system to performance well, it must be built using dataset which represents these variabilities.

We use Pytorch, and Kaldi-PyTorch toolkit to build ASR systems. The PyTorch-Kaldi collaboration seeks to bring Kaldi and PyTorch closer together. The toolkit uses PyTorch to train deep neural networks, while Kaldi handles data preparation and pre-processing. Several deep learning model implementations such as feedforward DNNs, CNNs, and RNNs versions are natively available in PyTorch-Kaldi. The toolkit provides options to parallelize the operations helping in faster system building by utilizing the processor and GPUs efficiently. The toolkit is designed to ensure a simple combination of user-defined acoustic models.

The reminder of the paper is organized as follows: literature review, the experimental setup, the experimental results, conclusion, and future work.

2 Literature Review

Many researchers did work on Bengali ASR and got state-of-the-art performance. Some of the very recent and quality works are described here. In 2020, Badhon et al. [1] did a literature review on Bengali automatic speech recognition. They went through 15 research papers and concluded that the largest dataset on Bengali is SHRUTI" open-access dataset contains 22,012 different words. The best accuracy got on this data is 99.06% on only 100 voices. The most used classification algorithms by the researchers are HMM and GMM. Mel-frequency cepstral coefficients and linear prediction coefficients are mostly used for feature extraction. And the toolkit used by the researchers is HTK, Sphinx, Julius, Kaldi. In 2019, Ravanelli et al. [2] proposed the Pytorch-Kaldi toolkit for speech recognition. The PyTorch-Kaldi design intends to resolve the gap between these two most common toolkits by combining Kaldi's efficiency with PyTorch's flexibility. PyTorch-Kaldi is more than just a basic bridge between this software; it also has many important features for creating new speech recognizers. By supporting different feature and label types as well as combinations of neural networks, Kaldi may use complex neural architectures. They used four datasets TIMIT, DIRHA, CHiME, and LibriSpeech. They used neural networks like MLP, CNN, RNN, LSTM, and GRU with this they also used some advanced architecture like Light-GRU, twin-regularized RNN. The studies consider a wide range of acoustic features, including 39 MFCCs (13 static + delta + delta-delta), 40 FBANKS, and 40 fMLLR features, all of which were calculated using 25 ms windows with a 10 ms overlap. The best WER they got with the FMLLR features, i.e., 14.2 with Li-GRU neural network. They also make some complex models by combining MLP + Li-GRU + MLP NN, and features like MFCC + FBANK + FMLLR to obtain lesser WER of 13.8.

Another researcher Amin et al. [3] proposed a continuous speech recognition system in the Bengali language based on deep neural network in 2019. They utilized the Kaldi toolkit to run DNN-HMM and GMM-HMM models on an open-source SRUTI dataset and obtained improved WER than the existing CMU-SPHINIx dependent approach to the continuous recognition of Bengali speech dataset. The research is carried on the SRUTI dataset which consists of 21.64 h of total audio in that 22,012 numbers of unique words, 13,025 utterances, and 34 different speakers out of which 26 males and 8 females. They split the dataset into 75% to train and 25% for validation, i.e., 25 speakers for training and 9 speakers for testing. They primarily used GMM-HMM and DNN-HMM mixed combination algorithms to achieve their target, and they used MFCCs for feature extraction. In addition, the features are combined with the use of MFCC over the LDA + MLLT + FMLLR. They used greedy layer-wise supervised training implementation technique, not deep belief network (DBN). And the features used in this experiment are MFCC, LDA, MLLT, SAT. The best WER was 0.92% on DNN-HMM and 2.02% on GMM-HMM. They also conclude that SPHINX GMM (sGMM) is very efficient for small datasets.

In 2018, Rahman Saurav et al. [4] carry their research on speech recognition for voice search on Bengali data. Researchers proposed various Gaussian mixture model-hidden Markov model (GMM-HMM) and deep neural network-based (DNN-HMM) models for speech recognition in Bangla language to construct a voice search module for a search engine called "pipilika." The dataset used for this experiment consists of 9 h of audio from 49 unique speakers. The corpus consists of 500 unique words. The speaker's age was between 17 and 35 years. And the dataset is split into an 80–20 ratio. Out of which 80% is for training, and 20% is for testing. The training file consists of 12,000 utterances of a total of 40 speakers out of which 35 male and 5 females. And test corpus consists of 2700 utterances of a total of 9 speakers out of which 7 male and 2 females. MFCC feature is used in this experiment after extracting MFCC features the researchers applied linear discriminant analysis (LDA) transform as well as maximum likelihood linear transform (MLLT) estimation and also applied speakers adaptation technique (SAT). The best WER they got was 3.96% for a GMM-HMM-based model and 5.30% for a DNN-HMM-based model.

In 2018, Hosain Sumit et al. [5] reported work on end-to-end deep learning technique for continuous Bangla speech recognition in noisy environments. Architecture of up to nine hidden layers was used, which included a convolutional layer, a recurrent layer, and a single fully connected layer in the network (3 CNN + 5 GRU + 1FC). Connectionist temporal classification (CTC) error was employed to train the system. In this experiment, two datasets, Babel [6], and CRBLP [7] were used. Babel dataset is used for both training and testing, and CRBLP dataset was employed only for testing. Babel dataset contains both noise and silent data, whereas CRBLP contains only noise-free data. The network was trained for 20 epochs, with the optimization function as SGD with 0.99 momentum and the size of the mini-batch is 32. On the

SOCIAN and BABEL datasets, the model achieves 9.15% and 34.83% CER, respectively, and on the SOCIAN and CRBLP datasets, it achieves 12.31% and 10.65% CER.

Ravanelli et al. [8] in 2018 proposed to eliminate the traditional form of GRU design into a more efficient and compact single gate model as well as replace hyperbolic tangent with ReLU activations. Researchers also used CTC method, for that 123-dimensional FBANK features (40 + energy + delta + delta-delta) are used. The experiment was carried out using four corpus TIMIT, DIRHA-English, CHiME, and TED-talk. The experiment results show that Li-GRU not only increases the accuracy but also reduces the computational complexity and over 30% of the time for training than traditional GRU.

In 2018, Guglani and Mishra et al. [9] proposed speech recognition using the Kaldi toolkit on the Punjabi language. From a Punjabi speech sample, the researchers extracted Mel-frequency cepstral coefficients (MFCC) and perceptual linear prediction (PLP) features. Using the *N*-grams model, they generated both a mono-phone and a triphone. The dataset consists of 240 h of speech of 24 different speakers of various age groups; the sampling rate is 16 kHz. They used LDA + MLLT features transformation over Deltas features. Also, decoding with MPE Aqustic model I bit more computational efficient when compared with boosted maximum mutual information (bMMI). They have concluded that LDA + MLLT is more accurate than Delta + Delta-Delta features, and also that LDA + MLLT + SAT is more effective than LDA + MLLT. Furthermore, the results suggest that increasing the linguistic model increases the complexity, which can degrade the accuracy of the speech recognition model. The best WER they got is 21.2 using MFCC features on a 3-g model and 22.7 using PLP features.

In 2018, another research team Ahmed Sumon et al. [10] proposed speech recognition using convolutional neural networks on Bangla short speech commands. They proposed three distinct convolutional neural network architectures and used MFCC features to carry out their experiment. They used only 10 Bengali words (e.g., আগেরটা, আস্তে, আট, বাবা, বামে যাও, বাড়ি, বাসা, বোন, বন্ধকর, বড়). In this paper, the researchers worked with the combination of two datasets. Their dataset consists of 1000 voices from 100 people divided into 10 groups for each speaker. They also used a dataset from Google called Google speech command, which is in English and contains 65,000 samples for a length of less than 1 s, where they discovered 30 words with 1000 variations. For the final result, they used three methods, and they were only getting started with the CNN in speech recognition using a small dataset. The accuracy they got in training using MFCC features is 85.44% and in testing is 74.01%.

In 2018, Syfullah et al. [11] carry out their research on character recognition from a speech on Bengali data, and they mainly worked on vowels and consonants in Bengali which means Sorborno and Byanjonborno. Some examples of vowels are অ,আ, ই, ঈ, উ, ঊ, ঋ, ঌ, এ, ঐ, ও, ঔ and some examples of consonants are ক, খ, গ, ঘ, ঙ, চ, ছ, জ, য়, ড়, ঢ়, ৎ, ঁ, ং, ঃ. They presented an algorithm called *k*-means LBG (*K*-means + Linde Buzo Gray). And the feature used for their experiment is MFCC. The dataset used for this experiment contains 900 audios of 20 different people, 45 Bengali characters for each speaker. To train the model, they used 60%, i.e., 12 speakers in the training set and 40%, i.e., 8 speakers in the testing set. And they got an accuracy of 81.61%.

Upadhyaya et al. [12] proposed a speech recognition model using the Kaldi toolkit on the Hindi language in 2017. From 1000 phonetically balanced Hindi speech data, they extracted MFCC and PLP features. AMUAV dataset used in this experiment consists of 100 speakers. For the acoustic model (AM), they used GMM-HMM. HCLG, which is derived from weight finite state transducers (WFSTs), was used for decoding. Every speaker narrates ten short Hindi sentences, two of which are shared by all speakers. As a result, a phonetically balanced 1000 continuous speech of Hindi database is developed. 900 sentences are used to train the model, with the remainder being used for evaluation. A 25 ms window that has been moved by 10 ms is used to extract the feature. The best WER they got using MFCC features on a 3-g model, i.e., 16.09. They also tested the result on a 4-g model and get a WER of 16.34. So, they conclude that the increase of the LM model increases the complexity vice versa decreases the performance of the model.

In 2011, Das et al. [13] researched developing a speech corpus for continuous speech recognition in the Bengali language. They demonstrate that the speech corpus plays a vital role in ASR. So, they develop speech corpus for two different age groups. One is the younger group, i.e., 20–40 years age and the older age group distributed as 60–80 years. They used HMM toolkit for aligning the speech data. The dataset used by them consists of 26 h of data, 7500 unique sentences, in those 19,640 unique words. The corpus is collected from 130 speakers out of those 70 males (30 young and 40 old) and 40 females (20 young and 20 old). For testing the model, they recorded 20 different sentences by 20 speakers (10 young and 10 old) which are not there in the training corpus. They used linear prediction cepstral coefficient (LPCC), MFCC, and PLP features. They used two types of MFCC features set one set consists of a 26-dimensional feature vector and another consist of 39-dimensional feature vectors. And normalization method used by them is MLLR. The results show that the 39-dimensional feature performs way better than the 26-dimensional feature. They conclude that the recognition rate is better for the young age group.

In 2011, Mandal et al. [14] developed a Bengali speech corpus for phone recognition. They demonstrate that choosing the proper text from the available text while maintaining a balanced phone distribution is a crucial step in designing a high-quality PR model. Anandabazar Patrika, Web-based blogs, common discussions, and editorial papers were used to assemble the corpus. It consists of 737,552 numbers of phones, and the sampling rate is 16 kHz. They selected 47 phonemes from the speech corpus, including 32 consonants, 7 common vowels, and 7 nasalized variations of those vowels. They combined several various HMM and GMM models. For GMM, the highest accuracy came from a three-state HMM with 16 mixture components. They used 12 speakers for the study (8 males and 4 females) and collected 92 words and the result shows that their method out from randomly selected method and greedy method. They concluded that their text filtering algorithm could be used to choose the best text for an automatic speech recognition application.

3 Proposed Approach

Deep learning-based approaches have shown promising results across all languages for ASR. Although Kaldi-based [15] systems are popular, deep learning architectures are increasingly gaining ground reporting better performance. It becomes important to compare the ASR performance for each technique on a given data to provide a baseline result for reference.

Apart from building an ASR, a Bengali grapheme to phoneme (G2P) was developed using recurrent neural networks using the available Google data [7]; the performance of which is also reported. G2P becomes important to convert the text data into component phonetic representations. The most Sanskrit-based languages have straight-forward rule-based phone representation in text which is easier to handle; however, Bengali vowels are context dependent. Sometimes, in Bangla, G2P conversion is influenced not just by orthographic information but also by POS and semantics, hence pose challenges in G2P [16].

3.1 Deep Learning-Based G2P for Bengali

To train the G2P, approximately, 56k words along with their ground truth representation are used to build a RNN model to output component phones. A pre-trained 3-layer transformer model having 256 hidden units and dimension of the conv-layer is 512 used to estimate grapheme-to-phoneme (G2P). The model was pre-trained on CMU English dictionary [17]. To get Bengali G2P, the model was trained on Google language source dataset [18], which contains more than 63,000 Bengali words along with their phonemes. The training was done up to 2 lakhs epochs; after that, there was no improvement on the training set. To get the phone error rate (PER), we split the 63k words into train and dev into 80:20 ratio. And for testing extra 150 words were extracted from the training dataset. The PER was found to be 28%.

The architecture of the model can be found in [19, 20] of the available ground truth data; a part was used for testing the performance of the system which was reported in Table 1, where * indicates the mismatch.

Table 1 Sequence-to-sequence model output

Words	Required output	Predicted output
পিত্তরস	p i t t o r O sh *	p i t t o * r sh O
বুঝতেও	b u jh t e O	b u jh t e * O^
ঈশ্বরদীতে	i ** sh o r d i t e	i SH sh o r d i t e
আক্রমণাত্মক	a k r o m o n a t t O k	a k r o m o n a t t O k

4 Experimental Setup

4.1 Dataset

The Bengali dataset used for this research was downloaded from https://openslr.org/. Though the dataset consists of 210 h of audio files, this experiment was carried out using a subset of data consisting of 9 h of audio. The selected subset contains 10,203 plus distinct utterances, with a vocabulary of 12,475 unique words. The dataset includes audios and TSV files. The TSV file contains audio ID, speaker ID, and the transcription of the audios, separated by a one-tab space. The audio ID is a ten-digit number. The second column's anonymized speaker ID was made up of five alphanumeric characters. The third column contained the transcription of that audio in the Bangla script. For instance, an audio file with the file ID 2a16205fb1 and a speaker ID of c749e was renamed to c749e-2a16205fb1. This file naming convention made it easier for us to organize the details required by the Kaldi toolkit and then separated the transcription with one blank space. After splitting the dataset into train and Val, 7140, and 3066 audio samples are used. Both male and female voices are included in the Bangla utterances. Since the speaker identity is anonymized, gender identity of speakers is lost. By listening tests to the dataset, it was identified that there are 508 unique speakers out of which 304 male and 204 female speakers. A set of 60 audios are used for testing purposes which was recorded by our team which contains 6 speakers (4 male, 2 female).

4.2 System Architecture

This project has been carried out using the Kaldi toolkit. Kaldi is a free and open-source toolkit for working with voice data. It is mostly used for speech recognition, as well as for other purposes in voice-related applications—such as speaker recognition and speaker dimerization. Kaldi can generate features such as mfcc, fbank, and fMLLR. As a result, one common use of Kaldi in recent deep neural network research is to pre-process raw waveforms into acoustic features for end-to-end neural models. There are three sections of Kaldi

- Data pre-processing and extraction of features
- Model
- The training process.

The Pytorch-Kaldi aims to bridge the gap between both the Kaldi and Pytorch toolkits to achieve Kaldi's performance and Pytorch's versatility. Not only is the Pytorch-Kaldi easy toolkit interface, but it also incorporates many practical features to improve advanced voice recognition systems.

Figure 1 shows the overall block diagram of the work carried out. First, the MFCC features are extracted from the audio using Kaldi toolkit. The output of the

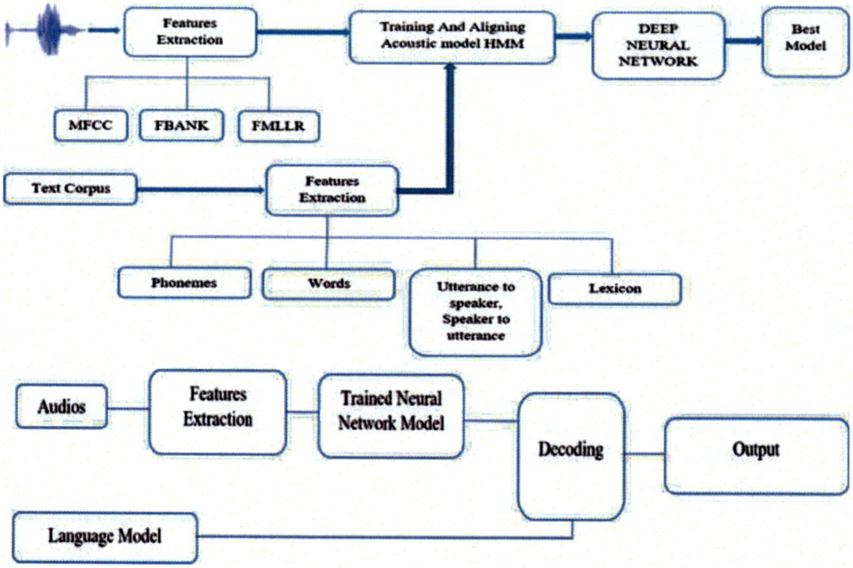

Fig. 1 Overall training and testing block diagram

extracted features stored in "ark" format. From text corpus, text features are extracted like phonemes, words, utt2spk, spk2utt, and Lexicon. Lastly, preparation of SRILM language model is done using text corpus. Before pushing the features to neural network, features extracted from the audio data and the transcription alignment are done in Kaldi (Monophone, Tri1, Tri2, Tri3, Tri4). A graph file will be generated after alignment which is further required to train DNN. After extracting all the features, the training is done using deep neural network. Finally, after decoding the model will give the decoded output.

4.3 *Data Preparation*

Kaldi requires data to be prepared in a special format to feed into the neural network. Also, the audio and text transcription files are required to be in a format fit for training and testing purposes. The OPENSLR data required cleaning since, for example, some of the audios were coded as numbers which were to be converted to the text format representation. Data preparation was started with cleaning the transcription file. All the numbers and words which are written in English have been changed into Bengali words in the transcription file. And all the special symbols were removed as described in Table 2.

Next, all the necessary folders are created inside the corpus like steps, utils, src, exp, conf, and data. Inside the data folder, three folders were created train, test, and language (lang). The train folder will contain files required for training

Table 2 Bengali transcription of words and numbers

Numbers/words	Transcription
১৯৯৩	উনিশ শ তিরানব্বই
.	দশমিক
National University	ন্যাশনাল ইউনিভার্সিটি

the NN, i.e., text, wav.scp, utt2spk, spk2utt. There are total 23,777 unique words extracted from the text file. Next, wav.scp file has been created. After this, utt2spk and spk2utt files are created. Each utterance mapped with its corresponding speakers is stored in utt2spk file. And speaker to utterance mapping is stored in spk2utt file. The language directory contains the following files silence_phones.txt, optional_silence.txt, extra_questions.txt, nonsilence_phones.txt, and lexicon.txt. The "lexicon.txt" file contains the phones for each Bengali word. The lexicon file created using sequence-to-sequence modeling [19]. The file nonsilence_phones.txt contains the phones found in the corpus, such as aa, umm, bh, and so on. Similarly, silence_phones.txt have the silence phones like "OOV" (out of vocabulary) "SIL", and optional_silence.txt simply contains a "SIL". After preparing all the files for the language directory, the data/language directory was prepared which contains phones.txt, words.txt, L.fst, L_disambig.fst, oov.txt, oov.int, and topo files. The finite state transducer version of the lexicon is stored in the file L.fst. On the input, there are phone symbols, and on the output, there are word symbols. The L disambig.fst file is the lexicon, but it has the #1, #2 discontingency symbols and even the #0 self-loop on it to "pass through" grammar's discontingent symbol. The file oov.txt contains just a single line. Lastly, the phones directory for the data/language folder was prepared. The majority of the files come in three different formats ".txt," ".int," and ".csl". The topology of the HMMs is defined here.

4.4 Feature Extraction

After creating all the necessary files, mel-frequency cepstrum coefficient (MFCC) features are extracted from the audio files using Kaldi features for Pytorch-Kaldi-based experiments. MFCC features are extracted using 25 ms frames shifted by 10 ms each time. After extracting 13-dimensional feature, MFCCs cepstral mean and variance are normalized (CMVN). CMVN helps to reduce audio data variability due to microphone change. A basic working sequence of the script to build ASR in Kaldi toolkit is given in Fig. 2.

To build the acoustic model, below tasks are performed sequentially.

- Train monophones: A simple monophone model is trained.
- Align monophones: Check alignment of monophone model with corresponding text.

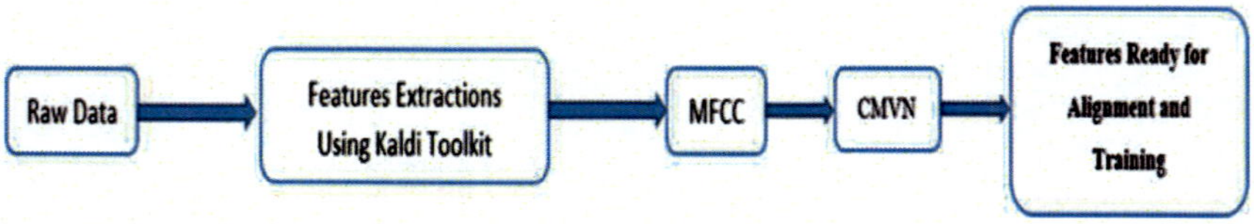

Fig. 2 Audio features extraction system

- Triphone training and alignment: Train and align triphone model.
- LDA-MLLT: Train triphone with LDA-MLLT features. For each input, LDA creates an HMM state with reduced feature space, and MLLT takes the reduced feature space from the LDA and extracts a specific transformation for each speaker. As a result, MLLT is a step toward speaker normalization because it reduces speaker difference.
- LDA-MLLT-SAT: Train triphone with LDA-MLLT-SAT. SAT is an acronym that stands for speaker adaptive training. It adapts to each speaker with a specific data transforms to perform speaker and noise normalization.

4.5 Deep Neural Network Architecture

After the alignment part using kaldi recipe, a graph folder will be created which contain "HCLG.fst" where H denotes HMM specifications, this symbol is used to indicate context-dependent phones, whereas transition IDs are used to encode the PDF ID, among other factors. C denotes context-dependency, L stands for Lexicon, and lastly, G is a grammar or language model acceptor. This HCLG file is required to train the Pytorch-Kaldi neural networks model. Now, as all the files are created, DNN model can be kept for training. This research is carried out with five NN models, i.e., multi-layer perceptron (MLP), recurrent neural network (RNN), long short-term memory (LSTM), light gated recurrent units (Li-GRU), gated recurrent unit (GRU). The architecture configuration for different neural networks is showed in Table 3.

Table 3 Neural network architecture configuration

Sl. No.	Neural networks models	Number of layers	Nodes in each layers	Activation function	Optimizer
1	MLP	6 Layers	1024	ReLU and Softmax	Stochastic gradient descent
2	RNN	4 Layers	550	ReLU	Adam
3	LSTM	4 Layers	550	Tanh	RMSProp
4	GRU	4 Layers	550	Tanh	RMSProp
5	Li-GRU	5 Layers	550	ReLU	RMSProp

5 Result Analysis

This experiment was carried out using five neural network models, those are MLP, GRU, Li-GRU, RNN, LSTM. Activation functions used are ReLU, tanh, and softmax. The optimizers used are SGD, RMSprop, and Adam.

From the triphone model, the features are pushed into the NN. Each of the NN is trained for 25 epochs and accuracy is noted. The accuracy graph and loss graph of all the models are given in Fig. 3.

From the above graph, it is inferred that the validation accuracy was not changing after a point in time. For this reason, we stopped training after 25 epochs as there is not much change in the accuracy.

The model's precision is measured in terms of word error rate (WER). The WER is calculated as

$$\text{WER} = \frac{S + D + I}{N}$$

"*S*" represents the number of substitutions, "*D*" represents the number of deletions, "*I*" represents the number of insertions, and "*N*" represents the number of words in the comparison. The model has been tested on 65 real-time audio samples recorded by four different speakers. And the WER of all the models is noted in Table 4.

From Table 4 it is cleared that Li-GRU gave the best WER compared to others models. The decoded output of the Li-GRU model is given in Table 5.

From Table 5, it can be referred that out of these seven sentenced only one word of the second sentence was misclassified. So, we can conclude that though validation accuracy was not that much good WER for all the models is very less. The words which are misclassified most of them have same kind of phonetic sound. And, also each word in the validation dataset must be there in the training corpus. As we cannot expect to generate new words from the model. It will only give the best phones match for the unknown words.

6 Conclusion

This paper's main goal was to examine the accuracy of an automatic speech recognition method using the Kaldi toolkit on a standard Bengali dataset. DNN-HMM model was used in the experiment with Kaldi feature extraction recipes as well as training approaches like MFCC with monophone and triphone also LDA + MLLT or LDA + MLLT + SAT. Also, based on the feature extracted from the Kaldi toolkit, five different neural networks models were developed like MLP, RNN, LSTM, GRU, and Li-GRU. The accuracy of all models was measured concerning word error rate. Li-GRU model gave the best WER of 4.16 which outperforms other models. It can be concluded from this experiment that if the full corpus of the Bengali dataset (i.e., 210 h of data) used for the experiment the accuracy will also increase as the number of words will increase.

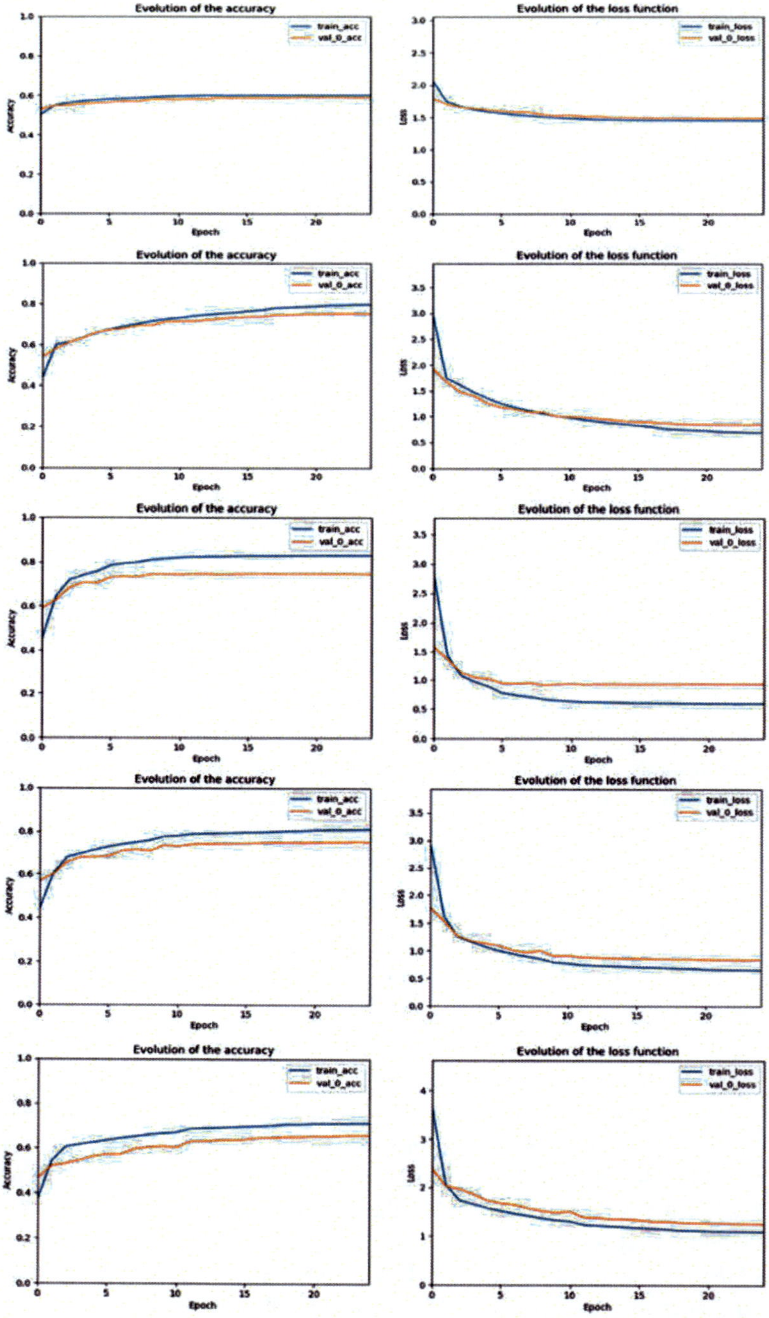

Fig. 3 **a** Accuracy and loss graph of MLP model. **b** Accuracy and loss graph of RNN model. **c** Accuracy and loss graph of LSTM model. **d** Accuracy and loss graph of GRU model. **e** Accuracy and loss graph of RNN model

Table 4 Word error rate table

Neural networks	Word error rate (WER)
MLP	5.14
RNN	5.24
LSTM	6.28
GRU	4.58
Li-GRU	**4.16**

Table 5 Expected and decoded output

Expected	Decoded
যখন চাঁদ পৃথিবীর কাছাকাছি পৌঁছায়	যখন চাঁদ পৃথিবীর কাছাকাছি পৌঁছায়
ঘুম থেকে জাগিয়ে	**ভুল** থেকে জাগিয়ে
ওর ধারণাই ঠিক	ওর ধারণাই ঠিক
একজন মারা গেছে বলে	একজন মারা গেছে বলে
ঢাকা সিটি কলেজ	ঢাকা সিটি কলেজ
সে হয়ে যায় মরা নদী	সে হয়ে যায় মরা নদী
একটা নতুন ভাড়াটে এসেছে	একটা নতুন ভাড়াটে এসেছে

So, for future work, we will run the model using the full dataset and get the result. Also, we will explore different activation functions and optimizers to improve the accuracy of the models.

Acknowledgements We would like to acknowledge Vikram Lakkavalli, Research & Development Head, Kaizen Secure Voiz Pvt. Ltd. for suggesting the problem and taking part in technical discussions without which this work would not have been possible. This work has been carried out at Kaizen Secure Voiz Pvt. Ltd. as a part of my thesis requirement.

References

1. Badhon SMSI, Rahaman MdH, Rupon FR, Abujar S (2020) State of art research in Bengali speech recognition. In: 2020 11th International conference on computing, communication and networking technologies (ICCCNT), Kharagpur, India, July 2020, pp 1–6. https://doi.org/10.1109/ICCCNT49239.2020.9225650
2. Ravanelli M, Parcollet T, Bengio Y (2019) The PyTorch-Kaldi speech recognition toolkit. In: ICASSP 2019—2019 IEEE International conference on acoustics, speech and signal processing (ICASSP), Brighton, United Kingdom, May 2019, pp 6465–6469. https://doi.org/10.1109/ICASSP.2019.8683713
3. Amin MdAA, Islam MdT, Kibria S, Rahman MS (2019) Continuous Bengali speech recognition based on deep neural network. In: 2019 International conference on electrical, computer and

communication engineering (ECCE), Cox's Bazar, Bangladesh, Feb 2019, pp 1–6. https://doi.org/10.1109/ECACE.2019.8679341
4. Rahman Saurav J, Amin S, Kibria S, Shahidur Rahman M (2018) Bangla speech recognition for voice search. In: 2018 International conference on Bangla speech and language processing (ICBSLP), Sylhet, Sept 2018, pp 1–4. https://doi.org/10.1109/ICBSLP.2018.8554944
5. Hosain Sumit S, Al Muntasir T, Arefin Zaman MM, Nath Nandi R, Sourov T (2018) Noise robust end-to-end speech recognition for Bangla language. In: 2018 International conference on Bangla speech and language processing (ICBSLP), Sylhet, Sept 2018, pp 1–5. https://doi.org/10.1109/ICBSLP.2018.8554871
6. IARPA Babel Bengali Language Pack. https://catalog.ldc.upenn.edu/LDC2016S08. Accessed Feb 2018
7. Alam F, Habib SM, Sultana DA, Khan M (2010) Development of annotated Bangla speech corpora. In: Spoken languages technologies for under-resourced languages
8. Ravanelli M, Brakel P, Omologo M, Bengio Y (2018) Light gated recurrent units for speech recognition. IEEE Trans Emerg Top Comput Intell 2(2):92–102. https://doi.org/10.1109/TETCI.2017.2762739
9. Guglani J, Mishra AN (2018) Continuous Punjabi speech recognition model based on Kaldi ASR toolkit. Int J Speech Technol 21(2):211–216. https://doi.org/10.1007/s10772-018-9497-6
10. Ahmed Sumon S, Chowdhury J, Debnath S, Mohammed N, Momen S (2018) Bangla short speech commands recognition using convolutional neural networks. In: 2018 International conference on Bangla speech and language processing (ICBSLP), Sylhet, Sept 2018, pp 1–6. https://doi.org/10.1109/ICBSLP.2018.8554395
11. Syfullah SM, Zakaria ZB, Uddin MdP, Rabbi MdF, Afjal MI, Nitu AM (2018) Efficient vector code-book generation using K-means and Linde-Buzo-Gray (LBG) algorithm for Bengali voice recognition. In: 2018 International conference on advancement in electrical and electronic engineering (ICAEEE), Gazipur, Bangladesh, Nov 2018, pp 1–4. https://doi.org/10.1109/ICAEEE.2018.8642994
12. Upadhyaya P, Farooq O, Abidi MR, Varshney YV (2017) Continuous Hindi speech recognition model based on Kaldi ASR toolkit. In: 2017 International conference on wireless communications, signal processing and networking (WiSPNET), Chennai, Mar 2017, pp 786–789. https://doi.org/10.1109/WiSPNET.2017.8299868
13. Das B, Mandal S, Mitra P (2011) Bengali speech corpus for continuous automatic speech recognition system. In: 2011 International conference on speech database and assessments (oriental COCOSDA), Hsinchu City, Taiwan, Oct 2011, pp 51–55. https://doi.org/10.1109/ICSDA.2011.6085979
14. Mandal S, Das B, Mitra P, Basu A (2011) Developing Bengali speech corpus for phone recognizer using optimum text selection technique. In: 2011 International conference on Asian language processing, Penang, Malaysia, Nov 2011, pp 268–271. https://doi.org/10.1109/IALP.2011.16
15. Povey D, Arnab G, Gilles B, Lukas B, Ondiez G, Nagendra G, Mirko H, Petr M, Yanmin Q, Petr S, Jan S, Georg S, Karel V (2011) The Kaldi speech recognition toolkit. In: 2011 IEEE Workshop on automatic speech recognition and understanding, Hilton Waikoloa Village, Big Island, Hawaii, US
16. Basu J, Basu T, Mitra M, Mandal SKD (2009) Grapheme to phoneme (G2P) conversion for Bangla. In: 2009 Oriental COCOSDA international conference on speech database and assessments, Urumqi, China, Aug 2009, pp 66–71. https://doi.org/10.1109/ICSDA.2009.5278373
17. https://github.com/cmusphinx/g2p-seq2seq.git
18. https://github.com/google/language-resources
19. Kaiser L (2017) Accelerating deep learning research with the Tensor2Tensor library. In: Google research blog
20. Vaswani A, Shazeer N, Parmar N, Uszkoreit J, Jones L, Gomez AN, Kaiser L, Polosukhin I (2017) Attention is all you need. arXiv preprint arXiv:1706.03762

Prediction of Priority to Individual for COVID Vaccine Distribution Using Soft Computing Techniques

S. Subbulakshmi, Adarsh R. Nambiar, A. K. Arun, Fawaz Al Faizi, and Vishnu Narayanan Harish

Abstract World is facing impact of COVID pandemic; every industry is facing corollaries to a great extent. With increase in number of COVID cases, it is necessary to control the spread of virus and wipe it out in the best possible manner. COVID-19, which have its rapid spread across the world, is a dangerous episode which entire community is cladding on. Proper treatment and efficient vaccine distribution are the major glitches of medical industry during this untoward outbreak. Priority in distribution of vaccine is extremely vital, to handle availability of vaccines. Our main objective is to develop methodologies with competing process which implements unsupervised k-means clustering and supervised naïve Bayes, decision tree and gradient boost classification algorithms to find the priority of individuals for distributing vaccines. Efficient pathway of distribution is implemented in layered manner based on individual's priority, assessed in par with impact level of disease on individuals in the community. Priority for vaccine distribution is finalized with classification algorithm with the highest accuracy. It helps to distribute the available vaccines optimally among the population, and it plays major role in reducing the impact of coronavirus on individuals and society.

Keywords COVID 19 · Data preprocessing · Label encoding · *K*-means clustering · Multi class classification · Naïve Bayes · Decision tree · Gradient boosting · Ensemble method

1 Introduction

Covid-19 pandemics have created a huge problem in whole world. It plunged national economies into deep recessions. The spread of corona virus through air and its rapid outbreak has increased the spread across the whole world, and necessities of vaccine for treatment have become more essential. Through proper prevention mechanisms,

S. Subbulakshmi (✉) · A. R. Nambiar · A. K. Arun · F. Al Faizi · V. N. Harish
Department of Computer Science and Applications, Amrita Vishwa Vidyapeetham, Amritapuri, India
e-mail: subbulakshmis@am.amrita.edu

S. Shakya et al. (eds.), *Proceedings of Second International Conference on Sustainable Expert Systems*, Lecture Notes in Networks and Systems 351,
https://doi.org/10.1007/978-981-16-7657-4_15

we can reduce the spread of disease and prevent hospitalization of person. The upward trend of COVID-19 victims has increased the demand for COVID-19 vaccines. It paves the basis for the need to distribute vaccine in an efficient manner. Vaccine distribution is also an important way to prevent the spreading of virus. If vaccine is distributed effectively, we can decrease the rising positive cases of COVID-19 virus. We have identified that there is requirement of creating methodologies, which will solve the vaccine distribution problem. Proper vaccine distribution is needed to handle demand and supply of vaccines, to render maximum benefit to the society [1].

Rise of COVID-19 lead to the complexity for vaccine distribution among people and have increased demand for vaccine across world. As it has a rapid spread, vaccination should be taken in such a way that its spread and severity of the problems due to this infection should be curtailed to a greater extent. Initially, it should be taken by persons who have the highest chance of getting infected due to commitment to his job or service. It should also include people who are more prone to disease due to reduction in metabolism with old age. First, category includes healthcare workers and older people. Distribution of vaccine is one of the greatest problems in this scenario as there is more and more people struggling to get the vaccine. It needs to be distributed in a proper manner so that we can reduce the crowding of people in vaccination centers. There should be a priority set for different categories of people, which is allotted to all individuals and they could be vaccinated according to priority [2].

Administrations of vaccines to general public have to consider different factors. Different option exists among people regarding vaccination, due to that many people fail to take vaccines. Most importantly, coronavirus has infected many people, and in due course of time, many powerful variants of the virus have emerged with crossover and mutations. It has resulted in second wave of infection in different countries, with unimaginable adverse effect on the health system of the community. Moreover, virus is becoming more powerful by continuous mutations with different genome sequences. Only possibility to control the spread of virus is to achieve herd immunity, which is possible only when 60–75% population is infected by the virus or vaccinated.

The more practical, feasible, and beneficial approach is to increase vaccination administration to maximum possible extent. It would help individual's health system to develop required immunity to handle coronavirus infection [3]. Since, vaccine availability in current stage is not proportional to population of country. It is highly essential to find a standardized model for the optimal distribution of vaccines, in order to provide protection and safety of the community as a whole. It is not practical to distribute vaccination to all individuals at the same time. Individuals in the population have to be categorized into five to six groups based on their exposure to virus. Our work aims to develop a multi-class classification model to address the issue with required solution approach. The work is divided into following modules—data set collection, data preprocessing, initial classification, splitting into train and test data, multiclass classification models with training data set, prediction for test data, and accuracy prediction of models. According to our models, we create our own dataset through the online survey forms. It consists of data like name, profession, age, if vaccinated, health condition, and prone to disease. Initially, before the start of vaccine distribution, data does not have priority labels, and it holds only features which are categorical and numeric datatypes. Preprocessing work is done with exploratory data analysis functions to perform (1) encoding for conversion of categorical data to numeric form, (2) normalizing data to bring them all to a specific range, (3) identifying outliers and handle them with appropriate functions.

Normalized preprocessed data without priority labels are used in unsupervised *k*-means clustering algorithm to classify them into different categories. On analysis of data in different clusters, they are assigned priority labels which could be used for vaccination distribution. If the real time dataset is labeled with priority numbers, in such cases, supervised machine learning of Naive Bayes, decision trees, and gradient boost algorithms is used to create models for classification. Naive Bayes algorithm creates model for classification based on probabilistic methods. In order to have more accurate results, gradient boost is used to create the prediction model in the form of an ensemble of decision tree. Decision trees are one of the weak prediction model and gradient boosting build trees one at a time, in which each new tree tries to correct errors made by the previously trained tree. So, the model is able to make prediction with high accuracy rate which is free from errors.

This document contains Sect. 2 Related Work, which includes details of major research articles, which forms the basis for our work, Sect. 3 Proposed System Design elaborates methodology with various components using the diagram, Sect. 4 System Implementation Results illustrate implementation details of the system. It also shows the implementation of different models, and Sect. 5 Conclusion and Future work concludes our work with suggestions for future enhancement.

2 Related Work

The effective way to prevent a pandemic is vaccination. Vaccinated individuals will have the immunity to prevent the disease, and it reduces the chance of spreading the virus to the unaffected population. Duijzer et al. [4] discuss the relation between the herd effect and vaccination fraction for the seminal SIR compartmental model which consists of an equation that describes the duration of a pandemic. They have identified the consequences of allocating limited vaccines to multiple non-interacting populations. Also, for each population, a unique vaccination fraction that is per dose and multiple doses of vaccine are administered. It is an efficient way to reduce the impact of the pandemic. An important finding of this paper is that the focusing vaccine on a limited number of populations can make a significant difference. That is the optimal vaccination allocation can reduce infection.

Generally, unsupervised learning algorithms are used for a dataset with features of numerical characteristics. It tries to group the data based on suitable similarity measure. Research work of Vani and Gupta [5] uses *k*-means clustering techniques for plagiarism detection. Focus of this paper is the automatic text categorization in plagiarism detection. It explores the clustering methods using different variations of

k-means algorithm and compares it with two methods, which are the general N-gram and vector space model. Finally, they analyzed and evaluated using a real dataset from pan 2013 and compared its performance based on precision, recall, and efficiency based on the time taken for algorithm execution. Based on this survey, we tried to implement k-means algorithm in order to find the initial set of priority labels for individuals in population.

For a dataset which have predefined multiple class labels based on the past behavior of various features of records, supervised multiclass classification model is used to predict the class label for a given record. Gradient boosting is such a classification algorithm that will classify the dataset with multiple class labels. In Lesion Localization methodology [6], gradient boosting is used for detecting tumor in the brain with its precise location and its characterization. For improved automatic lesion detection and characterization, an efficient machine learning algorithm is used which is the lee filtered bivariate correlative regression-based extreme gradient boosting. Here, gradient boosting technique is used to lesion characterization for the improved accuracy in certain features. Finally, this technique is able to produce better classification results, with high classification accuracy with respect to lesion time and peak signal ratio.

Naïve Bayes classifier is an efficient and simple algorithm in data classification, which use fast and suitable for real time predictions. Pandey et al. [7] have used a supervised multinomial naïve Bayes classifier which has good quality and better execution performance. They purely focused on statistical analysis of electronic business transaction which has increased since the outbreak of COVID pandemic. Motivation of work is to classify such voluminous business transaction data in an efficient way. Here, they used naïve Bayes classifier algorithm to classify product listings to anonymous marketplaces. Existing production is classified under some category based on the probabilistic score. Task of classifier is to automatically categorize a new product into existing categories.

Prioritizing vaccine can reduce the chance of spreading COVID-19 virus. Prioritizing the vaccine should not be limited only in individual's health factor and profession. It should also consider other aspects of individuals, so in strategic spatiotemporal vaccine distribution system is implemented in [8], which primarily prioritizes certain cities which increase survival rate of the population. Prioritization of region is based on new cases of COVID-19 during certain time and compares it with the standard vaccine distribution demographically. From the analysis, it is clear that result obtained by the system from a well-mixed population reduces the number of deaths. It also reveals that prioritizing for vaccine should not only be based on the individual factors but it also should consider other factors like region, which in turn can reduce death due to the virus.

Based on the information gathered through literature survey, we designed a methodology to find priority for vaccination using soft computing techniques, which uses both unsupervised and supervised algorithms. Initially, classification of preprocessed data is done through unsupervised algorithm to identify class labels. Later, for further predictions, supervised algorithms are used to classify new data, which shows higher accuracy rates.

3 Proposed System Design

Corona virus has taken over the whole world under its control. Spread happens mainly through person-to-person contact. Infections may be mild or a deadly one according to health condition and immunity levels of individuals. The upward rising of COVID-19 has increase the demand of COVID-19 vaccines. It has led to need for a system to distribute vaccine in an efficient manner. Vaccine distribution is also an important way to prevent spreading of virus. If vaccine is distributed effectively, we can decrease the rising COVID-19 virus positive cases. The designed methodology is depicted in Fig. 1. It shows flow of data to the system for classification, based on similarity of the content.

Architecture depicts major modules of the system, which starts with preprocessing of the raw data followed by initial clustering to assign class labels. Then, to create classification models, dataset is split into training and test data. Classification models are created with training dataset, and their predictions are verified with test data. Finally, accuracy of those models is evaluated.

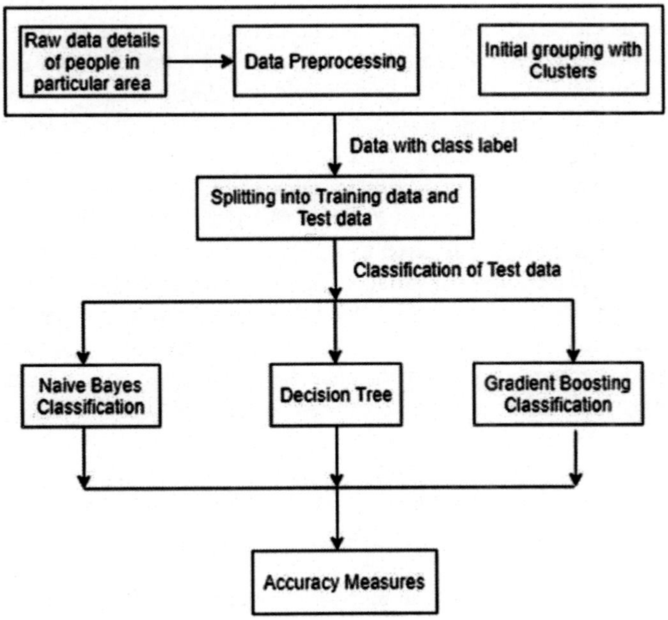

Fig. 1 System architecture

3.1 Modules

3.1.1 Data Preprocessing

Collection of data that is essential for the model is gathered from individuals in a particular locality. It includes name, age, profession, heart attack, allergies, and other health related problems, pregnancy/breastfeeding, locality, severely affected area, and most of the data are in string format. Those data are converted to numeric format, and it is normalized to change it to numeric values between 0 and 1, in order to enhance the data to be used in different models of soft computing techniques [9]. Preprocessed raw data hold several details of individuals in the community, but it neither has class labels nor it is grouped under certain categories. It represents details of population, which have to be grouped in order to assign priority. Next step is to assign initial classification by grouping them using unsupervised algorithm.

3.1.2 Initial Grouping with Clustering Algorithm

Data are not labeled and does not have any past categorization, as current pandemic situation is an unexpected happening occurred hundred years after swine flu. Spread of virus is subject to several attributes of individuals. Attributes are determined with respect to nature of spread, rate of spread, medium of transmission, health condition, and age of people getting affected by the deadly virus.

Dataset with attributes related to above-mentioned factors are given to unsupervised K-means clustering algorithm to group them into multiple groups. Here, the number of clusters taken as k is given a value six, in order to create six groups [10]. Similar ones are placed in a cluster according to its behavior [11]. This algorithm helps to label unlabeled data by classifying them into groups. It describes the best way to arrange data using centroids. K-means choose random k points as the first centroid and according to Euclidean distance; it will adjust all points and make them into groups [12]. Euclidean distance d between points is calculated using the formula:

$$d = \sqrt{\sum_{i=1}^{n}(x_i - y_i)^2} \tag{1}$$

where x and y represent points in Euclidean space, and n denotes the number of attributes considered. Clusters are formed by placing data point, which have minimum distance with cluster centroid into that cluster. Upon adding a member into the cluster, centroids are adjusted to get new centroid. This process continues until all data points are grouped and there is no change in cluster centroids. Thus, set of individuals in dataset are categorized into six groups. Data in each group are analyzed, and they are provided with priority numbers or labels according to nature of individuals in each cluster. Following is the detailed description of k-means clustering technique used for grouping the given dataset into a specified number of clusters.

K-Means Clustering Algorithm

Input: Unlabeled dataset with set of records holding multiple features*Input*: Unlabeled dataset with set of records holding multiple features

1. Set number of clusters $k = 6$ and obtain data points
2. Randomly select 6 centroids c_1 to c_6 one for each cluster
3. Repeat the following for fixed number of n iterations or until convergence for each record x_i in the dataset:
   - using Euclidean distance measure, find closest centroid c_j to x_i among the set of centroids (c_1 to c_6)
   - assign record x_i to the cluster of the centroid c_j for each cluster $j = 1 \ldots 6$
   - calculate new centroid value = mean of all data points in that cluster
4. End

Output: Labeled dataset with cluster labels.

One to six priority labels are given, where first priority will be given to health care workers, which include doctor, nurses, and police. First priority is also given to workers who have more interaction with COVID patients. Second priority is given to group of people above 60 years of age and people with heart problem and other major medical issues. Third category is allotted government officials and people with allergies or other diseases. People above 40 years of age and people residing in hotspot are given fourth priority. Fifth category is assigned to people above 20 years, and last priority 6 is to persons who cannot be vaccinated such as pregnancy/breastfeeding women and people below 20 years. Dataset with prioritized class labels will be used for classification model creation and prediction of new data pertaining different areas.

3.1.3 Priority Prediction Using Classification Models

Classification models are supervised algorithms, which identifies class labels based on existing dataset with class labels. Our problem is categorized as multiclass classification, since predicted value ranges from 1 to 6, which refers to different priority levels. Dataset includes records of different individuals in population. Each individual is described with set of attributes of varied datatypes. Each individual is also represented with the class label. Initially, class labels are defined to group the dataset into different categories based on their similarity after analyzing all its attributes. It is generally assigned based on some past behavior or by applying some unsupervised learning techniques.

The dataset with priority as class labels is split now into training and test data. Training data are used by different model to learn past behavior of all its attributes for prediction of the class labels. Only based on that understanding and behaviors of training data, test data are classified by the model.

Following section describes classification of data using different models, it also performs the comparative study of output of those models.

(a) ***Naïve Bayes Classification Model***

It is a probabilistic model, which will predict according to the priority by studying the given data. It shares a common principle that every pair of features being classified is independent of each other. Here, dataset is given to the model for prediction and will consider each column of dataset. This classification technique based on Bayes theorem it assumes presence of particular feature in the class is unrelated to the presence of other features [13]. Formula for Bayes theorem is given below:

$$P(A|B) = \frac{P(B|A) * P(A)}{P(B)} \tag{2}$$

where A, B are events which indicates class label and values of different attributes, respectively. $P(A|B)$ is posterior probability of A given B is true, and $P(B|A)$ is prior probability of B given A is true; $P(A)$ and $P(B)$ are the independent probabilities of A and B [14]. Posterior probability for every class label is calculated by using the above formula, and the label with the highest probability is chosen for prediction. Naive Bayes model is easy to build and calculate posterior probability mainly for larger datasets. It is well suited for multiclass classification problems.

(b) ***Gradient Boosting***

This technique is used for regression and classification of datasets. This machine learning technique uses a prediction model in the form of an ensemble of decision tree, which is one of the weak prediction models. Decision tree is like a flow chart like prediction of model using the data and predefined set of rules to find solution for the upcoming new data [15]. Here, attributes are represented using nodes, and the output of each branch is considered as the test outcome of the data that is considered in each node. Outcome of each node is considered as representation of class label that we get after computing all the attributes which is given to each node [16].

Multiclass classification problem is handled by using one-hot encoding in order to convert the problem into binary classifiers or directly we can adopt multiclass classification on original dataset. In both cases, loss function is implemented to assess the competency of model to make predictions. Loss function can be customized based on the problem and nature of prediction. For binary classification problem, log loss function uses the formula:

$$\text{log loss} = -\frac{1}{N}\sum_{i=1}^{N}\left[y_i \log p_i + (1 - (y_i))\log(1 - (p_i))\right] \tag{3}$$

where N represents number of observations, and y_i is actual output for ith observation, and p_i is the probability getting that desired output. Loss function for multiclass classifier called cross-entropy loss is given as:

$$\text{cross entropy loss} = -\frac{1}{N}\sum_{i=1}^{N}\sum_{j=1}^{M} y_{ij} \log p_{ij} \tag{4}$$

where M is the number of labels in the output classification and all other variables representation as similar to the Formula (3). Gradient boost algorithm adapts an iterative sequential process, which add weak learners gradually in every iteration. Moreover, it is a process where loss value gets reduced in each iteration which turns the model's prediction more closer to actual outcomes. Thus, gradient boosting is able to produce more accurate classification than any other models. Prediction accuracy is calculated using confusion matrix which take the numbers of true positive, false positive, true negative, and false negative predictions made by the models. Following algorithm explains the mathematical concepts adopted in this algorithm, which gradually removes the errors in prediction of weak learners,

Gradient Boosting Algorithm

Input: Training dataset $(\{x_i, y_i\})i = 1$ to n, Loss function $L(y, F(x))$, number of iterations M

1. Initialize model with a constant value:

$$F_0(x) = \arg\min \sum_{i-1}^{n} L(y_i, \Upsilon)$$

2. For $m = 1$ to M:

   (i) Calculate pseudo-residuals as given below:

$$r_{im} = -\left[\frac{\partial L(y_i, F(x_i))}{\partial F(x_i)}\right]_{F(x)=F_{m-1}(x)} \quad \text{for } i = 1, \ldots, n.$$

   (ii) Fit a weak learner—decision tree with scaling $h_m(x)$ to r_{im}

   (iii) Calculate multipliers as given below:

$$\Upsilon_m = \underset{\Upsilon}{\arg\min} \sum_{i=1}^{n} L(y, F_{m-1}(x) + \Upsilon h_m(x_i))$$

3. Update the model:

$$F_m(x) = F_{m-1}(x) + \Upsilon_m h_m(x)$$

Output: $F_M(x)$

Gradient boosting is an ensemble model, in which sequential learning process is adopted. Unlike other supervised classification techniques, ensemble method corrects the errors of previous models used for classification of the given dataset. Classification of successive models depends on classification errors of the predecessors. Thus, the sequential learning coupled with error correction adopted in gradient boosting produces more accurate classification of priority labels for vaccine distribution.

4 Implementation Results and Analysis

Implementation of the system starts with dataset collection, which includes the details of individuals like Aadhar id, name, age, profession, living in a hotspot or not, heart problems, other health problems including allergy, whether vaccinated or not, pregnancy/breastfeeding. Information related to all individuals in a particular locality is collected and stored in comma separated file format which is passed on to data processing module. Dataset includes records of 1500 individuals collected from the residents of southern portion of Kerala through google forms circulated among our friends, colleagues, and relatives. It includes details of various categories of individuals like health care front line workers, security personals, people working in agricultural sector, government employees, bank employees, teaching professionals, students, adults, children, teen agers, business people, and voluntary service providers. Sample dataset collected through online survey is shown in Table 1.

System is implemented in Python by invoking relevant application programming interface, after importing *numpy*, *scikit-learn* packages. This package is mainly used for machine learning-based classification and prediction.

Normalization is used for converting categorical data into continuous or discrete numeric values based on the attributes. Label encoding is used for fields—hotspot, heart disease, pregnancy, and vaccinated or not, which assigns 0 or 1 as they have two different values.

Attribute *age* uses label encoding to assign numerical values between 0 and 1 with 9 values depending upon the range. *Profession* is assigned values between 0 and 1, based on the nature of job. Sample data after normalization is shown in Table 2.

Table 1 Sample of original dataset

ID	Name	Age	Prof	Hotspot	Heart	Oth	Vac	Preg
1A001	Adish Nair	24	Health worker	No	Yes	No	No	No
1A010	Vineesh V.	32	Govt officials	No	No	Yes	No	No
1A430	Sneha K.	26	Private sector	No	No	No	No	No
1A002	Athulya Vijay	42	Private sector	Yes	No	No	No	No
1A509	Akshaya Anil	55	Govt officials	Yes	Yes	No	No	No
1A232	Krishnan K.	46	Govt officials	Yes	No	Yes	No	No
1A453	Salim Khan	33	Others	No	No	No	No	No
1A019	Sreelakshmi S.	36	Health worker	Yes	No	No	No	No
1A010	Sultan Safeer	24	Health worker	Yes	No	No	Yes	No
1A678	Jimmy Philip	23	Others	No	No	No	No	No

Table 2 Data after preprocessing

ID	Age	Prof	Hotspot	Heart	Oth	Vac	Preg
1A001	0.3	1	0	1	0	1	1
1A010	0.4	0.8	0	0	1	1	1
1A430	0.3	0.6	0	0	0	0	0
1A002	0.5	0.6	1	0	0	0	0
1A509	0.6	0.8	1	0	0	0	0
1A232	0.5	0.8	1	1	1	0	0
1A453	0.4	0.2	0	1	0	0	0
1A019	0.4	1	1	1	0	0	0
1A010	0.3	1	1	0	0	0	0
1A678	0.3	0.2	0	1	0	0	0

On analysis of dataset using exploratory data analysis using pair plot and heat map, to access the relationship between different attributes, it is identified that age is positively correlated with heart disease, other medical issues, profession have correlation with vaccinated or not. It is clear that age and profession have major impact in classfication and prediction of priority for distribution of vaccines.

Dataset is not labeled so unsupervised k-means algorithm is used to categorize the data with priority labels with k value as 6 whose result is shown in Table 3. Thus, K-means algorithm assigns priority labels to all individuals in the dataset [11].

Dataset is divided into 80% training and 20% test data, which is used for model creation and for further prediction. Data split is performed with API's in *train_test_split* library realized with *scikit-learn.* Training dataset is given as input to Naive Bayes and gradient booster classifier which creates model based on the data with class labels. Table 4 shows sample of the training dataset.

Table 3 Sample data with class labels based on *K*-means algorithm

ID	Age	Prof	Hotspot	Heart	Oth	Vac	Preg	Class
1A001	0.3	1	0	1	0	1	1	1
1A010	0.4	0.8	0	0	1	1	1	3
1A430	0.3	0.6	0	0	0	0	0	5
1A002	0.5	0.6	1	0	0	0	0	4
1A509	0.6	0.8	1	0	0	0	0	2
1A232	0.5	0.8	1	1	1	0	0	3
1A453	0.4	0.2	0	1	0	0	0	5
1A019	0.4	1	1	1	0	0	0	1
1A010	0.3	1	1	0	0	0	0	1
1A678	0.3	0.2	0	1	0	0	0	5

Table 4 Sample training dataset

ID	Age	Prof	Hotspot	Heart	Oth	Vac	Preg
IA392	0.3	1	0	1	0	0	0
IA671	0.3	0.6	0	0	0	0	0
IA791	0.6	0.8	1	1	0	0	0
IA453	0.4	0.2	0	0	0	0	0
IA010	0.3	1	1	0	0	1	0
IA551	0.5	0.6	0	1	1	0	0
IA890	0.4	0.4	0	0	1	0	0

Based on learning process with given training data, supervised algorithms are able to do prediction in best possible way. Naïve Bayes priority prediction is implemented by importing packages with *sklearn.naive_bayes import GaussianNB* [17]. Object instance is created, and predict() function is invoked with the relevant training and test dataset. For gradient boosting prediction, both decision tree classification and gradient boost classification are implemented. First relevant packages are imported with *sklearn.tree import DecisionTreeClassifier* and *sklearn.ensemble import RandomForestClassifier, AdaBoostClassifier, GradientBoostingClassifier.*

Decision tree classification is performed by creating instance of DecisionTreeClassifier() and calling predict() function. Gradient boosting predict() function is called after creating instance of GradientBoostingClassifier(). It is implemented over decision tree results, in order to rectify errors in decision tree prediction Log function is used to access errors, and it is corrected in every iteration. Finally, models are able to do prediction without errors. So, while creating instance of GradientBoostingClassifier(), it is required to specify number of estimations to be performed. Table 5 shows result of two models Naive Bayes and gradient boosting. Since in gradient

Table 5 Prediction resuls of three models

ID	Age	Prof	Hotspot	Heart	Oth	Vac	Preg	Naïve	Gradient	Decision
IA392	0.3	1	0	1	0	0	0	1	1	1
IA671	0.3	0.6	0	0	0	0	0	5	5	5
IA791	0.6	0.8	1	1	0	0	0	2	2	2
IA453	0.4	0.2	0	0	0	0	0	5	5	5
IA010	0.3	1	1	0	0	1	0	6	6	6
IA551	0.5	0.6	0	1	1	0	0	2	2	2
IA890	0.4	0.4	0	0	1	0	0	1	3	3
IA354	0.4	1	1	0	1	0	0	1	1	1
IA149	0.3	1	1	0	0	0	0	1	1	1
IA077	0.3	1	1	1	1	1	0	6	6	6
IA901	0.4	0.8	1	0	0	0	0	3	3	3
IA744	0.3	0.6	1	1	1	0	0	2	2	2
IA200	0.4	1	0	1	0	1	0	6	6	6
IA303	0.4	1	0	1	1	0	0	1	1	1
IA719	0.5	0.8	1	0	1	0	0	3	3	3
IA123	0.4	0.2	0	0	0	0	0	5	5	5
IA987	0.7	0.4	0	1	0	0	0	2	2	2
IA567	0.4	0.8	1	0	0	0	0	3	3	3
IA534	0.3	1	1	0	0	0	0	1	1	1

boosting classifiers, decision tree classification is performed, output of decision trees classifier is also displayed.

Graphical representation of *k*-means clustering is shown in Fig. 2 which represents individual ID and cluster labels assigned to a sample set of data. Figure 3 shows pictorial representation of class labels assigned to a sample set of individauls based

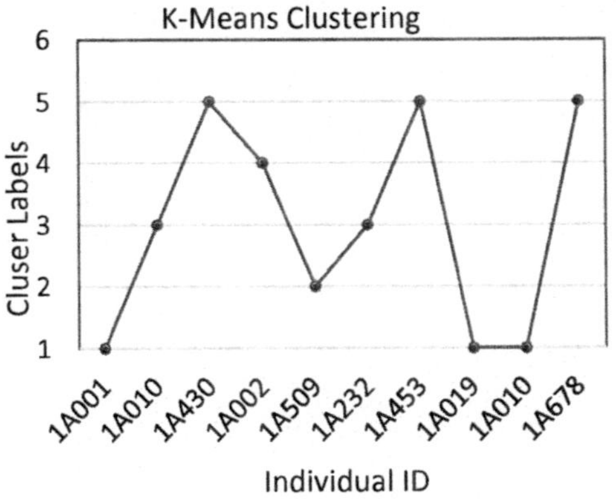

Fig. 2 *K*-means clustering-unlabeled data

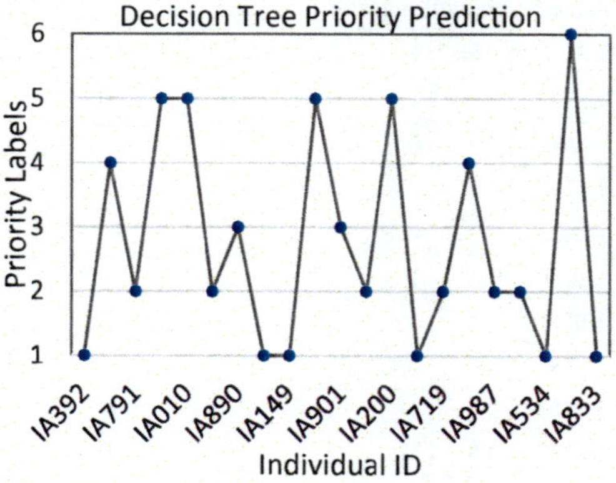

Fig. 3 Decision trees priority prediction

on the decision tree methodology. From the graphical representation, it is clear that decision tree classification is better than *k*-means algorithm.

Graphical representation of priority prediction by the two models Naive Bayes and gradient boosting is shown in Figs. 4 and 5, repectively. Figure indicates that are 4 error predictions in Naive Bayes when compared to original prediction. Comparison between the supervised classification model's priority predictions is shown in Fig. 6.

Prediction results show that Naive Bayes is subject to few wrong predictions when compared with gradient boosting algorithm. Since, decision tree prediction errors are assessed using loss function and they are corrected by performing multiple iterations in gradient boosting [18]. Prediction accuracy of the models is calculated with packages *sklearn.metrics import confusion_matrix*. accuracy_score() function is used on each prediction results in order to get the score. Table 6 shows the accuracy prediction scores of the three models [19]. From the above observation, it is clear that gradient boosting is able to predict priority labels with 100% accuracy. Priority

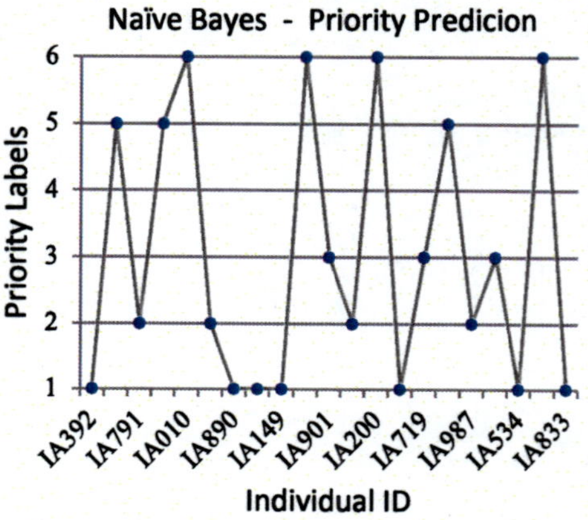

Fig. 4 Naïve Bayes priority prediction

Fig. 5 Gradient boosting priority prediction

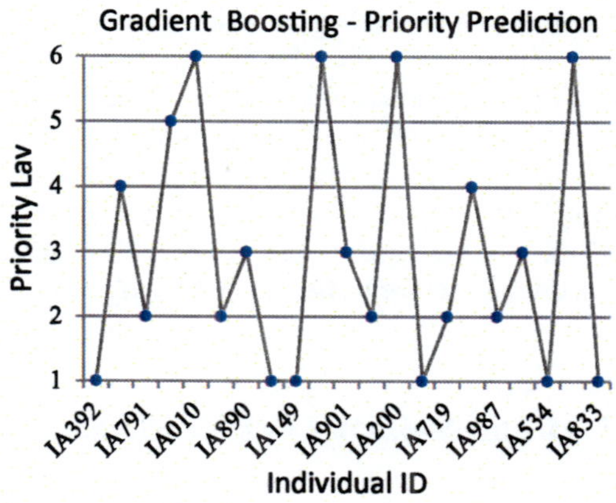

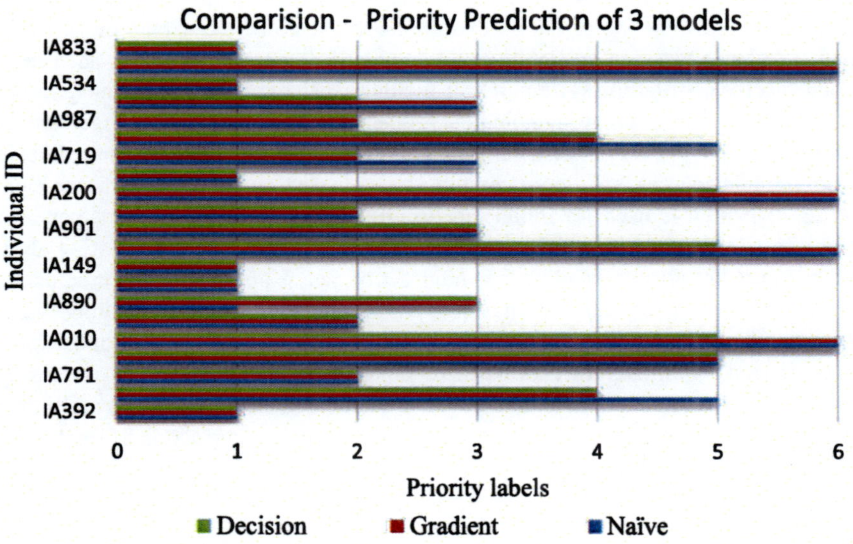

Fig. 6 Visualization of priority label prediction of models for sample data

Table 6 Prediction accuracy of three models

Model used	Naive Bayes	Decision trees	Gradient boosting
Accuracy rates	0.77333333	0.966666667	1.0

labels assigned to individuals will help for the efficient distribution of vaccines, which renders maximum benefit to the society.

For the initial phase of vaccine distribution with unlabeled data, *k*-means clustering algorithm is used as it is execution time is less, suitable for scalable data, creates clusters with closer data points and it creates clusters with any shape depending upon the nature of data. Then, for the dataset with priority labels, naïve Bayes algorithm is used as it can handle independent data features in the best possible manner [20]. It is very fast in performing classification when compared with other complex algorithms. In order to get rid of errors in classification of data, gradient boost algorithm is used which implements decision tree as weak learners in the process of classification. Thus, the final priority prediction with gradient boost ensemble techniques gives maximum accuracy when compared to other classification algorithms.

5 Conclusion and Future Enhancements

System implemented in this research plays a vital role in handling the pandemic situation of corona virus, which created a huge problem for the whole community. Vaccination drive is one of the best solutions to overcome the spread of virus. To have a balance between availability and demand for vaccine, there is a need for a good vaccine distribution system. It will avoid the rush of people toward hospitals to get vaccinated. This research implements supervised and unsupervised algorithm to provide an efficient way of distributing COVID vaccine. Initially, data are collected from reliable sources, and preprocessing of data is performed. Then, *k*-means clustering algorithm is used for initial labeling of data. Then, appropriate supervised classification models are implemented with Naive Bayes, gradient boosting algorithm to predict priority labels, when huge set of data is appended. The predictions are made based on the data, which is used for training the model. In addition, evaluation result shows that the gradient boosting algorithm provides maximum accuracy in prediction continuous correction of errors in gradient boosting leads to over fitting problems, so in future prediction could be implemented using multiclass multilayer perceptron models which uses adaptive learning with error correction process.

References

1. Deo S, Manurkar S, Krishnan S, Franz C (2020) COVID-19 vaccine: development, access and distribution in the Indian context. ORF Issue Brief No. 378, July 2020, Observer Research Foundation
2. Foy BH, Wahl B, Mehta K, Shet A, Menon GI, Britto C (2021) Comparing COVID-19 vaccine allocation strategies in India: a mathematical modelling study. Int J Infect Dis 103
3. Abebe H, Shitu S, Mose A (2021) Understanding of Covid-19 vaccine knowledge, attitude, acceptance, and determinates of COVID-19 vaccine acceptance among adult population in Ethiopia, vol 14, June 2021
4. Duijzer LE et al (2018) Dose-optimal vaccine allocation over multiple populations. Prod Oper Manage 27(1):143–159. https://doi.org/10.1111/poms.12788
5. Vani K, Gupta D (2014) Using K-means cluster based techniques in external plagiarism detection. In: 2014 International conference on contemporary computing and informatics (IC3I), pp 1268–1273

6. Siva Raja PM, Ramanan K (2020) Lesion localization and extreme gradient boosting characterization with brain tumor MRI images. In: Part of the Lecture notes on data engineering and communications technologies book series (LNDECT), vol 37
7. Pandey S, Muthuraman S, Shrivastava A (2018) Data classification using machine learning approach. In: Thampi S, Mitra S, Mukhopadhyay J, Li KC, James A, Berretti S (eds) Intelligent systems technologies and applications. ISTA 2017. Advances in intelligent systems and computing, vol 683. Springer, Cham
8. Grauer J et al (2020) Strategic spatiotemporal vaccine distribution increases the survival rate in an infectious disease like Covid-19. Sci Rep 10(1):21594. https://doi.org/10.1038/s41598-020-78447-3
9. Nawi NM, Atomi WH, Rehman MZ (2013) The effect of data pre-processing on optimized training of artificial neural networks. In: 4th International conference on electrical engineering and informatics, ICEEI 2013, vol 11
10. Li Y, Wu H (2012) A clustering method based on K-means algorithm. In: International conference on solid state devices and materials science, vol 25
11. Subbulakshmi S et al (2016) "Web services QoS prediction basedon dynamic non-functional quality factors and WS-security policy specification of web service". Second International Conference on Computer paradigms(ICCP 2016), International Journal of Control Theory and Applications, vol. 9. International Science Press., pp 4591–4601
12. Na S, Xumin L, Yong G (2010) Research on k-means clustering algorithm: an improved k-means clustering algorithm. In: 2010 Third international symposium on intelligent information technology and security informatics, pp 63–67. https://doi.org/10.1109/IITSI.2010.74
13. Mishan, MT et al (2017) An analysis on business intelligence predicting business profitability model using Naïve Bayes neural network algorithm. In: 2017th IEEE International conference on system engineering and technology (ICSET). IEEE
14. Harahap F, Harahap AYN, Ekadiansyah E, Sari RN (2018) Implementation of Naïve Bayes classification method for predicting purchase. In: 6th International conference on cyber and IT service management (CITSM)
15. Pandiangan N, Buono MLC, Loppies SHD (2019) Implementation of decision tree and Naïve Bayes classification method for predicting study period. J Phys Conf Ser 1569 (International conference on science and technology)
16. Bansal A, Kaur S (2018) Extreme gradient boosting based tuning for classification in intrusion detection systems. In: International conference on advances in computing and data sciences
17. Sathyadevan S, Sarath PR, Athira U, Anjana V (2014) Improved document classification through enhanced Naive Bayes algorithm. In 2014 International Conference on Data Science & Engineering (ICDSE), pp 100–104
18. Sungheetha A (2021) COVID-19 Risk minimization decision making strategy using data-driven model. J Inf Technol 3(01):57–66
19. Chen JI-Z (2021) Design of accurate classification of COVID-19 disease in X-ray images using deep learning approach. J ISMAC 3(02):132–148
20. Qin F, Tang X, Cheng Z (2012) Application and research of multi-label Naïve Bayes classifier. In: Proceedings of the 10th World congress on intelligent control and automation, pp 764–768. https://doi.org/10.1109/WCICA.2012.6357980

An Experimental Evaluation of Load Balancing Policies Using Cloud Analyst

N. Priya and S. Shanmuga Priya

Abstract With the rapid growth of cloud computing, the number of customers accessing the cloud systems and the workload of the servers has been increased tremendously. Resource management of infrastructure as a service (IaaS) is an important aspect in cloud computing that involves scheduling the resources among customers, but improper scheduling of cloud resources degrades the performance of the overall cloud systems. Load balancing is a technique in cloud computing that provides a solution to this problem through efficient resource management which improves the resource utilization, scalability, and overall performance of the cloud systems. In this paper, we made a comparison on various static and dynamic load balancing methods and implement them using a Cloud Analyst simulator to study and assess its performance, and the performance analysis is also presented, and the primary metrics used to measure these algorithms are response time, data center processing time, and cost.

Keywords Cloud computing · Load balancing · Virtualization · Static load balancing · Dynamic load balancing · Cloud analyst · Response time · Data center processing time

1 Introduction

Cloud computing is an emerging technology in the area of computer science and IT industry, and it offers enormous services like IaaS, SaaS, and PaaS to the small scale and startup large-scale industries. IaaS delivers customizable infrastructure on demand and allows the customers to outsource their IT infrastructures; i.e., it offers an umbrella range from single servers to entire infrastructures such as virtual machines, networking, processing power, storage, and servers through the Internet on a pay-as-you-go policy with low cost according to their needs [1, 2]. Since the cloud provides more benefits over an on-premise system, the number of customers using

N. Priya (✉) · S. S. Priya
Research Department of Computer Science, Shrimathi Devkunvar Nanalal Bhatt Vaishnav College for Women, Chennai, India

S. Shakya et al. (eds.), *Proceedings of Second International Conference on Sustainable Expert Systems*, Lecture Notes in Networks and Systems 351,
https://doi.org/10.1007/978-981-16-7657-4_16

the cloud resources increased immensely, and it is difficult to manage and process multiple requests simultaneously when the user requests increase day by day, which also down the performance of the cloud systems. Hence, resource management plays a vital role in efficiently allocating workload among various cloud resources which reduces the response time and increasing productivity with the same level of available resources [3].

2 Virtualization

A physical host having higher capacity processing power which are scattered in various locations can be able to run only one operating system at a time with only fewer connected users which leads to low resource utilization since some of the resources are idle. Further, they also process the job below their capabilities [4].

Virtualization is a technique that gives solutions to the above problem by combining multiple physical machines in a single data center and creates several virtual machines on top of each physical machine with different resource capabilities to make proper utilization of physical hardware [5]. A virtual machine is a machine that imitates the real machine, and each virtual machine contains a separate operating system called a guest operating system. Virtualization makes several operating systems to run simultaneously on a single physical machine which provides better utilization of physical resources [6].

The capacity of virtual machines is fully utilized with the help of the load balancing technique by evenly distributing the workloads among all available virtual machines. Virtualization is achieved with the help of a firmware called hypervisor (virtual machine monitor), and hypervisors like Microsoft Hyper-V, Citrix Xen Server, VMware Server, and Oracle VM act as an interface between the guest operating system and VMs and isolate the VM from the original physical hardware. Hypervisor supports multiple operating systems to run applications on a single physical hardware platform simultaneously. Figure 1 shows an overview of virtualization architecture [7, 8].

3 Load Balancing

Load balancing is an optimization technique that does efficient resource management by allocating users tasks and balancing the load among the resources in such a way that it gives higher performance, maximum resource utilization ratio, enhances the throughput of a system, reduces response time, avoids overloading of servers, and also reduces the amount of energy consumption. Resources offered by IaaS include CPU, memory capacity, virtual machines, servers. Figure 2 shows how a load balancer works in a cloud environment [9].

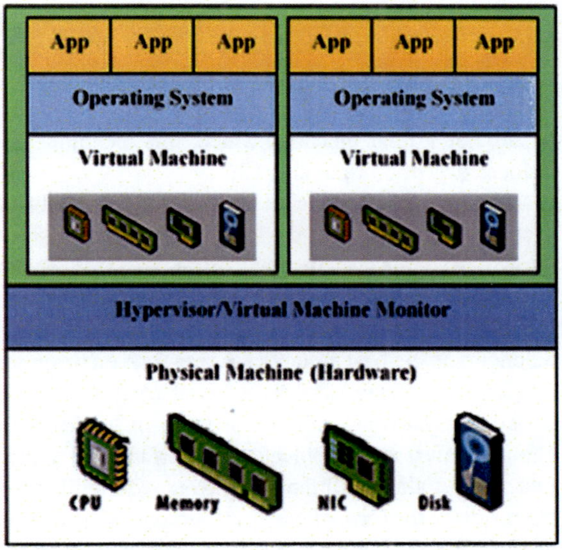

Fig. 1 Virtualization architecture

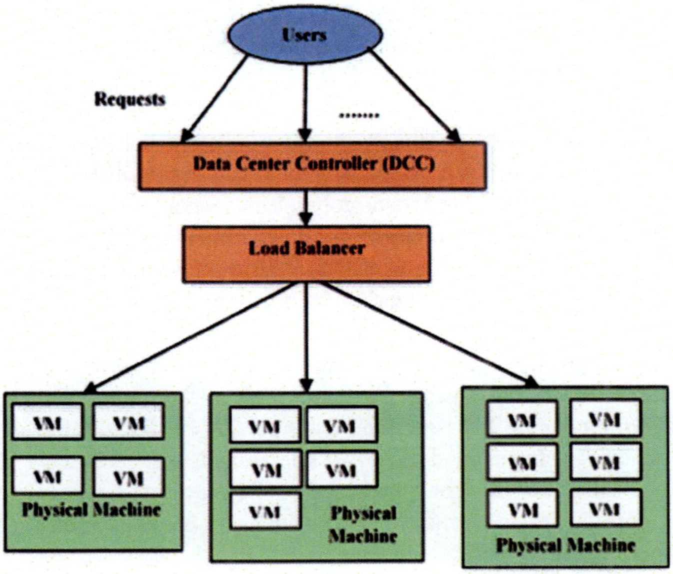

Fig. 2 Load balancing in cloud computing

There are two approaches for load balancing based on the allocation of tasks such as static load balancing and dynamic load balancing [10].

Static load balancing needs the performance of the virtual machine in advance before allocation. It does not consider the current state of the virtual machine while allocating tasks among them. Static algorithms are made for systems which have very low deviations in load since it divides loads equally to all the virtual machines regardless of its current capacity. Static load balancing techniques are non-preemptive in nature, i.e., load once allocated to the particular virtual machine cannot be relocated to another virtual machine. Since, this technique requires less context switching reduces execution time [11, 12].

Dynamic load balancing checks the present behavior of the cloud computing system in order to make the load assignment decision and does not require previous knowledge of the system. It can move tasks from one over-utilized virtual machine to an under-utilized machine dynamically. Although dynamic load balancing is complex in nature but can provide better fault tolerance and high performance, dynamic load balancing is the best suitable for real-time environments [13].

Cloud Analyst is a tool constructed to support and simulate the cloud environment. Cloud Analyst simulation tool is used for implementing load balancing algorithms. Cloud Analyst is a graphical user interface tool, and it has visualization capabilities that support almost all the parameters required for the evaluation process and show the results in the form of tables and graphs, which can be used to easily and accurately analyze the performance of the load balancing algorithms [14].

4 Literature Review

In this paper, the author examines the types of load balancers and also summarized the advantages and drawbacks of the existing load balancing algorithms such as round robin, least connection, weighted round robin, least response time, least bandwidth, and source hash. The paper also performs an experimental analysis of round robin and least connection algorithms in the Cloud Analyst tool according to the response time. After analysis, round robin algorithm performs better than the least connection algorithm by reducing the response time [15].

In this paper, the author proposed an improved genetic algorithm and compared it with the existing load balancing algorithms such as the traditional genetic algorithm, round robin, and FIFO algorithm. Optimization accuracy of the algorithm is improved by using selection and crossover operators. It compares these algorithms using a cloud simulation environment based on the execution time of the task. The proposed algorithm provides efficient load balancing of nodes and performs efficient scheduling [16].

In this paper, the author proposed an improved load balancing algorithm, and performance evaluation has been carried out for improved algorithm, round robin algorithm, and throttled algorithm in terms of response time, data center processing time, and overall throughput. Simulations were performed on the Cloud Analyst toolkit. The improved algorithm outperforms the existing approaches like RR and throttled algorithms by achieving a minimal response time, minimal data center processing time, and high throughput which leads better resource utilization on the cloud systems [17].

In this paper, the author developed a compound approach which is based on equally spread current execution (ESCE) and throttled algorithms for load balancing. This paper highlights the characteristics of cloud computing, load balancing, and various services that cloud computing offers. The simulation and result evaluation have been done using the CloudSim simulation toolkit. The average response time and makespan have been considered as evaluation metrics. The compound approach algorithms give better response time and processing time [18].

In this paper, the author proposed a new algorithm such as join minimum loaded queue based on the existing join idle queue algorithm, and the author reviewed the existing load balancing algorithms and concluded that the existing algorithms either maximize resource utilization or minimize response time but not both simultaneously. In this proposed algorithm, the author achieves maximum resource utilization with minimum response time [19].

In this paper, the author reviews various existing static and dynamic load balancing policies such as throttled, equally spread current execution, round robin, honey bee, min-min, and genetic algorithms concerning response time, fault tolerance, quality of service, and overall performance of a cloud system. The author also analyzed various simulation tools for executing load balancing algorithms in a cloud environment and suggested that the Cloud Analyst tool performs well than other tools. Finally, it has been concluded that more stable load balancing algorithms are required because most of the existing algorithms are not suitable for fault tolerance and virtual machine migration [20].

In this paper, the author develops an improved multitask scheduling algorithm also called an improved genetic algorithm which is based on a genetic algorithm and is compared with the existing min-min algorithm with respect to resource utilization and response time. Elitist selection strategy is used for this new algorithm in order to fix the fitness function model. CloudSim simulation tool has been used for simulation and analysis of this new algorithm. The author determines that the proposed algorithm outperforms than the genetic and min-min algorithms by optimizing the resource utilization and reduces the response time of the user's tasks [21].

In this paper, the author discussed three load balancing algorithms such as weighted round robin, throttled, and active monitoring load balancing algorithms based on time and cost. In weighted round robin method, weights are assigned to each virtual machine; i.e., higher weights are assigned to high-capacity virtual machines; higher loads are allocated to the higher weighted vm. In this way, it solves the problem of server overloading, but it does not consider the execution time of the tasks. The throttled algorithm maintains the table consisting of the status of each vm and assigns the request to the appropriate vm based on the size of the task. Active monitoring algorithm assigns the task to the vm with the least load. The author concluded that throttled algorithm outperforms than other two algorithms with minimum response time [22].

In this paper, the study represents issues and threats faced by the existing load balancing algorithms and helps the researchers to find more efficient algorithms to enhance performance, job scheduling, resource utilization, quality of service, and load distribution. The author differentiated load balancing algorithms into various categories like a task, hybrid, agent, and general-based load balancing algorithms [23].

In this paper, the author discussed the challenges related to the task scheduling and workload balancing in IaaS cloud environment. The author developed an improvised load balancing algorithm based on quality of service (QOS), energy efficiency, and resource utilization parameters. Here, the author outlined some of the existing load balancing algorithms such as SJF, FCFS and improvements of various min-min load balancing algorithms along with theirs advantages, drawbacks, and simulation tools used. Here, the author proved that the new improvised load balancing algorithm can work in a dynamic cloud environment and improves resource utilization and makespan than the existing algorithms [24].

5 Proposed Work

Numerous load balancing algorithms have been used for balancing the load of various resources. The proposed work illustrates the comparison and evaluation of three types of load balancing methods as round robin, throttled load balancing, and equally spread current execution algorithms.

5.1 Round Robin (RR) Algorithm

The RR algorithm is one of the simplest static load balancing algorithms. It assigns the user's requests to the pool of virtual machines in a circular manner. It randomly chooses one virtual machine from the list of virtual machines (VMs) and assigns the first request to it and continues assigning the subsequent requests to the remaining virtual machines in a circular fashion. This algorithm equally distributes the workload among available virtual machines without considering the capacity of the virtual machine. RR algorithm works well for homogeneous cloud; i.e., it works efficiently where all the virtual machines are having equal capacity. Figure 3 shows the working model of the RR algorithm in a cloud environment.

The procedure of RR algorithm:

(i) Start
(ii) Initially, keep the status of all the available virtual machines in the virtual machine state list.
(iii) The DataCenterController (DCC) receives the new request from each user base and queued it.
(iv) The DCC removes the request from the beginning of the queue and sends the request to the RR load balancer which arbitrarily finds the virtual machine from the vm state list based on the service broker policy given and allocates the request to that virtual machine.

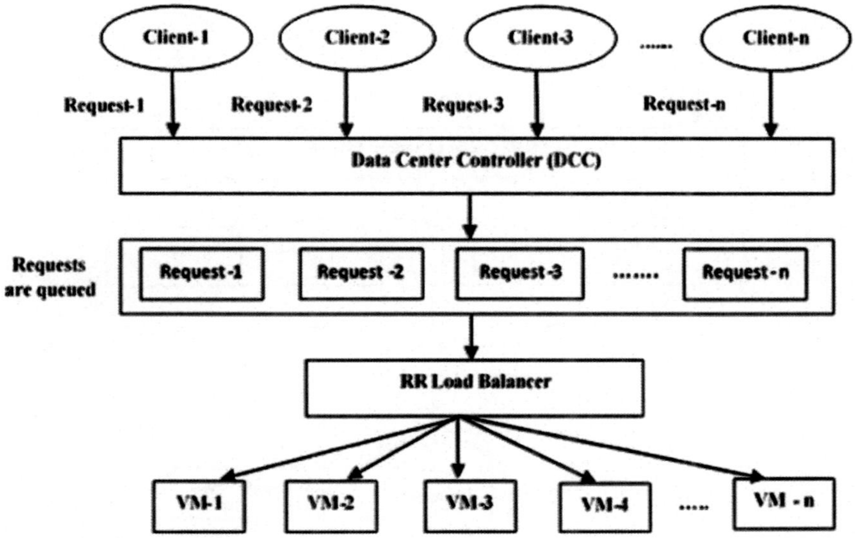

Fig. 3 Load balancing using round robin method

(v) Update the virtual machine state list based on the number of requests assigned.
(vi) The DCC allocates the subsequent requests to all the available virtual machines on a rotational basis and updates the state list.
(vii) If more requests are arrived on the user base, then go to Step iii and proceed with the remaining steps until all requests are assigned.
(viii) Stop.

5.2 Equally Spread Current Execution (ESCE) Algorithm

ESCE is a dynamic load balancing algorithm which keeps the information about each vm such as the availability of the virtual machine and the number of requests allocated to each virtual machine at present. Each time, it chooses the virtual machine with least number of current allocations for each request arrived. ESCE always finds the vm with the least load, but it does not consider the previous utilization of a particular vm which leads same VMs to be busy all the time. Figure 4 shows the working model of the ESCE algorithm in a cloud environment.

The procedure of ESCE algorithm:

(i) Start
(ii) ESCE keeps the status information of all the available virtual machines in the virtual machine (VM) state list, i.e., available (1) or not available (0), and also, it keeps the number of current allocations of each virtual machine at

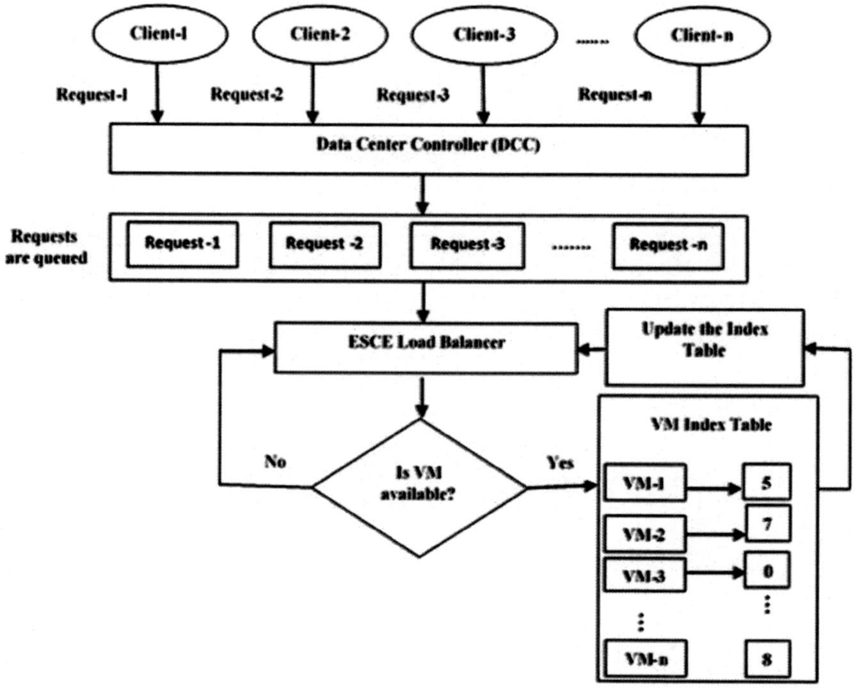

Fig. 4 Load balancing using ESCE method

present in the vm state list. Initially, all VMs are set to be zero for number of allocations.

(iii) The DataCenterController receives the new request from the user bases.

(iv) The DCC sends the received request to the ESCE load balancer which finds the first available virtual machine with least number of current allocations and assigns the request to that virtual machine from the VM state list.

(v) Now, update the virtual machine state list based on the number of current allocations done on the virtual machine.

(vi) Once the task has been completed by the virtual machine, then again, it updates the vm state list and the number of allocations.

(ix) If more requests are arrived in the user base, then go to Step iii and proceed with the remaining steps until all the requests are assigned.

(vii) Stop.

5.3 *Throttled Load Balancing (TLB) Algorithm*

Throttled load balancing (TLB) is a dynamic load balancing technique, which keeps all the available virtual machines in the virtual machine index table. It chooses the VM

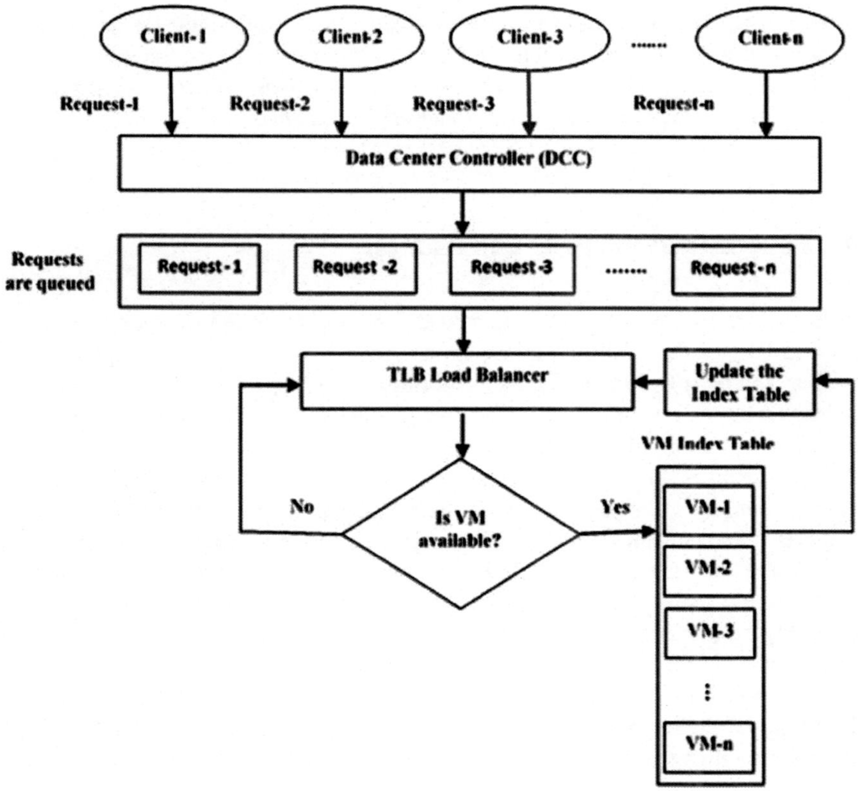

Fig. 5 Load balancing using TLB method

based on the size in such a way that it does the job easily and in a fast manner; thus, TLB provides fast processing of user's request and gives better response time than RR and ESCE algorithms. Figure 5 shows the working model of the TLB algorithm.

The procedure of TLB algorithm:

(i) Start
(ii) TLB maintains VM index table which keeps the status information of all virtual machines in the virtual machine state list, i.e., available (1) or not available (0).
(iii) The DataCenterController (DCC) receives the new request from the user base.
(iv) The DCC sends the received request to the TLB load balancer which scans the index table from the top and chooses the first available virtual machine appropriate for the load and assigns the request to that virtual machine and updates it in the index table.
(v) If the appropriate vm is not found for handling the request, then the load balancer has to wait until it finds the suitable VM.

(vi) If the virtual machine finishes the tasks, then again, it updates the status of the VM in the index table, and the response has been sent to the DCC.
(vii) If more requests have arrived in the user bases, then go to Step iii and proceed with the below steps until all requests are assigned.
(viii) Stop.

6 Experimental Results and Analysis

In the experiment, we have implemented and tested three load balancing algorithms such as RR, TLB, and ESCS in the Cloud Analyst tool, and also, we present the comparison table and graph for each algorithm. The primary performance measure of interest here is response time, datacenter processing time, and overall data transfer cost. Figure 6 shows the main screen of the Cloud Analyst tool.

Simulation Scenario: In each scenario, we have changed the number of data centers, number of virtual machines, number of user bases, and number of users' requests. The results acquired after simulation of various scenarios in view of RR, ESCE, and TLB algorithms are shown in Tables 1, 2, 3, 4, and 5. Figure 7 shows the average response time of each algorithm in different scenarios mentioned in the table. Figure 8 shows the data center processing time of each algorithm in different scenarios mentioned in the table.

Comparing with all the cases, the overall response time has been reduced in TLB than ESCE and RR algorithms. There is a little variation in the ESCE and RR algorithms with respect to the response time. ESCE algorithm gives better data center

Fig. 6 Cloud analyst—main screen

Table 1 Case 1: In the first case, we have created 1 data center with 25 virtual machines and 3 user bases with 50 requests in each user base

Algorithm	Response time (ms)			Data center processing time (ms)			Virtual machine cost ($)
	Avg.	Min.	Max.	Avg.	Min.	Max.	
RR algorithm	186.23	37.88	407.50	19.87	0.29	54.58	22.47
ESCE algorithm	185.95	37.89	407.50	19.57	0.29	53.63	22.47
TLB algorithm	176.47	37.89	407.50	10.13	0.29	45.82	22.47

Table 2 Case 2: In the second case, we have created 1 data center with 50 virtual machines and 5 user bases with 80 requests in each user base

Algorithm	Response time (ms)			Data center processing time (ms)			Virtual machine cost ($)
	Avg.	Min.	Max.	Avg.	Min.	Max.	
RR algorithm	218.21	55.18	654.46	14.10	0.14	50.46	358.75
ESCE algorithm	218.03	55.16	654.46	13.78	0.14	45.96	358.75
TLB algorithm	213.59	55.17	654.46	7.39	0.14	35.56	358.75

Table 3 Case 3: In the third case, we have created 2 data centers with 30 virtual machines and 5 user bases with 60 requests in each user base

Algorithm	Response time (ms)			Data center processing time (ms)			Virtual machine cost ($)
	Avg.	Min.	Max.	Avg.	Min.	Max.	
RR algorithm	189.31	37.62	812.13	20.54	0.04	54.17	28.62
ESCE algorithm	189.02	37.62	808.64	20.25	0.04	53.50	28.62
TLB algorithm	179.18	37.62	805.15	10.46	0.04	46.24	28.62

Table 4 Case 4: In the fourth case, we have created 2 data centers with 30 virtual machines and 10 user bases with 40 requests in each user base

Algorithm	Response time (ms)			Data center processing time (ms)			Virtual machine cost ($)
	Avg.	Min.	Max.	Avg.	Min.	Max.	
RR algorithm	150.33	35.33	836.51	0.49	0.00	1.01	44.47
ESCE algorithm	148.18	35.99	819.01	0.49	0.00	1.01	44.47
TLB algorithm	146.38	36.17	808.51	0.49	0.00	1.01	44.47

processing time than RR. While comparing all the three algorithms, Throttled load balancing outperforms than other algorithms by reducing the response time and also the data center processing time. The cost of virtual machines remains the same for all three algorithms since the same number of VMs is used for all three algorithms

Table 5 Case 5: In the fifth case, we have created 3 data centers with 35 virtual machines and we have created 6 user bases with 60 requests in each user base

Algorithm	Response time (ms)			Data center processing time (ms)			Virtual machine cost ($)
	Avg.	Min.	Max.	Avg.	Min.	Max.	
RR algorithm	108.41	37.65	542.83	42.05	0.02	89.03	28.89
ESCE algorithm	108.33	37.65	542.83	41.96	0.02	88.85	28.89
TLB algorithm	87.52	37.65	522.83	21.27	0.02	83.31	28.89

Fig. 7 Average response time of RR, ESCE, and TLB in all the cases

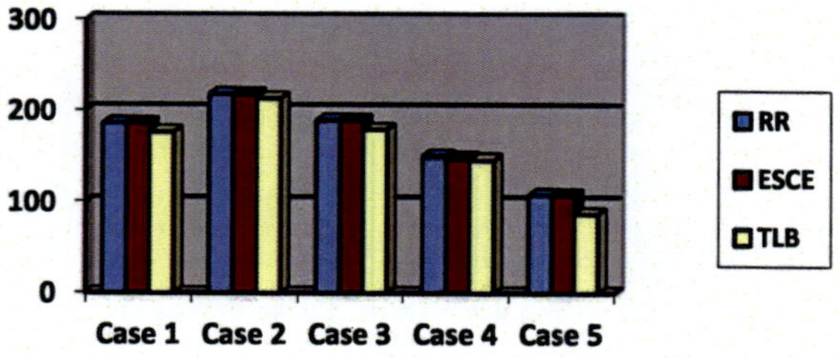

Fig. 8 Average data center processing time of RR, ESCE, and TLB in all cases

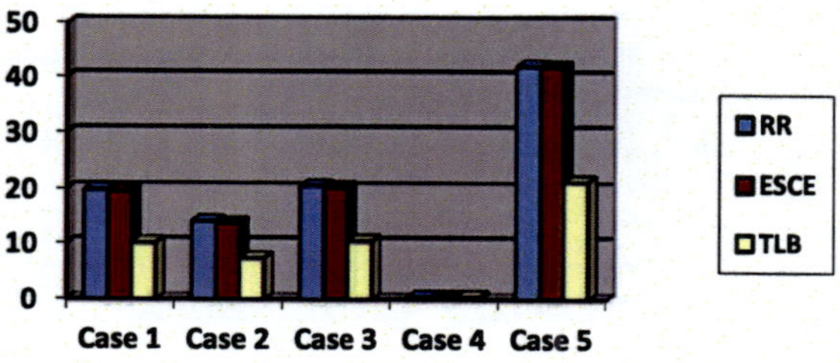

in each case. Moreover, TLB algorithm balances the load of the server in order to avoid overutilization or underutilization of the server.

The following chart represents the above cases as follows:

7 Conclusion

In cloud computing, load balancing is a major challenge that is required to distribute the workload in an efficient and scalable manner for efficient resource management. It is more important to find an algorithm that reduces substantially the response time and data center processing time. By comparing some of the significant load balancing algorithms, we tried to help make this deduction to choose the best load balancing algorithm in any given situation. The main aim is to balance the workload of the cloud infrastructure in order to efficiently handle the resources at hand and also to reduce the response time for the given number of tasks. The work can be further improved by analyzing the above mentioned algorithms and develops new algorithm which further reduces the response time and data center processing time and also develops an algorithm by applying different server broker policies and different performance metrics.

References

1. Kaur B, Singh R (2016) A comparison and analysis of load balancing algorithms in cloud computing. Adv Comput Sci Technol 9(1):51–59
2. Shakya S (2021) IoT based F-RAN architecture using cloud and edge detection system. J ISMAC 3(01):31–39
3. Madni SHH, Abd Latiff MS, Coulibaly Y (2017) Recent advancements in resource allocation techniques for cloud computing environment: a systematic review. Cluster Comput 20(3):2489–2533
4. Xu M, Tian W, Buyya R (2017) A survey on load balancing algorithms for virtual machines placement in cloud computing. Concurrency Comput Pract Exp 29(12):e4123
5. Patra MK, Patel D, Sahoo B, Turuk AK (2020) A randomized algorithm for load balancing in containerized cloud. In: 2020 10th International conference on cloud computing, data science & engineering (confluence), Jan 2020. IEEE, pp 410–414
6. Kong L, Mapetu JPB, Chen Z (2020) Heuristic load balancing based zero imbalance mechanism in cloud computing. J Grid Comput 18(1):123–148
7. Ala'Anzy M, Othman M (2019) Load balancing and server consolidation in cloud computing environments: a meta-study. IEEE Access 7:141868–141887
8. Mishra SK, Sahoo B, Parida PP (2020) Load balancing in cloud computing: a big picture. J King Saud Univ-Comput Inf Sci 32(2):149–158
9. Shah N, Farik M (2015) Static load balancing algorithms in cloud computing: challenges & solutions. Int J Sci Technol Res 4(10):365–367
10. Ali SA, Alam M (2016) A relative study of task scheduling algorithms in cloud computing environment. In: 2016 2nd International conference on contemporary computing and informatics (IC3I), Dec 2016. IEEE, pp 105–111
11. Elenin SA, Kitakami M (2011) Comparing static load balancing algorithms in grid. In: International conference on cooperative design, visualization and engineering, Sept 2011. Springer, Berlin, Heidelberg, pp 170–177
12. Shukla A, Kumar S, Singh H (2020) Fault tolerance based load balancing approach for web resources in cloud environment. Int Arab J Inf Technol 17(2):225–232
13. Jodayree M, Abaza M, Tan Q (2019) A predictive workload balancing algorithm in cloud services. Procedia Comput Sci 159:902–912
14. Edwin MST, A review: load balancing algorithm using cloud analyst environment. Miss. Rajeshwari Nema Mtech Schollar BTIRT, Sagar. E-Mail: reetunema04@gmail.com
15. Alankar B, Sharma G, Kaur H, Valverde R, Chang V (2020) Experimental setup for investigating the efficient load balancing algorithms on virtual cloud. Sensors 20(24):7342
16. Liu L, Qiu Z, Dong J (2017) A load balancing algorithm for virtual machines scheduling in cloud computing. In: 2017 9th International conference on modelling, identification and control (ICMIC), July 2017. IEEE, pp 471–475
17. Abdulkarim A, Muhammed I, Mohammed L, Babayaro A (2019) Performance analysis of an improved load balancing algorithm in cloud computing. Am J Netw Commun 8(2):47–58
18. Sachdeva R, Kakkar S (2017) A novel approach in cloud computing for load balancing using composite algorithms. Int J 7(2)

19. Sharma M, Kumar R, Jain A (2020) A proficient approach for load balancing in cloud computing-join minimum loaded queue: join minimum loaded queue. Int J Inf Syst Model Des (IJISMD) 11(1):12–36
20. Shafiq DA, Jhanjhi NZ, Abdullah A (2021) vLoad balancing techniques in cloud computing environment: a review. J King Saud Univ-Comput Inf Sci
21. Tang L (2018) Load balancing optimization in cloud computing based on task scheduling. In: 2018 International conference on virtual reality and intelligent systems (ICVRIS), Aug 2018. IEEE, pp 116–120
22. Sahu S, Pandey M (2019) Efficient load balancing algorithm analysis in cloud computing. In: 2019 International conference on communication and electronics systems (ICCES), July 2019. IEEE, pp 779–783
23. Balaji K (2021) Load balancing in cloud computing: issues and challenges. Turkish J Comput Math Educ (TURCOMAT) 12(2):3077–3084
24. Shafiq DA, Jhanjhi NZ, Abdullah A, Alzain MA (2021) A load balancing algorithm for the data centres to optimize cloud computing applications. IEEE Access 9:41731–41744

Classification of Ovarian Cyst Using Regularized Convolution Neural Network with Data Augmentation Techniques

N. Priya and S. Jeevitha

Abstract PCOS-polycystic ovary syndrome is one of the prevalent hormonal disorders which has currently affected women populations around the age group of 22–45, in their reproductive cycle. It has been widely observed that PCOS leads to infertility. Diagnosis of infertile has proceeded by using ultrasound images of follicles present in the ovary and further examined by the features like the size of the follicles, number of follicles, age group of patients, and the hormonal test. Based on the features, ovaries are classified into three categories like Normal ovary, Cystic ovary, and PolyCystic ovary. Usually, the diameter of a follicle is more than 2–9 mm, and the count of the follicles is more than 12, then it is considered polycystic ovary. In this paper, the classification of the ovarian cyst is implemented by using the regularized CNN method. In additionally, the justification of the classification process also improved with the data augmentation method and more droplet layer techniques for better accuracy. In the proposed algorithm, the performance of the combined procedure is evaluated with the objective type of metrics and shows the accurate detection of the follicle and leads to conclude the classification of ovarian cyst.

Keywords Convolution neural network (CNN) · Data augmentation · Segmentation · Image enhancement · Polycystic ovarian syndrome (PCOS)

1 Introduction

Polycystic ovary syndrome (PCOS) is the most endocrine disorder which produces more male hormones, which affects the reproductive cycle of women. Most of the issues like an improper menstrual period, ovarian cystic. It can lead to infertility problems. PCOS is a most universal common abnormality in women's reproductive

N. Priya (✉)
PG Department of Computer Science, Shrimathi Devkunvar Nanalal Bhatt Vaishnav College for Women, University of Madras, Chennai 600044, India

S. Jeevitha
Department of Computer Application, Shrimathi Devkunvar Nanalal Bhatt Vaishnav College for Women, University of Madras, Chennai 600044, India

S. Shakya et al. (eds.), *Proceedings of Second International Conference on Sustainable Expert Systems*, Lecture Notes in Networks and Systems 351,
https://doi.org/10.1007/978-981-16-7657-4_17

organs, and it causes crucial health issues including diabetes and heart diseases [1]. In recent studies, 5–10% of the women in the world population were affected by this disorder at reproductive age [2]. Recently different approaches are invented to diagnose this disease, and the most significant method is a pelvic ultrasound image. In general, ovaries can be divided into Normal Ovary, Cystic Ovary, and Polycystic Ovary based on the size and the number of follicles. Diagnosing the PCOS by eliminating many irrelevant test results or symptoms due to some lack of information and the complex pathological condition that occurs in the ovary and prime to take unnecessary imaging processes [3]. The detection of the follicle is about some of the criteria like polycystic ovary (PCO) and androgen excess (a common endocrine disorder at the reproductive age of women) [4]. Usually, PCOS leads the menopausal problems for adolescents who are affected in the age group 20–30 but now the problem is affected in the early age group like 11 years. PCOS women can also has a risk possibility of gestational diabetes, pregnancy issues and micro-vascular difficulties in the early stage [5, 6].

2 Related Works

Adiwijaya et al. [7] used two different schemas like determine the region of a follicle in the ovary and select the appropriate region of follicle with the number and size of the follicles included for finding PCOS in the ultrasound images. Untari et al. [8] proposed feature extraction and classification using Gabor wavelet and modified back-propagation method as a classifier to show the best accuracy level.

Hiremath et al. [9], explained different classification method like fuzzy ovarian classification and SVM algorithm (support vector machine algorithm) to classify the ovary in different categories as ovary which consist of more than 1–2 follicles greater than 28 mm in size they considered as cystic ovary, the ovary which consists 12 or more than follicles around the size of 10 mm, then considered as polycystic ovary and the ovary consist of 1–10 follicles along with 2–28 mm is considered as normal follicles. In most of the related works, classification can be done automatically using different approaches. Still, it is more challenging to predicate some manual extraction methods are also implemented. To avoid this situation automatically feature extraction is done by the system to ensure the quality of output [10]. Convolution neural network (CNN) is proposed for the classification of polycystic ovary or non-polycystic ovary because the CNN method is established as one of the best image classification methods. Some duplicate datasets or oversampling also used to adjust the unbalancing network to increase the performance [11]. Tajbakhsh et al. [3] demonstrated the pretrained dataset with a deep CNN model with the fine-tuning concept and given better results based on different medical images. Nii et al. [11] explained ovarian classification using two different proposed methods as convolution auto encoder for resolution of ultrasound images and other is layered convolution neural network for extracting features from the filtered layer. Sungheetha and Rajesh Sharma [12] implemented the CNN model to detect feature extraction and early diabetic retinopathy stages, further proved accuracy higher than the existing detection algorithm. Dewi et al. [13] used the CNN model to implement the classification of an ovary with feature extraction method (Gabor Wavelet Method) and concluded an accuracy of 80.84%. While creating the CNN model for a minimum number of a dataset is difficult to manage the output accuracy. Since here a minimal number of images have been used for the training section and to avoid the overfitting issues data augmentation has been applied along with the dropout regularization in the CNN model. PCOS can also detected by using image segmentation techniques and it is the challenging process to separate the foreground and background images due to noise presented in the ultrasound images. To overcome the issue, this paper has been proposed the regularized CNN model with the data augmentation and apply the preprocessing techniques for classifying the ovary.

2.1 Dataset Description

Dataset is collected from the infertility clinic from those who are all under the age group of 25–35. The dataset contains a total of 62 images, 48 PCOS ovary images, 4 normal ovary images, and 10 cystic ovary images. In overall consideration, PCO images are 48 and non-PCO is 14. The dataset can be partitioned into three stages for training as 75%, validation as 10%, and testing for 15%. Figure 1 represents the different images used in the dataset like normal ovary, PCOS, and cyst.

3 Methodology

The proposed system examines the robust deep learning classifiers called CNN classifiers in ultrasound images [10]. Generally, feature extraction can be done automatically by the system in the CNN model.

The model can be achieved and visualized for each input layer for classification purposes with different techniques. Eventually, a difference can be observed by the accuracy of the result obtained. Since there may be insufficient dataset images, the overfitting problem may occur. By introducing regularized CNN Model with more droplets, this problem can be solved with better accuracy exhibit and then applies the augmented dataset for further exploration.

The proposed system works more efficiently, comparing with the CNN Model, and hence it also accomplishes better accuracy in the ovarian detection and classification of ovarian cyst. Figure 2 represent the flow diagram for the proposed system that contains different step as follows

- Read the dataset of ovarian images for the proposed method.
- Built the CNN2d Model with different layers like (CNN2d layer, Maxpooling layer, Dense layer, other fully connected layers, etc.).

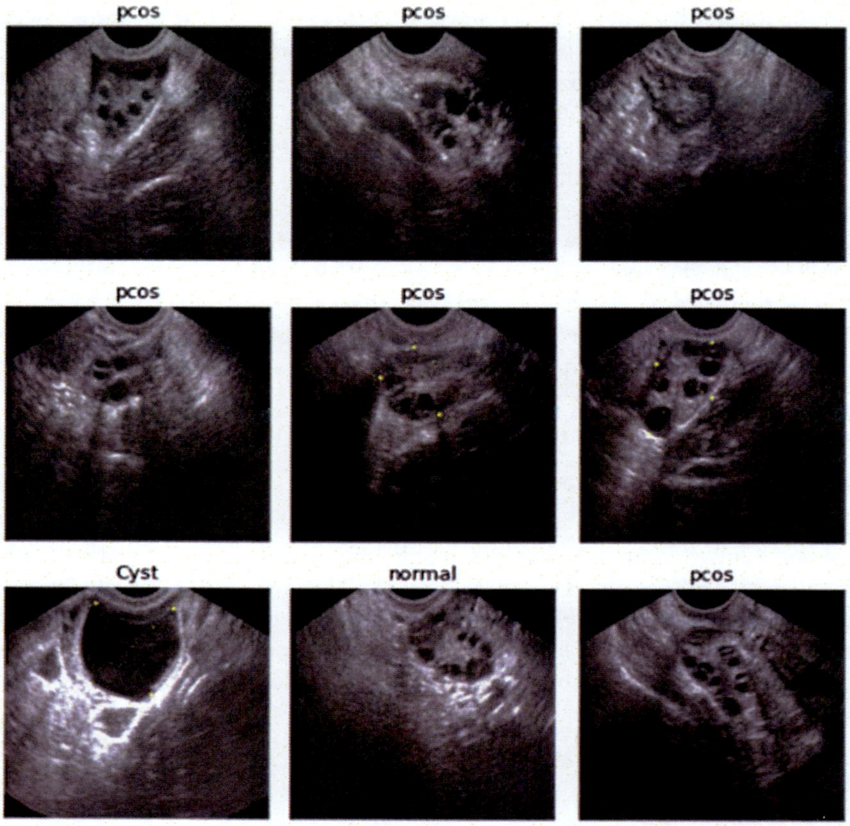

Fig. 1 Ovarian dataset with different types of cyst

- Calculate the output accuracy based on the CNN2d Model.
- Apply more Droplet Mechanism in CNN2d Model and examine the result.
- Apply the Data Augmentation to the CNN2d Model.
- Examine the accuracy of all the methods.
- Analyze the model on the test dataset and find the classification accuracy.
- Compare all methods for accuracy and classify the image.

3.1 CNN Model

Convolution neural network model is the most powerful model that is used for image classification for the past decade as known one. The model consists of the input layer, many hidden layers, and output layer. A fully connected layer (dense, flatten), Maxpooling layer with relu activation for converting translational invariance to reduce the size of an input image, all come under the hidden layer. CNN model

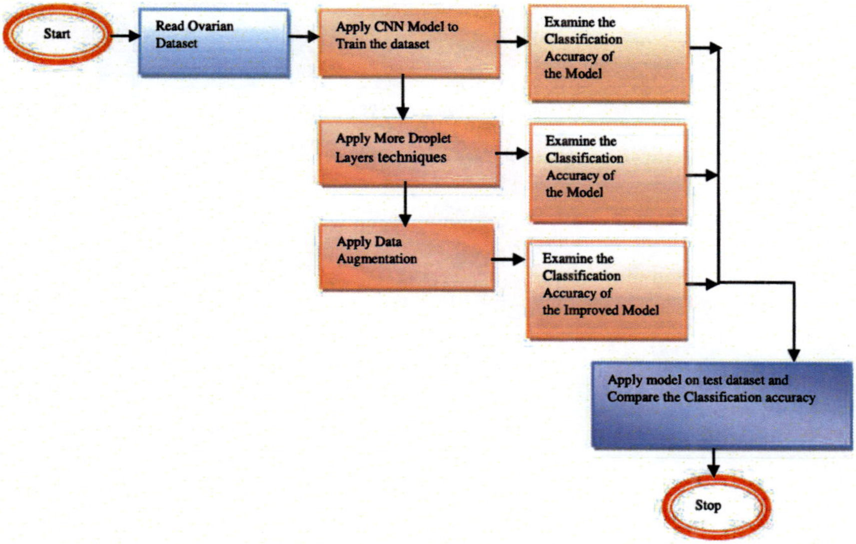

Fig. 2 Flow diagram for proposed system

allows us to represent maximum extraction of image content for classification, this takes raw image pixel data, then trains the model and extracts the features automatically for the superior classification model. CNN clean window through images then calculates the input and filter pixel values accurately, also which allow convolution neural network to emphasize the other features.

The proposed CNN Model

1. Create a 2dConvolutional layer for an input image.
2. Apply the Max pool layer with relu activation for converting translational invariance to reduce the size of the input image.
3. Flatten layer is used next, to flatten the image.
4. Dense layer is used, it is also called fully connected layer which is used to connect all the neurons in the before layer.
5. Train the model using the train dataset image.
6. Finding Accuracy Table 1. Represent the layers which is used in the CNN Model using Droplet and corresponding data complexity.

Increased the number of layers will help to build a deeper model, also increase the number of filter layers will help to perform more features. To regularizing the layer by adding dropout and dense layers. Figure 3 represents the proposed architecture of the CNN model. The output of the convolution layer is 75 with a corresponding feature map, then the flatten is used to flatten the output. Finally, the dense layer predicates the classification of ovaries. Filtering also applied to learn more features from the images Figs. 4 and 5. Figure 6 represent the feature map of CNN layer. By

Table 1 Proposed CNN model architecture details

Layers type	Data complexity (input and output)	
	Input	Output
Convolution2d_1 layer	200 × 200 × 1	200 × 200 × 2
Maxpooling layer_1	200 × 200 × 2	75 × 75 × 2
Dropout_1	75 × 75 × 2	75 × 75 × 2
Convolution2d_2 layer	75 × 75 × 2	37 × 37 × 4
Maxpoolling layer_2	37 × 37 × 4	18 × 18 × 4
Dropout_2	18 × 18 × 4	18 × 18 × 4
Flatten	20,736	
Dense	512	

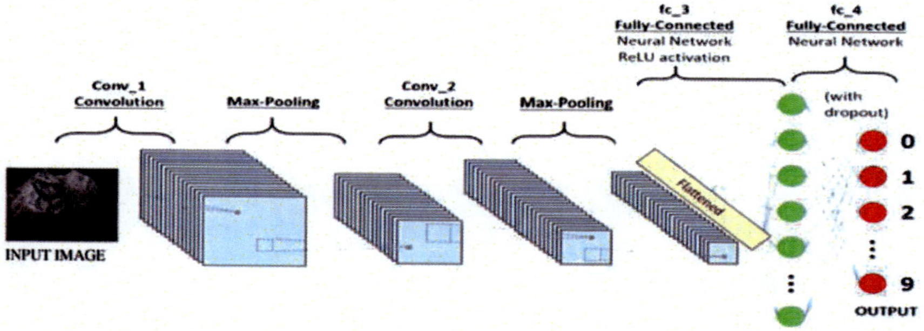

Fig. 3 Proposed architecture of CNN model

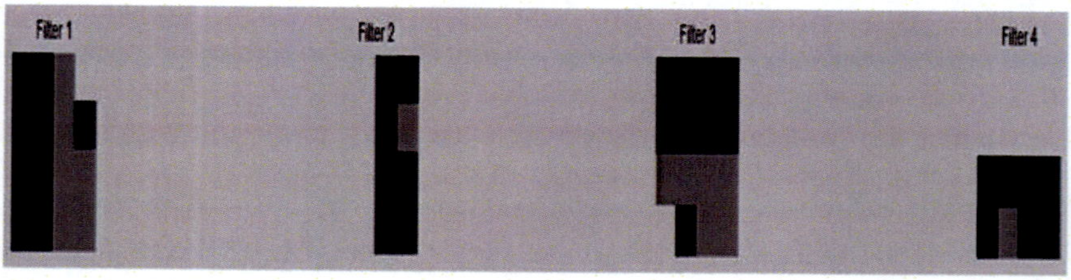

Fig. 4 Filter processing

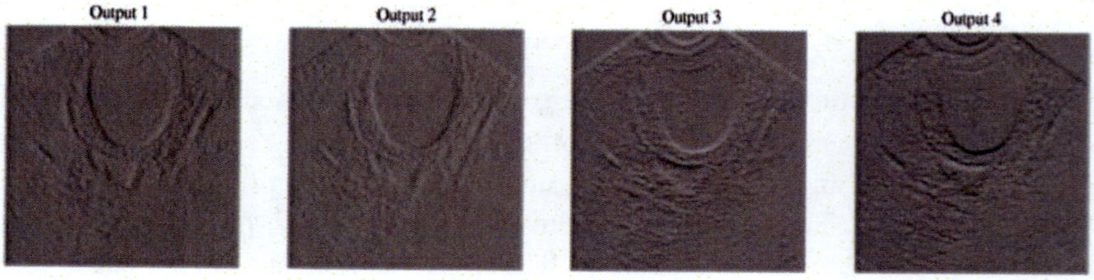

Fig. 5 Output of cyst image after filter processing

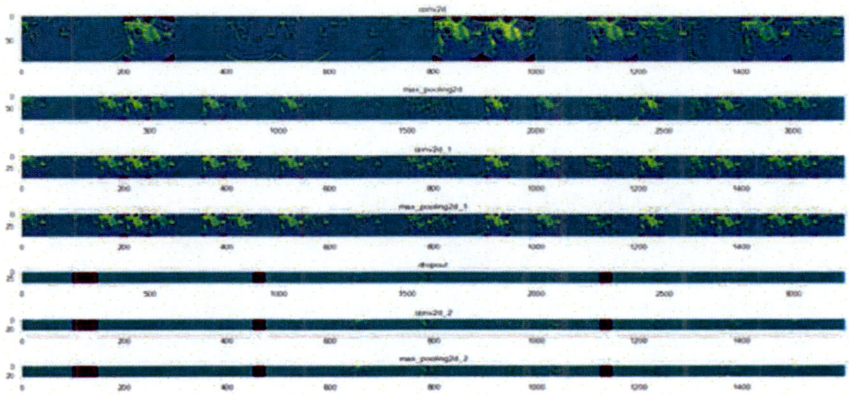

Fig. 6 CNN layers output using visualizing feature map

applying more layer to the fully connected network can achieve maximum accuracy of 81%.

3.2 *CNN with Droplet Regularization*

Droplet regularization is one of the alternative ways to regularize the convolution neural network. It is an inexpensive and secure way to regularizing the input layers in the model. Besides its blocks the model from overfitting issues. The overfitting occurs because of fully connected layers that occupy more parameters which will be overcome by the droplet and prevent the neural network from overfitting. It can be function by removing or dropping out input to a layer that consists of input variables in the data sample or from the activation data of the previous layer. Its regulation can be host at a large number of networks with different network structures. After few epochs, it seems the training model is overfitting because of the small training dataset. It additionally includes the enhanced droplet one from 0 to 0.2. In consideration of this, all the hidden layers were stimulated and get better results comparatively before the CNN model. By observing the performance and sampling reduction step by step maintained. The accuracy of the validation increases by 84%.

3.3 *Data Augmentation Technique for Overfitting*

Data augmentation concepts give more training images along with different positions in the ovary. The existing training images by augmenting them using random transformations which leads to giving better of the model. Because of this processing, will get more accuracy, not only used to avoid overfitting problem also maintained for good performance of models. Data augmentation techniques are classifieds into different types, which can be included in a dataset. They are as follows,

1. Geometric Transformation—normally used to rotate or transform, flip, a crop of the image.
2. Color Transformation—used to change the RGB colors in an image.
3. Filters—used to blur and sharpen an image.
4. Random Erasing—used to delete some part of an image.
5. Mixing Images—used to mix the images from one to another.

Vikas et al. [14], explained that the number of parameters required corresponds to the data complexity of the operation in the machine learning model must carry out by using this kind of augmentation process. Sometimes it might still go wrong even if we have the correct size of training set also. To regularized convolution neural network models, it can be upgrade the dataset by including more dataset images, using the data augmentation methods as shown in Fig. 7. Obtained 1:9 ratio of images by using the data augmentation method in the dataset from its performance, over-fitting has been gradually reducing similarly the accuracy found around 93% that is better than the previous CNN models.

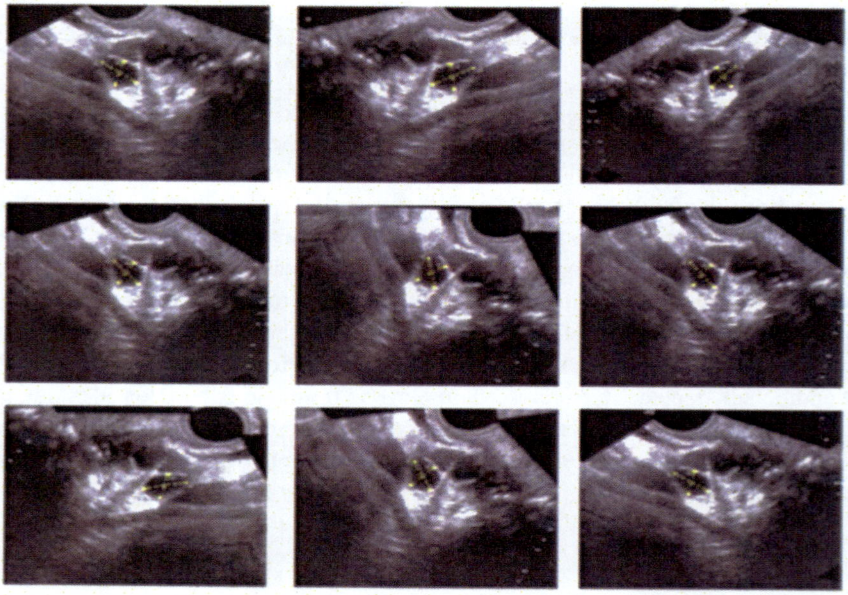

Fig. 7 Augmentation for train model

4 Result and Discussion

The above work is concluded with different deep learning methods and a good accuracy level obtained based on the results. Models build as per the corresponding train data set and tested with test set and classification can be done according to the classes like Normal, Cyst, and PCOS. Each of the CNN layer execution can be visualized using visualizing feature map, it can be generated by applying filter detector to the output of each layer in CNN. It shows the internal representation of the each input CNN layers. The technology used in the above paper performed well and the metrics values noted and shown in Table 2 and Fig. 8.

The above dataset is divided into two phase as training and testing dataset, respectively. It is clearly proved that each model has its gradually increasing the performance level as expected. By using the data augmentation finally achieved 93% of accuracy value. The confusion matrix is calculated by using the dataset as shown in Table 3.

Table 2 Accuracy obtained from different techniques

Techniques	Convolution neural network (*a*)	CNN with more droplet layers (*b*)	CNN with data augmentation (*c*)
Accuracy	81%	84%	93%

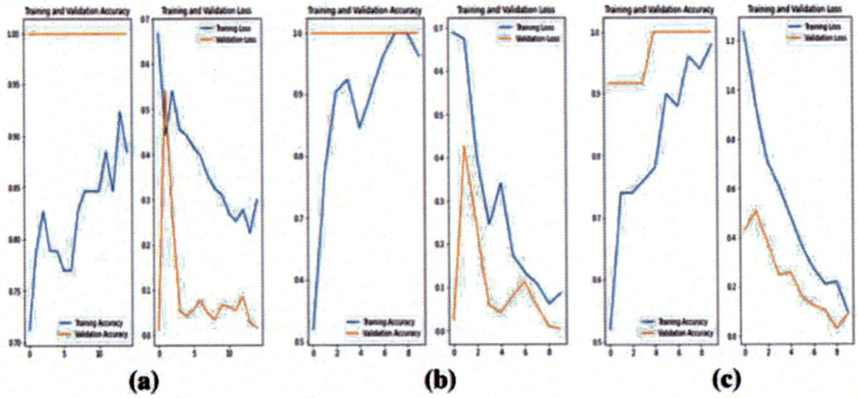

Fig. 8 Graph representation of different accuracies

Table 3 Confusion matrix for CNN model using data augmentation

Classes	Normal	PCOS
Normal (0)	90	18
PCOS (1)	18	423
Accuracy = TP + TN/TP + FP + TN + FN	93.44	

5 Conclusion

In the above work, deep learning model has been applied and conclude the classification of the ovary by analyzing PCOS and non-PCOS classes. This can be implemented with some necessary objective metrics like F_1-score, Precision, and recall while in the training and testing phase. In the proposed system, regularized convolution neural network method with the augmented dataset is used for classification and gives efficient accuracy and also solve the over fitting issues. In future, deep learning method is applied with the additional features to seek better accuracy and also aim to achieve some fine-tuning techniques to give better detection of follicle in the ovary. This work also provides the classification of ovarian and predicate the PCOS accurately for further processing.

References

1. Lee TT, Raush ME (2012) Polycystic ovarian syndrome: role of imaging in diagnosis. Radiographic.rsna.org 32:1643–1657. https://doi.org/10.1148/rg/326125503
2. Padmapriya B, Kesavamurthy T (2015) Diagnostic tool for PCOS classification. In: IFMBE Proceedings, vol 52. Springer International Publishing, Switzerland. https://doi.org/10.1007/978-3-319-19452-3_48
3. Tajbakhsh N, Shin JY, Gurudu SR, Todd Hurst R, Kendall CB, Gotway MB, Liang J (2016) Convolutional neural networks for medical image analysis: full training or fine tuning. IEEE Trans Med Imaging 53(5):0278–0062
4. Denny A, Raj A, Ram AAM, George R (2019) i-HOPE: detection and prediction system for polycystic ovary syndrome (PCOS) using machine learning techniques. IEEE. 978-1-7281-1895-6/19
5. Legro RS, Arslanian SA, Ehrmann DA, Hoeger KM, Hassan Murad M, Pasquali R, Welt CK (2013) Diagnosis and treatment of polycystic ovary syndrome: an endocrine society clinical practice guideline. J Clin Endocrinol 98(12):4565–4592. https://doi.org/10.1210/jc.2013-2350
6. Goodman NF, Cobin RH, Futterweit W, Glueck JS, Legro RS, Carmina E (2015) American Association of Clinical Endocrinologists American College of Endocrinology, and antrogen excess and PCOS society disease state clinic review: guide to the best practices in the evaluation and treatment of polycystic ovary syndrome. Part-2 Endocr Pract 21
7. Adiwijaya, Purnama B, Hasyim A, Septiani MD, Wisesty UN, Astuti W (2015) Follicle detection on the USG images to support determination of polycystic ovary syndrome. J Phys Conf Ser 622(2015):012027
8. Wisesty UN, Nasri J, Adiwijaya (2017) Modified backpropagation algorithm for polycystic ovary syndrome detection based on ultrasound images. In: Recent advances on soft computing and data mining. Advances in intelligent systems and computing, vol 549. Springer International Publishing AG. https://doi.org/10.1007/978-3-319-51281-5_15
9. Hiremath PS, Tegnoor JR (2013) Follicle detection and ovarian classification in digital ultrasound images of ovaries. In: Advancements and breakthroughs in ultrasound imaging. https://doi.org/10.5772/56518
10. Ciresan DC, Meier U, Masci J, Gambardella LM, Schmidhuber J (2011) Flexible, high performance convolutional neural networks for image classification. In: Twenty second international joint conference on artificial intelligence
11. Nii M, Kato Y, Morimoto M, Kobashi S, Kamiura N (2018) Ovarian follicle classification using convolutional neural networks from ultrasound scanning images. Int J Comput Vis Sign Process 8(1). ISSN: 2186-1390

12. Sungheetha A, Rajesh Sharma R (2021) Design an early detection and classification for diabetic retinopathy by deep feature extraction based convolution neural network. J Trends Comput Sci Smart Technol (TCSST) 03(02):2582–4104
13. Dewi RM, Adiwijaya, Wisesty UN, Jondri (2018) Classification of polycystic ovary based on ultrasound images using competitive neural network. In: International conference on data and information science. https://doi.org/10.1088/1742-6596/971/1/012005
14. Vikas B, Radhika Y, Vineesha K (2021) Detection of polycystic ovarian syndrome using convolutional neural networks. Int J Curr Res Rev 13(06)
15. Masko PHD (2015) The impact of imbalanced training data for convolutional neural networks. Degree project, in computer science, first level Stockholm, Sweden

Generating Game Indefinite Decision Tree in the Banking Sector Using Different Types of Algorithms

S. Sridevi and S. M. Chithra

Abstract In the banking sector, bank account holders play a crucial role in order to retain a advanced system that is fully functioning in the shortest or longest revolution. As a result, many queries about dedication based on rewards and lifetime maximization strategies have resulted in the integration of Game Indefinite Decision Tree [GIDT] idea. This GIDT technique identifies the association of GIDT with well-chosen fit-out and back to enhance its broad perspective on know-hows. The extensive technique is described by the arrangement of benefit rates, categories and time period. The advantage of the proposed method is higher than the traditional GIDT. Furthermore, as a direct objective of the squandering, all-encompassing model adaption research vocabularies have much lower variance for GIDT than for conventional game. It is discovered to be a crucial application offered to secure a client's bank account with a different algorithm. In addition, this study examines the group of account holders and advantages for regular setting out in order to enhance the benefits ratio and therefore minimize overall loss and enhance lifetime advantages.

Keywords Game theory · Game indefinite decision tree · Banking sector · Brute force algorithm · Next-door neighbour algorithm (NNA) · Repeated next-door neighbour algorithm (NNA)

1 Introduction

The roots of game theory are very primitive. The idea of game theory appeared throughout the history such as in the works of the Bible, the Talmud, Descartes and Sun along with the works of Charles Darwin. Nonetheless, some argue that the first actual learning of game theory began in 1700 with the work of the geometrician Daniel Bernoulli. Despite the fact that his drudgery 'Bernoulli's Principles' is the

---

S. Sridevi (✉)
Queens College of Arts and Science for Women, Punalkulam, Pudukkottai, Tamil Nadu, India

S. M. Chithra
RMK College of Engineering and Technology, Chennai, Tamil Nadu, India

S. Shakya et al. (eds.), *Proceedings of Second International Conference on Sustainable Expert Systems*, Lecture Notes in Networks and Systems 351,
https://doi.org/10.1007/978-981-16-7657-4_18

foundation for the jet device's design and process, he is delighted to have offered concepts regarding the projected usage and declining income. Another explanation for the cautious application was provided in England in the eighteenth century by Thomas Bayes, known as the 'Bayes theorem'. His drudgery entails by using possibilities as the focal point of a valid deduction. Nonetheless, the inside of up-to-date game theory can be restrained as the amplification of triple key works; Augustine Carnot proposed 'Research on the Mathematical Ethics of Wealth Theory' in 1838. It affords an inbuilt account of what is ultimately courteous as the Nash equilibrium and gives the group of actors the best retort to the arrangement of others. In 1881, Francis Y. Edge led into the idea of a no-nonsense two-person cheap.

Game indefinite decision tree uses the decision tree logic and creates profits in different account holders. After that, it selects the best one from all the trees that have been created. However, we have to be careful with that demonstration. The Game Indefinite Decision Tree [GIDT] is a quick end tree greenhorn that creates a strong end/profit tree by using information gain as a dividing scale. This is a set of values for numeric properties. So a lot of inquiries in terms of benefits have attracted to the notion of a sport indefinite end tree with a lifetime increases the system of devotion. Among the advantages, it is this syndicate technique of GIDT has used to determine the relevance of GIDT and to increase its broad vision knowledge while being appropriately and well-selected background. The foresight method is explained by the arrangement of benefits rate, categories and time. The benefits generated by this method are more profitable than conventional GIDT.

The Game Indefinite Decision Tree [GIDT] is one of the best ways to suggest a graphical description of a technique for classifying, predicting and evaluating the significance or concern. This is the easiest and most commonly used classification method. It deals with the game's indefinite decision tree analysis by using the structure of tree-like results and their associated potential effects. This is the easiest and most commonly used classification method. It deals with the game's indefinite tree analysis by using the structure of tree-like results and their associated potential effects. Each terminal has indefinite end tree nodes at each end and an attribute must be selected to divide the terminal events into sub-games.

The game indefinite decision tree accepts the input set of well-sorted data and the output shaft is provided. Each node (leaf) shows a result (a class) and each node (middle) shows a test. The most common algorithms used in the Game Indefinite Decision Tree are Brute Force Algorithm, Next-door Neighbour Algorithm (NNA), and Repeated Nearest Neighbour Algorithm (RNNA), which were adopted for the study and implemented in-game theory by using a fuzzy graph with the maximum package.

2 Literature Review

This section reviews the implementation of latest approaches to bank account holders with the goal line of maintaining a fully functional benefit system for the short and long term benefits. All relevant documents are systematically classified into two separate categories namely, non-cooperative game theory and cooperative game theory. GIDT concept approaches can be used to enhance node-level and network-level performance by using distributed decision-making capabilities.

The optimization outline distribution using GIDT was developed to analyse the benefits-allocation and interest allocation ratio of bank a/c holders Na et al. The proposed bank a/c holders target tax system based on GIDT and its objectives are aimed at reducing overhead and maximizing lifetime benefits. With views on economics and GIDT, Wu and Wei et al. wished for a mechanism proposal to hold the inducements of deliberate negotiators. Shami et al. developed a cooperative game in unfinished data for strewn GIDT.

The most collective set of rules used in the Game İndefinite Decision Tree are Brute Force Algorithm, Next-door Neighbour Algorithm (NNA), and Repeated Nearest Neighbour Algorithms (RNNA), which were adopted for the study and carry out in-game theory using a fuzzy graph with the integration of maximum package.

3 An Overview of Game Theory

Game Theory is a group of measured simulations, which are considered to study the struggle and mutual aid locations. It is disturbed with the verdict for achieving the best life out for distinct judgement producers in this status quo and identify the lasting moments.

3.1 Game Theory in the Fuzzy Decision Tree

Game theory is gradually enticing more care as a mechanism to resolve innumerable teething troubles in fuzzy decision tree (Kacher et al. [7], [14], Appiahen et al. [16]). Table 1 shows that the components of a Game and the respective elements present in a fuzzy decision tree.

Table 1 Constituents of a game indefinite decision tree (GIDT)

Constituents of a game	Elements of a GIDT
Players	Edge in the fuzzy graph
A set of strategies	A modulations scheme, coding rate, transmit weight level etc.
A set of payoffs (or) utility functions	Concert metrics in bank eg. Throughput, delay, SNR etc.)

3.2 Processes Involved in GIDT

A GIDT typically reflects all data as a root before preliminary. Then, under certain conditions, it begins to split through the branches or inner ends and decides to form a leaf effect. The imperative entity to know is that it condenses the uncleanness in the material goods and at the same time gets the statistics to succeed the right domino effect when the office block a tree.

Since the algorithm is unpretentious in the environment, it also encompasses some constraints that technologist consider essential to tell, as these factors decide how well a GIDT makes in the ending shop of an ideal.

3.3 Definition of the Problem

The present research mainly concentrates on Game Indefinite Decision Trees (GIDT) on fuzzy graphs and their application.

(a) Game Indefinite Decision Tree problem is examined on the grid with fuzzy edge weights.
(b) The problem under different conditions are overall game fuzzy amounts, a mixture intellectual set of rules is intended to resolve it.
(c) A fuzzy version of classical Prufer sequences to solve the Game Indefinite Decision Tree problem in the bank environment.
(d) Highlights with the application the effects of the changes in the provisioning of public bank services to make them more efficient and effective.
(e) To find a key application of the fund available for investment for the customer to the bank using different types of Algorithms.

3.4 Proposed Framework

The proposed structure intends to cast-off in order to ripen prophetic models. It lies (works) on three altered times: data gathering time (stage I), data pre-processing time (stage II), and calculation model change time (stage III). This implies that the records established for model enlargement spirits were completed three times, namely time I, time II, and stage III. During the data gathering time (Level I), the cause data gets serened from the banks. After data assembly, it is pre-processed in time II before taking place to the pre-sample increase time. Bond Versions, Optional/Regal Accounts, Senior Citizens Accounts, Women's Accounts, Youth Accounts, Children's Accounts, Current Accounts, Fixed Deposits RD, Senior Citizens FD.

4 Modelling of Indefinite Decision Tree

Definition 4.1 *A **directed (or) undirected Game Indefinite stroll** in a Game Indefinite diagram is a sporadic arrangement of crests and limits* $x_0, \varepsilon_1, x_1, \ldots \varepsilon_n, x_n$ *in which upper hand* ε_i is $x_{i-1}x_i$.

Definition 4.2 *A **Game Indefinite pathway** is an indefinite stroll in which fully points are divergent.*

Definition 4.3 *A barred Game Indefinite stroll in which on no account peak (except the primary and the closing peak) appears more than once is called a **Game Indefinite circuit.***

Definition 4.4 *A game indefinite graph having no cycles is called an **acyclic game indefinite gridiron**. A connected acyclic game indefinite gridiron is called **a game indefinite tree** (or) **indefinite course of action**.*

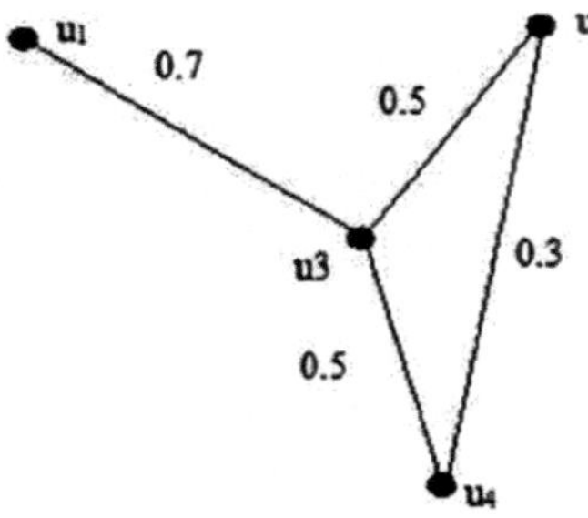

High quality data plays an important role in increasing the emporium/exploration efficiency of computational techniques. To produce an optimal solution, well-known graph algorithms are used. Finding direct paths on weighted maps is one of the archaeological problems faced the integrated optimization domain.

Finding short paths in weighted maps is one of the archaeological problems encountered in the integrated optimization domain. More recently, game indefinite tree weighted maps with the generalization of a method for finding optimal paths within them have evolved into an adequate modelling tool for restricted complex or naturally occurring methods that are used for path comparisons.

With the number of optimization criteria they impose, the realistic scenarios are often so complex that even the most sophisticated implementation becomes computationally unmanageable. The method based on the most popular centre of gravity sports anaesthesia has some drawbacks. This makes it irrelevant for the purposes of the game indefinite short path algorithm of a game blunt system, which is used as the basis for ranking obscure numbers. Game ambiguous weights are possible before the algorithm is implemented under some of the confusing operators.

The obvious use is that the minimal game indefinite decision tree can be used to almost solve the account holder's problem. A convenient way to define this problem is to find a short path that visits each point so that this problem can be solved by various methods. They have recently been developed on research topics. There are several better ways to do this, they are:

1. A random algorithm can solve it linearly at the expected time.
2. If the weights are large integers, it can be solved at the worst possible time.
3. Select the minimum weightage that is not used on the game blur map and highlight it.

Remark 4.1 The game indefinite tree has no journeys and it is fine to have peaked with a stage higher than dualistic.

Definition 4.5 *The main goal of the problem is the exaggerated goal of trivializing the problem of the game indefinite end tree. The goal should be specific to the problem at hand and the game should be one that can be properly explored by the tree makers for producing ambiguous results.*

For example, employing a new bank staff with a different account holder may have different objectives. This can lead to different consequences of their game indefinite end tree problem. Also, the goal must be single. That is, game indefinite end tree makers should not attempt to accomplish multiple goals within one problem. (In practice, many of these goals will be broken down into separate criteria as defined below). An example of a goal related to creating accounts. 'Determining the benefits that make me looks profitable' is something that can be determined by the specific, singular and game ambiguous end tree maker.

Definition 4.6 *A standard game ambiguity matrix is* ϱ_{ij} *(the admittance in the ith row and jth column),* $\varrho_{ij} = \varrho_{kj}/\varrho_{ki}$.

What this means in our matrix is that a standard matrix has some basic size. That is, each element has a set weight that does not change compared to another element. Now that we have the ratio team, notice that we can create an account of the following matrix equation:

$$\begin{bmatrix} G_1/G_1 & G_1/G_2 & G_1/G_3 & \cdots\cdots\cdots & G_1/G_n \\ G_2/G_1 & G_2/G_2 & G_2/G_3 & \cdots\cdots\cdots & G_2/G_n \\ \cdot & \cdot & \cdot & \cdots\cdots\cdots & \cdot \\ \cdot & \cdot & \cdot & \cdots\cdots\cdots & \cdot \\ \cdot & \cdot & \cdot & \cdots\cdots\cdots & \cdot \\ G_n/G_1 & G_n/G_2 & G_n/G_3 & \cdots\cdots\cdots & G_n/G_n \end{bmatrix} \begin{bmatrix} G_1 \\ G_2 \\ \cdot \\ \cdot \\ \cdot \\ G_n \end{bmatrix} = n \begin{bmatrix} G_1 \\ G_2 \\ \cdot \\ \cdot \\ \cdot \\ G_n \end{bmatrix} \quad (1)$$

We know that the matrix equation in 3 is valid because as the game goes through the ambiguous matrix multiplication we find that there is an $n \times 1$ matrix with inputs:

For the product matrix sequence, i is $G_1(G_i/G_1) + G_2(G_i/G_2) + \cdots + G_n(G_i/G_n)$. As can be seen, classes are cancelled, and we have left within G_i for row i. Therefore, we can express the product matrix with the measurement n as a factor, as in Eq. 1. N is an eigenvector of the $n \times n$ ratio matrix (this is called the matrix A) and the $n \times 1$ weight matrix (this is called the matrix W). Because of theorem 2, we know that the sum of the equivalent values of matrix A is simply its trace. First

$$T_r\left(\breve{G}\right) = G_1/G_1 + G_1/G_2 + G_1/G_3 + \ldots\ldots\ldots\ldots G_1/G_n$$

$$= 1 + 1 + \cdots + 1$$

We know that $T_r\left(\breve{G}\right) = n$. Therefore, we know that the quantity of the genetic values of the matrix $\breve{G}$ is like to n. Since we have already shown n as the eigenvalue of $\breve{G}$, and since $\breve{G}$ the matrix is in the ranking, we know by theorem 1 that n is only a non-zero eigenvalue. We now create the following definitions.

Theorem 4.1 *Any linear game indefinite decision tree that works out the Element Distinctiveness occupation has depth* ω ($n \log n$).

Proof

Let $x = x_1, x_2 \ldots x_n$ and $y = y_1, y_2 \ldots y_n$ be two flight footpaths with distinctive match up that are in hand in two poles apart orders. In certain, for some pair of alphabetical citation i and j, we have $x_i < x_j$ and $y_i > y_j$. Any continuous lane from x to y must contain a point z such that $z_i = z_i$ by the intermediary value proposal, but then we have $F(z) = 0$. Thus, arguments with poles apart permutations are in poles apart connected apparatuses of $F^{-1}(1)$, so $\#F1 \geq n!$ The worse bound now keep an eye on instantaneously.

Theorem 4.2 *Any linear game indefinite decision tree that works out the Immovable Near-term collected efficacy has severity* ω ($n \log n$).

Proof

For this badly behaved, the involvement is a point $(x, y) = (x, y) = (x_1, x_2 \ldots . x_n; y_1, y_2 \ldots\ldots y_n) \in \dagger^{2n}$ where the $x$and$y$—match up represent the two crowds $x$and$y$. If $x$and$y$ are disconnects, then we can be dividing wall $x$and$y$ into come apart subsets $x_1, x_2 \ldots x_n$ and $y_1, y_2 \ldots\ldots y_n$ that mollify the restricted order.

$$x_1 < y_1 < x_2 < y_2 < \ldots < x_k < y_k$$

Here, $A < B$ wealth that every constituent of A is less than every component of B, but the fundamentals indoors A are unparalleled. Whichever x_1 or y_k or both possibly will be unfilled. There are exactly n^2 such limited instructions where the $x$and$y$—rudiments substitute, that is, where each set x_i and y_i is a singleton, and $k =$

n. As in the aforementioned theorem, any two opinions that obey poles apart partial orders lie in poles apart associated mechanisms of $F^{-1}(0)$. Thus, $\#F_0 \geq n^2$ and the proposition follows instantaneously.

5 Types of Algorithm

Example 5.1 (see Fig. 1).

5.1 *Brute Force Algorithm*

Step 1: Gradient all likely game indefinite decision trees.
Step 2: Treasure trove the measurement of each game indefinite decision tree by tally the advantage weights.
Step 3: Top quality the game indefinite decision tree with insignificant total weight.

S. No.	Indefinite decision tree	Weights
1	$x_1\ x_2\ x_3\ x_4\ x_5\ x_6$	0.53
2	$x_1\ x_6\ x_3\ x_4\ x_2\ x_5$	0.89
3	$x_1\ x_2\ x_5\ x_6\ x_3\ x_4$	0.65
4	$x_1\ x_5\ x_3\ x_2\ x_6\ x_4$	0.79
5	$x_1\ x_5\ x_3\ x_6\ x_4\ x_2$	0.93
6	$x_1\ x_4\ x_5\ x_6\ x_2\ x_3$	0.76

There is a unique game indefinite decision tree on this game indefinite graph. All other game indefinite decision tree is the opposite of the registered ones or twitches

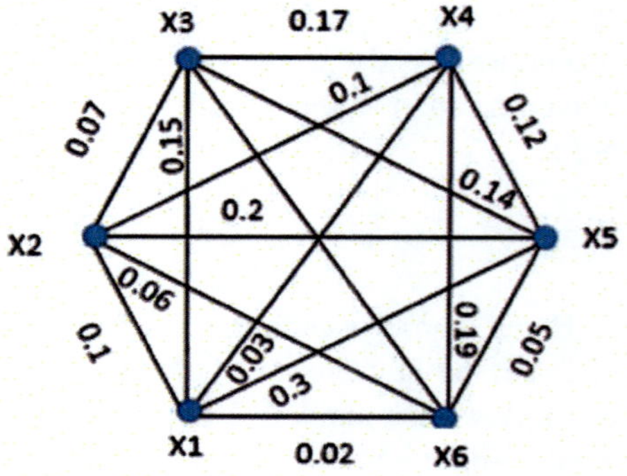

Fig. 1 Game indefinite graph

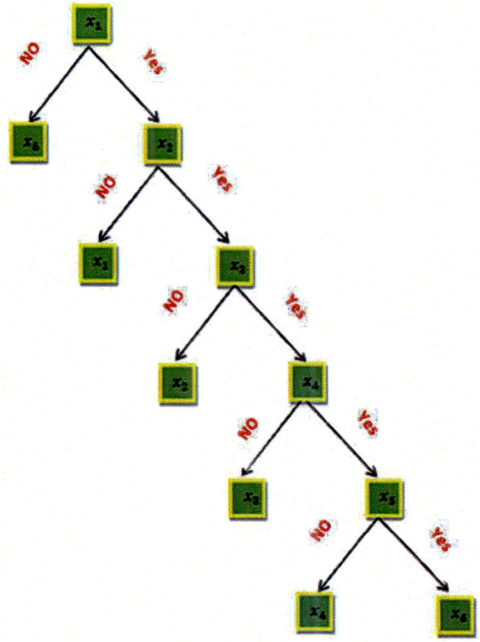

Fig. 2 Brute force game indefinite decision tree

at a poles apart vertex but end result in the same weights. From this, we can see that the first game indefinite decision tree, $x_1\ x_2\ x_3\ x_4\ x_5\ x_6\ x_1$, is the optimal game indefinite decision tree. İt the shortest term and get more benefits. A similar way to find another GIDT (Fig. 2).

5.2 *Next-Door Neighbour Algorithm (NNA)*

Step 1: Hand-picked a preliminary point.
Step 2: Interchange to the next-door unvisited uttermost (the verge with least load).
Step 3: Summary up to the game indefinite decision tree is all-embracing.

Consider Fig. 3 preliminary at vertex x_4, the next-door neighbour is vertex x_1 with a weight of 0.03. From x_1, the next-door neighbour is x_6 with a weight of 0.02. From x_6, the next-door neighbour is x_5 with a weight of 0.05. From x_5, the next-door neighbour is x_4 with a weight of 0.12 but it completes a circuit without visiting other fellow citizens. So, the next-door neighbour is x_3 with a weight of 0.14. From x_3, the only unvisited vertex x_2 with weight 0.07. The bring about game indefinite decision tree $x_4\ x_1\ x_6\ x_5\ x_3\ x_2$ is with a total weight of 0.31.

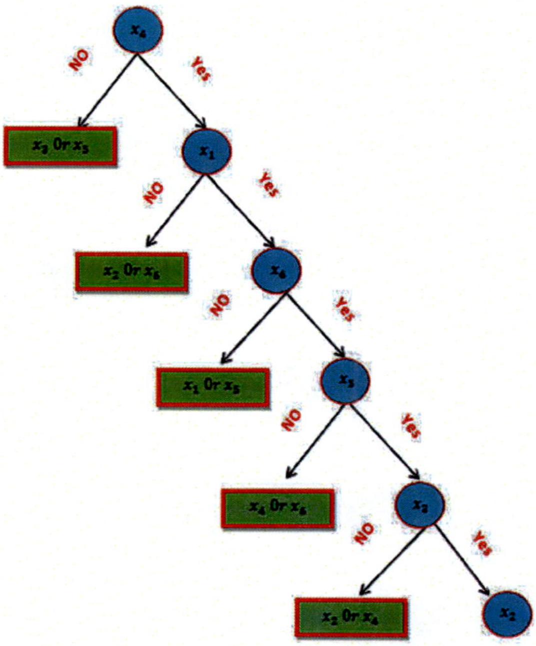

Fig. 3 Next-door neighbour game indefinite decision tree

5.3 *Repeated Nearest Neighbour Algorithm (RNNA)*

Step 1: Start at X. Apply nearest the neighbour algorithm.
Step 2: A repeated process using each of the other vertices.
Step 3: Of the game indefinite decision tree is obtained to use the cheapest.
Step 4: Of the nearest neighbour game indefinite decision tree is obtained to keep the best one.
Step 5: If there is a designated starting vertex, rewrite the indefinite decision tree using that vertex as the reference point.

Cogitate Fig. 3, first at vertex x_1, the next-door neighbour is vertex x_6 with a weight of 0.02. From x_6 the nearest neighbour is vertex x_5 with a weight of 0.05. From x_5, the nearest neighbour is vertex x_4 with a weight of 0.12. From x_4, the nearest neighbour is vertex x_3 with a weight of 0.17. From x_3, the adjoining neighbour is vertex x_2 with a weight of 0.07. The follow-on game indefinite decision tree x_1 x_6 x_5 x_4 x_3 x_2 is with a total weight of 0.43.

Similarly, starting at vertex x_2, the resulting game indefinite decision tree x_2 x_6 x_5 x_4 x_1 x_3 is with a total weight of 0.41. Starting at vertex x_3, the resulting game indefinite decision tree x_3 x_2 x_6 x_5 x_4 x_1 is with a total weight of 0.33. Starting at vertex x_4, the resulting game indefinite decision tree x_4 x_1 x_6 x_5 x_3 x_2 x_4 is with a total weight of 0.38. Starting at vertex x_5, the resulting game indefinite decision tree x_5 x_4 x_1 x_6 x_2 x_3 is with a total weight of 0.30. Starting at vertex x_6, the resulting game indefinite decision tree x_6 x_1 x_4 x_5 x_3 x_2 is with a total weight of 0.50 (Fig. 4).

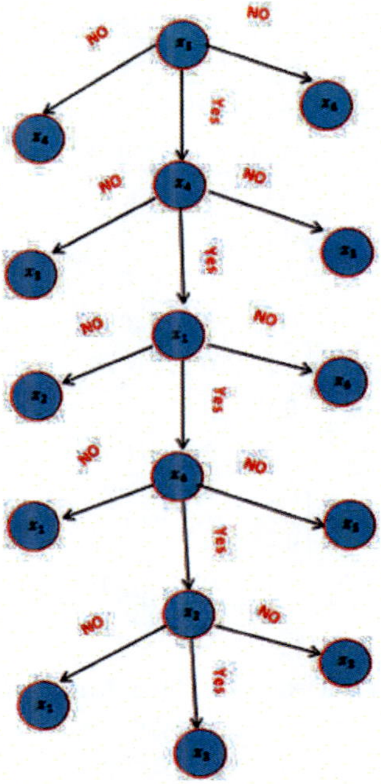

Fig. 4 Repeated nearest neighbour game indefinite decision tree

The minimum, resulting in the game indefinite decision tree, using the Repeated Nearest Neighbour Algorithm is vertex 4 with a total weight of 0.30 (Fig. 5).

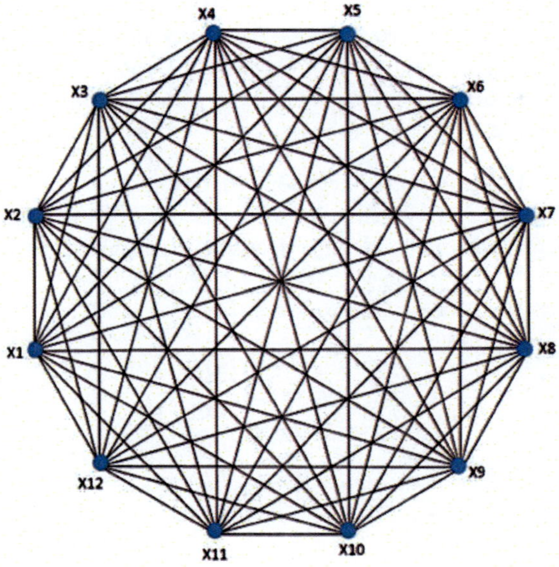

Fig. 5 Relations between bank account holders

Table 2 Relations between bank account holders amount

	x_1	x_2	x_3	x_4	x_5	x_6	x_7	x_8	x_9	x_{10}	x_{11}	x_{12}
x_1	–	0.24	0.17	0.28	0.35	0.25	0.2	0.12	0.06	0.1	0.21	01
x_2	0.24	–	0.13	0.06	0.08	0.27	0.16	0.07	0.19	0.43	0.18	0.01
x_3	0.17	0.13	–	0.15	0.34	0.12	0.03	0.11	0.25	0.05	0.27	0.31
x_4	0.28	0.06	0.15	–	0.22	0.08	0.15	0.33	0.27	0.1	0.47	0.09
x_5	0.35	0.08	0.34	0.22	–	0.3	0.14	0.09	0.23	0.48	0.19	0.1
x_6	0.25	0.27	0.12	0.08	0.3	–	0.06	0.21	0.13	0.08	0.1	0.2
x_7	0.2	0.16	0.03	0.15	0.14	0.06	–	0.11	0.02	0.22	0.12	0.17
x_8	0.12	0.07	0.11	0.33	0.09	0.21	0.11	–	0.38	0.27	0.13	0.05
x_9	0.06	0.19	0.25	0.27	0.23	0.13	0.02	0.38	–	0.19	0.04	0.09
x_{10}	0.1	0.43	0.05	0.1	0.48	0.08	0.22	0.27	0.19	–	0.43	0.06
x_{11}	0.21	0.18	0.27	0.47	0.19	0.1	0.12	0.13	0.04	0.43	–	0.03
x_{12}	0.11	0.01	0.31	0.09	0.1	0.2	0.17	0.05	0.09	0.06	0.03	–

5.2 Example In a private bank, there are connected with many services like Deposit Products, Loan Products, Investments, Products, Lockers, Foreign Currency Buy/Sell, CMS Services, Tax Collection, Internet Banking, etc. We take the deposit products to strategy and unite a proficient game indefinite direction for raising domestic revenue and to gear up through a dynamic credit policy to ensure that the fund available for investments to the customers it reaches to the all type account once and returns to the starting location (e.g.: Deposit Accounts). Consider vertex such as x_1—Deposit Accounts, x_2—Preferred/Royale Accounts, x_3—Senior Citizens Accounts, x_4—Women's Accounts, x_5—Youth Accounts, x_6—Kids accounts, x_7—Current accounts, x_8—Fixed Deposits RD, x_9—Senior Citizens FD, x_{10}—Super Saving Accounts, x_{11}—Tax Saving Deposits, x_{12}—Salary Accounts. The deposit measurements are exemplified as game indefinite weights in the horizontal column (Table 2).

Using Nearest Neighbour Algorithm (NNA) starting at x_8 with a fixed deposit RD. Observing in the row for x_8, the smallest distance is 0.05 to x_{12}. Succeeding that clue, our game indefinite decision tree minimum weightiness is will be:

S. No.	Game indefinite decision tree	Weights	S. No.	Game indefinite decision tree	Weights
1	x_8 to x_{12}	0.05	7	x_5 to x_7	0.14
2	x_{12} to x_2	0.01	8	x_7 to x_9	0.02
3	x_2 to x_4	0.06	9	x_9 to x_{11}	0.1
4	x_4 to x_5	0.08	10	x_{11} to x_6	0.08
5	x_3 to x_1	0.17	11	x_6 to x_{10}	0.05
6	x_1 to x_8	0.12			

Therefore, the total connecting length is 0.88. Now we consider 1 (approximately).

Using Repeated Next-door Neighbour Algorithm (RNNA) starting at x_1 with super saving accounts. Watching in the row for x_1, the smallest cold is 0.06 x_1 to x_6 Subsequent that hint our game indefinite decision tree least possible burden is will be:

S. No.	Game indefinite decision tree	Weights	S. No.	Game indefinite decision tree	Weights
1	x_1 to x_9	0.06	11	x_4 to x_6	0.06
2	x_9 to x_7	0.02	12	x_6 to x_{11}	0.1
3	x_7 to x_3	0.03	13	x_{11} to x_5	0.19
4	x_3 to x_{10}	0.05	14	x_5 to x_8	0.09
5	x_{12} to x_2	0.06	15	x_8 to x_1	0.12
6	x_2 to x_4	0.01	16	x_1 to x_9	0.06
7	x_9 to x_7	0.02	17	x_2 to x_4	0.01
8	x_7 to x_3	0.03	18	x_4 to x_6	0.06
9	x_3 to x_{10}	0.05	19	x_6 to x_{11}	0.1
10	x_{12} to x_2	0.06	20	x_{11} to x_5	0.19

Therefore, the total profit of the accounts is 0.87.Similarly, starting at vertex x_2 x_3 x_5 x_6 x_8 x_9 x_{10} x_{12}, the resulting game indefinite decision tree with the total profit is 1(approximately). Starting at vertex x_4, the resulting game indefinite decision tree with the total profit is 0.92. Starting at vertex x_7, x_{11}, the resulting game indefinite decision tree with the total profit are 0.83, 0.95 (Fig. 6).

Using the Repeated Nearest Neighbour Algorithm the minimum, resulting game indefinite decision tree is vertex x_7 with a total profit of 0.83.

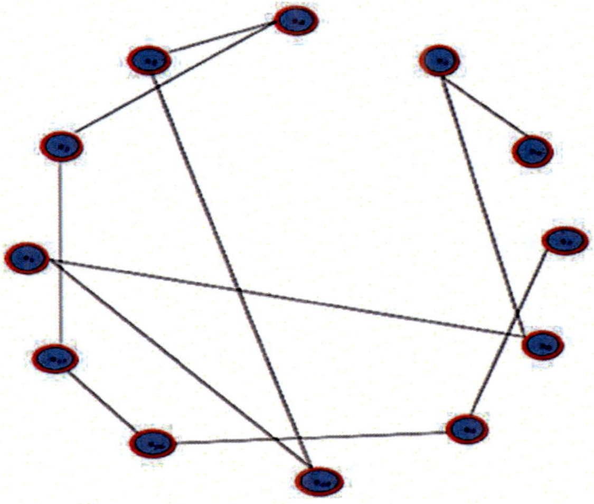

Fig. 6 Profits bank account holders

6 Applications

For the reason that GIDT is so meek in landscape and can be straightforwardly elucidated by any first-born running, they are used in a varied kind of diligences and parks.

(a) **In health-care industries**
In the health-care productiveness, GIDT can express whether an easy-going is distress from an infection constructed on period, bulk, sexual category and former features. Other tenders, such as influential the upshot of preparation based on features such as structure and time of creation, can be very operative in detecting a healthy loud noise.

(b) **In banking sectors.**
Eligibility for a loan tin be dogged at GIDT created on his or that girl pecuniary grade, kinfolk's affiliate, wages, etc. Other solicitations take in credit postcard fraud, financial transactions plans and agreements, and loan episodes.

(c) **In educational Divisions**
In universities, a scholar's closing list can be categorized built on his or her be fit cuts, turnout, and whole score. A GIDT can also fix the overall publicizing line of attack of capacity in institutions of higher education.

Rewards of GIDT

1. A GIDT model is self-same explicable and can be categorically denoted to first-born running and stakeholders.
2. No must pre-process files such as regulation and setting, which shrinks the struggle to craft a classic.
3. A GIDT set of rules can handgrip both top-secret and numeric data and is greatly well-organized equalled to other sets of rules.
4. Any disappeared rate in the data seems to not upset a GIDT, which is why it is well-thought-out a bendable procedure.

Drawbacks of GIDT

1. GIDT works off-colour once it emanates from getting worse.
2. A GIDT can now and then be unsound and unpredictable for the reason that a change in data can top to a GIDT bad arrangement, which can have an emotional impact on the truth of the exemplary.
3. If the data is not gagged in the approved manner, a GIDT set of rules may give inappropriate results and carry out poorly paralleled to other ways and means.
4. Teething troubles in working out arise when the domino effects are amalgamated and it can yield time while working out an archetypal.

7 Conclusion

In this work, we have extensively studied the game indefinite decision tree problem with game indefinite edge weights. The game indefinite decision tree problem on a graph, whose edge weights are assumed as a game indefinite decision variable, and the proposed notion of game indefinite decision tree are discussed. We also focus on the case when game indefinite decision tree weights have regular credibility distributions.

Game indefinite decision trees have direct applications in the design of the network, including transportation networks, the banking sector etc. We showed that the different data can be used in a unique network process, ease of use of information and tools to reduce the need to vacation industry for work and to profit local services. Matching the result with a heuristic algorithm for the problem that connects efficient game indefinite decision tree route, also accomplish the best algorithm to take on for these styles of network problem.

In this study, the authors linked the DEA to the algorithms of bank a / c holders for investigation and anticipated the act of bank branches. The DEA and its Two-Phase Build Hull set of rules are instigated in the most common set of rules used in GIDT are Brute Force Algorithm, Next-door Neighbour Algorithm (NNA), and Repeated Nearest Neighbour Algorithm (RNNA), was agreed for the revision and instigated in-game theory via fuzzy graph with the maximum array to weigh the effectiveness of the 44 bank kindling at the bond time and asset time. The whole neck and neck act of each bank is well-thought-out and regarded as capable (Class A) or infertile (Class B) by 0.8 units or 80% adopted 'cut-off point'. Pre-set by the CCR DEA, this arrangement lesson is rummage-sale as a reply variable.

Motivation

In recent years, sports indefinite end tree has been used mainly in presentations such as well-being, navy and environmentally friendly monitoring. This development was prompted by the widespread popularity of GIDT. However, there are limitations due to the increased benefits. Benefits are coded for a lifetime, due to level variation; so considerable effort is devoted to using it most effectively. This interesting conflict makes the game theory more useful in banks. To maximize network life, the loss must be minimized through a game-philosophical approach.

In recent times, more and more research is focussed on this zone and various researchers are presently concentrating on the strategy of benefits mindful of rules and methods for sensor links. To tally to this modus operandi, researchers are fixed on the exposure area of the measuring device nodes. These proposed algorithms state the tree of indefinite results, over time, the interest rate a/c will be restructured by any means to reduce the wastage of money by the holders.

Data Availability

The files for the revision stayed received from not the same banks in Vallam using their yearbook commercial announcements.

References

1. Bian F, Wang X (2021) School enterprise cooperation mechanism based on improved decision tree algorithm. J Intell Fuzzy Syst 40(4):5995–6005
2. Shibu D, Ayekpam IC (2018) A study on efficiency on Assam gramin vikash bank branches. Indian J Appl Res 7(5):115–118
3. Wu D (2006) Detecting information technology impact on firm performance using DEA and decision tree. Int J Inf Technol Manage 5(2–3):162–174
4. Aggelopoulos E, Georgopoulos A (2017) Bank branch efficiency under environmental change: a bootstrap DEA on monthly profit and loss accounting statements of Greek retail branches. Eur J Opera Res 261(3)
5. Grmanová E, Ivanová E (2018) Efficiency of banks in Slovakia: measuring by DEA models. J Int Stud 11(1):257–272
6. Sufian F, Kamarudin F, Nassir AM (2016) Determinants of efficiency in the Malaysian banking sector: does bank origins matter? Intellectual Econ 10(1):38–54
7. Kacher F, Larbani M (2006) Solution concept for a non-cooperative game with fuzzy parameters. Int Game Theory Rev 08(03):489–498
8. Zhang GN, Ye F (2019) Predicting financial distress in Norway using logistic regression and random forest models. Norwegian School of Economics, Bergen, Norway
9. Galka J, Jaciow P (2019) Speaker identification using fuzzy I-vector tree. J Intell Fuzzy Syst 37(4):4937–4949
10. Titko J, Stankevičienė J, Lāce N (2014) Measuring bank efficiency: DEA application. Technol Econ Dev Econ 20(4):739–757
11. Martin L, Sharma S, Maddulety K (2019) Machine learning in banking risk management: a literature review. Risks 7(1):29
12. Rzepecki L, Jaskowski P (2021) Application of game theory against nature in supporting bid pricing in construction. Symmetry 13:132
13. Kuhn M, Weston S, Coulter N, Culp M (2018) C5. 0: C5. 0 decision trees and rule-based models. CRAN, Germany. https://cloud.r-project.org/package=C50
14. Mu Y, Wang L, Liu X (2020) Dynamic programming based fuzzy partition in fuzzy decision tree induction. J Intell Fuzzy Syst 39(5):6757–6772
15. Hamid N, Ramli NA, Hussin SAS (2017) Efficiency measurement of the banking sector in the presence of non-performing loan. In: AIP conference proceedings, vol 1795
16. Appiahene P, Missah YAWM (2019) Predicting the operational efficiency of banks in the presence of information technology investment using artificial neural network. In: Proceedings of the academics world 132nd international conference, Florence Italy, pp 6–11
17. Appiahene P, Missah YM, Najim U (2019) Evaluation of information technology impact on bank' s performance: the Ghanaian experience. Int J Eng Bus Manage 11:1–10
18. Sridevi S, Chithra SM (2020) The system of harvest getting higher using game theory technique in wireless sensor networks. Int J Mech Prod Eng Res Dev (IJMPERD) 10(3):9653–9660 ISSN (P): 2249–6890; ISSN (E): 2249–8001
19. Sridevi S, Vinoba V (2016) A study on cooperative and non-cooperative game theory techniques in wireless sensor networks. Int J Math Arch 7(9):1–5 ISSN 2229–5046
20. Havidz SAH, Setiawan C (2015) Bank efficiency and non-performing financing (NPF) in the Indonesian Islamic banks. Asian J Econ Model 3(3):61–79
21. Sun J (2020) Power instability prediction method for wind turbine based on fuzzy decision tree. J Intell Fuzzy Syst 39(2):1439–1447
22. Addison YE (2018) 19 banks meet new capital requirement. https://www.graphic.com.gh/business/business-news/19-banks-meetnew-capital-requirement.html

23. Addison YE (2018) Banking crisis: BoG's roadmap for clearing UT, capital bank mess. https://citinewsroom.com/2018/08/14/banking-crisis-bogs-roadmap-for-clearing-utcapital-bank-mess/
24. Chen Z, Matousek R, Wanke P (2017) Chinese bank efficiency during the global financial crisis: a combined approach using satisficing DEA and support vector machines. North Am J Econ Finance 43

Detection of Pneumonia Using Convolution Neural Networks

Sai Hemanth Kumar Kotturi and S. V. N. Sreenivasu

Abstract *Pneumonia* is one of the severe contagious diseases affecting either on one or the both lungs of humans which is commonly caused by bacteria called *Streptococcus pneumonia* (Rajpurkar et al. in, 2017). World Health Organization reported that there is one in three deaths caused by pneumonia in India. Radiotherapists use the chest X-rays to evaluate and then diagnose pneumonia (Antin and martayan in Detecting Pneumonia in Chest X-Rays with Supervised Learning, 2017). So, developing an automatic system for detecting pneumonia would be beneficial for treating the disease without any delay particularly in remote areas (Das et al. in Micron 45:97,106, 2013; Poostchi et al. in Translational Research, 2018; Ross et al. in Medical and Biological Engineering and Computing 44(5): 427–436, 2006). Deep learning algorithms succeeded in analysing the medical images, and CNN also gained much attention for disease classification (Manoharan in J Artif Intell 2:201–206, 2020). Features learned by the use of pre-trained models of CNN are useful in image classification (Krizhevsky et al. in Advances in neural information processing systems, p 1097110, 2012). By the use of CNN and ANN with customizations in the structure of their layers in custom CNN models and modification of pre-trained models (Tripathi in Journal of Innovative Image Processing (JIIP) 3:100–117, 2021) is done in this study, we have built a binary classification model to predict whether the given chest X-ray image (Sathish in Journal of Soft Computing Paradigm 2(4):216–225, 2020;Tripathi in Journal of Innovative Image Processing (JIIP) 3:100–117, 2021; Adam and Sathesh in Journal of Innovative Image Processing (JIIP) 3(2):85–99, 2021) has pneumonia or not.

Keywords Pneumonia · Streptococcus · Chest x-rays · Convolution neural networks · Artificial neural networks · Radiotherapists · Deep learning · Image classification

S. H. K. Kotturi (✉) · S. V. N. Sreenivasu
Department of CSE, Narasaraopet Engineering College, Narasaraopet, A.P., India

S. Shakya et al. (eds.), *Proceedings of Second International Conference on Sustainable Expert Systems*, Lecture Notes in Networks and Systems 351,
https://doi.org/10.1007/978-981-16-7657-4_19

1 Introduction

PNEUMONIA is one of the dangerous diseases that affects people's lungs caused by bacteria known as Streptococcus pneumonia. It disturbs the air sacs of lungs of the person who is infected majorly. The disease is mainly evolved by various bacteria, fungi, or a virus that infects the air sacs of lungs which fills up with discharge fluids which ultimately takes up to chills, mucus cough, fever, and also breathing problems amongst the persons diagnosed. People who have weak immunity system, especially for children whose age is below five years and elder patients, there is a chance of possible attack of pneumonia disease. In the year 2018, the disease killed over a million. If the pneumonia is not detected or diagnosed, then it will become a life threatening disease. To detect pneumonia, we have different ways like radiography, CT-scan, or MRI. Radiography is used to check the patient's chest if they are infected with pneumonia or not.

Every year, 700,000 deaths occur due to pneumonia in children and also affecting the overall population by 7%. Chest X-rays are mainly used to treat the pneumonia. Sometimes, it is a challenging thing for experienced radiologists to recognise the pneumonia in early stages [1]. So, ultimately, there is the need to improve the accuracy of diagnosis. In this paper, to detect pneumonia, we used an efficient model using digital chest X-ray images [2], which will help radiologists in the decision-making process. Here, we introduced the approach that combines the weighted predictions [3] in an optimal way from the deep learning models such as ResNet50, CNN, AlexNet, GoogLeNet, and VGG16 [1–28]. It is a supervised learning process where the networks predict the result on the basis of dataset. To obtain a higher training and validation, accuracy transfer learning is used [10]. The CNN has an ability in image classification tasks and provides the output with higher accuracy and medical promising results in image analysis. The advantage of CNN [6] is to identify the image features and using these features to generate the accurate results [11, 13, 29]. To reduce the problem of overfitting and increase the accuracy, we majorly use ResNet50, GoogLeNet, and VGG16 on dataset training [11].

Our study developed some algorithms like optimized DL models of CNN [2] which can not only detect but also classify the pneumonia disease in efficient and quick manner [20]. This consists of an optimized convolution neural network models and the experimental analysis of each model towards the pneumonia diseases detection and classification. To increase the balanced way of training dataset, we use partial data augmentation techniques [16]. The methodology that we proposed is able to perform better than the all the available models [16]. The proposed model is considered and evaluated in the terms of accuracy and AUC score. So, the proposed model can be used to aid the radiologists in the diagnosis process and quick diagnosis of pneumonia.

2 Related Work

In the recent times, machine learning (ML) algorithms in the detection of bosom, bust, and peritoneum diseases have gained a very good attention in the medical image classification research area. Recently, Rajpurkar [27] in 2017 explored a method for detecting the pneumonia at a best level than radiotherapists and named their model as ChexNet [27] that uses the DenseNet architecture with 121 layers in it, which helps to detect a variety of diseases from a set of 112,200 images in the dataset. Antin [18] worked on the same dataset and proposed logistic regression model for the detection of pneumonia.

In 2017, Lakhani and Sundaram [26] proposed a method for pulmonary tuberculosis detection by following the GoogleNet and AlexNet architecture. Different methods are introduced to detect the pneumonia using images of chest X-rays. The parameters of deep layered CNN have been changed for pneumonia detection which can be used to obtain high accuracy in detecting the disease. In Yao [26] paper, they have used DilatedNet, R-CNN and for pre-processing they have used Soft-NMS [6] algorithm to ensure the quality. For the better exploration of ML in chest X-ray screening, a larger dataset of front chest X-rays dataset is released. Performance of variants of different CNNs for the abnormality prediction are also proposed.

3 Proposed Work Methodology

This section deals with the detailed description of the proposed methodology. In our proposed methodology, there are three steps involved, and they are as follows:

(a) Pre-Processing stage
(b) Feature-Extraction stage
(c) Prediction stage
(a) **Pre-Processing stage:** The main goal of maintaining the convolutional neural networks in all classification methodologies of images is to reduce the complexity of computation. In our model, for training, we are reducing all the images from their original dimensions to 220 × 220 pixels for the reduction of heavy combinations and for easy processing of the inputs. All the techniques used in this study are applied on the images which are reduced in this step.
(b) **Feature-Extraction stage:** In this step, we have used several neural network architectures to train the models and extract the features from them. All the techniques used in this study are explained detailed in the next section. Techniques used are:

    i. RESNET-50
    ii. ALEXNET
    iii. GOOGLENET
    iv. VGG-16

v. Customized CNN

(c) **Prediction stage:** Based on the training and accuracy of obtaining the features by the models which are trained, we will treat it as the best model and use that particular model for the feature selection process and prediction based on the outcome of feature extraction. According to our study, we are getting a good accuracy value in out customized CNN model with 90.67% accuracy rate.

4 Feature-Extraction Techniques

Classification comes under supervised learning, and it can perform on both structured and unstructured data. It categorizes the set of data into classes and then predicts output for class in data. Some of the algorithms that comes under classification are CNN, AlexNet, GoogleNet, and VGG16. These are used to classify the data and predict the results correctly (Fig. 1).

We have many classification algorithms, but it is not possible to say which one is superior to another. The way we conclude depends on the application and available dataset. Some of classification algorithms are:

1. CNN
2. RESNET-50
3. ALEXNET
4. GOOGLENET
5. VGG16

1. **CNN:** CNN's full form is convolutional neural network; it is a subclass of neural network. We send the images as i/p (input) and assign the weights to various objects for images and be able to differentiate with others. In comparison with other algorithms, CNN requires less pre-processing. There are three layers in CNN. They are input layer, hidden layer, and output layer. When the image

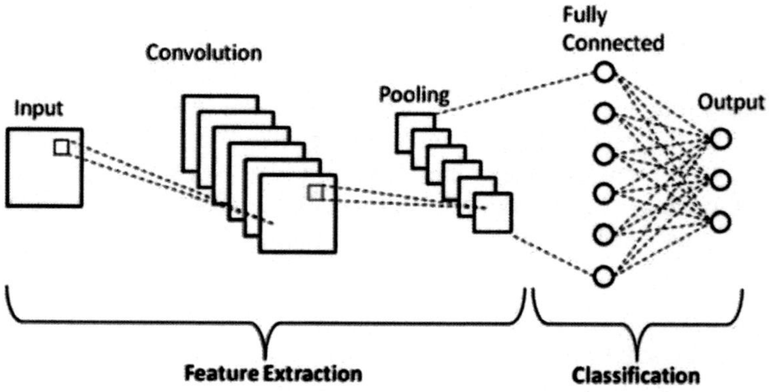

Fig. 1 CNN architecture

given as input, the CNN reduces it in a form which is easy to process it, and here, the image does not lose the features, which are used to get good prediction of image. In CNN, we are also having the pooling layer along with convolution layer. To reduce the computational power which is required to process the data, we use dimensionality reduction and extract the dominant features.

2. **RESNET-50:** Residual Network with 50 layers (ResNet-50). Due to increase of network depth degradation, problem will occur; RESNET is used to solve the problem. This network introduces new terminology that is residual learning. RESNET contains the 1000-way fully connected layer with softmax and global average pooling layer. The network divide into blocks and each block will be divided into either 2- or 3-layer deep. Between the input block and output block, residual mapping acts as bridge in this network. At the end of layers, the network learns several level features like low, mid, and high levels (Fig. 2).

3. **AlexNet:** AlexNet basically contains eight layers. In these eight layers, first five layers are for convolution layers with max pooling layers, and the last three are for fully connected layers. ReLU nonlinearity, multiple GPUs, overlapping pooling, data augmentation, and dropout are the features of AlexNet. By using the ReLU as activation function, it accelerates the training process speed by six times. Overfitting problem can be prevented by dropout layers. Error rate can be

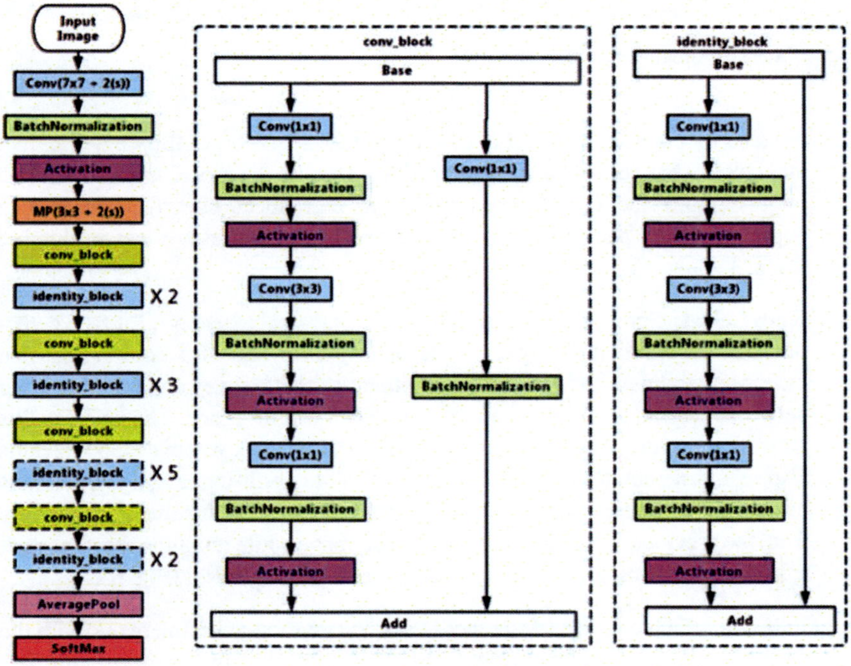

Fig. 2 RESNET-50 architecture

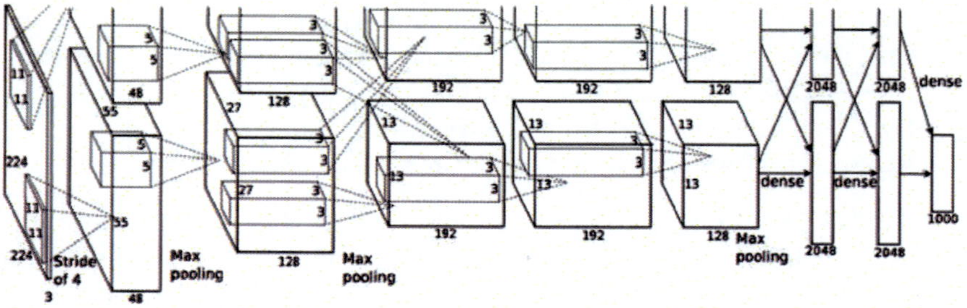

Fig. 3 AlexNet Architecture

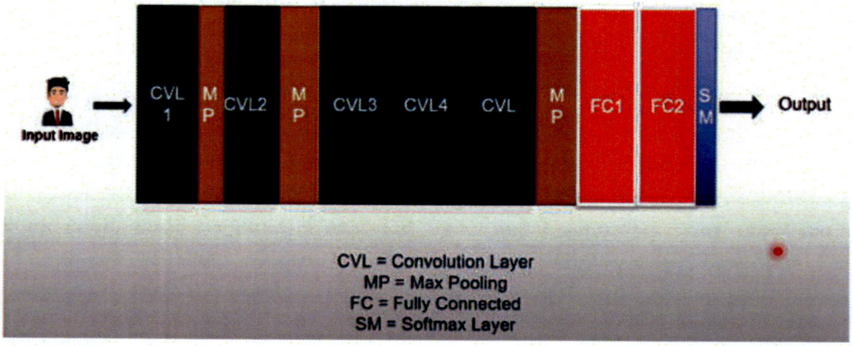

Fig. 4 AlexNet-Simplified Architecture

reduced by pooling. AlexNet added image augmentation of flipping, clipping, and colour changes for data augmentation. It makes the model more robust and larger sample size. Overfitting problem can be reduced effectively when data augmentation is performed for the classes of images that are less in number (Figs. 3, 4 and 5).

4. **GoogLeNet:** GoogLeNet architecture has 22 layers. Nowadays, for face detection and recognition, GoogLENet is mainly used and also consists of nine inception modules. To increase the computational efficiency compared to other networks created at the same time, GoogLENet was designed. The layers in this network are average pooling, droput, linear, and softmax where dropout layer is also regularization technique which can prevent overfitting; average pooling is used to reduce the width and height of input image these feature mappings are from the previous layers. Reduction of input image while retaining the important spatial information is the efficiency achieved by GoogLeNet (Fig. 6).

5. **VGG16:** VGG16 is one of the models of convolution neural network with the deep of 16 layers. The layers consist of layers like conv, max pooling, softmax, and pooling. It is majorly used in image classification problems of deep learning

	Layer	Feature Map	Size	Kernel Size	Stride	Activation
Input	Image	1	227x227x3	-	-	-
1	Convolution	96	55 x 55 x 96	11x11	4	relu
	Max Pooling	96	27 x 27 x 96	3x3	2	relu
2	Convolution	256	27 x 27 x 256	5x5	1	relu
	Max Pooling	256	13 x 13 x 256	3x3	2	relu
3	Convolution	384	13 x 13 x 384	3x3	1	relu
4	Convolution	384	13 x 13 x 384	3x3	1	relu
5	Convolution	256	13 x 13 x 256	3x3	1	relu
	Max Pooling	256	6 x 6 x 256	3x3	2	relu
6	FC	-	9216	-	-	relu
7	FC	-	4096	-	-	relu
Output	FC	-	1000	-	-	Softmax

Fig. 5 AlexNet Dimensions

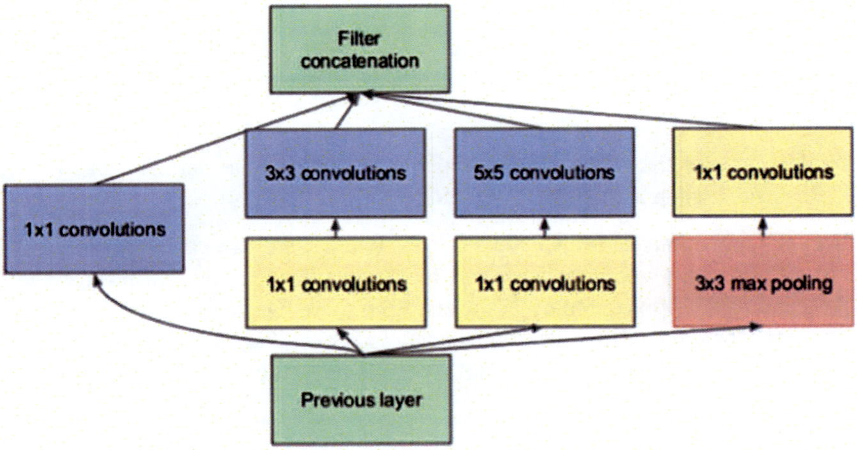

Fig. 6 GoogLeNet architecture

and also very easy to implement. The network of VGG16 uses the same techniques which are used in AlexNet in the training the image. The training process is carried out using mini batch gradient descent based on backpropagation. For first two fully connected layers of VGG16, dropout regularization was added to make sure that there is no loss of features from the extraction of images present in the dataset collected and used (Fig. 7).

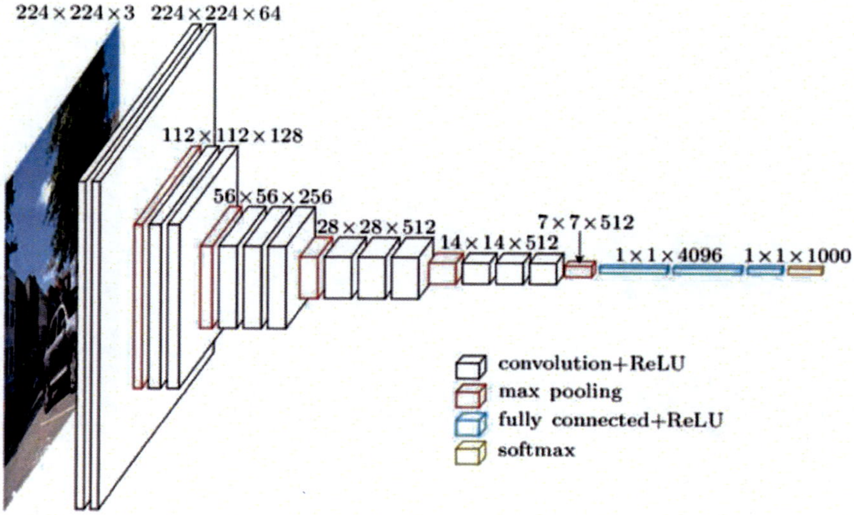

Fig. 7 VGG-16 architecture

5 Result Analysis

We can observe the below images which will show us the normal human lungs X-Rays and human lungs with PNEUMONIA (Figs. 8 and 9).

(A) **Result Analysis for Custom CNN Model:** Here, we constructed our own CNN model with eight layers in it. It consists of five convolution 2D layers each followed by max pooling 2D layers. After these five layers, model is flattened and then three dense layers are added with 1024, 512, and 1 as their features. All the layers, except the output layer, the activation function used is 'ReLU'

Fig. 8 Normal lungs X-ray

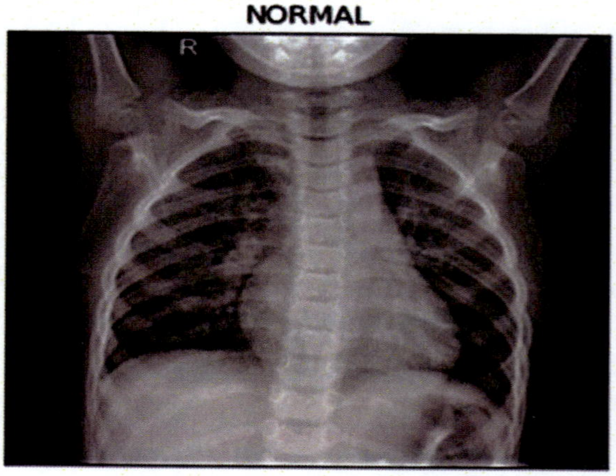

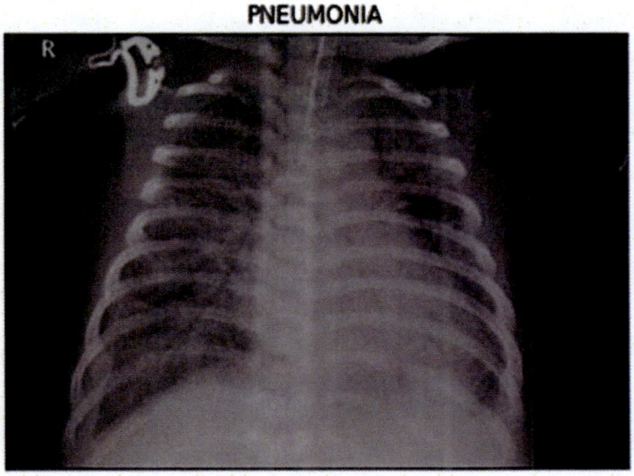

Fig. 9 Lungs with PNEUMONIA X-ray

and padding is marked as 'same'. The output layer's activation function is used as 'sigmoid'. Optimizer used here is RMSProp with a learning rate of 0.001. Binary cross entropy is used as loss factor, and accuracy, precision, recall are used as metrics. ImageDataGenerators are used to generate the images with width and height shift range of 0.2 each. Rescaling is done for the images with 1.0/255. A rotation of maximum of 30 is done along with zoom range of 0.2. All the images used here are in grayscale mode only. All the graphs related to custom CNN models training and validation accuracy, recall, precision, and loss are shown in the Figs. 10, 11, 12 and 13.

Our custom CNN model has provided the following results as metrics.
Accuracy: 0.9070512652397156
Loss: 0.6959954500198364
Precision: 0.8933649063110352
Recall: 0.9666666388511658

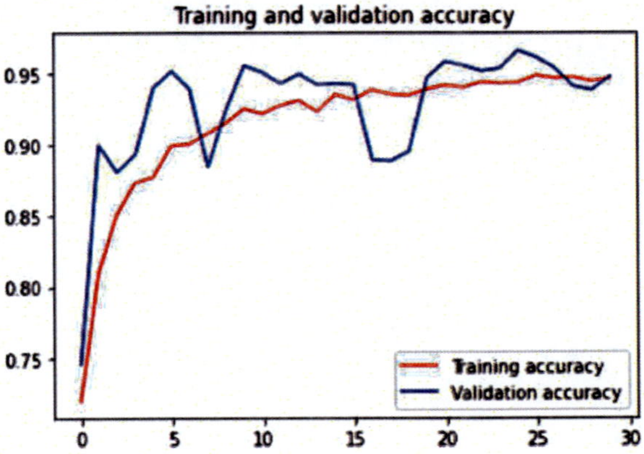

Fig. 10 Training and validation accuracy

Fig. 11 Training and validation precision

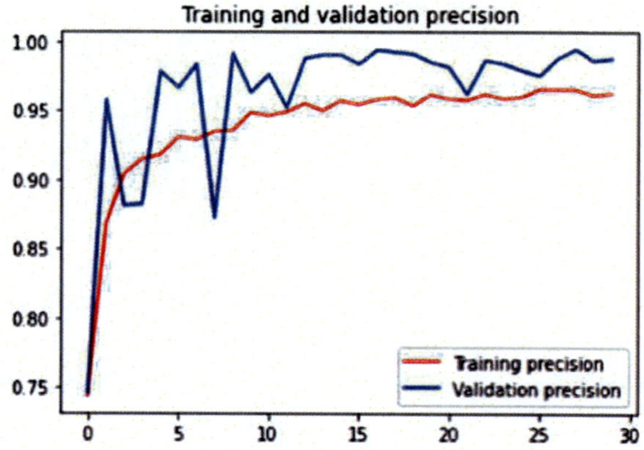

Fig. 12 Training and validation recall

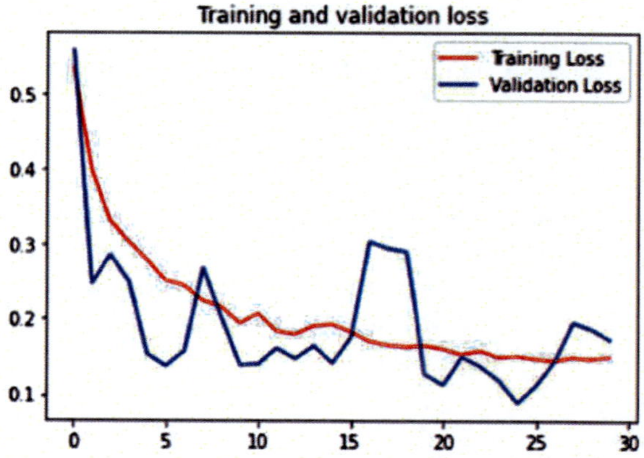

Fig. 13 Training and validation loss

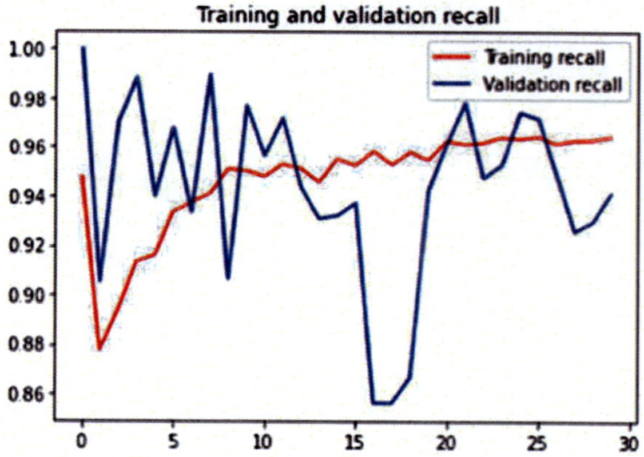

Result Analysis of ResNet-50: Here, we are using the ResNet-50 pre-trained model provided by keras and added some layers on top of it. The pre-trained model will take an input of 3-channel, but all our images are grayscale mode. So, in order to avoid the mismatch of the shape, we'll let our image data generator use the default color_mode as RGB instead of grayscale model with rescaling of 1.0/255, rotation range of 30, width and height shift range of 0.2 along with zoom_range of 0.2.

The output layer of the pre-trained model is taken and flattened before adding the new layers. Three dense layers are added on the top of ResNet-50 pre-trained model with 1,024,256 and 1 as output neurons. For the first two layers, 'ReLU' is used as activation function, and for the last layer, we are using 'sigmoid' activation function.

It is used along with RMSProp as optimizer with initial learning rate as 0.001 and binary cross entropy as loss factor along with accuracy, precision, and recall as metrics of its learning (Figs. 14, 15, 16 and 17).

Customized ResNet-50 model has given the following as metrics:

Fig. 14 Training and validation accuracy

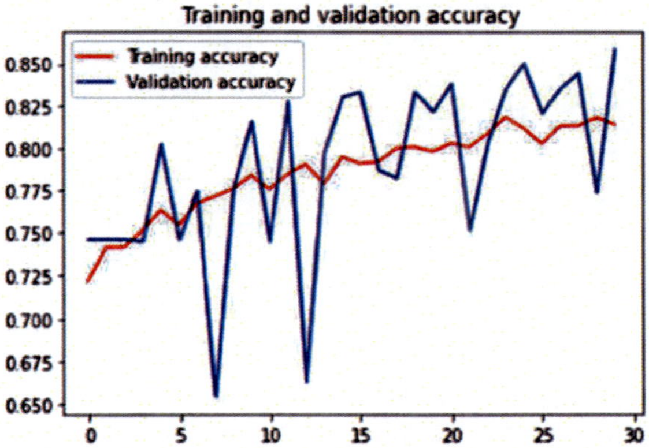

Fig. 15 Training and validation precision

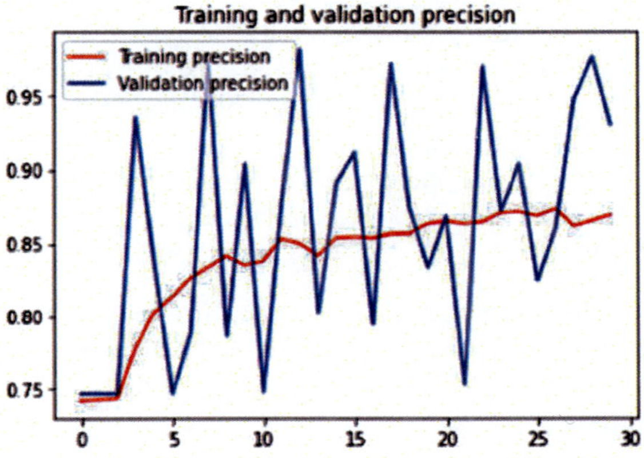

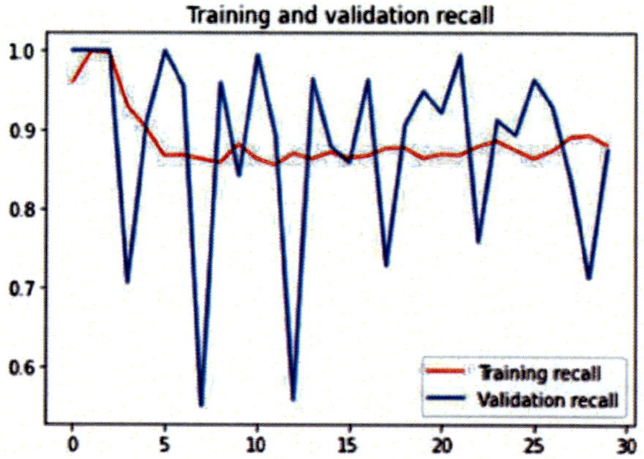

Fig. 16 Training and validation recall

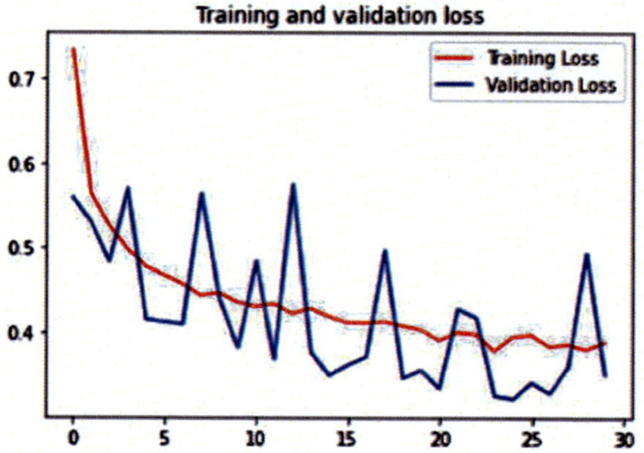

Fig. 17 Training and validation loss

Accuracy: 0.81891024112701142
Loss: 0.4258221983909607
Precision: 0.825882372403259
Recall: 0.8999999761581421
Result Analysis of Inception Model: Here, we are using the Inception pre-trained model provided by keras and added some layers on top of it. The pre-trained model will take an input of 3-channel, but all our images are grayscale mode. So, in order to avoid the mismatch of the shape, we'll let our image data generator use the default color_mode as RGB instead of grayscale model with rescaling of 1.0/255, rotation range of 30, width and height shift range of 0.2 along with zoom_range of 0.2.

The output layer of the pre-trained model is taken and flattened before adding the new layers. Three dense layers are added on the top of Inception pre-trained model with 1,024,256 and 1 as output neurons. For the first two layers, 'ReLU' is used as activation function, and for the last layer, we are using 'sigmoid' activation function.

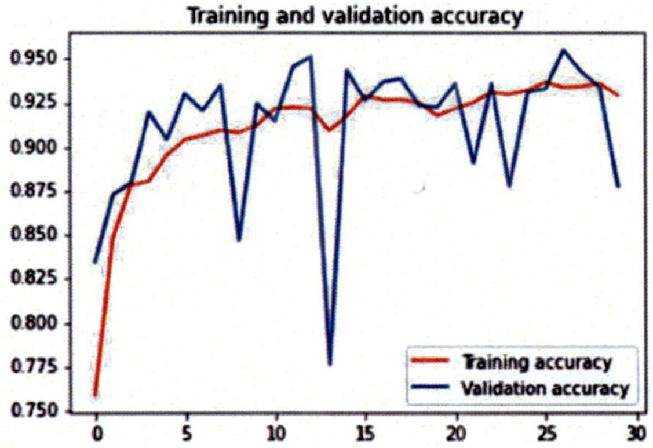

Fig. 18 Training and validation accuracy

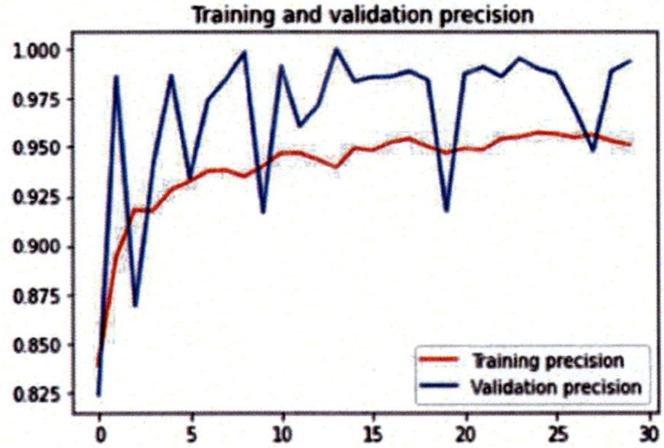

Fig. 19 Training and validation precision

It is used along with RMSProp as optimizer with initial learning rate as 0.001 and binary cross entropy as loss factor along with accuracy, precision, and recall as metrics of its learning (Figs. 18, 19, 20 and 21).

Customized Inception model has given the following as metrics:

Accuracy: 0.8605769276618958
Loss: 0.34215420484542847
Precision: 0.8955613374710083
Recall: 0.879487156867981

Out of all the three models mentioned here, our customized CNN Model has provided more accuracy than the other two models. It has given an accuracy of **0.9070512652397156**. We are still trying to improve the results to get more accuracy in very less time.

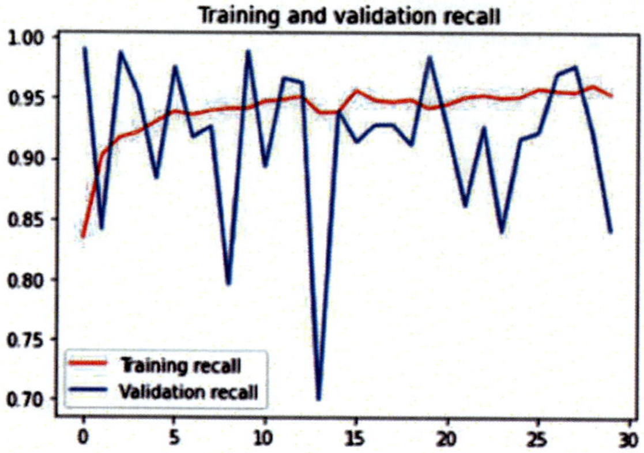

Fig. 20 Training and validation recall

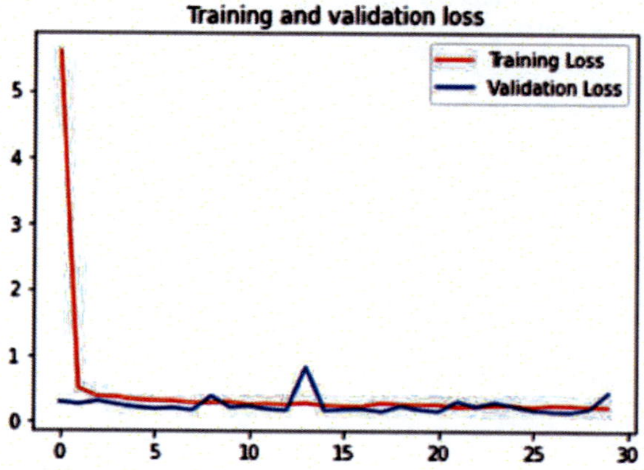

Fig. 21 Training and validation loss

6 Conclusion and Future Scope

To diagnose any kind of disease, presence of experts is one of the most necessary condition. Especially in the case of thoracic diseases, we need the expert radiologists. We are primarily aiming to improve the medical adaptiveness in the places where the radio therapists are still less and limited. Through our study, we can diagnose the pneumonia in the very early stage and prevent the consequences which are adverse including the death of the patient.

We are still in the development of our own customized CNN Model which provides more accuracy in predicting pneumonia. In future, we want to make this model to predict the percentage of lungs that is affected with pneumonia and also which aims to predict it in very less time.

7 Limitations

Since we are using the convolutions layers which automatically consumes more amount of memory to process, we need high computational capable systems so that the results are expected to be very fast and more accurate.

References

1. CDC URL: https://www.cdc.gov/features/pneumonia/index.html
2. Tripathi M (2021) Analysis of convolutional neural network based image classification techniques. J Innov Image Process (JIIP) 3(02):100–117
3. Hua K-L, Hsu C-H, Hidayati SC, Cheng W-H, Chen Y-J (2015) Computer-aided classification of lung nodules on computed tomography images via deep learning technique. OncoTargets Ther 8
4. Rajpurkar P, Irvin J, Zhu K, Yang B, Mehta H, Duan T, Ding D, Bagul A, Langlotz C, Shpanskaya K et al (2017) Chexnet: radiologist-level pneumonia detection on chest x-rays with deep learning. arXiv preprint arXiv:1711.05225
5. Manoharan JS (2020) Improved version of graph-cut algorithm for CT images of lung cancer with clinical property condition. J Artif Intell 2(4):201–206
6. Sathish (2020) Adaptive shape based interactive approach to segmentation for nodule in lung CT scans. J Soft Comput Paradigm 2(4):216–225
7. Krizhevsky A, Sutskever I, Hinton GE (2012) Imagenet classification with deep convolutional neural networks. In: Advances in neural information processing systems, p 10971105
8. Adam EEB, Sathesh A (2021) Construction of accurate crack identification on concrete structure using hybrid deep learning approach. J Innov Image Process (JIIP) 3(02):85–99
9. Simonyan K, Zisserman A (2014) Very deep convolutional networks for large-scale image recognition. arXiv preprint arXiv:1409.1556
10. Chollet F (2016) Xception: deep learning with separable convolutions. arXiv Preprint arXiv:1610.2357
11. He K, Zhang X, Ren S, Sun J (2016) Deep residual learning for image recognition. In: Proceedings of the IEEE conference on computer vision and pattern recognition, p 770778
12. Huang G, Liu Z, Van Der Maaten L, Weinberger KQ (2017) Densely connected convolutional networks. CVPR 1:3
13. Mohsen H, El-Dahshan E-SA, El-Horbaty E-SM, Salem ABM (2017) Classification using deep learning neural networks for brain tumors. Future Comput Inf J
14. Lakhani P, Sundaram B (2017) Deep learning at chest radiography: automated classification of pulmonary tuberculosis by using convolutional neural networks. Radiology 284(2):574582
15. Islam MT, Aowal MA, Minhaz AT, Ashraf K (2017) Abnormality detection and localization in chest x-rays using deep convolutional neural networks. arXiv preprint arXiv:1705.09850
16. Openi Dataset (2016). https://openi.nlm.nih.gov
17. Wang X, Peng Y, Lu L, Lu Z, Bagheri M, Summers RM (2017) Chestx-ray8: hospital-scale chest x-ray database and benchmarks on weakly-supervised classification and localization of common thorax diseases. In: 2017 IEEE conference on computer vision and pattern recognition (CVPR), IEEE, p 34623471
18. Antin B, Kravitz J, Martayan E (2017) Detecting pneumonia in chest x-rays with supervised learning
19. Kumar P, Grewal M, Srivastava MM (2018) Boosted cascaded convents for multilabel classification of thoracic diseases in chest radiographs. In: International conference image analysis and recognition, Springer, p 546552

20. Li Z, Wang C, Han M, Xue Y, Wei W, Li L-J, Fei-Fei L (2017) Thoracic disease identification and localization with limited supervision. arXiv preprint arXiv:1711.06373
21. 2018. Kaggle URL :. (2018). https://www.kaggle.com/nih-chestxrays/data
22. Raoof S, Feigin D, Sung A, Raoof S, Irugulpati L, Rosenow III EC (2012) Interpretation of plain chest roentgenogram. Chest 141(2):545558
23. Nijhawan R, Das J, Balasubramanian R (2018) A hybrid CNN + random forest approach to delineate debris covered glaciers using deep features. J Indian Soc Remote Sens 46:1–9
24. Nijhawan R, Verma R, Bhushan S, Dua R, Mittal A (2017) An integrated deep learning framework approach for nail disease identification. In: 2017 13th international conference on signal-image technology and internet based systems (SITIS), IEEE, p 197202
25. Nijhawan R, Rishi M, Tiwari A, Dua R. A Novel Deep Learning Framework Approach for Natural Calamities Detection. In Information and Communication Technology for Competitive Strategies 2019 (pp. 561–569). Springer, Singapore.
26. Yao S, Chen Y, Tian X, Jiang R (2021) Pneumonia detection using an improved algorithm based on faster R-CNN
27. Lakhani P, Sundaram B (2017) Deep learning at chest radiography: automated classification of pulmonary tuberculosis by using convolutional neural networks. Radiology 284(2)
28. Rajpurkar P, Irvin J, Zhu K, Yang B, Mehta H, Duan T, Ding D, Bagul A, Langlotz C, Shpanskaya K et al (2017) Chexnet: radiologist-level pneumonia detection on chest x-rays with deep learning
29. Razavian AS, Azizpour H, Sullivan J, Carlsson S (2014) CNN features off-the-shelf: an astounding baseline for recognition. In: Proceedings of the IEEE conference on computer vision and pattern recognition workshops, p 806813
30. Das DK, Ghosh M, Pal M, Maiti AK, Chakraborty C (2013) Machine learning approach for automated screening of malaria parasite using light microscopic images. Micron 45(2013):97106
31. Poostchi M, Silamut K, Maude R, Jaeger S, Thoma G (2018) Image analysis and machine learning for detecting malaria. Translational Res
32. Ross NE, Pritchard CJ, Rubin DM, Duse AG (2006) Automated image processing method for the diagnosis and classification of malaria on thin blood smears. Med Biol Eng Comput 44(5):427436

Blockchain Technology in Health Care Sector: An Experimental Approach

R. Jaichandran, H. Abdul Shabeer, Tony K. John, K. P. Karan, and Aravind Krishnan

Abstract Due to the innovation in the technology field, many researches are happening in the field of blockchain technology, and its scope in different industries are increasing. Blockchain technology is a chain of blocks which provides a secure environment between the users. To address the security, trust issues, many research studies are attracted toward blockchain in customizing and applying for their working process. In this paper, we propose how blockchain technology can be adopted in healthcare sector. Blockchain could provide secure storage of patient health information's. The security and privacy for patient's information are provided using proof of work (PoW). The blockchain network has been validated by miners, and cryptocurrency is used for transactions with security and transparency.

Keywords Healthcare · Blockchain · Patient health records

1 Introduction

As we are in the move of achieving digital India, the current healthcare sector demands to preserve patient health information digitally for easy access. Also, the major concern is the security and patient's identity preserving. Patient's privacy preserving would gain trust and confidentiality toward the system. To attain patient's confidentiality and trust, a secure proposed system invoking blockchain technology is needed. Also, aside lot of health devices are been implemented in which the patient's health data need to be securely transmitted and stored in a secure manner preventing intrusions [1].

In the current health record maintenance system, the patient's health records are stored and carried manually by the patients which always have the risk of theft, damages. Also, inspecting the health records which are maintained manually invoking much time for the doctors and usually they see the recent prescriptions,

R. Jaichandran (✉) · H. A. Shabeer · T. K. John · K. P. Karan · A. Krishnan
Department of Computer Science and Engineering, Aarupadai Veedu Institute of Technology, Vinayaka Mission's Research Foundation (Deemed to be University), Paiyanoor, Chengalpattu District, Tamil Nadu 603104, India

S. Shakya et al. (eds.), *Proceedings of Second International Conference on Sustainable Expert Systems*, Lecture Notes in Networks and Systems 351,
https://doi.org/10.1007/978-981-16-7657-4_20

medications only. Thus, the system providing historical-based approach of arranging patient's health records and consolidating in a historical-wise format would help the doctors to view the complete patient's history and their medicinal treatments undergone. Also, this system can be able to provide insight to the doctors about the allergic drugs within a short time. Thus, making the decision-making process for the doctors is easy. However, these patient's health data have to be stored and maintained securely, thus blockchain-based patient health record storage can be able to provide complete secure storage against unauthorized access with transparency [2].

The scope of blockchain technology has attracted many industries andacademic research. The main advantages of blockchain technology are distributed ledger in which complete transparency is maintained. The blockchain network would be public or private networks. The privacy of the patients is maintained by using public address. The security and user privacy are maintained by integration of smart contracts. The accessing rights and roles are predefined in the smart contracts. Smart contracts are defined as the set of rules which is defined in prior on mutual agreement by both the parties. Thus, blockchain technology with smart contracts, ledger, proof of work, provides secure data transmission and storage reducing the management cost and also rendering complete security, transparency [3]. Also, blockchain-based healthcare would eliminate the third parties' intermediates.

This paper focuses on performing an experimental result to adopt blockchain technology in healthcare sector to render security and preserve privacy of patients.

2 Related Works

In this paper, they have proposed utilization of blockchain technology in the healthcare sector. This paper states, this was the first study paper on explaining blockchain technology toward healthcare sector on a broad perspective. This paper was an explanatory paper majorly focusing on the blockchain technology functions and its advantages for the management and users [4].

This research paper explains how blockchain technology can be applied for the healthcare and biomedical sector for benefiting human lives. This paper also discusses about the cryptocurrencies and its advantages. This paper extends how blockchain technology can be adopted for health data storage, transmission management, patient insurance process, and research and development process in the health sector. This paper doesn't invoke practical approach [5].

This paper briefs about the integration of blokchain technology in the healthcare sector. This paper also explains how the patient's health records can be transmitted securely using proof of work concepts. It discusses about the hash value and block creation techniques invoked with the blockchain technology, but this paper missed to analyze how practically blockchain is adopted in the healthcare sector with results and tools [6].

This research paper elaborates a detailed approach of how blockchain can be integrated to the healthcare sector. In this, the authors discuss about blockchain technology and its advantages in the medicinal field. This paper invokes how blockchain can be able to automate the manual process rendering security in other medical verticals like drug identification process, supply chain, maintain health records, health insurance processing, and medicinal raw material procurement process. This paper missed to explain few important aspects like smart contracts, sharing of data in detail [7].

This paper explains adopting of new blockchain technology in the medicinal research areas like neuroscience and medicinal researchers to keep the research progress maintained securely. This paper also explains how blockchain can be able to prevent the external intrusions [8].

This paper explains attribute-based encryption policy along with blockchain technology [9]. Through the blockchian technique of nominating, the miners for the blockchain network could be able to validate the transactional requests and the patient, doctor nodes. This paper also explains prevention of collusion attack while group messaging. There are various science and applications uses diverge algorithms and systems ranging from blockchain, health, wireless networks, and data analysis [10–17].

3 Proposed Methodology

In the proposed methodology, we have used registration and login process for patients and doctors. In this step, we collect essential information about patients invoking personal and medical. Once the health information's are provided by the patients, hash value has been generated for the data block along with the time stamp are shown in Fig. 1. Thus, all patient records would be interconnected through the proof of work concept in blockchain.

In this proposed study, we use ganache-based public blockchain for experimental results. Once the block is created, it would be added to the blockchain after validation. Miners inside the blockchain network validate the request and accept nor reject the request. Accepted request gets added into the blockchain network, and transaction has been initiated. The hash value has been generated using SHA algorithm. This hashing technique generates unique hash values for different inputs are shown in Fig. 2.

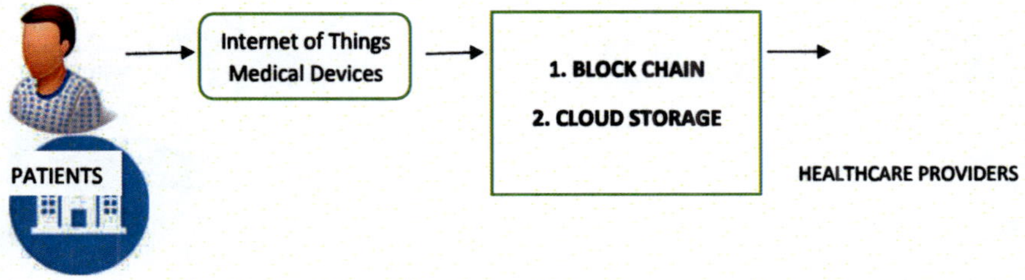

Fig. 1 Internet of things for medical applications

Fig. 2 Proof of work

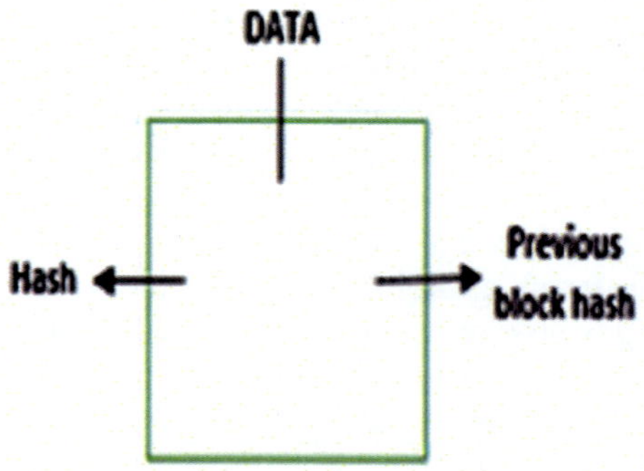

Another important feature in our proposed project study is that smart contract technique integration within our healthcare system to protect patient's privacy. We propose ethereum remix for writing the contracts. All blockchain users are connected with the smart contract using solidity language. Thus, before each transaction, the smart contract would be matched and validated. As said, we used ganache-based public blockchain for the local project development are shown in Fig. 3.

For the cryptocurrency, we used metamask as the middleware. Metamask is a cryptocurrency wallet to store and access patients cryptocurrencies. Ethers are deducted for each user transactions. Thus, our proposed study renders secure blockchain-based healthcare system invoking security and privacy for patient's health data.

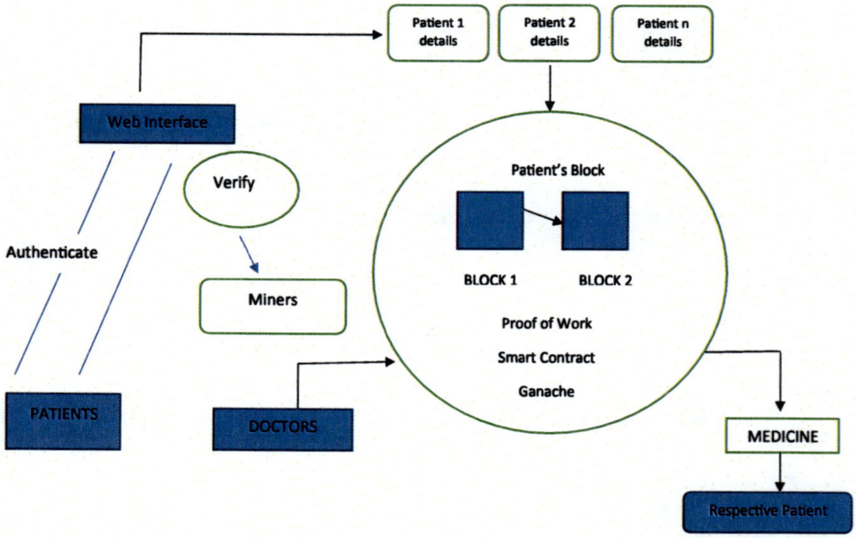

Fig. 3 Proposed architecture

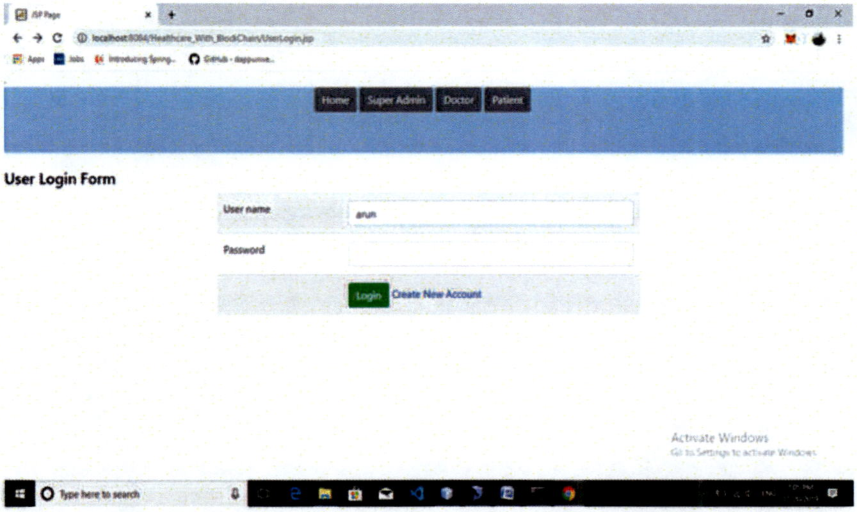

Fig. 4 Patient's login

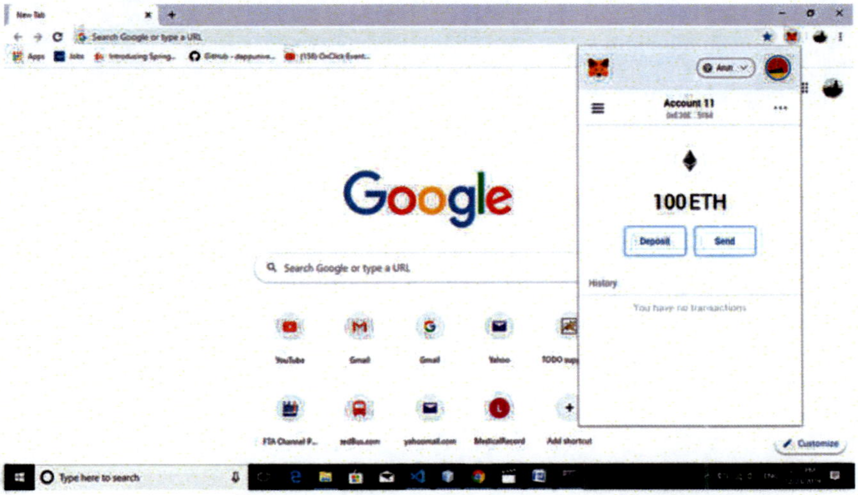

Fig. 5 Metamask: cryptocurrency wallet

4 Result and Discussions

In this section, the experimental session explains our proposed study approach on public blockchain-based healthcare system. Figure 4 explains the patient's login for authentication and once after authentication, the patients can provide their medical and personal information's. Figure 5 explains the cryptocurrency wallet which is used to store the free ethers. The free ethers can be bought by using social media-based technique using the networks available. Figure 6 explains the blockchain network accounts, and their privacy is preserved by identifying each with private keys. Figure 7 explains the block creation and hash value generation using SHA algorithm.

In Table 1 and Fig. 8, we have analyzed the block creation execution time in our Netbeans console for each patient's blocks. For performance analysis, we have analyzed the execution time for a block and 10 blocks during creation.

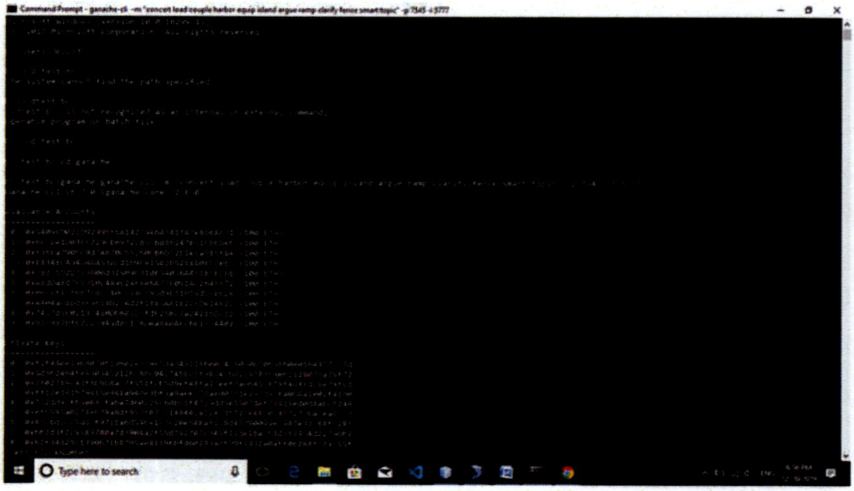

Fig. 6 Public blockchain: ganache

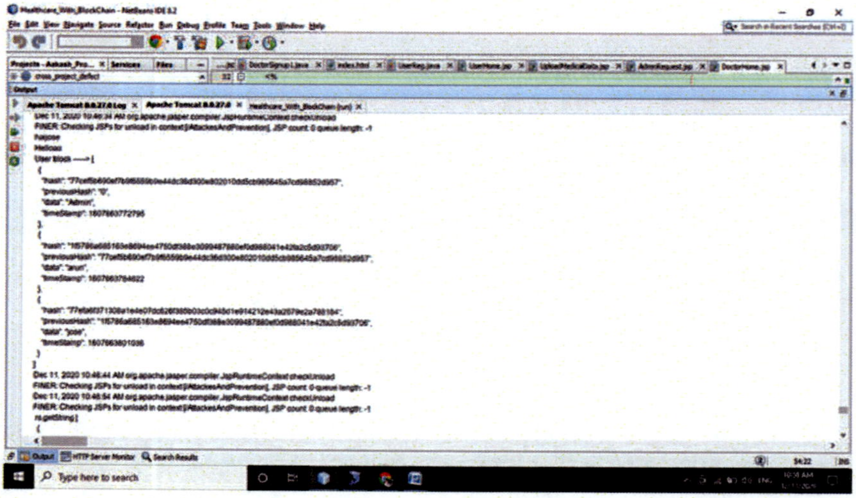

Fig. 7 Hash value generation for patients block

Table 1 Our proposed system block creation performance

Blocks	Execution time (s)
1	0.43
10	0.45

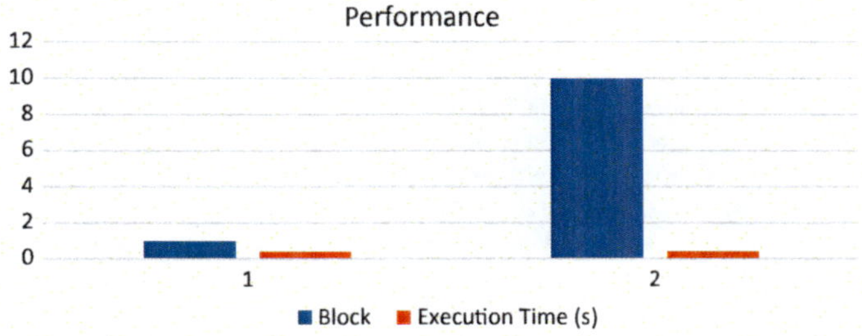

Fig. 8 Our proposed system block creation performance

5 Conclusion

This experimental paper explores study of blockchain technology in healthcare sector. Thus, we identified for security, privacy preserving many individuals, organizations, researches are migrating to blockchain technology. This experimental paper highlights how blockchain is applied in protecting patient's health data. The blockchain is made secure against intrusion, unauthorized data access through its important aspects like smart contracts, distributed ledger, decentralized approach, and proof of work. Few areas in the blockchain network such as flexibility, scalability need to be addressed. Once these aspects are being addressed, the blockchain network would be complexly adopted in the software products for enterprises. For experimental results, we used JavaScript for programming and Netbeans, MySQL as the software tool.

References

1. Al Ridhawi I, Aloqaily M, Kantarci B, Jararweh Y, Mouftah HTA (2018) Continuous diversified vehicular cloud service availability framework for smart cities. Comput Netw 145:207–218
2. Zhang P, Schmidt DC, White J, Lenz G (2018) Blockchain technology use cases in healthcare. In: Advances in computers, vol 111. Elsevier, Amsterdam, The Netherlands, pp 1–41
3. Casino F, Dasaklis TK, Patsakis C (2019) A systematic literature review of blockchain-based applications: current status, classification and open issues. Telemat Inform 36:55–81
4. Jiang S, Cao J, Wu H, Yang Y, Ma M, He J (2018) BlocHIE: a blockchain-based platform for healthcare information exchange. In: Proceedings of the 2018 IEEE international conference on smart computing (SMARTCOMP), Sicily, Italy, pp 49–56
5. Ji Y, Zhang J, Ma J, Yang C, Yao X (2018) BMPLS: blockchain-based multi-level privacy-preserving location sharing scheme for telecare medical information systems. J Med Syst 42:147

6. Uddin MA, Stranieri A, Gondal I, Balasubramanian V (2018) Continuous patient monitoring with a patient centric agent: a block architecture. IEEE Access 6:32700–32726
7. Mamoshina P, Ojomoko L, Yanovich Y, Ostrovski A, Botezatu A, Prikhodko P, Izumchenko E, Aliper A, Romantsov K, Zhebrak A et al (2017) Converging blockchain and next-generation artificial intelligence technologies to decentralize and accelerate biomedical research and healthcare. Oncotarget 9:5665–5690
8. Zhao H, Bai P, Peng Y, Xu R (2018) Efficient key management scheme for health blockchain. CAAI Trans Intell Technol 3:114–118
9. Cunningham J, Ainsworth J (2017) Enabling patient control of personal electronic health records through distributed ledger technology. Stud Health Technol Inf 245:45–48
10. Prakash UM, Kottursamy K, Cengiz K, Kose U, Hung BT (2021) 4x-expert systems for early prediction of osteoporosis using multi-model algorithms. Measurement 180:109543
11. Kottursamy K, Raja G, Saranya K (2016) A data activity-based server-side cache replacement for mobile devices. In: Artificial intelligence and evolutionary computations in engineering systems, Springer, New Delhi, pp 579–589
12. Raja G, Kottursamy K, Chaudhary SH, Hassan A, Alqarni M (2017) SDN assisted middle box synchronization mechanism for next generation mobile data management system. In: 2017 IEEE smart world, ubiquitous intelligence and computing, advanced and trusted computed, scalable computing and communications, cloud and big data computing, internet of people and smart city innovation (Smart World/SCALCOM/UIC/ATC/CBD Com/IOP/SCI), IEEE, pp 1–7
13. Ponnusamy V, Kottursamy K, Karthick T, Mukeshkrishnan MB, Malathi D, Ahanger TA (2020) Primary user emulation attack mitigation using neural network. Comput Electr Eng 88:106849
14. Raja G, Kottursamy K, Theetharappan A, Cengiz K, Ganapathisubramaniyan A, Kharel R, Yu K (2020) Dynamic polygon generation for flexible pattern formation in large-scale UAV swarm networks. In: 2020 IEEE globecom workshops (GC Wkshps), IEEE, pp 1–6
15. Mehra R, Pachpor K, Kottilingam K, Saranya A (2020) An initiative to prevent Japanese encephalitis using genetic algorithm and artificial neural network. In: 2020 International Conference on Computational Intelligence (ICCI), IEEE, pp 142–148
16. Saranya A, Kottilingam K (2021) A survey on bone fracture identification techniques using quantitative and learning based algorithms. In: 2021 International Conference on Artificial Intelligence and Smart Systems (ICAIS), IEEE, pp 241–248
17. Nallarasan V, Kottilingam K (2021) Spectrum management analysis for cognitive radio IoT. In: 2021 International Conference on Computer Communication and Informatics (ICCCI), IEEE, pp 1–5

Data Security Framework Using Block Chain Technology

R. Jaichandran, R. Shobana, S. Muthamil Selvan, Medagoni Rohan Sai, and Gunupati Lakshmi Sai Vikas Kumar Reddy

Abstract Information technology plays an important role in processing the information and securing the data. Many universities and industries with small scale and large scale are using this frame work for their implementations. The block chain with the hyper ledger framework helps to build the trust between the industries and universities. This paper provides the link of block chain with cryptography algorithms. Some companies are not effective to predict the fraud cases in terms of academic certificates, credits or percentages. This application finds out the fraud candidates and fraud bio data using advanced techniques.

Keywords Telugu language · FWT · HMM · DNN

1 Introduction

Industries are related with education. Applicable knowledge leads to good production in industries. The efficiency in the education is improved with proper practical knowledge. More over many educational intuitions are covering the enough practical knowledge but there is a gap in the live project implementations [1]. There is no clarity in the credit systems used by the universities there need to be an improvement done in sharing the credits. Companies can see the fraud in academics and as well as in their bio data. Some companies are not effective to predict the fraud cases in terms of academic certificates, credits or percentages. Due to this the companies or government agencies cannot obtain actual stuff of students. This cannot generate the trust between the passes out candidates and industries [2]. Technology like information technology helps to defend this situation. A reliable network can be generated for

R. Jaichandran (✉) · R. Shobana · S. Muthamil Selvan · M. R. Sai · G. L. S. V. K. Reddy
Department of Computer Science and Engineering, Aarupadai Veedu Institute of Technology, Vinayaka Mission's Research Foundation (Deemed to be University), Paiyanoor, Chengalpattu District, Tamil Nadu 603104, India

R. Shobana
e-mail: shobana@avit.ac.in

S. Shakya et al. (eds.), *Proceedings of Second International Conference on Sustainable Expert Systems*, Lecture Notes in Networks and Systems 351,
https://doi.org/10.1007/978-981-16-7657-4_21

the companies by using the advanced technology like block chains. Internet which was invented and used by many users helps the technology to interconnect.

Previously the World Wide Web is used by defence only but now every individual know the applications and usage. Information can be easily exchanged using the World Wide Web. The hi-tech world is created using the Internet. This has been advanced with the block chain technology. The technology is user friendly and easily operated by any user across the universe. The technology is very simple and feasible to operate from anywhere. Biodata fraud or certificates fraud can easily determine and which is very useful for the companies or government agencies. The accuracy need to high to determine the fraud cases. Starting only the prediction can be done. Transparency parameters are considered for determining the accuracy. There are many features which are generated but the useful parameters are only considered [3]. Block chain plays an important role here for obtaining the accuracy. Hyperledger block chain system or network is used for implementing this technology.

This hyperledger network is very flexible and provides promising results in terms of accuracy, operation speed and cost. Many universities and industries with small scale and large scale are using this frame work for their implementations. The block chain with the hyperledger framework helps to build the trust between the industries and universities. Factors like decentralization, transparency and immutability are used for generating the application [4]. Not only the higher educational institutions, even the school with international standards are following the same procedure to upgrade themselves. Many companies are selecting the candidates from the institutions which follow the protocols mentioned above. This scenario helps the organization or the world to improve perfectly without any loop holes [5].

2 Literature Review

Across organizations and industries data maintenance plays a vital role. In large level organizations the proposed technology is in use to maintain data security, eliminate vulnerability/redundancy of data, integrity [6]. These advantages are retained in this method by storing of data information in different nodes in and across the database [2], where relationship between them is a main aspect. In this method of technology data transparency is a major advantage for the companies so they have up-to-date information about any records where as in worldwide manufacturing industries, banking sectors, educational institutions and so on [8]. Though it is a high level technology it is a successful technology implemented for the major companies in an aspect of users' satisfaction and with return on investment.

3 Proposed System

The main objective is to computerize the results of institutional details and biodata for industries. This should be obtained as users can interact easily, with low cost, and high speed. All the analysis is done with the information technology using block chains. The relationships are very important in the network. So the main objective is to design an application based on blockchains to find the fraud cases in terms of educational qualifications.

Blockchain is a simple term; it is a time-stamped series of immutable record of data that is managed by cluster of computers not owned by any single entity. Cryptography is one of the major areas where the blockchains are integrated with advanced algorithms. Each of these blocks of data are secured and bound to each other using cryptographic principles as in Fig. 1. For an even more detailed explanation check out our guide on blockchain explained. The blockchain network has no central authority-it is the very definition of a democratized system [6]. Since it is a shared and immutable ledger, the information in it is open for anyone and everyone to see. Hence, anything that is built on the blockchain is by its very nature transparent and everyone involved is accountable for their actions [7].

The major properties used in the blockchain technology by considering all the factors are given below.

- Decentralization
- Transparency
- Immutability
- *Decentralization*

Many users heard about bit torrent and bit coin; this played a vital role in blockchain communication. The services are centralized in these applications. The information is stored and centralized [8]. The idea is simple and short. The data can be processed by a single user or n number of users. Many applications are there which used in centralized format. Banks are the best examples for decentralization as shown in Fig. 2 and Table 1. The transaction can be done easily from one user to other and

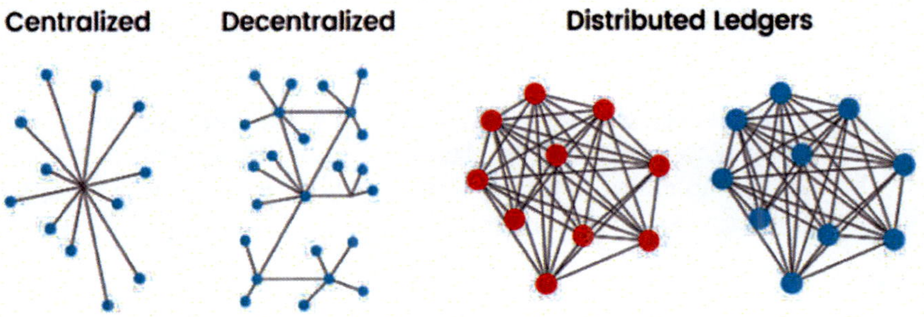

Fig. 1 Blockchain

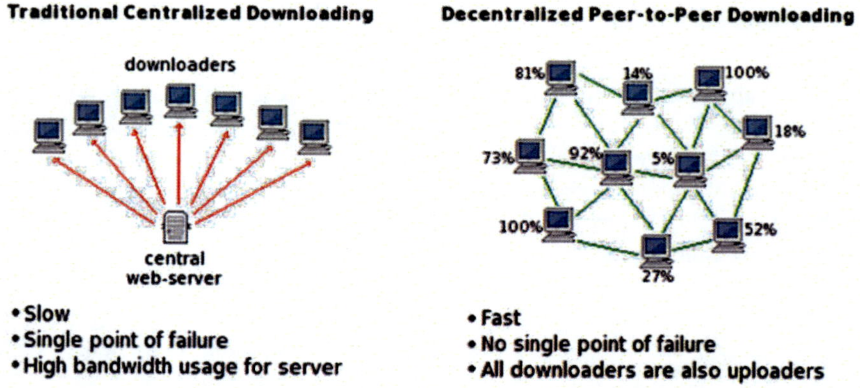

Fig. 2 Centralization and decentralization

Table 1 Centralized versus decentralized

Input	Hash
Hello	ftfg658hnty5by656b7bbhy5h156h4h1hg8h4h1h8yy7518h
Welcome to block brad nice to meet you	Hfg56h4gh14hhg1g4gh741ghghgfh4gh7hg4hgh7g8h4g8h74hfg89h7gh4g4r68g4r8grgrgrg4rg48dggg8rgewg4rhhjydfwaf44grg4g4dg4gr8g47htg44cf87fetef4fe7ff4d4f7fe7f7f7ef7ewrfretfdftrqetryjscsfdefeduiwhbdnbdcb4415tryrgr4184g4rg4rg4r84grgrgrg8rgrgr488rg4rrregregretre4576454ghtr5y89fberygfghjdsbdyc63bf7832983hefb7wetd383gr632geg
Thank you	Fh54hhfgh4h4h4h4f7hggwtyyiobcsds5544fdyuywdhtbbdgreyyddgrr5fdg4rg54155

they can store the money also. Google search is also one of the related examples [9]. All the information is stored in the Web and any user can extract whatever data they want. Hackers can easily track if centralization is there so advanced cryptographic algorithms are used for security. Eventually, the software will be updated continuously. The system is designed such that if any hacker tries to steal the information, then the system is locked directly by the algorithm. This helps the industries to lock the information easily [10].

- *Transparency*

Transparency is one of the interesting and confusing topics in blockchain technology. Blockchain uses private algorithms for obtaining security. Transparent blockchains are also available according to the applications. All the applications need to go through this process to obtain durability [11].

- *Immutability*

Immutability is once if some information is stored in the blockchain then it cannot be tampered. Especially in financial terms or networks this application helps a lot.

The blockchain uses this due to cryptographic algorithms and Hash functions used in the creativity. There are number of hash functions used in the algorithms [12]. In the context of crypto currencies like bitcoin, the transactions are taken as an input and run through a hashing algorithm (bitcoin uses SHA-256) which gives an output of a fixed length. Let us see how the hashing process works. We are going to put in certain inputs. For this exercise, we are going to use the SHA-256 (secure hashing algorithm 256) [13]. Blockchain can be used for various science and applications [14–19].

4 Methodology

There are many systems which are already available in the market. But those systems are not easy to operate. It is very hard to make over those applications. The retrieving the data is not an easy task for the system. Those systems are also not accurate. It takes much time so the new system is proposed. The proposed system architecture is shown in Fig. 3 does not have any redundancy problems and also easily operated. It has promising results in terms of speed and accuracy. The information is stored in external drives like hard disks; pen drives etc. which reduces the redundancy.

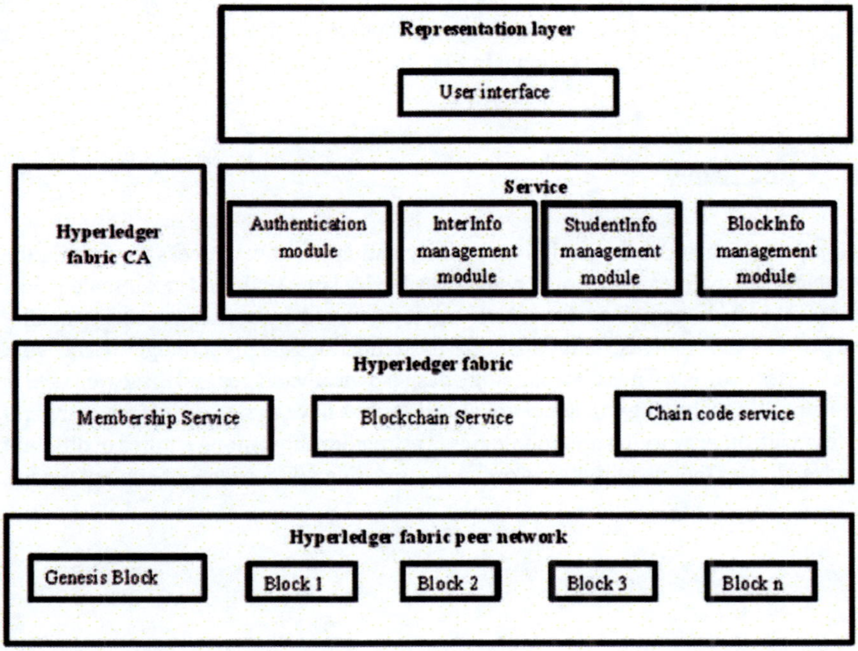

Fig. 3 Proposed architecture

- Blockchain technology is initiated for educational institutions
- Security is provide using advanced crypto algorithms
- Information tampering is not allowed using the analyzed application.
- All the student's data is inserted into the blockchains.
- Company authority can easily obtain the details of the students to recruitment.
- Less chances of occurrences of errors and fraud attainments are reduced

5 Testing

5.1 System Testing

System testing is done to visualize the errors. This is the basic step in the testing process. Testing is an important program to obtain the results. Various stages are considered for testing the system as given below.

- Unit Testing
- Integration Testing

At the development of the system this unit stepping is done. Unit stepping is also called as component testing. Every component or module is tested individually. This requires an esteemed platform to obtain perfect results. Next is integration testing where the integration between the components are tested. Finally the acceptance user testing is done. A module is shown in Fig. 4a, b, c, d, e and f.

6 Conclusion

This paper concludes the design of security and applicative process for storing the educational qualifications stored and accessible by large scale and small scale industries. A path is generated for many universities and industries to obtain quality employees and students. This paper uses hash functions and hyperledger frame work for this technology. Thus all technology has its own advantages and disadvantages in it, using this technology in educational institutions have more advantages comparatively with other working methodologies. The relationship among entities in different nodes all over the institutions makes easier updating and retrieval of student record.

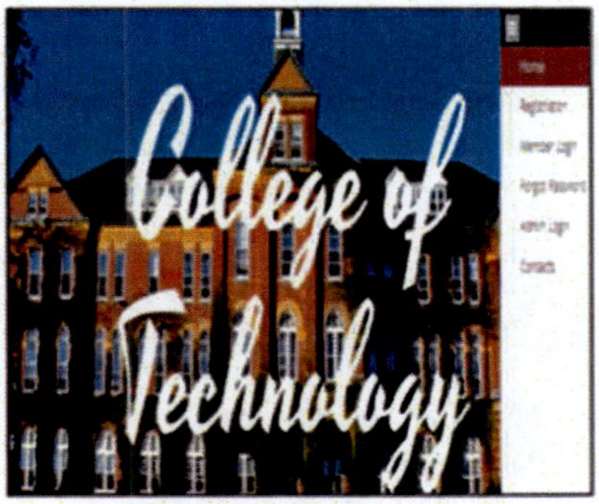

(a) Home page

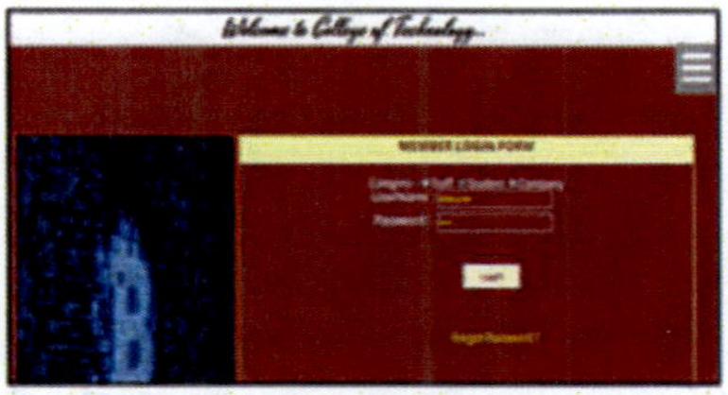

(b) Login page

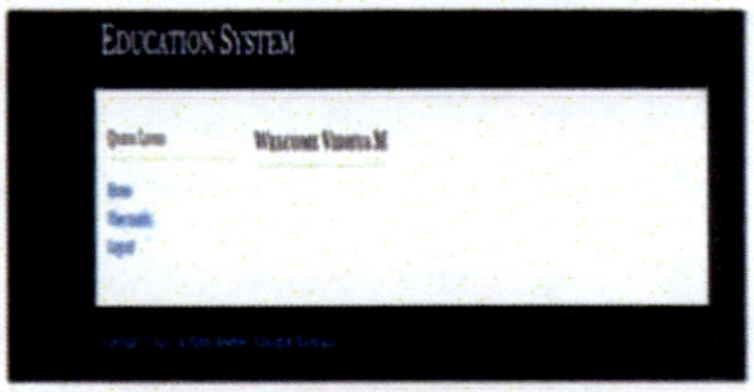

(c) Education system

(d) Student registration form

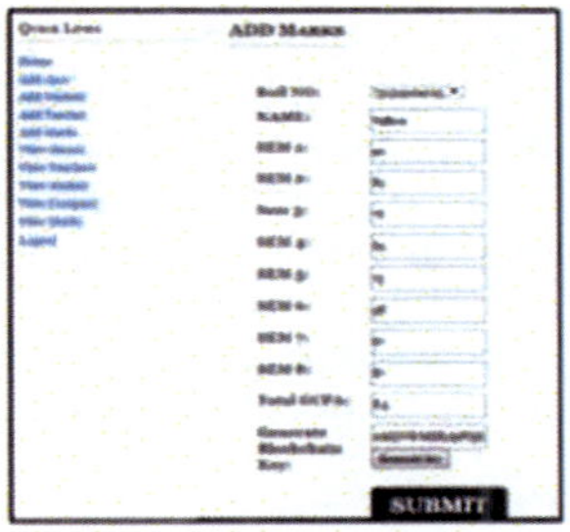

(e) Block chain key

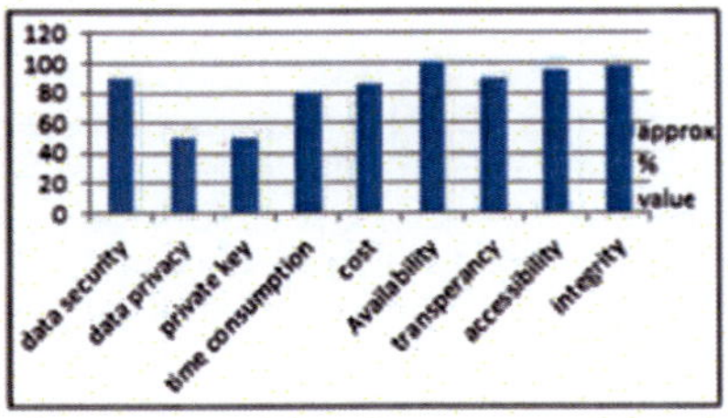

(f) Block chain features

Fig. 4 **a** Home of the module, **b** login page, **c** education system to find the employee details, **d** students registration form, **e** quick link to generate block chain key, **f** features of block chain technology

References

1. Collins R (2016) Blockchain: a new architecture for digital content. EContent 39:22–23
2. Memon M, Bajwa UA, Ikhlas A, Memon Y, Memon S, Malani M (2018) Blockchain beyond bitcoin: block maturity level consensus protocol. In: Proceedings of the 2018 IEEE 5th international conference on engineering technologies and applied sciences (ICETAS), Bangkok, Thailand, pp 1–5
3. Nakamoto S (2008) Bitcoin: a peer-to-peer electronic cash system. Available online: https://bitcoin.org/bitcoin.pdf
4. Monash University Library. Available online: https://guides.lib.monash.edu/subject-databases
5. Platt DS (2003) Introducing Microsoft.NET, 3rd edn, Microsoft Press
6. https://blockgeeks.com/guides/what-is-blockchain-technology/
7. https://blockgeeks.com/guides/blockchain-developer/
8. https://www.ibm.com/blockchain/what-is-blockchain
9. Banupriya S, Kottursamy K, Bashir AK (2021) Privacy-preserving hierarchical deterministic key generation based on a lattice of rings in public blockchain. Peer-to-Peer Netw Appl 1–13
10. Banupriya S, Kottilingam K (2021) An analysis of privacy issues and solutions in public blockchain (bitcoin). In: 2021 2nd international conference for emerging technology (INCET), IEEE, pp 1–7
11. Cengiz K, Sharma R, Kottursamy K, Singh KK, Topac T, Ozyurt B (2021) Recent emerging technologies for intelligent learning and analytics in big data. In: Multimedia technologies in the internet of things environment, Springer, Singapore, pp 69–81
12. Raja G, Kottursamy K, Theetharappan A, Cengiz K, Ganapathisubramaniyan A, Kharel R, Yu K (2020) Dynamic polygon generation for flexible pattern formation in large-scale UAV swarm networks. In: 2020 IEEE globe com workshops (GC Workshops), IEEE, pp 1–6
13. Mehra R, Pachpor K, Kottilingam K, Saranya A (2020) An initiative to prevent Japanese encephalitis using genetic algorithm and artificial neural network. In: 2020 international conference on computational intelligence (ICCI), IEEE, pp 142–148
14. Saranya A, Kottilingam K (2021) A survey on bone fracture identification techniques using quantitative and learning based algorithms. In: 2021 international conference on artificial intelligence and smart systems (ICAIS), IEEE, pp 241–248
15. Nallarasan V, Kottilingam K (2021) Spectrum management analysis for cognitive radio IoT. In: 2021 international conference on computer communication and informatics (ICCCI), IEEE, pp 1–5
16. Prakash UM, Kottursamy K, Cengiz K, Kose U, Hung BT (2021) 4x-expert systems for early prediction of osteoporosis using multi-model algorithms. Measurement 180:109543
17. Raja G, Anbalagan S, Kottursamy K, Aparna GS, Kumaresan J, Ihsan M (2019) Authorized arming and safeguarded landing mechanism. IEEE Trans Commun 67(7):5024–5036
18. Rajagopal RD, Murugan S, Kottursamy K, Raju V (2019) Cluster based effective prediction approach for improving the curable rate of lymphatic filariasis affected patients. Clust Comput 22(1):197–205
19. Nallarasan V, Kottursamy K (2021) Cognitive radio jamming attack detection using an autoencoder for CRIoT network. Wireless Pers Commun 1–17

Disease Detection Based on Human Eye Images and Analysis Using Multi-agent Systems with SPADE

R. Jaichandran, R. Magesh Sunder, K. L. Shunmuganathan, S. Leelavathy, and Vijay Harshan

Abstract This paper introduces a concept of identification and easy detection of diseases in human without any master health check-up, just by reading the image of the eye of the person. The vessel pattern of the sclera region in the eye is unique for each human along with the iris, retinal vessel pattern. We use this retinal vessel pattern for human identification as one of the biometrics. But, we have not concentrated much on the sclera vessel pattern; unlike retinal vessel pattern, this does not require biometric device to capture the image; we can just use the ordinary camera of eight megapixels and above. Some common diseases like diabetes, hypertension, cholesterol can be easy, and early detection can be made by the eye doctors itself by comparing the now captured image of the eye with that of the stored images of normal eye of a person, thus saving time and money for both doctors and patients. During this **COVID-19, pandemic** period, the doctors can remotely study the patients and diagnose easily.

1 Introduction

The diabetes, hypertension diseases are becoming quite common among the Indian public nowadays. The reason might be due to various factors, but not everyone is diagnosed at early stages itself. Again, this may be due to lot of reasons. The person getting affected does not feel the disease unless tested in most of the cases. As the early detection is lost, the disease gets aggravated and gets complicated. Also, most of the people are not aware of it also (Fig. 1).

Eye is the only organ in the body that provides features that can be used to indicate disease symptoms. The eye has two regions, namely sclera and retina; both

R. Jaichandran (✉) · R. Magesh Sunder · K. L. Shunmuganathan · S. Leelavathy · V. Harshan
Department of Computer Science and Engineering, Aarupadai Veedu Institute of Technology, Vinayaka Mission's Research Foundation (Deemed to be University), Paiyanoor, Chengalpattu District 603104, Tamil Nadu, India

S. Leelavathy
e-mail: leelavathy@avit.ac.in

S. Shakya et al. (eds.), *Proceedings of Second International Conference on Sustainable Expert Systems*, Lecture Notes in Networks and Systems 351,
https://doi.org/10.1007/978-981-16-7657-4_22

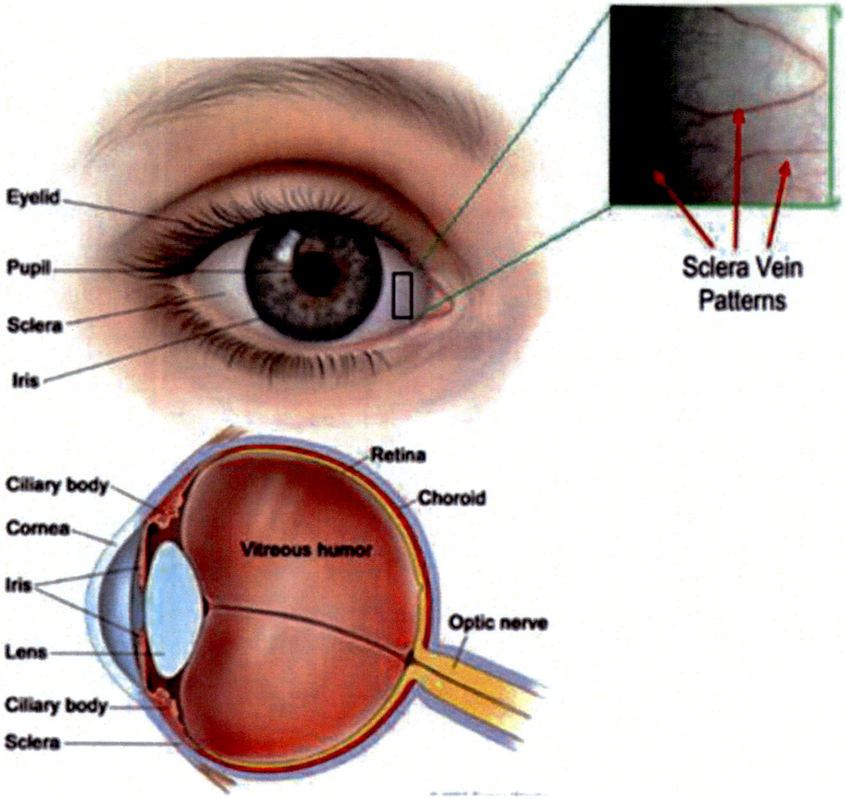

Fig. 1 Parts of the eye

have unique vessel patterns for every human. The retina vessel pattern capturing requires special biometric device and cooperation of the person to capture. But, the sclera vessel pattern does not require such devices or cooperation of the person to capture. A normal camera of minimum eight megapixels and above is enough.

A master database of every individual's normal eye images can be stored and then some periodic comparing with the current eye image with that in the database provides an easy and early detection of the disease. Whenever the disease affects a person, the sclera vessel pattern clearly indicates the changes, thus saving the time and energy and also lot of money for both doctors and patients.

In this pandemic times like COVID-19, the doctor and the patient need not come in physical contact also, as the person can capture his eye image easily with the help of mobile camera itself and send to the doctor for study. In turn, the doctor can feed the image into the software and can determine the nature of disease affected to the person easily. Thus, even the current mucormycosis disease commonly known as black fungus infection can also be prevented or detected at the early stages in a

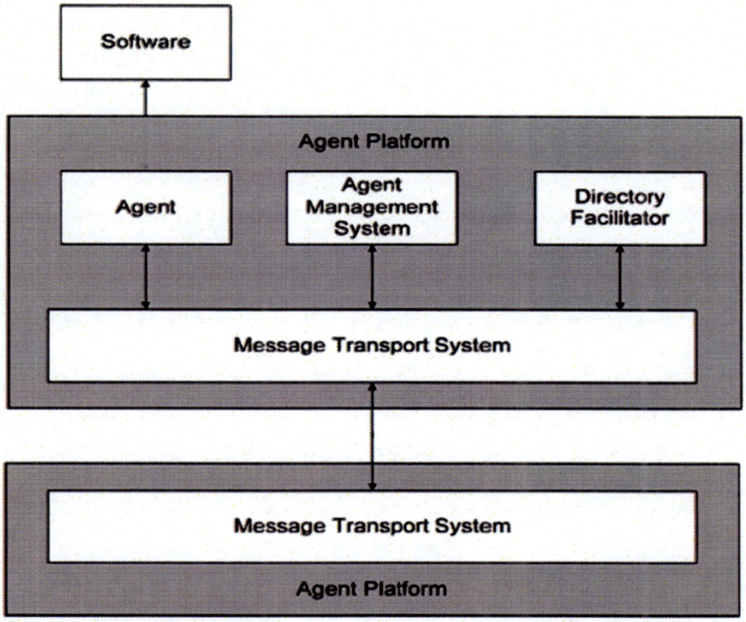

Fig. 2 Block diagram of agent system working

COVID-19-affected person. This disease affects person with hyperdiabetic and after cure of corona, but due to late treatments, few have lost sight and even eye itself.

Figure 2 shows the block diagram of agent system. Agent is a process that computes and implements sovereign behavior of an application. They work toward a goal in support of another entity, which can be either computational or human. The analysis part in the doctor's side can be speed up using the appropriate software tool aided with agent system. The algorithm for analysis can be embedded into the agent software for image processing, thus making the analysis faster and providing results quicker.

2 Literature Study

The main feature of this study is the classification of the disease severity based on the features of the sclera region of the eye. Literature study was done from medical text books like "Diseases of the eye" [1], referred journals, and articles published in various Web sites as mentioned in the reference. Literature survey was done in the area of image segmentation and enhancement techniques to extract the vessel patterns of the sclera region. Survey was also done in the area of obtaining suitable databases for carrying out the experiments.

As per the proposal of T. Vandarkuzhali in her 2013 paper "Detection of exudates caused by the appearance of diabetes in the fundus retinal image using complex K-neural channels" [2], we can automate the detection of abnormalities in our body due to diseases like diabetes using the eye image. The irregularities in the fovea are identified by fuzzy logic and neural networks. This is tested for standard and abstract images.

The research on sclera region requires examination of the anterior segment of the eye [1]. Several techniques are like.

(i) General inspection of the eye performed with the diffuse light of a torch or ophthalmoscope to acquire a gross picture of the eye and to identify possible areas of abnormality.
(ii) Examination in focal (or oblique) illumination using the slit-lamp or magnifying loupe.
(iii) Examination of the recess of the angle of the anterior chamber with a gonioscope.
(iv) Estimation of intraocular pressure by tonometry. All these techniques are dealt with in Parsons' diseases of the eye [1].

Holmes Keikobad, in an article entitled "Color Changes in Scleral Capillaries and Their Clinical Significance: Diagnosis and Treatment Protocol" published in acupuncture today (the sixty-first issue of Nan-Ching), discusses the importance and diagnostic value of color change occurring on the surface of the people. This text emphasizes the color changes that are commonly seen (and rarely) in the sclera capillaries, (interlocking vessel patterns seen in the sclera region of the eye) form an important part of the facial tissue, and bring the sea details to the diagnostician. We detected the symptoms of the disease using imaging techniques and detected the severity of the disease through neural networks [3].

A visible symptom of diabetic retinopathy is exudates that form a major cause of blindness in diabetic patients. Loss of vision can occur when exudates expand in the macular area. The paper entitled "A Study and Comparison of Automated Techniques for Exudate Detection Using Digital Fundus Images of Human Eye: A Review for Early Identification of Diabetic Retinopathy" discusses about the effectiveness of early detection of disease. Four methods were proposed and performed for diagnostic procedures for early detection of diabetic retinopathy. They are growing regional process, behavioral approach, K-mean integration process, and C-means clustering methods.

Sophark's article "Automatic exudates detection from non-dilated diabetic retinopathy retinal images using fuzzy C-means clustering" suggested a way to integrate FCM into outsiders. Comparison enhancement is used following the provision of data obtained from image elements to the hypothesis classification method using the FCM integration method as a first step. Image elements include standard deviation of power, number of pixels, intensity, and hue. The entropy element is used to identify the optical disk. For partitioning, the FCM integration algorithm is used. For obtaining the final result, the contrast image is set and reconstructed [4].

In the article entitled "Detection and classification of exudates using k-means clustering in color retinal images" [5], exudates are obtained using K-means clustering technique. This paper outlines how to identify exits and classify exits as soft and complex subjects. The retina image in the color space is treated first to remove the distortion. Next, to make the discovery and elimination of the optic disk easier, the blood network is removed. The Hough transform process was then used to eliminate the optic disk. Both hard and soft exudates were obtained using the K-means synthesis process. Finally, exudates are classified as hard and soft wood based on their edge strength and limit. The test results found encouraging. It also discusses about not using the Hough mutation to remove the optic disk from retina images, as they are not accurate if the position is not round.

A new logic based on fuzzy morphology is discussed in the paper [6] "Development of exudates detected by diabetes diagnosis using fuzzy morphology" for the diagnosis of diabetic retinopathy. Diabetic retinopathy is a very common disease in diabetic people. If exudates are found in the macular area, the disease is characterized. For developing exudates, the fuzzy morphology is used. The fundus image is first converted to grayscale image and then processed. Tests were performed on a database of various fundus images with and without infection. As a result, the tests led to convincing results, and the images were elevated to a rapid clinical trial.

In the survey of MAS from the perspective of ML, Stone et al. [7] explain the rules and regulations of conduct and comportment of agents in MAS. The survey focuses on highlighting the heterogeneous nature of agents and the communication and messaging between them. It explains that heterogeneous agents have the capacity to work alone or communicate with each other as and when required trough a common protocol or language. This communication is used for planning and negotiations, so as to attain a set goal and since the target is common, the agent system makes a deliberate commitment to complete the task in the most successful manner.

Luck et al. [8] explain the importance of agents in providing seamless computing that provides high performance with incessant communication and learning. According to them, agents are the future away from traditional computing. They go on to give the designs, roadmaps, technologies, and challenges in the development of agents.

Gregori et al. [9] put forward the idea of a Jabber XMPP-based agent platform that can be utilized for the development of multi-agent system. The platform is based on the instant messaging method and follows all the standards set by FIPA. Agents are *social*, *mobile,* and robust and are able to move with ease from one machine to the other. They are co-operative and support each other to attain their goal. The agents are modeled based on the FIPA standards and communicate with each other via FIPA ACL. The agents in this model are fashioned through supervised learning, and the threats are classified using machine learning (ML) techniques [10].

Van Der et al. [11], in his book on agents and multi-agents, explain that a society of intelligent agent systems proficient in adaptation to varying circumstances, interaction, and exchange of knowledge to achieve a common goal without external interventions is a MAS. Agents in MAS are autonomous with only an insular view of the system. Also, no agent in the MAS is assigned with governing or supervisory authority. MAS may encompass a combination of human and agents.

Bellifemine et al. [12] present JADE and all its advantages in the development of MAS and then move on to analyze the network neural element manager that has been built on a JADE platform.

Nikolay et al. [13] proposed that the actions of self-organizations of intelligent agents are modeled based on the observation of agent behavior. To achieve a goal, intelligent agents rationalize information and take appropriate action. The ability to reason and make decisions are requisites of an intelligent agent. The five levels of an intelligent agent are the sensors, philosophies, predilections, proficiencies, and actions.

3 Methodology

3.1 *Database Used for Experiment on Identification*

The UBIRIS database [14] contains color-coded iris images and is a public database. Previously, many details of the iris were obtained using NIR light. Images of 1877 are contained in the database (1214 in Session 1 and 663 in Session 2), and photographs were taken at two different times with 241 users. In Session 1, the image disturbance was reduced, especially those related to light, contrast, and reflection by taking pictures in a dark room. However, in Session 2, the location of the camera was changed to introduce a natural light element. This has enabled the emergence of bizarre images in terms of brightness, contrast, reflection, and concentration problems. The images obtained in this space often mimic those captured by a real visual system with or without the slightest interaction from the studies. In other words, the significant number of images in Session 2 is of very poor quality. In both databases, images are often cut in such a way that the eye region is much focused, and the eye is well cut out of the images. Various science and applications are contributing toward wider scientific platforms [15–23].

3.2 *Segmentation of Sclera Region*

There is a lot of research work being done to classify people based on eye parts. The sclera of the human eye is an open, white area that acts as a protective eye cover. The eye is surrounded completely by sclera. The vein patterns observed in the sclera region are unique and different for each individual in the visible wavelengths. It, therefore, can be used as a biometric identification tool. The size of the sclera varies with age. Elements extracted from a vein pattern as seen in the sclera region are used for diagnostic purposes. When sclera recognition is compared to other biometrics

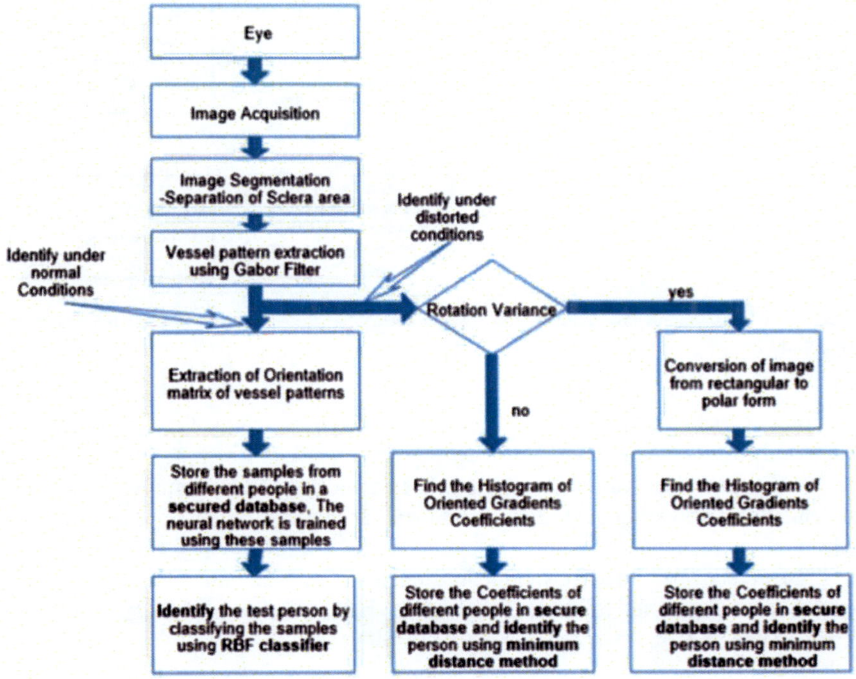

Fig. 3 Steps involved in the experimental identification based on vessel patterns as seen in the eye's sclera region

such as iris recognition, it has many benefits. Other benefits of sclera recognition are:

(1) Specialized cameras are not required for capturing the images. This reduces the imaging requirements, as a minimum eight megapixel camera can capture the image. If the camera is of higher resolution, images can be captured from longer distances.
(2) Frontal gaze images of the eye are not a requirement in the case of sclera recognition.
(3) Since we extract the features from either side vessel patterns of iris, the iris position is not very important; i.e., off angle segmentation and recognition are possible. Figure 3 shows the various steps involved in identification of vessel patterns in the eye's sclera region.

3.3 Classification of Illness Based on Scleral Features

Segmentation of the sclera area was done, and the vessel patterns were extracted.

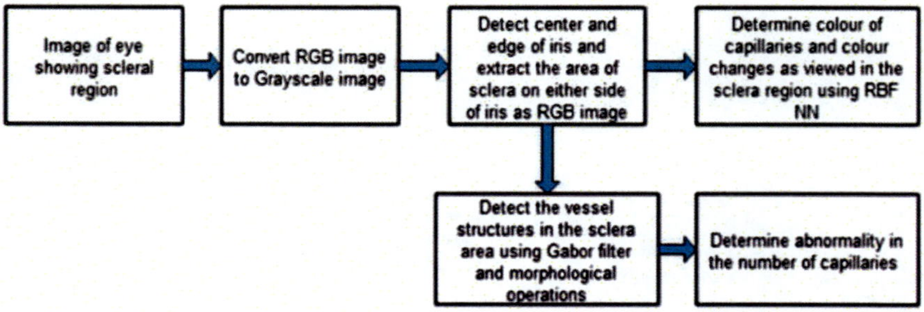

Fig. 4 Segmentation of the sclera region and the additional steps required for disease symptoms identification

Color estimates based on RGB values as viewed in the sclera of the eye region and the blood vessels color are known to identify related diseases symptoms. Few feature vectors based on the color observed in the sclera and color of conjunctival capillaries are calculated and assigned to the neural network to detect any color changes in the region of interest.

Figure 4 shows the feature vectors formed on the basis of color changes and other inferences are again fed to a neural network for finding the probable extend of illness a person can have.

3.4 *Classifications of Disease Severity Based on Symptoms of the Human Eye*

Table listing sclera and capillary changes and associated diseases [3]

Symptoms noticed in sclera region	Associated disease severity
Less number of capillaries and more red in color	Attack is recent, and illness is in an critical phase
More number of capillaries and somewhat of darker color	Condition and illness have become severe
Color of the capillaries are dark clear purple	Illness involves an element of heart and blood inactivity
Capillaries show a rupture and of blood extravasations into the sclera	Other than the root pathology, the spleen is also in short and cannot keep the blood within the vessels
Spilled out blood shows an yellow color	Spleen deficiency
Spilled out blood is reddish or pink	The breakage is recent
Spilled out blood is purplish or darkish	The breakage is of longer duration
Capillaries are narrower in some sectors	The spleen is in disarray, and the vessel walls have caved in

(continued)

(continued)

Symptoms noticed in sclera region	Associated disease severity
Cornea affected with capillaries	The illness is in more advanced stage and needs immediate attention

4 Results

To test the proposed system, images are got from the diabetic retinopathy dataset, DIARETDB0 and DIARETDB1, which is available in public. The fundus images are taken using TOPCON fundus camera TRC 50 DX with 50% of field vies. The proposed process is proved using the images of size 1500 × 1152 pixels using MATLAB. The exudates found are classified as hard ones using the fuzzy algorithm. We regarded the exudate with an output value > 0.25 as hard ones. This was arrived using by the membership function (MF) of linguistic variable of the Xout. In this sense, exudates are considered a weak exudate if the exudate output value is > 0.25. We found that the output from fuzzy algorithm was below but very close to 0.25 for very few hard exudates. Since small exudates are hard as well as soft exudates show similar intensity of color, it is very difficult to distinguish both types of exudates, and therefore, from time to time, this method identifies these very small exudates as non-hard exudates. In this study, better results were achieved using the Gaussian combination of MF combinations. Unusual output values of different overlapping points between MFs are examined, and it gave some inaccurate results for low levels of overlapping MFs. The unpredictable discharge rates of the most severe exudates are lower than the expected emissions rates at such low levels. MFs with less severe exudates are identified as less severe exudates. Non-aberrant excretion rates for non-severe exudates (excluding mild exudates) have increased, while non-aberrant excretion rates for weak exudates have decreased to lower levels of interaction between MFs. Therefore, indistinct MFs can produce inaccurate results that confirm the importance of having diffuse MFs in order to detect hard droplets.

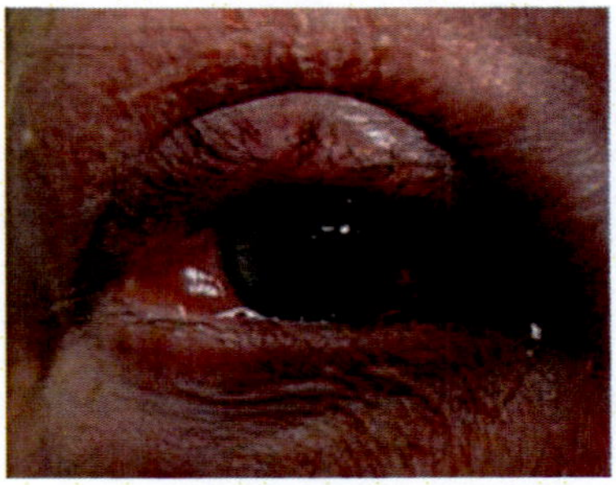

Fig. 5 Allergic conjunctivitis

Figure 5 shows eye inflammation caused by allergic reactions, which makes eyes red, itchy and watery. Figure 6 shows chinese pink eye caused by infection which is highly contagious. Figure 7 shows eye pus that leaves behind eye boogers that are are harmless, but changes in discharge from the eyes can give clues to health concerns. Figure 8 shows eye hemorrage caused by a small bleed behind the covering of the eye. It is usually harmless and clears within two weeks. Figure 9 shows subconjunctival hemorrage. A subconjunctival hemorrhage is a red spot on your eye caused by a broken blood vessel. Disorders related to blood clotting such as hemophilia or von Willebrand disease, make subconjunctival hemorrhage. Figure 10 shows stress induced eye hemorrhage. Stress induced eye hemorrhage is often accompanied by symptoms such as sensitivity to light, reduction of peripheral vision, blurred vision, dry eyes, twitching in the eyes, etc.

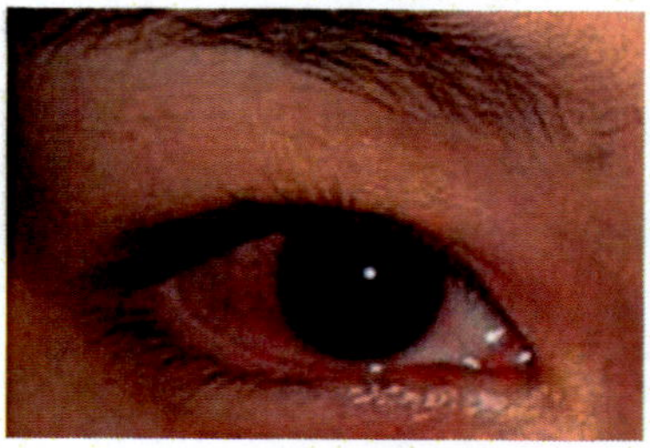

Fig. 6 Chinese pink eye

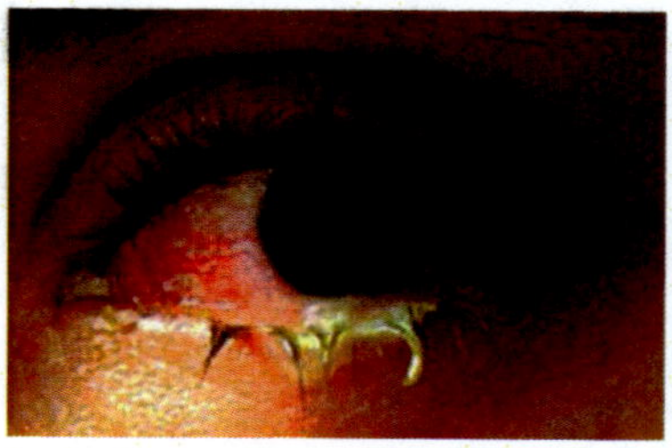

Fig. 7 Eye with pus

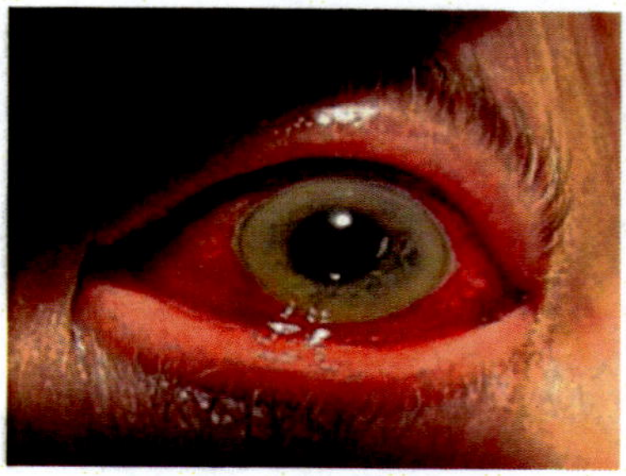

Fig. 8 Eye with hemorrhage

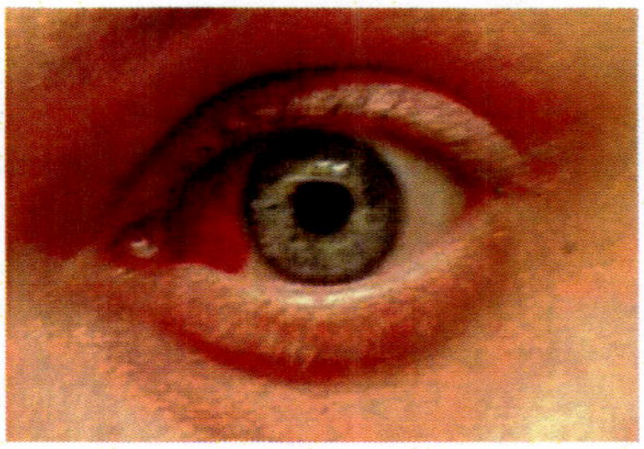

Fig. 9 Eye with subconjunctival hemorrhage

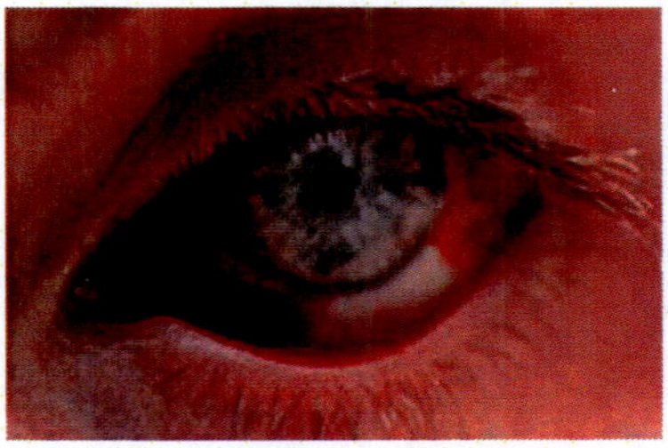

Fig. 10 Eye with stress induced hemorrhage

5 Conclusion

Thus, during this pandemic (COVID-19) times, both the doctors and patients are benefited by early detection of disease, and therefore, proper treatment can also be provided without much issues.

References

1. Sihota R, Tandon R, Parsons' diseases of the eye, 21 edn. Elsevier Publications, pp 3–29, 161–280
2. Vandarkuzhali T, Ravichandran CS, Preethi D, Detection of exudates caused by diabetic retinopathy in fundus retinal image using fuzzy K means and neural network. IOSR J Electr Electron Eng
3. Keikobad LH, Color changes in scleral capillaries and their clinical significance: diagnosis and treatment protocol. Acupuncture Today (61st difficult issue in the Nan-Ching)
4. Sopharak A, Uyyanonvara B, Barman S (2009) Automatic exudate detection from non-dilated diabetic retinopathy retinal images using fuzzy C-means clustering. J Sens 9(3):2148–2161
5. Rajput GG, Patil PN (2014) Detection and classification of exudates using k-means clustering in color retinal images. In: Fifth international conference on signals and image processing, 978-0-7695-5100-5/13 $31.00 © 2013 IEEE. https://doi.org/10.1109/ICSIP.2014.25
6. Mansoor AB, Khan Z, Khan A, Khan SA (2008) Enhancement of exudates for the diagnosis of diabetic retinopathy using fuzzy morphology. In: Proceedings of 12th IEE international multi-topic conference (INMIC)
7. Ramaswamy M, Anitha D, Priya Kuppamal S, Sudha R, FepslinAthishMon S, A study and comparison of automated techniques for exudate detection using digital fundus images of

human eye: a review for early identification of diabetic retinopathy. Int J Comp Tech Appl 2(5):1503–1516
8. Luck M, McBurney P, Preist C (2003) Agent technology enabling next generation computing (A roadmap for agent bassd computing)
9. Grégoire Y, Fisher R (2006) The effects of relationship quality on customer retaliation. Mark Lett 17:31–46. https://doi.org/10.1007/s11002-006-3796-4
10. Kotsiantis S (2007) Supervised machine learning: a review of classification techniques. Informatica (Ljubljana) 31
11. Van der Hoek W, Wooldridge M (2008) Multi-agent systems. In: van Harmelen F, Lifschitz V, Porter B (eds) Handbook of knowledge representation, Amsterdam, Elsevier, pp 887–928
12. Bellifemine F, Caire G, Poggi A, Rimassa G (2008) JADE: A software framework for developing multi-agent applications. Lessons learned. Inf Software Technol 50:10–21. https://doi.org/10.1016/j.infsof.2007.10.008
13. Tchernev N, Lacomme P, Larabi M (2013) Job-shop based framework for simultaneous scheduling of machines and automated guided vehicles. Int J Prod Econ 143:24–34. https://doi.org/10.1016/j.ijpe.2010.07.012
14. Proença H, Alexandre L (2005) UBIRIS: a noisy iris image database. Image analysis and processing. Image Anal Proces 3617:970–977
15. Banupriya S, Kottursamy K, Bashir AK (2021) Privacy-preserving hierarchical deterministic key generation based on a lattice of rings in public blockchain. Peer-to-Peer Netw Appl 1–13
16. Raja G, Kottursamy K, Theetharappan A, Cengiz K, Ganapathisubramaniyan A, Kharel R, Yu K (2020) Dynamic polygon generation for flexible pattern formation in large-scale UAV swarm networks. In: 2020 IEEE Globecom Workshops (GC Wkshps), pp 1–6. IEEE
17. Mehra R, Pachpor K, Kottilingam K, Saranya A (2020) An initiative to prevent Japanese encephalitis using genetic algorithm and artificial neural network. In: 2020 international conference on computational intelligence (ICCI), pp 142–148. IEEE
18. Saranya A, Kottilingam K (2021) A survey on bone fracture identification techniques using quantitative and learning based algorithms. In: 2021 international conference on artificial intelligence and smart systems (ICAIS), pp 241–248. IEEE
19. Nallarasan V, Kottilingam K (2021) Spectrum management analysis for cognitive radio IoT. In: 2021 international conference on computer communication and informatics (ICCCI), pp 1–5. IEEE
20. Prakash UM, Kottursamy K, Cengiz K, Kose U, Hung BT (2021) 4x-expert systems for early prediction of osteoporosis using multi-model algorithms. Measurement 180:109543
21. Cengiz K, Sharma R, Kottursamy K, Singh KK, Topac T, Ozyurt B (2021) Recent emerging technologies for intelligent learning and analytics in big data. In: Multimedia technologies in the internet of things environment, pp 69–81. Springer, Singapore
22. Raja G, Anbalagan S, Kottursamy K, Aparna GS, Kumaresan J, Ihsan M (2019) Authorized arming and safeguarded landing mechanism. IEEE Trans Commun 67(7):5024–5036
23. Rajagopal RD, Murugan S, Kottursamy K, Raju V (2019) Cluster based effective prediction approach for improving the curable rate of lymphatic filariasis affected patients. Clust Comput 22(1):197–205

Distributed Data Balancing on Multi Base Station Wireless Network to Solve the Routing Complexity Issues

R. Shobana, R. Jaichandran, V. Sivakumar, G. Devaraj, and A. Anbarasan

Abstract Routing protocol is one of the important parameters in wireless networks. Hybrid wireless networks have many advantages when the two streams of networks are combined. Here, we combine mobile adhoc network and infrastructure network. This combination is most commonly used by many platforms due to its high efficiency. Network should focus on capacity, scalability and durability. Adhoc and mobile modes of transmission are combined to generated high performance. This research uses DTR (Distribute three-hop routing protocol) which is integrated to the hybrid wireless networks. This helps to get the huge spreading of base stations. Generally, the data is segmented into packets of data and each packet is transmitted in the distribution method. This reduces the maintenance and easily route is discovered. Various parameters like throughput capacity, scalability and mobility resilience are calculated and compared where promising results are obtained.

Keywords Wireless networks · DTR · Adhoc networks

1 Introduction

Network is the major attraction in the communication world. For proper communication, network should be strong. There are two basic types of networks, wired and wireless [1–3]. Wired network have minimal issues in the communication but wireless networks have much complications in the design. So engineers use different protocols to get a feasible network. Node energy is one of the main factor in hybrid networks.

R. Shobana · R. Jaichandran (✉) · V. Sivakumar · G. Devaraj · A. Anbarasan
Department of Computer Science and Engineering, Aarupadai Veedu Institute of Technology, Vinayaka Mission's Research Foundation (Deemed to be University), Paiyanoor, Chengalpattu 603104, Tamil Nadu, India

R. Shobana
e-mail: shobana@avit.ac.in

S. Shakya et al. (eds.), *Proceedings of Second International Conference on Sustainable Expert Systems*, Lecture Notes in Networks and Systems 351,
https://doi.org/10.1007/978-981-16-7657-4_23

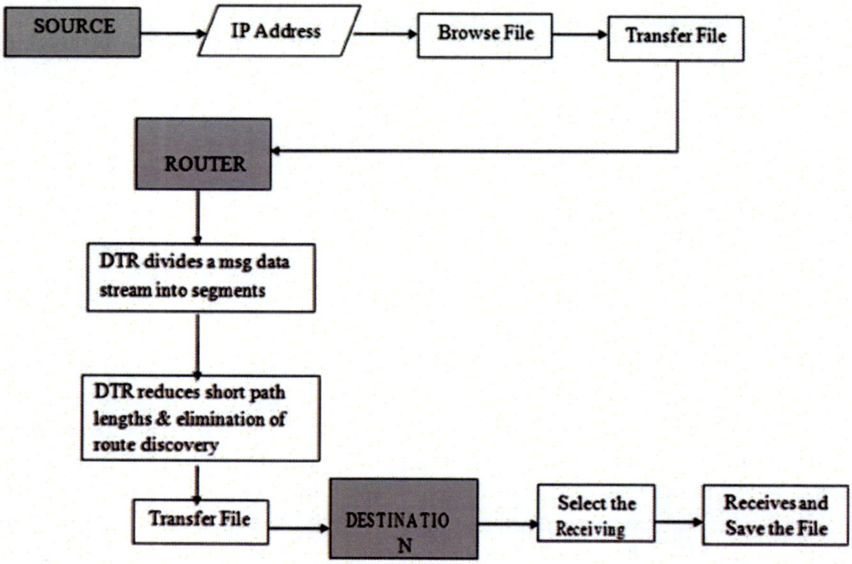

Fig. 1 DT block diagram

Hybrid wireless networks are one of the efficient network used for communication. Recent researchers are combining the some networks for higher efficiency. One of the best combinations is mobile adhoc network and infrastructure network [4–7]. The information or data is splitted in to number of packets. Each packet is distributed using distributed method. This helps in the reuse of the space, and the reuse can be done based on mobile interface. As many packets are sent to base station it will improve throughput which ultimately increases wide spread of base stations [8]. DTR is one of the best technique which consists of cognition control algorithm. DTR also helps to resolve the issues of overloading at the base stations. The DTR block diagram is shown in Fig. 1.

2 Literature Survey

Many authors presented their research on these topics but the related papers shown better results in the applications. Authors H. Y. Hsieh and R. Shiva Kumar proposed an algorithm named sphinx, which is used for peer-to-peer network. This is related to cellular network model. This sphinx removes pitfalls in the networks. Simulation is done on the cellular networks [9, 10]. This generates low-power consumption. Authors X. J. Li. B. C. Seat and P. H. J. Chong worked on multihop cellular network. They used single hop to generate multihop networks. All the technology is related to the economics. Very clear declaration is done between economics and technology.

This is very much useful in fourth generation [11–13]. There are various diverse applications are been used for different science and engineering process [14–21].

3 Methodolgy

System methodology is based on the architecture. This architecture is shown in Fig. 3. This is a concept-based design. Various sub systems and sub structures are available. Each structure is implemented, and overall output is generated. System architecture is the conceptual model that defines the structure, behavior and more views of a system. An system architecture can consists of system components and the sub systems developed that will work together to implement the overall system there have been efforts to formalize languages to describe system architecture, collectively these are called architecture description languages (ADLs) [14, 15].

Activity diagram is shown in Fig. 2 where the source and destination are connected with a router. Source used an IP address and browse the file and transfer the file. The transferred file is connected through the router. There it uses DTR technology which divides the data and transferred. The packets of data is selected by the receiver and finally it will save the data. The activity diagram flow is very simple, but it gives optimized results to perform any network simulations. This type of flow can be used for any type of nodes and any protocols [16]. Advanced protocols can be used for the interference of various networks. Energy consumption is very essential at the nodes. Various sensors are combined in wireless sensor networks. Many sensors in large area gives essential results. Error detection and error correction are easy in wireless sensor networks. The routing path can be replaced by seeing the sensors. At all the

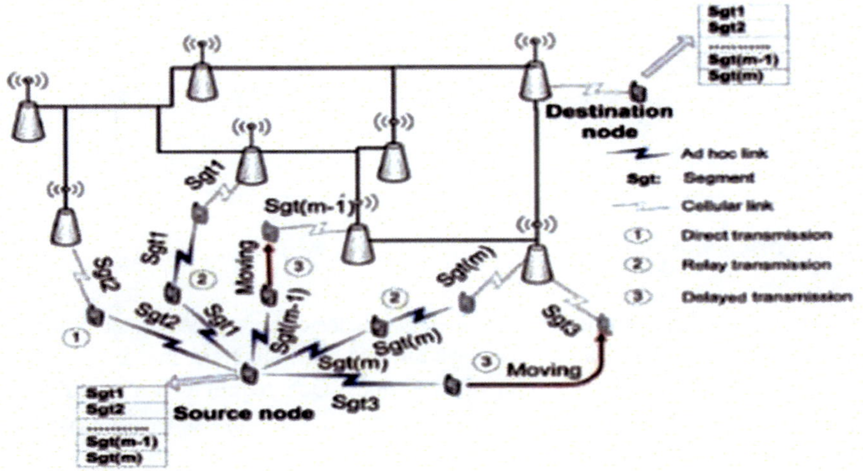

Fig. 2 Architecture of the design

Fig. 3 Activity diagram

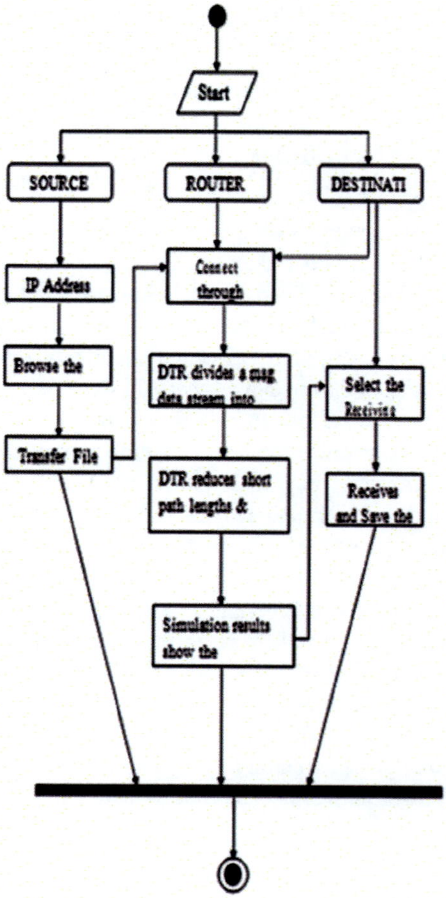

applications are huge and almost each and every network is related to wireless sensor networks [17] (Fig. 3).

4 Proposed System

As wide spread base stations are used there is chance of occurrence of encountering when the system is moving. This plays an equal advantage in the design. DTR helps in wide range of networking base stations. QoS requirement is considered and concerned packet is sent to that base station. Then, they choose two types of transmission as mentioned.

- Direct transmission
- Relay transmission.

In direct transmission, there is no consideration of capacity, any node can transfer any information to any nodes at the bases station. In relay transmission, the capacity is higher the forwarding principle is done on higher capacity node as compared to the capacity at the base station node. Three routing hops are used in DTR. The three hops are used by one in cellular transmission mode and two in adhoc transmission mode. Single hop is used to segment the data and distribute the data. The two hops maintain the capacity and networking through various nodes.

5 Testing

- Server
- Client
- Distributer three-hop router
- Load balancing.

Sever is a mechanical device at the source. As per the administration, the data is send from the server with proper acknowledgment. Server only processes the input but cannot indulge in to clients routing path and its property. Intermediate nodes are considered for connecting gateways. All the cellular mode is a dual mode. Exactly to say, it acts as WLAN radio interface and infrastructure network interface. Routing burden is major concern where the server helps to shift the routing burden from adhoc network to infrastructure network.

Client is the machine at destination, which receives the information from the server. A router is designed and developed such that it helps the client to receive the information from the server.

Distributer three-hop router is similar to the three hops as discussed one hop is used for cellular and two hos for adhoc communication. The uplink routing and down link routing are very important in this design.

Load balancing is the final chapter of this research where the load is balanced by algorithm named congestion control. Load balance is very important at the base stations. If load is properly balanced, then the data is transmitted with very high speed. After the overall design and simulation, the screen shots of the software generated are shown in Fig. 4a, b. The throughput and capacity are also shown.

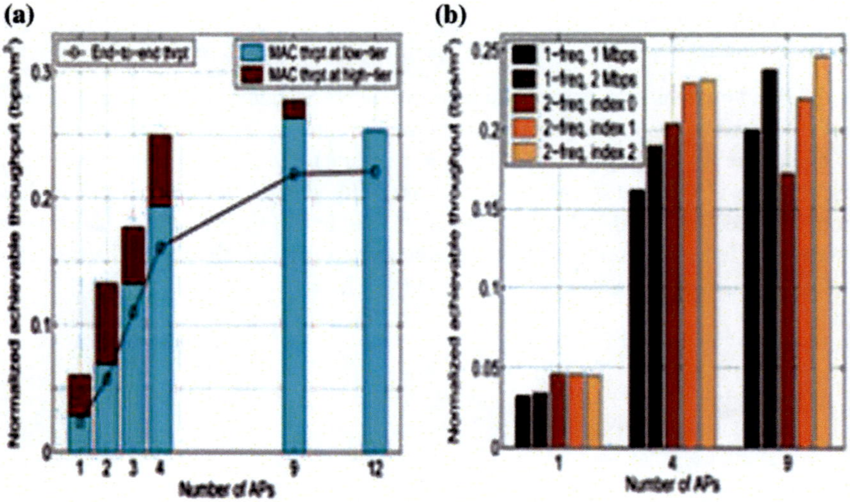

Fig. 4 **a**, **b** Throughput with MAC (low/high frequency)

6 Conclusion

HWN which have many applications has been designed with the help of DTR technique. The data transmission from server to client or source to destination is clearly seen Fig. 5a–d. The path of routing. The protocols used for routing the path has shown. DTR technique interfacing is shown with the flow diagrams. Distributed three-hop routing (DTR) data routing protocol that integrates the dual features of hybrid wireless networks in the data transmission process. In DTR, a source node divides a message stream into segments and transmits them to its mobile neighbors, which further forward the segments to their destination through an infrastructure network.

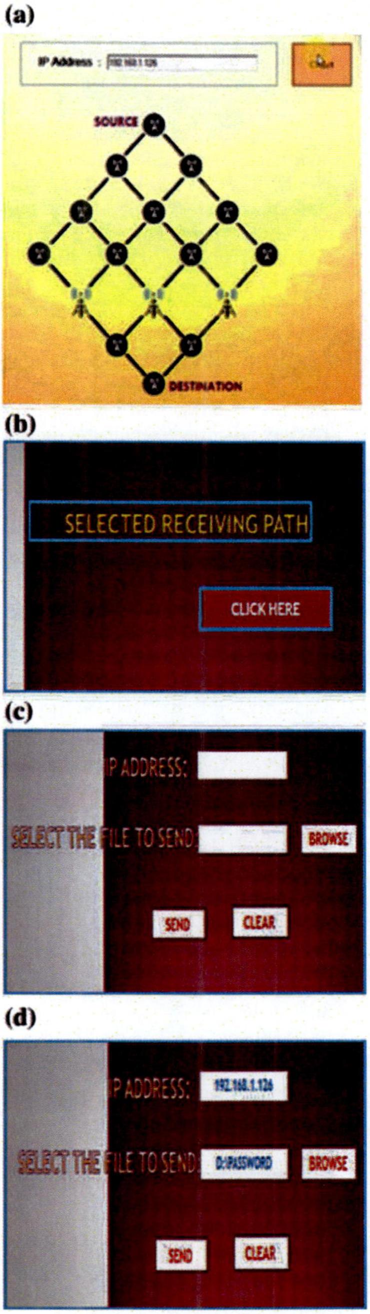

Fig. 5 **a** IP addresses of source and destination. **b** Identified path. **c** Searching file with IP. **d** Data transmission process

References

1. Luo H, Ramjee R, Sinha P, Li L, Lu S (2003) UCAN: a unified cell and ad-hoc network architecture. In: Proceedings of Mobicom
2. Mckinley PK, Xu H, Esfahanian AH, Ni LM (1992) Uncastbased multicast communication in wormhole-routed direct networks (TPDS)
3. Wu H, Qiao C, De S, Tonguz O (2001) Integrated cell and ad hoc relaying systems: Icar (J-SAC)
4. Tam YH, Hassanein HS, Akl SG, Benkoczi R (2006) Optimal multi-hop cellular architecture for wireless communications. In: Proceedings of LCN
5. Lin YD, Hsu YC (2000) Multi-hop cellular: a new architecture for wireless communications. In: Proceedings of INFOCOM
6. Oliver PT, Dousse HM (2002) Connectivity in ad hoc and hybrid networks. In: Proceedings of INFOCOM
7. Charles EP, Bhagwat P (1994) Highly dynamic destination sequenced distance vector routing (DSDV) for mobile computers. In: Proceedings of SIGCOMM
8. Perkins C, Belding-Royer E, Das S (2003) RFC 3561: ad hoc on demand distance vector (AODV) routing. Technical Report, Internet Engineering Task Force
9. Johnson DB, Maltz DA (1996) Dynamic source routing in ad hoc wireless networks. IEEE Mobile Comput
10. Park VD, Scott Corson M (1997) A highly adaptive distributed routing algorithm for mobile wireless networks. In: Proceedings of INFOCOM
11. Chang RS, Chen WY, When YF (2003) Hybrid wireless network protocols. In: IEEE Trans Veh Technol
12. Aggelou GN, Tafazolli R (2001) On the relaying capacity of NextgenerationGsm cellular networks. In: IEEE Personal Communications Magazine
13. Rouse T, Band I, Mclaughlin S (2002) Capacity and power investigation in TDD-CDMA based systems. In: Proceedings of ICC
14. Banupriya S, Kottursamy K, Bashir AK (2021) Privacy-preserving hierarchical deterministic key generation based on a lattice of rings in public blockchain. Peer-to-Peer Netw Appl 14:1–13
15. Vijayakumar KP, Kumar KPM, Kottilingam K, Karthick T, Vijayakumar P, Ganeshkumar P (2019) An adaptive neuro-fuzzy logic based jamming detection system in WSN. Soft Comput 23(8):2655–2667
16. Banupriya S, Kottilingam K (2021) An analysis of privacy issues and solutions in public blockchain (Bitcoin). In: 2021 2nd international conference for emerging technology (INCET), pp 1–7. IEEE
17. Raja G, Kottursamy K, Theetharappan A, Cengiz K, Ganapathisubramaniyan A, Kharel R, Yu K (2020) Dynamic polygon generation for flexible pattern formation in large-scale UAV swarm networks. In: 2020 IEEE Globecom Workshops (GC Wkshps), pp 1–6. IEEE
18. Mehra R, Pachpor K, Kottilingam K, Saranya A (2020) An initiative to prevent Japanese encephalitis using genetic algorithm and artificial neural network. In: 2020 international conference on computational intelligence (ICCI), pp 142–148. IEEE
19. Saranya A, Kottilingam K (2021) A survey on bone fracture identification techniques using quantitative and learning based algorithms. In: 2021 international conference on artificial intelligence and smart systems (ICAIS), pp 241–248. IEEE
20. Nallarasan V, Kottilingam K (2021) Spectrum management analysis for cognitive Radio IoT. In: 2021 international conference on computer communication and informatics (ICCCI), pp 1–5. IEEE
21. Raja G, Kottursamy K, Chaudhary SH, Hassan A, Alqarni M (2017) SDN assisted middlebox synchronization mechanism for next generation mobile data management system. In: 2017 IEEE SmartWorld, ubiquitous intelligence & computing, advanced & trusted computed, scalable computing & communications, cloud & big data computing, internet of people and smart city innovation (SmartWorld/SCALCOM/UIC/ATC/CBDCom/IOP/SCI), pp 1–7. IEEE

RPBK21 Technique Used to Secure the Data

K. Shantha Shalini, R. Shobana, Bhupalam Preetham, M. P. Shashank, and Bathula Pavan Kalyan

Abstract In the digital era, the information is generated by utilizing Internet media, which is highly ubiquitous and unrestricted. These characteristics may lead to significant challenges. To overcome these, this research work has proposed a ChaCha method to overcome this problem. The proposed method effectively steals data from software developers. The novel RPBK21 strategy is divided into eight parts, and they are: to find the mystery prime key S; to find the X_1 and X_2 values from prime numbers; to find the $\overline{X}_1$ and $\overline{X}_2$ values; to find the standard deviation values with the help of Step 2 and Step 3; to exchange the a and b esteems from left in a matrix; to locate the T-test values and pair them that numbers from left to right; to trade the undividable numbers and non-negative whole numbers in a given lattice; to exchange the M pair numbers. As illustrated in the results, the proposed RPBK21 strategy provides extraordinary security while appearing differently about ChaCha and RBJ25 techniques.

Keywords RPBK21 · Encryption · Decryption · Prime · T-test · ChaCha

1 Introduction

The large information time has been purchased with adequate freedoms for logical improvement in improving medical care, monetary development, improving the instruction framework, and different types of amusement. The examination of large information needs to go through numerous stages to acquire some significant qualities, which incorporate a few phases like information reconciliation, the information collection, data cleaning, data extraction, inquiry preparation, information demonstration, and translation. Each stage holds numerous difficulties like heterogeneity, practicality, intricacy, security, and protection. The significant challenges in massive frameworks such as large volume, speed, and diversity are security and protection. Despite the fact, there are primarily four characteristics of massive information security: security of the foundation and structure; information protection; information

K. Shantha Shalini (✉) · R. Shobana · B. Preetham · M. P. Shashank · B. Pavan Kalyan
Department of CSE, Aarupadai Veedu Institute of Technology, Vinayaka Mission's Research Foundation, Salem, India

S. Shakya et al. (eds.), *Proceedings of Second International Conference on Sustainable Expert Systems*, Lecture Notes in Networks and Systems 351,
https://doi.org/10.1007/978-981-16-7657-4_24

guidelines; vital and responsive security. The proposed ChaCha technique adequately hacks the data from the software engineers [1]. The developer has utilized a new hash idea for key speculating and ending conditions [2]. Also, the bricklayer assault has been presented for investigation of ChaCha [3]. They focused on the security of double-A [4]. They developed a new plan to securely implement a quick and adaptable calculation [5]. Further, SRB18 strategy is used to provide information security [6]. SRB21 Phase 1 and SRB21 Phase 2 strategies are also used to provide security to the information [7, 8]. CBB21 [9], CBB22 [10], CBB20 [12], and RJB25 techniques are used to leverage advanced information security [13]. Moreover, a novel RajaPrakashBagathKarthik (RPBK) strategy has been developed. As an implementation, various applications are used in performing different science and engineering process [11, 14–22].

2 Methods

This section discusses about the encryption and decryption methods. The proposed RPBK21 method has eight phases.

1. To find the mystery prime key S.
2. To find the X_1 and X_2 values from prime numbers.
3. To find the $\overline{X}_1$ and $\overline{X}_2$ values.
4. To find the standard deviation values with the help of Step 2 and Step 3.
5. To trade a and b esteems from left in a matrix.
6. To locate the T-test values and pair them that numbers from left to right.
7. To trade the undividable numbers and non-negative whole numbers in a given lattice.
8. To exchange the M pair numbers.

2.1 Encryption

- RPBK21 method mentioned in Table 1 is used for encryption.
- Input of the matrix (IP).

$$\mathrm{IP} = \begin{bmatrix} 401/4 & 402/4 & 403/4 & 404/4 & 405/4 \\ 406/4 & 407/4 & 408/4 & 409/4 & 410/4 \\ 411/4 & 412/4 & 413/4 & 414/4 & 415/4 \\ 416/4 & 417/4 & 418/4 & 419/4 & 420/4 \\ 421/4 & 422/4 & 423/4 & 424/4 & 425/4 \end{bmatrix}$$

- Prime Numbers—1, 3, 5, 7, 11, 13, 17, 19, 23
- $X_1 = 3, 5, 7, 11; X_2 = 13, 17, 19, 23$

Table 1 Encryption algorithm

Steps	Encryption procedure
I	To find a prime numbers in given matrix IP
II	Equally separate the two parts of prime numbers and apply those values in Eqs. (1)–(3)
III	T-test formula = $(\overline{X}_1 - \overline{X}_2) / \sqrt{((S_1^2/N_1) + (S_2^2/N_2))}$ (1)
IV	To find the values for Eq. (1) with the help of Eqs. (2) and (3)
V	$\overline{X}_1 = \sum X_1/N_1$ $\overline{X}_2 = \sum X_2/N_2$ (2)
VI	$S_1 = \sqrt{\sum (X_1 - \overline{X}_1)^2/(N_1 - 1)}$ $S_2 = \sqrt{\sum (X_2 - \overline{X}_2)^2/(N_2 - 1)}$ (3)
VII	To pair the T-test value from left to right
VIII	TEB = $E^n * M$ (4) where TEB is T-test encryption lattice *B*, *E* is an indivisible number, $n = 0$, and *M* is a non-negative whole number
IX	Recognize the conceivable number of indivisible numbers duplicate by the *M* for a request of a lattice
X	To trade an E and M in a lattice *B*
XI	To distinguish the quantity of conceivable pair of *M*
XII	TEC = $P_M * M$ (5) where TEC is T-test encryption grid *C*, *P* is outlined pair, and *M* is a non-negative number
XIII	To duplicate the *M* in the sets for just odd sets

- **Using Eqs. (2) and (9) and Table 2.**
- $\overline{X}_1 = \sum X_1/N_1$;
- $\overline{X}_1 = (3 + 5 + 7 + 11)/4$; $\overline{X}_1 = 26/4$; $\overline{X}_1 = 6.5$
- $\overline{X}_2 = \sum X_2/N_2$;
- $\overline{X}_2 = (13 + 17 + 19 + 23)/4$ $\overline{X}_2 = 18$

- **Using Eqs. (3) and (8)**
- $S_1 = \sqrt{\sum (X_1 - \overline{X}_1)^2/(N_1 - 1)}$ $S_2 = \sqrt{\sum (X_2 - \overline{X}_2)^2/(N_2 - 1)}$
- $S_1 = \sqrt{(35/(4 - 1))}$ $S_2 = \sqrt{(52/(4 - 1))}$
- $S_1 = \sqrt{(35/3)}$ $S_2 = \sqrt{(52/(3))}$
- $S_1 = 11.66$ $S_2 = 17.33$

Table 2 X_1 and X_2 values

X_1	$(X_1 - \overline{X}_1)$	$(X_1 - \overline{X}_1)^2$	X_2	$(X_2 - \overline{X}_2)$	$(X_2 - \overline{X}_2)^2$
3	−3.5	12.25	13	−5	25
5	−1.5	2.25	17	−1	1
7	0.5	0.25	19	1	1
11	4.5	20.25	23	5	25
	$\sum (X_1 - \overline{X}_1)^2$	35		$\sum (X_2 - \overline{X}_2)^2$	52

- **Using Eqs. (1) and (10)**
- T-test formula = $(\overline{X}_1 - \overline{X}_2) / \sqrt{((S_1^2/N_1) + (S_2^2/N_2))}$
- T-test formula = $(6.5 - 18) / \sqrt{((11.66^2/4) + (17.33^2/4))}$
- T-test formula = $11.5 / \sqrt{((135.95/4) + (300.32/4))}$
- T-test formula = $11.5 / \sqrt{((135.95 + 300.32)/4))}$
- T-test formula = $11.5 / \sqrt{(436.27/4)}$
- T-test formula = $11.5 / \sqrt{109.06}$
- "Pair the T-test value from left to right (1,1) (5,1) (0,9) (0,6) and swap it those numbers."

Step "1: (1,1)."

$$\text{TTE} = \begin{bmatrix} 401/4\ 402/4\ 403/4\ 404/4\ 405/4 \\ 406/4\ 407/4\ 408/4\ 409/4\ 410/4 \\ 411/4\ 412/4\ 413/4\ 414/4\ 415/4 \\ 416/4\ 417/4\ 418/4\ 419/4\ 420/4 \\ 421/4\ 422/4\ 423/4\ 424/4\ 425/4 \end{bmatrix}$$

where TTE is T-test encryption.

Step "2: (5,1)."

$$\text{TTE} = \begin{bmatrix} 401/4\ 406/4\ 403/4\ 404/4\ 405/4 \\ 402/4\ 407/4\ 408/4\ 409/4\ 410/4 \\ 411/4\ 412/4\ 413/4\ 414/4\ 415/4 \\ 416/4\ 417/4\ 418/4\ 419/4\ 420/4 \\ 421/4\ 422/4\ 423/4\ 424/4\ 425/4 \end{bmatrix}$$

Step "3: (0,9)."

$$\text{TTE} = \begin{bmatrix} 410/4\ 406/4\ 403/4\ 404/4\ 405/4 \\ 402/4\ 407/4\ 408/4\ 409/4\ 401/4 \\ 411/4\ 412/4\ 413/4\ 414/4\ 415/4 \\ 416/4\ 417/4\ 418/4\ 419/4\ 420/4 \\ 421/4\ 422/4\ 423/4\ 424/4\ 425/4 \end{bmatrix}$$

Step "4: (0,6)."

$$\text{TTE} = \begin{bmatrix} 407/4 & 406/4 & 403/4 & 404/4 & 405/4 \\ 402/4 & 410/4 & 408/4 & 409/4 & 401/4 \\ 411/4 & 412/4 & 413/4 & 414/4 & 415/4 \\ 416/4 & 417/4 & 418/4 & 419/4 & 420/4 \\ 421/4 & 422/4 & 423/4 & 424/4 & 425/4 \end{bmatrix}$$

- **Using Eqs. (4) and (6)**
- $EB = E^n * M \Rightarrow$ "$EB = 16$" $\Rightarrow$"$EB = 2^2 * 4$" $\Rightarrow$"$E = 2$," "$n = 2$," "$M = 4$"
- "$EB = 2^1 * 8$" $\Rightarrow$ "$E = 2$," "$n = 1$," "$M = 8$" $\Rightarrow$ "$EB = 3^0 * 16$"$\Rightarrow$ "$E = 3$," "$n = 0$," "$M = 16$" $\Rightarrow$ "Pair of numbers (2,4), (2,8), and (3,16)"

Step "5: (2,4)."

$$\text{TEB} = \begin{bmatrix} 407/4 & 406/4 & 405/4 & 404/4 & 403/4 \\ 402/4 & 410/4 & 408/4 & 409/4 & 401/4 \\ 411/4 & 412/4 & 413/4 & 414/4 & 415/4 \\ 416/4 & 417/4 & 418/4 & 419/4 & 420/4 \\ 421/4 & 422/4 & 423/4 & 424/4 & 425/4 \end{bmatrix}$$

Step "6: (2,8)."

$$\text{TEB} = \begin{bmatrix} 407/4 & 406/4 & 409/4 & 404/4 & 403/4 \\ 402/4 & 410/4 & 408/4 & 405/4 & 401/4 \\ 411/4 & 412/4 & 413/4 & 414/4 & 415/4 \\ 416/4 & 417/4 & 418/4 & 419/4 & 420/4 \\ 421/4 & 422/4 & 423/4 & 424/4 & 425/4 \end{bmatrix}$$

Step "7: (3,16)."

$$\text{TEB} = \begin{bmatrix} 407/4 & 406/4 & 409/4 & 417/4 & 403/4 \\ 402/4 & 410/4 & 408/4 & 405/4 & 401/4 \\ 411/4 & 412/4 & 413/4 & 414/4 & 415/4 \\ 416/4 & 404/4 & 418/4 & 419/4 & 420/4 \\ 421/4 & 422/4 & 423/4 & 424/4 & 425/4 \end{bmatrix}$$

- **Using Eqs. (5) and (7)**
- $EC = P_M * M \Rightarrow$ "$P_M = 4$" and "$M = 4$"
- $\mathbf{P_M = \{\}}$
- $P_4 = \{(), (1,2), (1,3), (1,4), (2,3), (2,4), (3,4), (1,2,3), (1,2,4), (1,3,4), (1,2,3,4)\}$
- "$EC = P_4 * M$" (only odd pairs)
- $EC =$ "{(4,8), (1,3), (4,16), (2,3), (8,16), (3,4), (4,8,12), (1,2,4), (4,12,16), (1,2,3,4)}"

- The request for 3 overlooks the first number from the second pair onward. The request for 4 is to shape a couple and check as of now exits, if exits discard the pair. "Pair of numbers are (4,8), (1,3), (4,16), (2,3), (8,16), (3,4), (4,12), (2,4), (12,16), (1,2)"

Step "8: (4,8)."

$$\text{TEB} = \begin{bmatrix} 407/4 & 406/4 & 409/4 & 417/4 & 405/4 \\ 402/4 & 410/4 & 408/4 & 403/4 & 401/4 \\ 411/4 & 412/4 & 413/4 & 414/4 & 415/4 \\ 416/4 & 404/4 & 418/4 & 419/4 & 420/4 \\ 421/4 & 422/4 & 423/4 & 424/4 & 425/4 \end{bmatrix}$$

Step "9: (1,3)."

$$\text{TEB} = \begin{bmatrix} 407/4 & 4017/4 & 409/4 & 406/4 & 405/4 \\ 402/4 & 410/4 & 408/4 & 403/4 & 401/4 \\ 411/4 & 412/4 & 413/4 & 414/4 & 415/4 \\ 416/4 & 404/4 & 418/4 & 419/4 & 420/4 \\ 421/4 & 422/4 & 423/4 & 424/4 & 425/4 \end{bmatrix}$$

Step "10: (4,16)."

$$\text{TEB} = \begin{bmatrix} 407/4 & 417/4 & 409/4 & 406/4 & 404/4 \\ 402/4 & 410/4 & 408/4 & 403/4 & 401/4 \\ 411/4 & 412/4 & 413/4 & 414/4 & 415/4 \\ 416/4 & 405/4 & 418/4 & 419/4 & 420/4 \\ 421/4 & 422/4 & 423/4 & 424/4 & 425/4 \end{bmatrix}$$

Step "11: (2,3)."

$$\text{TEB} = \begin{bmatrix} 407/4 & 417/4 & 406/4 & 409/4 & 404/4 \\ 402/4 & 410/4 & 408/4 & 403/4 & 401/4 \\ 411/4 & 412/4 & 413/4 & 414/4 & 415/4 \\ 416/4 & 405/4 & 418/4 & 419/4 & 420/4 \\ 421/4 & 422/4 & 423/4 & 424/4 & 425/4 \end{bmatrix}$$

Step "12: (8,16)."

$$\text{TEB} = \begin{bmatrix} 407/4 & 417/4 & 406/4 & 409/4 & 404/4 \\ 402/4 & 410/4 & 408/4 & 405/4 & 401/4 \\ 411/4 & 412/4 & 413/4 & 414/4 & 415/4 \\ 416/4 & 403/4 & 418/4 & 419/4 & 420/4 \\ 421/4 & 422/4 & 423/4 & 424/4 & 425/4 \end{bmatrix}$$

Step "13: (3,4)."

$$\text{TEB} = \begin{bmatrix} 407/4 & 417/4 & 406/4 & 404/4 & 409/4 \\ 402/4 & 410/4 & 408/4 & 405/4 & 401/4 \\ 411/4 & 412/4 & 413/4 & 414/4 & 415/4 \\ 416/4 & 403/4 & 418/4 & 419/4 & 420/4 \\ 421/4 & 422/4 & 423/4 & 424/4 & 425/4 \end{bmatrix}$$

Step "14: (4,12)."

$$\text{TEB} = \begin{bmatrix} 407/4 & 417/4 & 406/4 & 404/4 & 413/4 \\ 402/4 & 410/4 & 408/4 & 405/4 & 401/4 \\ 411/4 & 412/4 & 409/4 & 414/4 & 415/4 \\ 416/4 & 403/4 & 418/4 & 419/4 & 420/4 \\ 421/4 & 422/4 & 423/4 & 424/4 & 425/4 \end{bmatrix}$$

Step "15: (2,4)."

$$\text{TEB} = \begin{bmatrix} 407/4 & 417/4 & 413/4 & 404/4 & 406/4 \\ 402/4 & 410/4 & 408/4 & 405/4 & 401/4 \\ 411/4 & 412/4 & 409/4 & 414/4 & 415/4 \\ 416/4 & 403/4 & 418/4 & 419/4 & 420/4 \\ 421/4 & 422/4 & 423/4 & 424/4 & 425/4 \end{bmatrix}$$

Step "16: (12,16)."

$$\text{TEB} = \begin{bmatrix} 407/4 & 417/4 & 413/4 & 404/4 & 406/4 \\ 402/4 & 410/4 & 408/4 & 405/4 & 401/4 \\ 411/4 & 412/4 & 403/4 & 414/4 & 415/4 \\ 416/4 & 409/4 & 418/4 & 419/4 & 420/4 \\ 421/4 & 422/4 & 423/4 & 424/4 & 425/4 \end{bmatrix}$$

Step "17: (1,2)."

$$\text{TEB} = \begin{bmatrix} 407/4 & 413/4 & 417/4 & 404/4 & 406/4 \\ 402/4 & 410/4 & 408/4 & 405/4 & 401/4 \\ 411/4 & 412/4 & 403/4 & 414/4 & 415/4 \\ 416/4 & 409/4 & 418/4 & 419/4 & 420/4 \\ 421/4 & 422/4 & 423/4 & 424/4 & 425/4 \end{bmatrix}$$

2.2 *Decryption*

- RPBK21 method is Table 3 for decryption.

Table 3 Decryption algorithm

Steps	Decryption procedure
I	The information from scrambled information as a lattice A
II	$DA = D^n * M$ (6) where *DA* is unscrambling lattice *A*, *D* is an indivisible number, $n = 0$, and *M* is a non-negative integer
III	To recognize the number of potential sets of *M*
IV	$DB = P_M * M$ (7) where *DB* is unscrambling network *B*, *P* is outlined pair, and *M* is a non-negative whole number
V	To increase the *M* in the sets for just odd sets
VI	The request for 3 precludes the first number from the second pair onward
VII	The request for 4 is to shape a couple and check as of now exits, if exits exclude the pair
VIII	The subgroups should be traded values from right to left scrambled lattice
IX	Recognize the conceivable number of indivisible numbers duplicate by the *M* for a request of a lattice
X	To trade a *D* and *M* from right to left match in the scrambled framework *DB*
XI	To track down an indivisible number in framework TTD
XII	Similarly, independent the two pieces of indivisible numbers
XIII	$S_1 = \sqrt{\sum (X_1 - \overline{X}_1)^2/(N_1 - 1)} S_2 = \sqrt{\sum (X_2 - \overline{X}_2)^2/(N_2 - 1)}$ (8)
XIV	$\overline{X}_1 = \sum X_1/N_1 \quad \overline{X}_2 = \sum X_2/N_2$ (9)
XV	T-test formula = $(\overline{X}_1 - \overline{X}_2)/\sqrt{((S_1^2/N_1) + (S_2^2/N_2))}$ (10)
XVI	To consolidate the qualities Step XV from right to left
XVII	Step XVI values apply in the matrix

- **Using Eq. (7).**
- "Pair of numbers are (2,1), (16,12), (4,2), (12,4), (4,3), (16,8), (3,2), (16,4), (3,1), (8,4)"

Step "1: (2,1)."

$$\text{DB} = \begin{bmatrix} 407/4 & 417/4 & 413/4 & 404/4 & 406/4 \\ 402/4 & 410/4 & 408/4 & 405/4 & 401/4 \\ 411/4 & 412/4 & 403/4 & 414/4 & 415/4 \\ 416/4 & 409/4 & 418/4 & 419/4 & 420/4 \\ 421/4 & 422/4 & 423/4 & 424/4 & 425/4 \end{bmatrix}$$

Step "2: (16,12)."

$$\text{DB} = \begin{bmatrix} 407/4 & 417/4 & 413/4 & 404/4 & 406/4 \\ 402/4 & 410/4 & 408/4 & 405/4 & 401/4 \\ 411/4 & 412/4 & 409/4 & 414/4 & 415/4 \\ 416/4 & 403/4 & 418/4 & 419/4 & 420/4 \\ 421/4 & 422/4 & 423/4 & 424/4 & 425/4 \end{bmatrix}$$

Step "3: (4,2)."

$$\text{DB} = \begin{bmatrix} 407/4 & 4017/4 & 406/4 & 404/4 & 413/4 \\ 402/4 & 410/4 & 408/4 & 405/4 & 401/4 \\ 411/4 & 412/4 & 409/4 & 414/4 & 415/4 \\ 416/4 & 403/4 & 418/4 & 419/4 & 420/4 \\ 421/4 & 422/4 & 423/4 & 424/4 & 425/4 \end{bmatrix}$$

Step "4: (12, 4)."

$$\text{DB} = \begin{bmatrix} 407/4 & 4017/4 & 406/4 & 404/4 & 409/4 \\ 402/4 & 410/4 & 408/4 & 405/4 & 401/4 \\ 411/4 & 412/4 & 413/4 & 414/4 & 415/4 \\ 416/4 & 403/4 & 418/4 & 419/4 & 420/4 \\ 421/4 & 422/4 & 423/4 & 424/4 & 425/4 \end{bmatrix}$$

Step "5: (4,3)."

$$\text{DB} = \begin{bmatrix} 407/4 & 417/4 & 406/4 & 409/4 & 404/4 \\ 402/4 & 410/4 & 408/4 & 405/4 & 401/4 \\ 411/4 & 412/4 & 413/4 & 414/4 & 415/4 \\ 416/4 & 403/4 & 418/4 & 419/4 & 420/4 \\ 421/4 & 422/4 & 423/4 & 424/4 & 425/4 \end{bmatrix}$$

Step "6: (16,8)."

$$DB = \begin{bmatrix} 407/4 & 417/4 & 406/4 & 409/4 & 404/4 \\ 402/4 & 410/4 & 408/4 & 403/4 & 401/4 \\ 411/4 & 412/4 & 413/4 & 414/4 & 415/4 \\ 416/4 & 405/4 & 418/4 & 419/4 & 420/4 \\ 421/4 & 422/4 & 423/4 & 424/4 & 425/4 \end{bmatrix}$$

Step "7: (3,2)."

$$DB = \begin{bmatrix} 407/4 & 417/4 & 409/4 & 406/4 & 404/4 \\ 402/4 & 410/4 & 408/4 & 403/4 & 401/4 \\ 411/4 & 412/4 & 413/4 & 414/4 & 415/4 \\ 416/4 & 405/4 & 418/4 & 419/4 & 420/4 \\ 421/4 & 422/4 & 423/4 & 424/4 & 425/4 \end{bmatrix}$$

Step "8: (16,4)."

$$DB = \begin{bmatrix} 407/4 & 417/4 & 409/4 & 406/4 & 405/4 \\ 402/4 & 410/4 & 408/4 & 403/4 & 401/4 \\ 411/4 & 412/4 & 413/4 & 414/4 & 415/4 \\ 416/4 & 404/4 & 418/4 & 419/4 & 420/4 \\ 421/4 & 422/4 & 423/4 & 424/4 & 425/4 \end{bmatrix}$$

Step "9: (3,1)."

$$DB = \begin{bmatrix} 407/4 & 406/4 & 409/4 & 417/4 & 405/4 \\ 402/4 & 410/4 & 408/4 & 403/4 & 401/4 \\ 411/4 & 412/4 & 413/4 & 414/4 & 415/4 \\ 416/4 & 404/4 & 418/4 & 419/4 & 420/4 \\ 421/4 & 422/4 & 423/4 & 424/4 & 425/4 \end{bmatrix}$$

Step "10: (8,4)."

$$DB = \begin{bmatrix} 407/4 & 406/4 & 409/4 & 417/4 & 403/4 \\ 402/4 & 410/4 & 408/4 & 405/4 & 401/4 \\ 411/4 & 412/4 & 413/4 & 414/4 & 415/4 \\ 416/4 & 404/4 & 418/4 & 419/4 & 420/4 \\ 421/4 & 422/4 & 423/4 & 424/4 & 425/4 \end{bmatrix}$$

- **Using Eq. (6)**; "pair of numbers (16,3), (8,2), and (4,2)."

Step "11: (16,3)."

$$DA = \begin{bmatrix} 407/4\ 406/4\ 409/4\ 404/4\ 403/4 \\ 402/4\ 410/4\ 408/4\ 405/4\ 401/4 \\ 411/4\ 412/4\ 413/4\ 414/4\ 415/4 \\ 416/4\ 417/4\ 418/4\ 419/4\ 420/4 \\ 421/4\ 422/4\ 423/4\ 424/4\ 425/4 \end{bmatrix}$$

Step "12: (8,2)."

$$DA = \begin{bmatrix} 407/4\ 406/4\ 405/4\ 404/4\ 403/4 \\ 402/4\ 410/4\ 408/4\ 409/4\ 401/4 \\ 411/4\ 412/4\ 413/4\ 414/4\ 415/4 \\ 416/4\ 417/4\ 418/4\ 419/4\ 420/4 \\ 421/4\ 422/4\ 423/4\ 424/4\ 425/4 \end{bmatrix}$$

Step "13: (4,2)."

$$DA = \begin{bmatrix} 407/4\ 406/4\ 403/4\ 404/4\ 405/4 \\ 402/4\ 410/4\ 408/4\ 409/4\ 401/4 \\ 411/4\ 412/4\ 413/4\ 414/4\ 415/4 \\ 416/4\ 417/4\ 418/4\ 419/4\ 420/4 \\ 421/4\ 422/4\ 423/4\ 424/4\ 425/4 \end{bmatrix}$$

- Pair the T-test value from right to left (6,0), (9,0), (1,5), and (1,1) and swap those numbers.

$$TTD = \begin{bmatrix} 407/4\ 406/4\ 403/4\ 404/4\ 405/4 \\ 402/4\ 410/4\ 408/4\ 409/4\ 401/4 \\ 411/4\ 412/4\ 413/4\ 414/4\ 415/4 \\ 416/4\ 417/4\ 418/4\ 419/4\ 420/4 \\ 421/4\ 422/4\ 423/4\ 424/4\ 425/4 \end{bmatrix}$$

where TTD is T-test decryption.

Step "14: (6,0)."

$$TTD = \begin{bmatrix} 410/4\ 406/4\ 403/4\ 404/4\ 405/4 \\ 402/4\ 407/4\ 408/4\ 409/4\ 401/4 \\ 411/4\ 412/4\ 413/4\ 414/4\ 415/4 \\ 416/4\ 417/4\ 418/4\ 419/4\ 420/4 \\ 421/4\ 422/4\ 423/4\ 424/4\ 425/4 \end{bmatrix}$$

Step "15: (9,0)."

$$TTD = \begin{bmatrix} 401/4\ 406/4\ 403/4\ 404/4\ 405/4 \\ 402/4\ 407/4\ 408/4\ 409/4\ 410/4 \\ 411/4\ 412/4\ 413/4\ 414/4\ 415/4 \\ 416/4\ 417/4\ 418/4\ 419/4\ 420/4 \\ 421/4\ 422/4\ 423/4\ 424/4\ 425/4 \end{bmatrix}$$

Step "16: (1,5)."

$$TTD = \begin{bmatrix} 401/4\ 402/4\ 403/4\ 404/4\ 405/4 \\ 406/4\ 407/4\ 408/4\ 409/4\ 410/4 \\ 411/4\ 412/4\ 413/4\ 414/4\ 415/4 \\ 416/4\ 417/4\ 418/4\ 419/4\ 420/4 \\ 421/4\ 422/4\ 423/4\ 424/4\ 425/4 \end{bmatrix}$$

Step "17: (1,1)."

$$TTD = \begin{bmatrix} 401/4\ 402/4\ 403/4\ 404/4\ 405/4 \\ 406/4\ 407/4\ 408/4\ 409/4\ 410/4 \\ 411/4\ 412/4\ 413/4\ 414/4\ 415/4 \\ 416/4\ 417/4\ 418/4\ 419/4\ 420/4 \\ 421/4\ 422/4\ 423/4\ 424/4\ 425/4 \end{bmatrix}$$

3 Result and Discussion

The proposed algorithm RPBK21 encryption performance is compared with ChaCha, AES, and existing RBJ25. ChaCha concept is that the all diagonal values are moved to the first column. AES concept is the 16-round processes.

RBJ25 concept is quadratic with key and column operations. The three by three matrix has 24 bytes of fie size; the six by six matrix has 76 bytes of file size; the ten by ten matrix has 312 bytes of file size; the fifteen by fifteen matrix has 812 bytes of file size; the 20×20 matrix has 1531 bytes of file size; and the forty by forty matrix has 6580 bytes of file size as shown in Table 4.

From Fig. 1, the RPBK21 technique has compared the encryption speed in seconds. The RPBK21 gives more protection of the data; when compared to existing techniques.

Table 4 RPBK21 encryption performance

File size	ChaCha	AES	RBJ25	RPBK21
24	1.69	1.225	2.2	4.1
76	1.29	1.852	2.6	4.3
312	2.73	1.764	4.5	5.1
822	2.64	3.179	5.3	5.6
1531	3.4	2.039	5.5	6.5
6580	2.27	3.221	6.8	7.5

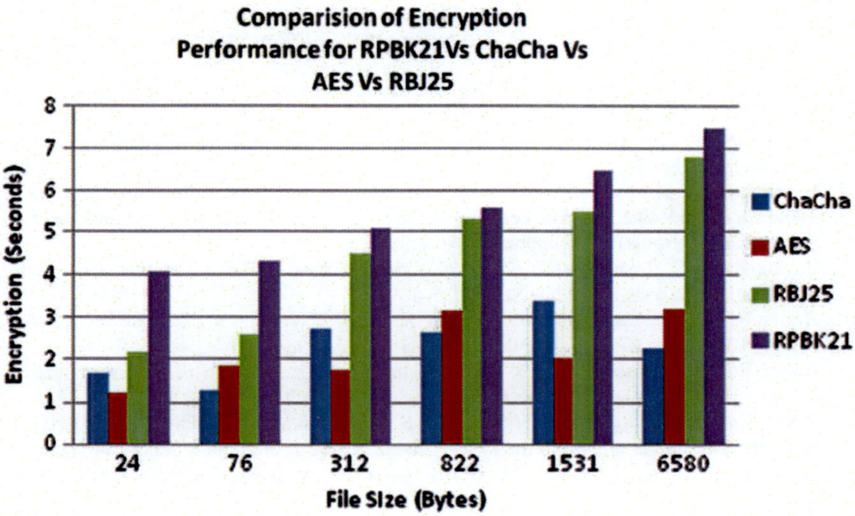

Fig. 1 Encryption performance

4 Conclusion

The proposed technique has successfully hacked the data from the software engineers. BBRP24 strategy has seven phases by finding the mystery prime key along with finding X_1, X_2, $\overline{X}_1$, and $\overline{X}_2$ values from prime numbers. Furthermore, the standard deviation values are obtained by considering the a and b values from left and T-test values pair from left. The quadratic equations are solved by utilizing the prime numbers. In a nutshell, it has been proven that the RPBK21technique leverages great security when contrasted and ChaCha, AES, and RBJ25 strategy. In the future, the prime variables tasks of the information security will be integrated.

References

1. Dilip Kumar SV, Patranabis S, Breier J, Mukhopadhyay D, Bhasin S, Chattopadhyay A, Baksi A (2017) A practical fault attack on ARX-like Cipherwith a case study on ChaCha20 (workshop on fault diagnosis and tolerance in cryptography (FDTC) 33:40
2. Arun Babu P, Jose Thomas J (2019) Freestyle, a randomized version of ChaCha for resisting offline brute-force and dictionary attacks. J Inf Secur Appl 102396
3. Adomnicai A, Jacques Fournier JA, Masson L (2017) Bricklayer attack: a side-channel analysis on the ChaCha quarter round. Progress in Cryptology—INDOCRYPT 2017. Lecture Notes in Computer Science, Springer, Cham, vol 10698, pp 65–84
4. Mazumdar B, Subidh Ali Sk, Sinanoglu O (2015) Power analysis attacks on ARX: an application to Salsa20. IEEE 21st international on-line testing symposium, pp 40–43
5. Watt C, Renner J, Popescu N, Cauligi S, Stefan D (2019) CT-Wasm: type-driven secure cryptography for the web ecosystem. In: Proceedings of the ACM on programming languages, pp 77:1–77:29
6. Bagath Basha C, Rajapraksh S (2020) Enhancing the security using SRB18 method of embedding computing. Microproces Microsyst
7. Bagath Basha C, Rajaprakash S (2019) Securing Twitter Data using phase I methodology. Int J Sci Technol Res 1952–1955
8. Bagath Basha C, Rajaprakash S (2020) Applying the SRB21 Phase II methodology for securing Twitter analyzed data. In: AIP conference proceedings of the international conference on mechanical electronics and computer engineering
9. Bagath Basha C, Rajaprakash S (2020) Applying the CBB21 Phase 2 method for securing Twitter analyzed data. Adv Math Sci J 1085–1091
10. Bagath Basha C, Rajaprakash S, Harish VVA, Krishna MS, Prabhas K (2020) Securing Twitter analysed data using CBB22 algorithm. Adv Math Sci J 1093–1100
11. Bagath Basha C, Somasundaram K (2019) A comparative study of Twitter sentiment analysis using machine learning algorithms in big data. Int J Recent Technol Eng 591–599
12. Bagath Basha C, Rajaprakash S, Muthuselvan S, Saisatishsunder P, Alekhya Rani SVL (2020) Applying the CBB20 algorithm for Twitter analyzed data. J. Phy. Conf. Ser. In: First international conference on advances in physical sciences and materials, Coimbatore, Tamil Nadu, India
13. Rajaprakash S, Bagath Basha C, Muthuselvan S, Jaisankar N, Pratap Singh R (2020) RBJ25 cryptography algorithm for securing big data. J Phys Conf Ser. In: First international conference on advances in physical sciences and materials, Coimbatore, Tamil Nadu, India
14. Prakash UM, Kottursamy K, Cengiz K, Kose U, Hung BT (2021) 4x-expert systems for early prediction of osteoporosis using multi-model algorithms. Measurement 180:109543
15. Banupriya S, Kottursamy K, Bashir AK (2021) Privacy-preserving hierarchical deterministic key generation based on a lattice of rings in public blockchain. Peer-to-Peer Netw Appl 1–13
16. Vijayakumar KP, Kumar KPM, Kottilingam K, Karthick T, Vijayakumar P, Ganeshkumar P (2019) An adaptive neuro-fuzzy logic based jamming detection system in WSN. Soft Comput 23(8):2655–2667
17. Cengiz K, Sharma R, Kottursamy K, Singh KK, Topac T, Ozyurt B (2021) Recent emerging technologies for intelligent learning and analytics in big data. In: Multimedia technologies in the internet of things environment, pp 69–81. Springer, Singapore
18. Raja G, Kottursamy K, Theetharappan A, Cengiz K, Ganapathisubramaniyan A, Kharel R, Yu K (2020) Dynamic polygon generation for flexible pattern formation in large-scale uav swarm networks. In: 2020 IEEE Globecom Workshops, GC Wkshps, pp 1–6. IEEE
19. Mehra R, Pachpor K, Kottilingam K, Saranya A (2020) An initiative to prevent Japanese encephalitis using genetic algorithm and artificial neural network. In: 2020 international conference on computational intelligence (ICCI), pp 142–148. IEEE
20. Saranya A, Kottilingam K (2021) A survey on bone fracture identification techniques using quantitative and learning based algorithms. In: 2021 international conference on artificial intelligence and smart systems (ICAIS), pp 241–248. IEEE

21. Nallarasan V, Kottilingam K (2021) Spectrum management analysis for cognitive radio IoT. In: 2021 international conference on computer communication and informatics (ICCCI), pp 1–5. IEEE
22. Raja G, Kottursamy K, Chaudhary SH, Hassan A, Alqarni M (2017) SDN assisted middlebox synchronization mechanism for next generation mobile data management system. In: 2017 IEEE SmartWorld, ubiquitous intelligence and computing, advanced and trusted computed, scalable computing and communications, cloud and big data computing, internet of people and smart city innovation, SmartWorld/SCALCOM/UIC/ATC/CBDCom/IOP/SCI), pp 1–7. IEEE

Classification of Extreme Reviews from Online Products Using RNN Model

Satuluri Naganjaneyulu, G. Tarun, Y. Sriram, B. Rama Devi, and P. Manoj

Abstract Online reviews on a particular product play an important role in the sales of a product. There exist genuine reviews where the customers review their opinion about the products out of joy or frustration. But, from the study, we came to know that some people get hired by certain brands to provide reviews which are not actually genuine. This reviewing is also done as group. Although the previous methods exist for detecting fake reviews, but they are just limited to identify the individual reviews. In this paper, the group reviews which are concentrating on a particular brand will be detected as extremist reviews. The characterizing is done using the algorithm RNN algorithm by which is also used to train data by itself. RNN is recurrent neural networks from TensorFlow which is same as the regular artificial neural networks, but the only difference is RNN consists of an additional memory for computations. Here, we used different layers of the network like LSTM, dense and dropout layers, and a softmax layer for classification of the reviews as extremist or moderate.

Keywords Characterizing · RNN · Reviews · Classification · Extremist · Moderate · Brand · Computation · LSTM · Neural networks

1 Introduction

E-commerce Web sites have been experiencing a lot of spam reviews in today's online world. Actually, the reviews will be delivered by the customers to express their opinion toward the product in both negative and positive reviews. We consider this kind of reviews as genuine reviews. On the other hand, there exist the reviews which are not genuine. These reviews come from the people who have been hired from the several brands to promote their brand through reviewing their products at

S. Naganjaneyulu (✉)
Department of IT, Lakireddy Balireddy College of Engineering, Mylavaram, Andhra Pradesh, India
e-mail: naganjaneyulu.s@lbrce.ac.in

G. Tarun · Y. Sriram · B. Rama Devi · P. Manoj
Lakireddy Balireddy College of Engineering, Mylavaram, Andhra Pradesh, India

S. Shakya et al. (eds.), *Proceedings of Second International Conference on Sustainable Expert Systems*, Lecture Notes in Networks and Systems 351,
https://doi.org/10.1007/978-981-16-7657-4_25

various e-commerce platforms. The same people also tend to attack on the other products by demoting the particular brand products by giving the negative reviews. Eventually, this process would cause the loss to other genuine brand products. This process is generally done in groups. Amazon has earlier observed such practices manually by taking the data of 923 group of reviewers. They have done the analysis on how the certain group is targeting a brand and characteristics of reviewers. But, doing the analysis manually on heavy datasets is a big challenge. So, we need an automated system.

Many systems have been developed to detect the spam reviewers so as to minimize their activity on the review of products. Most of the existing systems only concentrated on detecting an individual spam reviews but not the group reviews. Most of the algorithms used are unsupervised. The result of this eventually falls on the efficiency and accuracy. The traditional methods of the detection of spam reviews involves the sentimental analysis, classification algorithms, and naïve Bayes algorithm by which we can simply classify the review based on the data that have trained before.

In the proposed system, instead of detecting the individual spam review, we can be able to detect the extremist groups which are not exhibiting normal behavior. This also being an unsupervised system will train the data by itself many times until the perfect accuracy meets.

Algorithms used in this system are RNN and perceptron, where RNN is used to train the dataset by itself by remembering the previous outputs and give them as inputs to subsequent steps. This is how the datasets are being trained and unsupervised learning is achieved. And, perceptron algorithm is used as classifier which is used to differentiate the extremist group and moderate group. In functionality, when a desired review is provided in the form of dataset, then the system classifies the reviews as moderate ones and extremist groups. The accuracy and efficiency of this system are high when compared to other models [1–7].

2 Literature Survey

Detecting of extremist reviewer groups is done in this system. The previous papers that are taken into the consideration mainly deal with the individual reviews which mainly consist of sentimental analysis. Below are the papers and their brief that are taken for the literature survey and implementation [1].

This paper was proposed by Viresh Guptha, Aayush Agarwal, Tanmoy Chakraborty. In this system, they suggested SVM and random forest algorithms for classification of online extremist reviews. This system takes the data from the user and classifies the given review as moderate or extremist. This system makes use of supervised learning. The system gives accurate results about the review whether it is a moderate or extremist. But, the system gives low accuracy because of its classification algorithm and supervised learning used [2].

The paper proposed is by Mayur Gaikwad, Swati Ahirrao, Ketan Kotecha. This system uses the naïve Bayes algorithm to detect whether the review is extremist or not. This system involves the cleaning, visualization of data, and applying of machine learning algorithms. In this paper, the reviews from the Amazon product review site reviews are collected. It consists of 926 reviewer groups which are manually labeled. As the system involves in the usage of naïve Bayes, it produces the accurate results for detecting the extremist group, but only drawback that can find from this system is it is a supervised learning algorithm [3].

The paper is proposed by Shakeel Ahmad, Muhammad Zubair, Fahad M. Alotaibi, and Irfanullah Awan. This paper classifies the extremist-related tweets. In this system, the data are trained in a way such that the system can be able to differentiate test data into the extreme terms and moderate terms [4]. Based on the training data, the system can classify the input tweet into an extremist or moderate using the sentimental analysis. In this paper, CNN-LSTM is applied for differentiation of tweets where CNN is used to feature extraction and LSTM is used to obtain sequential correlation by taking the input as CNN output. As the algorithm used is CNN, the efficiency of system is high, but when comes to the accuracy, it is hard to obtain because of its sentimental usage [5].

This paper is proposed by Michael Crawford, Taghi M. Khoshgoftaar, Hamzah Al Najada [6]. This paper concentrates about detecting the spam reviews from the online product reviews. This process is done by extracting the relevant features from the reviews using NLP where NLP stands for natural language processing. Another advantage in this system is in addition to detecting the spam reviews, and this system also detects the reviewer based on their behavior on the product reviews [7]. This paper also concentrates on high comparative research on detecting spam with the help of different machine learning algorithms and shows path for methodology to conduct additional investigations.

3 Proposed System

In e-commerce sites, every customer has the facility of reviewing. But, fake reviews can lead to few more customers to make negative reviews. This influence can be minimized by detecting extremist groups that are concentrating on particular brand. In the proposed system, the extremist groups are detected using RNN and perceptron algorithm.

In the system, dataset is taken from the Amazon customer reviews, and preprocessing is done. As a part of preprocessing, tokenizing and stemming of data are done where the given reviews are converted into natural languages and further parts. In the next part, the data are trained which act as a reference to the test data. As the algorithm is an unsupervised learning perceptron, we have made to train the data by itself until the maximum accuracy meets.

RNN algorithm then classifies the reviews of the dataset as either extremist or moderate review based on the data that have trained earlier (Fig. 1).

A recurrent neural network is an artificial class neural networks in which connections between node come from directed graph with a temporal relation. This is the

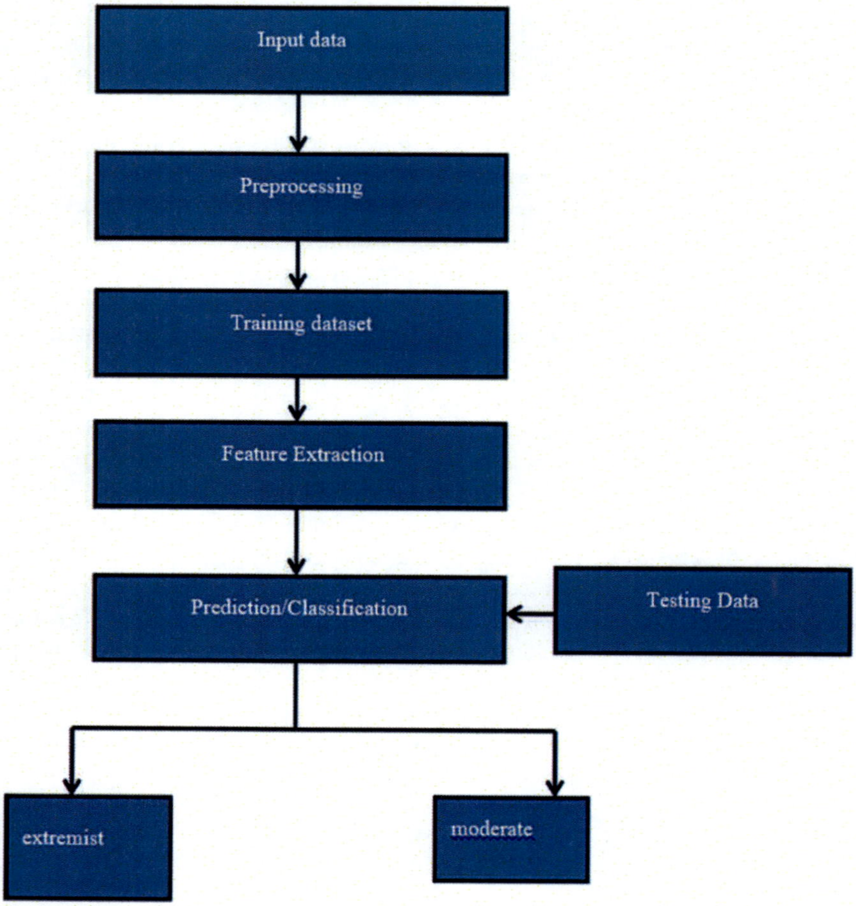

Fig. 1 Flow diagram of the proposed system

reason why this algorithm shows dynamic behavior. RNN algorithm takes input of any lengths and processes by using their internal memory to produce sequences. This makes the algorithm to be an unsupervised learning algorithm. In addition to this processing, the algorithm is also known for its accuracy and efficiency. RNN is mostly used in a situation where the temporal dynamics plays an important role more than the spatial context.

Perceptron algorithm is a linear classifier algorithm which is used to classify the input into desired category. It uses linear predictor function to done predictions. In this system, it takes the output from the RNN and takes it as reference for classifying the given input dataset as either extremist or moderate. This algorithm works on a large dataset as it can store array of 400 photocells. Another important advantage of the algorithm is its accuracy, and the algorithm makes updated after every iteration of training sets.

Table 1 Accuracy table

```
208/208 [==============================] - 1s 5ms/step - loss: 0.0467 - accuracy: 0.9838
Epoch 50/50
208/208 [==============================] - 1s 5ms/step - loss: 0.0473 - accuracy: 0.9824
Model Trained!

[39]: score = model.evaluate(x_test, encoded_y_test)

104/104 [==============================] - 1s 3ms/step - loss: 0.2967 - accuracy: 0.9244

[40]: print("Accuracy = " + format(score[1]*100, '.2f') + "%")

Accuracy = 92.44%
```

4 Results and Discussion

After training of data and applying of algorithms on datasets, the accuracy obtained is as in Table 1.

The confusion matrix that drawn between true label and predicted label is:

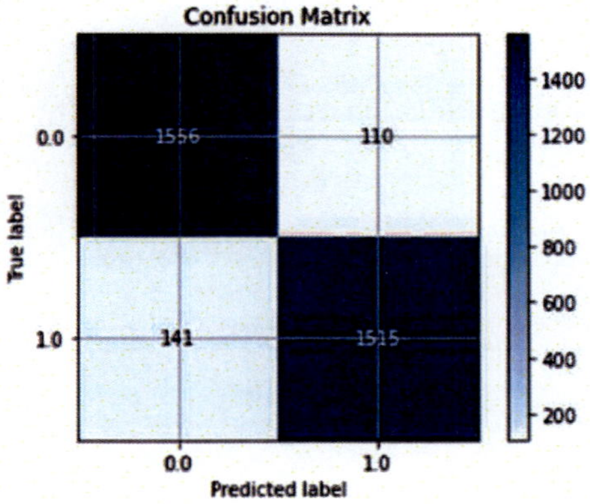

After training of data, algorithms have been applied on dataset, and the results for the Apple Iphone reviews dataset are as below shown.

	Reviewer	Rating	Review title	Review description	Result_category
0	Sanjay Tiwari	4.0 out of 5 stars	Balanced phone	Well nice product for daily use. Nice display ...	Moderate
1	Amazon customer	1.0 out of 5 stars	Not satisfied with this product. I do not want ...	I am not satisfied with this product, because ...	Extreme
2	Sukhjeet Sing	5.0 out of 5 stars	Nice phone under 25 k	Your browser does not support HTML5 video ...	Moderate

(continued)

(continued)

	Reviewer	Rating	Review title	Review description	Result_category
3	Gurumoorthy	3.0 out of 5 stars	Decent but with a doubtful battery	Delivers what is in the spec. Decent enough …	Moderate
4	Bhavesh Bhargav	5.0 out of 5 stars	Best	My luck is so bad because I receive 29th …	Moderate
5	Bijoy PK	2.0 out of 5 stars	V20 is a Dampner!! Considering the price it …	I exchanged my two year old vivo V9 pro to V20…	Moderate
6	Gopaal T	1.0 out of 5 stars	Waste of money	Average product… Vivo other mobiles are good…	Moderate
7	Pushpraj Upadhyay	1.0 out of 5 stars	Very much hanging issue	Please all friends purchase this phone carefully …	Extreme
8	Manish	5.0 out of 5 stars	What a camera!!!!	This review is after a week of using the phone …	Extreme
9	Md Altaf Hussain	3.0 out of 5 stars	Disappointing by product day by day	Mobile looks good, and camera is awesome but …	Extreme

5 Conclusion

This system produces the accurate results by using forward and backward propagation recursively. Being a system with unsupervised learning, it decreases the interaction of human. The system can be extended in the future by combining various platform reviewers and can be grouped.

References

1. Alatawi HS, Alhothali AM, Moria KM (2020) Detecting white supremacist hate speech using domain specific word embedding with deep learning and BERT. arXiv:2010.00357. [Online]. Available: http://arxiv.org/abs/2010.00357
2. PyTorch Documentation (2020) Embedding Layer PyTorch. Accessed: 10 Oct 2020 [Online]. Available: https://pytorch.org/docs/stable/generated/torch.nn.Embedding.html

3. Chatfield AT, Reddick CG, Brajawidagda U (2015) Tweeting propaganda, radicalization and recruitment: Islamic state supporters multisided Twitter networks. In: Proceedings of 16th *annual international conference* on digital *government research*, May 2015, pp 239–249. https://doi.org/10.1145/2757401.2757408
4. Berger JM (2016) Nazis vs. ISIS on Twitter: A comparative study of white nationalist and ISIS online socialmedia networks. George Washington University, Washington, DC, USA, Technical Report, 2016 [Online]. Available: https://extremism.gwu.edu/sites/g/_les/zaxdzs2191/f/downloads/Nazisv.ISIS.pdf
5. Lara-Cabrera R, Pardo AG, Benouaret K, Faci N, Benslimane D, Camacho D (2017) Measuring the radicalisation risk in social networks. IEEE Access 5:10892–10900. https://doi.org/10.1109/ACCESS.2017.2706018
6. Lara-Cabrera R, Gonzalez-Pardo A, Camacho D (2019) Statistical analysis of risk assessment factors and metrics to evaluate radicalization in Twitter. Future Gener Comput Syst 93:971–978. https://doi.org/10.1016/j.future.2017.10.046
7. De Bruyn PC (2020) Predicting behavioral profiles of online extremists through linguistic use of social roles.Behav Sci Terrorism Polit Aggression 1–25. https://doi.org/10.1080/19434472.2020.1775675; Benigni M (2017) Detection and analysis of online extremist communities, Ph.D. dissertation, Inst. Softw. Res., Carnegie Mellon University, Pittsburgh, PA, USA

A Novel Text-to-Speech Synthesis System Using Syllable-Based HMM for Tamil Language

J. Samuel Manoharan

Abstract A text-to-speech (TTS) is an intellectual system that converts the given language text into speech output. TTS system synthesizer can be evaluated using different aspects such as naturalness, ability of expressing intelligence, preference of synthetic speech, and some of the human perception factors like comprehensibility. TTS system is used in many application areas where the human assistance is to be removed. For example, the TTS program allows people with visual impairments or reading disabilities to listen to written words on a home computer. TTS system for Tamil language has the complexity of adaptive pronunciations that are dependent on the language constructs. This paper introduces a hidden Markov model (HMM)-based speech synthesis for Tamil language.

Keywords Text-to-speech conversion · Natural language processing · Human perception · Tamil language database · Hidden Markov model

1 Introduction

Natural language processing (NLP) is the technique of manipulating the natural language automatically. Natural language is the way, namely text and speech, through which the human beings communicate with each other. Natural language processing is the domain where the human languages can be understandable by computer and do the activities according to the instruction of human being. Text-to-speech synthesizer, automatic speech recognition, analysis of semantics, etc., are the various research fields that are related with the NLP. Text-to-speech (TTS) synthesizer is getting importance nowadays due to the requirement of automatic conversion of text to the corresponding voice in so many application fields such as reading emails

J. Samuel Manoharan (✉)
Department of ECE, Sir Isaac Newton College of Engineering and Technology, Nagapattinam, India
e-mail: drjsm1530@ieee.org

S. Shakya et al. (eds.), *Proceedings of Second International Conference on Sustainable Expert Systems*, Lecture Notes in Networks and Systems 351,
https://doi.org/10.1007/978-981-16-7657-4_26

aloud, providing services to hearing impaired people. TTS synthesizer is a computer-based system that converts the given text into speech. The general block diagram of conventional TTS system is shown in Fig. 1.

The working of TTS falls under two domains, one is natural language processing that deals with the processing of text analysis, and the second one is digital signal processing that deals with the speech generation and analysis. The general architecture of TTS synthesizer is shown in Fig. 2. It consists of two important phases, named as text analysis and speech synthesizers, for conversion of text into speech. The text analysis phase is the one that converts the given text into phonemic representation, and the speech synthesizer is the one that is used to convert the phonemic representation into waveforms and then these waveforms can be concatenated to obtain the speech representation of the given text. In view of the latter, second phase is also named as waveform synthesizer.

Fig. 1 Block diagram of TTS synthesizer

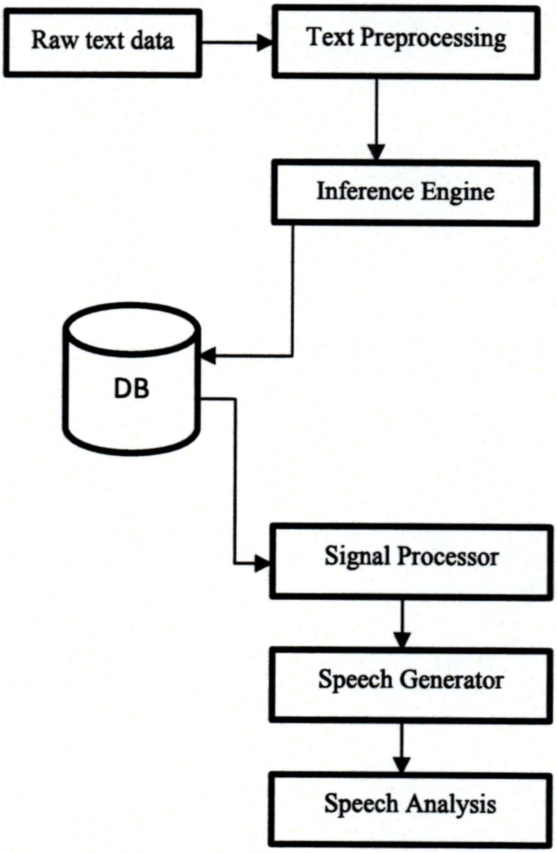

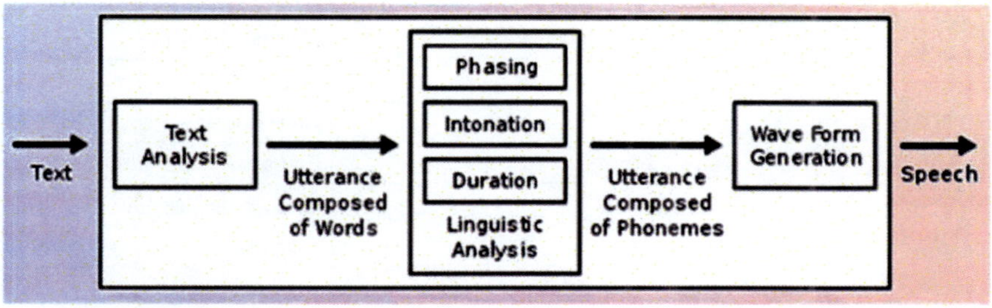

Fig. 2 Process flow of TTS synthesizer

1.1 Text Analysis

The first phase of TTS synthesizer is text analysis, and this phase is further classified into three steps called text preprocessing, phonetic analysis, and prosodic analysis. Text preprocessing is the method of preprocessing the text in such a way that the text is modified to a form that is easily predictable or analyzable for completing particular task. Lowercasing, stemming, lemmatization, stop words, and text normalization are the various techniques used for text preprocessing. Lowercasing is the process to lowercase all the text data. For example, the words "Welcome," "WELCOME" are preprocessed as "welcome." Stemming is the process of diminishing the given text to their root form by cutting off certain letters. For example, the word "troublesome" can be stemmed as "trouble." Lemmatization is the process of mapping the given word to its actual root form using rule-based approaches. For example, the word "best" can be mapped to "good." Text normalization methodology is getting importance among all other methodologies of text preprocessing. Text normalization is the technique of transforming the text into standard form. For example, the given text "good" and "gud" can be normalized as "good."

Social media comments, blog posts comments, etc., are encompassed with abbreviations, out of vocabulary words, and misspelling words. To deal with these noisy texts, text normalization mechanism is necessary. The text normalization is actually classified into two categories called sentence-level tokenization and word-level tokenization. In sentence-level tokenization, the document structure is analyzed, and the end of the sentence such as colon, double colon is identified in order to tokenize the sentence. In word-level tokenization, the given word is transformed to standard form as shown in Table 1.

Internal structure or processing of TTS consists of phases such as phonetic analysis, prosody analysis, articulatory synthesis, concatenative synthesis, and formant synthesis. Phonetic analysis is the branch of linguistics that determines the pronunciation of sound for each text produced by text normalization. The grapheme-to-phoneme (G2P) conversion technique is used for pronunciation. In prosodic analysis, the prosodic features such as intonation, tone rhythm, and stress for each phoneme are examined.

Table 1 Text normalization

Raw data instances	Normalized data
Gud	Good
Havent	Have not
8teen	Eighteen
Asap	As soon as possible
Tc	Take care

Speech synthesizer or wave form synthesizer can be categorized into three synthesis, named as articulatory synthesis, formant synthesis, and concatenative synthesis. The articulation of the speech is recognized by six major structures such as the jaw, velum, tongue body, tongue tip, lips, and hyoid bone. The articulator's position can be controlled manually for producing steady-state utterances or by means of a table of specification for dynamic productions. The amplitude of the glottal excitation, fundamental frequency, and the mode of representations can be controlled by the table of parameters. The amplitude of fricative excitation can also be specified. The segments of recorded speech are concatenated together to form speech which is called concatenative synthesis. It has the drawback of attaining audible glitches due to the difference experienced between nature of speech variations and automatic technique that are segmenting the waveform. Unit selection synthesis, diphone-based synthesis are the most commonly used concatenative synthesizers. The synthesized speech output is created by the combination of additive synthesis and an acoustic model which is known as the formant analysis. It is also named as rules-based synthesis; that is, the waveform of the artificial speech can be created using the rules. At the same time, parameters like fundamental frequency, voicing, and noise vary based on time to create waveforms.

Section 2 discusses the survey of the literature about TTS system for different languages, Sect. 3 discusses the proposed methodology for Tamil TTS system, Sect. 4 discusses the results, and Sect. 5 concludes.

2 Related Works

Text normalization can be applied by sorting the word into two categories, one is standard words which is in dictionary, and the second one is non-standard words (NSW). A new taxonomy for four different domains is proposed. The domains are newswire, a recipes newsgroup, hardware-product-specific newsgroup, and advertisements newsgroup. The corpora are a set of U.S. English words. The tools for normalizing these words are publicly available. These tools are applicable for different texts such as short messaging service [1], email [2], microblog text from Twitter [3]. "Do no harm approach" [4] has been introduced for attaining high-precision abbreviations. Heuristic-based normalization [5] has been developed for Internets and SMS text. After tokenizing the text, the non-standard words are detected by comparing the tokens against the word list, that is, the set of English words in the word list corpus and the set of alphabetic words in the brown corpus. After detecting the NSW, the words are classified under four general classes named as ALPHA for alphabetic tokens, NUMB for numeric tokens, SPLT for mixed token that needs further splitting, MISC for others. A new NSW taxonomy is proposed with the implementation in Python programming language [6]. Semi-supervised label propagation algorithm is proposed for ALPHA and NUMB tokens, and a rule-based mechanism is proposed for MISC tokens. The system consists of a dictionary of abbreviation for supporting domain-based activity. For classification of ALPHA, 13 domain-independent features and for NUMB, 26 domain-independent features are used. The system produces accuracy of more than 91%.

A hybrid approach based on N-gram model [7] is proposed for translating the short messages to English text. A corpus is maintained for storing the rule-based abbreviations. The combination of rule-based approach, direct mapping approach, and statistical machine translation approach is proposed for applying normalization over various types of text. The system produced 89.5%. A concatenative synthesis [8] approach with phoneme-to-grapheme is proposed. The given text is analyzed using bi-gram model and then phonetic conversion is implemented using dictionary. The prosody phasing is allocated for each token. The speech output for the given text is generated using the combination of unit selection concatenative synthesis, phoneme-based speech synthesis, and domain-specific speech synthesis.

A TTS that recognizes different font types of Indian languages has built [9]. A corpus is constructed for each font type of various Indian language Web sites. TF-IDF is used for recognition of font types, and base-map table has built for mapping the font to Akshara. Normalization is obtained by constructing word-level decision tree as a baseline. Acoustic-based phonetic speech synthesizer is used for generating the output. Moreover the normalization with syllable level is also constructed, and the performance is compared in which the syllable level shows better performance. A new model based on RNN, LSTM, RNN + FST [9] is proposed for normalization of English and Russian texts. The 16 NSW types of English and Russian texts are transformed into either plain English text or plain Russian text. The performance of all the model is compared. RNN model produces better accuracy than LSTM model. To avoid the wrong expansion, FST is added with RNN.

A deep learning-based text normalization techniques, context-aware classification (CAC) model for semiotic class, and sequence-to-sequence-based two-layer LSTM model for expansion of NSW are proposed [10]. The system gets better accuracy using these models, but sometimes, it leads worse when the expansion of NSW is complicated. To overcome this issue, the regular expression-based text normalization is proposed. An unsupervised model [11] that creates a lookup table for NSW is proposed. The unlabeled Twitter data-based corpus is constructed for text normalization. The model creates low-dimensional word embedding and semantic similarity for identifying NSW and also identifies OOV word for each English word. Moreover, a discriminative re-ranking for both word level and sentence level is proposed. TTS for all regional Indian languages [12] Marathi, Punjabi, Bengali, Kannada, Tamil, Hindi is proposed. The strategy for synthesizer, database creation and used, segmentation process for normalization, prosody analysis, and performance of the TTS system, etc., are compared.

Architecture for phrase-based translation scheme [13] for conversion of numerical element and abbreviations is discussed. The architecture has three modules, named as tokenizer module, phrase-based translation module, and post-processing module. The tokenizer module is used to split the given text into tokens graph using rule-based methodology. In phrase phase translation module, N-gram-based methodology is used to perform the translation of phrases with the help of parallel corpus, and post-processing module is used to remove the non-converted token. A probabilistic [14]-based TTS synthesizer is proposed. Text normalization for email data is implemented in three different levels such as paragraph level, sentence level, word level. The system has implemented as a two-step process: 1. text preprocessing and 2. tagging text. In text preprocessing, the given text is separated into paragraph and then tokens are separated. In tagging process, each token is tagged. For tagging, maximum likelihood estimation-based iterative algorithm is used.

A three-component-based TTS system based on naïve Bayes classifier [15] has introduced for token-to-token, SVM classifier for token-to-class, and regular expression-based grammar rules. The naïve Bayes classifier produces 98.97% accuracy, and SVM classifier produces 98.88% accuracy. The classification tree-based [16] text normalization technique is proposed for Croatian text. The classification tree contains NSW as the root node, and the number, character, and combination of both are used as child nodes. The algorithm is implemented using Perl language. The corpus contains Croatian data of size 111 K. The algorithm produces at least 80% of the tokens correctly. TTS system for Malayalam regional language [17] is proposed. Memory-based hierarchical model is proposed for text preprocessing. Memory prediction framework is used to display the hierarchical structure of the sentence level in which the sentence is used as root node, and the syllables of the sentences are used as child nodes. The concatenative synthesis is used to construct speech output for the given sentence using the syllable structure of (C) (C) (C) V (C) C where C denotes consonants and V denotes vowels. The Ziff's law is proposed for producing hyperbolic function of the natural language utterance from the corpus. N-gram model is proposed for correct semantic of the given word. This proposed methodology has introduced the concept of syllable-based N-gram model for natural language processing part and HMM-based speech synthesis in digital signal processing.

3 Proposed Methodology

Tamil TTS system is getting importance compared to other languages due to the factors such as language complexity, and the grammar faced by the language is also high. There is no standard Tamil TTS system that reduces the gap between the spoken language and the written language. The deviations are felt in understanding of Tamil phonetics and database creation of Tamil language. HMM is actively used

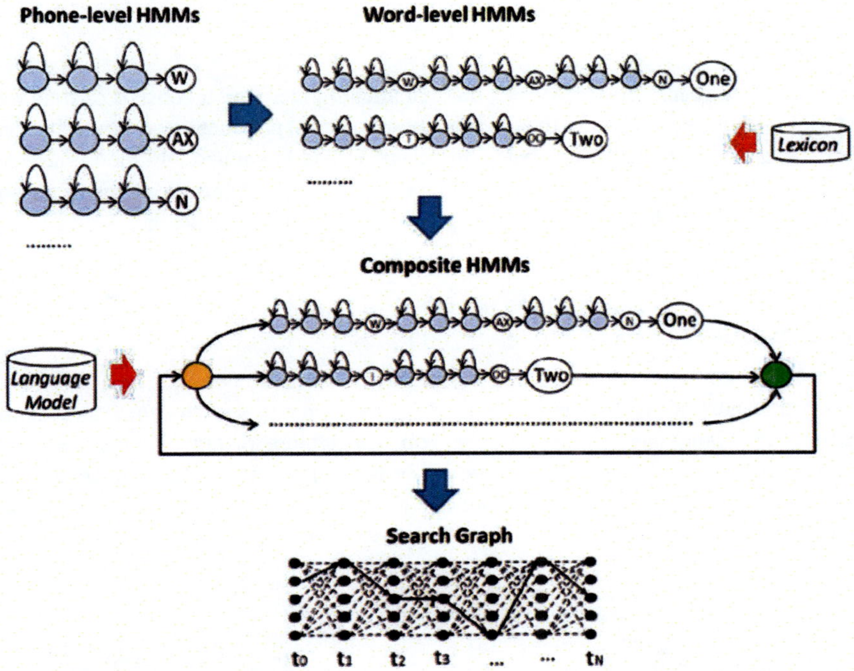

Fig. 3 General scheme of HMM-based speech modeling [18]

for speech generation and recognition of characters. A general scheme of HMM recognition model derived from Ravanelli, 2017, is depicted in Fig. 3.

This proposed methodology has introduced the concept of syllable-based N-gram model for phonetic analysis of natural language processing part and HMM-based speech synthesis in digital signal processing.

3.1 Syllable-Based N-gram Model

Phoneme is the basic linguistic unit of any language. Applying single phoneme is a common task for all the language, but it has the disadvantages of not being able to exhibit the specific information. A trigram-based syllable gives the best performance for Tamil language. The common syllable taken for the languages is C*VC* where C is a consonants and V is a vowel. In the training phase, the syllable with N-gram is extracted from the text, and their corresponding spectral and excitation parameters are stored.

3.2 HMM Speech Synthesis

Hidden Markov model (HMM) is a mathematical model which consists of training phase and synthesis phase. In the training phase, spectral parameter and the excitation parameter are extracted from speech database. HMM is applied and the frequency spectrum F0 of the speech being modeled is obtained by applying maximum criterion principle. This methodology produces good prosody without glitches.

4 Results

Festival framework is the one that is used for constructing speech synthesis system model. It is a language-independent platform. The application text-to speech conversion is implemented using C++ language. In festival, utterance plays an important role for the generation of speech synthesis as shown in Fig. 4. Table 2 denotes the details of input text data set taken for experimentation in this research work.

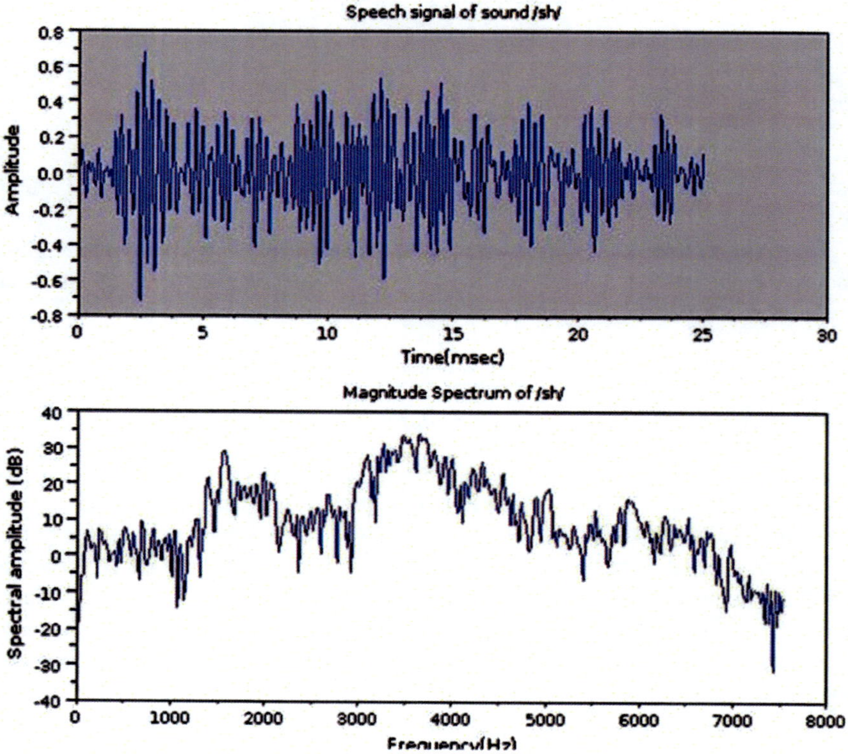

Fig. 4 Tamil speech output

Table 2 Experimental values

Parameters	Values/data
Clips used for testing	500
Total word count	15,000
Characters	42,150
Duration	30:00:00

Table 3 Preference score computation

Listener	Preference score (%)	Mean opinion score
1	20	4.1 ($F_o/2$)
2	10	4.2 ($F_o/1.5$)
3	15	4.1 (F_o)
4	10	4.0 ($F_o * 1.5$)
5	20	4.2 ($F_o * 2$)

The experimentation validation is carried out by performing the text generation from the Tamil script. Five listeners were used in the experimentation process, and preference scores are computed which are listed in Table 3.

The speech spectrum obtained is shown in Fig. 4.

5 Conclusion

TTS system is proposed and implemented using the powerful features of a HMM model for the given Tamil language. A novel TTS system based on syllable-based phonemic analysis and the HMM-based speech synthesis is proposed. The system is implemented with the help of festival framework, and result was analyzed. The method is quite simple and straightforward. Yet, it provides satisfactory confidence scores as well as MOS, thus validating the efficiency of the proposed work.

References

1. AiTi Aw, Zhang M, Xiao J, Su J (2006) A phrase-based statistical model for SMS text normalization. In: Proceedings of the COLING/ACL2006 main conference poster session. Association for Computational Linguistics
2. Moore S, Buchholz S, Korhonen A (2010) Annotating the enron email corpus with number senses. In: Proceedings of the seventh international conference on language resources and evaluation (LREC'10). European Language Resources Association (ELRA)
3. Han B, Baldwin T (2011) Lexical normalization of short text messages: makn sens a twitter. In: Proceedings of the 49th annual meeting of the association for computational linguistics. Association for Computational Linguistics

4. Roark B, Sproat R (2014) Hippocratic abbreviation expansion. In: Proceedings of the 52nd Annual Meeting of the Association for Computational Linguistics
5. Eisenstein J (2013) What to do about bad language on the internet. In: Proceedings of NAACL-HLT 2013. Association for Computational Linguistics
6. Flin E et al (2017) A text normalization system for non-standard English words. In: Proceedings of the 3rd workshop on noisy user-generated text, pp 107–115. Copenhagen, Denmark, 7 Sept 2017
7. Htun HM, Zin T, Tun HM (2015) Text to speech conversion using different speech synthesis. Int J Sci Technol Res 4(7):(2015)
8. Arokia Raj A et al (2007) Text processing for text-to-speech systems in Indian languages. In: 6th ISCA workshop on speech synthesis, Bonn, Germany, 22–24 Aug 2007
9. Sproat R, Jaitly N (2017) RNN approaches to text normalization: a challenge. arXiv preprint arXiv:1611.00068
10. Rohatgi S, Zare M, Norm D (2017) A deep learning approach to text normalization. arXiv preprint arXiv:1712.06994
11. Li C, Liu Y (2014) Improving text normalization via unsupervised model and discriminative reranking. In: Proceedings of the ACL 2014 student research workshop, pp 86–93
12. Mache SR et al (2015) Review on text-to-speech synthesizer. Int J Adv Res Comput Commun Eng 4(8):54–59
13. Lopez Ludeña V, San Segundo R, Montero JM, Barra Chicote R, Lorenzo J (2012) Architecture for text normalization using statistical machine translation techniques. In: Iber SPEECH 2012, pp 112–122. Madrid, Spain
14. Zhu C et al (2007) A unified tagging approach to text normalization. In: Proceedings of the 45th annual meeting of the association of computational linguistics, pp 688–695. Prague, Czech Republic
15. Lin C, (Sarah) Mu Q, Shao Y, iTalk A 3-component system for text-to-speech synthesis, unpublished
16. Beliga S, Pobar M, Martincic-Ipšic S (2015) Normalization of non-standard words in croatian texts. arXivpreprint arXiv:1503.08167v2, 30 Mar 2015
17. Gokul P, Thomas N, Thomas C, Gopinath DP (2015) Text normalization and unit selection for a memory based non-uniform unit selection TTS in Malayalam. In: Proceedings of the 12th international conference on natural language processing, pp 172–177, Trivandrum, India
18. Ravanelli M (2017) Deep learning for distant speech recognition

SOA-Based Middleware Framework for IoT Applications

M. Mallegowda, Pooja Sarashetti, and Anita Kanavalli

Abstract For the first time in 1999, the IOT definition was derived from the Automated ID (Auto-ID) laboratories. IoT highlights that each entity around us is connected by using a range of wireless communication technologies. It is marked by the Internet convergence of many real-life activities, with the aim of making high-level encounters with the physical world as straightforward as interactions with the virtual world today. In IoT, significant role is played by sensors and actuators. For the movement and regulation of mechanism, the actuators are used. These are nevertheless used for inferences with poor standards and essential utilities. But, there are millions and perhaps billions of sensors and actuators in the network. There are barriers to interoperability, scalability, and stability. So, at the middleware layers, we will have new solutions.

Keywords IoT · Middleware technology · Internet of technology

1 Introduction

The Internet of Things aims to link all entities around us through a spectrum of wireless communications technologies. Heterogeneous data fusion is often seen as an amazing and urgently necessary task in IoT data processing. In this research, the history of IoT design will first be discussed and then IoT architecture described in depth. And then, we develop a multisource data management middleware architecture (SOA). My study aims to provide a new research view on the scalable heterogeneous architecture of IoT data processing. Experiments are carried out using

M. Mallegowda · P. Sarashetti (✉) · A. Kanavalli
Department of Computer Science and Engineering, M. S. Ramaiah Institute of Technology, Bangalore, Affiliated to VTU, Bangalore, India

M. Mallegowda
e-mail: mallegowdam@msrit.edu

A. Kanavalli
e-mail: anithak@msrit.edu

S. Shakya et al. (eds.), *Proceedings of Second International Conference on Sustainable Expert Systems*, Lecture Notes in Networks and Systems 351,
https://doi.org/10.1007/978-981-16-7657-4_27

data collected from indoor environments for device validation. The SOA-based IoT data processing middleware provides a useful framework for the future integrating and interacting heterogenous platforms for various network information, decreasing integration process complexity, and enhancing complete system reusability.

The rapid advancement of information and communication systems, a single technology alone, cannot satisfy dynamic, contexts-conscious applications criteria, which involve external intervention in resource information. And, users need in real time and in the physical world, such as diverse sensory data collections and the analysis and interaction between humans and computers. IoT systems combine with IntelliSense identification software, widespread networking, and ubiquitous networks that are seen as an Internet and computer third wave. IoT is a key feature of today's technology of intelligence.

The foundation of IoT is still the Internet, and the Internet is continued and developed. By exchanging information and interaction through the Internet, the IoT is extending its client to all persons. Everything is linked to the Internet through IoT. For example, smart home systems, smart factory equipment, wearable health monitors. In order to secure data flow and interchange, the IoT combines RFID, wireless sensor, and ubiquitous terminal devices as a perceptual basis with a mix of wired or wireless network connection and Internet connectivity. Through cloud computing and hi-performance technology, we effectively provide input decision-making systems for closure monitoring in actual data processes, management, and strategic. This will potentially lead to a new wave of innovations in diverse industries such as intelligent housing, intelligent supervision, smart grid, and other new technologies, thanks to the rising achievement of IoT. IoT has been presented to the areas of smart production, smart agriculture, smart logistics, smart infrastructure, smart grid protection, environment, environmental safeguards, intelligent healthcare, intelligent home, and other demo techniques (Fig. 1). IoT is a leading technology company. In the new system, the limitations of the current system will be solved. The downside of the current system is that there will be no total protection for multi-applications. We

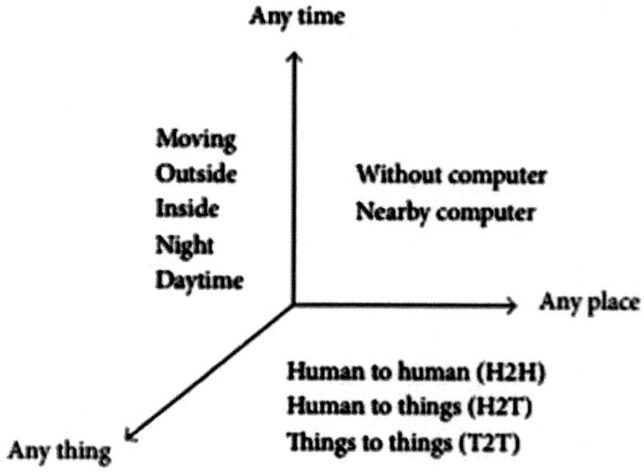

Fig. 1 Connections in the IoT

use the architecture of middleware to develop an application. It is used for the analysis of current IoT services and applications for middleware. Build an end-to-end solution for various fields—agriculture, health care, government, road traffic, etc., highly scalable wireless sensor network middleware solution design and support; using powerful machine learning and deep learning, companies and organizations can achieve sustainable growth by analyzing data.

2 Literature Review

The literature review was carried out during the entire project in order to acquire information and develop the skills required to complete this project. Previous similar ventures, academic thesis, and journals are the primary references for this project. This chapter focuses on the basic principles and all the fundamental ideas that apply to this project and the disadvantages of the new method.

IoT in middleware is very huge. Middleware serves as an intermediary between service providers who are IoT devices and sensors, etc., and service users who are analytical business applications. It is a layer of software between applications and objects. It is behaving like an intermediary. It makes it possible to communicate with stuff on the Internet. A proposed middleware is with M2M and M2M capabilities. Based on mapping techniques for heterogeneous devices, this is intended to push data from sensors to cloud-based infrastructure [1]. A paper offers a COTWare service middleware. The incorporation of COT and Fog computing is encouraged in large-scale IoT systems. Intelligent city traffic light management scheme to prevent congestion and reduce travel time increases the average speed of vehicles [2]. In addition with traditional historical data processing, middleware is used to analyze data in real time. Used various ML algorithms for predictive results. Random forest is used on the basis of the RMSE score [3]. Model-driven approach focuses on ECE-DDS middleware. The advantages of this mode are reusability, modularity, synchronization, and reduced complexity. It provides intuitive and higher-level abstraction for developers [4] in middleware and ubiquitous network engineering research history. Ubiquitous networking is the environmental delivery of wireless communication networks and technologies to enable continuous connectivity. The ubiquitous network can be split into the integration layer and service convergence layer [5]. Models, frames, and middleware are becoming increasingly common to fix important facets of the above-mentioned lens. Patterns codify reusable architecture, which offers time-tested solutions to popular software issues, which occur in specific circumstances and domains [6]. Suggested middleware farming architecture in Myanmar has embraced the method of plant cultivation and agricultural land. The architecture addresses current architectural challenges. IoT and web techniques are included in the extensible and configurable middleware [7] with the present global sensor network. Sun et al. [8], middleware IoT research has been performed to manage the laboratory's surroundings. They say, the method is highly accurate and efficient. It monitors the laboratory's light and sound. The IoT and middleware technology structure refers to the data gateways, the safety components, and the routing component. Amaral et al. [9], in this system, the software technology middleware for IoT is utilized to construct, manage, and integrate heterogeneous devices as well as applications in IoT settings. This study determines the research and development technology effect of this step in 5G's true maturity. This chapter's major contribution is to offer a future view on the IoT middleware technology as regards adjustments to suit the upcoming 5G communication paradigm. Terziyan et al. [10]: This paper contains agents to manage the roles and scenarios repository. Agents monitor data from a resource status adapter and decide according to the contents of the data. Other agents in the environment are also facilitated by their detection. This method enables the identification of services, use, and integration/composition of FIPA communications protocols. Agent-based layer contains an agent controlling the role and scenario repository represented in the language of the RDF-based semantic agent (S-APL). Agents handle atomic conduct repositories which are software components which agents load in accordance with the current situation and catalog which enables agents to be flexibly. Velmurugan et al. [11]: This paper explains the modeling and management of the IoT systems using dynamic orchestration middleware such as CHARIOT to help digitally increase intelligent grid operating by flexibly determining where the functionality of a grid hierarchy should be applied to effectively manage the risks resulting from volatility in the energy system area. HYDRA, UBIWARE, UBIROAD, and SMEPP offer context-conscious functionality. The middleware implements security and user privacy in its architecture. Tai-Yeon Ku et al. [12], in this paper, it is very vital to make effective use of energy at home, in hospitals, industries, etc. The authors have, therefore, proposed a method which improves the efficient use of energy. First, the system collects data on energy resources at home and analyzes energy use patterns, therefore decreasing energy waste. Sun et al. [13], middleware IoT has been performed to manage the laboratory's surroundings. They say, the method is highly accurate and efficient. It monitors the laboratory's light and sound. The IoT and middleware technology structure refers to the data gateways, the safety components, and the routing component.

The surveys are conducted on the basis of context-consciousness in [14, 15]. The survey in [15] takes architecture into account and offers the characteristics of a general context-aware. In the survey published in [15], many context-aware designs are assessed from a ubiquitous or general computing point of view based and on many important characteristics.

By reviewing, all previous ventures, academic thesis, and journals are the primary references for this project. This chapter focuses on the basic principles and all the fundamental ideas that apply to this project. There were used many middleware types like Hamra middleware, service middleware, GSN middleware, and MQTT middleware. These technologies are lagging in efficiency, speed, throughput, and security. The response time of middleware is not optimal, so we are creating a middleware framework which can overcome these challenges.

3 Proposed Methodology

In the new system, the limitations of the existing system will be solved. The downside of the existing system is that there will be no total protection for multi-applications. We use the architecture of middleware to develop an application. It is used for the analysis of current IoT services and applications for middleware. Build an end-to-end solution for various fields like agriculture and road traffic. It is highly scalability middleware that provides the solution design, and we can achieve sustainable growth by analyzing data. So, by introducing this method, we are trying to improve efficiency, decrease downtime, improve speed, reduce power, and increase throughput.

4 Design Methodology

This implementation is intended to achieve full user satisfaction by taking user preferences into account. It ensures that the project proposed is quick, safe, efficient, and cost-effective.

4.1 Problem Definition

In the existing system, there will be no total protection for multi-application. From the previously used method, i.e., solution of .net, we get to know that its accuracy and efficiency are low. There are many middleware types like Hamra middleware, service middleware, GSN middleware, and MQTT middleware that has been used in the past years. These technologies are lagging in efficiency, speed, throughput, and security. The response time of middleware is not optimal, so we are creating a middleware framework which can overcome these challenges.

4.2 Flowchart

See Fig. 2.

4.3 Algorithm

Step 1: Read the data from sensors.

Step 2: Sensor data are stored in database called lowdb.

Fig. 2 Flowchart

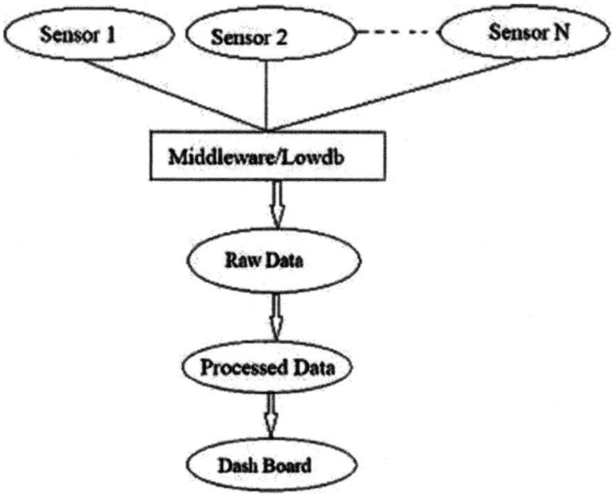

Step 3: Collecting data from lowdb and stored in raw data.

Step 4: The raw data are analyzed and stored in processed data.

Step 5: Visualization of data.

4.4 Dataset

For data collection, I have collected own dataset using Raspberry Pi model with a set of sensors. The data have been read and have been stored in cloud since 6 months and while project implementation fetched the dataset from cloud.

4.5 Methodology

At least 75 billion connected IoT devices will be located around 2024–2025, according to estimates. Most of them are heterogeneous, domain-specific, and cross-domain. Selecting middleware makes the application secure. Scalability, usability, high processing speed, versatility are the most relevant middleware parameters. This middleware often extends to public applications such as traffic management, hospitals, shopping places. All of these applications include analytics, cost, and security.

The design approach for wireless embedded and sensor networks is based on service-oriented architecture and agile development concepts.

IoT is very big in middleware, and in 2025, it is expected to hit billions of dollars. Middleware functions as an agent between IoT sensors and devices, etc., service

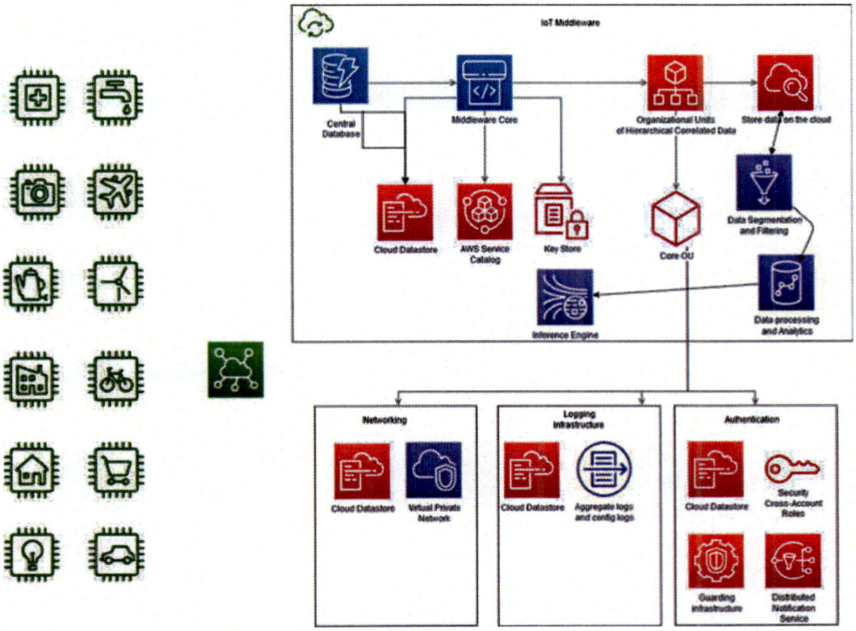

Fig. 3 Middleware solution for IoT

providers, and customers of services that are analytics business applications. It is a layer of software between apps and objects. It behaves like an intermediary. It allows for contact between things and the Internet (Fig. 3).

IoT Middleware Parts:

Middleware core: This component itself is middleware. It is referred to as entity management or object management. This is a distributed framework with middleware.

Key store: It stores sensitive information like password.

Organizational units of hierarchical correlated data: All middleware-processed data are stored here. It also clarifies what belongs to each case.

Cloud database: It is combination of distributed cash and distributed database.

Data segmentation and filtering: The information processed is clustered and data filtered. Huge data are stored in order to require segmentation.

AWS catalog: It can be used by many clients. It is deployed in middleware. It manages inquiries and replies. Middleware integrates with the catalog, so it is used by millions of clients.

Inference engine: It has analytics and offers responses.

Example 1: Lighting solution. When the light is on continuously for 12 h, the light can turn OFF automatically.

Example 2: If blood sugar rises per minute, the doctor, nurse, and hospital will be notified. Inferencing the insulin pumped into the bloodstream.

Core OU: The middleware manager performs operational tasks by creating requests to collect data from various store entities. It involves networking, logging infrastructure, authentication in the bottom part of the architecture. Cloud architecture central manager or distributed to communicate with them. The central manager handles items that are operational. Because of its extremely safe and scalable attribute, middleware offers an authentication service. It is a completely protected network supported by our private virtual network. The middleware for analysis is used. It was used for:

- How to reduce the flight delays
- How to reduce wastage of water
- How to conserve electricity
- Helps in smart shopping.

5 Advantages

This proposed system incorporates a new user framework and improves the heterogeneous reusability of data, allowing for better adaptation to the scenarios of IoT multisensory and multistream applications. It is easier to create a single framework for the processing of heterogeneous information. It provides better optimization of results. The cluster achieves better load balancing. This proposed study involves a detailed experimental assessment that shows for each application both efficiency and effectiveness.

6 Results

Middleware framework contains two dashboards. They are agriculture dashboard and traffic signal dashboard. In agriculture dashboard, we have 726 invocations. In traffic dashboard, we have 426 invocations. Comparing with the previous data, the present data are represented, i.e., how much improvement percentage can be done by using middleware technology. In agriculture dashboard and traffic dashboard, after all the analysis, 93.58 and 91.93% improvement can be done. Through graphs, we have represented the present and previous improvement (Figs. 4, 5, 6, 7, 8, 9, and 10).

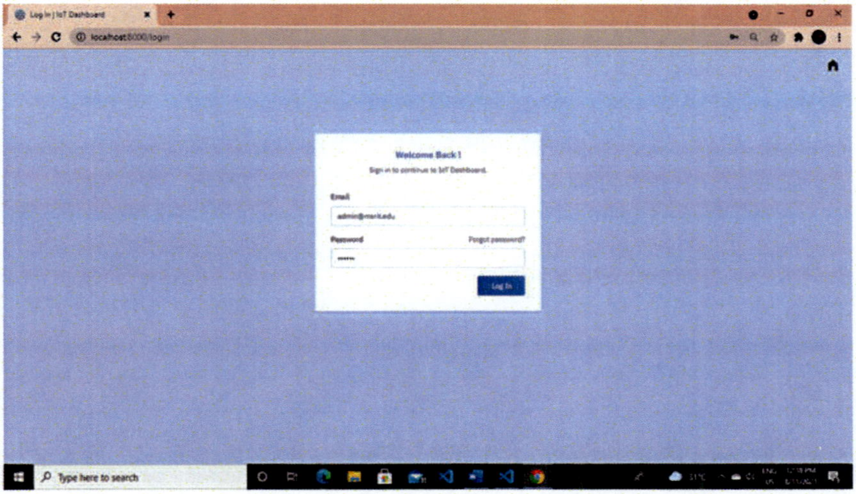

Fig. 4 Login page

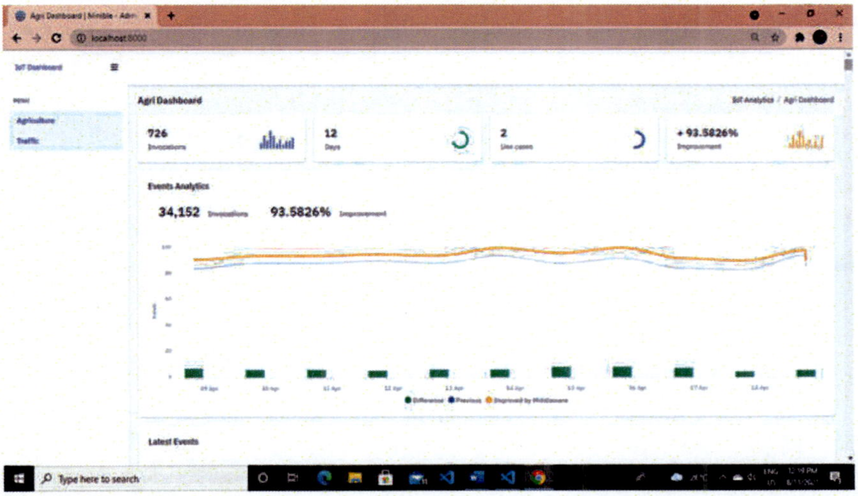

Fig. 5 IoT dashboard

7 Conclusion

Middleware model is based on SOA architecture for the integration of heterogeneous multisource data, based on the features of the agriculture system and traffic signal system. The present data are displayed in comparison with the prior data, that is to say, how much improvement can be achieved utilizing middleware technology. After all, 93.5 and 91.93% improvement is possible for the farm dashboard and

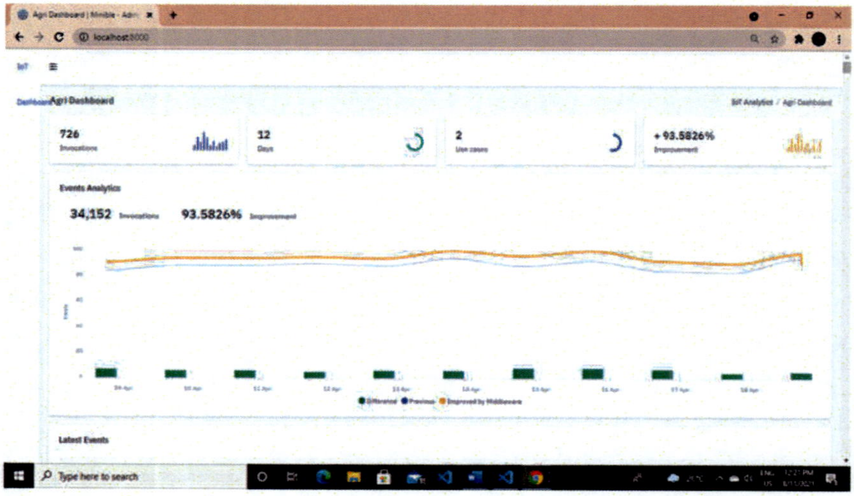

Fig. 6 Agriculture dashboard

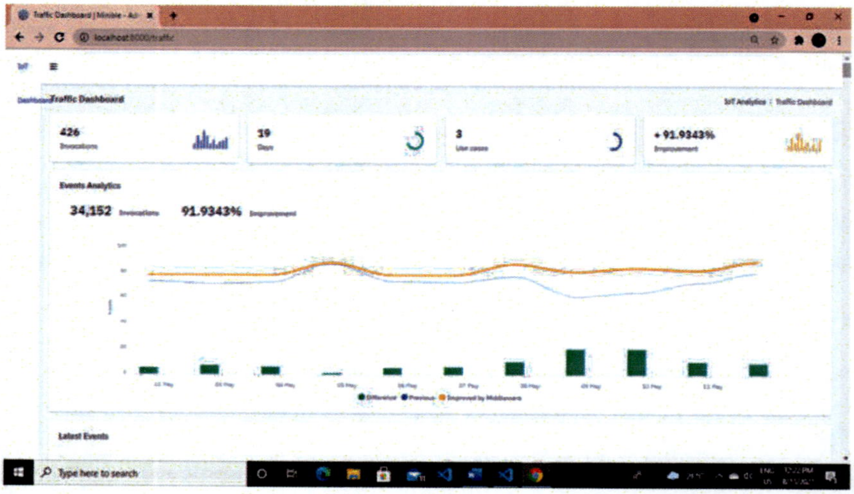

Fig. 7 Traffic signal dashboard

traffic dashboard. From this experiment, we can make certain conclusions. First, the SOA middleware processing architecture offers better reusability. Secondly, the test results show that an end-to-end system may be developed to accommodate for agricultural and road traffic areas. Third, distributed middleware implementation improves performance and provides massively scalable. In the future, we can integrate many IoT devices to middleware sensor, e.g., medical, inventory. We have used

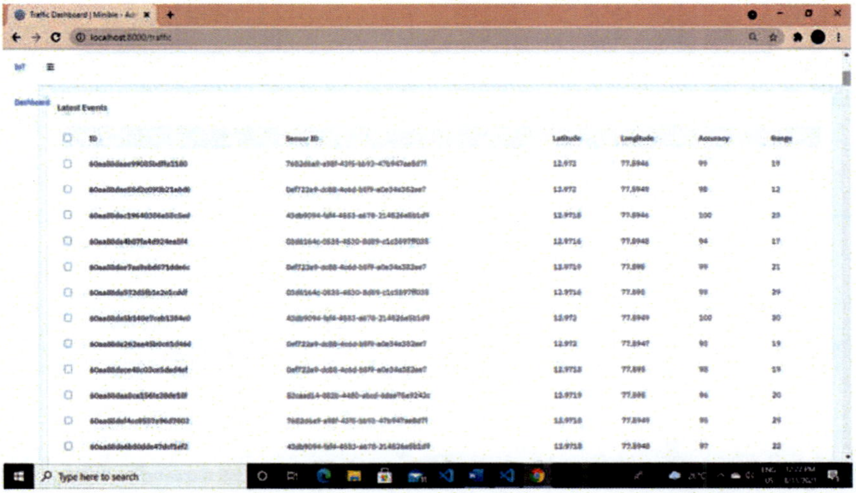

Fig. 8 Events captured from cloud

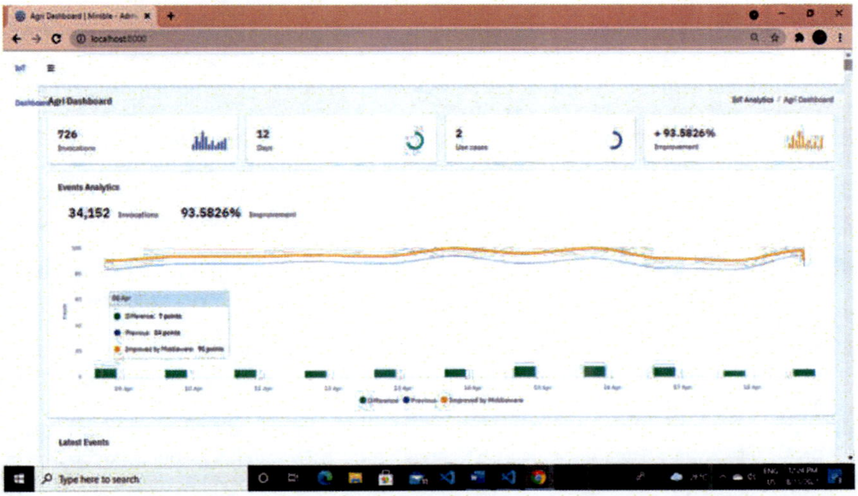

Fig. 9 Improved points by using middleware in agriculture dashboard

two applications for making middleware framework, but in the future, we can add many applications with more security and accuracy.

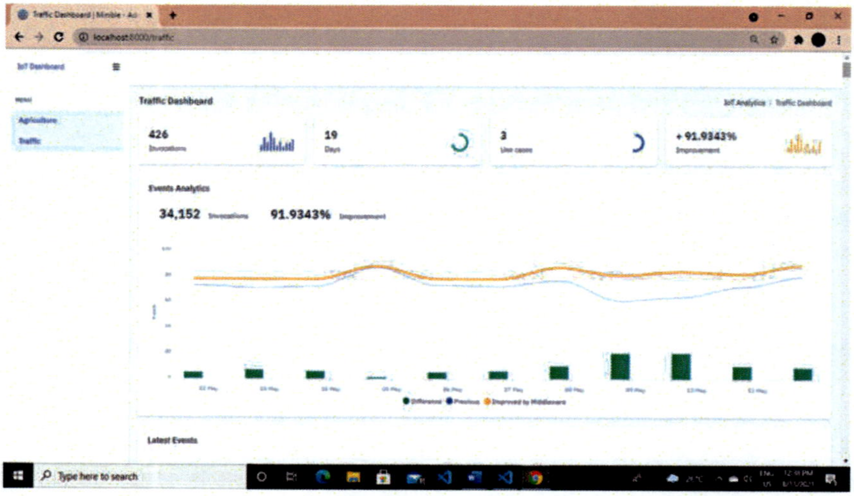

Fig. 10 Improved points by using middleware in traffic signal dashboard

References

1. Lomotey RK et al (2017) Middleware framework for IoT services integration. In: 2017 IEEE international conference on AI & Mobile Services (AIMS). IEEE
2. Al-Jaroodi J et al (2017) CoTWare: a cloud of things middleware. In: 2017 IEEE 37th international conference on distributed computing systems workshops (ICDCSW). IEEE
3. Ali MI, Patel P, Breslin JG (2019) Middleware for real-time event detection and predictive analytics in smart manufacturing. In: 2019 15th international conference on distributed computing in sensor systems (DCOSS). IEEE
4. Brummett T et al (2020) A model-driven middleware integration approach for performance-sensitive distributed simulations. In: 2020 IEEE 23rd international symposium on real-time distributed computing (ISORC). IEEE
5. Zhang J, Yan Z (2015) Middleware technology research and interface design in ubiquitous network. In: 2015 seventh international conference on measuring technology and mechatronics automation. IEEE
6. Htaik NM, Maung Maung NA, Zaw W (2018) Enhanced IoT-based interoperable and configurable middleware using semantic web techniques. In: 2018 15th international conference on electrical engineering/electronics, computer, telecommunications and information technology (ECTI-CON). IEEE
7. He W et al (2017) MODE: A context-aware IoT middleware supporting on-demand deployment for mobile devices. In: 2017 IEEE 23rd international conference on parallel and distributed systems (ICPADS). IEEE
8. Sun T et al (2018) Research on Internet of things middleware technology for laboratory environmental Monitoring. In: 2018 international conference on virtual reality and intelligent systems (ICVRIS). IEEE
9. Amaral LA, de Matos E, Tiburski RT, Hessel F, Lunardi WT, Marczak S (2016) Middleware technology for IoT systems: challenges and perspectives toward 5G. Springer International Publishing Switzerland
10. Terziyan V, Kaykova O, Zhovtobryukh D (2010) Ubiroad: semantic middleware for context-aware smart road environments. In: 2010 Fifth international conference on internet and web applications and services. IEEE

11. Velmurugan NB, Garoma L (2017) Role of middleware technologies for different applications: a study. Int J Creative Res Thoughts
12. Tai-Yeon Ku W-KP, Choi H (2017) IoT energy management platform for MicroGrid. In: 7th international conference on power and energy systems. IEEE, ISSN 978-1-5386-2429-6, pp 106–110
13. Tao S, Xu YJ, Li J, Zhang H (2018) Research on Internet of things middleware technology for laboratory environmental Monitoring. In: International conference on virtual reality and intelligent systems. ISSN 978-1-5386-8031-5/18, IEEE, pp 544–547
14. Kajer K (2007) A survey of context-aware middleware. In: 25th conference on IASTED international multi-conference: software engineering, pp 148–155. ACTA Press
15. Miraoui M, Tadj C, Amar C (2008) Architectural survey of context-aware systems in pervasive computing environment. Ubiquitous Comput Commun J 3(3)

The Future Warfare with Multidomain Applications of Artificial Intelligence: Research Perspective

M. A. Nikitha, B. S. Sai Swetha, Krishna Harika Mantripragada, and N. Jayapandian

Abstract We live in a period when historical fiction has become current reality. With our future being automated, using AI on a daily basis will only get more convenient. Making military weapons to detect, monitor, and engage a human being with attacks may all be done in the privacy of one's own garden. There is a plethora of AI software out there that can be readily integrated into combat weapons. The automobile industry is already incorporating AI into vehicles to assess driving circumstances and give augmented reality to drivers via heads-up displays in order to assist avert accidents. Similarly, artificial intelligence will be utilized to study the battlefield and give soldiers with augmented reality information via heads-up displays and weapon control systems. Since AI is not a single technology, it has been argued that it might be used by the military in a variety of ways. Intelligence, surveillance, and reconnaissance (ISR) activities, as well as processing and interpreting sensor data and geographic imaging analysis, are all examples of AI. Artificial intelligence has the potential to reduce human involvement in conflict, whether it is employed for combat robots or data analysis. AI has the potential to profoundly alter the nature of war. The article mainly focussed on warfare technologies and applications. The main aim of this review is to understand the current applications being used in armed forces and proposed technologies of artificial intelligence.

Keywords Artificial neural networks · Artificial ıntelligence · Cyber security · Machine learning · Robots · Navy application · Unmanned air vehicle

1 Introduction

The artificial intelligence is a subdivision of computer technology which is defined as something that makes machines work intelligently. AI demonstrates behaviours which mimic human intelligence such as planning, manipulation, problem-solving,

M. A. Nikitha · B. S. Sai Swetha · K. H. Mantripragada · N. Jayapandian (✉)
Department of Computer Science and Engineering, CHRIST (Deemed To Be University), Kengeri Campus, Bangalore, India
e-mail: jayapandian.n@christuniversity.in

S. Shakya et al. (eds.), *Proceedings of Second International Conference on Sustainable Expert Systems*, Lecture Notes in Networks and Systems 351,
https://doi.org/10.1007/978-981-16-7657-4_28

reasoning, motion, perception, creativity, and learning. It acts as a tool that completes tasks automatically with very little human intervention. In this rapidly growing technical era with new advancements every other day, artificial intelligence (AI) has become a reality, now more than ever [1]. Not to mention how AI has become a term that is fancied much. We have technologically evolved to an extent where our day to day interaction with AI is almost unrecognizable. AI in particular has been growing at a quick pace and also is growing widely by venturing into various domains. The progression of AI can be easily measured by looking around and noticing how handy the world has become. AI is now a part of almost all domains, from medical, engineering, warfare, visual arts, gaming, education, finance and management, cybersecurity, space industry, environmental protection [2]. Generally, AI uses algorithms along with constraints that help gear models towards processing, thinking, perception, and action. There are various ways to power AI such as deep learning, machine learning, or even simple lists of rules. The most basic AI system, to our surprise, can be extremely simple like a computer programme which performs a certain function or gives an output based on specific rules [3]. Discussed below is the rundown of technology that is implanted in today's AI picture recognition, prediction engines, and language processing. AI can only function when it is fed with data. Similar to our brain which takes in information in large amounts, AI takes in data and depends on it. Due to the result of the Internet of Things (IoT), multiple devices are connected and have become modes of communication, all around the globe [4]. Software which holds a layer for data collection is able to interact with the devices which hold information in regard to news or the weather. Such information is then collected and converted to human recognizable speech. When this data is collected, it requires a lot of storage space from where it can be accessed transiently. A cloud is created where storage can be scaled up or down. Machine learning, image recognition, language processing, emotion analysis are all processed. Third-party APIs are used to access these 'services'. Data outputs can be accessed for various reasons either to encourage machines to operate together more efficiently and effectively, or other revelations may be meant for humans to act on. In regard to the types of AI, there are four kinds. Some of them are considerably more advanced than others. There are four major AI kinds, according to the current categorization system: reactive, limited memory, theory of mind, and self-awareness. Reactive AI is a basic category of AI which receives input and provides a predictable output off of it. Limited memory AI takes lessons from the previous and perceives outputs from observing data. Theory of Mind AI is capable of making decisions identical to a human mind. Self-aware AI is very advanced and is quite capable in understanding its emotions and others around it.

The global impact of AI is boundless, as it has been playing a vital important role in human life since it has been discovered. From travelling to shopping to surfing the Internet, AI has taken over things. There was a prediction that the AI social media market may grow from \$633 million to \$2.1 billion. We have already been through the rise of social media, and what we are now observing is how AI helps social media. Social media uses AI to advertise more effectively [5]. The marketers are paid to talk about them, run ad campaigns, and without AI, they are only going to spend more time doing each of those separately. These ads are reduced to links, hash-tags, and a simple click. For example if a known brand has taken interest in a person who has been catching people's attention, they may ask that person to work for them. If that said brand has products, they will give that person a link or a code through which people can have a little discount. And, giving out discounts would attract more people. We also have digital assistants like Google Home, Alexa from Amazon[6]. These digital assistants are used for actions like: reminders, alarms, playing music, and scheduling things. They also automate home which allows us to control temperature, lights, and doors [7]. Seeing the things AI can do, one can only wonder how AI could be used in warfare. Tasks like image recognition, language translations, tracking, and setting the target on people are some of the tasks in which AI is used. Although the military has been trying to build killer robots, there are still some shortcomings. AI could also help in anomaly detection, prediction systems, and fraud detection. Massive datasets which are retrieved from computers and digital devices can be processed using AI. AI weaponization is being developed rapidly across aerial and terrain vehicles, deploying missiles, using systems to digitize and automate from personal systems to deployment of UAVs and more. In this article, we will be discussing how artificial intelligence is used in defence. Involving AI in defence would help make things more precise and at the same time a little easy. We will get to know the basic idea of AI and how it is used in the military, navy, and air force. Artificial neural networks, military logistics, AI robots, Remote Environmental Monitoring Unit System (REMUS), Integrated Submarine Imaging System (ISIS), Battle space Preparation Autonomous Underwater Vehicle (BPAUV), Sea hunter, drones, UAVs are the applications of AI in warfare that will be discussed in this article [8].

2 Military Applications

AI in warfare without a doubt will change the dynamics, shifting the fundamental principles and the traditional military thinking. Hearing AI in warfare brings up classic images like tonnes of drones and humanoid robots in the battleground, which is quite not how it is in reality. While these might be events belonging to the distant future, as of now, AI is majorly involved in improving human ability and productivity in warfare. It helps in advancing the nation's defence without replacing the humans, but rather by fortifying. Though warfare is not a well-established domain for AI with not so much of actually used applications, there are numerous researches and applications being done today most of which are in testing phases. That being said there are great chances for having minimal human involvement in the battlefields and at the same time a very distinct war approach. With the rapid advancements in technology, the defences department surely has to adopt and acclimate to the changes. Considering the ability of AI to perform complex missions is with high-level parameters such as ISR data integration, data analysis, unmanned vehicle operation, including unmanned undersea operations, nuclear attack clean-up. It surely would be a ground breaker in the defence domain. Further discussions on the usage of artificial intelligence will be probed below.

Figure 1 is to elaborate different applications in military. Artificial neural networks (ANNs) are comprehended by a strong learning set; deep learning. The structure of ANN is replicated after the neuronal structure of the human cerebral cortex. Artificial neural networks, like the human brain, include neurons (nodes) that are linked to one another in various levels of the networks. These interconnected neurons (nodes) are exhibited in three layers: input, hidden, and output. The retina has photoreceptors which are sensitive to light. These photoreceptors are represented by the input layers. The hidden layers behave similarly to the neurons that probe the retinal data received and transmit them to the visual cortex [9]. The output layer is now represented by the visual cortex. Simple ANNs use a single hidden layer. When ANNs with more than two hidden layers are present, it calls for deep learning. Intricate datasets are analysed using ANNs with more than two hidden layers. Impulse noise (IN) is a momentary noise which is made up of one or more bursts of sound energy.

The military is often exposed to such high levels of impulse noise which prompts a variety of hearing consequences including acoustic trauma. Although IN was perceived, it also resulted in a number of false positives. There was a lot of wind noise being detected which deteriorated the possibility of many events of interest. Therefore, ANN proved to be a driving force in receiving such high accuracy for the noise classifier by the military. The procedures, resources, and systems involved in creating, transporting, maintaining, and redeploying or reallocating material and troops are referred to as military logistics. The military might of a country determines its capacity to execute these duties. Particular soldiers are assigned specific tasks which would allow them to increase their efficacy in a variety of ways in the logistics branch. Supplies are always in demand at the battle stations. Supplies and real-time

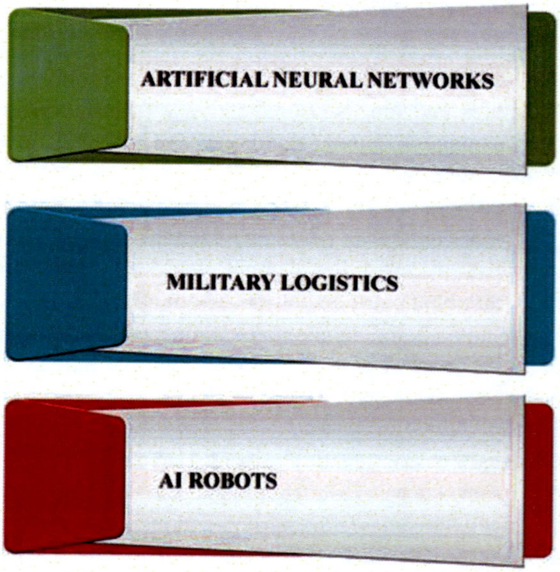

Fig. 1 Different military application

data need to be provided to the units who have requested it. But, this information is not present in the status quo. This would lead to immense perplexity among the units as they would have to check in with the supplies provider constantly. AI can be used to administer the needs of these units. This is done based on the supply pattern or through pre-ordered supplies. Such patterns can be designed using machine learning or process automation which will allow the units to receive their orders precisely on time. Machine learning is merely a subset of AI which enables a machine to learn automatically from previous data without explicit programming. This is also used in developing an automated infrastructure protocol for systems that are automated. Infrastructure developers frequently employ a testing framework, which enables the automation system to perform a set of tests. The direction of the application is determined by the results of each test. Metadata is received during the testing that is then redirected to these tests which are further acted on by the system on the results that are procured. Such a technology comes in handy when regulating if a threat received is forthcoming or just a false positive. Soldiers fighting on the battlefields are always putting their lives on the line. This would result in a loss of men and their health deteriorated. In order to protect the lives of these soldiers, companies are going to great measures in designing and creating humanoid robots. These robots would ultimately be able to fight on the battlefield, similar to a soldier. Boston Dynamics is the world leader in the field of mobile robots. Robots like Spot and Atlas are some of their mobile robots designed for a battle. Spot is a nimble quadruped (four legged) robot that was created for indoor and outdoor operations. It is operated hydraulically and runs on electricity. Spot is equipped with a sensor head that aids it in navigating and negotiating difficult terrain. Spot is around 160 pounds. Simultaneous localisation and mapping (SLAM) algorithms are frequently used in situations that require updating a map of an unknown environment while also monitoring a moving item, such as objects. Recognizing pedestrians or determining the location of unmanned aerial vehicles. This robot is capable of behaving autonomously or can be controlled using a remote [10]. Atlas is a humanoid whose height is 150 cm. This humanoid is capable of amazing athletic performances. It has actuators which are powered by a strong hydraulic system which is rather small. This technology provides the humanoid with frenzied power. This power enables the robot to do acrobatic jumps and flips. Since robots cannot get tired and do not require food or water to survive, it could become the new norm in future of the military across the world. Although, these kind of robots could be exposed to hackers, which would alter their functioning status. The major objective of future study in this field should be to ensure robot software dependability and resistance to hostile intervention.

3 Applications in Navy

Hydroid technical solutions provide Remote Environmental Monitoring Unit System (REMUS) which is a series of unmanned underwater vehicles. It is claimed to help the navy in threat detection using machine vision [11]. Machine vision basically is the use of cameras to identify and analyse an object automatically. This REMUS software can be integrated with the existing software used in the navy for surveillance and is capable of identifying and detecting threats to depths of about 6000 m underwater. It intercepts, locates, monitors, and records communication signals. REMUS 600 (one in the series) in specific is 4.3 m in length, 0.7 m in diameter, and around 320 kg [12]. It has a pencil beam, auxiliary battery pack, sonar and sensors for temperature, conductivity, and depth placed in the front most section, while the rear section has the main battery unit, communication systems, motor assembly, and system electronics. It is also designed with a tri-fin tail to improve its directional stability and control. The vehicle is capable of autonomous operation as a result of the pre-loaded mission programme which can be modified during the mission if needed. In standard REMUS configuration, the vehicle will be using long baseline acoustic positioning systems in order to track objects underwater and a GPS along with satellite-based augmentation systems to get definite positions. It has high-definition cameras fitted to capture images of the objects being detected. After it is launched, when it detects a threat, it immediately sends signals and data. The canister then gets released and prepares for the unmanned autonomous vehicle (UAV) to launch which further does surveillance of the threat and reverts back the images and data. This UAV travels at the speed of 2 m/s with a range of 150–200 km. Hydroid claims to have helped the navy in finding and destroying underwater mines. A case study also reveals that this has increased the navy's warfare capabilities by keeping adversaries 'guessing'.

The Integrated Submarine Imaging System (ISIS) is an electro-optical/infrared periscope sensor which is attached to the inboard software and hardware. Advanced processor build programmes are commonly used in submarine combat systems these days, and it provides improvements in the periscope photonic systems of the navy submarines. The image enhancement algorithms are tuned suitably for the environment, and the camera settings are properly set. Nonetheless, the current automatic cameras are not serving all the needs. Many maritime conditions are rarely caught in these commercial cameras. With ISIS, the operator can simply use a joystick to control and manipulate photonic masts by looking at digital video on monitors. In addition, it can share the video real time with the combat team on various displays. ISIS supports photonic mast and periscope applications and provides the submarine operators with extremely helpful resources such as analysis and image enhancement tools, options to store, recall and record data. Furthermore, the images can be transmitted from the submarine to other naval and joint forces using the system's infrared cameras. It is highly helpful for port security, surveillance, navigation, and to improve image quality in surface military vessels.

Battlespace Preparation Autonomous Underwater Vehicle (BPAUV) is a fast and small underwater robot which is used to detect changes in inshore conditions, to map the ocean bottoms near the shore, and to identify mines. It was developed and is manufactured by Bluefin Robotics. These BPAUVs can be operated from a ship or a boat, from launch, oceanographic vessels. They have a low drag fairing with articulated, single propellers. It is operated as entirely autonomous systems and did not communicate during the missions. It also lacks on board classification and identification capability. The missions are generally constructed using toolkits which

Fig. 2 Different application of navy

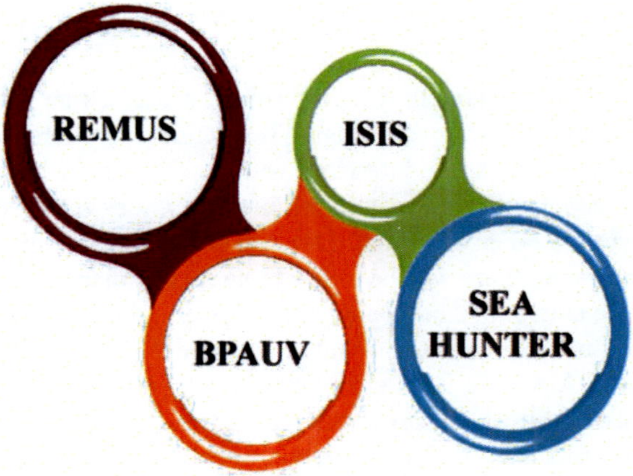

have a huge reliable database of behaviours that has been established. The reason behind flexible and powerful means of specifying mission objects by the system is the usage of behaviour-based control architecture. The BPAUV is around 4 m long and 0.5 m wide, weighing almost 800–1000 pounds. It is powered by lithium polymer batteries. Once the vehicle is launched, it moves autonomously along its present defined path that is loaded into its programme. Due to the presence of its side scan sonar, it is capable of detecting and classifying objects or mine like contacts ahead of the ship. Figure 2 is to elaborate different applications in navy.

Its compact size and easy launching and recovering process allow it to take advantage of the ships of opportunity. In addition, it also has greater deployment efficiency with swappable batteries and subsystems. BPAUV's self-contained navigation system and high performance help in reducing the logistical requirements of the mission. Sea hunter is one of its kind with the ability to hunt submarines without any crew onboard by using cameras and radars. It is capable of being in the sea for months together at a time. Furthermore, it costs around $20,000 a day to run which is much lesser in comparison to the crew ships. Its manufacturing costs are $20 million. Sea hunter is known for its capability to perform diverse missions and to carry various payloads. It is 132 ft. (40.2 m) long and has a top speed of 27 Knots. Weighing about 145 tonnes, it can travel approximately 10,000 nautical miles. Since it is autonomous, the crew are to only monitor the sensors and weapons when it is in action. And in most cases, a single crew is adequate to operate multiple sea hunters, as monitoring is not always necessary. Sea hunter is identified to be more cost-efficient and convenient.

4 Application in Air Force

The artificial intelligence in warfare, especially air force, can bring colossal changes in the way wars are fought. To use AI in warfare, it has to be decided when the system should be interfered with by the human and when not. For example when a country designs a weapon in such a way that it can set and engage targets, it should be ensured that the weapon targets only the rival. Using this autonomous weapon, there will be at least one less soldier killed in the combat [13]. AI being the emerging technology, it is only natural that we try it out in the fields. Countries all over the world now want to include AI in the military to improve their defences. The AI-enabled air defence system suggests using AI-enabled small, less expensive, and airborne sensors rather than expensive, limited in number sensors. These small sensors use AI edge computing with data which is processed partly. This data is sent through the cloud to a fusion centre and then to the command-and-control system. These sensors may be active, short-range emitters, but power constraints could limit such usage. These sensors are used to enhance the detection and possibly identify the targets early. It is observed that the present fighter jets are digitized to detect targets, categorizing said targets, tracking, and engaging them. Autonomous drones could allow operators to focus their attention on things that require their skills. A drone is built with various technologies like navigation, cameras, sensors, and controller. These drones are unmanned devices used for different purposes. At first, these drones were controlled manually. But now, using AI technology, they are automated. AI in drones helps us using the data from the sensors to collect and use that data. This data helps in making the operation much easier. The drones with AI technology mostly depend on computer vision. Using computer vision, we can detect objects and analyse them while recording the information at the same time [14]. This computer vision is high performance, and the images processing is done through a neural network. A neural network is something that is used to implement algorithms. They enable the drones to perform actions such as detecting the object, classifying them into categories, and tracking. The information collected throughout these processes is combined which in return helps drones to avoid collisions. The information from the drones is collected through sensors, which is the important component in AI-based drones. Machine learning models help in determining how the drone should respond after processing the data. It determines where the drones should fly to and places to avoid, what objects should it avoid, and etc. The data captured by the drone during the flight is called real-time data. This real-time data is processed in flight by AI, and it helps in getting constant feedback to the operators. These drones are needed to be trained with massive datasets for it to learn and detect different objects and obstacles. The self-navigated drones come with predefined GPS coordinates. They have the capability to reach the destination using the most optimal way without human intervention. The drones are used in different fields. In smart cities, the drones are used for surveillance which detects unwanted objects. It is also used in construction to get aerial views of the land. It is also used in traffic monitoring which can help us in knowing which route is the best at that particular moment. In agriculture, drones are used to monitor the plant's condition, monitor the crops, and figure out what is the best time to apply fertilizers at night. The police uses the drones to track humans using the camera in the drone to detect the face. Unmanned combat aerial vehicles are jets which are used for combat but with technology included. Several countries have been working on increasing this usage of AI in combat jets [15]. The AI-enabled UAVs are smaller, lighter, and less expensive compared to a full crew aircraft. These UAVs are commanded to engage the target. The UAVs are controlled by humans whether it is in the air or from the ground. AI is used to take UAVs to the next level by training them alongside humans. It needs to identify the threats, assist humans when they need to, and most importantly respond to the threats. AI in these jets provides the pilot with weapon guidance, trajectory, and surveillance among others. These UAVs are capable of communicating through a wide range. The AI-enabled jet could be operated from the ground alert interceptor (GAI) or in the combat air patrol (CAP). In the air, the UAV should be larger to show high endurance. With the simulations being run all over the world, it was observed that the AI-based jets have been improving. If the AI could not recognize the movement, it would not make the same mistake twice. AI helps in the decision-making process through monitoring every phase. UAVs can extend their time of operation by saving their battery power using perching. Scientists are also trying to build UAVs that are submersible and float on water. The UAVs act as communication nodes which offer mobile networks. AI compares the flight's performance and handling with current status. It also monitors where the error made by the pilot could affect the safety of the whole flight [16].

5 Research Challenges

The hype-surrounding AI is certainly increasing and will surely be the same in the coming years. With more AI systems being deployed, the world is entering into a phase where AI systems are becoming common in warfare. Although there are multiple consequences of initiation of AI in naval operations, foremost of them being resistance from conventional personnel who will oppose any change that has the possibility of replacing them. The fog-based smart surveillance system is also used [17]. At present, the functions that are to be performed by AI supported naval combat systems are done by personnel on the naval vessels or from stations at shore. Thus, replacing them with AI would transform multiple sectors of the naval force. Almost, all AI applications used in navy right now majorly perform analytical functions, and the final decision of engagement still rests in the hands of the human operator [18]. Henceforth, it requires a good understanding of the decision-making patterns between the human operator and the AI system. The security issues are also found in smart city environment [19]. Further, some autonomous systems based on deep learning neural networks do not provide complete transparency regarding input output operations causing barriers to man–machine teaming. Additionally, in situations where the GPS is either blocked by enemy forces or denied in certain environments, it raises serious concerns about its communication and working. Not to mention the possibilities for unmanned vehicles to be hacked and used against [20]. Just like any other domain in warfare, the navy sure does have its fair share. In addition, the some challenges that are being faced in the domain of military are mentioned below. Machine learning-based applications and its development in the military area are quite strenuous. Gathering and processing of data in the military, the platforms and training facilities are not suited for ML. The sensor networks and weapons are not suited for AI to develop. Data needs to be available for AI developers. Numerous AI systems are developed on the data basis and a system to classify this data. Corporations using manual approaches frequently categorize large datasets. Receiving and distributing this information are difficult. A military dataset comprising pictures is created by thermal-imaging equipment and annotated by specialists to identify the weapon systems identified in the image, if any. An AI that successfully leverages that set cannot be developed without sharing information with the developers. Humans are able to identify the enemy, their uniform, and vehicles. Based on this information, they are able to ploy the formations, weapons called for and predict the ultimate outcome of it. To be able to perform, each of these tasks is utterly impossible by an AI. Humans can perceive this information and plan ahead and make changes. But, enabling an AI to perform such tasks simultaneously is not possible.

Table 1 is providing recent pros and cons in the field of military. This comparison is taken with artificial intelligence technology. The closest proposition that can be decided is if there were multiple AI's performing separate tasks to recreate a close understanding of things as a human does on that battleground. When the results of machine learning are statistically significant, it means that there is some degree of connection between the input data and the model's output. When an AI-based system is loaded with statistics about two separate issues, it would fail to recognize the relationship between them. This kind of issue might arise in a military fraud prevention system that is given month-by-month data on purchases. A system like

Table 1 Pros and cons in AI-based military domain

Paramaters	Pros	Cons
Military operations	– To find or predict the opponent location – Reducing the man power and save the human life – Automated logistics to reduce the risk factor – To fix accurate target location during attacks	– In supply chain management facing risk – To minimize the strategic function – Opponent force easily traced autonomous vehicle
Training and management	– It is used to provide personalized and specific training – It is more reliable exercise and simulation training – Latest technology-based training is possible -Reduce the training cost	– Some physical environment is missing – Cultural and personal issues not able to solve

this may incorrectly suggest that fraud rises when expenditure rises, when it is really just a result of the year-end spending habits. In the AI and machine learning research field, adversarial machine learning, a technique that tries to mislead models using false data, is becoming an increasing danger. The most typical reason is to cause a machine learning model to malfunction. Such an attack can be labelled into—evasion, inference, poisoning, and extraction. Evasion attacks try to trick an AI engine in order to escape detection and hide a cyberattack. Inference attacks, which gather knowledge about the AI system and may be utilized to allow evasion assaults, are frequently preceded by evasion attacks. The following attack is the poisoning attack which has the ability to earmark an AI system while training to satisfy the need of their vindictive intents. Lastly, extraction makes use of the access provided to the AI interface and collects all the necessary data about the functioning of AI to build a parallel model of that AI system. It is no secret that some people are afraid of involving AI in the military, particularly the air force. It may be because of not needing pilots anymore or because of wrong decisions taken by then AI. There may be a chance of trusting the AI if it improves eventually. There is a concern that when humans let go of the wheel, the AI could spiral down gradually without someone there to control them. If the AI-enabled unmanned jet loses track of what it is supposed to do during aerial combat, it is really dangerous for humans to control it from the ground. There is also a concern that the jet could be hacked. Without the pilot controlling it, if the jet has been hacked, it can be turned against its country. If the AI fails to perform the task it has been assigned, it could lead to dangerous situations. Artificial intelligence has shown immense growth since the last few decades of its existence; in the fields of agriculture, medicine, and education. As for growth in warfare; AI will undoubtedly play a more significant role in future. Its usage in warfare although is very exciting presently, it is a field that is only going to grow and show ultimate promise in warfare. When its usage amplifies, it will result in enhanced productivity, lesser human workload, and quicker results. With such a high-paced growing field being utilized in warfare, it is unquestionably going to improve the capability and strength. Such a technology must grab the attention of the navy, airforce, and the military more than it is now.

Figure 3 is to elaborate the words search strategy in Google. This article shortly describes already existing technology that has been embraced by the navy, airforce, and the military. Such technology has to be popularized and must be spread out to the nations. To have good intentions, experimenting with such technology is salient. Research shows how people are agitated over AI taking over the world with its usage being predominant in every field. Such distress is understandable as there is no definite knowledge of what the future might turn out to be. In order to work forward to a better and convenient future, such bold steps must be taken to understand that these technologies can be just as beneficial. When a clear understanding of good and bad of AI is achieved, prominent and carefully thought-out actions will induce.

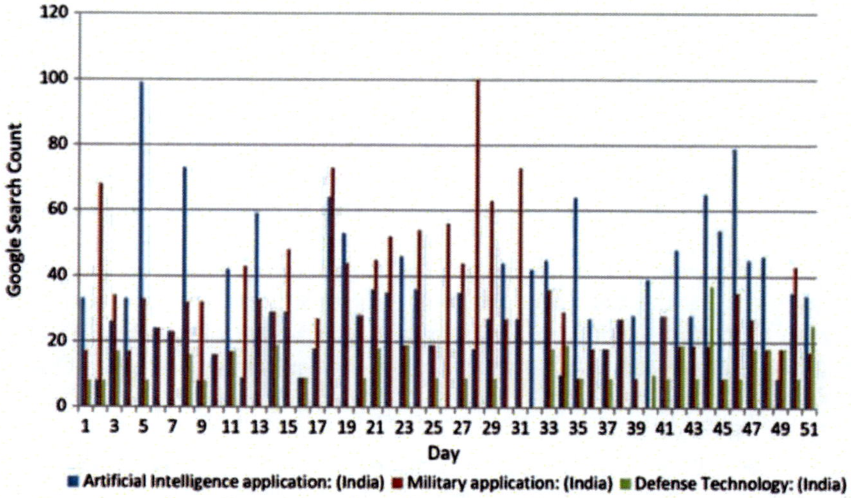

Fig. 3 Trending chart in Google

6 Conclusion

Artificial intelligence has shown immense growth since the last few decades of its existence; in the fields of agriculture, medicine, and education. As for growth in warfare; AI will undoubtedly play a more significant role in future. Its usage in warfare although is very exciting presently, it is a field that is only going to grow and show ultimate promise in warfare. When its usage amplifies, it will result in enhanced productivity, lesser human workload, and quicker results. With such a high-paced growing field being utilized in warfare, it is unquestionably going to improve the capability and strength. Such a technology must grab the attention of the navy, airforce, and the military more than it is now. This article shortly describes already existing technology that has been embraced by the navy, airforce, and the military. Such technology has to be popularized and must be spread out to the nations. To have good intentions, experimenting with such technology is salient. Research shows how people are agitated over AI taking over the world with its usage being predominant in every field. Such distress is understandable as there is no definite knowledge of what the future might turn out to be. In order to work forward to a better and convenient future, such bold steps must be taken to understand that these technologies can be just as beneficial. When a clear understanding of good and bad of AI is achieved, prominent and carefully thought-out actions will induce.

References

1. Ullah Z, Al-Turjman F, Mostarda L, Gagliardi R (2020) Applications of artificial intelligence and machine learning in smart cities. Comput Commun 154:313–323
2. Natarajan J (2020) Cyber secure man-in-the-middle attack intrusion detection using machine learning algorithms. In: AI and Big Data's potential for disruptive ınnovation, pp 291–316. IGI global
3. Xin Y, Kong L, Liu Z, Chen Y, Li Y, Zhu H, Wang C (2018) Machine learning and deep learning methods for cybersecurity. IEEE Access 6:35365–35381
4. Bhavana K, Nekkanti V, Jayapandian N (2019) Internet of things enabled device fault prediction system using machine learning. In: International conference on ınventive computation technologies, pp 920–927. Springer, Cham
5. Hayes JL, Britt BC, Evans W, Rush SW, Towery NA, Adamson AC (2021) Can Social Media listening platforms' artificial intelligence be trusted? Examining the accuracy of Crimson Hexagon's (now Brandwatch consumer research's) AI-Driven analyses. J Advert 50(1):81–91
6. Kepuska V, Bohouta G (2018) Next-generation of virtual personal assistants (microsoft cortana, apple siri, amazon alexa and google home). In: 2018 IEEE 8th annual computing and communication workshop and conference (CCWC), pp 99–103. IEEE
7. Pandey A, Singh Y, Malik M (2021) Design and development of home automation system. In: Intelligent computing and applications. Springer, Singapore, pp 245–254
8. Krishna K, Karumuri N, Christopher C, Jayapandian N (2021) Research challenges in self-driving vehicle by using ınternet of things (IoT). In: 2021 5th ınternational conference on ıntelligent computing and control systems (ICICCS), pp 423–427. IEEE
9. Lee H, Lee K, Yang H, Oh SY (2020, October) Applying FastPhotoStyle to synthetic data for military vehicle detection. In: 2020 20th ınternational conference on control, automation and systems (ICCAS), pp 137–140. IEEE
10. Du T, Zeng YH, Yang J, Tian CZ, Bai PF (2020) Multi-sensor fusion SLAM approach for the mobile robot with a bio-inspired polarised skylight sensor. IET Radar Sonar Navig 14(12):1950–1957
11. Liu D, Chen B, An J, Li C, Liu G, Shao J et al (2020) Wind-driven self-powered wireless environmental sensors for Internet of Things at long distance. Nano Energy 73:104819
12. Roper D, Harris CA, Salavasidis G, Pebody M, Templeton R, Prampart T et al (2021) Autosub long range 6000: a multiple-month endurance AUV for deep-ocean monitoring and survey. IEEE J Oceanic Eng 46:1179–1191
13. Jenkins PR, Caballero WN, Hill RR (2021) Predicting success in United States Air Force pilot training using machine learning techniques. Socio-Econ Plan Sci 101121
14. Jayapandian N (2019) Cloud enabled smart firefighting drone using internet of things. In: 2019 ınternational conference on smart systems and ınventive technology (ICSSIT), pp 1079–1083. IEEEl
15. Huh J, Park J, Shin D, Choi Y (2020) A Behavior optimization method for unmanned combat aerial vehicles using matrix factorization. IEEE Access 8:100298–100307
16. Hu J, Zhang H, Song L, Schober R, Poor HV (2020) Cooperative internet of UAVs: distributed trajectory design by multi-agent deep reinforcement learning. IEEE Trans Commun 68(11):6807–6821
17. Shrestha S, Shakya S (2020) A comparative performance analysis of fog-based smart surveillance system. J Trends Comput Sci Smart Technol (TCSST) 2(2):78–88
18. Ranganathan G (2020) Real life human movement realization in multimodal group communication using depth map information and machine learning. J Innov Image Proces (JIIP) 2(2):93–101
19. Jayapandian N (2019) Threats and security ıssues in smart city devices. In: Secure cyber-physical systems for smart cities, pp 220–250. IGI Global
20. Zhou H, Xu W, Chen J, Wang W (2020) Evolutionary V2X technologies toward the Internet of vehicles: challenges and opportunities. Proc IEEE 108(2):308–323

Evaluation of Various Classifier Models to Predict the Cardiovascular Disease

K. G. Rani Roopha Devi, R. Murugesan, and R. Mahendra Chozhan

Abstract CVD plays a major death all over the world. It happens at a sudden or prolonged illness; cardiovascular infection is a class of illnesses that include the heart or veins. It incorporates coronary vein infections (e.g. angina and myocardial localised necrosis (ordinarily known as a cardiovascular failure)). The sudden facial expression can be the key factor for diagnosis or identification of the CVD patient. It happens for much prolonged illness, and the proposed work concentrated and developed the monitoring system only for periodontal patient. In this paper, the proposed system has been evaluated with the many efficient related works done by various researchers. In this work, the optimised NN has been used to develop the model for CVD identification. In this regard, the patient data has been stored under the cloud environment in a secured manner to help the patient on-time caregiving to avoid sudden cardiovascular death. The proposed work achieves an efficient result with the accuracy of 96%, the sensitivity of 77%, and specificity of 97.84%. The result has been compared with other classifier models of NB, SVM, NN, and adaptive neuro-fuzzy inference system. The proposed work can also be evaluated with other performance metrics of F-measure, recall, MCC, FPR, FNR, NPV, and FRR.

Keywords Naïve Bayes · Support vector machine · Neural network · Optimised neural network · Adaptive neuro-fuzzy inference system

K. G. Rani Roopha Devi (✉)
Department of Computer Science, Madurai Kamaraj University, Madurai, India

R. Murugesan
Department of Computer Science, C.P.A College, Bodinayakanur, India

R. Mahendra Chozhan
Chozhan Dental Clinic, Periyakulam, Kodaikanal, Lakshmipuram, India

S. Shakya et al. (eds.), *Proceedings of Second International Conference on Sustainable Expert Systems*, Lecture Notes in Networks and Systems 351,
https://doi.org/10.1007/978-981-16-7657-4_29

1 Introduction

Unhealthy eating habits and other co-occurring illnesses are the major behaviour risk factors for cardiovascular disease and stroke. Increased pulse, increased blood glucose, high blood lipids, and obesity and heftiness are all symptoms of conduct hazard elements [1]. These "transitional dangers factors" can be calculated in hospitals and imply a higher chance of developing a coronary episode, CVA, cardiovascular breakdown, and other complications. Quitting tobacco, reduced salt intake, consumption of agricultural goods, traditional physical labour, and avoidance of harmful alcohol usage have all been shown to minimise the risk of cardiovascular disease. Furthermore, medications for diabetes, high blood pressure, and high blood lipids may be required to minimise cardiovascular risk and avoid respiratory failures and CVAs. People must be motivated to obtain and encourage positive conduct by well-being agreements that offer favourable opportunities for making sound decisions in a moderate and open manner [2].

In addition, there are also a variety of key factors that influence CVD risk or "the explanations for the triggers". Globalisation, urbanisation, and population maturation are some of the main factors influencing social, monetary, and social change. Destitution, tension, and innate elements are all determinants of CVDs [3, 4].

The most associated sickness adversary CVD is periodontal, and poor periodontal well-being has been emphatically corresponded with cardiovascular infection (CVD) in various observational studies; nonetheless, it is not clear if poor periodontal well-being causes CVD. Enormous proportions of affiliation revealed in early case-control and cross-sectional examinations assessing this affiliation have been ascribed to contemplate plan and insufficient control of puzzling; these affiliations are lessened yet persevere in meta-investigations of very much led accomplice studies. Additionally, in observational investigations, periodontal microorganisms have been related with atherosclerosis, hypertension, and dyslipidaemia [5].

There has been lots of techniques to identify CVD [6, 7]. In this work, the facial-based CVD has been identified with the facial expressions of the patient. The optimised neural network has taken as the classifier model to predict CVD.

2 Evaluation of Various Classifier Model

Periodontal microorganisms present in dental plaque show up in the circulatory system in direct extent to the seriousness and degree of periodontal disease and can affect removed destinations. Porphyrins gingivitis includes are higher in sub-gingival plaque in periodontal sickness, where it cause irritation and periodontal destruction [8, 9].

The creature has been appeared to enter the circulation system following tooth brushing and has been found in the intima of removed veins remembering for coronary arteries, where it multiplies, starting an incendiary course prompting apoptosis

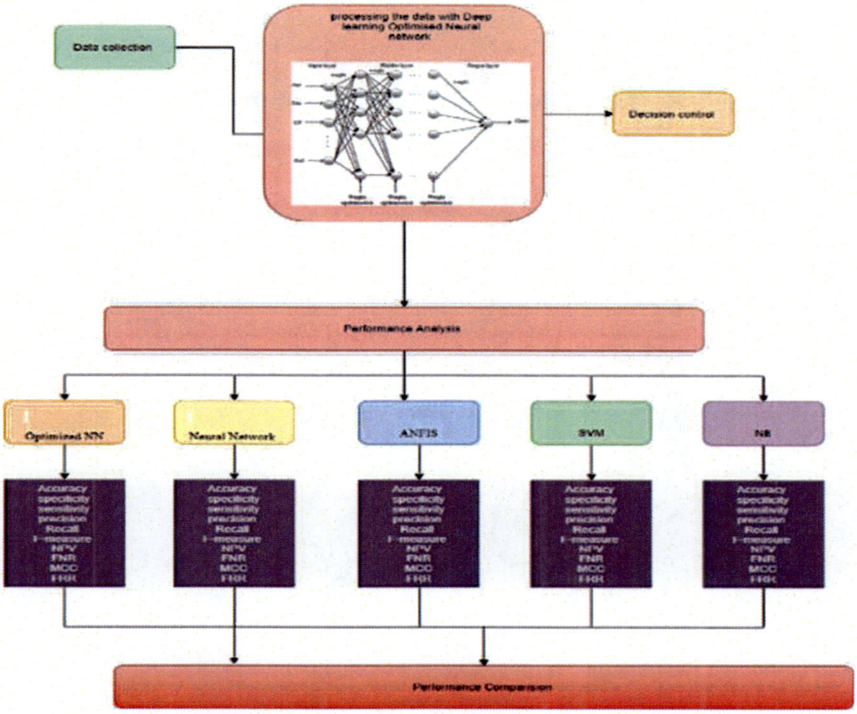

Fig. 1 Performance evaluation of various classifier model

and resulting endothelial dysfunction, which is related with hypertension. It likewise invigorates froth cell creation in the intima prompting atherosclerosis. *P. gingivitis* has been appeared to expand low-thickness lipoprotein (LDL) and absolute cholesterol levels by up-guideline of a protein which effects circling levels of LDL cholesterol in mice [10] (Fig. 1).

3 Proposed Optimised Neural Network

In this work, the deep learning optimised neural network has been employed. The deep learning neural network classifier has a convolutional layer, pooling layer, and fully linked-layer. It has weight and bias for the neural network learning structure; here, the weight has been applied based on the modification of the PSO algorithm. Figure 2 represents the optimised neural network.

There have been plenty of works carried out to measure the heart rate based on the facial video; in 2017, they reviewed the various research works based on heart rate

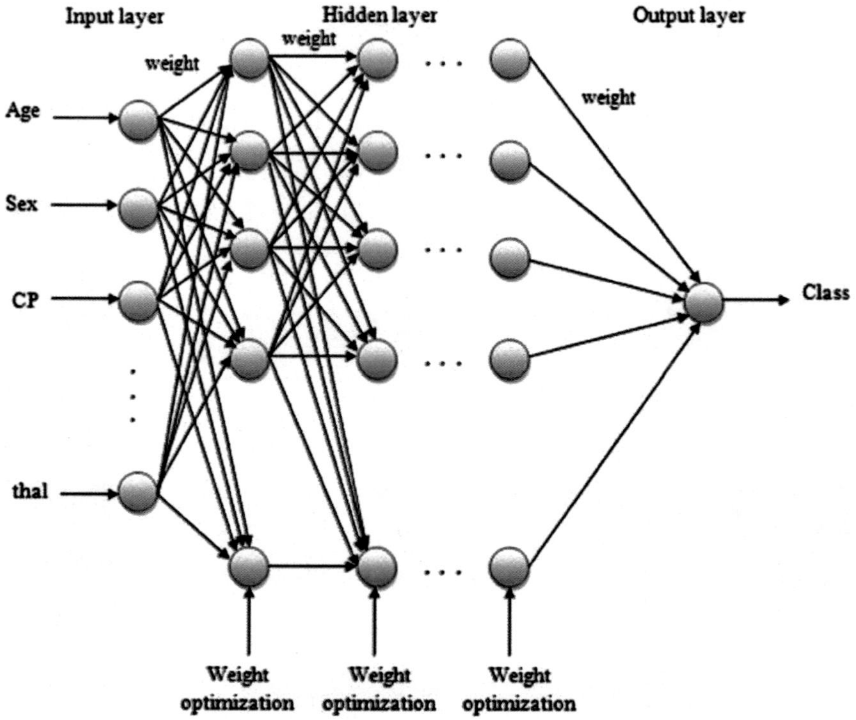

Fig. 2 Proposed optimised neural network

estimation using video, the most of the works are carried out using photoplethysmography and ballistocardiograph, and these are the methods to find the heart rate by human facial video.

4 Performance Metrics

In the previous section, we discussed classification and its algorithms. Here, we will look at a variety of success indicators that can be used to assess classification problem predictions.

5 Confusion Matrix

It is the simplest choice to assess the efficiency of a classification issue with two or more types of output. As shown below, an uncertainty matrix is a two-dimensional table: "real" and "predicted", as well as "true positives (TP)", "true negatives (TN)", "false positives (FP)", and "false negatives (FN)"' on both dimensions.

Below are the definitions of the terms related to the uncertainty matrix:

TP: If both data point of real and expected is 1.

TN: If both data point of real and expected is 0.

FP: If data point of real is 0 and expected is 1.

FN: If data point of real is 1 and expected is 0.

It is the commonly used efficiency metric for classification algorithms. The accuracy rate is the number of accurate predictions divided by the total number of predictions. Based on the result, the various performance measures have been applied.

6 Performance Analysis with Related Work

6.1 Naïve Bayes (NB)

The calculation must be simplified in order to use Bayes theorem (BT) for a model of conditional probability. In the BT, each input variable is assumed to be interdependent on the others. This adds to the calculation's sophistication. Remove this presumption and treat each input variable as if it was unrelated to the others [11].

Simple to construct, naïve Bayesian models are particularly useful for small- and medium-sized datasets (as the one used in this article demonstrates!). Naive Bayes is known to outperform even the most powerful classification algorithms due to its simplicity.

As a result, this classifier is employed in a variety of key fields, including disease detection, sentiment analysis, and email spam classifier development. For binary (two class) and multiclass classification issues, naive Bayes is a classification algorithm. Because the probability computations for each class are simplified to make their calculations tractable, it is known as naive Bayes or idiot Bayes [12].

Rather of attempting to predict the likelihood of each attribute value, the class value is regarded conditionally independent.

$$P(y_i|x_1, x_2, \ldots, x_n) = P(x_1, x_2, \ldots, x_n|y_i) * P(y_i) \tag{1}$$

The conditional probabilities of all variables with the same class label are then divided into conditional probabilities

$$P(y_i|x_1, x_2, \ldots, x_n) = P(x_1|y_i) * P(x_2|y_i) * \ldots P(x_n|y_i) * P(y_i) \tag{2}$$

This equation can be done with any of the class categories, with the category with the greatest probability being selected as the designation for the specific scenario [7].

7 Support Vector Machine

The support vector machine, or SVM, is a common supervised learning algorithm that can be applied to fix both classification and regression problems. However, it is mainly used in machine learning for classification problems. A hyperplane [13] is the right criterion boundary.

The goal of the help vector machine algorithm is to find a hyperplane that distinguishes between data points in an N-dimensional space (N—the number of features) [14].

In order to differentiate the two types of data points, you can select from a number of hyperplanes. Our goal is to find the plane with the largest margin or gap between data points from both categories. By increasing the margin difference, it becomes easier to distinguish possible data points.

8 Neural Networks (NN)

The term "neural networks" is one of those cool terms that is frequently used to give credibility to science. But, what does it mean to have a basic understanding of neural nets and convolutional NN, as well as the ability to code your own simple neural network model [6, 15].

The basic neural network model is shown in Fig. 3. Neural nets are focussed on the learning mechanism of human brains [16]. They are made up of a collection of parameters, or artificial functions, that allow the machine to learn and improve itself by analysing new data. Each parameter, also known as a neuron, is a function that produces an output after obtaining one or two inputs. The next layer of neurons

Fig. 3 Multilayer neural network

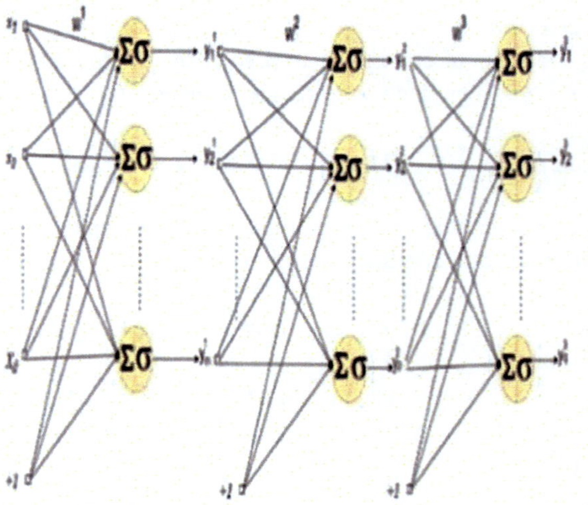

receives these outputs and uses them as inputs to their own functions, generating further outputs [17]. The outputs of these neurons are then transferred on to the next layer of neurons, and so on, until all of the neurons have been considered and the terminal neurons have received their input [11]. The model's final result is then output by those terminal neurons.

A multilayer perceptron (MLP) is a form of artificial neural network with multiple layers. It is made up of three layers: An input layer that links to input variables, one or two hidden layers, and an output layer that produces output variables. Figure 3 shows the different forms of layers in an MLP:

The visible layer is often referred to as the input layer because it contains the input variables.

$$\vec{y}^1 = \sigma\left(w^1\vec{x}\right) \tag{3}$$

$$\vec{y}^2 = \sigma\left(w^2\vec{y}^1\right) \tag{4}$$

$$\vec{y}^3 = \sigma\left(w^3\vec{y}^2\right) \tag{5}$$

Hidden layers are node layers that exist from the input layers to the output layers. One or two of these layers can exist.

The output variables are generated by a layer of nodes called the output layer.

Lastly, there are words to describe a neural network's form and capability, such as:

- The number of nodes in the model's size.
- The number of nodes in a layer is measured in width.
- The number of layers in a neural network is measured in depth.

After each piece of data is processed, the perceptron learns by adjusting relation weights based on the amount of error in the performance relative to the expected result. Backpropagation, a generalisation of the least mean squares algorithm in the linear perceptron, is used in this example of supervised learning.

$$b + \sum_{i=1} w_i x_i = w_1 x_1 + w_2 x_2 + \cdots + w_m x_m \tag{6}$$

To perform a multilayer activation function with m features in input X, m weights are needed. For performing dot products with n secret neurons in the hidden layer, n sets of weights (W_1, W_2 ... W_n) are needed. To get the secret output h, perform n dot products with one hidden layer: (h_1, h_2 ..., h_n).

9 Adaptive Neuro-fuzzy Inference System

In the form of MFs and fuzzy rule bases, fuzzy inference effectively explains the method by organising professional scientific information that is difficult to quantitatively express. Furthermore, since NN can learn, they can be configured in a variety of ways, and they are capable of exceptional parallel processing and fault tolerance. Theories of neuro-fuzzy systems and NN are being researched in a variety of fields [18].

In terms of optimising the nonlinear system model, the neuro-fuzzy system outperformed the model based on the current linear system. Based on the type of learning model or learning process, the performance of a neuro-fuzzy system that mimics human learning ability, decision-making, and other cognitive functions rather than a quantitative calculation technique may vary [19]. The adaptive neuro-fuzzy inference system (AFNIS) model rule generation technique can be divided into two approaches: grid-based rule generation and clustering-based rule generation.

ANFIS is a neuro-fuzzy system that is well-known. The ANFIS presumption is contingent on how the rule is written. The shape of the first equation stone, the TSK model [20, 21], is used to structure the conclusion portion.

10 Result and Discussion

The simulation is done with MATLAB environment for prediction and diagnosis of periodontal diseases. Confusion matrix is utilised for determining the performance of classification process by evaluating the performance metrics. The performance measurement is utilised for determining metrics of true positive (TP), true negative (TN), false positive (FP), and false negative (FN). The dataset is developed from the Chozhan group of dental clinics, Theni, Periyakulam, and Kodaikanal with the following information. There have been samples collected with real-time environment.

Data Sheet

S. no.	Parameters
1	Age
2	Sex
3	Cp
4	Trestbps
5	Chol
6	Fbs
7	Restecg
8	Exang
9	Oldpeak

(continued)

(continued)

S. no.	Parameters
10	Slope
11	Ca
12	Thal

Table 1 represents the sensitivity results of the five classifier models, and Fig. 4 evaluated the sensitivity.

Table 2 represents the specificity results of the five classifier models, the proposed optimised NN produces the efficient specificity results of 0.97842, and Fig. 4 evaluated the specificity.

Table 3 represents the accuracy results of the five classifier models, the proposed optimised NN produces the efficient accuracy of 0.96, and Fig. 5 evaluated the accuracy results of classifier model.

Table 1 Performance analysis of sensitivity

Methods	Sensitivity
Optimised NN	0.72727
NN	0.45455
ANFIS	0.90909
SVM	0.63636
NB	0.90909

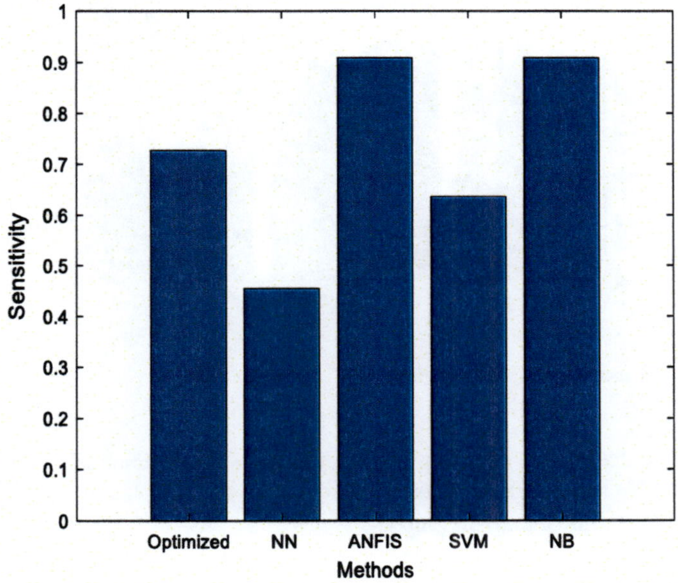

Fig. 4 Performance evaluation of sensitivity

Table 2 Performance analysis of specificity

Methods	Specificity
Optimised NN	0.97842
NN	0.96043
ANFIS	0.86691
SVM	0.93885
NB	0.91007

Table 3 Performance analysis of accuracy

Methods	Accuracy
Optimised NN	0.96
NN	0.92333
ANFIS	0.87
SVM	0.91667
NB	0.91

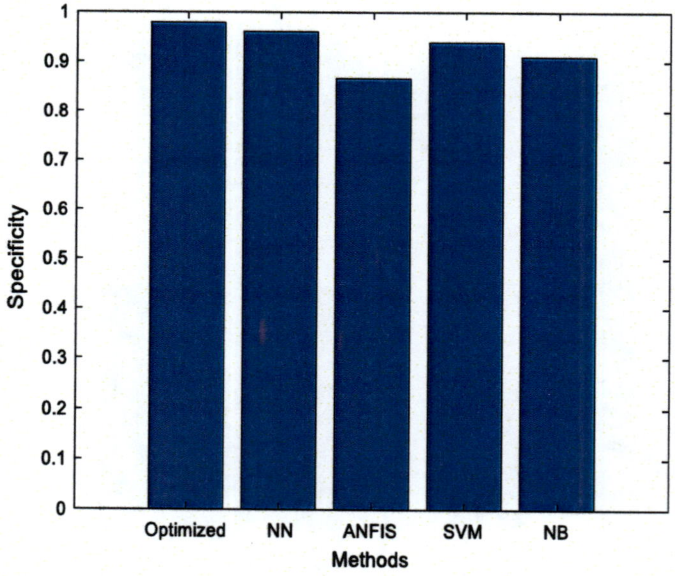

Fig. 5 Performance evaluation of sensitivity

Table 4 represents the precision results of the five classifier models, the proposed optimised NN produces the efficient precision rate of 0.72727, and Fig. 6 evaluated the precision results of classifier model.

Table 5 represents the *F*-measure of the five classifier models, the proposed optimised NN produces the efficient precision rate of 0.72727, and Fig. 7 evaluated the precision results of classifier model (Fig. 8).

Table 4 Performance analysis of precision

Methods	Precision
Optimised NN	0.72727
NN	0.47619
ANFIS	0.35088
SVM	0.45161
NB	0.44444

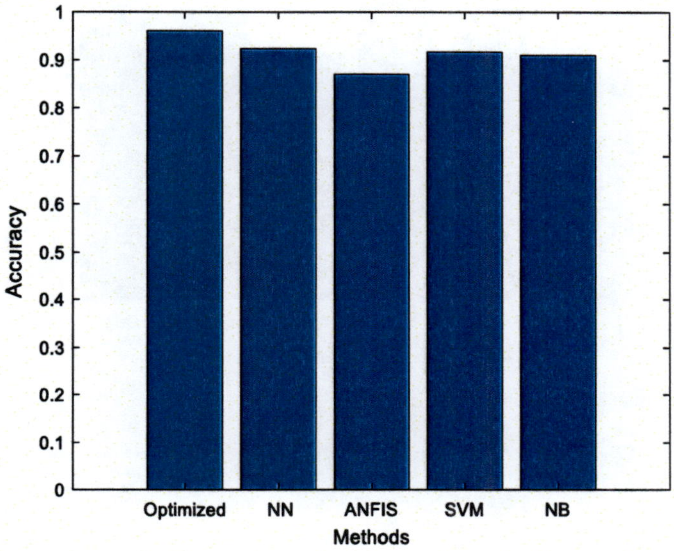

Fig. 6 Performance evaluation of sensitivity

Table 5 Performance analysis of *F*-measure

Methods	*F*-measure
Optimised NN	0.72727
NN	0.46512
ANFIS	0.50633
SVM	0.5283
NB	0.59701

Table 6 represents the overall performance metrics of five various classifier models for detection of cardiovascular disease. The proposed optimised NN produces the efficient results (Fig. 9).

From the overall performance, there have been various performance metrics that produced the efficient result for proposed optimised NN. The work has also been

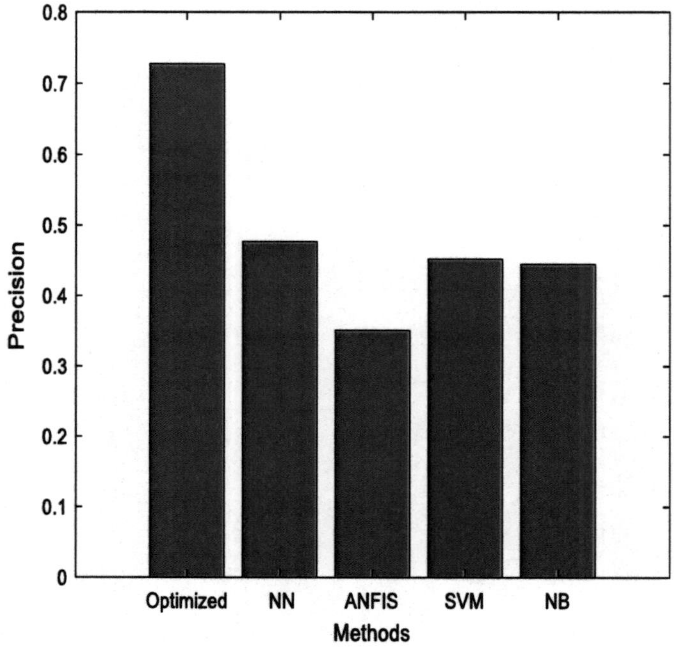

Fig. 7 Performance evaluation of sensitivity

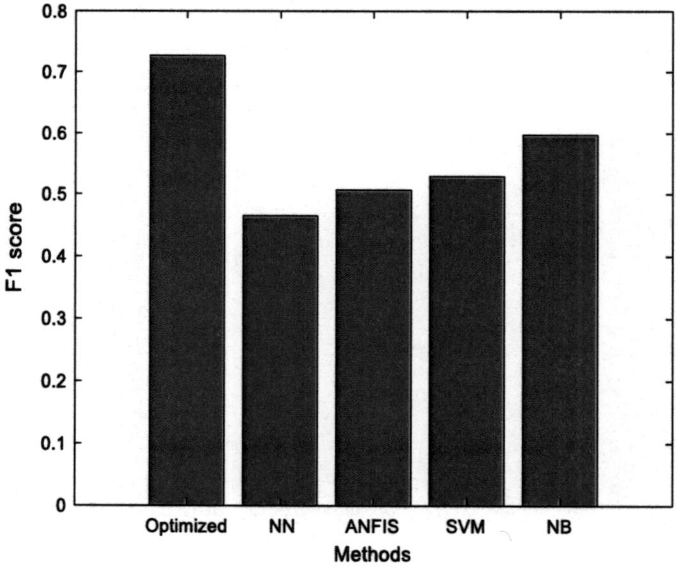

Fig. 8 Performance evaluation of sensitivity

Table 6 Overall performance analysis

Methods	Optimised NN	NN	ANFIS	SVM	NB
Sensitivity	0.72727	0.45455	0.90909	0.63636	0.90909
Specificity	0.97842	0.96043	0.86691	0.93885	0.91007
Accuracy	0.96	0.92333	0.87	0.91667	0.91
Precision	0.72727	0.47619	0.35088	0.45161	0.44444
Recall	0.72727	0.45455	0.90909	0.63636	0.90909
F-measure	0.72727	0.46512	0.50633	0.5283	0.59701
NPV	0.97842	0.95699	0.99177	0.97026	0.99216
FPR	0.021583	0.039568	0.13309	0.061151	0.089928
FNR	0.27273	0.54545	0.090909	0.36364	0.090909
MCC	0.70569	0.42398	0.51565	0.49261	0.59804
FRR	0.27273	0.54545	0.090909	0.36364	0.090909

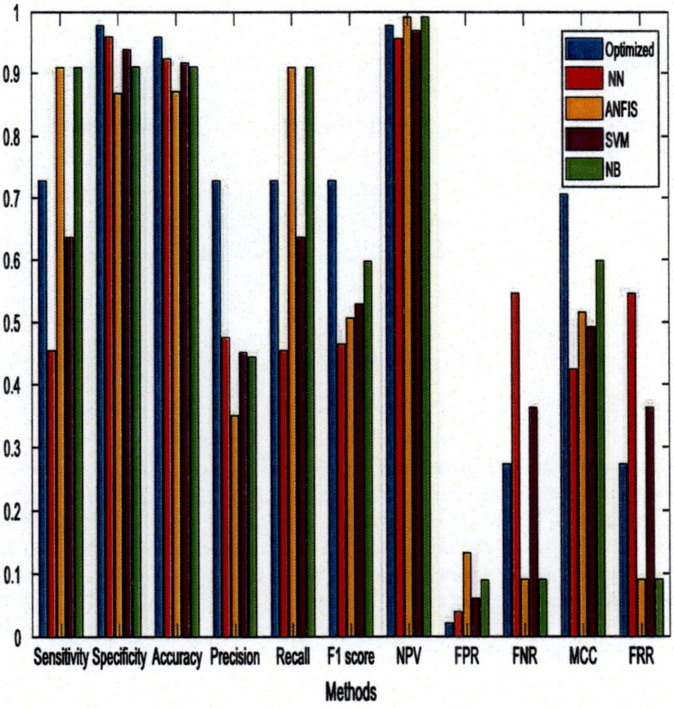

Fig. 9 Overall performance evaluation

compared with the other conventional algorithm. The proposed work produces the higher accuracy rate of 96%.

11 Conclusion

Based on the evaluation results, the proposed optimised NN approaches outperformed alternative methods for cardiovascular disease identification with the performance metrics of specificity, accuracy, *F*-measure, precision, and MCC. CVD is a leading cause of death all over the world. Cardiovascular infection is a group of illnesses that affect the heart or veins which can occur suddenly or over time. Coronary artery disease is a form of cardiovascular disease (ordinarily known as a cardiovascular failure). The sudden facial expression can be a crucial factor in deciding whether or not someone is ill. The patient's sudden facial expression may be a crucial factor in diagnosing or identifying them. Since it is a long-term disease, the proposed research focussed on developing a monitoring system specifically for periodontal patients. The proposed framework has been tested with a variety of efficient classifications using the proposed optimised NN approach in this article.

The optimised NN was used to create the CVD recognition model in this study. In this situation, the medical data was safely stored in the cloud to assist the patient in accessing prompt treatment and preventing sudden cardiovascular death. In comparison to other similar work, the proposed work produces an effective result. In future, direction of the work will design an efficient model with IoT-based CVD prediction on real-time scenario.

References

1. Aminoshariae A, Kulild J, Fouad AF (2020) The impact of cardiovascular disease and endodontic outcome: a systematic review of longitudinal studies. Clin Oral Invest 1–7
2. Li X, Guan B, Su T, Liu W, Chen M, Waleed KB, Guan X, Gary T, Zhu Z (2020) Impact of cardiovascular disease and cardiac injury on in-hospital mortality in patients with COVID-19: a systematic review and meta-analysis. Heart 106(15):1142–1147
3. Carbone S, Kirkman DL, Garten RS, Rodriguez-Miguelez P, Artero EG, Lee D-C, Lavie CJ (2020) Muscular strength and cardiovascular disease: an updated state-of-the-art narrative review. J Cardiopulm Rehabil Prev 40(5):302–309
4. Anselmi G, Gagliardi L, Egidi G, Leone S, Gasbarrini A, Miggiano GAD, Galiuto L (2020) Gut Microbiota and cardiovascular diseases: a critical review. Cardiol Rev
5. Lee JH, Oh JY, Choi JK, Kim YT, Park YS, Jeong SN et al (2017) Trends in tooth extraction incidence due to periodontal disease: results of a 12-year longitudinal cohort study in South Korea. J Periodontal Implant Sci
6. Rajpurkar P, Hannun AY, Haghpanahi M, Bourn C, Ng AY (2017) Cardiologist-level arrhythmia detection with convolutional neural networks. arXiv e-print 2017: arXiv:1707.01836
7. Ramotra AK, Mahajan A, Kumar R, Mansotra V (2020) Comparative analysis of data mining classification techniques for prediction of heart disease using the Weka and SPSS modeler tools. In: Smart trends in computing and communications, pp 89–97. Springer, Singapore

8. Lee JH, Lee JS, Park JY, Choi JK, Kim DW, Kim YT et al (2015) Association of lifestyle-related comorbidities with periodontitis: a nationwide cohort study in Korea. Medicine (Baltimore)
9. Lee JH, Choi JK, Kim SH, Cho KH, Kim YT, Choi SH et al (2017) Association between periodontal flap surgery for periodontitis and vasculogenic erectile dysfunction in Koreans. J Periodontal Implant Sci
10. Lee JH, Kweon HH, Choi JK, Kim YT, Choi SH (2017) Association between periodontal disease and prostate cancer: results of a 12-year longitudinal cohort study in South Korea. J Cancer 2017
11. Danish S, Aamer S, Kharde SP, Gadekar SS (2020) Review on chronic kidney disease using Naïve Bayes algorithm
12. Khan AAR, Shajun Nisha S, Mohamed Sathik M (2020) Researcher reviews on data mining techniques for medical dataset: a survey
13. Nilashi M, Ahmadi N, Samad S, Shahmoradi L, Ahmadi H, Ibrahim O, Asadi S, Abdullah R, Abumalloh RA, Yadegaridehkordi E (2020) Disease diagnosis using machine learning techniques: a review and classification. J Soft Comput Decis Support Syst 7(1):19–30
14. Manonmani M, Balakrishnan S (2020) review of optimization-based feature selection algorithms on healthcare dataset. In: Emerging research in data engineering systems and computer communications, pp 239–245. Springer, Singapore
15. Litjens G, Kooi T, Bejnordi BE, Setio AA, Ciompi F, Ghafoorian M et al (2017) A survey on deep learning in medical image analysis. Med Image Anal
16. Kim TS, Obst C, Zehaczek S, Geenen C (2008) Detection of bone loss with different X-ray techniques in periodontal patients. J Periodontol
17. Ronneberger O, Fischer P, Brox T (2015) U-net: convolutional networks for biomedical image segmentation. In: International conference on medical image computing and computer-assisted intervention, pp 234–241. Springer, Berlin
18. Jerlin Rubini L, Perumal E (2020) Efficient classification of chronic kidney disease by using multi-kernel support vector machine and fruit fly optimization algorithm. Int J Imaging Syst Technol 30(3):660–673
19. Sethy PK, Behera SK, Ratha PK, Biswas P (2020) Detection of coronavirus disease (COVID-19) based on deep features and support vector machine
20. Kandel I, Castelli M (2020) How deeply to fine-tune a convolutional neural network: a case study using a histopathology dataset. Appl Sci 10(10):3359
21. Bandyopadhyay S, Dutta S (2020) Adaptive neuro fuzzy inference system used to classify the five major types of brain tumors

Blockchain Advancement for a Safe and Transparent COVID-19 Vaccination

S. Vinushree and S. Thejaswini

Abstract The recent years are witnessing the phantom of coronavirus infection. Even though the disease is emerged in Wuhan, it has spread across the globe since people used to travel and spread all over the world. Remarkable spreads as well as significant obliteration rates are the two main consequences of the SARS-CoV-2 infection that put the whole world on lockdown. The emerging monetary crisis along with increased mortality has placed the world in a very difficult scenario. The unusual demand on the clinical concept structure revealed different breakdown in this sector as well as numerous regions, resulting in a huge load of new difficulties for medical experts and specialists. In these circumstances, we might take an interest in the development of numerous invulnerable response up-and-comers in an unimaginably shorter amount of time. Later, the vaccinations were clinically tried to provide immunization. Vitality, security, straightforwardness, and noticeable quality are critically demanded in this situation. To assure a competent vaccination campaign, this research work proposes a blockchain-based framework to handle the enrollment, collection, and distribution of antibodies.

Keywords Blockchain · COVID-19 · Clinical idea

1 Introduction

The World Health Organization (WHO) declared coronavirus as a global pandemic on January 30, 2020. An examination into the spread of COVID-19 disease revealed that the most countries have persevered despite a huge increase in COVID-19 disease cases. As a result of the nation's adoption of total lockdown as a measure to limit the spread of the illness, which prompted the growing levels of physical and mental abusive conduct at home among various other forms of social maltreatment: Inoculation is an important element of medical services and is also seen as a fundamental freedom. Investing in immunization is essential and yields tremendous

S. Vinushree (✉) · S. Thejaswini
Siddaganga Institute of Technology, Tumkur, India

S. Shakya et al. (eds.), *Proceedings of Second International Conference on Sustainable Expert Systems*, Lecture Notes in Networks and Systems 351,
https://doi.org/10.1007/978-981-16-7657-4_30

Fig. 1 COVID-19 vaccine

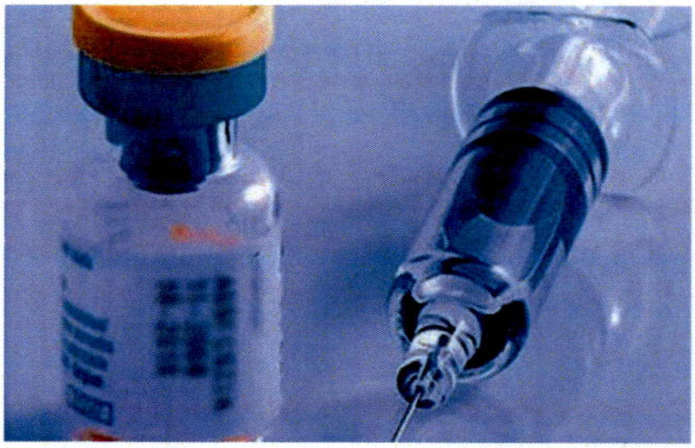

results. Moreover, antibodies are vital for the anticipation and control of irresistible sickness episodes. Compelling and safe antibodies can comprise of an extremely useful asset in forestalling COVID-19 cases and end the pandemic. Numerous nations and associations throughout the planet are currently working together in the process of exploration. Furthermore, the development of a COVID-19 antibody and currently few COVID-19 vaccinations such as Pfizer BioNTech antibody and Moderna's antibody, as illustrated in Fig. 1, is available.

The fundamental point is the openness of a utilitarian and direct beginning-to-end creation organization and collaboration structures [1, 2]. On one hand, it works to ensure the vaccination, stock organization, and afterward the careful temperature control in the infection chain [3].

Blockchain can assemble the efficiency and straightforwardness of COVID-19 immunizer movement by ensuring the perceptibility and the exhaustive survey of the limit and transport conditions. As we might want to think, blockchain-based game plans may give a modernized execution of data duty and provenance continuing in counteracting the agent movement, which will enable the coordination of different information storage facilities additionally had and regulated by different kinds of accomplices in all the dispersal chain. Self-executing smart arrangements may assure the visibility of the COVID-19 inoculation stock affiliation, particularly the infection component. To continue, the chain in which the immunizer is kept at inconceivably low temperature.

The cooler temperature should be −70 °C, and most prominent storing capacity of the cooler is 30 days. The resultant perspective is the straightforwardness and accuracy in the selection and the directives of the keeping things under control for immunization. This overview on data is not just delicate, but it also demands accuracy, avoidance of duplication, security, and immutability.

Finally, the third point of view is to increase faith in neutralizers by performing a direct and public uncovering of potential coincidental effects, such as adjusting the backup vaccination package level.

WHO declared coronavirus as a global pandemic on January 30, 2020 [4]. Most countries are coping with an increasing spike in COVID-19 defilement cases according to a study on the COVID-19 spread [5]. Countries tolerating comprehensive lockdown as an action to limit the illness's fundamental spread as a response to slow down the spread of disease that causes physical and mental health issues [6].

Coronavirus illness results in a fundamental socioeconomic disaster. Han et al. [7] present an examination of the affected locations.

In [3], blockchain and AI-based designs were suggested as ways to combat the COVID-19 crisis. Scene tracking, customer security, consistent workouts, clinical creation organization, and gift tracking are among the five essential blockchain-based remedies to solve the problems with centers:

An all-encompassing blockchain model arrangement was developed in [8] to monitor different times of the counteracting agent store organization.

IBM collaborated with the partnership at various stages of the drug development process to show how data from a blockchain association might be utilized for coronavirus vaccination [2], according to a separate article.

The use of blockchain as a solution to affiliation and the main group of industry supply chains has been recommended [9]. For pharmacy networks, where temperature detection or counterfeit pharmaceutical expectations are of the biggest concern, IoT and blockchain-frameworks may be a wise line of action [1]. The leaders of a blockchain drug creation network are brought together with an AI proposal in [10]. In Groin, the use of blockchain in the medical data stream has been proposed to make clear medicine trade data, wherever the unit that is locked in with the drug creation organization can partake in comparative opportunity to prevent phony of meds.

In a pandemic case, an immune response can be the magnificent elective for performing a countermeasure. Individuals are by and large centered on getting the neutralizer subject to their risk profile and transmission rates. A trustworthy strategy is essential for the success of vaccine communication, which must consider a compelling use of vaccination resources on a consistent basis. The technique in use is assessed by using the leading mathematical models. Coincidentally, vaccination data is required to provide a basis for precise assessment and essential planning. Getting such data can be a challenge for various countries. A significant concern is enlisting inoculations on a piece of paper for each individual, with no accompanying documentation. The main difficulties in immunization measures are perceptibility, information non-appearance, and model non-appearance. This research work presents an adaptive approach for assisting a general population's vaccination pattern by enlisting inoculation data in blockchain. **HOW TECHNOLOGY OF BLOCKCHAIN IS USED IN COVID-19 VACCINE.**

Vaccine injections are biomedical substances that cannot be directly exposed to temperatures inside their threshold value depending on freezing temperature, which cannot be sustained with light and so eliminates its effects. Vaccine monitoring is extremely demanding and complex for vaccine distribution, storage, and administration. Here, this section will discuss about how blockchain will improve the management and process of the vaccine so that it simplifies the operation and makes it transparent to the other part of the members and keeps the data secure.

This is the technique in which no one has control over encrypted information [11]. It is entirely based on a node-to-node network with all nodes interacting with one

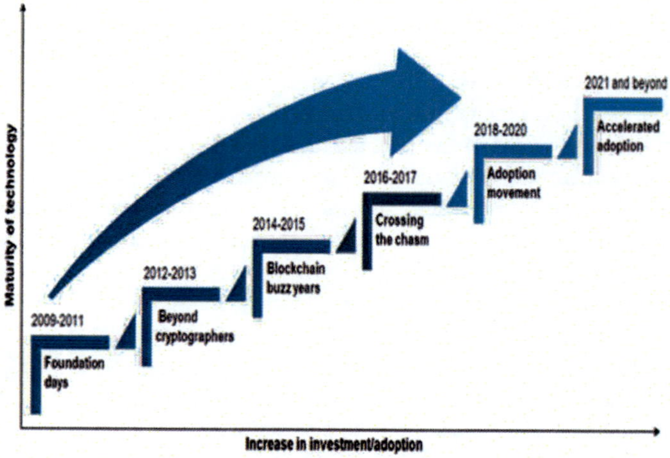

Fig. 2 Evolution of blockchain

another. At the most basic level, blockchain enables hubs with a similar organization to record exchanges in the organization's shared record. The emergence of Bitcoin in 2008 has introduced the world to a new concept that had a significant social influence. It had the potential to have a significant impact on any industry. Blockchain was the central concept.

The proposed technology gives the necessary features and structure for vaccine distribution camp from the pharmacy to the patients [12]. The use of blockchain technology will result in a transparent record of how the parties shared the supply chain. To ensure the quality of the vaccination vials and to be careful in the event of quality concerns, such as antibody temperature exceeding the permitted range, precise agreements that are used to check the correctness and similarity of collected data. Figure 2 depicts the evolution of blockchain.

By using blockchain, data is not moved, and also, it becomes independent to verify the records, so that the process of the vaccination will be much clearer. This process of blockchain is showed in Fig. 3.

2 Main Contributors of the Platform

The contributors who worked on blockchain technology to overcome the scam of the coronavirus vaccine are:

- Vaccine manufacturer
- Vaccine distributor
- Beneficiaries/patients
- Doctors
- Medical center

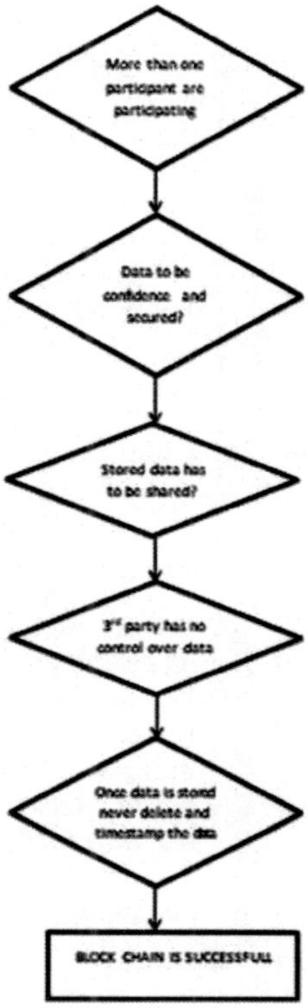

Fig. 3 Flowchart of blockchain

Roles of all the contributors are mentioned in Fig. 4.

A **Vaccine Manufacturer**
The main objective of recipient enrollment for immunization through the suggested blockchain-based innovation is to ensure the security of human life while avoiding pantomimes. As shown in Fig. 2, prior to a secret key, that will be overseen off-chain and, therefore, utilized to show his/her person. To maintain individual information confidentiality and namelessness on-chain in mind, a Markel proof is employed to allow recipients to display their individual without enamoring it. The root neighborhood holds the hash of the recipient's particular key and the hash of its single identification number (Fig. 5).

$$\mathrm{P\_HASH} = \mathrm{Hash}(\mathrm{Hash}(\mathrm{PI}), \mathrm{Hash}(\mathrm{SK}))$$

B **Vaccine Distributor**
The current goal is to determine the extent to which the described parameters for immune response storage and control are met throughout the appointment chain [13]. This is refined by utilizing rapid arrangements that compare the data obtained from sensors attached to limit units or connected to transportation coolers to the represented conditions/regulations indefinitely.

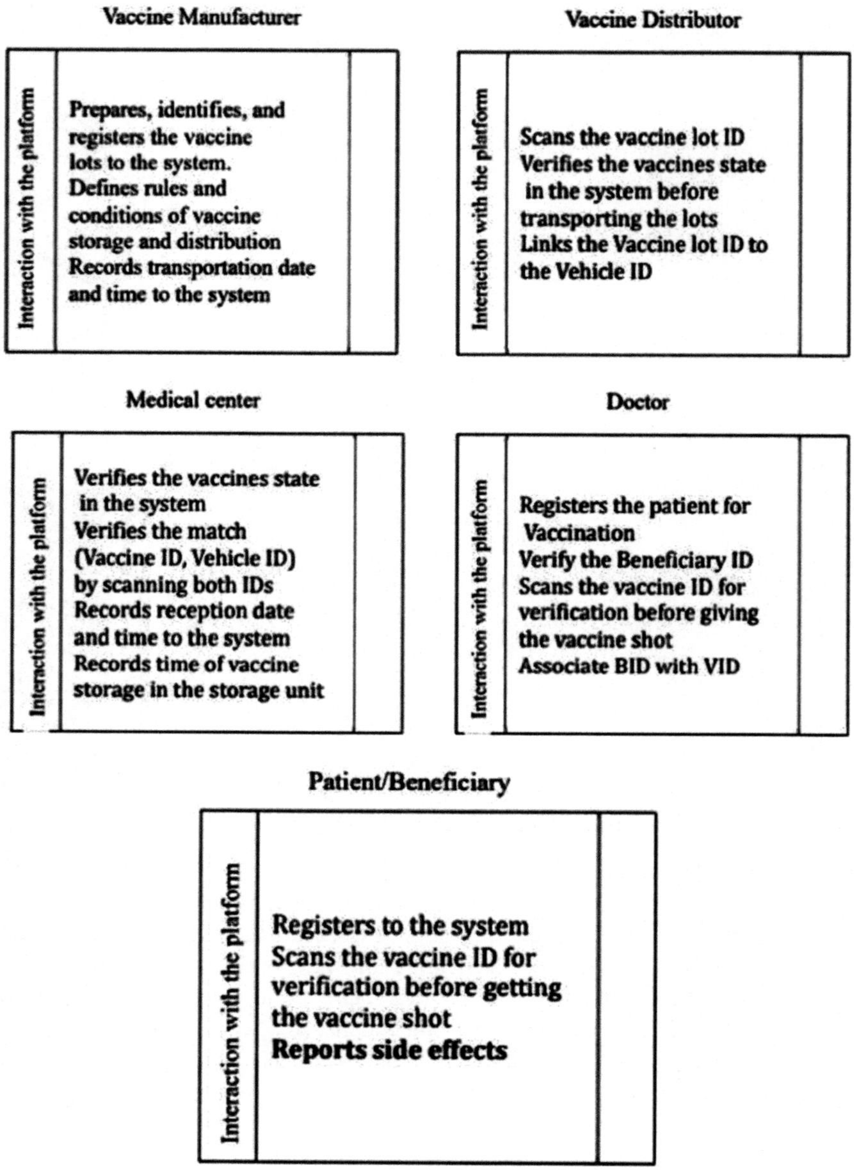

Fig. 4 Roles of contributors

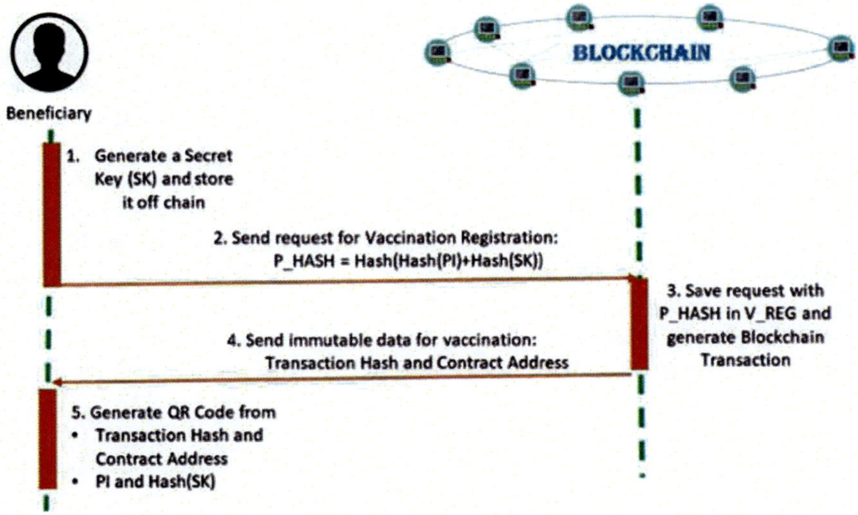

Fig. 5 Registration process of the vaccine through the blockchain

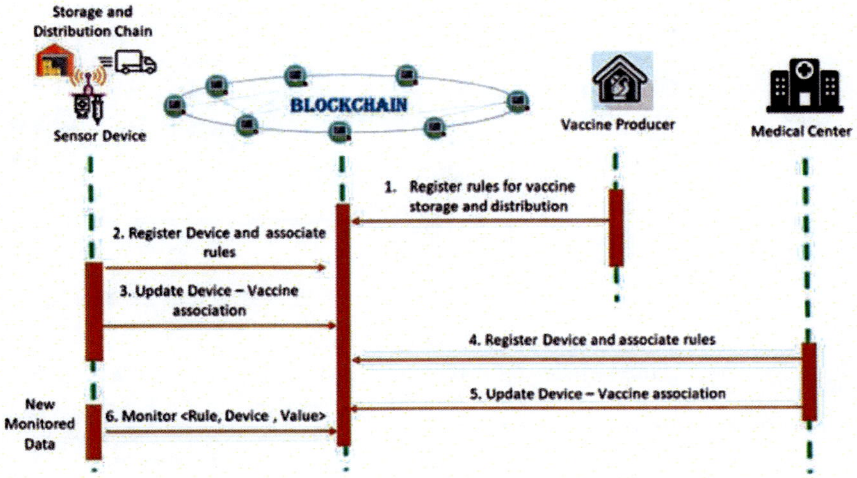

Fig. 6 Distribution of the vaccine using the proposed technology

C **Beneficiary/patients**

D **Managing vaccine delivery is a difficult task. As illustrated in Fig. 6, the following stages must be managed in this operation: Examine the vaccine administered to each beneficiary, the conditions of immunizer movement and handling in accordance with the standards described by the creator, and the beneficiary's relationship with the inoculation to be administered, including the further specification of possible accidental consequences. Doctors/beneficiary**

Everyone has the registration code, which includes the blockchain technology's hash key, the contract's address, a PIN code that is encrypted with hash and secret key. Using the hashes acquired from the QR code, the specialist executes an on-chain character confirmation to confirm the enrollment. During the recipient registration step, the hash of the two characteristics is compared to the root stored in the blockchain by using the Markel proof.

The vaccine QR code is scanned after the beneficiary's identity has been validated in order to obtain relevant vaccination data from the blockchain such as the smart contract address and vaccine lot ID, as well as tracking transit conditions. Once the vaccination has been provided, a two-step locking procedure involving the signatures of the doctor and the recipient is employed to mark it on the blockchain, as seen in Fig. 7. The vaccine is marked as administered by using the hash present in the blockchain.

After the beneficiary person affirmation is done, the counteracting agent QR code is checked to isolate material inoculation information set aside on the blockchain, for instance, the splendid arrangement addresses and the vaccination parted by noticing transportation conditions. Following the administration of the counteracting agent, a two-adventure locking device is used to check the vaccination on the blockchain

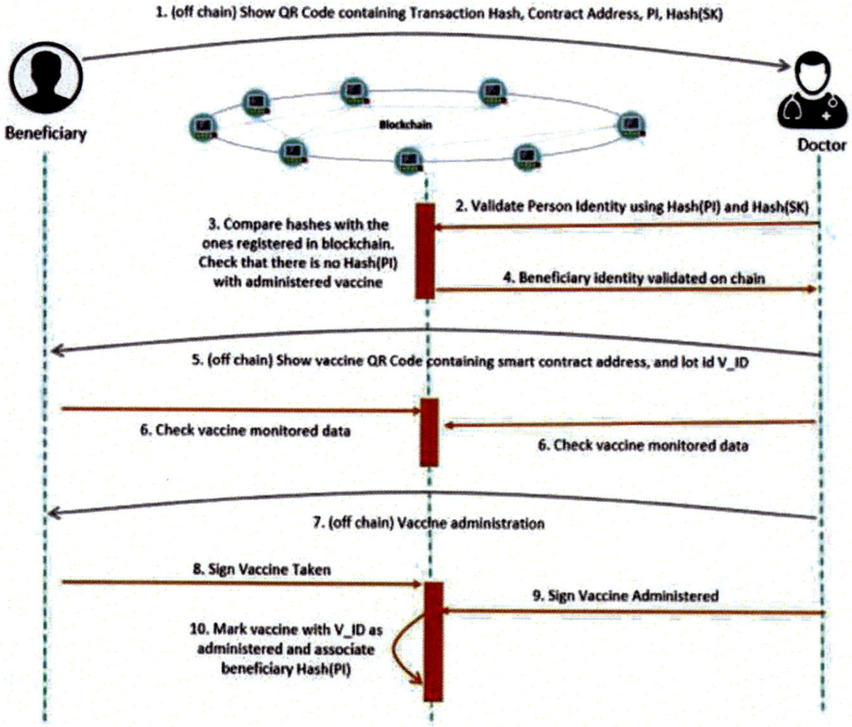

Fig. 7 Vaccine administration sequence diagram

by using the characteristics of subject matter expert and beneficiary. The counteracting agent is administered by using the hash (PI) of the recipient, who received the immunization in the blockchain.

3 Results that Are Considered

A COVID-19 vaccination approach is designed based on the parameters with safe transit and cutoff provided in [10] to evaluate the proposed blockchain-based progress. We hypothesized that the spread affiliation sends the immunization components to the enthralled clinical idea unit with transportation coolers that should arrive at their destination in no more than ten days, and the neutralizer ought to be kept at a temperature between −80°C and −60 °C.

3.1 Tracking of the Vaccination Distribution

A model has been attempted out by Ropsten [14] with a public Ethereum test group. The results obtained by each on-chain action can be endorsed on the Etherscan [15] by using the trade hash. They display on-chain exercises as well as the associated blockchain transaction receipt with the performer by representing the highlighted deal.

3.1.1 Attributes and Rules Enlistment in the Vaccination Program

We start by giving the savvy understanding that has been used to enlist the essential entertainers of the inoculation crusade on the blockchain mentioned below. The vaccine manufacturer should use its Ethereum address to sign the transaction that initiates the sharp understanding transmission.

- **Operation:**
  Signing Address: 0xD2796dE988975DD292e8aC981c4011B23E801DCd (To issue the vaccine).
  Receipt of Transaction:
  Transaction_hash:0xaed1e3267afa17f22d5a125cda449945407c2485b299a8a 15d9c 0ff2beec6738.
  From:0xD2796dE988975DD292e8aC981c4011B23E801DCd.
  To: Vaccine Registry. (constructor).
  Then, the immunization maker should sign exchanges for enlisting the beneficiary clinical consideration units as the recipient of things to deliver antibody parcels.

- **Operation**

  Signing Address: 0xD2796dE988975DD292e8aC981c4011B23E801DCd (Issue vaccine).

  Receipt of the Transaction:

  Transaction_hash:0 x563ddcd6fba93dbb3c05d6cb6b366a9289efcf8dd209d5 51.

  6dbee66e86a8bbc6.

  From:0xD2796dE988975DD292e8aC981c4011B23E801DCd.

  To: VaccineRegistry.registerDoctor(address)

  0x536798D9D1f0507C1a8600d9A475d410a90D5A0A.

  decoded_input: {"address_doctor":"0xF3A1846C82c74EA5D5d32a9BB8 759A 8093C26eFa"}.

The inventor of the neutralizer should choose the rules for safe transportation and inoculation limits. In the depicted scenario, two trades should be enlisted: one to arrange transportation regulations and the other to oversee limit restrictions.

Operation	Register Tracking Rules
Signing Address	0xD2796dE988975DD292e8aC981c4011B23E801DCd (Vaccine Issuer)
Rule	The vaccine should be transported at a temperature between -80 and -60 Celsius degrees for maximum 10 days.
Transaction Receipt (Transportation Rule)	transaction hash: 0xbc8b46345096d4a69faccd7300cf3450f1fc66d1e238236d511dce1771aacf9f from: 0xD2796dE988975DD292e8aC981c4011B23E801DCd to: VaccineRegistry.registerTrackingRules(string, int32, int32, uint256) 0x536798D9D1f0507C1a8600d9A475d410a90D5A0A transaction cost: 216219 gas input: 0x4a2...00000 **decoded input** { "string rule": "transport-temperature", "int32 minValue": -80, "int32 maxValue": -60, "uint256 timeDelta": {"type": "BigNumber", "hex": "0x337f9800" } }
Rule	Vaccine should be stored at a temperature between 2 and 8 Celsius degrees for maximum 5 days.

Any time after the sending of savvy contract, any recipient can buy the pausing list for the antibody. This is conceivable by marking an on-chain exchange. The location of the marking recipient will be put away in the holding up list on-chain. Upon membership, the antibody recipient should likewise give a hash of his/her data.

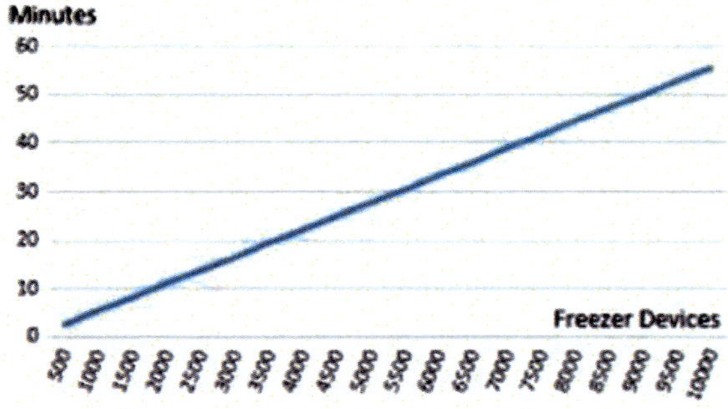

Fig. 8 Vaccine distribution time is taken and tracking transaction

Operation	Subscribe Beneficiary
Signing Address	0xFfb4b11D94CFbbBA8665f7682D4d3B76261EAacC (Beneficiary)
Raw Personal Data	PI: 20-10563145-8 Secret: my-super-secret
Hashed-Personal Data	Hash (PI): 0xa3f6550e5420ddda304a6b22772eb70b48ada3c7eb14648e321bb65387c8cfab Hash (Secret): 0x820371900007448f4a8d909327870ece84168bf90f1de8dddc0b6c7473c44b40
Patient Hash	Merkle Root: 0xfe08609620228b43d9eb80125dfab7a1686e9c3cd7ea5326aa1c5abf7e689b87

Transaction Receipt	transaction hash: 0x76e5a604b3ca803e7738947dbe8616435d5fa410208e382a6d51b58d86b0374c from: 0xFfb4b11D94CFbbBA8665f7682D4d3B76261EAacC to: VaccineRegistry.registerPatient(bytes32) 0x536798D9D1f0507C1a8600d9A475d410a90D5A0A transaction cost: 84808 gas input: 0x8eb...89b87 **decoded input {** "bytes32 patientHash": "0xfe08609620228b43d9eb80125dfab7a1686e9c3cd7ea5326aa1c5abf7e689b87" }

3.1.2 Vaccine Tracking and Administration:

When the vaccine is complete, the manufacturer can register vaccine lots on the blockchain system, revealing the number of vaccine samples in each lot as well as the lot ID.

Representation of the transaction mining is shown in below graph (Fig. 8).

4 Open Hardships

One of the major issues we confront is the authenticity and good leadership of structural individuals. In the event that a limit unit breakdown is not addressed or a power unsettling effect is not addressed, unable vials of vaccine may result.

5 In Situations Where IoT Devices Cannot Be Employed, We Should Be Vulnerable to the performer's Good Lead. CONCLUSION

This paper has thoroughly investigated how the emerging blockchain progression components and benefits can be used for battling the COVID-19 pandemic. We investigated the possible blockchain applications from a clinical benefits emergency standpoint, focusing on the crucial role that blockchain can play during the COVID-19 pandemic. This research study has recognized the importance of collaborative efforts in developing blockchain-based systems for clinical idea emergency relationships to combat the COVID-19 pandemic. The advantages of blockchain advancement in terms of liberal trust, security, perceptibility, and straightforwardness can greatly assist experts in devising solutions for the upcoming COVID-19 pandemic. Performance of modernized contact following courses of action hugely depends upon the total and speed of accumulated information related to the region, travel history, and COVID-19 test delayed consequences of individuals.

- Blockchain development is relied upon to give an accommodating, mindful, and local area arranged environment for individuals that are locked in with the stock organization collaborations of PPE or immunizer.

References

1. Kartoglu UH, Moore KL, Lloyd JS (2020) Logistical challenges for potential SARS- CoV-2 vaccine and a call to research institutions, developers, and manufacturers. Vaccine 38(34):5393–5395
2. Rahim M, Sharma A (2020) Mass vaccination against COVID-19 may require replays of the polio vaccination drives, EClinicalMedicine. Lancet 25(100501)
3. Haque A, Pant AB (2020) Efforts at COVID-19 vaccine development: challenges and successes. Vaccines 8:739
4. Rauf A, Abu-Izneid T, Olatunde A, Ahmed Khalil A, Alhumaydhi FA, Tufail T, Shariati MA, Rebezov M, Almarhoon ZM, Mabkhot YN, Alsayari A, Rengasamy KRR (2020) COVID-19 pandemic: epidemiology, etiology, conventional and non-conventional therapies. Int J Environ Res Public Health 17:8155
5. Andersen KG, Rambaut A, Lipkin WI et al (2020) The proximal origin of SARS-CoV-2. Nat Med 26:450–452
6. WHO Coronavirus Disease (COVID-19) Dashboard, https://covid19.who.int/. Accessed Dec 2020
7. Han E, Mei Jin Tan M, Turk E, Sridhar D, Leung GM, Shibuya K, Asgari N, Oh J, García-Basteiro AL, Hanefeld J, Cook AR, Hsu LY, Teo YY, Heymann D, Clark H, McKee M, Legido-Quigley H (2020) Lessons learned from easing COVID-19 restrictions: an analysis of countries and regions in the Asia Pacific and Europe. Lancet 396(10261):1525–1534
8. Kaur SP, Gupta V (2020) COVID-19 vaccine: a comprehensive status report. Virus Res 288:198114
9. Race for a COVID-19 vaccine (2020) EBioMedicine Editorial, The Lancet, Volume 55, 102817, May 01, 2020

10. COVID-19 Vaccine U.S. Distribution Fact Sheet, November 20, 2020, Pfizer, https://www.pfizer.com/news/hottopics/covid_19_vaccine_u_s_distribution_fact_sheet. Accessed 5 Dec 2020
11. Eisenstadt M, Ramachandran M, Chowdhury N, Third A, Domingue J (2020) COVID-19 antibody test/vaccination certification: there's an app for that. IEEE Open J Eng Med Biol 1:148–155. https://doi.org/10.1109/OJEMB.2020.2999214
12. Shih H-I, Wu C-J, Tu Y-F, Chi C-Y (2020) Fighting COVID-19: a quick review diagnoses, therapies, vaccines. Biomed J 43(4):341–354. ISSN: 2319-4170
13. Zafar S (2020) COVID-19: Current developments and further opportunities in drug delivery and therapeutics. Pharmaceutics 12:945
14. "Oxford-AstraZeneca vaccine is cheaper than Pfizer's and Moderna's and doesn't require super cold temperature, the Conversation," Nov.2020. Accessed: 20 Dec 2020. [Online]. Available: https://theconversation.com/oxford-astrazeneca-vaccine-is-cheaper-than-pfizers-and-modern-and-doesn't-require-supercold-temperature-150697
15. Chamola V, Hassija V, Gupta V, Guizani M (2020) A comprehensive review of the COVID-19 pandemic and the role of IoT, Drones, AI, blockchain, and 5G in managing its impact. IEEE Access 8:90225–90265. https://doi.org/10.1109/ACCESS.2020.2992341

Learning Conversational Stance from Multiple Views of Social Graph Using Multitask and Other Approaches

Bhaskarjyoti Das, Anvitha Poosarla, V. Mahima, and Chetana Kulkarni

Abstract Social conversations happen in multiple contexts on social graphs, and these contexts can be visualized as a multi-view graph. Inside the views, there can be tasks to be learnt with homogeneous or heterogeneous labels. In the most general case, this presents a dual challenge of view as well as task alignment. This work presents a prototype of such a multi-view multitask scenario for the problem of conversational stance detection. This work investigates and analyses the three possible approaches in such a multi-view Twitter graph. Firstly, cumulative conversational content across views is considered in lieu of separate views of the graph. Secondly, each view separately is considered with two view alignment strategies of relational graph convolution and multitask multi-view learning. Thirdly, merging of views for the special case of homogeneous task label across tasks in views is considered. So far, the problem of multi-view learning has been researched only in domains other than computational social science and those domains mostly deal with euclidean dataset unlike social graphs. Also usage of multitask learning as a glueing strategy has been rarely attempted. This work successfully examines and presents an approach framework for multitask multi-view learning problems in multi-view social graphs.

Keywords Multi-view learning · Multitask learning · Multi-view multitask learning · Multi-view social graph · Graph representation learning · Graph convolution network · Relational graph convolution network · graphSAGE · Laplacian eigenmap · Node2Vec · Conversational stance

1 Introduction

Multi-view learning is learning from different views of the same event. An example of multi-view learning can be a learning problem using multiple modalities in a multi-modal conversation. For example, emotion can be assessed using two views,

Supported by PES University.

B. Das (✉) · A. Poosarla · V. Mahima · C. Kulkarni
PES University, 100 Feet Ring Road, Banashankari Stage III, Bengaluru, Karnataka 560085, India

S. Shakya et al. (eds.), *Proceedings of Second International Conference on Sustainable Expert Systems*, Lecture Notes in Networks and Systems 351,
https://doi.org/10.1007/978-981-16-7657-4_31

i.e. facial expression and physiological signals. An example of views on social graph can be different explicit and implicit relationships between nodes where each view may have a particular type of edge along with a particular characteristics of the nodes. Merging of views makes sense only if the views are homogeneous and non-complementary in nature.

Multitask learning, on the other hand, is a technique where learning from related tasks are used to augment the learning from an individual task. This is typically the case of heterogeneous labels across different tasks. An example in social media is the task of learning stance from the same dataset by taking help from related learning tasks of sentiment, emotion and hate. Multiple related tasks are typically addressed in a single view.

The most general case will be a way of combining these two approaches of learning from complementary views as well as related tasks. Such kind of learning tasks can be further categorized into classification and clustering problems. The work described in this paper analyses the different considerations and approaches for such a multi-view multitask classification problem on a social media dataset.

In social conversations about an issue, people normally express their stance by agreeing, disagreeing or staying neutral to the opinion expressed by others. This gets noticed once the expressed stance acts as maximizing [27] factor behind an influence. Stance detection on social media has been traditionally a learning problem using linguistic features. This is a single-view single task approach. Researchers have also addressed this as single-view multitask learning when sentiment, emotion, stance etc. are considered related learning tasks. Conversations happen on social media across different views of the social graph and the views consist of explicit as well as implicit relations. If these separate views or relations on social graph are considered, then stance detection can be also positioned as a multi-view learning problem with homogeneous task labels of stance across views. The work described in this paper examines the different options in such a scenario.

Graph data in particular is non-Euclidean in nature and traditional machine learning techniques can only be applied on a set of derived features or metrics of the graph. This approach fails to encapsulate the propagation characteristics in the graph. Recently, emerged graph representation learning techniques and algorithms attempt to address this shortcoming. Hence, the work described in this paper investigates such graph representation learning algorithms for a multitask multi-view scenario.

2 Existing Work

2.1 Multitask Learning

Multitask learning (MTL) can be considered a special case of transfer learning, i.e. simultaneous transfer learning across related tasks. This use case of MTL can be extended across related tasks in the same dataset and same task across different dataset of the same domain. MTL indirectly augments the training data, introduces regularization and brings down the bias in any representation learning. This approach of joint optimization is considered better than just concatenating task specific representation as that may cause a loss of task specific learning and over-fitting on small dataset. Traditionally joint optimization is approached in two ways:

- In hard parameter sharing, generalization is achieved by sharing a common layer while keeping the task-specific layers separate.
- In soft parameter sharing, the regularization is by optimizing the distance between the learning parameters of the tasks instead of having a shared layer.

Hard parameter sharing is the most common method of MTL that has also been adopted in this work. The typical challenge in MTL is the case of negative transfer [29] when the tasks do not aid each other in learning and joint learning performance becomes worse than the individual learning from individual tasks. Multitask learning has been used in disinformation research to learn from related tasks, i.e. to detect emotion [1], offensive language [6], rumour [18], fake news [36] and even stance [19]. However, the application of MTL in this field so far has been limited to single-view multitask learning.

2.2 *Multi-view Learning*

In multi-view learning, a learning problem can be addressed by learning from different views. This can be achieved by using the same model or different models across views. Multi-lingual text categorization, citation prediction, Web recommendation, music classification, multi-modal emotion prediction etc. can be structured as multi-view learning problem. If the data is graphical, each view will signify a different proximity relationship between nodes. Just like the case in multitask learning, concatenation of multiple views into a single view may cause loss of view specific properties and an optimization approach between multiple views can help. Traditionally multi-view learning approaches address this optimization challenge in three ways:

- Co-training that essentially builds on assumptions such as complementary views and conditional independence between views where views consist of different feature set for different learners.
- Latent common subspace that essentially translates each view into a common subspace and subsequent fusion of the individual views for joint learning. However, this approach may suffer from the shortcoming of limited representation of the fused views.
- Multiple kernels for multiple views and then subsequently combination of view-specific learning.

In both the co-training and multi-kernel approach, there is a challenge of appropriately combining learning from different views. The other limitation of multi-view learning is the fact that it attempts to learn a task from each view individually. For example, comprehending the emotion from a multi-modal meme is kind of learning together process of human mind, and one is expected to get a sub-optimal learning from individual views or simple combination of learning from multiple views. Multitask learning over views can be a strategy for achieving the same goal of effective joint learning.

The problem of multi-view learning on a graph in the social network domain has been earlier attempted as a multi-layer network [4] problem trying to extend the existing body of algorithms and heuristics for a single layer social graph into a multi-layer social graph. However, the core issue of alignment of representations of various layers remained. This line of approach happened before the graph representation learning techniques evolved and extended the machine learning body of knowledge into the graph domain in recent years. In this work, the approaches adopted for multi-view learning are structured around the recently emerged machine learning techniques on graph.

2.3 *Combining Multi-view and Multitask Learning*

Learning from multiple tasks and multiple views is referred to as multi-view multitask learning (MVMT) in this paper. There are several ways the tasks and views can be combined:

1. **Joint learning from multiple related tasks when each learning task individually is a multi-view learning task**: In this most general case, the learning involves addressing alignment challenge at both view and task levels where the tasks are heterogeneous. Huang et al. [15] used the different related tasks of node and link prediction to learn node representation from different views of graph using graph convolution network. They tried to address the alignment challenge among views and related tasks by two specific attention mechanisms, i.e. view attention and task attention. Jin et al. [16] attempted to address this by a two-step process of first converting view specific feature space into an intermediate latent feature space and then using an extended version of classical LDA (linear discriminant analysis) algorithm, converted this intermediate latent feature space representation into a discriminant feature space. The nearest neighbour algorithm is then used to classify in this discriminant feature space.
2. **Joint learning from multiple related tasks when each learning task individually is a single-view learning task**: This is a relative simplification of the previous case where view alignments are not needed for a specific task. Depending on the type of task labels in different views, this category can be further divided into two:

   – **homogeneous tasks**: Here, the learning tasks on multiple views have the same set of labels. For example, detecting emotion from a multi-modal dataset where each mode is having same set of labels for emotion. Conceptually, it maps to MTL between same task from different views of the same dataset.

- **Heterogeneous tasks**: Here, the learning tasks in different views have different labels. For example, in a music recommendation problem for a dataset of songs in different languages, the labels in each language may signify different genre. Conceptually, it maps to MTL between different tasks from different views of the same dataset.

There are very few works where multiple homogeneous tasks from multiple views are glued together. Liu et al. [20] used multitask learning in water distribution network to align learning tasks from different views, i.e. water quality in the local spatio-temporal view of water station vs. water quality in spatio-temporal view across water stations. In a recent work, Wang et al. [34] have used MTL to align the representations learnt from multiples views of graph to solve a Web-scale recommendation problem. This work uses graph representation learning method to derive node representation in a graph view on a recommendation dataset. In the work described in this paper, a similar approach has been used.

2.4 Graph Representation Learning

For graph-based learning, there have been two key milestones, i.e. representation learning and graph convolution. The representation learning [7] addresses a key challenge of making non-Euclidean graph data amenable to modern machine learning approaches by capturing the graph structure information into a fixed length embedding. This method uses a mapping function that can come up with a low dimensional vector representation for the nodes preserving the properties and neighbourhood information of the nodes. Once the node embedding is obtained, it can be further used in a machine learning problem. Laplacian Eigenmaps (LE) [3] is the classical embedding approach that essentially does a spectral decomposition of the graph Laplacian into eigen components with contribution of each decided by the eigen values so that each node gets represented by ranked eigenvectors. The graph representation learning algorithms are loosely divided [5] into several categories by researchers:

1. Algorithms that use random walk and skip-gram, i.e. Node2Vec [12] and algorithms derived from Node2Vec such as Edge2Vec [32] and Graph2Vec [22]. These approaches essentially capture graph structure and are shallow encoding methods that are essentially transductive in nature.
2. Algorithms that use matrix factorization on graph adjacency matrix such as HOPE [23].
3. Algorithms that are auto-encoder-based such as SDNE [33]. They follow an unsupervised approach to learn encoding and can be used to generate embedding on unknown graphs. However, this approach has certain limitations. A recent work that attempts to build the embedding based on structural roles of the nodes into a

continuous vector representation is GraphWave [10] leveraging concept of spectral graph wavelet [21] diffusion.

4. Graph neural network based algorithms that extends the idea of convolution in neural network into graph [9], i.e. spectral convolution approach such as graph convolution network (GCN) [17] and spatial convolution approach such as graphSAGE. graphSAGE [13] also enables inductive learning by focusing on learning the local aggregation function in the GCN work flow. Recently introduced Attribute2Vec [35] is a claimed improvement over graphSAGE. Graph attention network (GAT) [30] extends the concept of neural attention into graph representation learning. For a multi-graph such as knowledge graph, relational graph convolution network (R-GCN) [26] extends the GCN into scenarios where nodes have multiple types of relationships. If each view is based on a type of edge, R-GCN can be used for a multi-view graph. These algorithms utilize both structure and attribute information of the nodes.

Essentially, a suitable representation learning method can be used to learn embedding for nodes, edges or graph as a whole based on the particular learning task. Once a suitable algorithm is identified, edge and nodes are extracted with their specific attributes and embeddings are generated. The logical next steps are the typical learning steps followed in machine learning. The investigations described in this paper follow the same approach for a view-specific learning task.

2.5 *Stance Detection*

Stance detection in a social graph has been primarily investigated as a supervised learning problem using the textual content of conversation, i.e. for against and neutral about a topic. The problem of learning stance is seen to be benefited using MTL techniques by learning emotion, hate, topic etc. as they accompany stance often. However, it is still challenging to predict user stance without knowing the users in their social contexts. The social context can be captured in a global manner by capturing the follower–followee relationships. The social context in a local perspective is the propagation context or the conversational context such as retweet, reply, mention graph on the Twitter. For example, people who have similar stances may follow each other or retweet each other's tweets on Twitter. With passing time, as the propagation context grows, a joint learning approach will become more effective compared to content only method. The graph-based joint learning approach using both conversational context and text content as used in this work is more suitable for such a scenario.

Stance datasets are very few due to privacy issues [14] and pose a leaning challenge associated with voluminous [24] data. Almost all of them are made for the content-based approach and usually social context is ignored in stance detection. In this work, a recently published conversational stance dataset [31] is used where the goal is to detect the conversational stance of the users on Twitter.

3 Dataset and Pre-processing

3.1 Dataset

The dataset [31] has been collected for four events, i.e. "Santa Fe Shooting", "Student Marches", "Iran Deal" and "General Terms". The tweets for each of the events have been collected with certain hashtags pertaining to that event or some contentious terms in the tweets. The dataset related to "general terms" were chosen purely based on contentious terms and not for any particular event. Hence, as shown in Table 1, for the case study described in this work, the conversations of this category are not used.

The dataset is that of Tweeter conversations in the form of target and response. As per the terminology in the dataset, target tweet is the source tweet, whereas response tweet is the response tweeted by another user. It has the target tweet text and response tweet text along with the tweet IDs, stance label, interaction type, the dates on which the tweets were created. There are two types of interaction, i.e. quote and reply. The "label" has six different values for the response, i.e. explicit denial, explicit support, implicit support, implicit denial, comment and queries. The labels comment and queries do not reveal the user stance on the event clearly whereas the labels Explicit_Denial, Explicit_Support, Implicit_Support, Implicit_Denial express the stance of the responding user. Hence, only the samples with explicit and implicit support or denial labels are used. The dataset provides the stance of only the responding users. The stance labels for the target users are derived as per the heuristics below as a first step in framing the problem as a supervised learning problem:

- If the responding user is labelled as deny, the target user is labelled as support as they essentially have opposite stances.
- If the responding user is labelled as support, the target user is labelled as support as they essentially have similar stances.

Table 1 Number of tweet conversations for each event

Dataset		
Event	Total no of tweets	Total no of tweets after pre-processing
Student Marches	996	730
Santa Fe Shooting	1595	1259
Iran Deal	1648	1230

3.2 Pre-processing to Construct Views of User Conversation Graph

Extraction of tweet meta data The dataset was first pre-processed to get the tweet metadata such as user information for the collected tweets. This was done using an open-sourced hydration tool.[1] The collected metadata was used to form a dictionary with tweets as keys and values as user IDs as a step towards forming a user ID-based graph. Out of these tweets, few were unauthorized and the user IDs could not be accessed. Also, few of the tweets IDs could not be hydrated and such tweets were hydrated by directly using Twitter API.
Determination of node features Users are the nodes in the user conversation graph for each event in the dataset. Since user stance detection is the goal of the exercise, text of the response tweet is taken as the user attribute in the user–user interaction graph. Also, stance is better interpreted from the context of the conversation. Hence, the response user's attribute for every conversation is derived from the concatenated target tweet and the response tweet. It was not necessary for the target users who started the conversation. This was done for every event irrespective of the reply or quote kind of conversation.

The tweet text is then pre-processed to remove stop words and all emojis present in the tweet text are converted into appropriate text. This is done as emotion is an important marker for user stance. The words in the pre-processed text are then converted to embedding of dimension 1 cross 300 using Glove pre-trained embedding. The average of the embeddings of the words across each considered text attribute is calculated, and these vector embeddings are used to represent the node attributes in the user conversation graphs.
Extraction of multiple views Same user can interact in multiple modes of communication on the twitter social graph, and these can be considered as local views of the social relations. These are retweet, reply or mention relationships, and the dataset provides only the mention (quote) and reply kind of conversations. On social graph, implicit relationships are often more significant [2, 11, 25, 28] than explicit ones. For the dataset used, such explicit modes of relationships are mention and reply, whereas implicit relationship between users can be derived based on common hashtags present in the tweets. These views of user interaction have been extracted from the dataset in the form of undirected graphs.

1. **Hashtag view**: Hashtag view is obtained by creating connections between users if they mention the same hashtags in their tweets. For this purpose, a list of hashtags which are most representative of the event is chosen and that explains the relatively less number of edges in this view. From the dataset, a dictionary of hashtag to user ID was created for each event where key is the hashtag and value is the list of users with that hashtag. From this dictionary, the edges were then obtained by

[1] `Documenting the Now. (2020). Hydrator [Computer Software].` Retrieved from: https://github.com/DocNow/hydrator.

making combinations of users using a particular hashtag. The top five hashtags of the three events in our dataset are:

(a) **Santa Fe Shooting**: NRA, GunControl,SantaFe,Texas and SantaFeHigh-School.
(b) **Student Marches**: MarchForOurlives, FakeNews, NRA, NeverAgain and LoveAMuslim.
(c) **Iran Deal**: Iran, IranDeal, Israel, IranNuclearDeal and MAGA.

2. **Reply view**: The reply view is created such that there is an edge from the target user to the response user who replied to the target user's tweet. To construct a reply view event wise, a dictionary with key as target tweet ID and value as the list of reply tweet IDs was constructed.
3. **Mention view**: This view is created such that there is an edge between the target user to the user who mentioned the target user's tweet. To construct a quote view event wise, a dictionary with the key as target tweet ID and value as the list of mentioning tweet IDs was constructed.

Table 2 lists the number of edges between users in each view for each event being considered and Table 3 shows the user overlap or participation across views. It is clear that only some users are more active, i.e. have participated more than once announcing their stances. The number of users common to all three views is very few. This clearly illustrates the opportunity in multi-view learning.

Table 4 lists various graph metrics extracted from the interaction graphs for each event after the views are merged. Low average betweenness centrality indicates the

Table 2 Number of edges in each view for each event

Number of edges in different views			
View	Iran Deal	Santa Fe Shooting	Student Marches
Hashtag	2127	460	714
Reply	192	279	186
Quote	329	323	133

Table 3 Number of edges in each view for each event

User overlap across different views			
View	Iran Deal	Santa Fe Shooting	Student Marches
Reply-Hashtag	55	39	39
Reply-Quote	38	46	19
Quote-Hashtag	78	48	55
Users in only one view	745	791	386
Users participating in all views	7	36	3

Table 4 Network metrics for each event after merging all views

Network metrics—DC: Degree centrality BC: Betweenness centrality CC: Closeness centrality									
Event	Nodes	Edges	Av. degree	Isolate	Av. DC	Av. BC	Av. CC	Global efficiency	Transitivity
Iran Deal	906	2540	5.6	4	0.0003	0.006	0.032	0.03	0.8
Student Marches	495	877	3.54	2	0.0002	0.007	0.021	0.02	0.8
Santa Fe Shooting	906	1022	2.25	2	0.0003	0.002	0.02	0.02	0.46

existence of relatively larger number of non-bridging nodes. Low average closeness centrality means that nodes are not really close to each other. Low average degree centrality and moderate value of average degree mean that nodes are not really too inter-connected. The global efficiency indicates the average of multiplicative inverse of the shortest path distance between all pairs of nodes. Since this value is not very high, it indicates short conversation chains (lack of sufficient edges) in the dataset. There are also few isolates in the dataset as the edges could not be constructed. Overall, as the conversation structure are not very elaborate, the metrics indicate that difference in approaches based on structure only, structure and attribute combined and conversational content (if response text and target text are concatenated together as node attribute) will not be appreciable. Also, in the student marches event, since the number of edges is minimal, conversational content approach will outperform graph approach.

4 Methodology

A multi-view graph $G = (V,E)$ consists of vertices V_i that exist on all layers and edges L_j that exist on jth view. Assuming there are n nodes, the adjacency matrix for jth view will be A_j where $A_j \in R^{n \times k_t}$ and the node V_i is suitably represented by an attribute vector or embedding of the relevant conversational text in jth view $x_i \in R^{k_t}$. Let T_j be the learning task on the jth view. The learning task in this case is to learn the stance of the user nodes.

In the specific case of the dataset used, in each view of the graph, the learning tasks are the same, i.e. user stance detection. This is a case of homogeneous labels in every task in each view. Following approaches are considered for this learning task:

1. **Merging views effectively transforming the learning problem into a single task single-view learning problem**: This is feasible as not only the tasks, the views are also homogeneous for the dataset and learning task chosen. In each each view, the same set of users are participating in the conversation and the user attributes are also the same, i.e. the conversation content. In this approach, user node embeddings are derived first followed by supervised learning using classical machine learning algorithms such as support vector machine (SVM)

and logistic regression. Two sub-approaches can be followed after making event specific conversation graphs by merging views:

(a) Algorithms that focus primarily on the structure, i.e. Laplacian Eigenmaps, Node2Vec.
(b) Algorithms that utilize both structure and node attributes, i.e. GCN, graph-SAGE (both transductive and inductive), GAT, Attribute2Vec and Graph-Wave.

2. **Pure content-based learning using conversational context**: This is an indirect approach to merging of views as for the event graph views constructed, node attributes are defined as conversational content collected from all views for the nodes concerned.
3. **Avoiding merging of views by treating each view separately so that it becomes a single task learning problem**: An algorithm such as relational graph convolution network (R-GCN) can be used. This method offers view alignment but does not offer task alignment.
4. **Multi-view multitask learning**: Conversational stance of user for each event is detected for each of the three views, and these three learning tasks are then optimized jointly using hard parameter sharing approach. This is a more general approach that can cater to heterogeneous task labels across views. This method offers task alignment across views and is the most general approach.

Implementation is done in Python language using the following libraries, i.e.

1. Graph Machine learning library, i.e. StellarGraph [8] and Node2Vec.
2. MTMV implementatio is done in Pytorch using reference implementation [17] of GCN.
3. ScikitLearn for classical Machine learning algorithm and Networkx for network analytics.

5 Results and Discussion

The results are displayed in Table 5 for the approaches which essentially merged the views. Table 6 lists the accuracy obtained using R-GCN (in this, the views were not merged). Table 7 lists the accuracy obtained using multi-view multitask approach with the significant results shown in bold. Following conclusions can be arrived at from the results:

1. Learning from single view proved inferior compared to all other approaches considered here. Though the comparison is using GCN-based model, similar result is expected using other algorithms.
2. For the conversation content approach, concatenating the response and target text indirectly achieves what the graph-based algorithm offers with merged views. This is feasible only when views are homogeneous. This approach, as evident in

Table 5 Stance detection using explicitly merged views

Obtained accuracy on test data using merged views				
Method	Iran Deal	Student Marches	Santa Fe Shooting	Model details
Conversation content	0.68	**0.63**	0.63	Glove 300, SVM with 70:30 train-test split
Laplacian eigenmaps	0.66	0.47	0.54	Embedding size 32, SVM with 70:30 train-test split
Node2Vec	**0.72**	0.51	0.61	Random walks = 50, walk length = 3, $p = 0.5$, $q = 2.0$, Logistic Regression with 70:30 split
GCN	0.70	0.48	0.58	2 GCN layers with 16 units each and ReLU activation and 50% dropout, 20 epoch, adam optimizer, learning rate 0.005, 70:30 split
GraphWave	0.68	0.51	0.62	Degree = 20, scales = [5, 10], SVM with 70:30 train-test split
graphSAGE transductive	0.7	0.48	0.58	Encoder with [32, 32, 16], no of samples in the three layer as [10, 7, 3], adam optimizer, learning rate 0.03, dropout 50%, 70:30 train-test split
Attribute2Vec	0.68	0.48	**0.68**	No of walks = 4; length = 5, layer size = 128, edge embedding method 'ip', epoch 20, Logistic Regression with 70:30 split
graphSAGE inductive	0.68	0.46	0.61	Model parameters same as transductive but 20% of data removed and then model trained on 5% of remaining. Final prediction done on held out data
GAT	0.70	0.48	0.58	Two GAT layers with 8 units in the first and 2 in the second; attn heads = 8, dropout = 0.5, attn dropout = 0.5, activation = [ReLU, softmax], 70:30 split

Table 6 Stance detection using separate views

Obtained accuracy on test data using R-GCN				
Method	Iran Deal	Student March	Santa Fe Shooting	Model details
R-GCN	**0.68**	**0.55**	**0.65**	Same model parameters as GCN transductive method in Table 5

Table 7 Stance detection using multi-view multitask

Obtained accuracy on test data in MTMV and single view using GCN					
Event	Hashtag view	Reply view	Quote view	Multi-view multitask	Model details
Student Marches	50	58	50	**60**	1. 3 layer GCN [32, 32, 32] with ReLU activation for each view
Santa Fe Shooting	58	52	58	**62**	2. GCN embeddings are concatenated and passed through shared dense layer [16]
Iran Deal	43	43	50	**54**	3. 3 linear layers with RelU, 50% Dropout and finally softmax. The loss function over three views are optimized together

the Student Marches event, will outperform the graph-based approach as number of edges is very small in the views.

3. R-GCN as a view alignment strategy has put up competitive performance and is a feasible option when non-homogeneous views cannot be merged.
4. The dataset used and the learning problem investigated is a special case of multitask multi-view learning where all views are homogeneous and all tasks have homogeneous labels. Given this, all the four approaches used should perform similarly and results obtained validates that expectation.
5. Multi-view multitask learning provided near comparable performance to other methods proving that it works and should be pursued as the preferred method for a general case of view and task alignment.

6 Conclusion and Future Work

Existing work on social graph using machine learning techniques are mostly focused on single view with or without multiple tasks. In multi-view scenario, if the views are complementary, there can be a view alignment challenge. In the most general case, if the learning tasks on each view are heterogeneous labels, it presents a task alignment challenge as well. This work presents an approach framework in such a general scenario and examines them with a conversational stance dataset on Twitter.

Four approaches were outlined and results presented. The first approach is a content only approach that indirectly merges the views by using full conversational content from all views for each user. The second approach uses merged views by various representation learning algorithms (it can work well if the views are homogeneous in nature). The third approach addresses each view separately by two methods, i.e. relational graph convolution network that offers view alignment and multitask multi-view learning using graph convolution network as representation learning mechanism offering both view and task alignment.

The relative efficacy of each approach will depend on two factors, i.e. nuances in the dataset and conversational context available. In the dataset chosen, the views are homogeneous and the learning task in each view is having homogeneous labels. So, all approaches should offer roughly equivalent results and the results obtained support that.

This work successfully presents the important first step of presenting a systematic approach framework of a multi-view multitask learning problem in computational social science domain. The logical next step will be to use the same framework with additional dataset in increasing levels of complexity, i.e. non-homogeneous views presenting view alignment challenge and non-homogeneous task labels presenting task alignment challenge. R-GCN for view alignment with MTL for task alignment also should be investigated along with additional investigation of better tuning of the hyper parameters of each model.

References

1. Akhtar MS, Chauhan DS, Ghosal D, Poria S, Ekbal A, Bhattacharyya P (2019) Multi-task learning for multi-modal emotion recognition and sentiment analysis. arXiv preprint arXiv:1905.05812
2. Al-Sabaawi AMA, Karacan H, Yenice YE (2020) Exploiting implicit social relationships via dimension reduction to improve recommendation system performance. PLoS One 15(4):e0231457
3. Belkin M, Niyogi P (2001) Laplacian eigenmaps and spectral techniques for embedding and clustering. In: NIPS, vol 14, pp 585–591
4. Bindu P, Thilagam PS, Ahuja D (2017) Discovering suspicious behavior in multilayer social networks. Comput Hum Behav 73:568–582
5. Chami I, Abu-El-Haija S, Perozzi B, Ré C, Murphy K (2020) Machine learning on graphs: a model and comprehensive taxonomy. arXiv preprint arXiv:2005.03675
6. Chen PC, Huang HH, Chen HH (2020) Ntu_nlp at semeval-2020 task 12: identifying offensive tweets using hierarchical multi-task learning approach. In: Proceedings of the fourteenth workshop on semantic evaluation, pp 2105–2110
7. Cui P, Wang X, Pei J, Zhu W (2018) A survey on network embedding. IEEE Trans Knowl Data Eng 31(5):833–852
8. Data61 C (2018) Stellargraph machine learning library. https://github.com/stellargraph/stellargraph
9. Defferrard M, Bresson X, Vandergheynst P (2016) Convolutional neural networks on graphs with fast localized spectral filtering. Adv Neural Inf Process Syst 29:3844–3852
10. Donnat C, Zitnik M, Hallac D, Leskovec J (2018) Learning structural node embeddings via diffusion wavelets. In: Proceedings of the 24th ACM SIGKDD international conference on knowledge discovery & data mining, pp 1320–1329
11. Frey D, Jégou A, Kermarrec AM (2011) Social market: combining explicit and implicit social networks. In: Symposium on self-stabilizing systems. Springer, pp 193–207

12. Grover A, Leskovec J (2016) node2vec: scalable feature learning for networks. In: Proceedings of the 22nd ACM SIGKDD international conference on knowledge discovery and data mining, pp 855–864
13. Hamilton WL, Ying R, Leskovec J (2017) Inductive representation learning on large graphs. In: Proceedings of the 31st international conference on neural information processing systems, pp 1025–1035
14. Haoxiang W, Smys S (2021) Big data analysis and perturbation using data mining algorithm. J Soft Comput Paradigm (JSCP) 3(01):19–28
15. Huang H, Song Y, Wu Y, Shi J, Xie X, Jin H (2020) Multitask representation learning with multiview graph convolutional networks. IEEE Trans Neural Netw Learn Syst
16. Jin X, Zhuang F, Xiong H, Du C, Luo P, He Q (2014) Multi-task multi-view learning for heterogeneous tasks. In: Proceedings of the 23rd ACM international conference on conference on information and knowledge management, pp 441–450
17. Kipf TN, Welling M (2016) Semi-supervised classification with graph convolutional networks. arXiv preprint arXiv:1609.02907
18. Li Q, Zhang Q, Si L (2019) Rumor detection by exploiting user credibility information, attention and multi-task learning. In: Proceedings of the 57th annual meeting of the association for computational linguistics, pp 1173–1179
19. Li Y, Caragea C (2019) Multi-task stance detection with sentiment and stance lexicons. In: Proceedings of the 2019 conference on empirical methods in natural language processing and the 9th international joint conference on natural language processing (EMNLP-IJCNLP), pp 6299–6305
20. Liu Y, Zheng Y, Liang Y, Liu S, Rosenblum DS (2016) Urban water quality prediction based on multi-task multi-view learning
21. Manoharan S (2019) Study on hermitian graph wavelets in feature detection. J Soft Comput Paradigm (JSCP) 1(01):24–32
22. Narayanan A, Chandramohan M, Venkatesan R, Chen L, Liu Y, Jaiswal S (2017) graph2vec: learning distributed representations of graphs. arXiv preprint arXiv:1707.05005
23. Ou M, Cui P, Pei J, Zhang Z, Zhu W (2016) Asymmetric transitivity preserving graph embedding. In: Proceedings of the 22nd ACM SIGKDD international conference on knowledge discovery and data mining, pp 1105–1114
24. Pandian AP (2019) Review of machine learning techniques for voluminous information management. J Soft Comput Paradigm (JSCP) 1(02):103–112
25. Roth M, Ben-David A, Deutscher D, Flysher G, Horn I, Leichtberg A, Leiser N, Matias Y, Merom R (2010) Suggesting friends using the implicit social graph. In: Proceedings of the 16th ACM SIGKDD international conference on knowledge discovery and data mining, pp 233–242
26. Schlichtkrull M, Kipf TN, Bloem P, Van Den Berg R, Titov I, Welling M (2018) Modeling relational data with graph convolutional networks. In: European semantic web conference. Springer, pp 593–607
27. Sivaganesan D (2021) Novel influence maximization algorithm for social network behavior management. J ISMAC 3(01):60–68
28. Srivastava V, Ladda V, Shanmugam V, Das B (2019) Assessing impact of explicit and implicit relationships on user's interests. In: 2019 9th international conference on advances in computing and communication (ICACC). IEEE, pp 103–109
29. Standley T, Zamir A, Chen D, Guibas L, Malik J, Savarese S (2020) Which tasks should be learned together in multi-task learning? In: International conference on machine learning. PMLR, pp 9120–9132
30. Veličković P, Cucurull G, Casanova A, Romero A, Lio P, Bengio Y (2017) Graph attention networks. arXiv preprint arXiv:1710.10903
31. Villa-Cox R, Kumar S, Babcock M, Carley KM (2020) Stance in replies and quotes (SRQ): a new dataset for learning stance in twitter conversations. arXiv preprint arXiv:2006.00691
32. Wang C, Wang C, Wang Z, Ye X, Yu PS (2020) Edge2vec: edge-based social network embedding. ACM Trans Knowl Discov Data (TKDD) 14(4):1–24

33. Wang D, Cui P, Zhu W (2016) Structural deep network embedding. In: Proceedings of the 22nd ACM SIGKDD international conference on knowledge discovery and data mining, pp 1225–1234
34. Wang M, Lin Y, Lin G, Yang K, Wu XM (2020) M2GRL: a multi-task multi-view graph representation learning framework for web-scale recommender systems. In: Proceedings of the 26th ACM SIGKDD international conference on knowledge discovery & data mining, pp 2349–2358
35. Wanyan T, Zhang C, Azad A, Liang X, Li D, Ding Y (2020) Attribute2vec: deep network embedding through multi-filtering GCN. arXiv preprint arXiv:2004.01375
36. Wu L, Rao Y, Jin H, Nazir A, Sun L (2019) Different absorption from the same sharing: sifted multi-task learning for fake news detection. arXiv preprint arXiv:1909.01720

Facial Emotion Recognition Using ML Algorithms

V. V. S. S. Ramya, Shaik Afifa Reshma, Afrah Samreen, and U. Chandrasekhar

Abstract A facial expression is nothing but the movement of muscles under the skin covering the areas of the face. It can be used for non-verbal communication. Facial emotion recognition is considered as a process of identifying human emotions from various facial expressions. The facial expression for detecting an emotion has forever been a challenging task to achieve through computer algorithms. Facial emotion recognition (FER) is an active research area in the field of artificial intelligence, with several recent works and studies. The recent advancement in the area of computer vision and machine learning has made it possible to detect emotions from the images. This automatic emotion detection which is done based upon facial expression is an interesting research area that has presented and can be applied in several areas such as e-learning, safety, gaming, health, automobile and in machine–human interactions. Human emotions which were only possible to be recognized by a human brain can be detected and identified using a software as well. This particular technology is becoming more and more accurate and, hence, will eventually be able to detect emotions with a precision that human brains do. In this paper, we propose a technique called facial emotion recognition using convolutional neural networks and Haar cascade classifier.

Keywords Facial emotion · Convolutional neural network · Haar cascade classifier algorithm · Background elimination

1 Introduction

Emotions are helpful in understanding and interacting socially and can be considered as the universal language. They can be used and understood by each and every person very easily irrespective of the regions where they live in. Since these emotions do not contain any boundaries, they are also helpful to understand the cultural diversities.

V. V. S. S. Ramya · S. A. Reshma · A. Samreen · U. Chandrasekhar (✉)
Department of Computer Science and Engineering, BVRIT Hyderabad College of Engineering for Women, Hyderabad, India
e-mail: chandrasekhar.u@bvrithyderabad.edu.in

S. Shakya et al. (eds.), *Proceedings of Second International Conference on Sustainable Expert Systems*, Lecture Notes in Networks and Systems 351,
https://doi.org/10.1007/978-981-16-7657-4_32

Facial expressions play an important role in conveying the information. Human feelings can be identified through facial expressions, because it relates to the emotions. Most of the times, emotions are used in non-verbal communication, and it can be contemplated in many applications as an evidence to understand whether a person is guilty or not in case of criminal investigations.

Without just confining the emotion recognition to research laboratories, various geometric and machine learning-based algorithms are being developed to achieve the required purpose in real time [1, 2]. Hence, developing such a system that is capable of not only detecting the face but also recognizing the emotion has always been quite challenging and an essential research area. The facial expressions consist of different categories, and the purpose of facial emotion recognition is to identify and understand these categories to gain the required insights into human perception [3, 4]. This technique is used to determine the emotion of humans based on their expressions. The current dataset focusses on seven essential facial expression classes reported, which are happy, sad, surprise, fear, neutral, disgust and anger [3].

2 Proposed System

Facial emotion recognition is essential to recognize the emotions in many cases like in criminal investigation, e-learning, gaming and psychology. In criminal investigation, it is used to identify the criminals based on their facial expressions. Similarly, in e-learning, this can be used to conclude whether a student is understanding a concept or not, whereas in other applications, it ensures the understanding of a person's emotional condition.

The classification of the facial expressions was performed using convolutional neural network and Haar cascade classifier algorithm [5–7]. The image is uploaded which is then converted as a greyscale image. The face is detected using Haar cascade classifier which is then classified by the model that displays the image along with its emotion as the output.

3 Implementation

3.1 Architecture

The following are the steps involved in classification (Fig. 1).

Initially, the dataset has been acquired in the first step after which preprocessing is performed on all the images in the dataset. A CNN model has been created which is then trained with this data in order to achieve some accuracy and predict accurately [8, 9]. The model is tested by providing an image as input. This image is at first converted into a greyscale image. The entire background is eliminated from the image, and the

Fig. 1 Flowchart

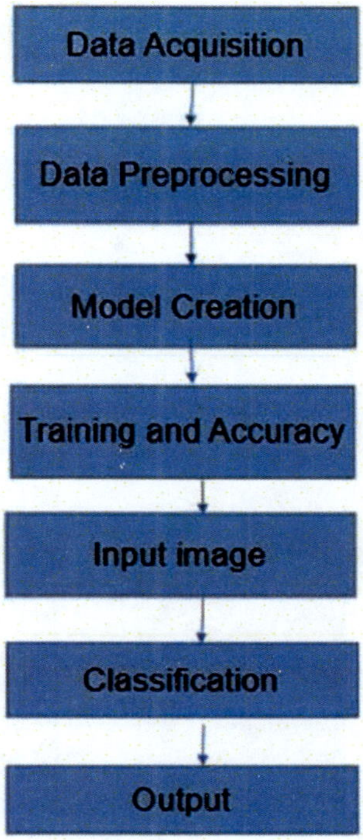

face is detected using the Haar cascade classifier after which it is fed into the model to predict the appropriate emotion. In the classification part, the images are tested according to the given categories and produce the final predicted emotion [10, 11].

3.2 Dataset

Proper and large dataset is required for the classification during the training and the testing phase. The dataset for the experiment is downloaded from the Kaggle which contains different facial images and their respective labels. A dataset containing approximately 35,887 images of seven classes is downloaded. Each image size is 48 * 48 pixels.

These seven classes are shown in Fig. 2.

Classification of each face into the mentioned seven categories is the task at hand.

The dataset is a CSV file with a categorical class labelled from 0 to 6, representing the seven emotions, and second is the "pixels", which is a string having an image enclosed. The string is nothing but space-separated pixel values in row major order. The rows with the column value as test consists of only the "pixels" column where the task is to predict the value of the emotion column (Fig. 3).

Fig. 2 Seven classes of the dataset

S.No.	Emotion	Total images
1	Angry	4953
2	Disgust	547
3	Fear	5121
4	Happy	8969
5	Sad	6077
6	Surprise	4002
7	Neutral	6198

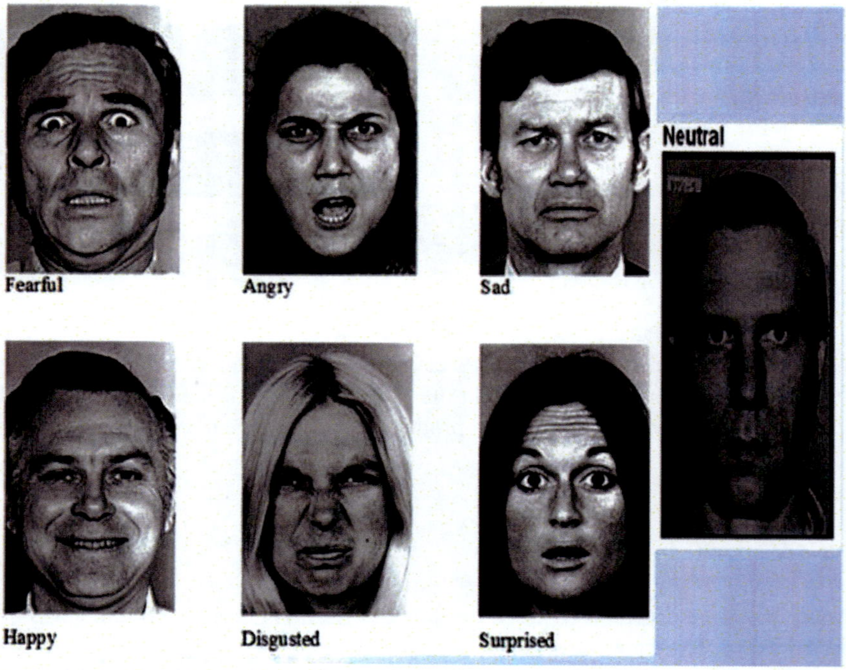

Fig. 3 Dataset screenshot

3.3 Convolutional Neural Network (CNN)

CNN architecture varies according to the type of the problem at hand. The proposed model consists of three convolutional layers each followed by a maxpooling layer. Rectified linear unit (ReLU) activation function is applied to the output of every convolutional layer and fully connected layer [8, 9].

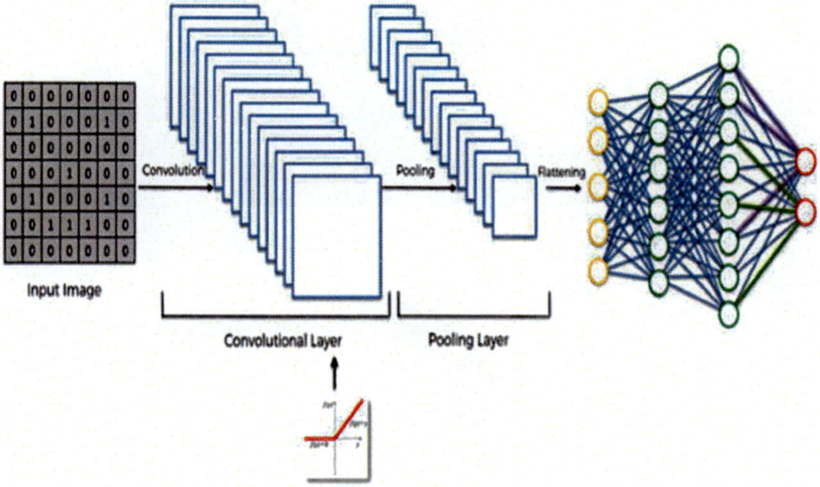

Fig. 4 Convolutional neural network

The first convolutional layer attempts to filter the input image with 64 kernels of size 3×3. After applying maxpooling, the output is given as an input to the second convolutional layer with 64 kernels of size 3×3. The last convolutional layer has 128 kernels of size 3×3 followed by a fully connected layer of 1024 neurons. The output of this layer is given to softmax function which produces a probability distribution of the seven output classes. The model is trained using stochastic gradient descent (SGD) with batch size of 16 for 100 epochs (Fig. 4).

There are five CNN algorithm steps:

(a) Convolution
(b) Maxpooling
(c) Dropout
(d) Flattening
(e) Full connection

3.4 *Haar Cascade Classifier Algorithm*

Haar cascade classifier is considered as an effective object detection approach which is used for boosting the implementation process [12, 13]. This approach was proposed by Paul Viola and Michael Jones in their respective paper, namely "Rapid Object Detection using a Boosted Cascade of Simple Features" in the year 2001. Based on the training, it is then used to recognize the objects in the given images (Fig. 5).

The algorithm can be explained in four stages:

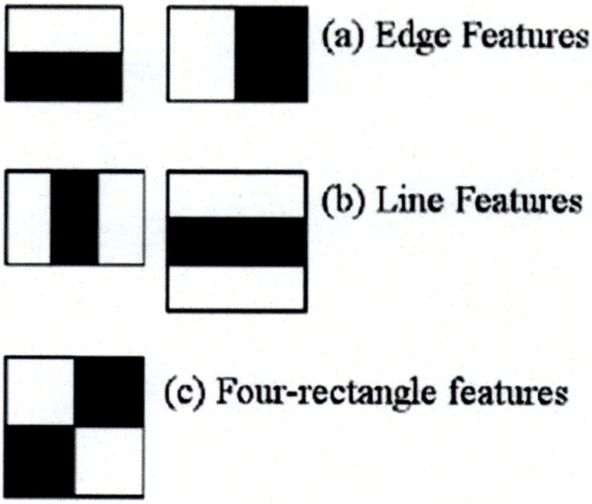

Fig. 5 Calculating Haar features

(a) Calculating Haar features
(b) Creating integral images
(c) Using AdaBoost
(d) Implementing cascading classifiers

3.5 *Results and Discussions*

The class distribution gives an overview on the number of images present in each individual class of emotions (Fig. 6).

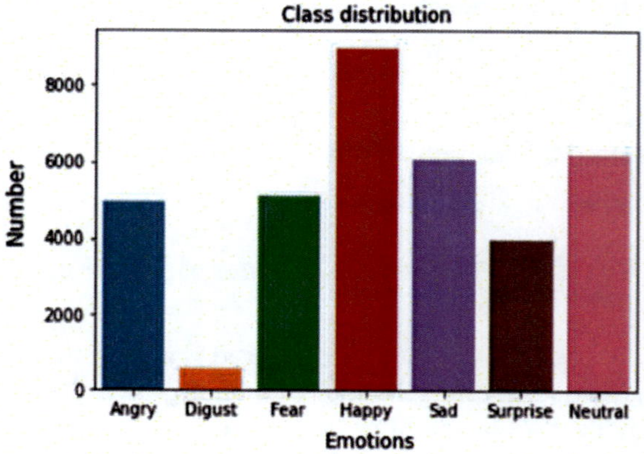

Fig. 6 Class distribution graph

Accuracy is one of the methods for evaluating the classification models. It is the fraction of predictions a model got right. Formally, accuracy has the following definition: number of correct predictions/total number of predictions. So, the accuracy of the model obtained is 60.30%.

The next step is visualizing the performance of the model. This helps us in making sense of the output that is not exactly in human understandable form. Hyperparameter tuning is possible for machine learning models from these visualizations in a simpler manner (Fig. 7).

The above graph shows the comparison between training and validation accuracies. The validation accuracy is the accuracy which is calculated on the dataset that cannot be used for training, whereas training accuracy is the accuracy calculated only on the training set.

Figure 8 shows a comparison between training loss and validations loss.

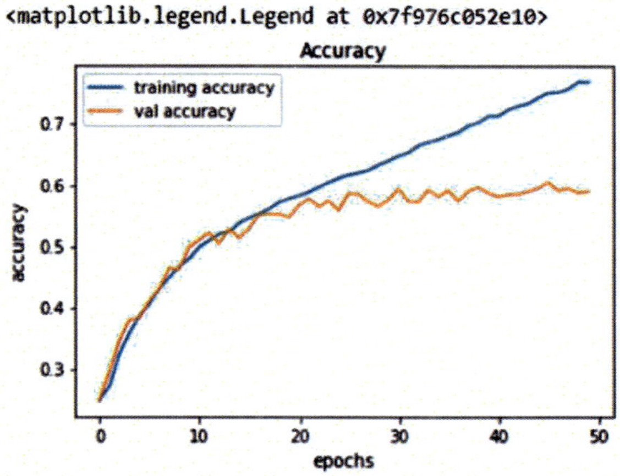

Fig. 7 Accuracy grap

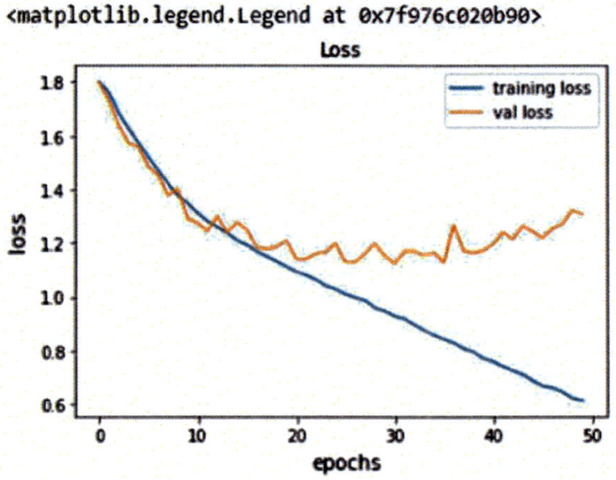

Fig. 8 Loss graph

A confusion matrix can be defined as a table that is often used in order to describe the performance of a model built for performing classification taking a set of test data where true values are known for this set. It can also be used to visualize essential predictive analytics such as recall, accuracy and precision (Fig. 9).

Initially, the model is saved and loaded whenever required to predict the required emotion accurately (Fig. 10).

There are numerous facial images available which can be used to test the model whether it is predicting accurately or not. So, any random image is given as input to the model which has been loaded to predict the emotion accurately. This uploaded

Fig. 9 Confusion matrix

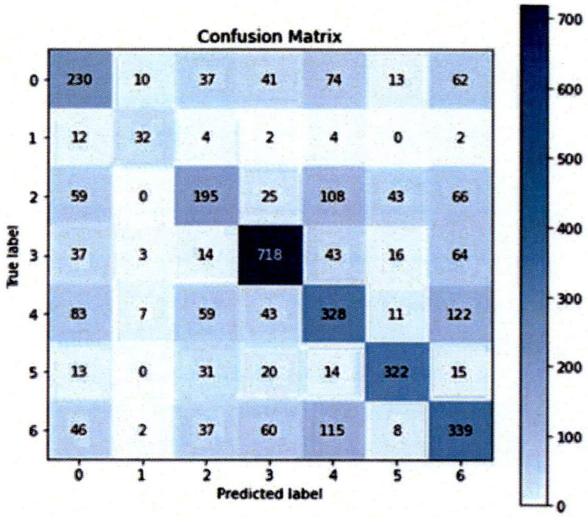

Fig. 10 Input image

Fig. 11 Converted greyscale image

image is firstly converted into a greyscale image before proceeding for background elimination. The greyscale image will be displayed as shown (Fig. 11).

The next step is to eliminate the background from the given image which can be done by using the background subtraction (BS) technique. As the name itself suggests, background subtraction is used to calculate the foreground mask by performing a subtraction between the given current frame and a background model which is present in the image. This background model can either be a static part of the scene or, in general, everything that can be taken as background considering the characteristics of the scene [14, 15].

This technique can be understood with the example (Fig. 12).

Background modelling consists of two steps, namely background initialization and background update.

In the background initialization step, an initial model of the background is calculated, whereas in the second step, that particular model is updated so as to adapt to all the possible changes in the scene.

Hence, this technique helps in eliminating the background from the uploaded images easily (Fig. 13).

The image after eliminating the background will be displayed in the following manner (Fig. 14).

The face is detected using the Haar cascade classifier after which it is fed into the model to predict the appropriate emotion [5, 6]. This prediction will be displayed as shown in Fig. 15.

With this, the task of classifying the face image based on the emotion displayed in the expression is completed. So, the image is tested according to these given categories by the model which then produces the final predicted emotion.

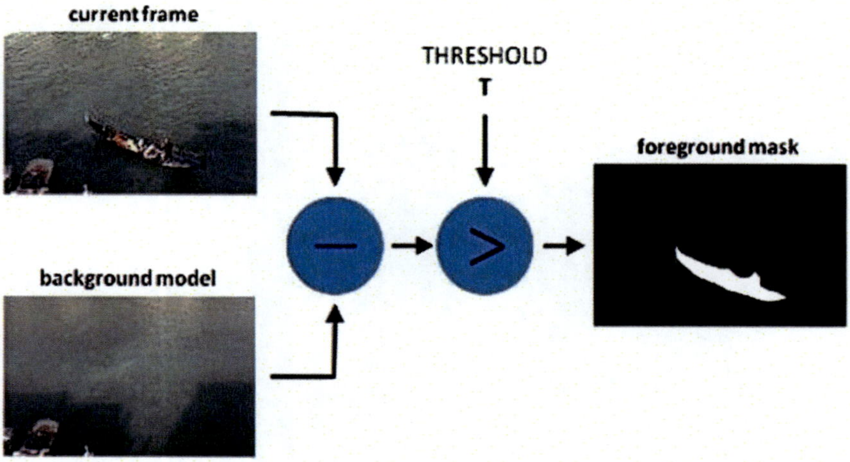

Fig. 12 Background subtraction

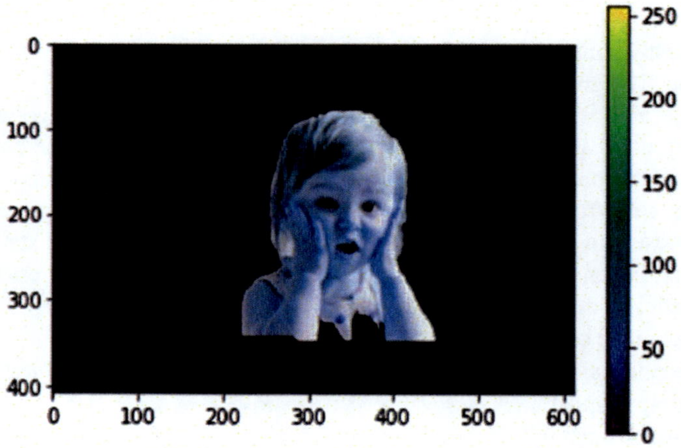

Fig. 13 Elimination of the background

The accuracy obtained is 60.30%. This accuracy is found to be improving if the variance among the data of different classes is high. The images are misclassified in some cases when mixed emotions exist in the uploaded image.

This misclassification can be observed in Fig. 16:

The given image can be considered as consisting of two different emotions which can be viewed as neutral and sad. So, the model tries to give any one of these emotions as the final predicted emotion.

Fig. 14 Background eliminated image

Fig. 15 Final output

4 Conclusion

The purpose of this work is not to get the required conclusive results but to bear out some of the main challenges and difficulties which are involved in detecting the emotion from various facial expressions. The paper discusses the technique which helps in identifying the different emotions of a person, namely angry, disgust, happy, sad, fear, surprise and neutral. These facial emotions can be mainly used in e-learning,

Fig. 16 Misclassified image

criminal investigation, online gaming, psychology and many other applications [16]. Using a public dataset of train and test samples for seven different classes of human emotions, we build and train a convolutional neural network model and use it along with Haar cascade classifier to produce the final predicted emotion of an uploaded image.

5 Applications and Future Scope

The facial emotion recognition is a technique which can be used in various applications. In e-learning [17], the images of the students can be taken in order to analyse whether the student is understanding properly or not based on the facial expressions. Similarly, this technique can be applied in the areas of psychology [16, 18] along with gaming in which the mood of the gamers is analysed to better understand and develop the product efficiently.

This technique is of great use in automobiles as well. It can be used to detect the behaviour and drowsiness of a driver so that they can be further given some sort of alerts in order to prevent accidents. The approach has an increasing importance in marketing too. Though various companies have employed different types of verbal surveys to know about the customer satisfaction, they may not always express it correctly. So, in these cases, knowing their feedback based on facial expressions is quite useful [18, 19]. This technology will be soon found even in future car dashboards, refrigerator doors and also in the conference room walls—or we can say essentially any surface will become a possible means for recognition of emotions.

The future enhancements for this approach include:

- Increasing the accuracy
- To predict the compound emotions

This work can be carried out by others in order to convert it into a model for predicting the emotions of a person by using the real-time capturing of the face. Similarly, the other advancement of this project may include the extraction of images from the videos and predicting the emotions from those images.

References

1. Rowley H, Baluja S, Kanade T (1998) Neural network-based face detection. IEEE Patt Anal Mach Intell 20:22–38
2. Chivers RC, Ramalingam VV, Pandian A (2018) Facial emotion recognition system—a machine learning approach facial emotion recognition system ± a machine learning
3. Du S, Tao Y, Aleix M (2014) Martinez1 Department of Electrical and Computer Engineering, and Center for Cognitive and Brain Sciences, The Ohio State University, Columbus, OH 43210. In: Heeger DJ (ed), New York University, New York, and approved 28 Feb 2014
4. Lindquist KA, Wager TD, Kober H, Bliss-Moreau E, Barrett LF (2012) The brain basis of emotion: a meta-analytic review. Behav Brain Sci 35(3):121–143
5. Sharifara A, Mohd Rahim MS, Anisi Y (2014) A general review of human face detection including a study of neural networks and Haar feature-based cascade classifier in face detection. In: 2014 International symposium on biometrics and security technologies (ISBAST), pp 73–78. https://doi.org/10.1109/ISBAST.2014.7013097
6. Cuimei L, Zhiliang Q, Nan J, Jianhua W (2017) Human face detection algorithm via Haar cascade classifier combined with three additional classifiers. In: 2017 13th IEEE international conference on electronic measurement & instruments (ICEMI), pp 483–487. https://doi.org/10.1109/ICEMI.2017.8265863.
7. Zhenxing G, Guo B (2021) Facial expression recognition with LBP and ORB features. Comput Intell Neurosci 2021
8. Rzayeva Z, Alasgarov E (2020) Facial emotion recognition using convolutional neural networks. IEEE Xplore. Auckland University of Technology
9. Mendel N (2020) Facial emotion recognition using convolutional neural networks (FERC)
10. Tian Y-I, Kanade T, Cohn JF (2001) Recognizing action units for facial expression analysis. IEEE Trans Pattern Anal Mach Intell 23(2):97–115
11. Du S, Martinez AM (2011) The resolution of facial expressions of emotion. J Vis 11(13):24
12. Bartlett MS et al (2005) Recognizing facial expression: machine learning and application to spontaneous behavior. IEEE Comp Vis Pattern Recog 2:568–573
13. Ding L, Martinez AM (2010) Features versus context: an approach for precise and detailed detection and delineation of faces and facial features. IEEE Trans Pattern Anal Mach Intell 32(11):2022–2038
14. Parks DH, Fels SS (2008) Evaluation of background subtraction algorithms with post-processing. In: 2008 IEEE fifth international conference on advanced video and signal based surveillance, pp 192–199. https://doi.org/10.1109/AVSS.2008.19
15. Mohamed SS, Tahir NM, Adnan R (2010) Background modelling and background subtraction performance for object detection. In: 2010 6th International colloquium on signal processing & its applications, pp 1–6. https://doi.org/10.1109/CSPA.2010.5545291
16. Chibelushi CC, Bourel F (2003) Facial expression recognition: a brief tutorial overview. CVonline: On-Line Compendium Comput Vis 9

17. Savva A, Stylianou V, Kyriacou K, Domenach F (2018) Recognizing student facial expressions: a web application. In: IEEE Global Engineering Education Conference EDUCON, vol 2018, pp 1459–1462
18. Izard CE (2009) Emotion theory and research: Highlights, unanswered questions, and emerging issues. Annu Rev Psychol 60:1–25
19. Russell JA (2003) Core affect and the psychological construction of emotion. Psychol Rev 110(1):145–172. 9

Green Initiative Low-Cost Electric Vehicle

K. Gopala Reddy, H. K. Pooja, Saima Zulfa, M. N. V. R. Shwetha, and J. Nanditha

Abstract In order to reduce the usage on oil and to reduce pollution, the advancement of electric vehicles has been stimulated in many countries. The execution of EVs especially battery electric vehicles is observed as the solution to the energy calamity and environmental subject. In non-conventional energy system, batteries play an important role, but bit expensive part of the system. In addition, lithium-ion batteries not only are high priced but also involve high risk factor in case of overcharging. Hence in order to overcome this problem, a battery management system (Fotouhi et al. in Renew Sustain Energy Rev 56:1008–1021, 2016 [1]; Du et al. in J. Power Sour, 2016 [2]) should incorporate with a lead–acid battery which also aids in improving the battery life. Even though the popularity of electric vehicles is growing, there are certain obstacles to overcome. Specifically, the cost of an electric vehicle ranges anxiety and charging time (Lagraoui , et al. in International renewable and sustainable energy conference (IRSEC), Oct 2014. IEEE, pp 636–641, 2014 [3]; Shen et al. in IEEE Trans Veh Technol 67:92–103, 2017 [4]). Conventional car emissions are now a significant contributor to increasingly severe environmental challenges. Furthermore, the current energy crisis, as well as the low energy efficiency of conventional automobiles, provides an excellent opportunity for the implementation of electric vehicles. The rapid development of pure electric vehicles is aided by increased realization of environmental issues and the energy crises. Low cost, range, and greater performance are three sectors served by this project. In this paper, we are focused on building a compact electric vehicle with cost-effectiveness. The vehicle top speed is 60 kmph, and on a single charge, we can travel up to 70 km. Comparing

K. Gopala Reddy (✉) · H. K. Pooja · S. Zulfa · M. N. V. R. Shwetha · J. Nanditha
Electrical and Electronics Engineering Department, Vidyavardhaka College of Engineering, Mysore, Karnataka, India
e-mail: gopal.reddy@vvce.ac.in

S. Shakya et al. (eds.), *Proceedings of Second International Conference on Sustainable Expert Systems*, Lecture Notes in Networks and Systems 351,
https://doi.org/10.1007/978-981-16-7657-4_33

the EV to conventional vehicle, on a single we can drive up to 70 km which requires one unit of electricity to charge that costs rupees 3, whereas the conventional vehicle to cover the same distance requires 3 L fuel costs around rupees 300 making EV economical. The key highlights of our work include low maintenance cost as it is battery operated, and as an electrical student, our prior duty is to protect the circuit; in view to this, we have placed a circuit breaker along with the ignition key technique, a battery charging indicator to indicate the percentage charge, and a vital aspect in EVs (Buchman and Lung in 2018 IEEE 24th International symposium for design and technology in electronic packaging (SIITME), Oct 2018. IEEE, pp 382–385, 2018, [5]). Unlike in internal combustion engine vehicles a hand brake is used to park in sloppy roads, in our design we have used "leg brake" for this purpose.

Keywords Electric vehicle · Battery management system · Battery life

1 Introduction

Batteries are electrochemical devices which, through redox reactions, convert chemical energy into electrical energy. There are three major categories of electric vehicles, defined by the degree to which electricity is used as their energy source: BEVs, hybrid electric vehicles, and plugged-in hybrid electric vehicles. Only BEVs can charge at a level 3, DC quick charge. In this paper, we concentrate primarily on electric battery vehicles and their features. Examples of BEV that can charge fast chargers on DC level 3 are tesla model 3, BMW i3, Chevy bolt, etc. Three battery states, namely open-circuit voltage, charge, and discharge, are available. The current flows between the battery cathode and anode and through the load connected across the battery terminals in the discharge state. Because the battery is not now in the open-circuit mode, the battery voltage is lower than OCV, due to the battery's internal impedance falling [6–8].

It is an eco-friendly vehicle venture, with the goal of creating a prototype that is compact, cost-effective, and minimal size and weight. If these compact vehicles are offered to the market at a time when fuel prices are rising, EV sales will skyrocket. However, the expenses of traveling the same distance will reflect a huge difference [9–11]. Hence, usage of EVs has rapidly started growing which will bring a sustainable transport and will thereby reduce the fuel expenses reducing the cost making it cost-effective plus reducing the harmful emissions and increasing health benefits leading to better air quality and putting a curb on noise pollution [12, 13]. These vehicles are cheaper to maintain, and the electricity required to charge the vehicle is also about one-third as much per kilometer to purchase fuel for vehicle. Thus as we know, the EV concept is unique and is bringing betterment in the lives of people leading to a huge increase day by day. Global warming, on the other hand, is hurting people's lives and health as a result of greenhouse gas emissions, with the transportation sector being one of the main contributors to GHG emissions (90%). Global CO_2 emissions reached 32.3 gigatons in 2015, according to figures from the International Energy

Agency, with the transportation sector accounting for 24% of overall emissions and the road sector accounting for three-quarters [14]. In reality, according to a well-to-wheel analysis, GHG emissions from electricity or hydrogen production are highly reliant on main energy sources [15]. The foregoing arguments imply that, with the development of technology and legislation, PEVs can entirely outperform ICEVs in terms of reducing GHG emissions.

This paper is structured as follows: Sect. 1 Introduction to E vehicles, Sect. 2 Methodology, Sect. 3 Conclusion, and Sect. 4 References.

2 Characteristics and Models of Energy Sources for PEVs

There are a variety of energy systems for storing electrical energy (batteries, super capacitors, and superconducting magnets), generating electrical energy (FCs, photovoltaic cells, and wind turbines), and storing mechanical energy (flywheels, pumped hydroelectric plants, and hydraulic accumulators) [16]. However, some of them are unsuitable due to the difficulties or limits associated in their installation. The focus of this research is on one of the energy storage devices known as batteries. It is well known that it is the most widely used storage device in power system applications because of its ability to convert from electrical to mechanical. Lead–acid batteries, nickel-based batteries, and lithium-based batteries are the three main types of batteries used in road transportation today. In this experiment, we used a lead–acid battery instead of the other batteries mentioned above. The electrochemical device can be utilized in both domestic and business settings. Low initial costs, excellent energy efficiency, quick responsiveness, and small self-discharge rates of 2% of rated capacity per month are significant advantages of lead–acid batteries.

3 Problem Solving and Challenges

Pure electric vehicles are the automobile industry's ultimate development aim; however, there are still several obstacles to overcome in the development process. For starters, PEVs have short driving ranges, and those powered by a single energy source have poor dynamic performance [12]. Integrating at least two different energy sources into a single powertrain of PEVs can meet specific requirements and maximize the dynamic performance of the entire system, while extending driving ranges [13]. However, in order to solve the problem of range anxiety, infrastructure for charging and replenishing PEVs must be constructed. Another issue is that charging stations and hydrogen refueling stations is now sparse, necessitating the installation of a large number of them on public land. The third issue is that these PEVs are still expensive [14]. The cost is mostly determined by energy storage technologies and is difficult to estimate because it is influenced by a variety of parameters such as storage type, application requirements, size, and so on. The capital cost of energy storage, on the other hand, can be assessed in a variety of methods, including cost per kW, cost per kWh, and cost per kWh per cycle.

4 Working Technology

Specifications:

Particular	Specification
Battery pack	12 V, 32 AHr (4 nos.)
Motor type	PM BLDC motor
Motor power	1 kW
DC driver controller	40 A
Charging time	30 min
Mileage	70 km on a single charge

At the start, the battery needs to be charged where it requires a time period of 30 min for full charge as we have used a lead–acid battery of 32 AHr capacity, but if a higher capacity battery is used, then probably the charging time will in turn reduce even more. So, the output of the battery is given to the driver controller to obtain voltage for the requirement of speed. Then, that particular output is given to the motor that is connected to wheels that in turn makes the wheels to rotate. Thus, this brings in the final movement of the vehicle. The electric vehicle once charged can go up to a range of 60 km per hour providing additional features like less noise pollution, compact in size, light weight, and a safer drive with absolutely no emissions. This vehicle can also be used in safari places to go deep inside the forest as there is no noise pollution and will not disturb the animals living there giving a smoother and comfortable ride.

The overall weight of the vehicle is 135 kg. The battery has a lifetime of about 4–6 years. The main specialty of our vehicle is it is back wheel drive where a single back wheel is driven. This is a unique concept as if two wheels would be driven, then torque would be more and the friction would also be more so by driving the back wheel more flexibility can be achieved with less losses and there will not be any problems while turning.

Our vehicle is powered by a 12 V, 32 AHr four lead–acid batteries connected in series to boost up to 48 V is connected to 1 kW, 48 V PM DC motor connected to rear axle of the vehicle, and a controller to control the current and voltage parameters. This electrical power is converted to mechanical power for the vehicle movement and a charging time of 30–40 min [17].

The vehicle top speed is 60 kmph, and on a single charge, we can travel up to 70 km. Comparing the EV to conventional vehicle, on a single we can drive up to 70 km requires one unit of electricity to charge which costs rupees 3, whereas the conventional vehicle to cover the same distance requires 3 L fuel costs around rupees 300 making EV economical. The key highlights of our work include low maintenance cost as it is battery operated; as an electrical student, our prior duty is to protect the circuit, and in view to this, we have placed a circuit breaker along with the ignition key technique and a battery charging indicator to indicate the percentage charge, a vital aspect in EVs. Unlike in ICE vehicles a hand brake is used to park in sloppy roads, in our design we have used "leg brake" for this purpose [18].

Block diagram of working mechanism

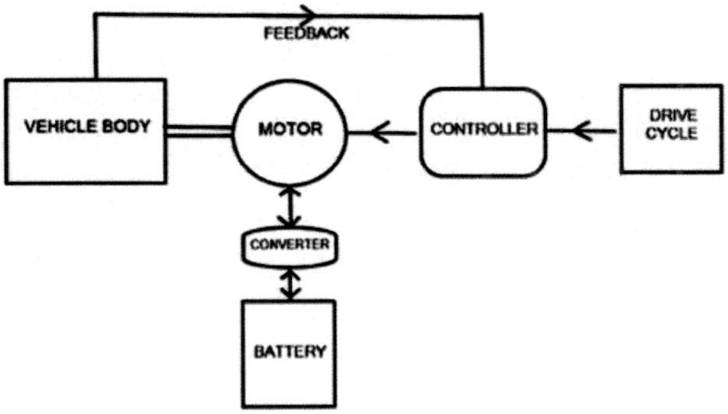

Specification	Details
Length * width * height	10798 mm * 762 mm * 1341 mm
Kerb weight	120 kg
Ground clearance	180 mm
Turning radius	2 m
Seating capacity	1
Transmission	Motor drive
Steering type	Rack and pinion
Front suspension	NA
Rear suspension	Swing arm with hydraulic shocks
Brakes	Drum
Boot capacity	Occupied by four batteries
Battery pack capacity	12 V, 32 h

Salient features of the vehicle:

- Unlike the ordinary cars, they have hand brake, but the speciality of our vehicle is it has "leg brake" included that is as an additional safety.
- It is eco-friendly and does not emit out any kind of emissions [15, 16] and provides a clean environment resulting in a healthy living.
- Additional solar panels can be added on to charge the vehicle when it has been parked.
- The vehicle is compact and has a smoother ride that occupies very less space for parking.
- It provides full comfort and can be used by aged persons who are willing to drive.
- Any person who does not have driving skills can learn using this vehicle and improve their skills.
- A layman can also effectively use the vehicle.
- It can be easily used in safari places as it is a noiseless vehicle that simply runs smoothly and will not disturb the animals in the forest. Hence, it provides a "noiseless operation."
- It can be taken on narrow roads very flexibly and has ease while turning or reversing.

It can effectively be used during "wildlife photography" due to the noiseless operation of the vehicle.

5 Results

The performance of the electric vehicle is tested, and the results are shown in the following table (Figs. 1, 2, 3, and 4).

Specifications	EV	Specifications
Travel range	60 km	Travel range
Charge condition	Full charge	Charge condition
Expenditure	Rupees 5	Expenditure

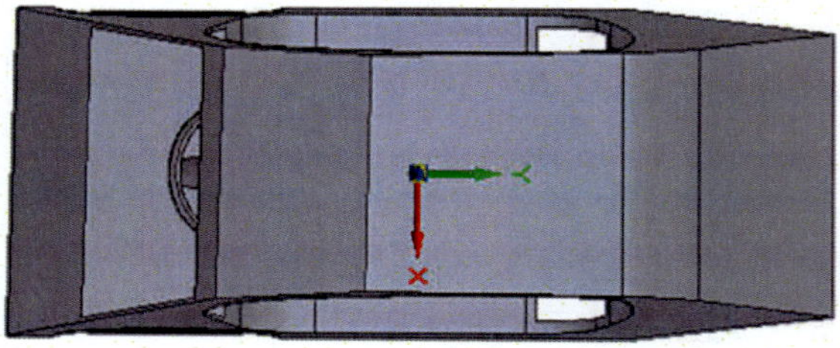

Fig. 1 Top view of the EV

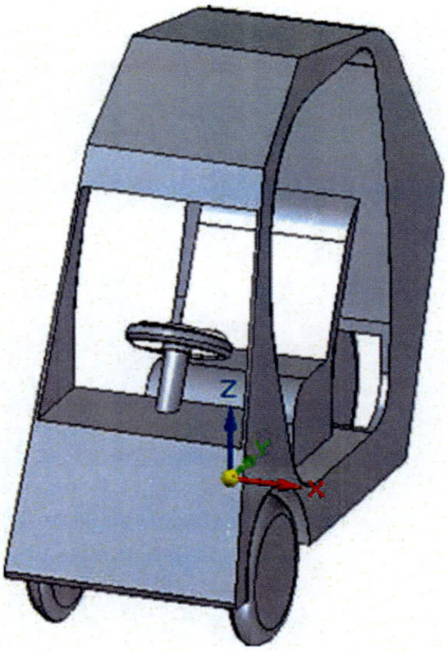

Fig. 2 Front view of the vehicle

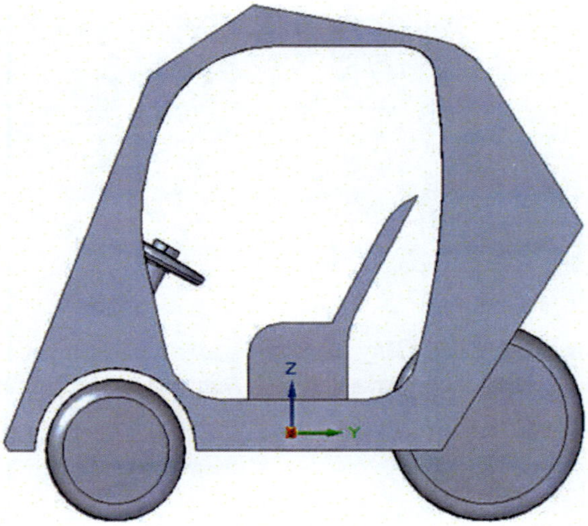

Fig. 3 Side view of the vehicle

6 Conclusion

The electric vehicle that is compact and cost-effective is the major attractive aspect described, and as the main constraint of electric vehicle would be the cost and the range anxiety, hence the proposed design and implementation will be the solution for the upcoming EV models. The experience of the ride in this compact vehicle makes it more effective as the automobile industry is shifting toward the eco-friendly aspect. The vehicle's main concept lies in back wheel driven where it makes the

Fig. 4 Team members with the EV

friction considerably less and the user finds it flexible to drive. Further research and developmental works can be made to develop and find the new designs that make EV more demandable in the market.

References

1. Fotouhi A, Auger DJ, Propp K, Longo S, Wild M (2016) A review on electric vehicle battery modelling: from lithium-ion toward lithium–sulphur. Renew Sustain Energy Rev 56:1008–1021
2. Du R, Hu X, Xie S, Hu L, Zhang Z, Lin X (2020) Battery aging- and temperature-aware predictive energy management for hybrid electric vehicles. J Power Sour
3. Lagraoui M, Doubabi S, Rachid A (2014) SoC estimation of lithium-ion battery using kalman filter and luenberger observer: a comparative study. In: 2014 International renewable and sustainable energy conference (IRSEC), Oct 2014. IEEE, pp 636–641

4. Shen P, Ouyang M, Lu L, Li J, Feng X (2017) The co-estimation of state of charge, state of health, and state of function for lithium-ion batteries in electric vehicles. IEEE Trans Veh Technol 67(1):92–103
5. Buchman A, Lung C (2018) State of charge and state of health estimation of lithium-ion batteries. In: 2018 IEEE 24th International symposium for design and technology in electronic packaging (SIITME), Oct 2018. IEEE, pp 382–385
6. 1 Million Pure EVs Worldwide: EV Revolution Begins! Available online: https://cleantechnica.com/2016/11/22/1-million-ev-revolutionbegins/. Accessed on 22 Nov 2016
7. Electric and Plug-In Hybrid Cars Whiz Past 3m Mark Worldwide. Available online: https://www.theguardian.com/environment/2017/dec/25/electric-and-plug-in-hybrid-cars-3m-worldwide. Accessed on 25 Dec 2017
8. Global Top. Available online: http://ev-sales.blogspot.com/2019/01/global-top-20-december-2018.html. Accessed on 20 Dec 2018
9. Matthews L, Lynes J, Riemer M, Dilator T, Cloet N (2017) Do we have a car for you? Encouraging the uptake of electric vehicles at point of sale. Energy Policy 100:79–88
10. Das HS, Tan CW, Yatim AH (2017) Fuel cell hybrid electric vehicles: a review on power condition in gun it sand topologies. Renew Sustain Energy Rev 76:268–291
11. Hydrogen. Available online: https://www.iea.org/tcep/energyintegration/hydrogen/. Accessed on 15 May 2019
12. Hannan MA, Lipu MSH, Hussain A, Mohamed A (2017) A review of lithium-ion battery state of charge estimation and management system in electric vehicle applications: Challenges and recommendations. Renew Sustain Energy Rev 78:834–854
13. Pelegov D, Pontes J (2018) Main drivers of battery industry changes: electric vehicles—a market overview. Batteries 4(4):65
14. Choi H, Shin J, Woo JR (2018) Effect of electricity generation mix on battery electric vehicle adoption and its environmental impact. Energy Policy 121:13–24
15. Santos G (2017) Road transport and CO_2 emissions: what are the challenges? Transp Policy 59:71–74
16. Vidhi R, Shrivastava P (2018) A review of electric vehicle lifecycle emissions and policy recommendations to increase EV penetration in India. Energies 11(3):483
17. Dhar S, Pathak M, Shukla PR (2017) Electric vehicles and India's low carbon passenger transport: a long-term co-benefits assessment. J Clean Prod 146:139–148
18. Davidov S, Pantoš M (2018) Planning of electric vehicle infrastructure based on charging reliability and quality of service. Energy 118:1156–1167

Vision Transformer Framework Approach for Yellow Nail Syndrome Disease Identification

Vikas Kumar Roy, Vasu Thakur, and Rahul Nijhawan

Abstract Yellow nail syndrome disease is an infrequent disorder distinguish by abnormalities affecting the fingernails and toenails. Yellow nail syndrome disease is sometimes associated with fluid collected in the lungs' membranes and lining the chest cavity. This disease usually affects older adults. The treatment of yellow nail syndrome is geared toward the symptoms available in each person. Here, the author proposes a novel deep transformer-dependent convolutional network to classify yellow nail syndrome disease through images. The framework uses Vision Transformer for feature extraction. A novel dataset is constructed for our framework for good accuracy because there was a lack of good quality of the dataset. This work is tested on our dataset and is compared with other state-of-the-art algorithms (SVM, ANN, KNN, Tree, RF, Adaboost), resulting in an excellent performance in feature extraction.

Keywords Vision Transformer (ViT) · Convolutional neural network (CNN) · Classification · Yellow nail syndrome disease · Feature extraction · VGG-16 · Inception V3

1 Introduction

A nail is a claw-like plate at the tip of the fingers and toes in most primates and is made of alpha-keratin. Alpha-keratin is a fibrous structural protein, which is a polymer and found in the hair, claws, and horns of vertebrates [1]. Our body's health is directly reflected by the changes in nail-like discoloration or thickness. The nail disease can occur at any age. Around fifteen percent of humans suffer belongs to the age group 60–70.

V. K. Roy (✉) · V. Thakur
Computer Science and Engineering, RIT Roorkee, Roorkee, India

R. Nijhawan
UPES Dehradun, Dehradun, India
e-mail: rahul.nijhawan@ddn.upes.ac.in

S. Shakya et al. (eds.), *Proceedings of Second International Conference on Sustainable Expert Systems*, Lecture Notes in Networks and Systems 351,
https://doi.org/10.1007/978-981-16-7657-4_34

The most severe and dangerous nail diseases among the global population is "yellow nail syndrome," which is a rare condition. This disease can happen to anyone but usually occurs in adults over the age of 50. It can be detected by observing the changes in the nail, such as the loss of the cuticle, "which is the part of the protective skin covering the nail," the nails that curve nail that stop growing, nail that separates from the nail bed, and sometimes nail losses.

Nail disease spotting has begun to congregate consequence, and after numerous studies, many researchers have begun to spectacle consideration in this field. The disease is possibly detected from the images because many techniques can categorize diseases from the image of the nail [2]. In deep learning, there is a Vision Transformer (ViT) architecture in which the input images are split into patches and then sequenced as linear embeddings as the input of the Transformer. Here, image patches are treated as tokens, which will be used to train the network in a supervised manner. This deep learning model is used by many researchers to classify and detect various types of diseases and abnormalities in the medical field. Vision Transformer has been used for diagnosis of COVID-19 and expands the model to estimate with clinical specifications [3]. Different CNN models have trained on a dataset, and these are Inception V3, VGG-16, VGG-19, and vision transformers. All of these have very remarkable accuracy, and ViT has more precision than others.

The main difficulty confronted by our structured work is the shortage of a fine and best dataset of images. We require the proper and structured dataset for feature extraction.

Considering the apparent difficulties and problems, the author has proposed a vision transformer-based work that uses a corps of convolutional neural nets. The proposed approach has outperformed the state-of-the-art deep learning and hand-crafted-based architectures. We have given a detailed disparateness with state-of-the-art techniques. Several statistical measures have been computed to prove the efficacy of our proposed approach.

Before 2004, none of any remarkable research had been done for recognizing nail disease [4]. But now, a large number of researchers are initiating to work in this area [5]. The nail unit is used to predict and detect psoriatic arthritis and color analysis to detect disease, [6] early disease detection and related work have been done.

In this study, the researcher used the Vision Transformer (ViT) to demonstrate the detection of yellow nail syndrome disease and performed the comparative study of various deep learning models.

The paper is organized as follows: Introduction (Sect. 1), Literature review (Sect. 2), Dataset (Sect. 3), Methodology (Sect. 4), Convolutional neural network (Sect. 5), Proposed vision transformer approach (Sect. 6), Results and discussion (Sect. 7), and Conclusion (Sect. 8).

2 Literature Review

For the image classification of nail disease, many researchers have put effort into deep learning concepts in the past few years.

Indi and Gunge [7] have used techniques of image processing on the human nail to create a disease detection system. Specifically, the nail pigment feature that gave an accuracy of 65%. Liao [8] has devised a deep learning approach for skin disease classification, using 23 top-level categories of the dermnet taxonomy, with nail diseases being one of the categories with minimal images per class for nail diseases. Another usage of the nail unit has been seen in computer-aided nailfold laparoscopy image analysis, performed by Paradowski and Markowska [9], in which they performed an enhancement on the images obtained from nail capilaroscopy; this method is used for assessment of the condition of the blood capillaries of the nailfold which finds its application to detect diseases such as scleroderma, and Raynaud's phenomenon. A color histogram methodology to analyze psoriasis is used in [10], and HSV color space is used to extract features and the average accuracy accomplished is 92%.

Park et al. [3] developed an algorithm to estimate COVID-19 through chest-X-ray (CXR) by ViT architecture that utilize low-level CXR feature obtained from a backbone network that extracts common CXR findings. Karimi et al. [11] used the CNN-free deep network for medical image segmentation to improve the precision of unlabelled images. Duong et al. [12] proposed an expert system that classifies chest-X-ray (CXR) images and introduced a solution for the detection of tuberculosis from CXR images by using a computer vision algorithm and achieved the precision of 97.72% with an AUC of 100%. Mondal et al. [13] have proposed the deep learning technique on chest radiography such as X-ray and CT to detect COVID-19 and employed a multi-stage transfer learning using Vision Transformer. Kamran [14] used Vision Transformer generative adversarial network (GAN) that comprised of feature fusion for discriminators and proposed a disease progression monitoring system to predict the increasing rate of the disease. Dalmaz et al. [15] introduced the model for image synthesis that incorporates transformers in a generator network and weight sharing strategy to decrease model complexity and computation. Balasubramaniam [16] proposed the detection of melanoma by artificial intelligence algorithm with SVM classification and introduced the melanoma classifiers to enhance diagnosis. Dai and Gao [17] applied transformers to medical image classification and improved the precision and efficiency, and proposed a novel multi-model image fusion strategy, which can be leveraged to capture manual information from images of different modalities in a more efficient way. Yuan et al. [18] developed novel progressive tokenization for ViT and demonstrated its advantages over the simple tokenization approach by ViT by proposing a T2T module that can encode the important local structure for each token. Ranftl et al. [19] introduced an architecture that leverages vision transformers in place of convolutional networks as a backbone for dense prediction tasks and observed an improvement of up to 28% in relative performance when compared to a state-of-the-art fully-convolutional network and showed that the architecture could be fine-tuned on smaller datasets.

Here, we have used the Vision Transformer, and the top five achieved accuracy is 99% which is much more than the CNN models. Following different CNN models are used. VGG-16 architecture is already trained on the ImageNet dataset [20] and achieved an accuracy of 93.57%. Due to this architecture of VGG-16 has gained maximum accuracy than that of VGG-19 (accuracy 90%) and Inception V3 (accuracy 92.1%).

Therefore, the author put forward a deep learning framework of ViT architecture for the detection of nail disease with an accomplished accuracy of 99% at an early age.

3 Dataset

The most accessible problem for deep learning to solve is the most challenging work, which is the fair data in the correct structure. The essential section of the deep learning approach is data collection. The data needs to be arranged according to expected results. Deep learning needs a robust dataset to be adequately trained for correct results. The model is trained finer for a good outcome. The deficiency of good quality of dataset with a large range of images, author create a novel dataset, keeping 200 images belongs to the corresponding classes. We chose the good images that represents the normal and diseased parts (Fig. 1).

All the dataset is further divided into training as well as testing dataset. About 81.951% of images are selected randomly for training purposes, and the rest is set for testing purposes (Table 1).

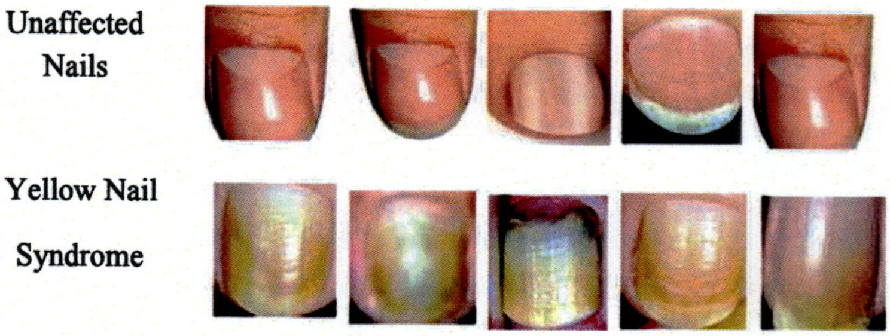

Fig. 1 Sample of unaffected and affected nails

Table 1 The table shows the total number of images present in the dataset

Types of nails	Number of nails images
Unaffected nails	100
Affected nails	100

Table 2 The architecture of VGG-16 [24]

Layers	Kernel filters	Filter size
1 and 2	64	3 by 3
3 and 4	124	3 by 3
5, 6 and 7	256	3 by 3
8–13	512	3 by 3

4 Methodology

4.1 Convolutional Neural Network (CNN)

Convolution neural network (CNN), often called ConvNet, can discover in a better way in comparison with networks with fully connected layers. It gets high features and can recognize objects effectively. CNN's are extensively applied in different fields due to their astonishing performance, such as image classification. It is also used to determine convoluted complications [21]. CNN has shown magnificent achievement in numerous computer vision and machine learning complications.

4.1.1 Inception V3

Inception-V3 is a convolutional neural network that focuses on less computational power and makes various enhancements, including factorized convolutions [22], parallelized computations, and the utilization of an additional classifier to generate label details.

4.1.2 VGG-16

VGG-16 is a convolutional network for extensive identification of images that accomplishes an accuracy of 92.7% in the top five tests in ImageNet (dataset of 15 million images belongs to 22,000 classes) [23] (Table 2).

4.1.3 VGG-19

VGG-19 has six structures in which every structure comprises convolutional layers. It is a preprocessing model and uses different convolutional layers and non-linear activation layers [25].

We used this CNN architecture to understand the crucial features needed for the classification of nail disease. For the classification of nail disease and we have also used the different machine learning algorithm and comparison of accuracy obtained by the different algorithm are performed.

4.2 Ensemble Methods

In this, procedures are meta-algorithms that integrates various machine learning approach into a single predictive prototype to reduce variance, bias or improving indications.

4.2.1 Stacking

It is a learning method that integrates multiple classifications or regression models via a metaclassifier. In this paper, different algorithms used for stacking are logistic regression, random forest, neural network, and CNN.

4.3 Logistic Regression

It uses a logistic function to prototype for the binary dependent variable. It predicts the criterion of the logistic model in the form of binary regression. It describes data and correlates with dependent and independent variables.

These are the different algorithm and method which is used for classification, and obtained accuracy is compared with other model and the most accurate technique are used for classification.

The machine on which we have executed our proposed constitute 8 GB RAM, AMD Ryzen 7 4700U with Radeon Graphics.

5 Convolutional Neural Network (CNN)

Convolution neural network (CNN), also called ConvNet, can derive more finely in comparison with networks with fully connected layers. It can learn conceptual features and recognize things effectively. CNN is extensively used in different fields due to its exceptional achievement like image classification, speech recognition, etc. [21]. We used this deep learning for the feature extraction of images. The output is used as an input to a logistic regression model, which classifies the yellow nail syndrome disease of the normal nail and the diseased nail.

6 Proposed Vision Transformer Approach

Over the past decade, deep CNN has been the go-to method to solve computer vision problems [24]. Compared to traditional machine learning algorithms, the significant advantage of CNNs is to apply numerous convolution operators on images to learn hierarchical spatial features, and CNN's have been successful in terms of performance [24]. From the last few years, CNN-based architectures such as VGG-16, VGG-19, and Inception-V3, etc., have been solving complex image recognition challenges on ImageNet.

However, CNN has many disadvantages. Since convolution operates on a fixed size window, the model is more focused on detecting local information than long-range spatial relationships between parts of the images and the images as a whole. Additionally, max-pooling can lead to a loss of data [24].

In the NLP field, attention-based architectures such as Transformers have been developed to tackle different language-related tasks more efficiently, such as language translation and text classification [24].

Large performance gains have been achieved using transformers and, as a result, have sparked great interest in the computer vision community to use similar self-attention models for vision tasks. There has been multiple computer vision works focused on combining self-attention with CNN-like architectures. A few works have attempted to replace the convolution entirely by self-attention mechanisms. However, these models have not yet been able to scale efficiently on modern hardware accelerators due to specialized attention patterns.

The Vision Transformer (ViT) was recently proposed as the first deep neural network algorithm for large-scale computer vision datasets without any convolution operators. The ViT uses the original Transformer developed in Vaswani et al. [26] for NLP tasks with a few changes. First, the input images are split into patches and then sequenced as linear embeddings as the Transformer's input. From an NLP point of view, image patches are treated as tokens which then will be used to train the network in a supervised manner [24]. In contrast to CNN's which encode prior knowledge about the spatial domain, transformers operate on vectors and need more data to discover knowledge from high-dimensional and large-scale datasets.

The architecture resembles language transformers, although in this model, the result of the transformer encoder is fed directly to a Multilayer Perceptron (MLP) head, and there is no decoder arm. A 2D flattening is completed on image patches to reshape them into a vector which is the required input of the Transformer. Then, vectorized image patches are converted to a linear patch embedding. At the same time, a position embedding is injected into each of these embedding patches to make the network able to retain the positional information of each embedding patch. Similar to BERT's method, a class token ([cls] which stands for classification) is added to the beginning of the sequence of embedded patches as a learnable embedding.

The transformer encoder in the prominent architecture consists of multiple encoder blocks, and each block has numerous layers of multiheaded self-attention mechanisms. Layer normalization is applied on the embedded patches before being fed into the multi-head self-attention and second before the MLP blocks. On the right side of Fig. 2, the general mechanism of an encoder is shown. The comprehensive tool of how an encoder, multi-head attention, and scaled-dot product attention works are explained in Fig. 2.

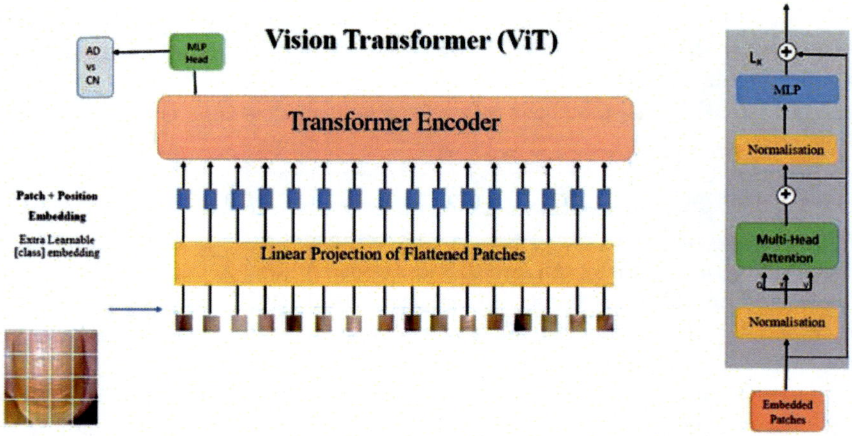

Fig. 2 ViT model architecture: illustration of our proposed study for yellow nail syndrome detection using ViT. The input image is split into equal-sized patches and embedded using linear projection. Position embedding is added, and the resulting sequence is fed to a transformer encoder

7 Results and Discussion

This segment represents the ultimate results derived by the Vision Transformer architecture suggested in this paper. It is discovered that the precision of our proposed way is the highest at 97.3% in comparison with another advanced prototype. The accomplishment of ViT is observed better and more accurately than CNN and other models. The results are derived using models preliminarily trained on different datasets. Overall, we observed four frameworks entirely.

In scenario 1, we proposed the Vision Transformer approach, which completely removes recurrence and convolution. Transformers use self-attention, a highly efficient technique that makes it possible to update the embeddings of every word in the input text in parallel. With Vision Transformer architecture, we divide an image into patches, and these patches are treated as a token and then sequenced as linear embeddings as the input and supplied the series of linear embeddings of these patches as an input to a Transformer. We achieved a remarkable accuracy of 97.3%.

In scenario 2, with an Inception V3 convolutional deep neural network which uses an auxiliary classifier to transmit label information towards the lower network to improve the convergence of a very deep network. Using this architecture achieved an accuracy of 88.6% with SVM, 91.4% with Logistic Regression, 81.4% with Random Forest, 88.6% with Naive Bayes, 64.3% with AdaBoost, and 94.3% with KNN.

In scenario 3, we have used the VGG-16 architecture and achieved the highest accuracy. A 13-convolutional layer, accompanied by three fully connected layers, achieved an accuracy of 88.8% with SVM, 93.6% with Logistic Regression, 91.4% with Random Forest, 88.6% with Naive Bayes, 64.3% with AdaBoost, and 88.6% with KNN.

In scenario 4, we proposed the VGG-19 deep architecture. A 19 layers deep model consists of 16-convolution layers, three fully connected layers, 5 max pool layers, and 1 softmax layer. Fixed-size (224 * 224) RGB images were inserted as input and used kernels of (3 * 3) sizes with a stride of 1 pixel. The dimensional padding used to structure the image's resolution and max-pooling over a (2 * 2) pixel with a stride of 2 is performed, followed by a rectified linear unit (RELU) non-linearity to make the model better to classify and improve computational time. Using this architecture achieved an accuracy of 75.7% with SVM, 90% with Logistic Regression, 82.8% with Random Forest, 94.9% with Naive Bayes, 74.3% with AdaBoost, and 91.4% with KNN.

It is observed that the ViT achieved an accuracy of 97.3%, and the VGG-19 deep convolutional neural network has the second highest accuracy of 94.9% with Naïve Bayes. For the feature extraction, it is observed that VGG-16 takes more time than another model.

The accuracy assessment table of all feature extractors of CNN and ViT is given. The AUC, accuracy, and precision are shown in Table 3.

In Table 3, the accuracy assessment of feature extraction of different convolutional neural networks (VGG-19, VGG-16, Inception-V3) and our proposed approach Vision Transformer are shown. It is observed that ViT has the highest accuracy of (97.3%) which is far better than the following mentioned CNN models. The second highest accuracy of (94.9%) is obtained by VGG-19 with the Naïve Bayes classifier, and the third-highest accuracy of (94.3%) is obtained by Inception-V3 with the KNN classifier.

Binary classification results of yellow nail syndrome disease are shown in Fig. 3. This shows the accuracy of the proposed approach model Vision Transformer and different convolutional neural network architecture (VGG-19, VGG-16, and Inception V3) using a bar graph. Various machine learning algorithms, SVM, KNN, logistic regression, Random Forest, Ada boost, etc., are used in advance CNN. *Y*-axis shows the accuracy infraction, and the results of all four scenarios with the corresponding model are represented.

Why is Vision Transformer better than CNN's Models?

The Vision Transformer proposed approach has more accuracy of (97.3%) as compared with following CNN model; the reason behind this is because ViT uses self-attention, which allows integrating information across the entire image, even in the lowest layer. ViT provides high accuracy with less computation time for training. The variable-sized information is handled by the model with transformer architecture using stacks of self-attention instead of CNNs. Applying self-attention between the image pixels in the Transformer without the participation of CNN is a big challenge. ViT has overcome this problem by splitting images into small patches (16×16). The Transformer encoder module comprises a multi-head self-attention (MSA) layer and a Multilayer Perceptron (MLP) layer. The multi-head self-attention layer split inputs into several heads so that each head can learn different levels of self-attention. The outputs of all the heads are then concatenated and passed through the Multilayer Perceptron. While CNN works on ideas where image consists of pixels, one

Table 3 Accuracy assessment table of feature extractor [VGG-19, VGG-16, Inception-V3, and Vision Transformer]

Feature extractor	Models	AUC	Accuracy	Precision
VGG-19 [25]	SVM	0.959	0.757	0.964
	Random forest	0.912	0.828	0.840
	AdaBoost	0.728	0.743	0.722
	Naive Bayes	0.911	0.949	0.868
	KNN	0.947	0.914	0.853
	Tree	0.718	0.757	0.716
	Stack	0.964	0.928	0.890
	Neural network	0.955	0.900	0.900
	Logistic regression	0.977	0.900	0.900
	CN2 rule inducer	0.710	0.686	0.686
VGG-16 [23]	Random forest	0.961	0.914	0.914
	Neural network	0.975	0.907	0.907
	SVM	0.953	0.878	0.888
	Naive Bayes	0.939	0.907	0.908
	KNN	0.959	0.886	0.887
	AdaBoost	0.814	0.814	0.817
	Tree	0.807	0.793	0.793
	Stack	0.957	0.928	0.929
	Logistic regression	0.981	0.936	0.936
	CN2 rule inducer	0.664	0.636	0.667
Inception-V3 [22]	KNN	0.946	0.943	0.835
	Random forest	0.907	0.814	0.891
	Naive Bayes	0.967	0.886	0.925
	SVM	0.978	0.886	0.925
	AdaBoost	0.686	0.643	0.703
	Neural network	0.968	0.886	0.912
	Tree	0.726	0.757	0.746
	Stack	0.973	0.914	0.914
	Logistic regression	0.976	0.914	0.927
	CN2 rule inducer	0.701	0.600	0.627
Transformer (PA)	Vision Transformer	0.986	0.973	0.972

pixel is dependent on neighboring pixels, and the next pixel is dependent on the next neighbor pixels. CNN uses filters on the patches of the image to pull out important features [13]. And the model is trained only on the extracted feature but not on the all feature of each pixel of images. However, if the entire image data is trained on the model rather than just training on the important extracted features, the chances

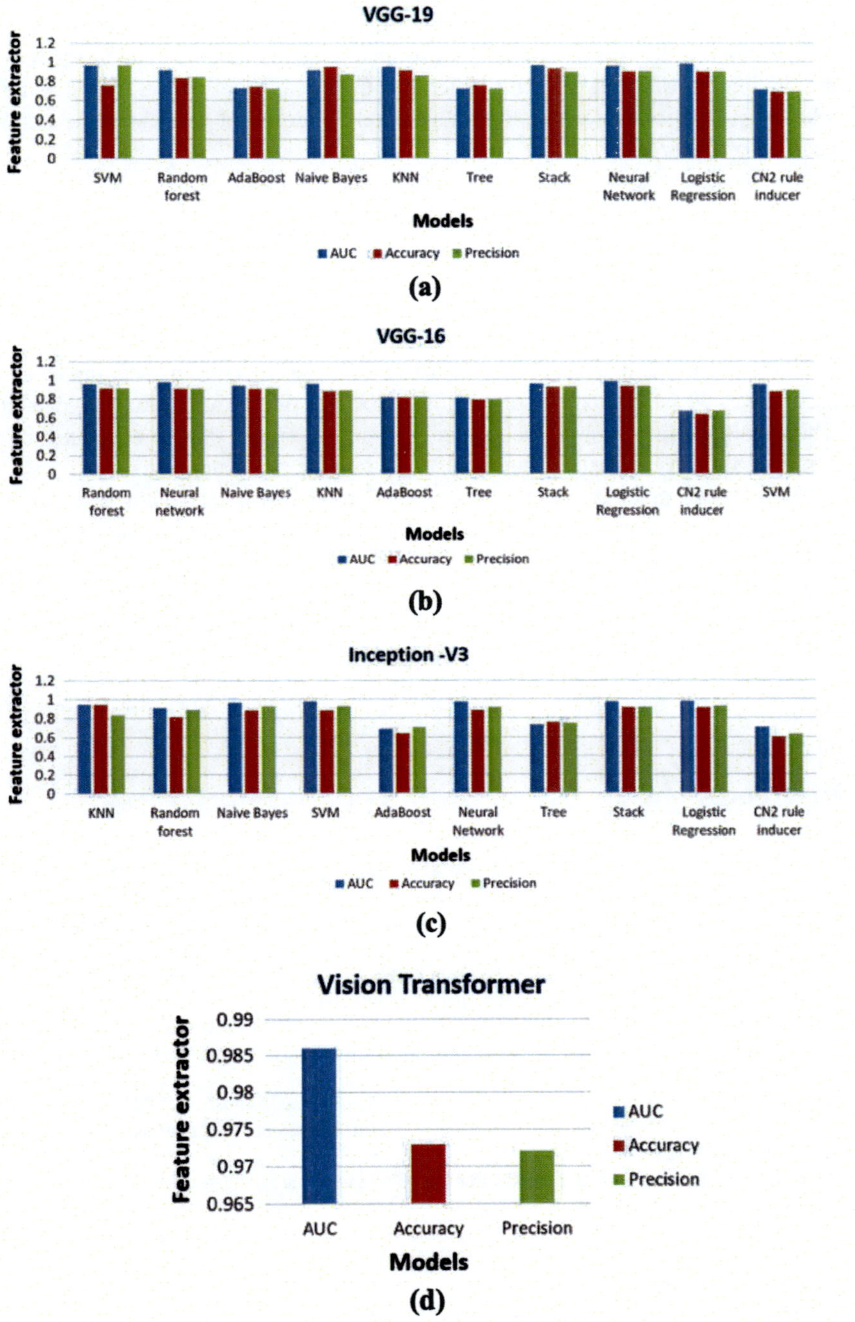

Fig. 3 Binary classification results of yellow nail syndrome disease. **a** VGG-19, **b** VGG-16, **c**Inception-V3 and **d** Vision-Transformer

of getting model performance are higher. This is happening in the case of a Vision Transformer. CNN fails to encode relative spatial instruction. This can be the reason that Vision Transformer performs better than CNN. CNN only looks for the features whether it is present or not in the image pixels. It does not care about the relative position of features.

8 Conclusion

Our proposed approach used Vision Transformer, logistic regression, and a hybrid of CNNs to classify and detect the yellow nail syndrome diseases and accomplished an accuracy of 99%. As a consequence of the lack of a preliminarily existing database with considerable pictures of yellow nail syndrome disease, the project gained more complications. Nevertheless, there must be a range for an extension soon.

We have an objective to create a general model to detect other diseases like hyperpigmentation and onychomycosis, which are other nail diseases and are difficult to detect.

Our research could be of great usage to detect yellow nail syndrome disease, which could acquire the necessary medical care. Our work could be extended and could be a usage of dermatologists to analyze the victim information and recognize the disease with fewer individual associations.

References

1. Nijhawan R et al (2017) An integrated deep learning framework approach for nail disease identification. In: 13th international conference
2. Goel A, Nijhawan R (2020) A deep learning approach for classification of onychomycosis nail disease. In: Proceedings of ICETIT 2019. Springer, Cham, pp 1112–1118
3. Park S et al (2021) Vision transformer using low-level chest X-ray feature corpus for COVID-19 diagnosis and severity quantification. arXiv arXiv:2104.07235
4. Fletcher CL, Hay RJ, Smeeton NC (2003) Observer agreement in recording the clinical signs of nail disease and the accuracy of a fungal and non-fungal nail disease clinical diagnosis. Br J Dermatol 148(3):558–562
5. Langenbruch A, Radtke MA, Krensel M, Jacobi A, Reich K, Augustin M (2014) Nail involvement as a predictor of concomitant psoriatic arthritis in patients with psoriasis. Br J Dermatol 171(5):11231128
6. Sharma V, Ramaiya M (2015) Nail color and texture analysis for disease detection. Int J Bio-Sci Bio-Technol 7(5):351–358
7. Indi TS, Gunge YA (2016) Early stage disease diagnosis system using human nail image processing. Int J Inf Technol Comput Sci (IJITCS) 8(7):30
8. Liao H (2016) A deep learning approach to universal skin disease classification. University of Rochester, Department of Computer Science, CSC
9. Paradowski M, Markowska-Kaczmar U, Kwasnicka H, Borysewicz K (2009) Capillary abnormalities detection using vessel thickness and curvature analysis. In: International conference on knowledge-based and intelligent information and engineering systems. Springer, Berlin, Heidelberg, pp 151–158

10. Dhandra B et al (2013) Color histogram approach for analysis of psoriasis skin disease. In: International conference on multimedia processing
11. Karimi D, Vasylechko S, Gholipour A (2021) Convolution-free medical image segmentation using transformers. arXiv arXiv:2102.13645
12. Duong LT, Le NH, Tran TB, Ngo VM, Nguyen PT (2021) Detection of tuberculosis from chest X-ray images: boosting the performance with vision transformer and transfer learning
13. Mondal AK, Bhattacharjee A, Singla P, Prathosh AP (2021) xViTCOS: explainable vision transformer based COVID-19 screening using radiography. TechRxiv. Preprint
14. Kamran SA, Hossain KF, Tavakkoli A, Zuckerbrod SL, Sanders KM, Baker SA (2021) VTGAN: Semi-supervised retinal image synthesis and disease prediction using vision transformers. arXiv preprint arXiv:2104.06757
15. Dalmaz O, Yurt M, Çukur T (2021) ResViT: residual vision transformers for multi-modal medical image synthesis
16. Balasubramaniam V (2021) Artificial intelligence algorithm with SVM classification using dermascopic images for melanoma diagnosis. J Artif Intell Capsule Netw 3(1):34–42
17. Dai Y, Gao Y (2021) TransMed: transformers advance multi-modal medical image classification
18. Yuan L, Chen Y, Wang T, Yu W, Shi Y, Tay FE, Feng J, Yan S (2021) Tokens-to-token ViT: training vision transformers from scratch on ImageNet
19. Ranftl R, Bochkovskiy A, Koltun V (2021) Vision transformers for dense prediction
20. ul Hassan M (2018) VGG16-convolutional network for classification and detection. Consulta: 10 Apr 2019. Disponible en: https://neurohive.io/en/popular-networks/vgg16
21. Indolia S, Goswami A, Mishra SP, Asopa P (2018) Conceptual understanding of convolutional neural network—a deep learning approach
22. Szegedy C et al (2016) Rethinking the inception architecture for computer vision. In: Proceedings of the IEEE conference on computer vision and pattern recognition
23. Krizhevsky A, Sutskever I, Hinton G (2012) Imagenet classification with deep convolutional neural networks. Adv Neural Inf Process Syst 1097–1105
24. Rezvantalab A, Mitha S, Khademi A (2021) Alzheimer's disease classification using vision transformers
25. Xiao J, Wang J, Cao S, Li B (2020) Application of a novel and improved VGG-19 network in the detection of workers wearing masks. J Phys Conf Ser 1518:012041. https://doi.org/10.1088/17426596/1518/1/012041
26. Vaswani A, Shazeer N, Parmar N, Uszkoreit J, Jones L, Gomez AN, Polosukhin I (2017) Attention is all you need. In: Advances in neural information processing systems, pp 5998–6008

An Efficient Fuzzy Logic Cluster Formation Protocol for Data Aggregation and Data Reporting in Cluster-Based Mobile Crowdsourcing

Check for updates

M. Renukadevi, E. A. Mary Anita, and D. Mohana Geetha

Abstract Crowdsourcing is a procedure of outsourcing the data to an abundant range of individual workers rather than considering an exclusive entity or a company. It has made various types of chances for some difficult issues by utilizing human knowledge. To acquire a worldwide optimal task assignment scheme, the platform usually needs to collect location information of all workers. During this procedure, there is a major security concern; i.e., the platform may not be trustworthy, and so, it brings about a threat to workers location privacy. Recently, many distinguished research papers are published to address the security and privacy issues in mobile crowdsourcing. According to our knowledge, the security issues that occur in terms of data reporting were not addressed. Secure and efficient data aggregation and data reporting are the critical issue in Mobile Crowdsourcing (MCS). Cluster-based mobile crowdsourcing (CMCS) is the efficient way for data aggregation and data reporting. In this paper, we propose a novel procedure, the efficient fuzzy logic cluster formation protocol (EFLCFP) for cluster formation, and use cluster cranium (CC) for data aggregation and data reporting. We recommend a couple of secure and efficient data transmission (SET) protocols for CMCS, (i) SET-IBE uses additively homomorphic identity-based encryption system and (ii) SET-IBOOS uses the identity-based online/offline digital signature system, respectively. Then, we have widen the features of cluster cranium by increasing the propensity to achieve aggregation and reporting on the data yielded by the requesters without scarifying their privacy. Also, considering query optimization using cost and latency.

Keywords Mobile crowdsourcing · Cluster cranium · Data aggregation · Data reporting · Security · Query optimization

---

M. Renukadevi (✉)
Department of CSE, S.A. Engineering College, Chennai, India

E. A. Mary Anita
Department of CSE, School of Engineering and Technology, CHIRST University, Bengaluru, India

D. Mohana Geetha
Department of CSE, Sri Krishna College of Engineering and Technology, Coimbatore, India

S. Shakya et al. (eds.), *Proceedings of Second International Conference on Sustainable Expert Systems*, Lecture Notes in Networks and Systems 351,
https://doi.org/10.1007/978-981-16-7657-4_35

1 Introduction

Howe invented crowdsourcing in the year 2006 [1]. Later on, with the brisk raise of mobile and communication technologies, the smartphones are becoming as the part of the human life, and mobile crowdsourcing is becoming an emerging paradigm in the current era. In mobile crowdsourcing, the end users are using mobile phones for collecting and sharing the data [2]. Today, things are getting better by using mobile phones for communications and lots of things like texting, education, maps, navigation, banking by using Internet or Wi-Fi connections. In advance to this, the sensors are also embedded in smartphones to sense scientific and industrial sectors using Internet of Things (IoT).

Crowdsourcing is a platform which is used to make the human-related task simple by outsourcing a task to preferred people. To toughen the platform, we are using mobile phones as the end users for data collections and processing by various capabilities like sensing, communication, and computation. Mobile phones are open environment where anyone can act as workers. Generally, the tasks are classified into two types: explicable and inexplicable. The tasks which can be solved by machines are called as explicable tasks, whereas the tasks that cannot be solved by machines are called as inexplicable tasks. The task which is inexplicable by the machines is posted to the crowdsourcing platform. To solve this kind of tasks, the platform consists of a group of workers known as crowd. This crowd of workers will take the tasks, solve and give the results to the end users known as requesters. Generally, in this platform, they have a group of workers specialized in various fields. The tasks which are inexplicable by the machines are posted to the platform in form of a query. The MCS will receive abundance of queries per day, and these queries may or may not be relevant to the platform. The queries from the requesters are structured based upon the input given by the requester, and it is allocated to the most preferable workers based upon Gale-Shapley matching game selection [3]. To reduce the burden of MCS, a declarative crowdsourcing is designed [4], in which a CROWDOP algorithm is used to optimize the query. Due to this algorithm, the repeated and irrelevant queries are rejected and an optimized query is given to the platform. The main objective of request-response method is to get the best answer for a particular query. Therefore, various levels of inspection are done such as linguistic inspection along with extra attempts is made. The tasks are further classified into micro-task and submitted to the workers with the help of the HIT manager [5]. Selecting a worker for a particular task is forever issue in MCS, since a task assigned to incompetence worker may result in undesirable consequences. A matrix factorization framework [6] was created to utilize for task recommendation in crowdsourcing systems. A database is required to post and receive a task from the platform [7]. In recent years, many applications are developed such as smart road [8] and public safety [9] by using mobile crowdsourcing and are used for various kinds of monitoring such as communications monitoring, ecological monitoring, and civic safety monitoring (Fig. 1).

In spite of greater benefits in MCS, it faces certain challenges in security and privacy. Due to openness of the platform, anyone can enter into the MCS as workers

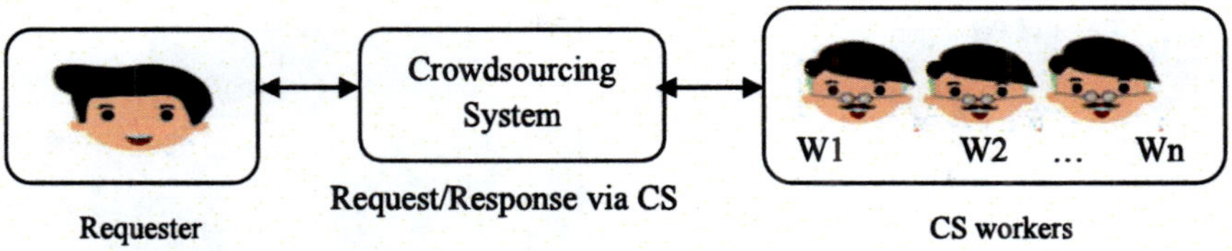

Fig. 1 Crowdsourcing systems

without proper authorization and can threat the entire platform. Also, there is no proper idea about whether the task is assigned to the expertise of the particular domain; hence, the integrity is lost. To overcome these issues, we propose a cluster-based mobile crowdsourcing (CMCS) for effective data transmission and data reporting.

Recently, most of the research focused on selection of cluster cranium [10, 11] whereas in our proposed work, we focused on cluster formation. The rest of the paper is discussed with the structure of the system various challenges in MCS and related work of crowdsourcing systems. Specifically, the contributions of this paper can be summarized as follows.

(1) We examine with the existing research work and recommend a novel procedure, the efficient fuzzy logic cluster formation protocol (EFLCFP) for cluster construction, and use cluster cranium (CC) for data aggregation and data reporting.
(2) We propose a pair of secure and efficient data transmission (SET) protocols for CMCS, namely, SET-IBE and SET-IBOOS.
    a. SET-IBE uses additively homomorphic identity-based encryption method.
    b. SET-IBOOS uses the identity-based online/offline digital signature system for reducing the computational cost.
(3) Also, we a considered query optimization in CMCS with CSELECT and CJOIN.
(4) As a result, we show the feasibility in security aspects using the proposed EFLCFP protocol with respect to data aggregation and data reporting.

1.1 Structure of the System

The MCS requesters and workers play a vital role in the structure of the MCS. The MCS requesters post the task to the platform which is not solvable by the machine. They acquire the knowledge of the human intelligence known as MCS workers by posting the task in the form of a query. Abundant tasks are posted, but only the optimized tasks are posted to the MCS platform. Further, those tasks are classified into micro-tasks by the HIT manager, and these micro-tasks are assigned to the workers who are expertise about the task. From the structure of a MCS system, we

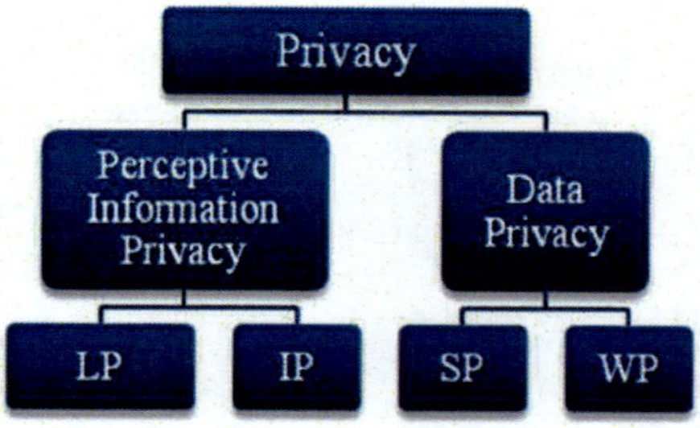

Fig. 2 Types of Privacy in MCS

conclude that it consists of a service provider, MCS requester, and MCS workers [12].

(a) **Service Provider**
The prominent role of service provider is to accept the request, select the proper workers, and assign relevant task to them. After receiving the relevant results, it will aggregate the results and report the same to the requesters.

(b) **MCS Requester**
The requesters will post the task to the platform to acquire the knowledge of human intelligent for complex tasks by fixing certain incentive for the tasks.

(c) **MCS Workers**
The MCS consists of group or cluster of workers to perform the task posted by the requester.

1.2 Various Challenges in MCS

(a) **Privacy**
The concept of privacy (Fig. 2) is to ensure both task and users privacy while assigning a task. They are categorized into two types known as perceptive information privacy and data privacy. Generally, before a task is published into a MCS, both requester and workers need to give certain information like their name, location, mobile number, mail id, state, and many which are sensitive. These types of data are known as perceptive data [13]. In any working atmosphere, the details of workers will be collected such as their specialization and environmental data may probably cause data leakage which consists of private information.

(b) **Security**
The data collected in the MCS environment should be secure. When a requester benevolent their details in an open environment, the risk factors are more with respect to security. Various security threats occur [12] when we furnish our data such as Sybil attack, false riding attack, impersonation attack, collusion attack, false personal information uploading attack, free-riding attack, and false data uploading attack. These types of attacks are harmful for both task assignment and reporting in Table 1 [12].

Table 1 Various types of security attacks [12]

Attacks	Description
Sybil attack	A malicious worker may transfer bogus information by making numerous identities for themselves to meddle the judgment of the service provider
False data uploading	A worker may transfer bogus information to the service provider
Free-riding attack	Worker may not show interest to complete a specific task after receiving the installment
False personal information uploading	Workers may give bogus information while registering in the MCS platform
Impersonation attack	The malicious workers who urge another genuine user to transfer an invalid information or data
Collusion attack	One crowd in MCS scheme with another to accomplish certain objective
Eavesdropping attack	Taking the data over the network. It is also known as snooping or sniffing attack

(iii) **Data Aggregation**
The most important component to get the statistics of data in processing is known as data aggregation [12]. We must make sure the privacy of end users while aggregating the data since; the SP cannot be trusted fully while data transmitting; therefore, it is essential to guarantee the data privacy of workers against the attackers. The privacy of data aggregation can be ensured by using the most popular techniques known as homomorphic encryption and by adding random noise/perturbation to the data.

(iv) **Data Reporting**
Data reporting [12] is the process that occurs from worker side after a successful completion of a task. The data generated by the workers will transmit to the service provider through various WNS. It faces several problems while reporting the data such as security, privacy, and trust. It also suffers from several attacks such as eavesdropping and data tampering attacks. Therefore, the process should guarantee from security and privacy while reporting. SP should authenticate the provenance of data.

2 Related Work

The mobile crowdsourcing network (MCS) is the platform using numerous number of people to work on a particular task. The Amazon Mechanical Turk is the example for crowdsourcing platform which is the way for most of the company to affordable outsources the task and to make money. The major part to be focused in MCS is to preserve privacy of the requester and workers at the time of assigning the task. We can able to preserve the data by using homomorphic encryption [14]. Based on the existing survey, a homomorphic encryption can be categorized into three types: additive homomorphic (AHME), multiplicative homomorphic (MHME), and mixed homomorphic encryptions (MiHME). The first encryption algorithm to evaluate the statistical information is known as AHME. Günther et al. [15] proposed an additive homomorphic IBE scheme in which the collusion attack was addressed. This paper proposes a framework model known as privacy-enhanced participatory sensing infrastructure for collusion resistance (PEPSICo) which includes mobiles hubs, questions, a specialist organization, and an enrollment authority with indistinguishable jobs as in the PEPSI model. The versatile hub senses and reports the information to the SP, which should be scrambled with the private key of workers. Afterward, they can aggregate the information without knowing the substance of it. This encryption algorithm guarantees collusion attack, confidentiality, and authentication. The author [16] addresses security issues faced by task owners and mobile users in spatial crowdsourcing. An architecture and a data aggregation scheme (PTAA) were proposed to give security assurance. Paillier cryptosystem [17] was the most popular cryptosystem developed by Pascal Paillier. The author uses asymmetric algorithm for public key cryptography which belongs to an additive homomorphic encryption.

The user privacy can also be prevented by using homomorphic and data cloaking [14]. A novel privacy-aware and trustworthy data aggregation protocol (PATA) is proposed [18] for mobile sensing. This paper merges the digital signature and homomorphic encryption to check the security issues. A research based on query optimization [19] is performed using the range and KNN query results with the help of whale optimization algorithm which is used to minimize the computation time as well as communication overhead. To minimize cost and latency in crowdsourcing, using EFGSA is elucidate [20]. Few papers focused on pushing and pulling the correct tasks for the workers [21]. Many researchers have investigated about the security and privacy issues that faced during data aggregation. Several papers are proposed for clustering strategies in WSNs [22, 23]. Recently, Zhang et al. [24] proposed a novel bilayer collaborative clustering (BLCC) method for label aggregation in crowdsourcing. Also, many researches are done for selection of cluster cranium [25]. In our proposed work, we extend a narrative procedure, the efficient fuzzy logic cluster formation protocol (EFLCFP) for cluster configuration, and use cluster cranium (CC) for data aggregation and data reporting (Table 2).

3 Proposed Approach

Currently, the association between computations based on machines and humans is materialized by mobile crowdsourcing which leads to the next level of information dispensation. Mobile crowdsourcing can be an effective way to solve a complex task with the help of human intelligence. The queries can be collected from the requester and will be sent to the workers via the crowdsourcing platform. We are using the recommender system [28] for collecting the queries from the requesters. This system will control the remote devices with the help of smart phones. Even though the MCS

Table 2 Comparison of existing systems

Existing system	Description	Algorithm	Cluster	Disadvantage
CROWDOP [4]	A cost-based optimization in crowdsourcing platform	Optimization algorithm (select, join, and complex queries)	Clustering is not addressed	Security not discussed
Smart road [8]	A crowd-sourced road sensing system	Classification algorithm	Spectral clustering	Security not discussed
QoE [26]	A crowdsourceable framework to quantify the QoE of multimedia content	TSR algorithm (quantify the individual consistency of a participant)	Clustering is not addressed	Security not discussed
LIRIS-ACCEDE [27]	A large video database LIRIS-ACCEDE	Quick sort algorithm	Clustering is not addressed	Does not provide security
TSCM [9]	A framework where IoT can enhance public safety by crowd management via sensing service	Outlier detection algorithm—trustworthiness of the user	Clustering is not addressed	Does not provide security
PTAA [16]	Privacy-aware task allocation and data aggregation scheme for bilinear pairing and homomorphic encryption	Aggregation algorithm is used	Clustering is not addressed	HMAC-SHA1 for security which is vulnerable to collision attacks
WOA, FGSA [19]	Optimal query plan selection using metaheuristics algorithms	Whale optimization algorithm	Clustering is not addressed	Security not discussed
TOP-K-T, TOP-K-W [21]	Effective algorithms used for pulling and pushing the right tasks to the right workers	TOP-K-T and TOP-K-W recommendation algorithm	Clustering is not addressed	Security not discussed

has abundant benefits, it has pitfall such as security issues in data aggregation and data reporting. To address these issues, in our proposed work, we are implementing clustering methodology in MCS. Initially, the cluster cranium is elected based on the LEACH Protocol [29]. The cluster cranium is used to send the aggregated queries from the requester to the workers and report the results to the same. To ensure the security of requester, we use a couple of data transmission protocol, namely SET-IBE and SET-IBOOS, for efficient data aggregation and reporting. The proposed concept is explained in the following section.

3.1 Cluster Formation

Clustering is the errand of partitioning the data into a number of clusters that are similar or dissimilar. Clustering plays a vital role in mobile crowdsourcing since the MCS workers are categorized into the various cluster based upon their domain of interest. Several studies are reviewed for selection of cluster cranium [30] in which energy level and lifetime of cluster-based MCS (CMCS) is not addressed. In our proposed approach, we use a novel strategy; the fuzzy logic cluster formation protocol (FLCFP) which uses fuzzy logic inference system (FIS) for cluster formation [29]. The clustering approach adds elasticity to achieve energy-efficient operation, lifetime of CMCS, and reduce the number of workers that communicate with service provider (SP). Our clustering formation can be done via a couple of iterations, namely system phase and stable phase. The fuzzy logic is verified using the conditional statements derived using Mamdani-type fuzzy rules [31].

The fuzzy rules are represented in Table 3 [31], in which L—low, BM—below minimum, M—medium, EL—extremely low, H—high, Le—least, BN—below normal, N—normal, AN—above normal, VS—very strong, AM—above margin, respectively. In our proposed work, the EFLCFP is used for cluster formation of worker by FIS. It is used for formulating the input and output by using the fuzzy rules. The queries received from the receivers are aggregated and sent to the workers

Table 3 Fuzzy rules

Iterations (i)	Estimation (E)	Output
L	L	BN
L	M	EL
L	H	Le
M	L	BN
M	M	N
M	H	AN
H	L	H
H	M	VS
H	H	AM

via the CC. After the completion of work, the results are reported to the requester via CC. We use the following equation to compute for effective data aggregation and reporting:

$$\mathrm{Eff}_{\mathrm{DA}}(q, \mathrm{dist}) = E_{\mathrm{en}} * q + \alpha * q * \mathrm{dist}^{\beta} \tag{1}$$

$$\mathrm{Eff}_{\mathrm{DR}} = E_{\mathrm{en}} * q \tag{2}$$

where the effective data aggregation can be represented as $\mathrm{Eff}_{\mathrm{DA}}$, effective data reporting can be represented as $\mathrm{Eff}_{\mathrm{DR}}$, q is the size of query in bits, and dist is the distance between the requester and worker. By using Eqs. (1) and (2), we are able to identify the query intention and are able to aggregate it in the cluster cranium. With the help of cluster cranium, the query will be shared to the, respectively, cluster or to the worker expertise, and the data reporting is done vice-versa.

3.2 Protocol Objective

To represent the parameter, we use triangular and trapezoid membership functions as shown in Eqs. (3) and (4), which are identified in MATLAB fuzzy logic toolbox.

$$M1(x) = \begin{Bmatrix} 0, & x \le x1 \\ \frac{x - x1}{y1 - x1}, & x1 \le x \le y1 \\ \frac{z1 - x}{z1 - y1}, & y1 \le x \le z1 \\ 0, & z1 \le x \end{Bmatrix} \tag{3}$$

$$M2(x) = \begin{Bmatrix} 0, & x \le x2 \\ \frac{x - x2}{y2 - x2}, & x2 \le x \le y2 \\ 1, & y2 \le x \le z2 \\ \frac{s2 - x}{s2 - z2}, & z \le x \le s2 \\ 0, & s2 \le x \end{Bmatrix} \tag{4}$$

where the values of $x1, x2, y1, y2, z1, z2$, and $s2$ used are described as energy level, distance to the SP, and distance between the CC and the node.

Consider cluster-based mobile crowdsourcing in which the service provider (SP) will be acting like an intermediate node for collecting the queries from the requester and responding for the same. We believe that the service provider is the trusted authority. At the time of registering the MCS platform, the requester needs to provide their sensitive information like address, locations, phone numbers. Almost, all the sensitive information of the requester as well as workers will be maintained in the SP. Occasionally, the service provider might get hacked by the attacker, and there is chance of misusing these sensitive information which in turn leads to security and privacy issues for the requester and workers. To overcome these issues, in CMCS, each and every worker is grouped into a cluster based upon their domain interest such as big data, artificial intelligence, marketing, medical, engineering, respectively. These clusters will have a cluster cranium (CC), which is nominated autonomously. The non-CC will join a cluster based upon the queries received from the requester. The CC performs data collection from the SP, aggregates the data, sends the data to the respective cluster, and finally, the results will be reported to the SP. In addition to this, timestamp is allotted for SP and all the worker nodes in the cluster. If the result is not responded within the timestamp, then the query will be passed to the next worker.

Several clustering-based protocols such as LEACH protocols [32] are highly secured for data aggregation, cluster head selection, and data communication in WSNs. Many researchers are proposed based on clustering strategy such as SecLEACH [33], GS-LEACH [34], and R-LEACH [35]. The universal loom used for security in these protocols is symmetric key management which leads to the orphan node problems. To overcome the orphan node problem, we use digital signature which is one of the most significant security service in asymmetric key management systems to provide security in data aggregation and data reporting.

3.2.1 CMCS with IBE and IBOOS

In this section, we propose a secure and efficient data transmission (SET) protocols, known as SET-IBE and SET-IBOOS [36]. The major inspiration of both the protocol is to provide a secure data aggregation and data reporting in MCS. To provide an authenticated platform in MCS, we created a cluster-based mobile crowdsourcing (CMCS) in which the data aggregation as well as reporting will be secured. Initially, the cluster formation is done in CMCS rather than selection of cluster cranium (CC). Later, the workers in MCS are grouped into a cluster based upon their domain of interest, and each cluster is allotted with CC autonomously. The major goal of CC is to gather the data from the service provide (SP), group the data, and send the same to the relevant cluster along with the timestamp. The security of sensitive information with respect of both requester and workers needs to be ensured with the help of key management protocol. We propose identity-based encryption for trustworthiness of reporting between the requester and the SP. The IBE is an additive homomorphic encryption scheme [37, 38] which is based upon elliptic curve cryptography. Equation (5), in IBE system, gleans from bilinear map $\mathcal{B}$.

$$g_1 * g_1 \rightarrow g_2 \tag{5}$$

where g_1, g_2 are the two groups from any bilinear map on elliptic curve. In IBE, there are two essential keys, namely public key and a private key. The public key can be calculated from any sequence, and no certification authority is required for it, whereas the private key is calculated by the trusted authority such as service provider (SP). The IBE scheme consists of four basic strategies for ensuring the

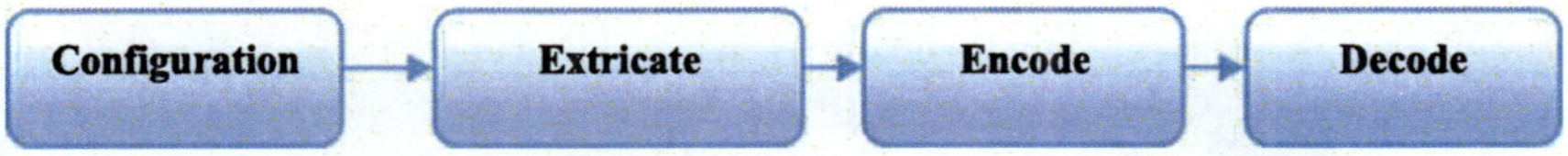

Fig. 3 Basic strategies of IBE

security as shown in Fig. 3. The configuration is the first strategies which is done by the trustworthy authority (TA) such as service provider (SP). Secondly, the extricate which is accomplished at the SP or at end user. Finally, encoding and decoding of data using public key and private key are made.

The security in IBE depends upon the bilinear Diffie–Hellman problem (DHP) in the pairing domain.

We also propose IBOOS for authenticating the encrypted data by using digital signature in which the security relies on hardness of discrete logarithm. The IBOOS is designed for secure and efficient transmission for CMCS. It follows certain strategies for authentication such as configuration, abstraction, signature, and verification. In the configuration phase, a master key and public parameters for private key are generated, and all are given to the CC. In the second phase, an ID string is provided in which each node generates a private key associated with the ID using master key. In third phase, the signature is provided with a message, timestamp, and signing key. IBOOS has two types of signing: offline signing and online signing in which offline signing is given with public parameter and timestamp, and the CC will generate the offline signature, whereas online signing generates online signature from leaf node. Finally, verification is done, and it results as "accept" if signature is valid and "reject" if signature is invalid (Fig. 4).

The IBOOS scheme depends upon the discrete logarithmic problem (DLP) in multiplicative clusters and proposes a secure data transmission protocol with IBOOS specifically for CMCS (Fig. 5).

As we mentioned earlier, the four strategies such as configuration, abstraction, signature, and verification are proposed. Equation (4) describes about the abstraction as,

$$K = (L, c_i) \tag{6}$$

where

$$L = h^{\text{au}} \tag{7}$$

$$c_i = \text{au} + G(L, \text{ID}_n)y \bmod z$$

In the abstraction, master key and identity are to be generated for the secret key. The offline signature creates the offline sign (β_j) associates through generated time period g_p for sending and saves the intellectual information for signing online sign while transmitting the information. Eventually, the offline sign is generated through

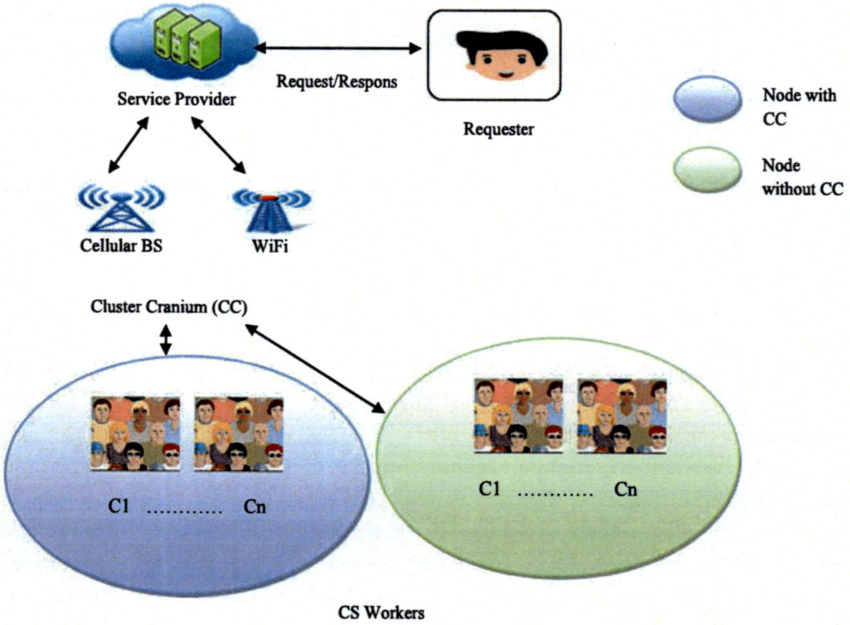

Fig. 4 Cluster-based mobile crowdsourcing architecture for efficient data aggregation and data reporting

sensor region or by the authorized external agency. Example, the RM or the DI sensor region.

$$\begin{aligned}&\text{Let } Y = h^{y}, \text{ then,}\\&h^{\text{au}} = h^{\text{au}}h^{G(L,\text{IDn})y \bmod z} = LY^{G(L,\text{IDn}) \bmod z}\\&\beta_j = h^{-tn}\end{aligned} \tag{8}$$

In online signature, each node generates the online sign (ρ_i, s_i) based on the data which are encrypted and based on offline signing.

$$\begin{aligned}&g_i = G(C||\delta_i)\\&x_i = \delta_i + g_i\text{au} \bmod z\\&\delta_i = h^{\delta i}\end{aligned} \tag{9}$$

Based upon the receiving queries, the verification is done along with the timestamp. Finally, for checking the correctness, an equation is shown below.

$$\begin{aligned}&\delta^{i} L^{gi} Y^{giG(L,\text{IDn}) \bmod z}\\&\quad = h^{\delta i} h^{rgi} h^{ygi\,G(L,\text{IDn}) \bmod z}\end{aligned}$$

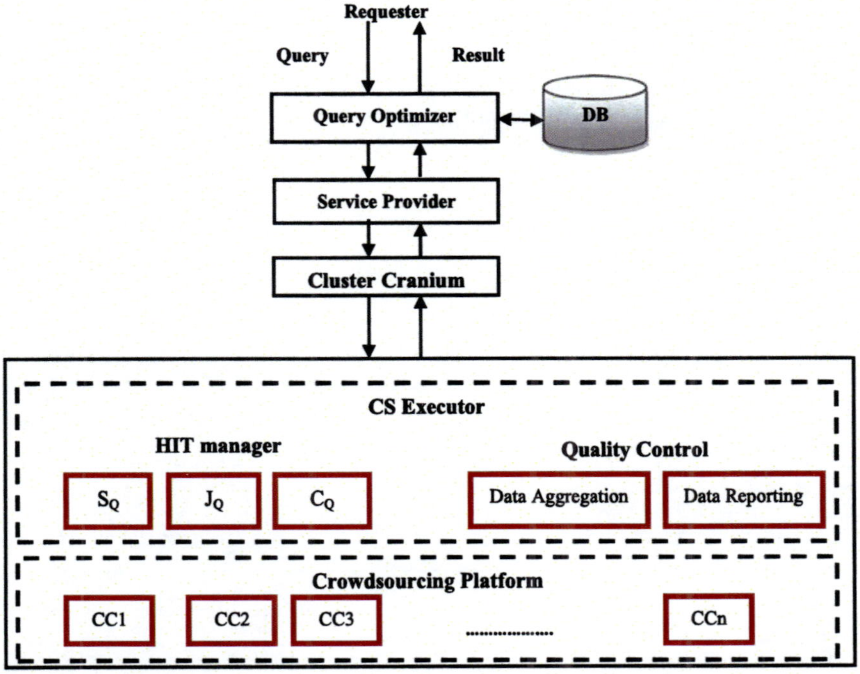

Fig. 5 Query optimizer in CMCS

$$= h^{\delta i + gi(r + (G(L, \mathrm{IDn})y \bmod z))}$$
$$= h^{\delta i + gici \bmod z} = h^{zi} \tag{10}$$

The data reported are said to be authenticated, if the above Eq. (10) is equal to the received message.

4 Query Optimization in CMCS

The query optimizer is used to optimize the query which is been sent by the requester. The crowdsourcing query optimizer [39] will choose the optimized query and send it to the cluster cranium. Generally, the query plan can be classified into CSELECT and CJOIN [4]. These query plans are used for evaluation of each query which in turn used to select an optimized query.

4.1 Various Query Languages

The query Qu used in CMCS is an SQL query over specified relations. The evaluation of query with its semantics is done in the crowdsourcing which gives the result of Qu. The query can be categorized with three types: selection query, join query, and selection-join query.

4.1.1 Selection Query (S_Q)

A selection query is a most commonly used for data manipulation. It is used to retrieve one or more tuples from one or more relations (Fig. 6). Generally, this query is used for viewing the data and which is used for filtering data and for finding certain items. An example for selection query is shown below:

T1: Vehicle

Id	Brand	Variant
x1	Maruti	Petrol
x2	Tata	Diesel
x3	Nissan	Diesel
x4	Hyundai	Petrol

T2: Preview_Vehicle

Img_preview	Brand	Variant	V_Color	V_Performance

Fig. 6 Example with two relations: vehicle and Preview_Vehicle

Qu1

SELECT T2.image

FROM Preview_Vehicle T2

WHERE Brand = "Maruti" and Variant = "Petrol" and V_color = "Red" and V_Performance = "High".

4.1.2 Join Query (J_Q)

A join query is used to combine one or more relations based upon certain join condition using human intelligence. In real-world scenario, a database in which data are structured into a large number of relations, and there is a need to combine these multiple relations based used upon a logical relationship between them. An example for selection query is shown below:

Qu2

SELECT T1.*, T2.img_preview

FROM Vehicle T1, Preview_Vehicle T2

WHERE T1.Brand = T2.Brand JoinFilter T1.Variant = T2.Variant.

4.1.3 Complex Query (C_Q)

A complex query is the combination of selection query and join query.

$$C_Q = S_Q + J_Q \tag{11}$$

These queries are used by the requester to send complex queries in crowdsourcing platform. In our proposed work, we are using complex queries in crowdsourcing platform.

4.2 CS Executor

The CS executor is used for executing the operators iteratively by a bottom-up traversal of the tree plan. An operator opr is elected based upon the algorithm whose input is already located. These opr are generated by human intelligence tasks (HITs), and the results are published in the CS platform. Once after completing all the operators, the answers A(opr) are communicated to the parent operator, and the algorithm gets expire (Fig. 7).

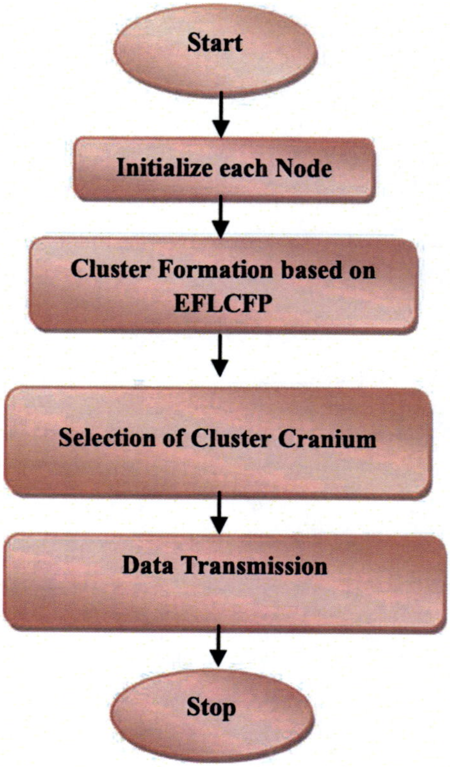

Fig. 7 Flow diagram of EFLCFP

4.3 *Performance Metric in Query Optimization*

Monetary Cost

Here, the monetary cost of each query is denoted as $C(\mathrm{Qp})$. The reward paid for overall query plan is denoted as,

$$C(\mathrm{Qp}) = \sum_{\mathrm{opr} \in O} C(\mathrm{opr}) \tag{12}$$

where $C(\mathrm{opr})$ represents the cost of execution of each operator. Initially, the crowd in the CS platform will have penalty for each and every query for the generated operator which is known as unit cost (u_c). The unit cost will differ for individual operators based upon the workload given for crowd. As such, the larger the workload, the higher the unit cost u_c. The variable $Ç_s$ is used as set of selection query for the workload evaluating condition with respect to the unit cost u_s. The enormous the $Ç_s$, the higher the unit cost u_s. Likewise, the more the number of join condition used $Ç_j$, the higher the unit cost u_j.

$$\text{Qp}^* = \arg_{\text{Qp}} \min(C(\text{Qp})) \tag{13}$$

where Qp query plan is used for minimizing the monetary cost.

Latency

Generally, we use latency to qualify the speed of query evaluation, in which the prediction and optimization are non-trivial. In crowdsourcing platform, there are many uncertainty, such as the workers may join and leave the CS platform anytime. Since there are only inadequate workers in CS platform such as AMT, they can choose the HITs of their own interest which has high reward. In this paper, the latency L(Qp) of query plan is calculated iteratively. Each iteration refers to the level of the query plan tree, and the latency L(Qp) is equal to height of the query tree. There are various uncertainty in the real-world crowdsourcing platform such as Amazon Mechanical Turk (AMT).

$$\text{Qp}^* = \arg_{\text{Qp}} \min(L(\text{Qp})) \tag{14}$$

where Qp is the query plan, budget cost βc. We can find minimum latency with the above Eq. (12) with respect to $C(\text{Qp}) \leq \beta c$.

5 Experimental Analysis

In order to provide security in the MCS, we proposed a hierarchical scheme known as EFLCFP for cluster formation. This is used to create the cluster cranium for data aggregation and data reporting. This is efficient protocol when compared to LEACH, C-LEACH, and SecLEACH. In FLCFP, for cluster formation, we use deduction scheme (DS) for fuzzy logic in cluster formation. This is formulated based on 3 inputs, 1 output with 27 fuzzy rules. The fuzzy rules are derived from Mamdani method. In EFLCFP, we are creating the cluster in an efficient way for secured data aggregation and reporting. For secured data transmission, we used IBE and IBOOS. The performance was measured by various metrics such as privacy, security, authentication, confidentiality, and computation cost. In the proposed system, the security is ensured by additive homomorphic encryption and digital signature with verification. We used MATLAB along with a real-time dataset such as Auto1 dataset. For simulation, we used real-time environment. Therefore, we use certain factors for simulation such as MCS size, service provider location, number of participating workers, and the possibility of flattering CC (Fig. 8).

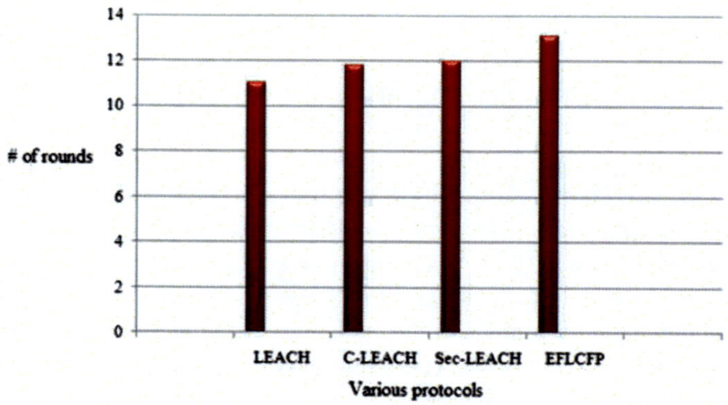

Fig. 8 Performance of various protocols

6 Conclusion

The MCS is the trendy platform used to outsource the data. Day-by-day numerous number of queries were posted by the requesters, and it is difficult to solve with tiny workers. In our paper, instead a selecting cluster cranium, we proposed EFLCFP for cluster formation. For ensuring the security in data aggregation and data reporting, we use an efficient protocol, namely SET-IBE and SET-IBOOS. In our proposed system, we also considered the query optimization in CMCS with CSELECT and CJOIN. The performance is measured based upon the monetary cost and latency. Our proposed work improves performance with respect to data transmission in MCS. In this work, we have also considered query optimization in terms of cost and latency for data retrieval.

References

1. Howe J (2006) The rise of crowdsourcing. Wired Mag 53(10):1–4
2. Fuchs-Kittowski F, Faust D (2014) Architecture of mobile crowdsourcing systems. In: Baloian N, Burstein F, Ogata H, Santoro F, Zurita G (eds) Collaboration and technology. Lecture notes in computer science. Springer, Cham, pp 121–136
3. Abououf M, Singh S, Otrok H, Mizouni R, Ouali A (2018) Galeshapley matching game selection—a framework for user satisfaction. IEEE Access 7:3694–3703
4. Fan J, Zhang M, Kok S, Lu M, Ooi BC (2015) Crowdop: query optimization for declarative crowdsourcing systems. IEEE Trans Knowl Data Eng 27(8):2078–2092
5. Parameswaran A, Park H, Garcia-Molina H, Poluzotis N, Widom J (2012) Deco: declarative crowdsourcing. In: CIKM. ACM, pp 1203–1212
6. Yuen M-C, King I, Leung KS (2012) Task recommendation in crowdsourcing systems. In: CrowdKDD'12, Beijing, 12 Aug 2012
7. Franklin MJ, Kossmann D, Kraska T, Ramesh S (2011) Crowd DB: answering queries with crowdsourcing. In: SIGMOD, pp 61–72

8. Hu S, Su L, Liu H, Wang H, Abdelzaher TF (2015) SmartRoad: smartphone-based crowd sensing for traffic regulator detection and identification. ACM Trans Sens Netw 11(4):1–27. https://doi.org/10.1145/2770876
9. Kantarci B, Mouftah HT (2014) Trustworthy sensing for public safety in cloud-centric Internet of Things. IEEE Internet Things J 1(4):360–368. https://doi.org/10.1109/JIOT.2014.2337886
10. Ran G, Zhang H (2010) Improving on LEACH protocol of wireless sensor networks using fuzzy logic. J Inf Comput Sci 7(3):767–775
11. Gupta I (2005) Cluster head election using fuzzy logic for wireless sensor networks. Halifax, Nova Scotia, Mar 2005
12. Feng W, Yan Z (2018) A survey on security, privacy, and trust in mobile crowdsourcing. IEEE Internet Things J 5(4)
13. Basu A, Vaidya J, Corena JC, Kiyomoto S, Marsh S, Guo G, Zhang J, Miyake Y (2014) Opinions of people: factoring in privacy and trust. SIGAPP Appl Comput Rev 14(3):7–21
14. Blasco S, Bustos-Jimenez J, Font G, Hevia A, Prato MG (2015) A three-layer approach for protecting smart-citizens privacy in crowdsensing projects. In: Proceedings of international conference of the Chilean computer science society (SCCC), Santiago, pp 1–5. https://doi.org/10.1109/SCCC.2015.7416585
15. Günther F, Manulis M, Peter A (2014) Privacy-enhanced participatory sensing with collusion resistance and data aggregation. In: Proceedings of international conference on cryptology and network security (CANS), Heraklion, pp 321–336. https://doi.org/10.1007/978-3-319-12280-9_21
16. Wu H-Q, Wang L, Xue G (2019) Privacy-aware task allocation and data aggregation in fog-assisted spatial crowdsourcing. IEEE Trans Netw Sci Eng. https://doi.org/10.1109/TNSE.2019.2892583
17. Paillier P (1999) Public-key cryptosystems based on composite degree residuosity classes. In: Eurocrypt
18. Fan J, Li Q, Cao G (2015) Privacy-aware and trustworthy data aggregation in mobile sensing. In: 2015 IEEE conference on communications and network security (CNS), Florence, pp 31–39. https://doi.org/10.1109/CNS.2015.7346807
19. Bhaskar N, Mohan Kumar P (2020) Optimal processing of nearest-neighbor user queries in crowd sourcing based on the whale optimization algorithm. Soft Comput. https://doi.org/10.1007/s00500-020-04722-0
20. Bhaskar N, Mohan Kumar P, Arokia Renjit J (2020) Evolutionary fuzzy-based gravitational search algorithm for query optimization in crowdsourcing system to minimize cost and latency. Comput Intell. https://doi.org/10.1111/coin.12382
21. Safran M, Che D (2016) Real-time recommendation algorithms for crowdsourcing systems. Appl Comput Inform. https://doi.org/10.1016/j.aci.2016.01.001
22. Kim J, Park S, Han Y, Chung T (2008) CHEF: cluster head election mechanism using fuzzy logic in wireless sensor networks. In: 10th international conference on advanced communication technology, ICACT 2008, pp 654–659
23. Gheisari M, Abbasi AA, Sayari Z, Rizvi Q, Asheralieva A, Banu S, Awaysheh FM, Shah SBH, Raza KA (2020) A survey on clustering algorithms in wireless sensor networks: challenges, research, and trends. In: International computer symposium (ICS). https://doi.org/10.1109/ICS51289.2020.00065
24. Zhang J, Sheng VS, Wu J (2019) Crowdsourced label aggregation using bilayer collaborative clustering. IEEE Trans Neural Netw Learn Syst 30(10)
25. Adil M, Khan R, Ali J, Roh B-H, Ta QTH, Almaiah MA (2020) An energy proficient load balancing routing scheme for wireless sensor networks to maximize their lifespan in an operational environment. IEEE Access 8
26. Wu C-C, Chen K-T, Chang Y-C, Lei C-L (2013) Crowdsourcing multimedia QoE evaluation: a trusted framework. IEEE Trans Multimedia 15(5):1121–1137. https://doi.org/10.1109/TMM.2013.2241043
27. Baveye Y, Dellandréa E, Chamaret C, Chen L (2015) LIRISACCEDE: a video database for affective content analysis. IEEE Trans Affect Comput 6(1):43–55. https://doi.org/10.1109/TAFFC.2015.2396531

28. Joe CV, Raj JS (2021) Location-based orientation context dependent recommender system for users. J Trends Comput Sci Smart Technol (TCSST) 3(01):14–23
29. Mhemed R, Aslam N, Phillips W, Comeau F (2012) An energy efficient fuzzy logic cluster formation protocol in wireless sensor networks. In: The 3rd international conference on ambient systems, networks and technologies (ANT), 1877-0509 © 2012. Published by Elsevier Ltd. https://doi.org/10.1016/j.procs.2012.06.035
30. Heinzelman W, Chandrakasan A, Balakrishnan H (2002) An application-specific protocol architecture for wireless microsensor networks. IEEE Trans Wireless Commun 1(4):660–670
31. Mamdani EH (1977) Application of fuzzy logic to approximate reasoning using linguistic synthesis. IEEE Trans Comput C-26:1182–1191
32. Pithva B, Pattani K, Christian A (2014) Optimization of leach protocol in wireless sensor network. Int J Comput Appl 93(12)
33. Oliveira LB, Ferreira A, Vilaca MA et al (2007) SecLEACH—on the security of clustered sensor networks. Signal Process 87
34. Banerjee P, Jacobson D, Lahiri S (2007) Security and performance analysis of a secure clustering protocol for sensor networks. In: Proceedings of IEEE NCA
35. Zhang K, Wang C, Wang C (2008) A secure routing protocol for cluster-based wireless sensor networks using group key management. In: Proceedings of WiCOM
36. Lu H, Li J, Guizani M (2014) Secure and efficient data transmission for cluster-based wireless sensor networks. IEEE Trans Parallel Distrib Syst 25(3)
37. Galindo D (2005) Boneh-Franklin identity based encryption revisited. In: Caires L, Italiano GF, Monteiro L, Palamidessi C, Yung M (eds) Automata, languages and programming. ICALP 2005. Lecture notes in computer science, vol 3580. Springer, Berlin, Heidelberg. https://doi.org/10.1007/11523468_64
38. Castelluccia C, Mykletun E, Tsudik G (2005) Efficient Aggregation of encrypted data in Wireless Sensor Networks. The Second Annual International Conference on Mobile and Ubiquitous Systems: Networking and Services. http://doi.org/10.1109/MOBIQUITOUS.2005.25
39. Renukadevi M, Mary Anita EA, Mohanageetha D (2021) An efficient privacy-preserving model based on OMFTSA for query optimization in crowdsourcing. Concurr Comput Pract Exp. https://doi.org/10.1002/cpe.6447
40. Azad P, Sharma V (2013) Cluster head selection in wireless sensor networks under fuzzy environment. ISRN Sens Netw 2013. https://doi.org/10.1155/2013/909086

Utilization of Adaptive K-Harmonic Means Clustering and Trust Establishment in VANETs

K. M. Jini, J. Senthilkumar, Y. Suresh, and V. Mohanraj

Abstract A Vehicular Ad-hoc Network (VANET) can be portrayed as an exceptional kind of Mobile Ad-hoc Network (MANET), which utilizes vehicles to make a versatile system. Vehicular Ad-hoc Networks (VANETs) are a gathering of smart vehicles intends to deal with transportation issues and decrease inadvertent proportion. Because of the high portability and dynamism, steering the messages to their last goal in VANETs is a difficult assignment. VANET is the eventual fate of the transportation framework as every one of the vehicles in system speaks with one another and consequently evades a sheer measure of mishaps on streets. The principal disadvantage of VANET is the system flimsiness, which lessen the system productivity. These issues can be tended to by clustering procedures. Clustering is a strategy utilized in system directing to improve the execution and save the system assets. Bunching alludes to the characterization of vehicles in gatherings so as to limit the traffic overhead by choosing a dependable hub called cluster head. Huge numbers of the VANET grouping calculations are gotten from Mobile Ad-hoc Network (MANET). Development of stable clusters withstanding the portability example of the vehicles is an essential key factor for grouping calculations in vehicular systems. To help productive information collaboration among vehicles, bunching based topology can be connected which bunches vehicle hubs in a topographical region together and bolsters the direct cooperation inside one group and between group information communication through cluster heads (CHs). Both K-Means and K-harmonic means (KHM) calculations are ordinarily utilized clustering calculations for remote sensor systems, in any case, these calculations can not be connected to VANETs legitimately because of the particular attributes of VANETs. The adaptive K-harmonic means clustering is proposed. Because of its extraordinary advantages, it is exceptionally helpless against different assaults and security in VANET ought to be mulled over. Foundation of trust is among the most basic parts of any framework's security. For any system, trust alludes to a lot of connections among the elements taking an

K. M. Jini (✉)
Anna University, Chennai, India

J. Senthilkumar · Y. Suresh · V. Mohanraj
IT, Sona College of Technology, Salem, Tamil Nadu, India

S. Shakya et al. (eds.), *Proceedings of Second International Conference on Sustainable Expert Systems*, Lecture Notes in Networks and Systems 351,
https://doi.org/10.1007/978-981-16-7657-4_36

interest in the system activities. The apprehension utilizing static framework is the inaccessibility of fixed foundation in certain areas. The primary goal of this work is to portray clustering and trust foundation approaches for VANET.

Keywords VANET · Clustering · Mobility · Trust establishment

1 Introduction

The fast increment in street traffic causes a great deal of street mishaps. With the quick developing of cooperative and transformative advances in the current computerized age, connected is a popular expression today. The headways in data innovation and the wide utilization of remote correspondences have pulled in much consideration of the specialists towards information transmission between the vehicles and the road side units (RSUs), which is known as Vehicular Ad-hoc Networks (VANETs). Vehicular Ad-hoc Network (VANET) includes vehicular hubs associated by means of remote connections in an Ad-hoc structure. The Internet-associated cars are aftereffect of another developing innovation called Vehicular Ad-hoc Network (VANET). VANETs can be broadly utilized in security, traffic data, and other business applications, for example, a street side eatery promotion, advanced amusement, and so forth [1].

VANET can likewise give driver-less vehicles which will be more secure, causing less mishaps and less clog. In VANET, all vehicles in the system transform themselves into a remote hub. A vehicle is a vigorous and stable stage which can undoubtedly bear the cost of the sort of preparing, detecting, stockpiling, and vitality capacities important to help nontrivial remote applications. In this manner, it offers extensive points of interest contrasted with other MANET stages, in light of the fact that the applications are not going to be obliged by the vitality prerequisites or weight of the essential gear. Hubs can distinguish each other with the assistance of sensors inbuilt in vehicles and being able to recognize vehicles around them.

At the point when vehicles are in the system, they send messages with respect to their own data, street data, unnatural exercises on the street, and so forth to different parts of the system. In VANET, vehicles are outfitted with an On-Board Units (OBUs) that can speak with one another, which called vehicle-to-vehicle (V2V) interchanges, or/and with stationary street foundation units, which called vehicle-to-infrastructure (V2I) correspondences. VANET has a few attributes that make it not quite the same as MANET, for example, high hub portability, unsurprising and limited versatility design, fast system topology change, and boundless power supply. Notwithstanding, when vehicles become inaccessible to other people or leave the system, different vehicles can join the system bringing about a versatile system with hubs as vehicles associating with each other [2].

In VANET vehicles trade information to encourage route arranging, street well-being, and other non-security applications. These interchanges are done through the remote medium given by the WiFi's. A vehicle can speak with another vehicle straightforwardly or vehicle can impart to road side unit. VANET demand agreeable correspondence with neighboring hubs in an activation domain of high versatility and powerfully evolving topology. These difficulties can be tended to by cluster arrangement. Go through the difficulties of clustering topology along with the executive's plan to be connected over VANET. It can grouped together nearby vehicles according to the same cluster. Implementing transmission node within the layered design, the transfer of data from one node to another in the VANET perhaps raised [3].

As a standout amongst the most generally utilized clustering calculations, K-Means calculation utilizes the aggregate inside cluster difference preferentially complete mean-square quantization error as for gauge decency based on clustering. The K-Means calculation is the disadvantage so as to execution about calculation be delicate before reinstatement as for centroids. Stretch this disadvantage, K-Means strategy is proposed and it is called KHM calculation, these figures effective execution work be subject to symphonious midpoints about separations of every node before focuses, also the goal that the execution of the calculation is basically coldhearted to the introduction of the centroids. Subsequent to clustering, VANET security is significant in light of the fact that an ineffectively planned VANET is powerless against system assaults and this bargains the wellbeing of drivers. Security frameworks ought to guarantee that transmission originates from a confided in source and is not altered in transit by different sources. Trust and security are two reliant ideas that can not be de-isolated. Subsequently, trust esteems assurance assumes a vital job for the development of system security and unwavering quality. Trust foundation procedure can be parceled into two classes: infrastructure-based trust and self-sorted out trust.

2 Related Work

Scientists across the globe proposed many clustering plans for VANET. These grouping systems center around different points and even use bunching plans proposed for MANET for group development. Alongside the bunching technique, security issues are considered and the arrangements are given for those issues.

Tao et al. [4] proposed a Cluster-Based Directional Routing Protocol (CBDRP) for the highway situation, in which the current CH chooses new CH that moves a similar way. The headers which are nearest to the middle coordination of the clusters are chosen as CH. This convention demonstrates noteworthy improvement contrasted with AODV and identical with GPSR as far as transmission execution. They talked about a system where cluster head is chosen relying upon the path having the most extreme traffic stream. Vehicles have the learning of the path of traffic on street and they communicate this data to the adjacent vehicles. The mobility dataset generation method of social vehicles traveling based on floating car data and social functional areas [5].

Khan et al. [6] overviewed a strategy to recognize mischief or malevolent hubs. It makes a dependable and secure condition for correspondence. Jerbi et al. [1] proposed an improved insatiable traffic-mindful directing convention which is a crossing point based geological steering convention. It utilizes clustering idea between nearby convergence to advance the information parcel, and it likewise thinks about the separation to group focus to choose the CH. The CH is utilized to assemble a portion of the traffic insights for current cluster to figure the general expense for the whole fragment between each nearby convergence [7]. Advanced randomized authentication system protocol prevents vehicles from being tracked by any single party including peer vehicles, service providers, authentication servers, and other infrastructure [8]. The clustering algorithm is provided according to the highway road and the algorithm searches to increase the network stability by recognising the cluster head. Than using a recommended variable range of cluster head to be a measure of finding the CH for each new vehicle.

Raya [2] exhibited a strategy to shield the security of vehicles from different vehicles by creating the counterfeit personality of vehicles. Chuang and Lee [9] proposed a plan TEAM to upgrade the confirmation among the vehicles in less time and figure the keys by utilizing hashing systems. B. They exhibited a CBR called CBVANET for basic expressway correspondence. This model has concentrated on the improvement of clustering system for correspondence among the VANET hubs. This model diminished the idleness in VANET, by decreasing the clustering arrangement time, CH race time, and cluster exchanging time. The hub with least speed was picked as the CH. The proposed convention beat existing plans as far as the creation time and exchanging time. Umer et al. [10] proposed multi-hop communication system via dual ring message delivery scheme on the behalf of clustering head based routing system to enhance network coverage, and connectivity systems. Wu et al. [11] used a fuzzy logic algorithm in the initial part of clustering to generate a stable cluster head nodes while taking into account vehicle velocity, vehicle distribution and link quality between vehicles. Than other one clustering to tune the number of gateway nodes in order to get the high overall network performance under different network conditions [12]. Introduces the concept of finding traffic congestion and the shortest path for data transmission between nodes.

Rong et al. [3] proposed another clustering-based calculation for VANET, clustering-based plan is connected for information transmission in VANET, and the proposed CH determination calculation mutually thinks about hub degree (the quantity of legitimately associated neighbors), the accessibility assets of the applicant CHs and others cluster members (CMs), and the speed distinction between hopeful CHs and different CMs in a given group. They gave the essential idea of the framework and it was set up by making the period of key through and through, for example, TA to vehicles. At the point when vehicles communicate the messages with endorsements, the message must be approved among vehicles by authentications. Monir and Abdel [13] proposed an instrument to build up the trust by utilizing trust among the hubs. Each hub is checked and after that ascertains the trust estimation of that hub. Hortelano et al. [14] proposed a plan to identify the conduct of the neighbor hub by watching them consistently. It utilized interruption identification of a watch dog to separate the pernicious hub. Kandali and Bennis [15] proposed routing scheme protocol based on K-harmonic means clustering algorithm to enhance the lifespan of the node in between the vehicles and to improve the efficiency of the vehicular networks. The impact of the different VANET size when changing VANET topology during the collaborative learning is identified during the study of privacy preserving technique [16]. Chen et al. [17] proposed to overcome the local maximum congestion problem, and improve the network efficiency.

3 Materials and Methods

A prevalent procedure of presenting stable structures in generally shaky systems, including VANETs, is to enable the versatile hubs to self-arrange into groups. Those clusters at that point decrease the weight of more elevated amount conventions on the worldwide populace of individual hubs. For instance, they give a chain of command inside the system which assists with steering. Furthermore, clustering can be utilized as a way to build hub coordination and lessening the number of hubs meddling with one another, and furthermore to alleviate the shrouded terminal issue. Clustering alludes to the arrangement of vehicles in gatherings so as to limit the traffic overhead by choosing a capable hub called cluster head. In this work, another group based directing convention for VANETs are presented. The convention is isolated into three phases. The main stage is identified with the development of bunch/gatherings of vehicles [18].

KHM calculate that the details of execution work subject to consonant mid-points of separations from every point to the focuses, with the goal so as execution before calculation like basically inhumane as for reinstatement as for centroids. K-harmonic means calculation can not a connected via VANETs legitimately by number of groups ought to a resolved ahead of time and ought to been fixed cluster, that is illogical of VANET applications. Moreover, the remote transmission scope of hubs is not specified in KHM, in this manner the productive information collaboration linking cluster members and CHs in VANETs can not be ensured. These work, concentrate that the specific qualities in VANETs, and focusing on the issues in KHM calculation in Vehicular Ad-hoc Network, AKHM grouping technique in VANETs is proposed.

Because in a substantial number of self-governing system individuals and the nearness of human factor, the misconduct of hubs in future vehicular systems can not be precluded. A few sorts of assaults have been distinguished and grouped based on layer utilized by the assailant. At the physical and connection layers, an aggressor can exasperate the system framework by overburdening the correspondence channel with garbage messages. The assailant can infuse false messages or rebroadcast an old message moreover. A few aggressors can corrupt with an On-Board Unit (OBU) or annihilate road side unit (RSU). At the system layer, an assailant can embed false directing messages or over-burden the framework with steering data. Protection of drivers can be unveiled by uncovering and following the situation of drivers. Consequently, trust foundation must be made for correspondence subsequent to clustering process.

3.1 AKHM Algorithm for VANET

To accomplish ensured transmission execution, the cluster heads group comprises the vehicle-to-vehicle communication data transfer capacity requirement is shaped and the underlying estimations in different groups than places at every centroid to be grouped, at that point there is an unequal separation among vehicles to centroids has characterized, in view of which the target work is planned and the ideal CHs are resolved. When group oriented system design, cluster heads generally function in hand-off vehicles for cluster members (CMs) inside each cluster [3].

Cluster members advance information gets through clusters that is one cluster head wants the value for some measure on assets. When the transmission bandwidth plays an important role in determining service transmission execution services determined by transmission bandwidth and than huge bandwidth along with vehicles to recommend maximum while selecting cluster heads, forwarding packets of one another vehicle. The underlying centroid appropriation is given in Fig. 1.

K-harmonic means calculation when different groups along with its situation among the centroids ought to become resolved ahead of time. There is no loss in sweeping statement, vehicles are approximately straight in the street. Consequently, different groups K_{Ini} are evaluated as pursues:

$$K_{\text{Ini}} = \max\left\{\left\lceil \frac{L}{2R} \right\rceil \cdot \left\lceil \frac{N}{N_{\max}} \right\rceil\right\}$$

Hence, L signifies length and R means the transmission span of vehicles, and $N_{\max}$ indicates each cluster having high quantity cluster members. when length on street L, total K clusters which is equivalent to the quantity of beginning centroids, consistently disseminated among street by picked along with centroids, when separation link with two adjoining centroids and to be equivalent at L/K. In Fig. 1, demonstrates the situation of starting centroids on a street with the length being L and the underlying number of clusters being K.

The target capacity of A_{KHM} bunching is given as,

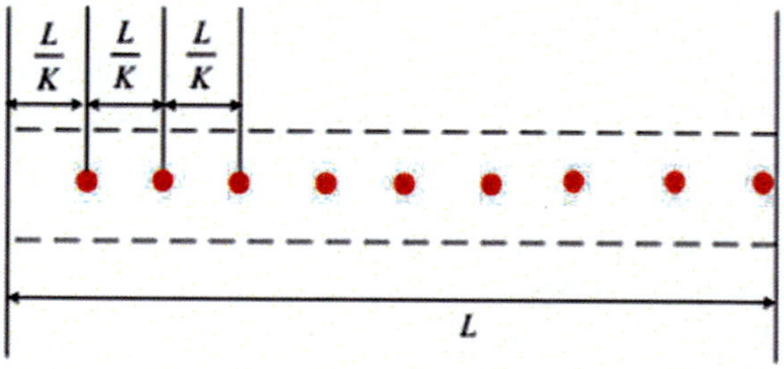

Fig. 1 Initial centroid distribution

$$f_{\text{AKHM}} = \sum_{i=1}^{N} \frac{K}{\sum_{k=1}^{K} \frac{1}{D_{ik}^2}}$$

To determine the ideal position of the centroids, halfway subordinate of f_{AKHM} is taken as for the situation of the centroids,

$$\frac{\partial f}{\partial x_k^c} = -K \sum_{i=1}^{N} \frac{4\left((x_i - x_k^c) + (v_i - v_k^c)\right)}{D_{i,k}^3\left(\sum_{k=1}^{K} \frac{1}{D_{i,k}^2}\right)}$$

The optimal position of *k*th centroid is given by,

$$x_k^c = \frac{\sum_{i=1}^{N} \frac{x_i + v_i}{D_{i,k}^3\left(\sum_{k=1}^{K} \frac{1}{D_{i,k}^2}\right)^2}}{\sum_{i=1}^{N} \frac{1}{D_{i,k}^3\left(\sum_{k=1}^{K} \frac{1}{D_{i,k}^2}\right)^2}}$$

Like the conventional KHM calculation, the worldwide ideal centroid position may resolved via cycle, when cluster heads would capable to be picked by vehicles and the nearest of centroids, vehicles may related for adjoining cluster heads in like manner. Be that as it may, while conventional KHM calculation will not take any transmission scope for hubs and these problem must take an arranging vehicles into specific CHs for down to earth VANET applications. For the situation that the *K* centroids neglect to cover every one of the vehicles, new clusters must be shaped.

3.2 Trust Establishment

There are different strategies for foundation by trust in specially appointed systems. For instance, a hub can build up trust intermittently or on interest premise. Direct trust is set up with one-jump neighbors and aberrant trust is set up with different hubs as in specially appointed systems. In certain methodologies, few substances is required. One methodology centers on confiding in messages as opposed to their characteristics. Figure 2 demonstrates the scientific classification of trust foundation strategies.

3.2.1 Infrastructure Based Trust

Believed frameworks can keep any sort of assault on VANET. The framework utilize secure and confided in correspondence foundation that can fulfill a lot of security necessities: confirmation, respectability, accessibility, no revocation, and protection.

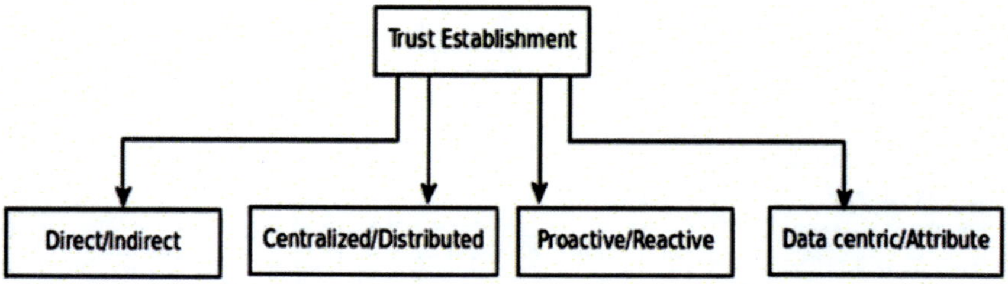

Fig. 2 Trust establishment

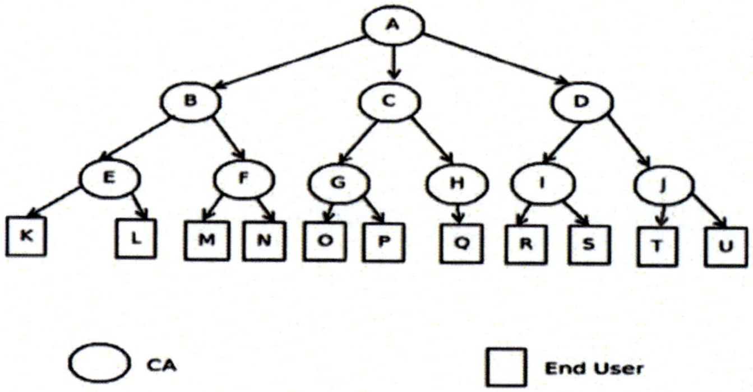

Fig. 3 Infrastructure based trust

Fundamentally various leveled trust demonstrate is followed for this situation as appeared in Fig. 3.

Trust choices must be made independently in light of the fact that fixed security framework is not ensured consistently in VANET. Choices must be founded on the fractional data gathered for a brief span from obscure hubs [13].

3.2.2 Self Organized Trust

Self sorted out trust foundation is required as appeared in Fig. 4 in view of non-accessibility of framework and shared worldwide information among the participating hubs. Self organizing trust foundation component can be immediate, aberrant and half breed. In direct instrument, trust is set up by direct correspondence of hubs, for example, the hubs trade trust esteems with every one of the hubs inside its correspondence go. Trust connections are transitive in backhanded system. For instance hub is a trading trust relations with its one jump neighbor B. Presently while correspondence of hubs B and C, B is additionally trading the trust relations of hub A separated from its own trust relations. So, this trust foundation component is roundabout. Cross breed instrument can likewise be utilized which is a mix of both immediate and roundabout trust foundation components.

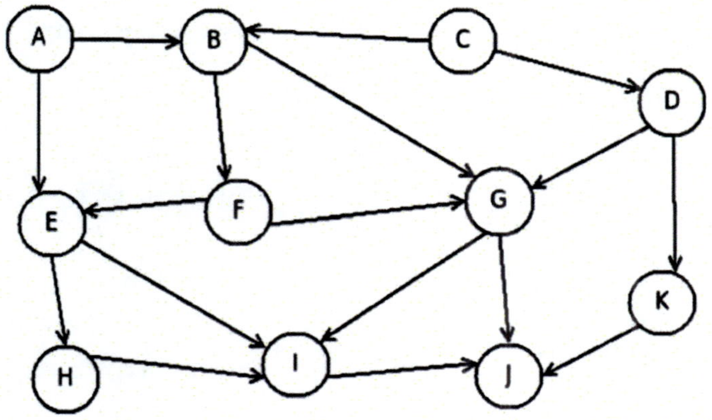

Fig. 4 Self organized trust

4 Results and Discussion

The above method is used to improve the scalability and routing in the network. The delay and throughput is determined and the results are shown in Tables 1 and 2 as well as Figs. 5 and 6.

The results show that the throughput rate is higher in adaptive K-means harmonic clustering method and when that is combined with trust establishment. Similarly, the average delay is quite less when the security procedure is integrated with AKMH method.

Table 1 Average throughput

S. No	Time (s)	Average throughput	
		AKMH + trust	KHM + trust
1	12	0.42	0.32
2	14	0.31	0.22
3	16	0.22	0.12

Table 2 Average delay

S. No	Time (s)	Average delay (ms)	
		AKMH + trust	KHM + trust
1	12	14	17
2	14	15	19
3	16	17	22

Fig. 5 Average throughput

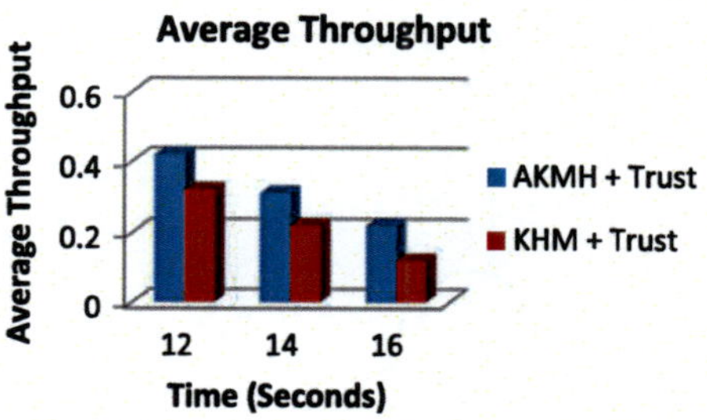

Fig. 6 Average delay

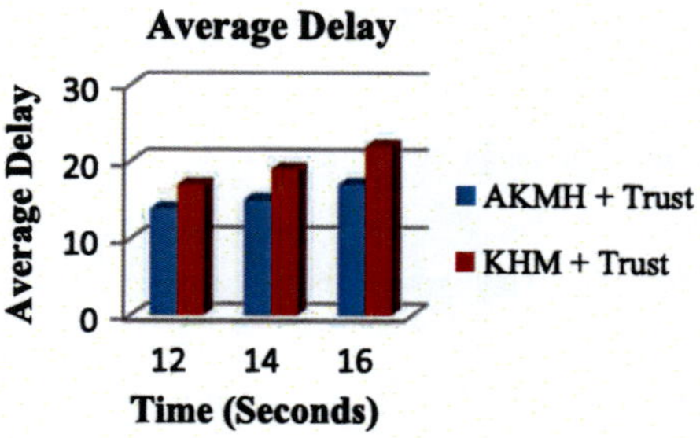

5 Conclusion

Hence the work, focusing on disadvantages for K-harmonic means calculation, modern cluster calculation has proposed in VANET and it is mutually take that versatility attributes for vehicles, then transmission capacity for competitor cluster heads. The outcomes demonstrate when proposed calculation executes superior to conventional K-harmonic means calculation particularly regarding being obtuse for underlying different groups, beginning places for centroids than provides 100% inclusion rate. VANET trust system depends on appropriated and secluded design. Every part of believed structure lives in each hub and performs set of activities to assess the notoriety of different hubs. It fuses self confirmations, suggestions, and experience measurements for assessing the trust dimension of different hubs. This structure is utilized to distinguish pernicious, narrow minded and temperamental conduct of hubs speaking with one another. In this system, it gives input to different hubs in the system. Trust foundation structure for VANET is lightweight since it does not perform broad hazard and conduct examination for dependability appraisal.

References

1. Jerbi M, Senouci SM, Rasheed T, Ghamri-Doudane Y (2009) Towards efficient geographic routing in urban vehicular networks. IEEE Trans Veh Technol 58
2. Raya M (2005) The security of vehicular ad-hoc networks. SASN, Alexandria, pp 11–21

3. Rong C, Bin Y, Lifan L, Xiao S, Qianbin C (2013) Clustering-based data transmission algorithms for VANET. In: International conference on wireless communications & signal processing (WCSP), pp 1–6
4. Tao S, Weiwei X, Tiecheng S, Lianfeng S (2010) A cluster-based directional routing protocol in VANET. In: 12th IEEE international conference on communication technology (ICCT), pp 1172–1175
5. Kong X, Xia F, Ning Z, Rahim A, Cai Y, Gao Z, Ma J (2017) Mobility dataset generation for vehicular social networks based on floating car data. IEEE Trans Veh Technol 1–13
6. Khan U, Agrawal S, Silakari S (2015) A detailed survey on misbehavior node detection techniques in vehicular ad hoc networks. In: Advances in intelligent systems and computing, vol 339, pp 11–19
7. Kang J, Lin D, Jiang W, Bertino E (2018) Highly efficient randomized authentication in VANETs. Pervasive Mob Comput 31–44
8. Hasson ST, Abbas AT (2021) A clustering approach to model the data dissemination in VANETs. In: IEEE Babylon international conference on information technology and science (BICITS), pp 337–342
9. Chuang M-C, Le J-F (2013) TEAM: trust-extended authentication mechanism for vehicular ad-hoc networks. IEEE
10. Umer T, Afzal MK, Munir EU, Alam M (2018) A dual ring connectivity model for VANET under heterogeneous traffic flow. Wireless Pers Commun 105–118
11. Wu C, Yoshinaga T, Chen X, Zhang L, Ji Y (2018) Cluster-based content distribution integrating LTE and IEEE 802.11p with fuzzy logic and Q-learning. IEEE Comput Intell Mag 41–50
12. Rathore NPS, Yadav A, Chauhan A (2020) Multi-agent based stable clustering and collision detection in VANET. In: International conference on intelligent computing and control systems (ICICCS), pp 1147–1150
13. Monir M, Abdel A (2013) A categorized trust based message reporting scheme for VANETs. In: Advances in security of information and communication networks, pp 65–83
14. Hortelano J, Ruiz JC, Manzoni P (2010) Evaluating the usefulness of watchdogs for intrusion detection in VANETs. In: IEEE international conference on communications workshops (ICC), pp 1–5
15. Kandali K, Bennis H (2020) A routing scheme using an adaptive K-harmonic means clustering for VANETs. In: Intelligent systems and computer vision (ISCV). IEEE, pp 1–5
16. Zhang T, Zhu Q (2018) Distributed privacy-preserving collaborative intrusion detection systems for VANETs. IEEE Trans Signal Inf Process Netw 1–14
17. Chen J, Rajib P, Choi Y-J (2021) An efficient neural network-based next-hop selection strategy for multi-hop VANETs. In: IEEE international conference on information networking (ICOIN), pp 699–702
18. Runkler T (2011) Partially supervised k-harmonic means clustering. In: IEEE symposium on computational intelligence and data mining (CIDM), pp 96–103

Design and Development of Blockchain-Based Security and Privacy-Preserving System

Nilima V. Pardakhe and V. M. Deshmukh

Abstract Blockchain offers the potential for creating the latest business models and trust issues can be solved in a structured way. It conducts many business innovations and research opportunities. Industry and academia present many solutions of blockchain root software, including the extensive reach of domains. A system design is shown in the paper where blockchain technology will be present in the healthcare system. In this, the essential information among the hospitals is going to be shared concerning medical analysis. Patients define research institutes and medical clinics root on approach policy. For the protection of confidential data, 2 types of chains are used by our solutions—a private one. It is also known as a sidechain that stores a patient's accurate ID data. The public one is also known as the mainchain that keeps patients' temporary ID health marked data. With the use of the Hyperledger framework, we have developed a design to test it. The present result of the experiment shows the better performance of a system concerning the given metrics: Nikoloudakis et al. (Peer-to-Peer Netw Appl J 2019 [1]) to recognize the patient's medical data time required by observing the turnaround time. It shows that the proposed approach gives better performance and security.

Keywords Blockchain · Security · Medical data · Healthcare

1 Introduction

Blockchain is the latest technology that allows value sharing. We have seen that it is used by various areas from some of the years, especially the financial areas. Blockchain can be said as a digital ledger in that all implemented transactions are kept. Blocks refer to squint networks for constructing an ongoing list of recorded orders which is used by blockchain. Each block restrains signed transactions set and authenticated itself by a network with the help of the consensus mechanism. In a network, blockchain copies are issued on every node participating. Blockchain

N. V. Pardakhe (✉) · V. M. Deshmukh
P.R.M.I.T.&R., Badnera, Amravati, India

S. Shakya et al. (eds.), *Proceedings of Second International Conference on Sustainable Expert Systems*, Lecture Notes in Networks and Systems 351,
https://doi.org/10.1007/978-981-16-7657-4_37

is perhaps examined as the permanent database; alteration is made to prevent the executed algorithms in previously kept information. For any system, in charge of storing and handling medical data, users' rights are taken into account that legislation in force imposed like personal data protection. The entity which proceeds health/personal data GDPR applies to it. Even though it was founded in the European Union or founded outward, the European Union moves people's data from the EU. It is a challenge to implement these rules in the blockchain, bearing in mind that the fact was created as an immutable ledger implemented in most blockchains.

Healthcare IT system innovative model is suggested in this that is root on privacy-preserving. They can donate namelessly to determined statistics. For building timestamped, append-only, immutable content chain, the blockchain uses cryptography of public keys. The design of our system consists of 2 types of blockchain—a private sidechain and a public mainchain. The hinge on the nodes types untrusted or trusted; everyone has a duplicate of both the blockchains. Because of the generation of abundant data and privacy reasons by every engaging device and patient. The node's content is set up by links to other auxiliary, permissions, and health data information.

The organization of the article is as follows. The following section related work overview is presented in the extent of solutions based on blockchain for healthcare. The design of the suggested system is described in Sect. 3. Section 4 describes the results of experiments to access the presentation of the proposed approach. In the last, the article concludes with the final section.

2 Related Works

In healthcare, we have proposed countless solutions based on blockchain to integrate patients, doctors, and clinics that target to provide better quality services on time.

Within [2], an application is granted by the writer authorizing MedRec. Proof-of-work is a mining algorithm, and for the contribution in accumulate undisclosed mining data, medical collaborators are awarded as the reward for mining. In real life, if we use such a system where anonymous personal data is used as a motive apart from the original one, it regulates the clients for not using such services. Additionally, the patient's identity can be revealed when it cannot be used as an organized anonymization process. Then the personal data protection is not met.

In [3] a blockchain-based architecture is presented. In that, the critical public infrastructure is used by the system. The identities are generated for facilitating data sharing for clinicians and for making preferable decisions for the patients. However, it violates personal data protection.

In [4], it describes an attribute-based signature system with many jurisdictions. In that, a message is endorsed by the patient without relieving any data following the attribute. The plan's design considers succeeding entities—patients, data verifiers, many *N* authorities, and a server clinic node generates the storage transactions by different firms in the medical system like medical insurance companies, hospitals, medical research institutes, etc. and there is a role of registration acceptation and medical data exchange of patients. Moreover, keeping medical data for all the patients fails when an organization is not joined to the network if we use a single server.

Within [5] suggested that offers sharing of medical data amidst custodians of big medical data in the suspicious environment. The key motive of blockchain is to support an immutable database where the data request is stored and actions are linked to delivery. The writer launched a specific type of child block, joined with the upper block. Its character is to retain the logs generated with the appeal of distinct organizations by intelligent contracts. The user signed every data request, and the authenticator checks it. When it seems to be valid then requested, the requestor receives information. It is operated with decryption of data, and data will be monitored. Consensus and processing nodes receive data handling information attached to the compatible parent block, which holds the specific data requests as a side block. A blockchain-based system.

Rifi et al. [6], the blockchain-based healthcare system is proposed by it combines hospitals, doctors, medical sensors, and patients. For increasing the presentation, data is kept outside the blockchain. On the other hand, the use of blockchain is to keep data parts only or its pointer. Another node name is the gateway with sufficient energy and power and can be mobile phones and laptops. A medical sensor transmits the information with the help of a gateway in the system. The use of smart contracts is to conserve access policy like doctor–patient relationship, but the way of accomplishment is not described.

In [7], the author proposed the faith is the root of network consensus that is in accord with semantic and structural interoperability evidence. At the time of the process, proof of interoperability, the pitman examines that inspection is interoperable with structural and semantic constraints. That's why the patient's medical analysis is envisaged by other organizations apart from medical clinics where there is the performance of the research. The author is using intelligent contracts like security policies unaccompanied by many details.

Kavitha and Vidhya Saraswathi [8] address various methods of encryption and decryption of images in this paper. Today, image security is extremely important. Various encryption methods are researched and analyzed to support the encryption recital. The original picture is incorporated and encrypted in all processes and then sent to the recipient. Every algorithm is special, the method is uniquely implemented. The new technique of encryption is changing every day. High-security encryption methods are still working out.

Tahir and Javaid [9] the data is the most precious commodity, they claim. Internet markets are increasingly exchanging data. These markets allow data owners to publish their information sets and data users to find suitable services. Data is a special commodity, however, rather than conventional items such as clothes and food. It is very difficult to protect copyright and privacy for emerging data exchange markets. Moreover, the management of data resources requires specific IT techniques, something that is difficult for many organizations, including hospitals, government agencies, planetariums, and banks, that have big datasets. A decentralized way to share big data is proposed in this paper. This approach aims to build an environment where all participants will participate in the peer-to-peer sharing of data. The main part is to use blockchain technology to archive transaction records and other important documents. Our solution does not need third parties, unlike current data exchange markets. It also provides data owners with a simple way to verify the use of data and protect the rights and privacy of data. They will discuss the environment and the technological problems and solutions.

Bandara et al. [10] with Mystiko, a high transaction capacity, high scalability, and high availability blockchain has been developed. The distributed storage is used as the background storage framework by Apache Cassandra. We introduced full text search functionality to Mystiko by indexing transactions and blocks in Elasticsearch. Our micro-service architecture from Docker and Kubernetes makes scaling simpler and more scalable. Mystiko is a storage system with these features that supports data. It has been demonstrated with empirical tests in terms of scalability and transaction throughput. Mystiko has been integrated in the banking and financial fields of development graduation applications. The implementations are Mystiko as an ideal Big Data and Cloud Storage blockchain network.

Sokouti et al. [11], the Goldreich-Goldwasser-Halevi (GGH) algorithm is used in the numerical frames for the encryption of medical images. Thus, both algorithms and confidential information may be used. In addition, the GGH algorithm does not increase the image size and thus its complexity remains as straightforward as O(n2). However, the chosen-ciphertext attack is one of the drawbacks of using the GGH algorithm. This deficiency in the GGH algorithm was resolved and improved in our strategy by using padding (i.e., snail tour XORing) before the GGH encoding process. Three measurement parameters are considered for measuring their efficiency, including (i) the number of pixels changing rate (NPCR), (ii) unified average intensity change (UACI) and (iii) the effect of avalanche. The tests on three different imagery sizes showed that the GGH padding strategy, as opposed to the regular GGH algorithm, has improved by approximately 100, 35, and 45%. The findings also allow the padding of the GGH resist the cipher text, the cipher text selected and statistical attacks. However, the improvement of more than 50% in the avalanche effects is a positive achievement compared to the more complicated encryption and decryption processes of the proposed system.

Ramani et al. [12], the author discusses secure mechanisms for access to information, which can provide patients with access only by authorized entities. This paper therefore considers that blockchain technology safeguards data on health care systems as a distributed approach. This work proposes that patients and doctors in a given medical system have a safe and efficient data accessibility mechanism based on blockchain. A program that can also protect patient's privacy is also suggested. Our security review shows that it can withstand common attacks and maintain system integrity. In order to check the viability of our proposed program, an Ethereum-based implementation was also used.

Explanting EMRs through decentralized hospitals, this is hampering knowledge exchange and jeopardizing the privacy of patients [13, 14]. In order to solve these problems, we suggest a data sharing network known as BPDS based on blockchain privacy. In BPDS, the initial EMRs are safely stored in the cloud and the indexes are held in a blockchain consortium which is manipulated. This will dramatically reduce the risk of medical data leakage while, at the same time, ensuring that blockchain indexes cannot be unilaterally updated throughout the EMRs. According to the predefined patient access privileges, the secure data sharing can be done automatically via the intelligent blockchain contracts. Furthermore, the CPABE-based framework for access controls and the Content Extraction Signature Scheme ensure that data sharing maintains good privacy. Security analysis shows that BPDS is a secure and efficient way to share EMR data.

3 Proposed Work

The literature review has not provided concrete information in the section. The medical analysis is known as personal data, which is accessible only to patients and established organizations. The patient's ID is stored by reviewed proposals in openly attainable blockchain transactions. Hence, make patients unsafe for tracking the records of medical services.

The certainty is provided to the patient by our approach that the information will stay safe and their privacy. The patient has been put at first in the proposed solution so that the security policy will retrieve the medical data. The transactions of policy are attainable by the blockchain, which is available publicly. Although the public blockchain carries the only temporary ID of the patient, it results in that it is not at all permitted to track the patient's history of medical treatment. Further, patients are reviewed as to their medical analysis owners in our designed system. Thus, they are having complete control over it, although they are kept in clinics' datacenters.

3.1 Working of Proposed System

We represent the design in the section of the security system that guarantees the authenticity and confidentiality of patients' medical data. The data is contained by the system like organizations that have approaches to data and their types, the patients and doctors in the design, medical information of patients, and obtain policies of these types of data. In this, patients are said to be possessors of their medical data, which results in complete control over them and could apply for approach policy at any moment.

The proposal is made based on the blockchain network for the e-healthcare record-sharing strategy. Therefore, an EHR system sharing design blockchain-based is presented. Different network block transfer techniques and settings are used. In the suggested technique, the EHR may be disseminated to several other users on the blockchain through a shared symmetry key and private key.

System architecture

The best EHR sharing techniques are indeed examined for smooth functioning and reduced communication time. The chapter defines the blockchain EHR system design. The arrangement as indicated comprises 4 respondents: system admin, doctor and patient. Different assets or intelligent contracts, not all of which are restricted to creating patient records, providing record access, granting doctor access, and revoking access, should be described in this platform. The workflow of the system is easy to use. Through the respondent applied for an entry certificate, the members enroll the certificate authority via the subscription network operator. The certification authority shall then provide a new ID to register the user by mail and certification and private key. Every operation is dispersed through the blockchain network of ledger fabric. Members have distinct responsibilities in the system, so they can acquire only logs of access. Using the client application, patients may add records that call up the chain code to conduct a network transaction.

The revised transactions will be disseminated throughout the network post submitting the action into the blockchain network. This assures, how every network operation is shared amongst every member in the program, and so no unauthorized patient may modify or remove any transaction. Transfers with a timestamp are now only appended to the past key, such that the network is completely safe. Data on the blockchain network will be refreshed and available to all users. The suppliers can request the necessary data over the network such as the students and sub administrative personnel.

The security mechanism purpose is to protect patient's identities and keep personal data confidential. The methods that the system use are temporary IDs and blockchains for the guarantee of anonymity of patients. Furthermore, the consent of the patient for data publication is constituted by a permitted password (Fig. 1).

Proposed security mechanism

For the association of medical data authorized password is used with the doctors who have made it. After all, the non-trusted organizations accessed the mainchain, and all the kept information distinctive to every entry that patient attributed like electronic signature and ID. The algorithms are presented by the following is for generating storage transaction that includes patient's generation for the signature of medical analysis and public–private keys. This mechanism root in cryptographic pair of keys used for message signing and authentication. This paper uses a public-key cryptosystem [14] known as the most desired cryptography algorithms.

The pair of keys (PKI, ski) is created at the patient wallet application. For every medical analysis, every key pair is distinctive. Static identity was founded in the blockchain, which patients have and are well known by trusted nodes. The result is received by the patient of medical analysis, which is signed by doctors and clinics. On a dedicated server, we can share the public keys that are why the patient can substantiate signatures of both for accessing data. Patients sign medical analysis which implicit that they concur for bringing out the mainchain data. When medical

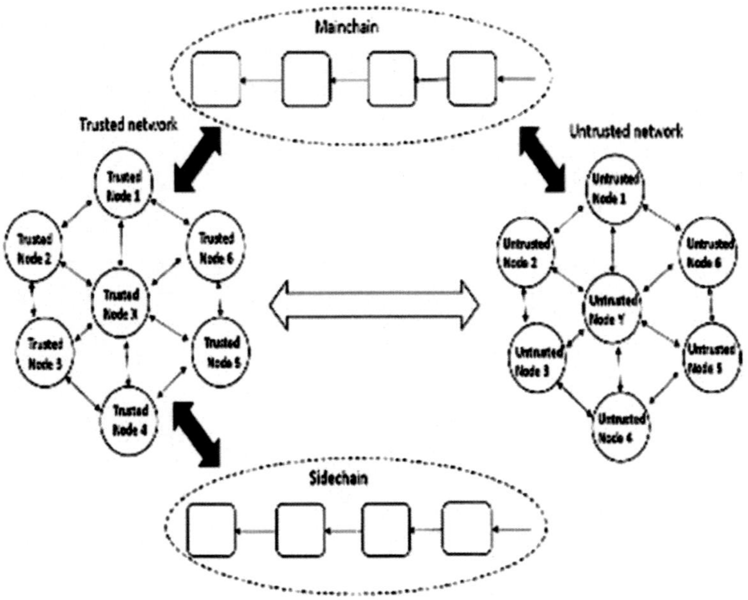

Fig. 1 Flow of proposed work

research of patient accepts it, the storage transaction is generated by a clinic node and to verify it, they distribute it to other nodes.

4 Implementation of Proposed System

Proposed algorithms

The EHR sharing system includes Algorithm 1 shows precisely how the admin is executed on a blockchain network. The certificate of registration of the administrative authorities should be sought. The administrator has complete permissions to the device, involving writing, reading, updating, and deletion. When patient and administrative staff, i.e., doctor or laboratory worker are authentic, then every participant is enabled by the admin or patient to provide an appropriate ID for blockchain access. If the attitude of the participant means unauthorized access is discovered, then the admin/authorized person may with a statement on the Hyperledger blockchain network eliminate/reject that person for accessing store dataset. The table includes all the algorithms employed used while accessing store information with proper working.

The doctor node asks for a private connection key for the network management. The doctor has several powers, such as reading and revoking writing records when they has accessed the blockchain store records. This method uses its ID on the blockchain network only for doctor sites. Unless the user has a legitimate node it is possible to browse or find student and subdirectory staff data through the network. Only authorized person can access the dataset otherwise secure massage send to admin about unauthorized access.

Algorithm 1: Admin or patient side algorithm

Input: Enrolment Certificate (EC) requested from Certification Authority (CA)
Output: Access to P HR , D HR and L HR transactions for all (P HR , C HR , L HR) _ B N **Initialization:** N Admin should be valid node. N Admin can Write/Read/Update/Remove nodes P ID ,D ID , L ID
procedure Admin (P ID , D ID , L ID)
while (True) **do**
if (P ID is valid) **then**
Add_Sub admin to the blockchain Network
Add_ Sub admin (B N , P ID)
Grant_access(P ID , U Name , P K)
else
Not_exist(P ID)
end if
if (P ID is valid) **then**
Add Patent to the blockchain Network
Add_Patient(B N , P ID)
grant_access(P ID , U Name , P K)
else
Not_exist(P ID)
end if
if (L ID is valid) **then**
Add Books to the blockchain Network
Add_Book(B N , L ID)
grant_access(L ID , U Name , P K)
else
Not_exist(L ID)
end if
end while
int N; {0 means bad behaviour, 1 means good behaviour}
for all (,,) **do**
if (behaviour_node(N) **then**
Not update(P ID ,D ID , L ID)
else
Remove_update(P ID ,D ID , L ID)
end if

```
end for
//Doctor Accessing records
procedure Student ( D ID )
while (True) do
if ( D ID _ B N ) then
if ( D REC_I not B N ) then
Create_records( D ID ,D REC_I , B N )
else
Update_records( D ID , D REC_I , B N )
Read_records(P ID , P REC_I , D ID , L ID , B N )
end if
else
Not_exist( D ID )
end if
end procedure (D ID)
end procedure (P ID)
```

Abbreviations

P HL	Patient hyperledger
D HL	Doctor hyperledger
L HL	Laboratory hyperledger
N Admin	Network admin
B N	Blockchain network
P ID	Patient ID
D ID	Doctor ID
L ID	Laboratory ID
P K	Private key
P REC _ I	Patient records
M PID	Laboratory records

Following are the working process of our system from registration to how to access the store dataset.

4.1 Registration

The registration module helps the doctors and patients to register their details.

In this module, we are collecting the all details regarding patient for future usage. In the case, each doctor in the system belongs to any one of the healthcare providers. In the case of patients, they have to submit their name, date of birth, address, mobile number, and details about their health insurance (Fig. 2).

For authentication purpose doctor and patient need to verify by using proper and valid password. The patients also note the details of the disease which the patient

Fig. 2 Registration screen proposed work

suffers from with their detailed doctor prescription and other laboratory details. All these details are converted into a block. These details cannot be viewed by anyone. Even the doctors cannot view the details of the patients until the patients permit the doctors (Fig. 3).

4.2 *Smart Contract Creation*

A smart contract is a self-operating program that executes automatically when the specific condition is satisfied. The proposed approach uses the data collection with code verification is done. As user can interact with system by using the smart authentication with proper interface. With the help of blockchain, the user can communicate with the contractor and the doctor can send the prescribed medicines to the patients. Doctors can also make a one-by-one session of the reports. The same report will be reflected on the patients which make their communication easier in the future (Fig. 4).

As well by using the proposed system patient and doctors can store and retrieve the database securely with proper automatic notification about the regular health checkup to reduce the disease.

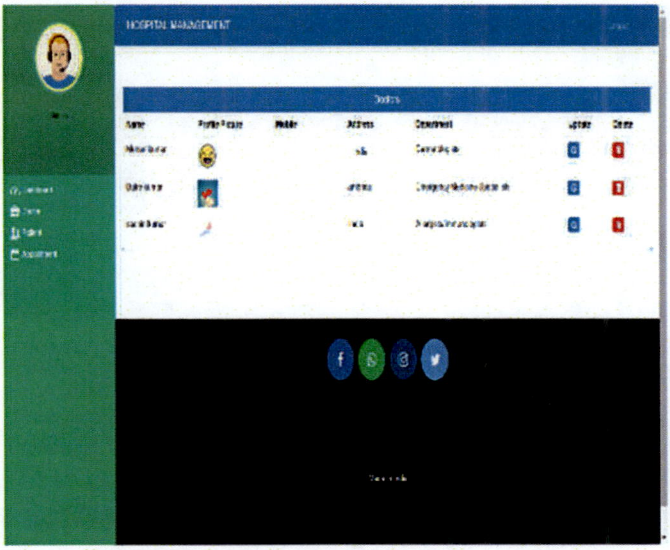

Fig. 3 Adding report of a patient in database

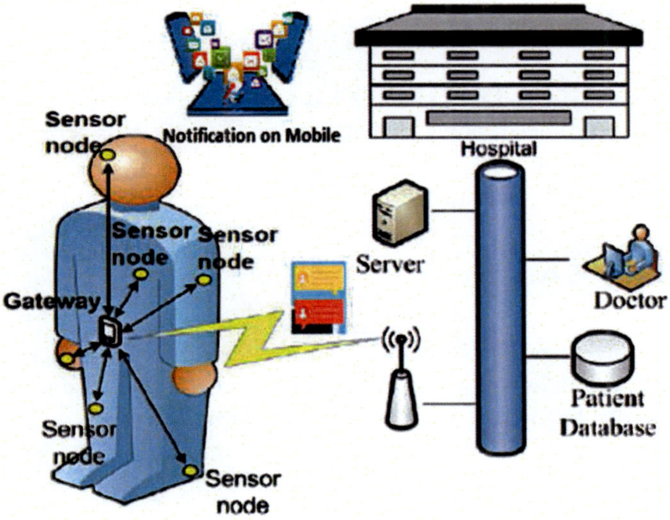

Fig. 4 Automatic notification of regular medical checkup

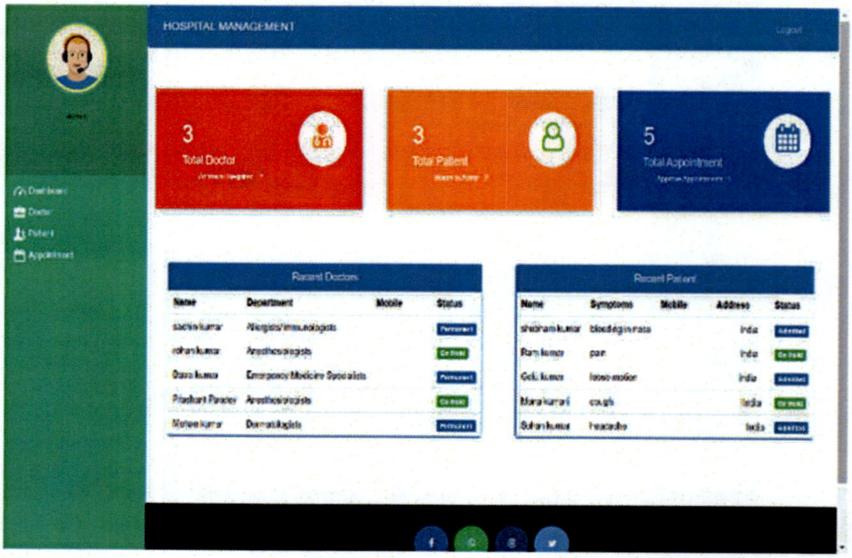

Fig. 5 Notification of appointment

4.3 *Block Creation*

Patients receive EHR notification which has to be approved by them. This way of approval avoids double-spending kind of attacks. After all clarification, the block is generated in EHR details (Fig. 5).

4.4 *Privacy-Preserving*

Blockchain technology acts as a complete solution for data privacy and security. The block is connected in form of a chain [4]. Each block contain the all basic information of each block connected to it for future use with proper data access.

5 Result and Discussion

In this arrangement of tests, we concentrate on how the framework performs under various traffic conditions when the data transmission is restricted to 4 Mb/s and the organization dormancy is set to 3 ms. Specifically, we scale the traffic focusing on the requesting administration as far as TPS from 2.5 to 13 TPS (exchanges each second). Results show that start to finish delay (demand full circle time) both for the end-client and the REST administration is not affected for the experiments up to

Table 1 Throughput with 4 ms bandwidth

Submission rate per s	Average commit rate per s	Average roundtrip time in s
3	2.54	1.43
6.35	6.21	1.2
7	6.8	0.93
8	7.67	0.48
10	9.76	0.57
12.5	11.32	5.25

8 TPS. Truth be told, the normal reaction time changes somewhere in the range of 1 and 1.3 s. This pattern is affirmed by the REST administration reaction time that pursues a similar direction. Table 1 shows the outlines the full circle time (Fig. 6).

Network latency in this arrangement of tests, we examine what network inertness can mean for administration execution. Specifically, the organization transmission capacity is restricted to 4 Mb/s and the organization idleness for all the associated customers is designed to various qualities as displayed in Table 2 (Fig. 7).

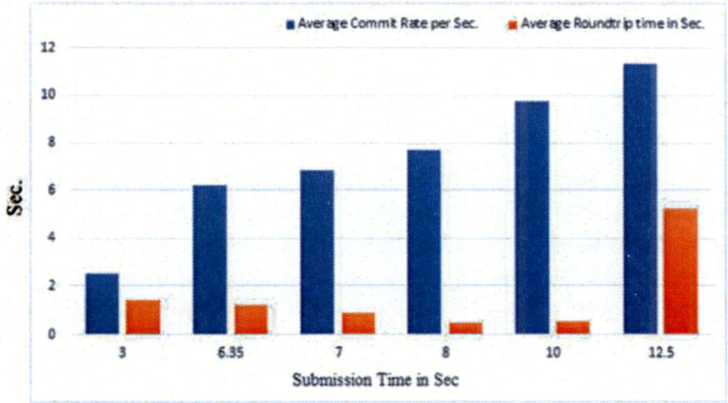

Fig. 6 Performance analysis throughput with 4 ms bandwidth

Table 2 Network latency distribution between blockchain nodes

Delay in ms	Number of nodes
50	2
60	4
80	5
100	3
120	6
150	8

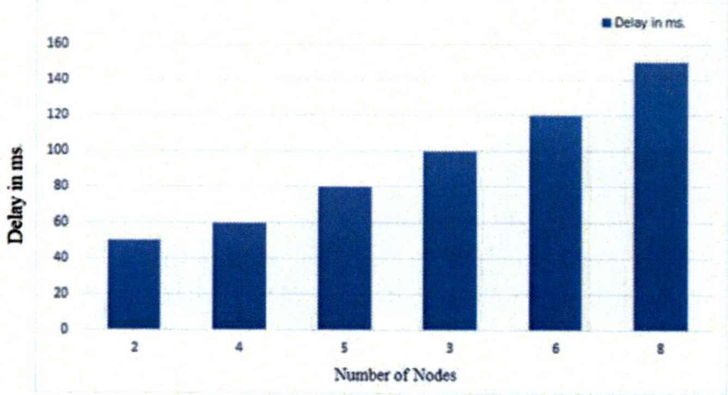

Fig. 7 Delay between blockchain nodes

6 Conclusion

Another name of medical analysis is personal data which is attainable to well-established organizations and the patients. A collective system is organized by medical data sharing among ancillary doctors and medical organizations. With public keys and blockchain cryptography, the system's design guarantees personal data confidentiality and acknowledges the patient as his data owner. The requirement is met by the proposed system obtrude by directives of protection of personal data. Data access is transparent that's why we can access health information by ancillary organizations root in security policy only. The workflow of application guarantees the creation of medical data by the medical unit that other reliable organizations argue. In real life, such protocol implementation makes numerous benefits to every actor involved. The patients efficiently managed medical analysis. The doctor predicted the hospital doctor could give medical services of better quality using a health database. In our experimental research, it is better shown that the time taken is less, and performance will be improved.

In our future work, we will center on proposed system integration with the storage infrastructure of actual medical data by which international standards are implemented to exchange healthcare data.

References

1. Nikoloudakis Y et al (2019) Vulnerability assessment as a service for fog-centric ICT ecosystems: a healthcare use case. Peer-to-Peer Netw Appl J. https://doi.org/10.1007/s12083-019-0716-y
2. Ekblaw A, Azaria A, Halamka JD, Lippman A (2016) A case study for blockchain in healthcare: 'MedRec' prototype for electronic health records and medical research data. In: 2nd international conference on open and big data

3. Zhang P, White J, Schmidt DC, Lenz G, Rosenbloom ST (2018) FHIRChain: applying blockchain to securely and scalably share clinical data. Elsevier
4. Guo R, Shi H, Zhao Q, Zheng D (2018) Secure attribute-based signature scheme with multiple authorities for blockchain in electronic health records systems. IEEE Access
5. Xia Q, Sifah EB, Asamoah KO, Gao J, Du X, Guizani M (2017) MeDShare: trust-less medical data sharing among cloud service providers via blockchain. IEEE Access
6. Rifi N, Rachkidi E, Agoulmine N, Taher NC (2017) Towards using blockchain technology for eHealth data access management. In: Fourth international conference on advances in biomedical engineering (ICABME)
7. Peterson K, Deeduvanu R, Kanjamala P, Boles K (2016) A blockchain-based approach to health information exchange networks. In: ONC/NIST use of blockchain for healthcare and research workshop
8. Kavitha PK, Vidhya Saraswathi P (2017) A survey on medical image encryption. In: ICASCT2501 I ICASCT I Mar–April, vol 3(5), pp 01–08
9. Tahir J, Javaid N (2017) Bootstrapping a blockchain based ecosystem for big data exchange. IEEE. 978-1-5386-1996-4/17 $31.00 © 2017 IEEE. https://doi.org/10.1109/BigDataCongress.2017.67
10. Bandara E, Ng WK, De Zoysa K, Fernando N, Tharaka S, Maurakirinathan P, Jayasuriya N (2018) Mystiko blockchain meets big data. IEEE. 978-1-5386-5035-6/18/$31.00 © 2018 IEEE
11. Sokouti M, Zakerolhosseini A, Sokouti B (2016) Medical image encryption: an application for improved padding based GGH encryption algorithm. Open Med Inform J 10:11–22
12. Ramani V, Kumar T, Braeken A, Liyanage M, Ylianttila M (2018) Secure and efficient data accessibility in blockchain based healthcare systems. https://doi.org/10.1109/GLOCOM.2018.8647221
13. Liu J, Li X, Ye L, Zhang H, Du X, Guizani M (2018) BPDS: a blockchain based privacy-preserving data sharing for electronic medical records. IEEE. 978-1-5386-4727-1/18/$31.00 © 2018 IEEE
14. Tasatanattakool P (2018) Blockchain: challenges and applications. IEEE. 978-1-5386-2290-2/18/$31.00 © 2018 IEEE
15. Zheng Z, Xie S, Dai H, Chen X, Wang H (2017) An overview of blockchain technology: architecture, consensus, and future trends. In: 2017 IEEE 6th international congress on big data
16. Dubovitskaya A et al (2017) Secure and trustable electronic medical records sharing using blockchain. In: Proceedings of the AMIA 2017, American Medical Informatics Association annual symposium, Washington, DC, 4–8 Nov 2017
17. Amofa S et al (2018) A blockchain-based architecture framework for secure sharing of personal health data. In: 2018 IEEE 20th international conference on e-Health networking, applications and services (Healthcom), Ostrava, pp 1–6. https://doi.org/10.1109/HealthCom.2018.8531160
18. Liang X, Zhao J, Shetty S, Liu J, Li D (2017) Integrating blockchain for data sharing and collaboration in mobile healthcare applications. In: 2017 IEEE 28th annual international symposium on personal, indoor, and mobile radio communications (PIMRC), Montreal, QC, pp 1–5. https://doi.org/10.1109/PIMRC.2017.8292361
19. Rahman MA et al (2018) Blockchain-based mobile edge computing framework for secure therapy applications. IEEE Access 6:72469–72478. https://doi.org/10.1109/ACCESS.2018.2881246
20. Markakis E et al (2019) Acceleration at the edge for supporting SMEs security: the FORTIKA paradigm. IEEE Commun Mag 57(2):41–47
21. Gajewski M et al (2019) Two-tier anomaly detection based on traffic profiling of the home automation system. Comput Netw 158:46–60
22. Tsampi K et al (2018) Extending the Sana mobile healthcare platform with features providing ECG analysis. In: Mobile big data. Springer, Cham, pp 289–321
23. Moriarty K et al (ed) (2016) PKCS #1: RSA cryptography specifications version 2.2. In: RFC 8017, IETF, Nov 2016

24. NIST special publication 800-57 part 3, revision 1. Available online: https://doi.org/10.6028/NIST.SP.800-57pt3r1
25. Hyperledger fabric—open source blockchain framework. Project webpage https://www.hyperledger.org/projects/fabric
26. Kuo TT, Kim HE, Ohno-Machado L (2017) Blockchain distributed ledger technologies for biomedical and health care applications. J Am Med Inform Assoc 6:1211–1220
27. Mettler M (2016) Blockchain technology in healthcare: the revolution starts here. In: Proceedings of 18th IEEE international conference on e-Health networking, applications and services, Sept 2016
28. Gordon W, Catalini C (2018) Blockchain technology for healthcare: facilitating the transition to patient-driven interoperability. Comput Struct Biotechnol J 224–230
29. Dubovitskaya A, Xu Z, Ryu S, Schumacher M, Wang F (2017) Secure and trustable electronic medical records sharing using blockchain. In: Proceedings of AMIA annual symposium proceedings, pp 650–659
30. Holbl M et al (2018) A systematic review of the use of blockchain in healthcare. Symmetry
31. Jiang S, Cao J, Wu H, Yang Y, Ma M, He J (2018) BlocHIE: a blockchain-based platform for healthcare information exchange. In: Proceedings of IEEE international conference on smart computing (SMARTCOMP)
32. Lo'ai AT, Mehmood R, Benkhlifa E, Song H (2016) Mobile cloud computing model and big data analysis for healthcare applications. IEEE Access 4:6171–6180

PSO- and GA-Based Hybrid Algorithm for Feature Selection

S. Sandeep Kumar, N. Raghavendra Sai, and M. Jogendra Kumar

Abstract One of the rule problems of artificial intelligence and data mining is to domesticate a substantial, multifaceted version to reduce the estimates drawn with the multifaceted computational nature of the representation. The ability to break down the computational time factor shortens the length of the facts and brings together the precision and breadth of the AI's unambiguous work. Because of their similarity to general smoothing strategies, two metaheuristics were used to determine the FS problems. Therefore, mutts metaheuristics help develop the tracking speed and the affiliation of the basic calculations. One nation of the work of art will ensure that the computation becomes a member of the two estimates; this document performs innate estimation (GA) and particle swarm optimization (PSO) to overcome search limits. The sufficiency of our proposed approach is specified in a motion of diversion stages, the use of the UCI gaining knowledge of display as a benchmark dataset.

Keywords Hybrid approach · Feature selection · Genetic algorithm · Particle swarm optimization · Evolutionary computation · Metaheuristic

1 Introduction

Features choice is a critical issue for the achievement of an intrusion detection system. It is essential for excessive variety facts mining and is an essential facts processing step with inside the education segment previous to transferring to the subsequent stage (testing) [1]. There are one-of-a-kind strategies for choosing capabilities, including the wrapper, the filter, and the embedded strategies. Other strategies are bio-stimulated metaheuristic [1–3]. Determining the pleasant wide variety of capabilities will enhance the achievement charge and performance. This study's define and most important contribution may be summarized with the subsequent

S. Sandeep Kumar · N. Raghavendra Sai (✉) · M. Jogendra Kumar
Department of Computer Science and Engineering, Koneru Lakshmaiah Education Foundation, Vaddeswaram, Andhra Pradesh, India
e-mail: nallagatlaraghavendra@kluniversity.in

S. Shakya et al. (eds.), *Proceedings of Second International Conference on Sustainable Expert Systems*, Lecture Notes in Networks and Systems 351,
https://doi.org/10.1007/978-981-16-7657-4_38

points: using PSO and GWO in a brand new emergent method, choosing the intersection of their (PSO and GWO) capabilities, then inspecting the ensuing maximum often repeated capabilities, a good way to constitute the pleasant set.

Part identification can be a critical phase of information preprocessing in information mining. For a long time, a variety of metaheuristic methods were used to get rid of superfluous and insignificant elements, but similar to what one can expect from high-dimensional datasets. Often, independent of the connection between many selected highlights, the guarantee of options can be a bit of a confusion as to the fact that usable halftones are possible. The great local demand and the live incremental backup are awesome. The circle of inquiries is growing steadily as to whether it relates to the proportion of the characteristics of the illuminated list [4]. After that, an absolute search is essentially useless most of the time. To solve this problem, various methods of dynamic filtering are used, such as B. reverse serial selection (SBS) [5].

In any case, these processes for the selection of storage also face a number of problems, for example, the optimal development of the neighborhood and the high costs of the machine. Having a compelling general search system is critical to having a penchant for warranty issues that (EG) are withheld for their general application limit. PSO could be a fairly new CT technique that focuses on swarm data [6, 7]. Ascent is the right gift day, as alternative estimates of the EC as gas. Reasons for including the PSO in this area, as well as a group of options, is a key methodology [1, 8, 9]. This method was used by Eberhart and Kennedy [2] as a specialized, mostly random smoothing technique based on the general population. PSO creates a variety of unpleasant animals, a celebration of birds, or the social leader of a school of fish. PSO offers safe equivalents to exceptional laptop advances such as innate computations and simulated neural associations. It comes downtown with unpredictable people and updates ages to channel the optimum. Approval to the rest of the region is something like the swarm of ants. In a high-dimensional evaluation area, the large direct range is shown in particles with optional positions and speeds. In hyperspace, these particles are fed by the dynamics of their region and speed. Spurred on by PSO and GA, an additional innovative cross-manifold computation (HEA) is created by changing the integrity of the potential gains of the two estimates to give the most plausible properties to determine within the selected feature set. To achieve this, the large direct number in particles is represented with regions and velocities in a high-dimensional research environment. They penetrate hyperspace through the dynamics of their own position and direction.

2 Literature Survey

Element alternative was a research region working in information mining, style recognition, and knowledge networks. The entire research area is to track down the most important and non-repeating elements as well as each of the conceivable elements.

Subset is 2*n*, where *n* is the number of distinctive parts. Complete hunt guarantees that the correct elements are found, but normally, this is not computationally meaningful for medium-sized datasets anyway. Since the evaluation of all potential subsets is exorbitant, a solution must be sought that is computationally practical and valuable as well as qualitative. Several component selection techniques use meta-heuristic computations so as not to add complexity to the method. These computations may really want to update the alternative component problem with the correct truth within a satisfactory time. Breeding strategies obsessed with the world and the state of insect sweetener (ACO), inheritance stone (GA), reproduced hardening (SA), no chase (TS), and progress swarm molecules (PSO) have recently been used in. In fact, mixed search techniques are used that consolidate nearby coverage and channel strips. In the concept of a race channel, [* fr1] was used, which includes the computation of the determination of sets obsessed with the PSO for the grouping of support vector machines (SVMs).

Also, some current methods think about that association of elements of their research methods. For example, in an improved inheritance, calculation was projected to better determine a subset of elements from a multiple character set, documented categories to kill invalid chromosomes. Over the course of several years, numerous organic process calculations are based primarily on calculations such as the genetic formula (GA), the particle swarm amplification (PSO), the pepper colony amplification (ACO), and the artificial bee colony (ABC). In the SI-based calculation, genetic was used profitably in the alternative theme of the elements to create a high-dimensional dataset. Therefore, the probability of selecting an excess rate can be increased. To overcome these drawbacks, this post presents a hereditary calculation based mainly on the area for determining dispositions called CGAFS. The country cluster placement strategy is used in the projected approach to separate the highlights in completely different sessions, for insurance clusters of elements of abuse. A recently handed over progeny is being assembled to require supplies related to posterity.

In contrast to the previous hereditary calculation, which is mainly based on the component alternative that uses channel and skin models in the application, the native spatial recognition methodology is integrated into the GA-based coverage model, and the quantity could also be determined accordingly. Completely different improvement calculations are proposed to address confusing progression problems because this wide variety of estimates have different direct differences. Organize it in three major orders: Innovative estimate that depends on swarm and materials science. The formative estimate depends on the regular progression that the Darwinian extraordinary norms organize. They are for the better known half (GA). Options for action are initially created at discretion, and people are modified with each estimate of the cycle of action. To adequately search for the simplest approach, GA uses crossover and alternate managers. Since the ideal situation can undoubtedly be involved in the creation of new successors, the new strategy is forecasted to be higher [2]. In addition, the swarm is ready for the movement of hordes, ovens, universities, or animal waves in the wild. The activities of the law enforcement agencies are examined through the total, social and knowledge generation. It is inspired by birds flocking social behavior.

Use the simplest general technique to validate the analysis and the best location for abuse too. ACO is another standard estimate for this class. Third, materials science focuses on the motives of royal law. Hardening (SA) and HS proliferation stones are the fastest estimates focused on materials science.

Feature selection choosing options hope to select a feature extension from the variety of underlying data, which can advantageously clarify the input file by reducing the effect of upheavals and dissidents.

1. Identification of a set the recognition phase for the development of candidate subsets. This step is the amount of research, which is achievable within the initial place all items or a subset of limitless self-affirmation variables…
2. Set examination: The premise is used to record the participating quality level of the completed subassemblies.
3. Waiting state: The execution of the calculation begins with understanding the stopping condition and ends until the stopping state has been solved. The condition of isolation is still in the air the subject of a work of examination.
4. Validity: The release of the chosen set is the subject of this point. This accreditation is made on mistreatment imitated from the bases of knowledge (or authentic instruction sheets), testing, and separating the consequences of the projected calculation using past methods.

Particle swarm optimization (PSO) is also an evolutionary computational version that is primarily based totally on swarm intelligence. PSO is advanced with the aid of using Eberhart and Kennedy [2] who have been stimulated with the aid of using the studies of the synthetic livings. Similar to GAs, PSO is likewise an optimizer primarily based totally on population. The device is initialized first of all in a hard and fast of randomly generated capability solutions, after which plays the look for the most effective one iteratively. Whereas the PSO does now no longer own the crossover and mutation methods utilized in GAs, it reveals the most effective answer with the aid of using swarms following the great particle.

3 Related Work

Existing position underwriting calculations could also be broken down into two arrangements: hedging approaches and channeling methods. Hedging approaches use computation to learn/plan within the assessment cycle, while channel approaches do not, usually machine-controlled [10], while covers can largely convey the results generated [11]. For the selection of options, some estimates have been suggested over the past few years. This section describes a typical estimate of the limit guarantee. The final performance of the two associations shows that the wave network is indeed more fundamental than MLP, in the case of pronounced psychological standard problems (Fig. 1).

The PSO is very extensive in the area of FS. A sigmoid cutoff was used to increase the partner's understanding of regional issues regarding the likelihood of receiving

	Population	F1	F2	F3	F4	F5	F6	F7	F8	Accuracy	WF1	WF2	WF3	WF4	WF5	WF6	WF7	WF8
GA	GA 1:	1	1	0	1	0	0	1	0	0.98	0.98	0.98	0	0.98	0	0	0.98	0
	GA 2:	1	1	0	0	0	1	1	0	0.93	0.93	0.93	0	0	0	0.93	0.93	0
	GA 3:	1	0	0	1	0	1	1	0	0.87	0.87	0	0	0.87	0	0.87	0.87	0
	GA 4:	1	1	0	1	0	1	1	1	0.75	0.75	0.75	0	0.75	0	0.75	0.75	0.75
	GA 5:	1	0	0	1	0	0	1	1	0.89	0.89	0	0	0.89	0	0	0.89	0.89
	GA 6:	0	1	0	1	0	1	1	0	0.98	0	0.98	0	0.98	0	0.98	0.98	0
	GA 7:	0	1	1	1	0	0	1	1	0.95	0	0.98	0.98	0.98	0	0	0.98	0.98
	GA 8:	1	1	0	1	0	1	1	0	0.79	0.79	0.79	0	0.79	0	0.79	0.79	0
	GA 9:	1	0	0	1	1	0	1	0	0.82	0.82	0	0	0.82	0.82	0	0.82	0
	GA 10:	0	1	0	1	0	0	1	1	0.91	0	0.91	0	0.91	0	0	0.91	0.91
PSO	PSO 1:	0	0	1	1	0	1	1	0	0.97	0	0	0.97	0.97	0	0.97	0.97	0
	PSO 2:	1	0	1	1	0	1	0	0	0.82	0.82	0	0.82	0.82	0	0.82	0	0
	PSO 3:	0	0	1	1	1	1	1	0	0.98	0	0	0.98	0.98	0.98	0.98	0.98	0
	PSO 4:	1	1	0	0	0	1	0	1	0.72	0.72	0.72	0	0	0	0.72	0	0.72
	PSO 5:	1	0	1	1	0	1	1	1	0.93	0.93	0	0.93	0.93	0	0.93	0.93	0.93
	PSO 6:	0	1	1	1	0	0	1	0	0.85	0	0.85	0.85	0.85	0	0	0.85	0
	PSO 7:	1	1	1	1	0	1	1	1	0.96	0.96	0.96	0.96	0.96	0	0.96	0.96	0.96
	PSO 8:	0	1	0	1	0	1	0	0	0.74	0	0.74	0	0.74	0	0.74	0	0
	PSO 9:	1	1	0	1	1	0	1	1	0.89	0.89	0.89	0	0.89	0.89	0	0.89	0.89
	PSO 10:	1	1	0	1	0	1	1	1	0.95	0.95	0.95	0	0.95	0	0.95	0.95	0.95
Feature importance											11.3	11.43	6.49	16.06	2.69	11.39	15.43	7.98

Fig. 1 Feature importance measurements

options for pooled PSO [12]. For fast events, all known constant and different PSOs with extended SVM and cutoff limits were used [13]. The assumed care assistant structure was used for the PSO calculations by Jack's performing arts, and therefore, the SVM design and testing were done by the buyer. The PSO was modified to indicate the numerical PSO in [14], develop the best venues and the best in the world with these trained professionals before the professionals created dynamic social gatherings. The estimate was used for quality assurance. The use of an evil set has been added to perform FS slight change to PSO.

Chung and Wahid [15] developed a new hybrid model for intrusion detection and used intelligent dynamic smart swarm-based rough set (IDSRS) for function selection. The proposed model must select as many relevant features as possible that characterize the traffic pattern, tested in simplified swarm optimization. The authors show that the proposed model can improve performance. In addition, the experiments are applied using the KDDCup99 dataset and show that the projected model can achieve an accuracy of approximately 93.3°.

Çavuşoğlu [16] used selection methods to provide a maximum number of accuracies. The data record is reduced by selecting two different characteristics (CfsSubsetEval, WrapperSubsetEval). By choosing a better collection of insightful features, the proposed hybrid model enhanced the efficiency of the k-means clustering algorithm.

Buczak and Guven [17] reported on a survey on data mining and machine learning methods for cybersecurity and intrusion detection. The complexities of machine learning and data mining are addressed, and the crucial aspect of this cybersecurity study is the importance of a dataset for training and testing. The authors mention that machine learning and data mining cannot work without data representation, and it is difficult and time-consuming to obtain a dataset. It also gives recommendations on how to use a particular method.

Chitrakar and Huang [18] suggested a hybrid learning model that combines NB with the clustering technique based on kMedoids. The authors suggest that it is better to use KMedoids grouping techniques followed by NB classification method for more accurate results. The results show that the planned model improves precision and indicates false positive results.

4 Proposed Method

For real datasets, there are innumerable surplus and insignificant components which {can} significantly worsen the representation of the learned model and, thus, the learning rate of the models. Part identification could be a fundamental step in information preprocessing in information mining to remove excess components from a particular dataset [19]. Different innovations can undoubtedly destroy the intangible determinations of opposing techniques for determining the amounts of elements, but they cannot manipulate the components excessively. According to GA, our structure is applied, and every move has pushed the final population to new and higher responses [20]; in value work evaluates, the expenditure of someone from the general population, maintaining the pace of the general population and a short time later changes Gbest and Pbest {every} half, which bothers the new cost. st implies the simplest response that any general population has cultivated in recent times. Certainly, it has been revised that GA might not abuse, and therefore, the mere institution of management of GA could be a change that causes almost no abuse, physical disorder.

In any case, GA will do an amazing mission room survey with cross-ventures [21]. Taking the PSO into account, however, it was found that the PSO has a strong ability to find space, but cannot bring a good exam home with its ankles. Abuse and disclosure developments encourage you to place your disclosures for viable and usable results. GA and PSO are two different types of FS estimates to be specifically innovative and comprehensive. GA is useful for shifting vital limits starting at age so that PSO has the advantage of having a full look through the cabin with particles related to information from the other skills [22]. These advantages of PSO and GA are coupled in an affordable way with manipulation and exploration.

This section can represent our fully calculation as follows: Step one confirms the number of options. Step two initializes the population. Step three updates the positions of the particles. Step four is philosophical rules of the present procedure. Step five is modification of the best. Step six is the k-nearest neighbor (KNN). From the beginning, all options in the information set are temporarily deactivated. To achieve this, the electrode lengthening technique is used [23]. The calculation provides for the selection of features that have the first influence in each category as a separating limit with regard to the attachments of the selection tree. The lengthening of the lead could be a change in the data obtained that reduces the distinction between the attributes of the upper branches [24]. When choosing a property, on the other hand, the profit margin evaluates the variety and size of the branches.

Computation 1. The pseudocode of the projected computation.

(1) Each part of GRI is calculated
(2) Select 75% of the components with GR high
(3) Select the first group of people
(4) Analysis of fitness physically
(5) Data format of the components and health assessment
(6) Repeat
(7) Better choice of people
(8) Crossover and change of use for the flaming new age of humans
(9) New human health assessment
(10) Updates of rate and position
(11) Updates of Pbest and Gbest
(12) Analysis of the suitability of the components
(13) Integrating the simplest advantages of people, speed, and positions
(14) The population involved is organized according to their assessment of well-being
(15) Selection of the population in better way
(16) Solutions extraction in a Best defined population
(17) Classifier: better take notes (KNN).

5 Design of Experiment

Parameter Setting and Datasets

In our recovery tests, seven perceptive records are followed to verify the adequacy of our fine-grained prompt estimates and practices. These insightful records are taken from the UCI learning device [25] shown in Table 1. The insightful records contain many variables, social events, and patterns that allowed a full study of the projected approach [26] and, therefore, the procedures used. Maybe, the foremost notable ANN calculations were accustomed to calculate the capability to act. *k* matches five during this preliminary (5NN). The instructive records utilized in the basics examined

Table 1 Description of datasets

Dataset	No. of features	No. of classes	No. of examples
Wine	13	2	178
German	24	3	1000
Lung	56	2	32
Hill valley	100	3	606
Musk-I	166	3	476
WBCD	30	3	564
Ionosphere	34	3	351

within the UCI store for AI are in Table 1. The cutoff amounts for PSO-based multi-goal assurance calculations are scrutinized (13–166), social aspects (2–3) associate degreed events (32–1000). In every information display, the two cases are willy-nilly separated into two sessions throughout the assessments: 70% as a take a look at set and 30% as an assessment set. (e) Contains a set of boundaries during the look part.

6 Analysis and Results

The max. no. of iterations and also the population play an essential role in the characterization of the actions of an agent on the basis of the knowledge of other agents and the progressive progress of the agents among one another, when the multi-agent evolutionary algorithm is spoken of.

6.1 Comparison

The model is determined by applying it to different datasets. The efficiency of the model is based on the performance of LFFS, LFS, and our proposed approach in Table 2. In this section, you will find the related material on the experiments when talking about the evolutionary multi-agent algorithm.

7 Conclusion

Additionally, in this article, we propose a cream philosophy for FS that is subject to a standard metaheuristic called PSO and GA estimation. The proposed computation has been different with respect to ten higher-level metaheuristic, and metaheuristic FS methods using seven common ICU records show that the proposed calculation is superior to several techniques. So, we can explain that our technique for dealing with FS issues is a reckless connection. In the event that we do notice any disclosures, our estimate provides a way to deal with PSO disclosure and abuse restrictions. Issues with this assessment are that the proposed calculation can be applied to other common and interesting subjects related to exploration as a potential degree of work, similar to the facial expression.

Table 2 LFFS and LFS accuracy rates with our proposed method

Dataset	LFFS	LFS	Our proposed
Wine			
Average size	8	8	6.44
Average acc	74.07	74.07	80.30
German			
Average size	6	6	5.08
Average acc	66.33	66.33	78.17
WBCD			
Average size	11	12	10.09
Average acc	91.18	92.35	9.48
Ionosphere			
Average size	8	8	10.85
Average acc	90.48	90.58	91.37
Lung			
Average size	5	6	30.25
Average acc	66.67	66.77	70.54
Hill-valley			
Average size	8	9	30.25
Average acc	53.85	55.49	78.53
Musk-I			
Average size	13	13	90
Average acc	79.29	80.71	90.06

References

1. Fourie PC, Groenwold AA (2002) The particle swarm optimization algorithm in size and shape optimization. Struct Multidiscip Optim 23:259–267
2. Eberhart RC, Kennedy J (1995) A new optimizer using particles swarm theory. In: Proceedings of the sixth international symposium on micro machine and human science (Nagoya Japan), pp 39–43
3. Sungheetha A, Sharma R (2020) A comparative machine learning study on IT sector edge nearer to working from home (WFH) contract category for improving productivity. J Artif Intell 2(04):217–225
4. Guyon, Elisseeff A (2003) An introduction to variable and feature selection. J Mach Learn Resour 3:1157–1182
5. Marill T, Green D (1963) On the effectiveness of receptors in recognition systems. IEEE Trans Inf Theory 9(1):11–17
6. Yu L, Liu H (2003) Feature selection for high-dimensional data: a fast correlation-based filter solution. In: Proceedings of the twentieth international conference on machine learning (ICML 2003), Washington DC, vol 3, pp 856–863
7. Sadollah A, Bahreininejad A, Eskandar H (2013) Mine blast algorithm: a new population based algorithm for solving constrained engineering optimization problems. Appl Soft Comput 13(5):2592–2612

8. Smitha TV (2021) A study on various mesh generation techniques used for engineering applications. J Innov Image Process 3(2):75–84
9. Chen LF, Su CT, Chen KH, Wang PC (2012) Particle swarm optimization for feature selection with application in obstructive sleep apnea diagnosis. Neural Comput Appl 4:2087–2096
10. Ghosh A, Datta A, Ghosh S (2013) Self-adaptive differential evolution for feature selection in hyperspectral image data. Appl Soft Comput 13:1969–1977
11. Vicsek T, Czirók A, Ben-Jacob E, Cohen I, Shochet O (1995) Novel type of phase transition in a system of self-driven particles. Phys Rev Lett 75(6):1226–1229
12. Kumar PJS, Devi PR, Sai NR, Kumar SS, Benarji T (2021) Battling fake news: a survey on mitigation techniques and identification. In: 2021 5th international conference on trends in electronics and informatics (ICOEI), pp 829–835. https://doi.org/10.1109/ICOEI51242.2021.9452829
13. Raghavendra Sai N, Bhargav J, Aneesh M, Vinay Sahit G, Nikhil A (2021) Discovering network intrusion using machine learning and data analytics approach. In: 2021 third international conference on intelligent communication technologies and virtual mobile networks (ICICV), Tirunelveli, pp 118–123. https://doi.org/10.1109/ICICV50876.2021.9388552
14. Vijaya N, Arifuzzaman SM, Raghavendra Sai N, Manikya Rao Ch (2020) Analysis of Arrhenius activation energy in electrically conducting casson fluid flow induced due to permeable elongated sheet with chemical reaction and viscous dissipation. Front Heat Mass Transfer (FHMT) 15. ISSN: 2151-8629
15. Chung YY, Wahid N (2012) A hybrid network intrusion detection system using simplified swarm optimization (SSO). Appl Soft Comput 12(9):3014–3022
16. Çavuşoğlu U (2019) A new hybrid approach for intrusion detection using machine learning methods. Appl Intell 49(7):2735–2761
17. Buczak AL, Guven E (2016) A survey of data mining and machine learning methods for cyber security intrusion detection. IEEE Commun Surv Tutor 18(2):1153–1176
18. Chitrakar R, Huang C (2012) Anomaly based intrusion detection using hybrid learning approach of combining k-medoids clustering and Naïve Bayes classification. In: 8th international conference on wireless communications, networking and mobile computing, Shanghai, pp 1–5
19. Sai NR, Cherukuri T, Susmita B, Keerthana R, Anjali Y (2021) Encrypted negative password identification exploitation RSA rule. In: 2021 6th international conference on inventive computation technologies (ICICT), Coimbatore, pp 1–4. https://doi.org/10.1109/ICICT50816.2021.9358713
20. Kumar MJ, Kumar GVSR, Krishna PSR, Sai NR (2021) Secure and efficient data transmission for wireless sensor networks by using optimized leach protocol. In: 2021 6th international conference on inventive computation technologies (ICICT), Coimbatore, pp 50–55. https://doi.org/10.1109/ICICT50816.2021.9358729
21. Raghavendra SN, Jogendra KM, Smitha CC (2020) A secured and effective load monitoring and scheduling migration VM in cloud computing. IOP Conf Ser Mater Sci Eng 981. ISSN: 1757-899X
22. Jogendra Kumar M, Raghavendra Sai N, Chowdary CS (2020) An efficient deep learning approach for brain tumor segmentation using CNN. IOP Conf Ser Mater Sci Eng 981. ISSN: 1757-899X
23. Pavan Kumar A, Gajjela L, Raghavendra Sai N (2020) A hybrid hash-stego for secured message transmission using stegnography. IOP Conf Ser Mater Sci Eng 981. ISSN: 1757-899X
24. Chowdary CS, Edamadaka G, Raghavendra Sai N, Jogendra Kumar M (2020) Analogous approach towards performance analysis for software defect prediction and prioritization. IOP Conf Ser Mater Sci Eng 981. ISSN: 1757-899X
25. Edamadaka G, Chowdary CS, Jogendra Kumar M, Raghavendra Sai N (2020) Hybrid learning method to detect the malicious transactions in network data. IOP Conf Ser Mater Sci Eng 981. ISSN: 1757-899X
26. Katakam MSK, Devineni K, Kanagala P, Raghavendra Sai N (2020) Analysis of artificial neural networks based intrusion detection system. Int J Adv Sci Technol 29(5s):928–935. Accessed 30 July 2021. http://sersc.org/journals/index.php/IJAST/article/view/7832

System for Verification Using Blockchain: Demonstration of PAN Cards, and College Transcripts Verification

Akshay Kale, Prathviraj Walke, Omkar Wanave, Kaushal Bhandari, and S. A. Deshpande

Abstract Verification is one of the most important factors of security and trust in this modern society, whether it is verification of college transcript for admissions, PAN cards for banks, Aadhar cards for voting, etc. Today's online verification processes are complicated and prone to cybersecurity attacks like data tampering and server hacking, and due to all these issues, trust between two entities can be compromised during the verification process. In this paper, we propose a new Blockchain- based architecture for the verification of the data in cyberspace. Blockchain technology is a modern disrupting technology that can be used to solve this issue. We also demonstrate a prototype of the system for the Indian PAN Card and college transcript.

Keywords Blockchain · Smart contract · Ethereum · Verification · DApp

1 Introduction

1.1 Background

Modern society needs a trusted, secure, and easy system for verification. Blockchain technology has those parameters built in them. The system we intend to build will have a simple and holistic way to authenticate multiple data together. As security is of paramount importance in verification and the current system has various security loopholes in them, a Blockchain is a viable option for mitigating most of the security issues.

In today's world, the need to verify the credentials and documents is very important. Data tampering in online systems is a very real problem facing the age of the Internet. Verification of data or document credentials is very important in various fields like finance, public offices, forums, banking, and academia. Nowadays, a gen-

A. Kale (✉) · P. Walke · O. Wanave · K. Bhandari · S. A. Deshpande
Department of Information Technology PVG's COET, Savitribai Phule Pune University, Pune, India
e-mail: sad_comp@pvgcoet.ac.in

S. Shakya et al. (eds.), *Proceedings of Second International Conference on Sustainable Expert Systems*, Lecture Notes in Networks and Systems 351,
https://doi.org/10.1007/978-981-16-7657-4_39

eral system for verification is missing; a one-stop platform where different types of data verification and document credentials verification can be done. Few online websites that provide such services are not general and are prone to data tempering and various online security concerns.

1.2 Motivation

Creating a system that can be used to do verification of data using blockchain is the main motivation behind this paper. As the need for this system is worldwide for authentication of college transcripts, PAN cards, Aadhar cards, Passport, etc., the scope of this system is vast. The architecture we develop will help create a verification system that is secure and reliable as it uses Blockchain technology. Various applications of the system include citizen identity authentication, documents verification, etc., under one system.

1.3 Objective

1. To create an architecture for a system that can be used for any type of verification securely and reliably.
2. Reducing complexity in the verification process.
3. To demonstrate the working of the system by creating a system with different authentication domains.

1.4 Blockchain Technology

The invention of Blockchain happened as the technology which is the base of Bitcoin [13]. A Blockchain is distributed ledger (file) that keeps a record of all the transactions that have taken place, in such a way that they are secure and immutable. It can be used for the transfer of money as well as storing records securely and verifying data by comparing data present within the distributed immutable ledger without requiring an intermediary or third party. Ethereum is a decentralized blockchain that uses Ether as its currency and is capable of running smart contracts [4]. It is open-source and has been online since July 2015. Thanks to the smart contracts ethereum has a wide range of applications like NFT's (non-fungible token), distributed applications (Dapps), etc.

2 Literature Review

In [6], Comparison of Blockchain-Based Solutions to Mitigate Data Tampering Security Risk, the paper does a deep analysis of data tampering issues and explores the mitigation of the issues using Blockchain. Paper explores various Blockchain tools like Ethereum Blockchain, Hyperledger, etc., and their uses for data tampering security.

In [1], Blockchain-Based Academic Records Verification in Smart Cities, the system proposed uses a smart contract-based approach to academic records verification. The authors detail the use of a decentralized platform for digital academic record verification. In this paper, the prototype is created using the Hyperledger framework.

In [11], Blockchain-Based Identity Verification Model, A hyper ledger Approach is taken for solving the verification problem. The authors used the Blockchain approach to solve the problem of identity verification.

In [8], A Blockchain-based aadhar is proposed. The proposed system has a single smart contract for a digitally distributed aadhar. The system does not take a general approach to verification and is limited in scope.

In [7], A Blockchain-based DBT method is proposed for social welfare benefits distribution in India. The paper reviews the current approach and proposes a Blockchain-based solution for a DBT.

In [3], Identity solutions for verification using Blockchain technology, the paper reviews the use of blockchain technology for verification in schools and certificates. The paper reviews different types of blockchain solutions like Bitcoin, Blockchain in healthcare and smart contracts, etc.

In paper [9], the authors provide a system using a hyper ledger blockchain framework to solve the problem of CCTV data integrity. Using Blockchain's distributed ledger and trustworthiness, they provide a better solution than the conventional systems.

Paper [14] proposes a Blockchain-based system for location verification. The system uses several connected mobile devices and a blockchain framework to solve the problem of location verification.

Paper [10] provides a blockchain-based solution to do a distributed machine learning application in a privacy-preserving manner. Paper uses Etherium Test Blockchain and smart contract to make ML model data accessible and give penalties to the rule-breakers.

Paper [5] considers problems in supply chain management (SCM) like fraud, tampering of products, delay, etc., and provides the blockchain-based mechanism integrated with SCM to build an effective supply chain. This paper provides a blockchain-based scheme for information sharing securely in the SCM with smart contracts and consensus mechanisms.

Paper [12] proposes a methodology where the hash values of all the original documents of a job application submitted by organizations are computed and then saved on a consortium blockchain. The verification of the details of job applicants is done during the process of hiring by comparing the hash value of documents with the hash value of the document present on the immutable distributed ledger of the blockchain.

In paper [15], the author purposes a data-driven trust mechanism based on blockchain (DDTMB) to detect and mitigate attacks on IoT devices. The system uses the data obtained from IoT devices to exemplify the resultant communication. The paper concludes that blockchain-based authentication and security during data transfer can be incorporated in future systems.

As seen in paper [6], the data tampering issue can be mitigated by using Blockchain technologies and papers [1, 2, 7, 8, 11, 16] various technologies of Blockchain are applied to some small to medium scope problems.

3 Problem Definition

As discussed in Sect. 1, the paper intends to target a few problems in general. Blockchain has its own limitations, so does our approach to solve the verification problem (as discussed in Sect. 5). Hence, it is important to discuss the scope and objectives of this paper. The problems this paper focuses on are:

1. Trust and reliability of online verification system.
2. Mitigate Data tampering in today's online systems.
3. System that can efficiently handle and process large amounts of transactions simultaneously.
4. Loss of data due to system failure.

4 Research Methodology

4.1 Use Case

As shown in Fig. 1, use case diagram has two actors that we have defined as follows:

1. Users—Users are the people or organizations that will use the verification system. A user can be a single individual to University or bank, anyone that would like to avail verification facility on the system.
2. Admin—Admins are trusted entities and organizations whose services will be available through the system. Admins will be the only actors who are allowed to add and update the data on the blockchain.

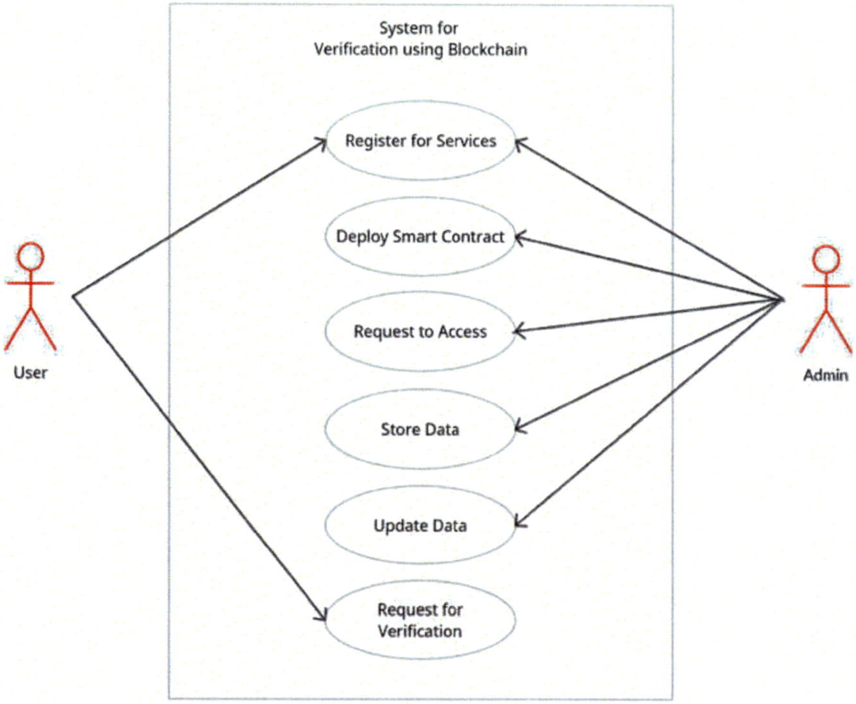

Fig. 1 Use case diagram

4.2 *Architecture*

As shown in Fig. 2, the system has three main components, i.e., WebApp, Smart contracts, and the Ethereum blockchain. The [data flow diagram] and the [architeure diagram] together show end-to-end data flow and working to the application.

1. Web App—The Web app will be used to interact with the blockchain. User and admin both can log in through the Web app to perform their tasks as mentioned in the [use case diagram]. Web3 interface is used for the communication between blockchain and the Web app. Web3.js helps to send/receive ethers and interact with the smart contract on the blockchain.
2. Smart contract/s—Smart contracts are the heart of the system. We have all the logic for inserting and verifying blockchain records with access modifiers for the admin in smart contracts.
3. Blockchain—We have used Etherum blockchain for our system. Ethereum Blockchain gives us the freedom to deploy smart contracts that are essential to the working of the system.

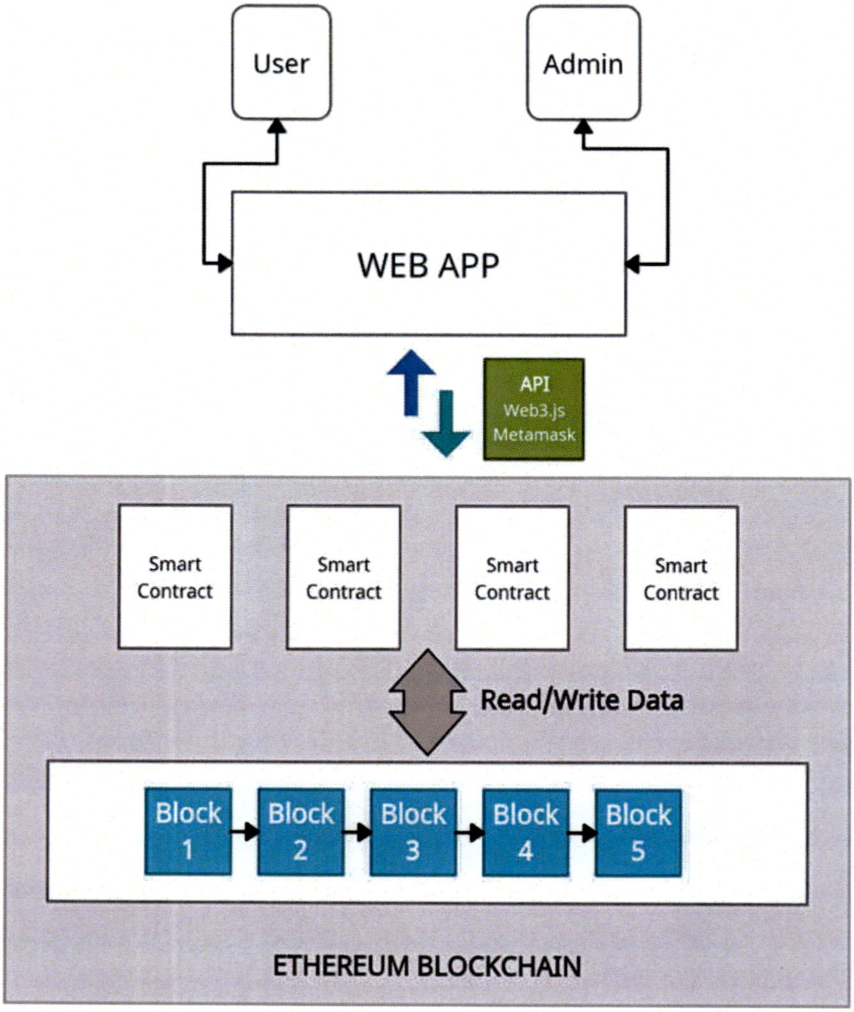

Fig. 2 Architecture diagram

4.3 Data Flow Diagram

Figure 3, shows how the data flows through the system.

1. As mentioned in the use case, diagram two actors are accessing the system, i.e., user and admin. Admin can update or insert data on the blockchain while the user can only verify the credentials.

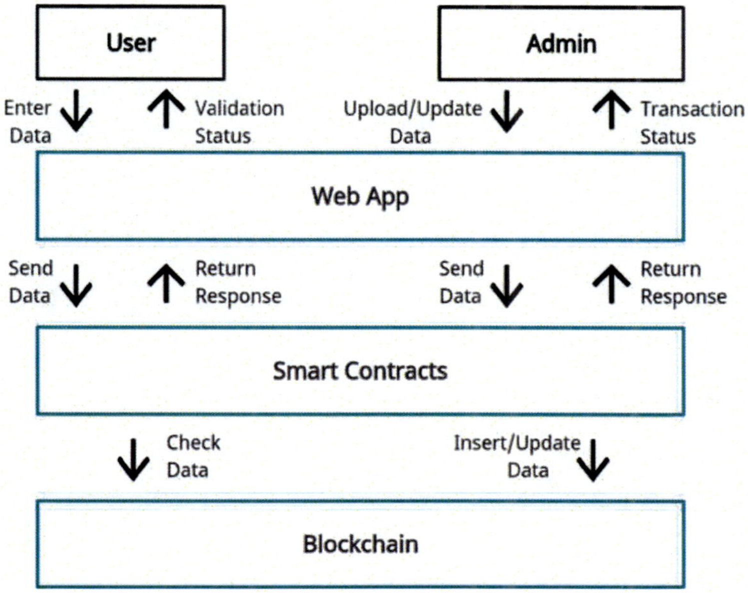

Fig. 3 Data flow diagram

2. User or admin sends a request through a Web app to the smart contract on the blockchain through web3.js. The request is executed on the blockchain with the help of a smart contract where all the instructions (Business logic) are written.
3. Once the request is executed, it sends back the response to the Web app to display it to the user.
4. If an unauthorized user makes an admin access request, the function will not execute and give an error message.

5 Prototype

5.1 Blockchain Environment

We used Ganache that is Ethereum blockchain an open-source decentralized public and permissionless blockchain-based platform providing computer applications to run on top of it. We are developing our application on it which is DApp. Below is Ganache Ethereum local blockchain which contains ten accounts each containing 100.00 ETH fake ethers.

We used the truffle framework, which is a development and testing framework and enables asset pipeline for Ethereum that makes Ethereum development faster and structured. We used the metamask plugin which is an extension for having access to the smart contract deployed in Ethereum/Dapps in the browser. This extension injects the Ethereum Web3 APIs into any website's javascript contents, which results in communication between DApps and the blockchain.

5.2 *Smart Contract*

The implementation of the business logic required for the verification system is done by writing the smart contracts on the immutable blockchain ledger. The smart contract is programmed in solidity, which is an object-oriented programming language for smart contract development. The development and deployment of two smart contracts

1. Smart Contract for PAN Card Verification
2. Smart Contract for College Transcripts Verification

is done to demonstrate the working modules of the project.

1. Smart Contract for PAN Authentication
   - Algorithm
     (a) Declare structure for PAN Card with all the PAN Card Variable details
     (b) Write the Function uploadPan() which adds the PAN Card in immutable distributed ledger.
     (c) Write the function verifyPan() which is used to Authenticating or verifying one's PAN in an immutable chain of data.
   - Code

```
\noindent
// Pan_Validity.sol
contract Pan_Validity {
  string public name = "Pan Validity Smart Contract";

  address public minter;

   constructor() public {
        minter = msg.sender;
    }

  uint itemcount = 0;

  mapping(uint => Pan_Struct) public pan_list;

  struct Pan_Struct{
    uint id;
    uint pan_no;
    string name;
    string dob;
    string gender;
  }
  function uploadPan(uint  _pan_no,string memory _name,string memory _dob,
  string memory _gender) public {
```

```
        require(msg.sender == minter);
        pan_list[itemcount] = Pan_Struct(itemcount,_pan_no,_name,_dob,_gender);
        itemcount++;
    }

    function verifyPan(uint  _pan_no,string memory _name,string memory _dob,
    string memory _gender) public view returns(uint){
        uint f = 0;
        for(uint i=0;i<itemcount;i++)
        {
          if(pan_list[i].pan_no == _pan_no &&
          keccak256(bytes(pan_list[i].name)) == keccak256(bytes(_name)) &&
          keccak256(bytes(pan_list[i].dob)) == keccak256(bytes(_dob)) &&
          keccak256(bytes(pan_list[i].gender)) == keccak256(bytes(_gender)))
            {
              f=1;
              break;
            }
        }
        return f;
    }
}
```

2. Smart Contract for College Transcripts Authentication
   - Algorithm
     (a) Declare structure for College Transcript with all the Transcript Variable details
     (b) Write the Function uploadTranscript() which adds the College Transcript in immutable distributed ledger. distributed ledger.
     (c) Write the function verifyTranscript() which is used to Authenticating or for verifying one's record in an immutable chain of data.
   - Code

```
// College_Transcript.sol
contract college_transcript{
    string public name = "Collage Transcript verify smart contract";

    address public minter;

   constructor() public {
        minter = msg.sender;
    }

    struct transcript_struct{
        uint id;
        string unique_reg_no;
        string name;
        string gender;
        string university;
        string cgpa;
        string degree;
        string year_of_passing;
    }

    uint itemcount = 0;

    mapping(uint => transcript_struct) public transcript_list;
```

```
    function uploadTranscript(string memory _unique_reg_no,string memory _name,
    string memory _gender,string memory _university,string memory _cgpa,string
    memory _degree,string memory _year_of_passing) public {
        transcript_list[itemcount] = transcript_struct(itemcount,_unique_reg_no,
        _name,_gender,_university,_cgpa,_degree,_year_of_passing);
        itemcount++;
    }

    function verifyTranscript(uint _id,string memory _unique_reg_no,
    string memory _name,string memory _gender,string memory _university,
    string memory _cgpa,string memory _degree,
    string memory _year_of_passing) public view returns(uint){
        uint f=0;
        for(uint i=0;i<itemcount;i++)
        {
          if((transcript_list[i].id == _id ) &&
         (keccak256(bytes(transcript_list[i].unique_reg_no )) ==
          keccak256(bytes(_unique_reg_no))) &&
         (keccak256(bytes(transcript_list[i].name)) ==
          keccak256(bytes(_name))) &&
         (keccak256(bytes(transcript_list[i].gender)) ==
          keccak256(bytes(_gender))) &&
         (keccak256(bytes(transcript_list[i].university)) ==
          keccak256(bytes(_university))) &&
         (keccak256(bytes(transcript_list[i].cgpa)) ==
          keccak256(bytes(_cgpa)))  &&
         (keccak256(bytes(transcript_list[i].degree)) ==
          keccak256(bytes(_degree))) &&
         (keccak256(bytes(transcript_list[i].year_of_passing)) ==
          keccak256(bytes(_year_of_passing))) )
            {
                f=1;
                break;
            }
        }
        return f;
    }
}
```

As soon as we deploy the smart contracts on the immutable distributed ledger, as the ledger is being updated, it is considered as a transaction and it costs some fee, i.e., gas (smaller unit of ETH) which will be deducted from the deployer's account. In such a way, each operation that modifies the ledger costs a fee. The deployment of these smart contracts is shown in Fig. 4.

Now the system for verification is ready, and the communication with the system can be done using the truffle console as shown in Figs. 5 and 6.

5.3 Web App

To seamlessly connect the distributed backend with a front end, we have created a Web app that communicates with the Blockchain using a meta mask. The Web app is developed using ReactJS. As shown in the figure, the Web app removes all Blockchain complexities from the end-user making it very user-friendly and easy to access using the Internet (Figs. 7 and 8).

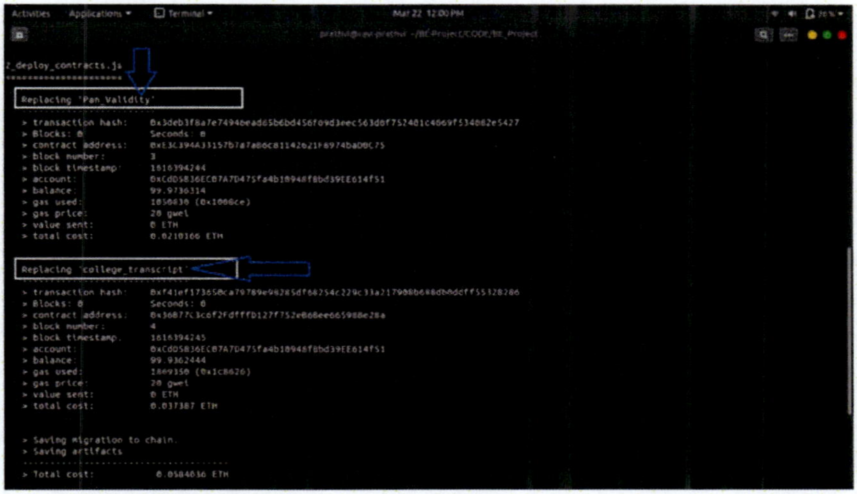

Fig. 4 Deployment of smart contract

Fig. 5 Pan validity smart contract

6 Results and Discussion

6.1 Results

Through the use of our proposed architecture following results are observed:

1. Multiple transactions can be processed simultaneously.

Fig. 6 College transcript smart contract

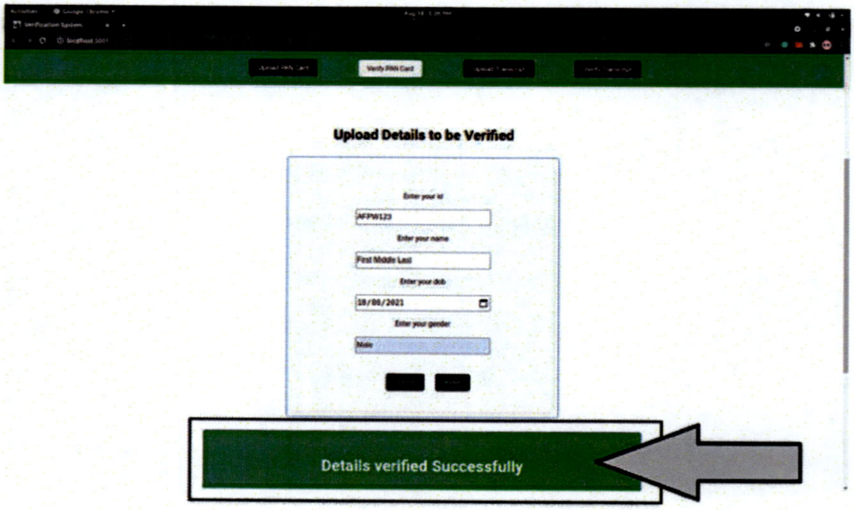

Fig. 7 Verified information

2. As it is a decentralized system, therefore, data is secured even after a system failure.
3. Trustworthy and Reliable: Blockchain is not immune to hacking, but being decentralized gives blockchain a better line of defense. To alter a chain, a hacker needs to control more than half of all the computers in the same distributed ledger. It is nearly impossible, and therefore, it is trustworthy and reliable.

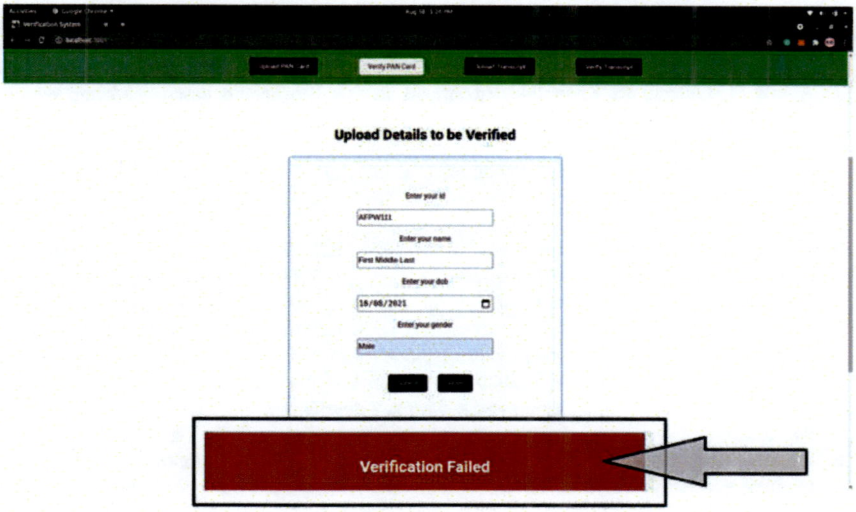

Fig. 8 Information not-verified

4. Data Tampering: Each block is connected to all the blocks before and after it. This makes it hard to tamper with a single record because an intruder would need to change the block/s containing that record as well as those linked to it to avoid detection. Network participants have their private keys that are assigned to the transactions they make and act as a personal digital signature. If a record is altered, the signature will become invalid and the peer network will know right away that something has happened. Therefore, data is secured and there cannot be any tampering of data.

7 Conclusion

As society becomes increasingly technological and Internet-centric, a better form of digital verification is necessary and current systems of verification are becoming increasingly insecure. A major overhaul of these systems is inevitable and the blockchain is a possible avenue to explore in solving these problems. Thus, the proposed system will overcome all these issues using the distributed immutable ledger and will transition everything from the current insecure system into the more secure, reliable, and the best way of verification.

References

1. Aamir M, Qureshi R, Khan FA, Huzaifa M (2020) Blockchain based academic records verification in smart cities. Wireless Pers Commun. https://doi.org/10.1007/s11277-020-07226-0
2. Adam EEB, Sathesh (2021) Survey on medical imaging of electrical impedance tomography (EIT) by variable current pattern methods. J ISMAC 2(2):82–95. https://doi.org/10.36548/jismac.2021.2.002
3. Belurgikar DA, Kanak Kshirsagar J, Dhananjaya KK, Vineeth N (2019) Identity solutions for verification using blockchain technology. In: 2019 1st international conference on advanced technologies in intelligent control, environment, computing communication engineering (ICATIECE), pp 121–126. https://doi.org/10.1109/ICATIECE45860.2019.9063802
4. Buterin V (2015) On public and private blockchains. In: Ethereum blog. https://ethereum.org/
5. Dwivedi SK, Amin R, Vollala S (2020) Blockchain based secured information sharing protocol in supply chain management system with key distribution mechanism. J Inf Secur Appl 54:102554
6. Iqbal M, Matulevičius R (2019) Comparison of blockchain-based solutions to mitigate data tampering security risk. In: Di Ciccio C, Gabryelczyk R, García-Bañuelos L, Hernaus T, Hull R, Indihar Štemberger M, Kő A, Staples M (eds) Business process management: blockchain and central and eastern Europe forum. Springer International Publishing, Cham, pp 13–28
7. Jaffer SA, Pandey S, Mehta R, Bhavathankar P (2020) Blockchain based direct benefit transfer system for subsidy delivery. In: 2020 international conference for emerging technology (INCET), pp 1–6. https://doi.org/10.1109/INCET49848.2020.9154178
8. Juno Bella Gracia SV, Raghav D, Santhoshkumar R, Velprakash B (2019) Blockchain based aadhaar. In: 2019 3rd international conference on computing and communications technologies (ICCCT), pp 173–177. https://doi.org/10.1109/ICCCT2.2019.8824892
9. Khan PW, Byun YC, Park N (2020) A data verification system for CCTV surveillance cameras using blockchain technology in smart cities. Electronics 9(3). https://doi.org/10.3390/electronics9030484, https://www.mdpi.com/2079-9292/9/3/484
10. Lee KM, Ra I (2020) Data privacy-preserving distributed knowledge discovery based on the blockchain. Inf Technol Manag 21(4):191–204
11. Malik G, Parasrampuria K, Reddy SP, Shah S (2019) Blockchain based identity verification model. In: 2019 international conference on vision towards emerging trends in communication and networking (ViTECoN), pp 1–6. https://doi.org/10.1109/ViTECoN.2019.8899569
12. Marella V, Vijayan A (2020) Document verification using blockchain for trusted CV information
13. Nakamoto S (2008) Bitcoin: a peer-to-peer electronic cash system. https://bitcoin.org/bitcoin.pdf
14. Nosouhi MR, Yu S, Zhou W, Grobler M, Keshtiar H (2020) Blockchain for secure location verification. J Parallel Distrib Comput 136:40–51
15. Sivaganesan D (2021) A data driven trust mechanism based on blockchain in IoT sensor networks for detection and mitigation of attacks. J Trends Comput Sci Smart Technol 3:59–69. https://doi.org/10.36548/jtcsst.2021.1.006
16. Sungheetha A (2021) 3D image processing using machine learning based input processing for man-machine interaction. J Innov Image Process 3(1):1–6. https://doi.org/10.36548/jiip.2021.1.001

A Review on Data Hiding Mechanism for Enhancing Embedding Capacity

Priyanka V. Deshmukh, A. S. Kapse, and V. M. Thakare

Abstract Reversible data hiding is used to embed information or data in the encrypted domain. In RDH, an input image is converted into unreadable format with encryption technique by using secret key and while transmission, additional information is embed in the encrypted image, with no knowledge of image content. At the time of decoding, extraction of embedded data can be performed and the input image can be reconstructed is the original image. Reversible data hiding allows hiding large amount of information, whereas existing methods does not allow hiding large amount of information. Here, a new reversible method with a very high payload that means enhancing embedding capacity is proposed. In the proposed research approach, intention is to use image as a carrier object for hiding data along with cryptography technique. Combination of these two will provide more security to the sensitive data. Lots of techniques have been studied and conclusion made that combination of image steganography and cryptography will provide peak level of security to the data. Steganography hide data behind image and cryptography completely convert that image into another meaningless image. This system allows hiding large amount of information, and it also proposes high-embedding rate (Payload) with good reformation of image quality (PSNR). For evaluating the performance of system, parameters like PSNR, MSE and Cross Correlation will be considered.

Keywords Image encoding · Image decoding · Steganography · Cryptography · Image security

P. V. Deshmukh (✉)
Department of CSE, SSGMCE, Shegaon, India

A. S. Kapse
Department of IT, AEC, Chikhali, India

V. M. Thakare
Department of CSE, SGBAU, Amravati, India

S. Shakya et al. (eds.), *Proceedings of Second International Conference on Sustainable Expert Systems*, Lecture Notes in Networks and Systems 351,
https://doi.org/10.1007/978-981-16-7657-4_40

1 Introduction

Security of the digital image plays an important role in the fields of confidential areas such as military and medical worlds. By using the growth of cloud computing, the growth in information technology has taken to different security problems whereas confidentiality, authentication and integrity are constantly threatened, through the illegal parts like the gaining of unauthorized access to data, copying data or malicious use of our data. Encryption focuses on the data privacy by randomizing the content of original input image. At the time of transmission of encrypted images, it is essential to analyze the image with no knowledge of content of the image and also secret key used during encryption [1]. Data hiding uses the data where hiding of data inside image is performed and also widely used for copyright protection and index based authentication. It embeds data in image, and then hidden data in the image is sent to the receiver. Input may include image, text, audio, videos. In some cases such as military, medical and legal applications, distortion are intolerable [2].

Data hiding has given more importance to the process in the field of secure data transmission over wireless network. There are too many techniques existing that provide security to the sensitive data especially when it is transmitted over network. It is found that available techniques are not enough to protect data because it shows the existence of data over network. Intention of this research work is to completely hide the existence of sensitive data behind any object called it as carrier object. An existence of sensitive data creates a challenges for an intruder to find out decode which is quite impossible. There are lots of techniques to hide the data behind carrier object like image steganography, audio and video steganography. Text steganography is also existing but not too much popular due to its complexity and low-data heading rate. Reversible data hiding is use to insert data in the encrypted domain. In RDH, encryption of input image is performed by using secret key and while transmission, additional information is inserted in the encrypted image, with no knowledge of content of the image [3]. At the time of decoding, extraction of the embedded data can be performed and the original input image can be restored. Reversible data hiding allows hiding large amount of information, whereas existing methods do not allow hiding large amount of information. Reversible data hiding methods are designed for enrichment of data and authentication. The important part in reversible data hiding is payload that is nothing but data embedding capacity and the quality of the restored image. Many methods have developed but, the embedding capacity is very less, and quality of image is poor [4].

In data hiding approach, it uses image as a carrier object for hiding data along with cryptography technique. Combination of these two will definitely provide more security to the sensitive data. Lots of techniques have been studied and conclusion made that combination of image steganography and cryptography will provide the security of the data, whereas steganography hides data behind image. Cryptography includes encryption and decryption, and it completely convert the image into another meaningless image. Reversible data hiding techniques are categories into difference expansion, histogram shifting and lossless compression. Actually in reversible data hiding, it is having two steps. The first step is generation of image with low entropy. In the second step, it is use to insert additional data into image [5]. In all the existing methods, the embedding capacity of image is low and also quality of the image is not good. Some methods provide high-embedding capacity that is payload, but the extracted image is change if comparison is performed with the original image. In many papers, existing methods used are data hiding by LSB (least significant bit) substitution. In LSB, substitution embedding capacity is not more than 1 bit per pixel, and image quality is also low.

2 Literature Review

Qian et al. [6], the author presented and addressed the previous issues and introduced separate embedding and extraction of reversible data hiding procedure in JPEG bit pattern. In recent years, research has given more attraction toward the signal processing. Also in cloud computing when users are not ready to display the content of the signal that time encrypted copy may send to the remote server. The signal processing in the encrypted domain can be accomplish by the server. Reversible data hiding gives the permission to server to insert the data into encrypted image and which is then uploaded by the content of the owner, also it gives the guarantee that contents cannot be alter at the time of decryption. Generally, embedding capacity that is nothing but payload is related to the reversibility. As existing techniques are used to develop for uncompressed images, this paper addressed a separate embedding and extracting reversible data hiding method in JPEG bit pattern. The author has compared work with the existing; the existing method has divided into three parts. First, separate methods are required for extracting the data and image recovery, while assuming both blocks are realized jointly. Second, to avoid the reveal of image content steps for rearrangement and enciphering are proposed, make sure that method should be secure. Third, image compression and reconstructing algorithm are proposed to hide data. Author introduced an encryption algorithm on JPEG, which resize image into smaller part. The workflow having three parties such as owner of the content, to whom who hides the data, receiver. Content owner generates cipher text by performing encryption, data hider adds the data inside the image and in receiver side extraction of data is performed from received bit stream.

Pu et al. [1], the author introduced two bit MSB prediction data hiding in reversible manner. Development in cloud computing and cloud storage is performing very fast; for personal data more importance is given to the privacy and security problems. To provide, the privacy encryption is used. Encryption encrypt the original image and convert it into unreadable. In order to share file with other receiver which are related to multimedia, sender or content owner convert the message into unreadable format before sending then it is possible to embed the data. After decryption original message is recovered and extraction of data is performing at receiver side. In the existing method, no one is considering relation between neighboring pixels and data hiding is done only LSB substitution. Here, author has explored neighboring pixel concept that is data hiding is performed by taking two MSB substitutions. In this paper, encryption of original image is done by using XOR operation, and it uses prediction error concept by using error location binary map.

Qian et al. [2], the author addressed progressive recovery based reversible data hiding. They summarize the existing methods and found that the idea of reversible data hiding is used in cloud storage and medical systems application. Encryption of image is performed by the content owner to preserve the privacy and after encryption; encrypted data is stored on cloud in cloud storage. While managing the encrypted data, additional message like labels, any information can be embedded. The existing system having three parties such as owner of the content, to whom who hides the data, receiver. Content owner generate cipher text by performing encryption, data hider insert the data inside the image using LSB substitution and in receiver side extraction of data is performed from received bit stream. LSB substitution is not secure and does not provide that level of privacy. Embedding capacity that is payload is low in the LSB substitution. After the analysis, it is found that improvement is done by using progressive recovery. The reversible data hiding progressive recovery gives better prediction way in the LSB substitution.

Yin et al. [3], the author introduces multi-MSB prediction based reversible data hiding. Author has summarized existing methods, data hiding technique is use to embed the data, and extracted embedded data should not content error. Existing methods are divided into two different parts; these are reserving room (RRBE) and vacating room (VRAE). In VRAE, encryption of image is performed using encryption procedure then data can be embedded by doing some modification in the pixels of encrypted image. RRBE method is different from VRAE method, before encryption, RRBE method is use to reserve the room for embedding data. It provides the privacy protection in cloud. At the decoding, extraction of data should be error free.

Puteaux et al. [7], the author have introduced EPE based reversible data hiding for enhancing the embedding capacity in encrypted images. While transmitting the encrypted image, it is important to analyze the process so that content of image and also secret key will not be leak after encryption phase. Actually, RDH based on EPE taken more research attention and developed different methods in the image processing. Data enrichment and authentication of image are used in methods of reversible data hiding. In decoding phase, extraction of data should be error free and recovery of image is necessary. The author has proposed the method in that it focuses on the payload that is near about 1 bit per pixel but not more than 1. Firstly, encryption of image is done by using encryption techniques, and embedding data inside the image is possible. Author also uses prediction errors for the detection purpose. For the prediction, it takes neighboring pixels so that information regarding PE location can be calculated. If first bit plane in encrypted image is indicated by $I[k]$ then next bit indicated by $I[k + 1]$.

Yi et al. [4], the author presented parametric binary tree labeling based separate method for data hiding and extraction of data in reversible manner. Encryption is procedure to convert the image into unreadable format image so that no unauthorized user can access this image. Encryption can be done by using secret key and at the decryption receiver can see the original image. Hiding of data in reversible data hiding is done by changing the particular pixels value. The parametric binary tree labeling is used for labeling the pixels. It makes the parts of image into small blocks and then inserts data inside the image.

Ja et al. [8], the author introduced 3D object reversible data hiding method and this method worked on encrypted image. This method develop the marked 3D object which is same as the original image at the time of recontruction. As it works on the 3D objects so this objects are based on the paillier cryptosystem, with the help of this method reconstructed image quality is increased.

3 Research Issues

There are various research issues identifies such as privacy, originality and integrity of the data is one of the major issue in reversible data hiding. In many of the existing methods embedding capacity is very low, also there are problems regarding imperceptibility and distortion. Constructing high-performance multiple histograms and protecting the users image data in both academia and industry, this is also the issue in data hiding. As hiding of data is performed but the focus is on the reliability of the hidden data against the small changes in the image such as image compression. In existing methods, embedding capacity of the image is less that is limited embedding capacity is there so it can be extended by using histogram of prediction errors and MSB substitution. Extraction of data and restoration of image is also important phase in this, data and image should not be altered. Another issue is security of the image at the time of transmission and exchange of image.

4 Existing Methodologies and Analysis

4.1 Joint Method

The author uses AES for encrypting the input image and for embedding data the bits are selected randomly in 4 * 4 block pixels. At the time of decoding the image, author analyzes the standard deviation. In this method, the embedding capacity that is payload is somewhat small which is near to 0.0625 bits per pixel [7].

4.2 XOR Operation

The author performed encryption using XOR operation. Then, after encryption, this input image is being divided into number of blocks and each and every block having two sets. In set, it uses three bit LSB substitution in which each pixel were used to get out of room for embedding the data. In decoding procedure, process is followed to recover the image and data is extracted. Also embedding capacity that is payload is less which is 0.1 bits per pixel [9].

4.3 Side Match Techniques

Side match technique is addressed by the author and uses formula for smoothness evaluation. Here, after reconstructing, the image quality is not good this is near to 30 dB, whereas the payload is 0.518 bits per pixel which is high if compared with the other techniques [10].

4.4 Histogram Shifting

To enhance the performance of histogram shifting, while embedding data, data hiding on reference pixel and selection of block method is introduced by the author. Actually, it gives low payload and it leads to degradation of the image quality also. Algorithms used in histogram shifting gives high-image quality but with less embedding capacity [11].

5 Scope

The data hiding mechanism will work efficiently for enhancing embedding capacity and reconstruction of image quality. MSB substitution based reversible data hiding will be used to cover the data inside image. The performance parameters consider for system is peak signal to noise ratio (PSNR), mean square error (MSE), min value, max value and cross correlation. Also, this system is to provide security to the data irrespective of its sensitiveness over network, and it can be used anywhere where there is a need for sharing sensitive data or information like password, account number and other details which are not easily shared. As steganography completely hide the existence of data, there is no need of other additional scenarios used to protect data. Data hiding capacity of proposed system can be improved further by incorporating some robust techniques. Here, prediction of future pixels depends on the past pixels.

6 Methodology

In the research approach, intention is to use image as a carrier object for hiding data along with cryptography technique. Combination of these two will provide more security to the sensitive data. Lots of techniques have been studied, and conclusion made that combination of image steganography and cryptography will provide peak level of security to the data. Steganography hide data behind image and cryptography completely convert that image into another meaningless image. This data hiding mechanism is use for enhancing capacity using reversible data hiding technique. It works on encrypted input image based reversible data hiding and hiding data inside image. This system provides higher payload that is embedding capacity which is more than 1 bit per pixel and after reconstruction image quality will be good.

This system works as;

- Sender select image as a carrier object.
- Sender performs visual cryptography on the image.
- Sender extracts pixels from carrier image.
- Sender hides input secret data in carrier image.
- Receiver extracts data from pixel component of image.
- Receiver performs visual cryptography on cryptic image to decode an image with the public key.

Here, it is the overview of encoding method, encoding phase includes three main steps as; MSB prediction error detection, encryption and error consideration, data hiding by MSB substitution. Aim is to encrypt the original image or source image I by using secret key Ke and embed the message by using data hiding key Kw without knowlegde of Ke. Then, marked encrypted image Iew is obtained (Figs. 1 and 2).

The secret message is embedded by MSB substitution, and then original MSB value is lost after data hiding. So, during decoding phase, it is able to predict this

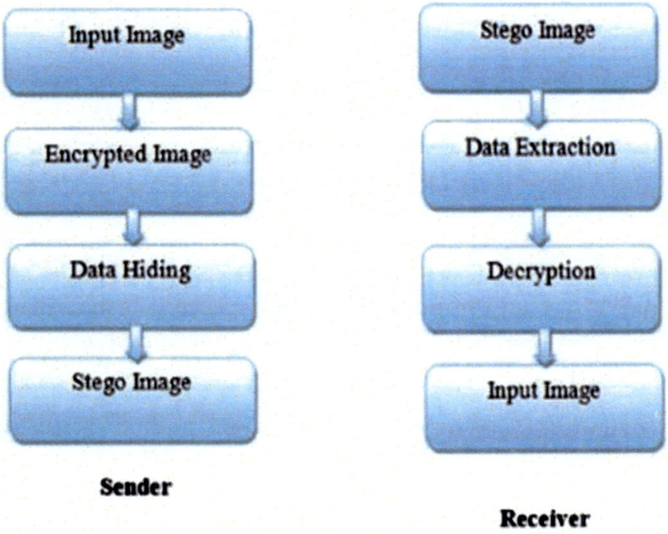

Fig. 1 Data hiding and data extraction process

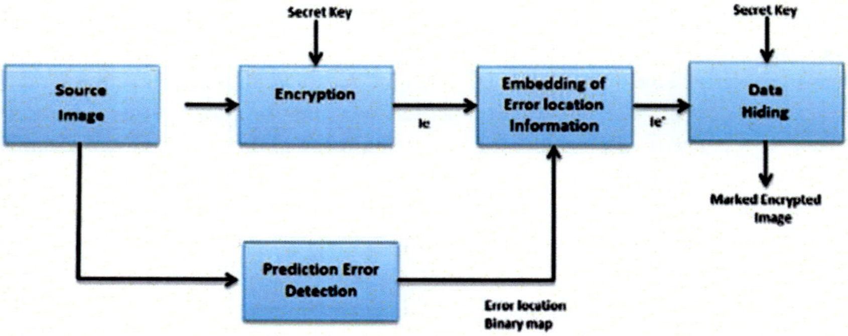

Fig. 2 Overview of encoding method

without any errors. In order to reconstract the original image, previous pixel is required to predict the current pixel values. At the time of encryption, by using chaotic generator, a sequence of pseudo-random bytes is obtained and with the help of XOR operation, encryption can be done. In the data hiding, pixels of encrypted image are scanned from left to right and then from top to bottom and MSB of each available pixel is substituted by one bit. During prediction error detection, the location of the prediction error is stored in the error location binary map.

7 Conclusion

Reversible data hiding is used to insert data in the encrypted domain. By substituting bit values of pixels in the encrypted image, a larger data can be covered and in decoding phase, the original input image can be recaptured. This system provides good security level, and it can be used to preserve original image content confidentiality. The performance is calculated on the basis of the parameters like PSNR, MSE and cross correlation. This system measures the accuracy and different performances depending on extracted bits, the payload that is embedding capacity and the reconstructed image quality after data extraction. Merits of this system are to enhance the embedding capacity as compared to the other other system, existing system having less payload which is less that 1 bits per pixels. Appilcation of this sytem is to hide the confidential data, also in the QR code, in military and medical field.

References

1. Puyang Y, Yin Z, Qian Z (2019) Reversible data hiding in encrypted images with two-MSB prediction

2. Qian Z, Zhang X, Feng G (2016) Reversible data hiding in encrypted images based on progressive recovery. IEEE Signal Process Lett 23(11):1672–1676
3. Yin Z, Xiang Y, Zhang X (2018) Reversible data hiding in encrypted ımages based on multi-MSB prediction. IEEE Trans
4. Yi S, Zhou Y (2019) Separable and reversible data hiding in encrypted ımages using parametric binary tree labeling. IEEE Trans Multim 21(1)
5. Zhou J, Sun W, Dong L et al (2016) Secure reversible image data hiding over encrypted domain via key modulation. IEEE Trans Circuits Syst Video Technol 26(3):441–452
6. Qian Z, Hang Z, Zhang W (2018) Separable reversible data hiding in encrypted JPEG bitstreams. IEEE Trans Dependable Secur Comput 15(6)
7. Puteaux P, Puech W (2018) EPE based huge capacity reversible data hiding in encrypted ımages. IEEE Int Workshop Inf For Secur (WIFS)
8. van Rensburg BJ, Puteaux P, Puech W, Pedeboy J-P (2021) Homomorphıc two tıer reversıble data hıdıng ın encrypted 3D objects. In: IEEE ınternational conference on ımage processing (ICIP).978-1-6654-4115
9. Huang F, Huang J, Shi YQ (2016) New framework for reversible data hiding in encrypted domain. IEEE Trans Inf For Secur 11(12):2777–2789
10. Hong W, Chen T, Wu H (2012) An improved reversible data hiding in encrypted images using side match. IEEE Signal Process Lett 19(4):199–202
11. Zhang R, Lu C, Liu J (2019) A high capacity reversible data hiding scheme for encrypted covers based on histogram shifting. J Inf Secur Appl. Elsevier
12. Tai W-L, Yeh C-M (2009) Reversible data hiding based on histogram modification of pixel differences. IEEE Trans Circuits Syst Video Technol
13. Zhang X (2011) Reversible data hiding in encrypted images. IEEE Signal Process Lett 18(4):255–258
14. Tai W-L, Yeh C-M, Chang C-C (2009) Reversible data hiding based on histogram modification of pixel differences. IEEE Trans Circuits Syst Video Technol 19(6):906–910
15. Yu J, Qian Z (2015) Reversible visible watermark embedded ın encrypted domain
16. Nikolaidis A (2015) Reversible data hiding in JPEG images utilising zero quantised coefficients. IET Image Processing
17. Hassan FS, Gutub A (2016) Novel embedding secrecy within images utilizing an improved interpolation-based reversible data hiding scheme. J King Saud Univ-Comput Inf Sci
18. Jiang R, Zhou H, Zhang W, Yu N (2018) Reversible data hiding in encrypted 3D mesh models. IEEE Trans Multimedia 20(1)
19. Ni Z, Shi Y-Q, Ansari N, Su W (2006) Reversible data hiding. IEEE Trans Circuits Syst Video Technol 16(3):354–362
20. Cai S, Gui X (2013) An efficient reversible data hiding scheme based on reference pixel and block selection. IEEE Trans Circuits Syst Video Technol
21. Qian Z, Zhang X (2016) Reversible data hiding in encrypted ımages with distributed source coding. IEEE Trans Circuits Syst Video Technol
22. Manikandan VM, Bini AA (2019) An ımproved reversible data hiding through encryption scheme with block prechecking. In: Third ınternational conference on computing and network communications (CoCoNet'2019)
23. Kumar R, Jung K-H (2020) Enhanced pair wise IPVO-based reversible data hiding scheme using rhombus context. J Inf Sci
24. Guan B, Xu D (2020) An efficient high-capacity reversible data hiding scheme for encrypted images. J Image Commun Image Representation
25. Wu X, Yang C-N, Liu Y-W (2020) A general framework for partial reversible data hiding using hamming code. J Signal Processing
26. Dzwonkowskı M, Rykaczewskı R (2021) Reversible data hiding in encrypted dıcom ımages using cyclic binary golay (23, 12) code. IEEE Access 9

Deep Convolutional Neural Network for Brain Tumor and Skin Cancer Detection Over Traditional Neural Networks

T. Ashish Narayan, Ch. Anudeep, Pavan Nageswar Reddy Bodavarapu, and P. V. V. S. Srinivas

Abstract According to the skin cancer foundation, skin cancer is detected in one out of every three malignancies, and skin cancer affects one out of every five Americans at some point in their lives. This research work considers a total of 24,530 malignant brain or spinal cord tumors (13,840 men and 10,690 females). The statistics would be substantially higher if benign (non-cancerous) tumors were included. The proposed model was trained separately with the datasets obtained from skin cancer (SC) and brain tumor (BT) with an overall accuracy of about 0.95 and 0.72, respectively. As a consequence, Resnet-50, VGG-19, and VGG-16 were used to compare the proposed model. On a brain tumor dataset, the suggested model will have 0.95 accuracy rate.

Keywords Brain tumor · Skin cancer · Deep learning · Convolutional neural network

1 Introduction

Skin cancer classification using a convolutional deep neural network, Here, classified the skin cancer using some techniques. The data includes different images of skin cancers. After implementation, it has shown different results which includes basal cell carcinoma, and squamous cell car cinoma has 95.1% and 91.8%, respectively [1]. Taken a dataset from a hospital in Stanford. The dataset consists of different images which includes blurry images, for-away images. These images can cause a lot of changed the noise is removed. Here, CNN is used which turns out to 93.33% accuracy [2]. Collected image data set which contains 1,300 biopsied images. Images that contain both body and skin are visualized and checked for various area in skin where there are spots, yellowish areas. CNN is applied to this to get high accuracy. Spots suspicious points on skin to determine the cancer [3]. Took data based on MRI scans. Removed noise from the data. Look overed the overfitting of CapsNets. Used CNN in implementing since it has more learning capacity. It performs less for small

T. Ashish Narayan (✉) · Ch. Anudeep · P. N. R. Bodavarapu · P. V. V. S. Srinivas
Department of Computer Science and Engineering, Koneru Lakshmaiah Education Foundation, Guntur, India

S. Shakya et al. (eds.), *Proceedings of Second International Conference on Sustainable Expert Systems*, Lecture Notes in Networks and Systems 351,
https://doi.org/10.1007/978-981-16-7657-4_41

datasets. So, capsule networks are used since these network handles small datasets. This will improve the accuracy [4, 5].

Focused on different group of tumors within the CNS. Analyzed how the brain forming in different ways. classified brain tumors into different types in it. Took data from US and classified by percentages. Classified ways of getting the brain tumor [6]. Used deep learning models for cancer diagnosis. Mostly in every execution, big problem is execute any algorithm on the dataset but in this paper all access links for the dataset which present in the work. Also, CNN different models are described in manuscript. In performance of algorithms in brain, skin, breast, and lung cancer are center of attraction is manual script. The implementation is measured provided by lot of studies which are dissimilar [7, 8]. Decrease the draw bags for classification of feature introduced deep cues. It is a deep learning using CNN model to tell the features of the DNA data. That DNA date of infection classification Deep Cues is applied to date from TCGA analysis different types which of seven major cancers in this we get 77.6% accuracy [9, 10].

The goal is to identify of cervical cancer automatically. So, in these 668 samples, 4 target variable and 30 attributes are used. Soft max classification with collected auto encoder this auto encoder collection applied to the dataset. In this data, sample was divided into two phase one is training for 70%, and another phase is testing of 30% and few machine learning methods are used. For the first time, the autoencoder was used applied for cervical cancer 97.8% is the rate of correct identification other than machine learning [11]. HTS was used to identify the urine tumor DNA and this called as UtDNA CAPP-seg. It detects the BLCA because Cytology missed 82% this UtDNA CAPP-seq applied to 67 normal healthy people and applied to 188 illness people of urine whose collected prior for the approach which identified 6 variations per BLCA having people [12]. The date was used to classify is histological images used by Gleason grading. Epithelial network is used by RCNN for multitasking prediction. Using fivefold cross validation, 99.07% epithelial cell identification accuracy is 89.40%-pixel accuracy [13]. CNN based of cancer subtype classification method by attaching multi-instance and multi scale to reduce real-time problems 196 people data collected by different hospitals. This CNN-based cancer is more accuracy and better performance than standard CNN and different conventional methods [14]. The proposed CNN system is more effective than existing ML models and traditional CNN model designs, as shown in Tables 1 and 2. And also model uses lesser trainable parameters when compared to other existing conventional models which results in lesser epoch time, i.e., overall execution time of the model will be reduced.

2 Related Work

Oliveira et al. [15] here, the author addressed about analyzing different patterns in the collected data. Data consists of images that has noise. This paper mainly focused on macroscopic and dermoscopic images. The author hopes to create more

Table 1 Accuracy and loss comparison of various models

S. No.	Cancer type	Model	Training accuracy	Testing accuracy	Training loss	Testing loss
1	Brain tumor	Own model	0.99	0.95	0.02	0.51
		VGG-16	0.97	0.94	0.05	0.42
		VGG-19	0.97	0.94	0.06	0.39
		Resnet-50	0.93	0.81	0.29	1.15
2	Skin cancer	Own Model	0.88	0.72	0.26	0.64
		VGG-16	0.89	0.68	0.34	1.19
		VGG-19	0.90	0.68	0.32	1.59
		Resnet-50	0.89	0.69	0.30	0.77

Table 2 Performance metrics comparison of various models

S. No.	Cancer type	Model	Class	Precision	Recall	F1-Score
1	Brain tumor	Own model	Benign	0.99	0.93	0.96
			Malignant	0.93	0.99	0.96
		VGG-16	Benign	0.98	0.94	0.96
			Malignant	0.94	0.98	0.96
		VGG-19	Benign	0.96	0.94	0.95
			Malignant	0.94	0.96	0.95
		Resnet-50	Benign	0.78	0.91	0.84
			Malignant	0.88	0.73	0.80
2	Skin cancer	Own model	Benign	0.78	0.63	0.70
			Malignant	0.69	0.82	0.75
		VGG-16	Benign	0.74	0.68	0.71
			Malignant	0.62	0.69	0.65
		VGG-19	Benign	0.68	0.78	0.73
			Malignant	0.66	0.55	0.61
		Resnet-50	Benign	0.71	0.80	0.73
			Malignant	0.70	0.59	0.60

efficient and effective expert systems in future. SVM classifier is also used because the generalization and simplification are good in non-linear data. Teck Yan Tan et al. [16] the author addressed about using the particle swarm optimization (PSO) algorithm for the optimal solution. It is an evolutionary algorithm. Feature extraction is applied to the data. It takes iterations to get the accurate data. Here, the maximum number of iterations are 500. This will be helpful to get to know about the skin more. H. R. Mhaske et al. [17] with the help of neural network and SUM going to identify the skin cancer because melanoma skin cancer is very dangerous it also can kill the human beings in this image processing is used supervised, and unsupervised algorithms are introduced.

Ilias Maglogiannis et al. [18] visual techniques to identify the skin cancer in this data set is digital images. To extract the characteristics from digital pictures operations techniques that methods are color and texture and border identification in this paper, explain the statistics and output of the executions that are in the papers and compare the implementations at different classifiers of the skin cancer. Mahamudul Hasan et al. [19] author introduced machine learning techniques and image processing to identify skin cancer artificially. The damaged skin calls feature is taken out for dermoscopic images. Deep learning technique is CNN, and it is used for standard feature extraction. 89.5% is an accuracy, and 93.7% is training accuracy. Khalid M. Hosny et al. [20] proposed skin lesion identification method is deep learning automated told. In this deep learning network of pre-trained and also used transfer learning, these models are proposed because they have ability to three classes of only. Major challenge is going to control by pre-trained Alex Net.

Achim Hekler et al. [21] the CNNs was used and in this dermatologists and artificial intelligence are work together to identify the skin cancer classification in this 11,444 dermoscopic images are used 112 dermatologists belongs to German universities of 13. Deep learning of five diagnostic classes. Pradeep Kumar Mallick et al. [22] the author used Deep Wavelet Auto encoder which has mage analysis as well as replication property of autoencoder. Brain-image data is collected which has noise in it and applied future extraction. Before with using DNN, the overall accuracy improved to is 91%. After using DWA, it has improved to 96%. By using DKA-DNN, a lot of terms has improved such as accuracy, sensitivity. Amin Zadeh Shirazi et al. [23] classified brain cancers into four classes brain cancer patients. Collected brain-image dataset and classified into four classes. Checked survival rates of each classes for few months on one class. Accuracy precession after this has got around 0.9 and 0.8. By using DeepSurvNet, accuracy the author got is 99% in testing phase.

Emanuele Torti et al. [24] for brain cancer detection, by using K-means algorithm, it fixes the boundaries and limits of image data. It improved the quality of images. By this, the map is generated that is unsupervised map. Image data is collected that has 100,000 subjects it includes MRI and CT Scans. K-Means defines with high accuracy and boundaries. Certain iteration is performed to increase the accuracy. Javaria Amin et al. [25] on MRI images, we use Long Short Term Memory (LSTM). The actual method got DSC (Dice similarity Coefficient) as 1,0.95, 0.99, 0.95 on each year whereby using LSTM, it achieved 0.97 DSC scorex. LSTM accurately classified low and high tumors than the previous methods. This model achieved 98% accuracy. Yalanati Ayyappa et al. [26] When compared to other classifiers such as PNN and SVM, a Gaussian Filtering with MLP-Based BPNN delivers higher accurate results in identifying the brain tumor with an accuracy rate of 93%. M. Jogendra Kumar et al. [27] the goal of this project might be to create a completely new tumor division mechanism using convolutional neural networks. On the test set of patches, a fix wise division process will be used, with 98% accuracy. Ramachandro Majji et al. [28] skin cancer is quickly becoming one of the most dangerous diseases that people may have. Skin cancers include basal, melanoma, carcinoma, and squamous cell carcinoma, with melanoma being the most unexpected. As a result, early identification of skin cancer is critical for successful treatment.

3 Proposed Work

3.1 About Dataset

Skin and breast cancer datasets are considered for experimentation and evaluation of the model. Both the datsets consists of two differnt classes namely Benign and Malignant. All the images in the datsets are converted to .jpg format, normalized by min-max normalization technique and resized to the dimensions 224 × 224. After resizing the images, they are stored in their respective directories. The suggested model is then trained on the training set and tested on the test set.

3.2 About Model

A convolutional neural network with six fully connected layers. Three connected layers, one flatten layer, and two thick layers were built for testing and assessment. A convolutional layer, a relu activation function layer, and a max pooling layer make up each fully linked layer. All six fully connected layers are connected in order, with the flatten layer receiving the output of the final fully connected layer, dense layer 1 and dense layer 2 following. The flatten layer and dense layer 1 consist of 512 hidden layers and dense layer 2 consists of two neurons to classify the output as benigm or malligant. Relu activation function is used in fully connected layers and dense layer 1 where softmax activation function is used in dense layer 2 where final output is taken. A cancer image size with dimensions 224 × 224 is given as input to the model. The description of the model architecture is given in Fig. 1 given below. The first fully connected layer consists of a conv2d_1 with 16 filters, activation_1 and maxpooling2d_1 with 16 filters, the input image given as input to the the conv2d_1 layer and relu activation function is applied on it and was fed to a maxpooling 2d_1 layer that resulted in dimensions 111 × 111. The first completely connected layer's output is delivered to the second fully connected layer, which is made up of the conv2d 2 layer with 32 filters, activation_2 and maxpooling2d_2 with 32 filters which resulted in an output dimensions of 54 × 54 that was given as input to the third fully connected layer with a conv2d_3 layer with 64 filters, activation_3 and maxpoolingg2d_3 with 64 filters resulted in an dimensions of 26 × 26 that was fed as input to the fourth fully connected layer with a conv2d_4 layer with 128 filters, activation_3 and maxpooling2d_3 with 128 filters resulted in an output dimensions of 12 × 12 and given for the fifth fully connected layer with conv2d_5 nad maxpolling_5 with 256 filters and resulted in output of 5 × 5 whose output is

```
Layer (type)                 Output Shape              Param #
=================================================================
conv2d_1 (Conv2D)            (None, 222, 222, 16)      448
_________________________________________________________________
activation_1 (Activation)    (None, 222, 222, 16)      0
_________________________________________________________________
max_pooling2d_1 (MaxPooling2 (None, 111, 111, 16)      0
_________________________________________________________________
conv2d_2 (Conv2D)            (None, 109, 109, 32)      4640
_________________________________________________________________
activation_2 (Activation)    (None, 109, 109, 32)      0
_________________________________________________________________
max_pooling2d_2 (MaxPooling2 (None, 54, 54, 32)        0
_________________________________________________________________
conv2d_3 (Conv2D)            (None, 52, 52, 64)        18496
_________________________________________________________________
activation_3 (Activation)    (None, 52, 52, 64)        0
_________________________________________________________________
max_pooling2d_3 (MaxPooling2 (None, 26, 26, 64)        0
_________________________________________________________________
conv2d_4 (Conv2D)            (None, 24, 24, 128)       73856
_________________________________________________________________
activation_4 (Activation)    (None, 24, 24, 128)       0
_________________________________________________________________
max_pooling2d_4 (MaxPooling2 (None, 12, 12, 128)       0
_________________________________________________________________
conv2d_5 (Conv2D)            (None, 10, 10, 256)       295168
_________________________________________________________________
activation_5 (Activation)    (None, 10, 10, 256)       0
_________________________________________________________________
max_pooling2d_5 (MaxPooling2 (None, 5, 5, 256)         0
_________________________________________________________________
conv2d_6 (Conv2D)            (None, 3, 3, 512)         1180160
_________________________________________________________________
activation_6 (Activation)    (None, 3, 3, 512)         0
_________________________________________________________________
max_pooling2d_6 (MaxPooling2 (None, 1, 1, 512)         0
_________________________________________________________________
flatten_1 (Flatten)          (None, 512)               0
_________________________________________________________________
dense_1 (Dense)              (None, 1)                 513
_________________________________________________________________
activation_7 (Activation)    (None, 1)                 0
=================================================================
Total params: 1,573,281
Trainable params: 1,573,281
Non-trainable params: 0
_________________________________________________________________
```

Fig. 1 Proposed CNN model architecture

given to tha last fully connected layer with conv2d_6, maxpooling_6 layers with 512 filters each and an activation_6 layer that resulted in an output dimensions of 1 × 1. The output of last fully connected layer is fed to a flatten layer with 512 neurons, and the resultant output is supplied to a dense layer where the output is categorized using the softmax function in the final dense layer.

3.3 *Algorithm*

Input: Breast Or Skin Cancer Image Dateset

Output: Image Classified as benigm or malligant

CNN Model For Cancer Detection(Img Dataset)

```
Begin
    if len(Img Dataset = Null)  then
        for all images in Img Dataset

            Convert(Img Dataset, .jpg)

            imgst = resize(Img Dataset, 224 X 224)

            imgst = MinMax Normalize(imgst)

            limgst = Assign Lab(imgst, benigm, maligant)

        End for loop

      Tr_st, Ts_st = Split(limgt, 80:20)

    End if

      Prop CNN Model = Train(Prop CNN Model)

      Evaluation_Metrics = Evaluate(Prop CNN Model()

        return (Evaluation_Metrics)

  End
```

3.4 *Model Architecture*

4 Experiment and Results

Across all datasets, our model consistently VGG16, VGG19, and Resnet-50 were all outperformed. The train and test accuracy were 0.99 and 0.95, respectively, when we used the brain tumor dataset in our own model, with train and test losses of 0.02 and 0.51, respectively. When using similar datasets, train and test accuracy in the VGG16 model are 0.99 and 0.94, respectively, whereas train and test loss are 0.03 and 0.39, respectively. When VGG19 was applied, train and test accuracy were

0.99 and 0.94, respectively, while train and test losses were 0.04 and 0.37. Train and test accuracy were 0.92 and 0.81, respectively, the train and test losses were 0.26 and 1.05, respectively, when Resnet-50 was employed. With greater train and test accuracy and lower train and test loss, the model we presented significantly beat the others. When we used the skin cancer dataset in our own model, the train and test accuracy were 0.88 and 0.72, respectively, but the train and test loss were 0.26 and 0.64, respectively.

In the VGG16 model, the train and test accuracy were 0.90 and 0.68, respectively, when the identical datasets were utilized, but the train and test loss were 0.35 and 1.16. The train and test accuracy are 0.92 and 0.68, respectively, when VGG19 is employed. The train and test losses, on the other hand, are 0.31 and 1.60, respectively. The train and test accuracy were 0.90 and 0.69, respectively, when Resnet-50 was installed, whereas the train and test loss were 0.30 and 0.75. As can be seen in the graph below, the aforementioned model outperforms the previous models in terms of test accuracy, while requiring little training and causing little test loss. The performance measures accuracy, precision, recall, and F1-score are provided for each model and dataset since the classification was binary, i.e., benign or malignant. Because it was a binary categorization, the model we demonstrated had superior accuracy, recall, and F1-score. The model exhibited a precision of 0.99 and 0.93 for benign and malignant classifications, respectively, when using the cancer dataset, whereas VGG16 had a precision of 0.98 and 0.94 for benign and malignant classes.

Similarly, VGG19 had a precision of 0.96 and 0.94 for benign and malignant and Resent-50 had a precision of 0.78 and 0.88 for benign and malignant. Our model outperforms the Resnet-50 model by a significant margin. The model was 0.78 and 0.69% accurate for benign and malignant classes, respectively, when utilizing the skin cancer dataset, whereas VGG16 was 0.74 and 0.62% correct for benign and malignant classes. The benign and malignant VGG19 scores are 0.69 and 0.67, respectively, whereas the benign and malignant Resnet-50 scores are 0.70 and 0.68, respectively. Resnet-50 is clearly outperformed by our model. The model accuracy graph in Fig. 2 demonstrates that as the number of epochs rose, so did the accuracy when the proposed model was applied to the brain tumor dataset. The accuracy forecasts for the training and test sets were almost identical. Similarly, the loss decreased as the number of epochs increased. There were significant variations in train set and test set loss between epochs 20 and 60.

There were minor variations between them beyond the 60th epoch, indicating that the suggested model performed flawlessly on both the train and test datasets. As seen, Fig. 3 shows how the model accuracy varies between the train and test datasets. The difference between the train and test datasets is shown in Fig. 4 accuracies was lowest at the 100th epoch. As the number of epochs increased, the model's accuracy improved. VGG16's loss was also outstanding since the model performed exceptionally well at the conclusion of the 100th epoch on both the train and test datasets. The accuracy of the VGG19 model improved after the 20th epoch, as seen in Fig. 5, as the model performed nearly identically on the train and test datasets. The model loss in VGG19, which is almost non-existent after the 20th epoch, produces comparable findings. The suggested model upgraded its accuracy as the epochs count

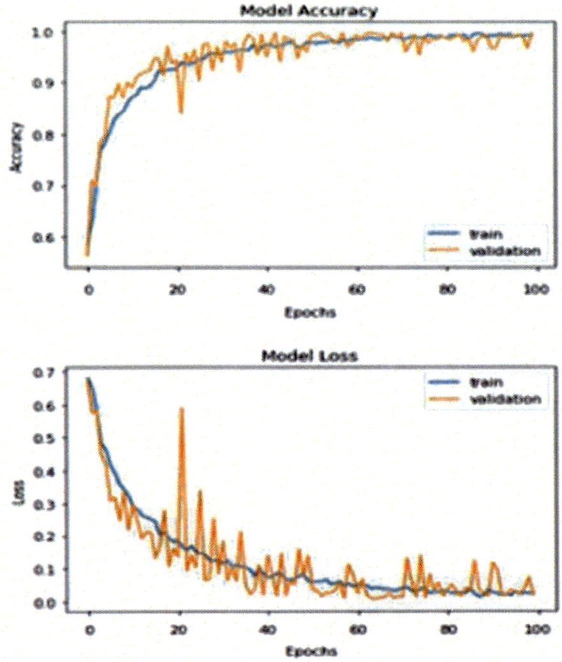

Fig. 2 Accuracy and loss of proposed model on brain tumor

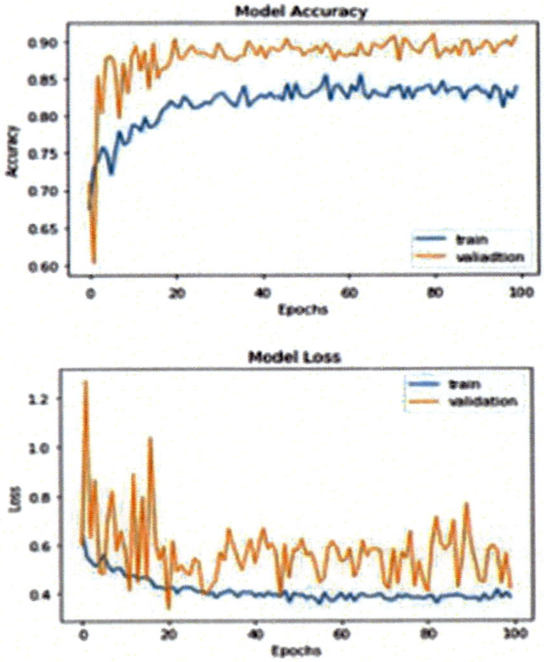

Fig. 3 Accuracy and loss of ResNet-50 model on brain tumor

Fig. 4 Accuracy and loss of VGG-16 model on brain tumor

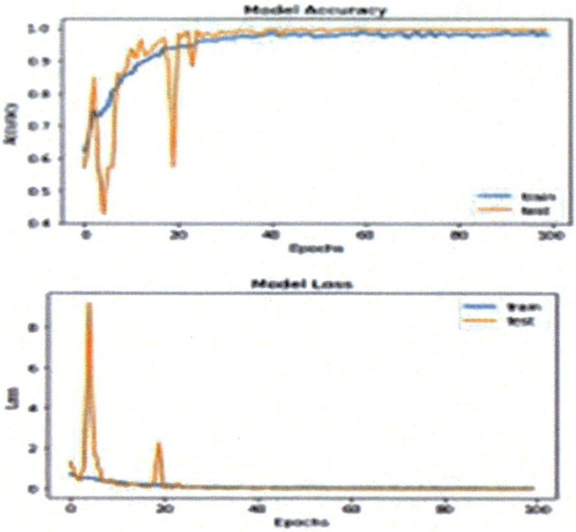

Fig. 5 Accuracy and loss of VGG-19 model on brain tumor

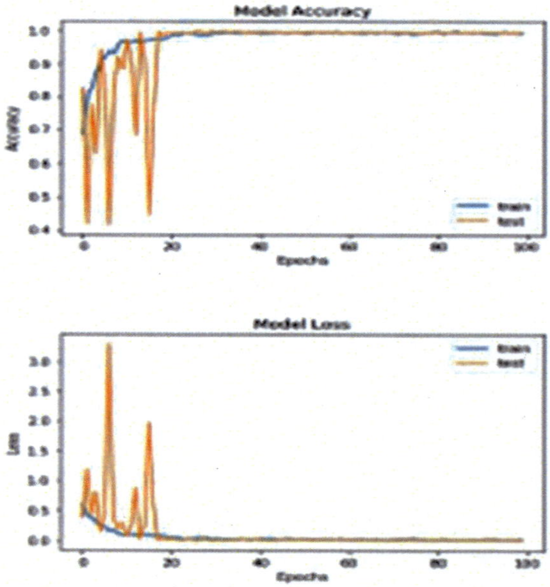

increased in Fig. 6, and the accuracies achieved throughout the train and test datasets were minimum.

1. Graphical Representation

2. Experiment and Results Table

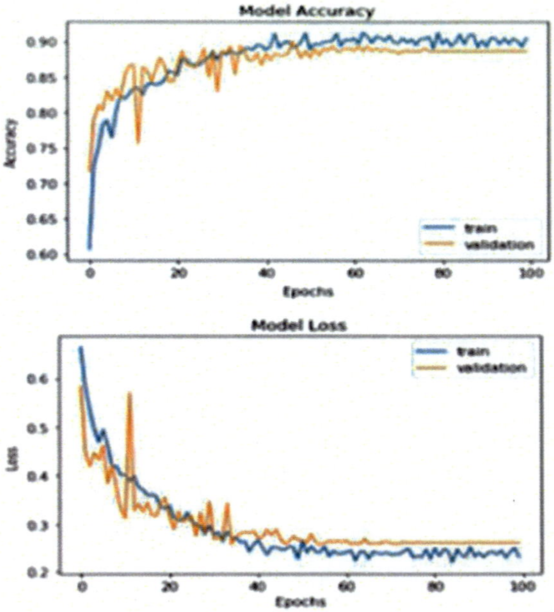

Fig. 6 Accuracy and loss proposed model on skin cancer

5 Conclusion

A convolutional neural network with six convolutional layers and a dense layer are proposed in this model. To calculate the maximum value, each pooling layer employs a max pool technique. Because it aids binary classification, the sigmoid activation function is used in the output layer. For training and testing, datasets are split 80:20. The train and test accuracy were 0.99 and 0.95, respectively, when we used the brain tumor dataset in our own model, while the train loss and test loss were 0.02 and 0.51, respectively. ResNet-50 model accuracy and loss in skin cancer (Fig. 7) are similar. VGG-16 model accuracy and skin cancer loss (Fig. 8) VGG-19 model accuracy and skin cancer loss (Fig. 9). Train accuracy, test accuracy, train loss, and test loss were all 0.92, 0.81, 0.26, and 1.05 when Resnet-50 was used. On the training set, vgg16 and vgg19 perform better, but on the testing set, they perform poorly. The main disadvantages of vgg16, vgg19, and resnet-50 are their excessively significant test losses as compared to the recommended model. We may deduce that our model is significantly better than the others.

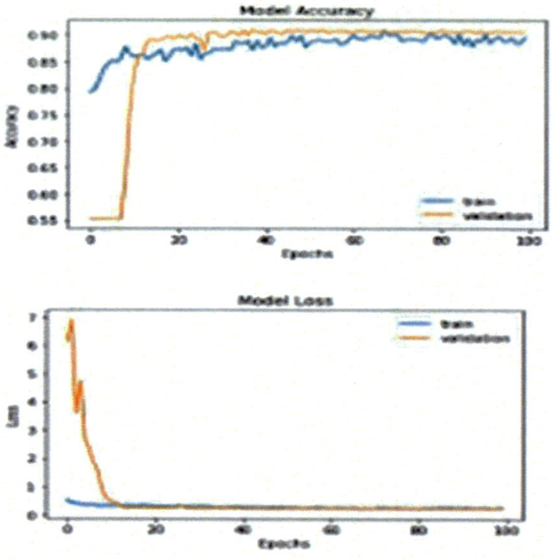

Fig. 7 Accuracy and loss on ResNet-50 model on skin cancer

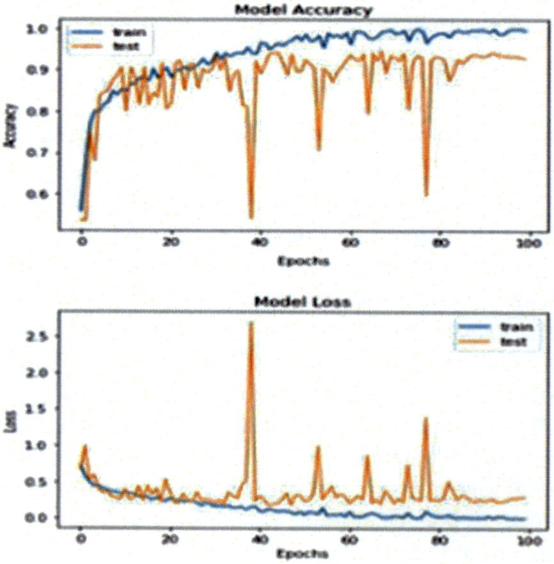

Fig. 8 Accuracy and loss of VGG-16 model on skin cancer

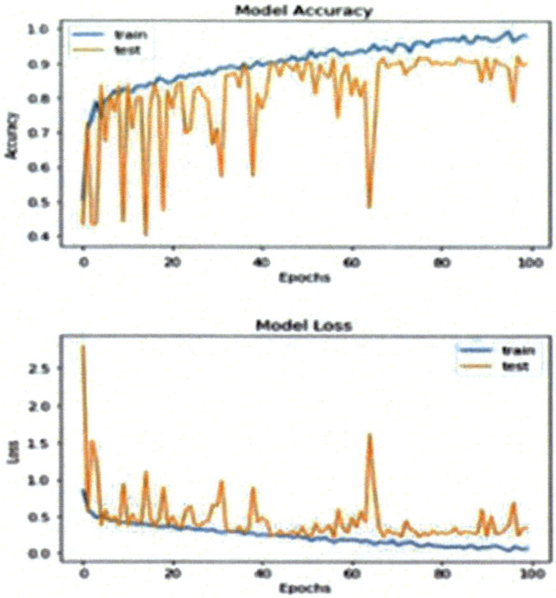

Fig. 9 Accuracy and loss of VGG-19 model on skin cancer

Acknowledgements We sincerely thank P.V.V.S Srinivas for helping in finding the right datasets required for this research work and guiding us throughout the process.

References

1. Dorj U-O, Lee K-K, Choi J-Y, Lee M (2018) The skin cancer classification using deep convolutional neural network. Multimedia Tools Appl 77(8):9909–9924
2. Esteva A, Kuprel B, Novoa RA, Ko J, Swetter SM, Blau HM, Thrun S (2017) Dermatologist-level classification of skin cancer with deep neural networks. Nature 542(7639):115–118
3. Li Y, Esteva A, Kuprel B, Novoa R, Ko J, Thrun S (2017) Skin cancer detection and tracking using data synthesis and deep learning. In: Workshops at the thirty-first AAAI conference on artificial intelligence
4. Afshar P, Mohammadi A, Plataniotis KN (2018) Brain tumor type classification via capsule networks. In: 2018 25th IEEE international conference on image processing (ICIP). IEEE, pp 3129–3133
5. Babu K, Indira ND, Vara Prasad K, Shameem S (2021) An effective brain tumor detection from T1w MR images using active contour segmentation techniques. J Phys: Conf Ser 1804(1):012174
6. Lapointe S, Perry A, Butowski NA (2018) Primary brain tumours in adults. The Lancet 392(10145):432–446
7. Babu KR, Sai Deepthi U, Sudha Madhuri A, Sai Prasad P, Shameem S (2019) Comparative analysis of brain tumor detection using deep learning methods. Int J Sci Technol Res 8(12):250–254
8. Tripathi M (2021) Analysis of convolutional neural network based image classification techniques. J Innov Image Process (JIIP) 3(02):100–111

9. Zeng Z, Mao C, Vo A, Nugent JO, Khan SA, Clare SE, Luo Y (2019) Deep learning for cancer type classification. bioRxiv 612762
10. Bodavarapu PNR, Srinivas PVVS, Mishra P, Mandhala VN, Kim H (2021) Optimized deep neural model for cancer detection and classification over ResNet. In: Smart technologies in data science and communication. Springer, Singapore, pp 267–280
11. Adem K, Kilicarslan S, Comert O (2019) Classification and diagnosis of cervical cancer with softmax classification with stacked autoencoder
12. Dudley JC, Schroers-Martin J, Lazzareschi DV, Shi WY, Chen SB, Esfahani MS, Trivedi D et al (2019) Detection and surveillance of bladder cancer using urine tumor DNA. Cancer Discov 9(4):500–509
13. Li W, Li J, Sarma KV, Ho KC, Shen S, Knudsen BS, Gertych A, Arnold CW (2018) Path R-CNN for prostate cancer diagnosis and gleason grading of histological images. IEEE Trans Med İmaging 38(4):945–954
14. Hashimoto N, Fukushima D, Koga R, Takagi Y, Ko K, Kohno K, Nakaguro M, Nakamura S, Hontani H, Takeuchi I (2020) Multi-scale domain-adversarial multiple-instance CNN for cancer subtype classification with unannotated histopathological images. In: Proceedings of the IEEE/CVF conference on computer vision and pattern recognition, pp 3852–3861
15. Oliveira RB, Papa JP, Pereira AS, Tavares JMRS (2018) Computational methods for pigmented skin lesion classification in images: review and future trends. Neural Comput Appl 29(3):613–636
16. Tan TY, Zhang L, Neoh SC, Lim CP (2018) Intelligent skin cancer detection using enhanced particle swarm optimization. Knowl-Based Syst 158:118–135
17. Mhaske HR, Phalke DA (2013) Melanoma skin cancer detection and classification based on supervised and unsupervised learning. In: 2013 International conference on circuits, controls and communications (CCUBE). IEEE, pp 1–5
18. Maglogiannis I, Doukas CN (2009) Overview of advanced computer vision systems for skin lesions characterization. IEEE Trans Inf Technol Biomed 13(5):721–733
19. Hasan M, Barman SD, Islam S, Reza AW (2019) Skin cancer detection using convolutional neural network. In: Proceedings of the 2019 5th ınternational conference on computing and artificial ıntelligence, pp 254–258
20. Hosny KM, Kassem MA, Foaud MM (2018) Skin cancer classification using deep learning and transfer learning. In: 2018 9th Cairo ınternational biomedical engineering conference (CIBEC). IEEE, pp 90–93
21. Hekler A, Utikal JS, Enk AH, Hauschild A, Weichenthal M, Maron RC, Berking C et al (2019) Superior skin cancer classification by the combination of human and artificial intelligence. Eur J Cancer 120:114–121
22. Mallick PK, Ryu SH, Satapathy SK, Mishra S, Nguyen GN, Tiwari P (2019) Brain MRI image classification for cancer detection using deep wavelet autoencoder-based deep neural network. IEEE Access 7:46278–46287
23. Shirazi AZ, Fornaciari E, Bagherian NS, Ebert LM, Koszyca B, Gomez GA (2020) Deep-SurvNet: deep survival convolutional network for brain cancer survival rate classification based on histopathological images. Med Biol Eng Comput 58(5):1031–1045
24. Torti E, Florimbi G, Castelli F, Ortega S, Fabelo H, Callicó GM, Marrero-Martin M, Leporati F (2018) Parallel K-means clustering for brain cancer detection using hyperspectral images. Electronics 7(11):283
25. Amin J, Sharif M, Raza M, Saba T, Sial R, Shad SA (2020) Brain tumor detection: a long short-term memory (LSTM)-based learning model. Neural Comput Appl 32(20):15965–15973
26. Ayyappa Y, Bekkanti A, Krishna A, Neelakanteswara P, Basha CZ (2020) Enhanced and effective computerized multi layered perceptron based back propagation brain tumor detection with Gaussian filtering. In: 2020 Second ınternational conference on ınventive research in computing applications (ICIRCA). IEEE, pp 58–62

27. Kumar MJ, Raghavendra Sai N, Smitha Chowdary C (2020) An efficient deep learning approach for brain tumor segmentation using CNN. IOP Conf Ser: Mater Sci Eng 981(2):022012
28. Majji R, Prakash PGO, Cristin R, Parthasarathy G (2020) Social bat optimisation dependent deep stacked auto-encoder for skin cancer detection. IET Image Proc 14(16):4122–4131

Design and Development of the Efficient RF Energy Harvesting System

P. Kalpana Devi and T. Yuvaraj

Abstract In today's life, mobile usage causes an increase in the radiation levels in the atmosphere which affects all the living and exponentially increases the risk of diseases. Energy harvesting means acquiring signals and gathering energy from various sources which are deemed to be unused for any applications. These kinds of remaining energies are generally released as wasted potential to the environment. But the wasted potential can be used to provide power for the wireless sensors. The battery source is normally used for powering up these devices. To avoid the expensive repeated battery replacement in wireless devices, energy harvesting system is proposed to receive the energy from RF source, and the power is transferred from the antenna to the rectifier circuit. The impedance matching circuit is required to transfer the maximum power gain from antenna tower to the successive stages. The rectifier circuit convert the incoming RF signal to DC signal for efficient rectification to improve the power level which is further fed into the battery. This process in turn reduces the atmospheric radiation level and to maintain the sustainable environment.

Keywords Rectenna · Super capacitor · Energy harvesting · Rectifier · Energy source · Environment

1 Introduction

In recent years, wireless devices are used in most of the applications. This increases the radiation level in the environment as shown in Fig. 1. This in turn creates the need of increase in energy storage batteries for wireless devices [1]. Hence, research was diversified in reducing the consumption of devices and the process of recycling the wasted energy in the environment. The charging of wireless devices specifically

P. Kalpana Devi (✉)
Veltech Rangarajan Dr. Sagunthala R&D Institute of Science and Technology, Chennai, India
e-mail: drkalpanadevip@veltech.edu.in

T. Yuvaraj
Prathyusha Engineering College, Thiruvallur, India

S. Shakya et al. (eds.), *Proceedings of Second International Conference on Sustainable Expert Systems*, Lecture Notes in Networks and Systems 351,
https://doi.org/10.1007/978-981-16-7657-4_42

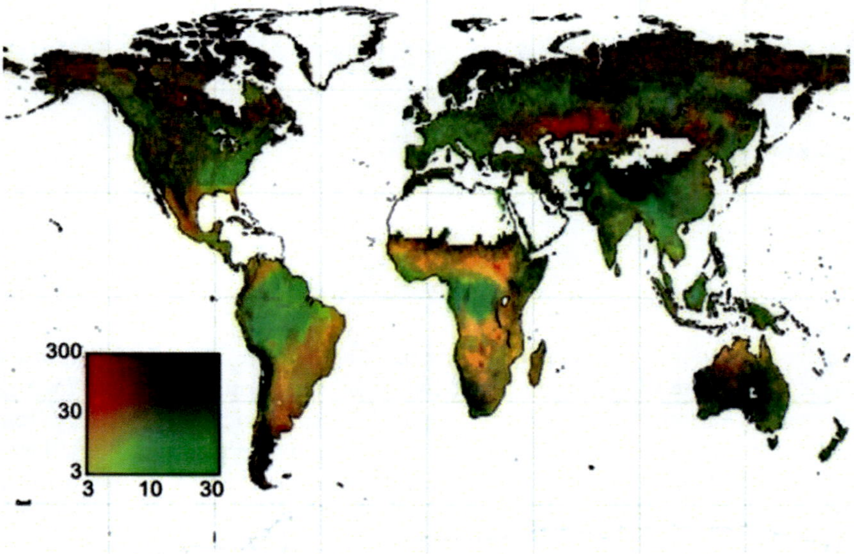

Fig. 1 Intensity of radiation levels across the world

wireless sensor nodes which are located in remote places remains a major challenge in communication.

For the increase in use of wireless devices, energy requirement for charging these devices is also increased. The rectification of RF energy in to DC signal was proposed in 1950s for powering the helicopter and solar power satellite system [2].

In general, 98% of total power can be captured by the antenna from mobile frequency bands and 2% contributed as FM, Wifi WiMax power bands. Hence, it is proposed to design a rectenna in PCB with miniaturized to get operated in multiple bands for receiving RF energy. Energy scavenging mechanisms are used to recharge batteries either through solar energy or vibration harvesting techniques.

Efficient energy management methodologies at the sensor nodes are designed to extend the network lifetime as maximum. Energy is a limited resource in every wireless device [1], compared to the regular task complexity in sensor networks [2], the sensor nodes are very small in size to place the high-capacity power backups.

2 Energy Efficient RF Harvesting System

RF energy harvesting system includes impedance matching section, rectification and energy storage load as shown in Fig. 2. The received signal is set to matched with the rest of the blocks in the system by matched transmission stub. This effective matching can reduce the losses and further conditioned by multiplier circuit [3]. The

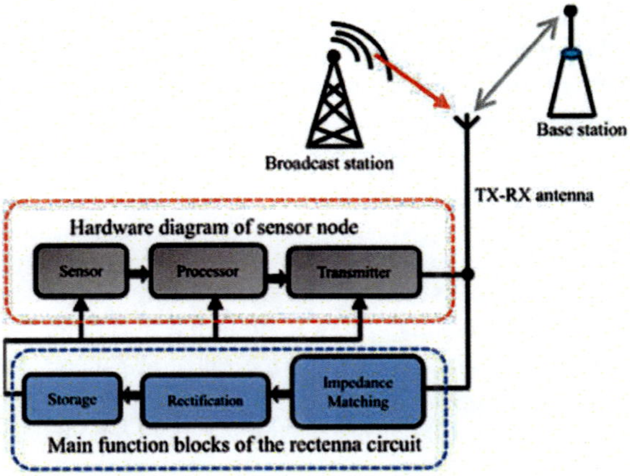

Fig. 2 Block diagram of RF energy harvesting system

output voltage from multiplier to be maintained by its peak values by using surface mount RF schottky barrier diodes in different combination for multiple frequencies. Double-stage rectifier circuit with super capacitors can store the energy which can powering the sensor nodes in the system.

Circularly Polarized Rectenna: The antenna required to be multiband such that it can receive the RF energy present in all frequency sources in the environment [3]. The reason is that the power levels are in the range of microwatts but more energy is needed for powering up the sensor nodes [4].

Adaptive Matching stub design: Adaptive matching stub design is achieved by connecting the microstrip line in a specific pattern for impedance matching. This in turn processes the rectifier to operate at multiple bands.

Adaptive reconfigurable rectifier: Depletion-mode field-effect transistor switch can make the rectifier to adapt to the required input power level, it will enhance the RF-to-DC power conversion to extend the operating power range significantly when compared to the existing rectifiers.

Rectenna: The incident RF signal fall on antenna, followed by rectifier circuit. This RF wave is converted into DC power by the diode-based converter [5]. A high-frequency filter performs the impedance matching between the antenna and the diode at 2.45 GHz for optimum power conversion factor. The DC–DC converter filter smoothens the output DC.

3 Proposed RF Energy Harvesting System—R-Ger

The proposed block diagram deals with for the effecient energy harvesting mechanism for acquiring the energy from most freely available RF source in environment.

The impedance matching is implemented to improve the gain from an incoming RF signal. Further RF signal received is converted into DC signal that is fed into battery for an efficient rectification which improves the output power as represented in Fig. 3 which consists of antenna, matching circuit, rectifier, charging circuit and battery. The proposed system design improves the overall performance of RF energy harvesting system.

Antenna
In general, the radio frequency that is commonly available in the range of 2.4 GHz. Hence, the antenna design is preferred in this frequency band. Microstrip antenna is designed with CST tool in the range of S-band. Slots various shapes of slots are selected for opereating in various resonant frequency in L band range [6].

S-Parameter describes the input–output relationship between ports in an electrical system. Electrical properties such as VSWR, reflection coeffecient, gain, return loss can be simulated for determining the component characteristics (Fig. 4).

Matching Circuit
The impedance matching charcteristics are designed with the circuit of LC components. The performance is analyzed suitable for RF energy harvesting that is most opted for known probability density function. Then, the simulated graph shows that 280 mV matching network gives the best power conversion effeciency system, and the high value of output harvested power is obtained as shown in Figs. 5 and 6.

RF to DC converter: The RF signal received is acting as a source in RF to DC converter. The RF power is transferred from the rectenna which is so called the combination of rectifier and antenna and impedance matching is performed to provide more power gain from source and the rectifier circuit. The efficient rectification improves the output power which will be stored in the battery. Hence, the RF energy from the radiation is converted to dc for optimum power level. Super capacitors will be an alternate source for batteries for energy storage applications (Fig. 7).

DC to DC buck boost converter
A self oscillating DC–DC converter is required to convert the energy harvested from a radio-frequency rectifier into a charge sources. A linear regulator is used to scale

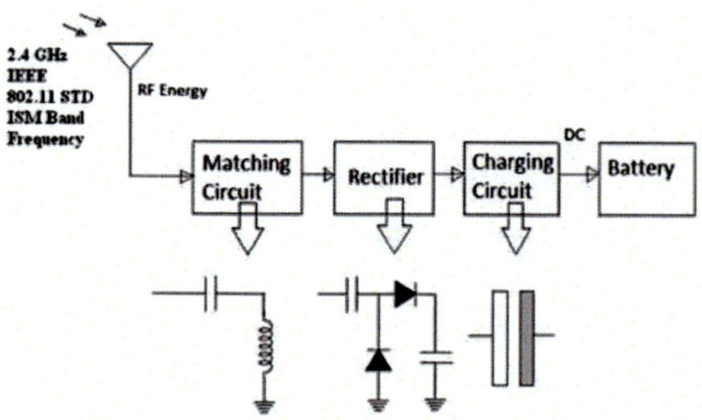

Fig. 3 RF energy harvesting system

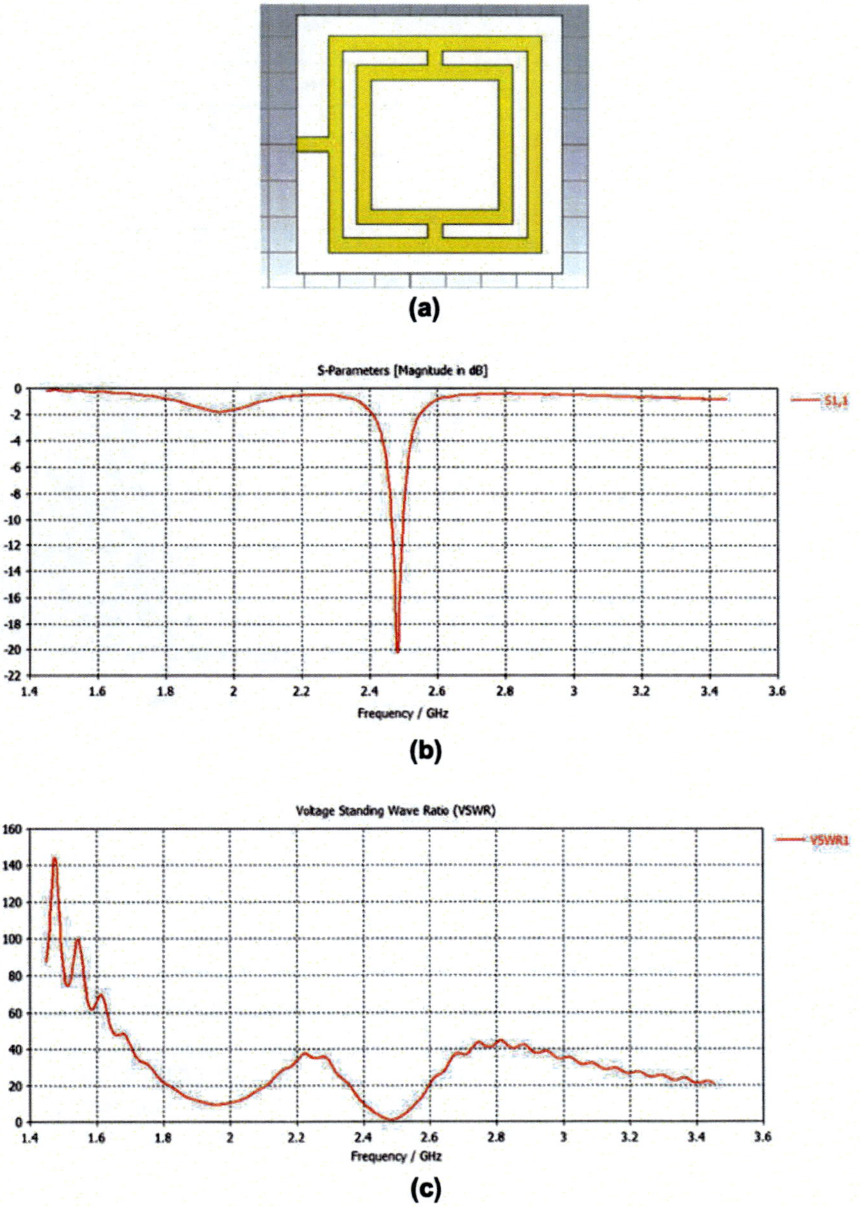

Fig. 4 **a** Design of microstrip antenna. **b** S-Parameter. **c** VSWR

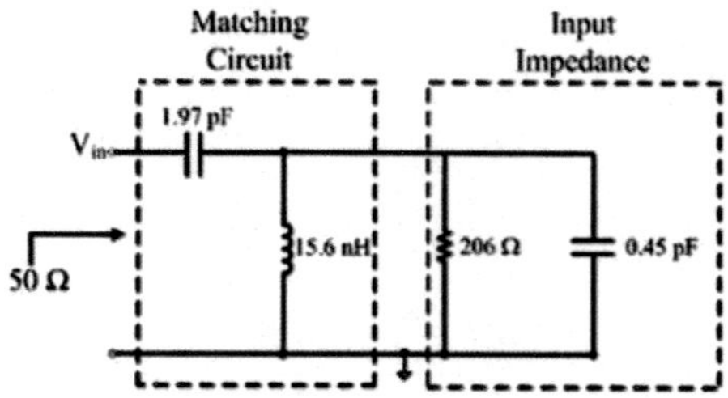

Fig. 5 280 V matching circuit

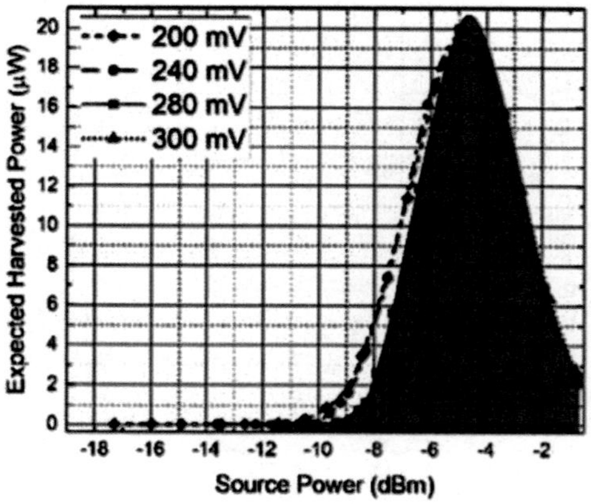

Fig. 6 Matching circuit PCE

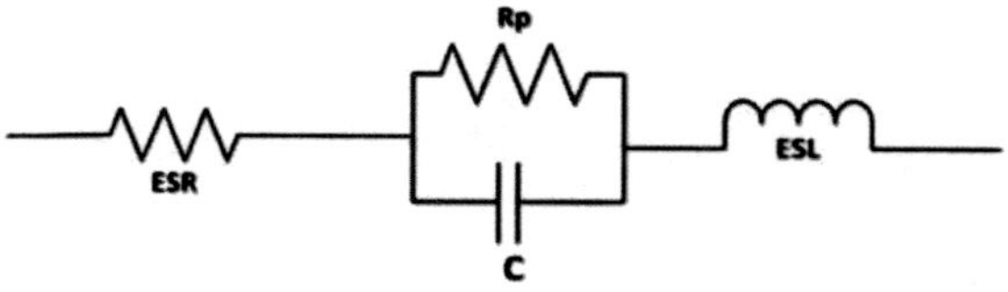

Fig. 7 Equivalent circuit model

the voltage supply for a suitable value of C used for sensing and conversion circuit in wireless sensor nodes.

The oscillator is designed with JFET to amplify the gate input control signal. 2-coupled inductors and gate source capacitor comprised the oscillator return loop. Voltage value can be raised by step up transformer from very low-source voltage level. The rectifier is preferred to convert gate oscillating voltage. If the state oscillating voltage is greater than the diode threshold voltage, then the diode will be in ON condition, and load is supplied with input voltage (Fig. 8).

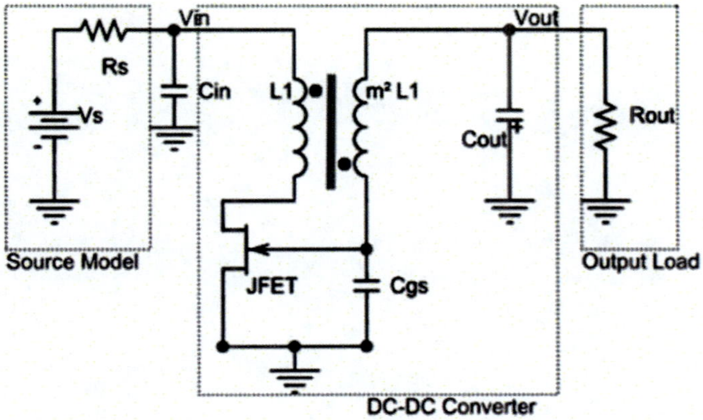

Fig. 8 Block diagram of the buck-boost converter

4 Measurement of RF Energy Harvesting System for the Optimized Design

RF energy harvesting is a predominant technology since it is an alternative source of energy for sensor nodes. The optimized RF harvesting system design involves the efficiencies of indivdual blocks and together constitute the integrated design describes the overall efficiency. As per the requirement of operation, the antenna design to be matched with gain, frequency and size. The rectifier element is to be selected appopriate DC to DC converter and corresponding power conversion efficiency factor to be analyzed. Measured received power error are measured with respect to distance for the operating frequency of 2.45 GHz free unlicensed ISM band is shown in Fig. 9.

In general, energy calculated is extremely low from available RF electromagnetic sources. The measured value states that the power calculated is in the range of microwatts (Fig. 10).

5 Conclusion

Systematic methodology for designing of RF energy harvesting circuits to maximize the harvested energy for a defined range of input power levels is proposed in this paper. The defined system consists of antenna, matching circuit, rectifier circuit, DC to DC converters. The converted energy value is stored in battery for future applications rather than the radiation of RF signal wasted in the environment.

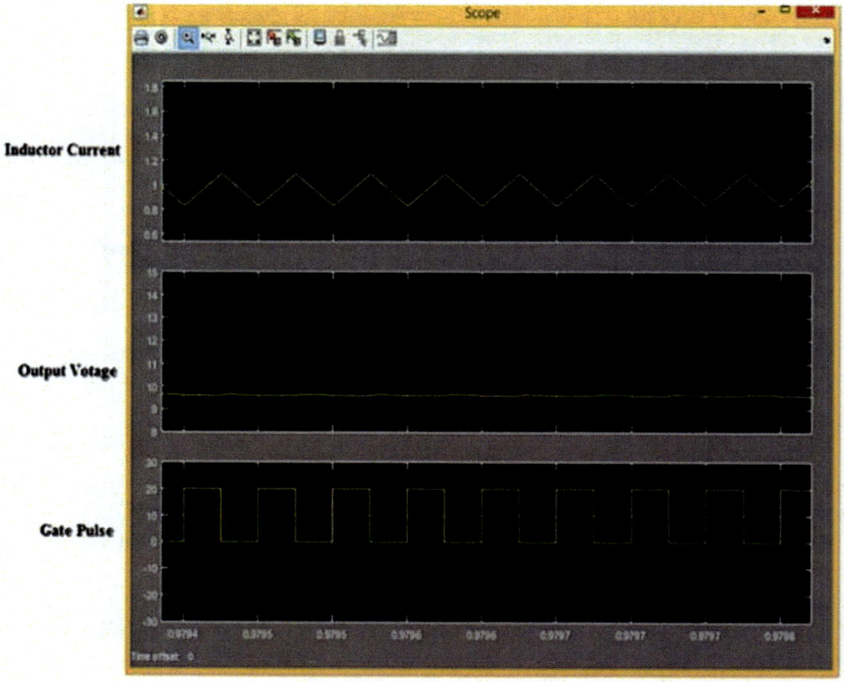

Fig. 9 DC to DC converter simulation graph

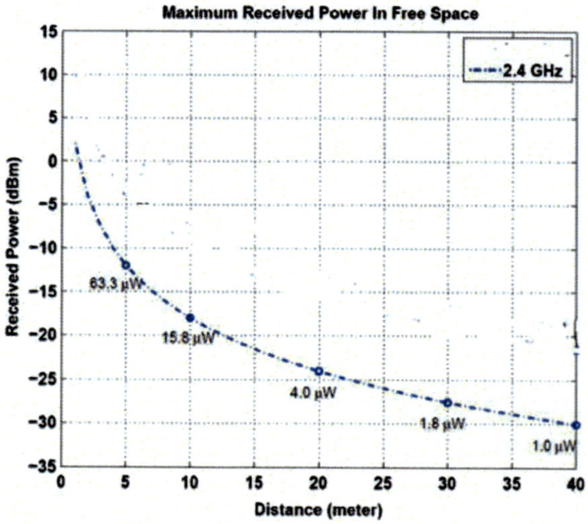

Fig. 10 Received power versus distance

References

1. Song C et al (2016) A novel six-band dual CP rectenna using improved impedance matching technique for ambient RF energy harvesting. IEEE Trans Antennas Propag 64(7):3160–3171
2. Jabbar H, Song YS, Jeong TT (2010) RF energy harvesting system and circuits for charging of mobile devices. IEEE Trans Consumer Electron 56(1):247–253
3. Shafique K et al (2018) Energy harvesting using a low-cost rectenna for Internet of Things (IoT) applications. IEEE Access 6:30932–30941
4. Din NM, Chakrabarty CK, Ismail AB, Devi KKA, Chen W-Y (2012) Design of RF energy harvesting system for energizing low power devices. Prog Electromagn Res 132:49–69
5. Sun H et al (2012) Design of a high-efficiency 2.45-GHz rectenna for low-input-power energy harvesting. IEEE Antennas Wireless Propag Lett 11:929–932
6. Devi PK, Sujatha M, Prasad JA (2020) Design of reconfigurable integrated patch antenna in ISM band for IoT applications. In: 2020 7th international conference on smart structures and systems (ICSSS), pp 1–4
7. Shen S et al (2019) An ambient RF energy harvesting system where the number of antenna ports is dependent on frequency. IEEE Trans Microw Theory Tech 67(9):3821–3832

A Combined K-Mean Semantic Approach for the Implicit Document Clustering

R. S. Rehna

Abstract In this study, an effective document clustering algorithm is introduced. The suggested algorithm comprises of existing K-mean algorithm with semantic approach. The PSO clustering algorithm is applied to assess the performance of the developed document with proposed hybrid algorithm. The present study focuses on assessing the proposed algorithm using reports on four different document dataset like 20NG, Reuters, WebKb. Finally, this newly proposed algorithm is compared with the clustering algorithm like Hierarchical to improve the factors such as efficiency, scalability and quality of clustering.

Keywords K-mean · PSO algorithm · Clustering · Semantics

1 Introduction

The grail of a document clustering method is to reduce the intra-cluster distances within the materials. The similarity meed slander at the navel of the document clustering. The large heterogeneity of documents makes it almost non-viable to turn out a general algorithm that can masterstroke in the event of all kinds of datasets.

The threats are [1]:

1. Determine significant attributes of the documents that have to be worn for clustering.
2. Determine the significant resemblance meed between documents.
3. Determine the significant clustering process employ the exceeding similarity meed.
4. Applying the clustering algorithms in a more realistic way to optimise the memory and CPU resources.

R. S. Rehna (✉)
Department of Computer Science and Engineering, LBS Institute of Technology, Poojapura, Thiruvananthapuram, Kerala, India
e-mail: rehna@lbsitw.ac.in

S. Shakya et al. (eds.), *Proceedings of Second International Conference on Sustainable Expert Systems*, Lecture Notes in Networks and Systems 351,
https://doi.org/10.1007/978-981-16-7657-4_43

5. Detecting routine of appraising the peculiarity of the complete clustering. Additionally, with moderate to bigger document collections, the number of term-document relations is sufficiently overhead and the ciphering complexity of the algorithm has to do with. If an impenetrable matrix is contriving to constitute term-document relations, then this matrix could assuredly flatter enormous memory. Whether the vector model is put in, then the capacity of the emerging vector space will correspondingly be fairly at the peak. This signifies elementary operations such as finding the Euclidean distance between two documents in the vector space, become time-consuming tasks [2].

Particle Swarm Optimization (PSO) is a functional method for unceasing nonlinear function expansion that recreates communal practices. The proposed methodology is bind to bird flocking, fish schooling, and broadly speaking swarming theory. It is an immensely potent algorithm for optimizing expansive functions. The main intuition of the algorithm is to sustain a set of potential solutions particles.

2 Partitional Clustering Algorithms

Partitional Clustering Algorithms uses the values of similarity that optimizes certain criterion. These type of algorithms are meant to improve certain criterion.

The most commonly used algorithms are detailed here: Partitioning Around Medoids, which is also called PAM is the one among them. Clustering LARge Applications, commonly named as CLARA and Clustering LARge ApplicatioNS (CLARANS) are also widely employed. Out of the common practise, K-means algorithm seems to be relevant in this field. Hence for the present research, K-means algorithm is used.

3 System Analysis

The demerits of an existing algorithm such as PSO and ways of improvement of its performance are addressed here [3]. In the present study, clustering algorithm that uses semantic similarity measure is adopted and it is presented in this section.

3.1 Existing System

The main issues in the existing systems are

- Does not capture the meaning behind the words.
- The number of clusters should be known a priori A priori, which is basically represents the figure of clusters.

- Need to calculate the weight of each dimension in each cluster, leads to decrease in performance.
- High dimensional representation of data.

3.2 *Proposed System*

With the increasing amount of information, document clustering is applied to the information for easy recognition of the relevance of content. The present study focuses on a clustering algorithm that uses semantic similarity measure. Thus, a methodology to interpret and cluster knowledge documents using ontology is presented. The Particle Swarm Optimization (PSO) clustering method can be adapted to the illustrated documents. We further evaluate the performance of using ontology with the classical K-means and Particle Swarm Optimization clustering algorithm [4]. The proposed work is evaluated using reports on four different document dataset like 20NG, Reuters and WebKb. The scope of this work also extends to the performance comparison of clustering algorithm like Hierarchical with the newly proposed algorithm to improve the factors such as efficiency, scalability and quality of clustering. Also, the proposed system when applied to the distributed environment will show good efficiency, scalability and clustering quality.

4 Methodology

To improve the clustering results, a new methodology is applied in this work. The four different document datasets like 20NG, Reuters and WebKb are taken as input. After the pre-processing, the semantic relations between the words in a sentence is described and generates the feature vectors [5]. Then, the PSO clustering method is used to the illustrated documents and further apply the classical K-means algorithm to the document. Then, the Particle Swarm Optimization (PSO) clustering algorithm can be applied to the annotated documents and further apply the classical K-means algorithm to the document. Nevertheless, the traditional K-means clustering algorithm is nearly precise to the initial cluster core, and the final clustering result is extremely relative on the initial core. So as to achieve a more accurate clustering result, we put forward a K-means clustering algorithm centred on semantic strengthening.

5 Overview of the Process

- Document dataset like 20NG, MEDLINE, TREC and WebKb is taken as input
- Pre-processing is run on the dataset
- The weight measure is recalculated based on the semantic similarity.
- Particle Swarm Optimization is run on the semantic documents.

- Finally, K-mean algorithm is applied on the optimized document set.

5.1 PSO Algorithm

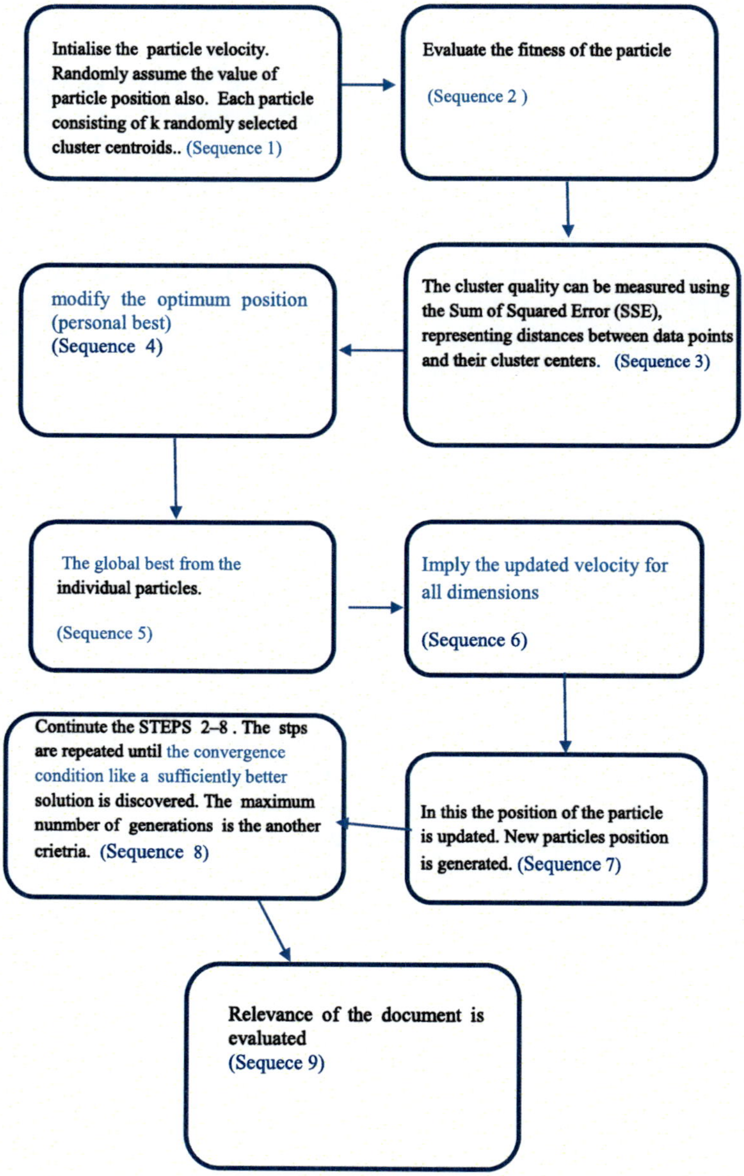

The present section illustrates the different steps involved in PSO algorithm. The sequence of operation involves nine steps, starting from initialization of the particle velocity as given in the subsequent section.

5.2 *K-Mean Algorithm*

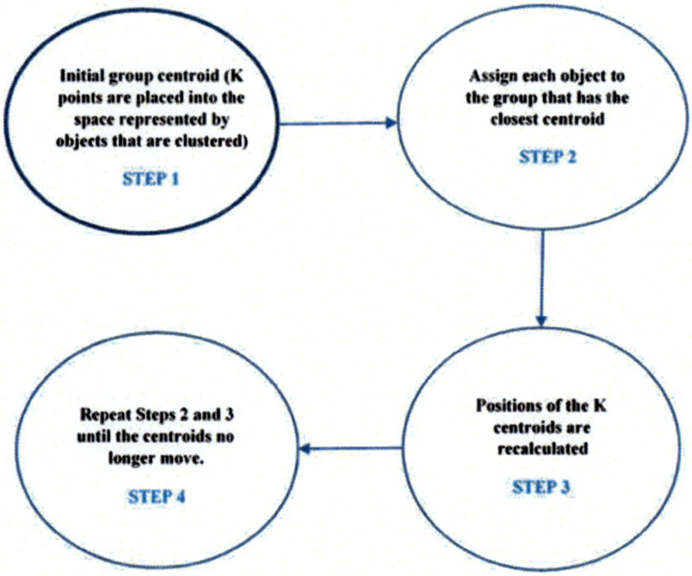

6 System Design

The document clustering scheme utilizes an approximate distance measure between the documents to maximize the inter-cluster distances. The distance measure or sometimes termed as the similarity measure is the importunate parameter in the document clustering [6]. Stop word removal, filtering and stemming are the most widely employed among them. The advantage of Particle Swarm Optimization (PSO) method is being it is capable of generating fine cluster centroids for the K-means. The method relies on population-based stochastic method and implies to obtain the optimal, near-optimal solution to a specified numerical or quantitative problems. Also, the clustering algorithm would suffer from the curse of dimensionality. In these cases, the assumption of equal importance for all features is not effective. The reason is the negligence in the consideration of the semantically related words which

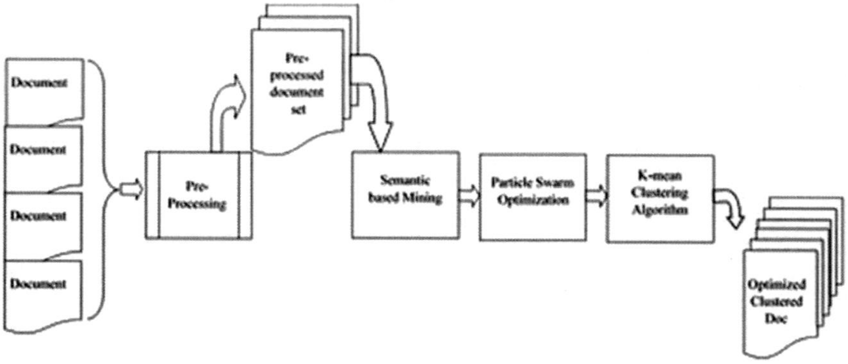

Fig. 1 System architecture

can cause problems. So in this work, semantically related words are considered after the pre-processing steps [7].

6.1 System Architecture

The system architecture consists of four modules shown in Fig. 1. They are

- Datasets
- Pre-processing steps
- PSO optimization
- K-mean clustering

6.2 Datasets

The three datasets used are 20 newsgroups, Reuters and WebKb.

20 newsgroups
The 20 newsgroups is considered to be the most commonly accepted dataset for evaluation of many text applications. The use of dataset covers in data mining methods, machine learning methods, etc. The dataset uses 18,828 unique documents in 20 categories, and the number of unique words after removing the stop word is found to be 71,830.

Reuters-21578
Reuters-21578 is found to be the most widely employed technique for the evaluation of document categorization and clustering. It uses 19,715 documents in five categories.

WebKb

This uses 62,239 number of documents in 69 categories with 336,656 unique words after removing the stop words.

6.3 Pre-processing

The text data normally has number formats, date formats and most common words like prepositions, articles and pronouns [8]. Hence, the text data has to be subjected a pre-screening before the actual processing. The database is collected from a heterogeneous source. The database contains massive volume of the data for the assessment. Due to this heterogeneity, real-world data tends to be inconsistent and noisy. The consistent data is imperious in data mining process. If the data is inconsistent, the process gives a wrong interpretation in the data. Hence, the pre-processing of the data is required to obtain accurate and consistent data. Such process can further reduces the complexity of the mining process as well. The different pre-processing steps are shown in Fig. 2.

- Stop word removal;
- Stemming;
- Filtering.

Term Filtering

Term filtering stipulates the elimination of stop words from the dataset. Further filtering includes elimination of terms with low document frequencies [9]. Term filtering indicates the removal of stop words from the dataset. The other filtering terms includes removal of terms with low document frequencies.

Stemming

As the name indicates, 'Stemming' represents the elimination of the unwanted term and finally shaped to their individual stem or root form. Let us consider the words such as 'eats', 'eating' and 'ate'. These are considered to be the different forms of a same word. But for the study, these different forms should be represented in a single phrase.

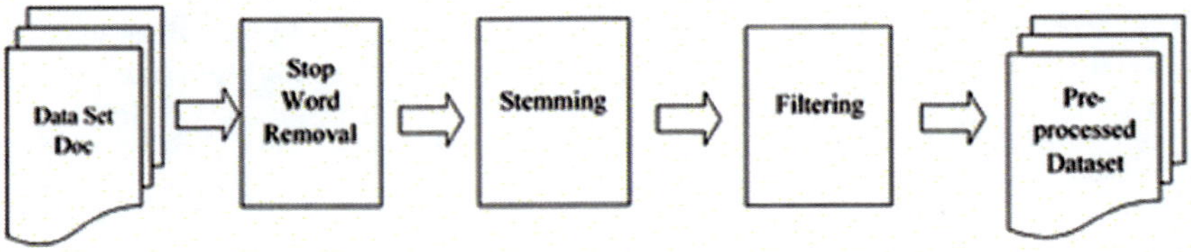

Fig. 2 Pre-processing steps

6.4 Semantic-Based Document Mining

The semantic approach document mining purely depends on the analysing of the text in documents [5]. The optimum way of clustering relies in the use of combined phrases analysis and word analysis to explore better ways of document representation. Syntax structural descriptions of the document like parts of speech tags, phrasal chunks and parse trees are extracted. Pool of words used for the clustering is often not considering the relationship between the decisive terms that interdict arise literally. To solve these problems, we integrate core ontologies as background knowledge into the process of clustering text documents. Semantic approach assigns weights to both words and phrases. The assigned weights plays a vital role in augmenting the document clustering. These weights generally improve the accordance between the terms and terminologies in the document and gather the semantic information of the monument.

6.5 Particle Swarm Optimization

The algorithm is a population-based stochastic optimization technique used to find the solution to a numerical and qualitative problem [10]. In PSO system, particles move through the search space. The particles position in the swarm represents solution of clustering the documents. The PSO algorithm for the document clustering and each swarm represents several candidate clustering solutions for the document collection [11]. The method used is updation of position of individual particle based on P_{bset} or G_{best}. The terms Pbest and Gbest represent particle best and global best, respectively. The movement of the particle is directed by the velocity. The method involves specifying the position and velocity of the individual particles. The position of ith particle of the swarm is illustrated with $Yi = (Yi1, \ldots, Yin)$ and velocity $Ui = (Ui1, \ldots, Uin)$. The velocity and position of particle updates by the expressions as shown below.

$$\mathrm{Uid}(t+1) = \mathrm{wUid}(t) + \mathrm{C1r()(rid - Yid)} + \mathrm{C2s()(rg - Yid)} \quad (1)$$

$$\mathrm{Yid}(t+1) = \mathrm{Uid}(t+1) + \mathrm{Yid}(t) \quad (2)$$

rid represents rbest position found by the ith particle, and rg the Gbest position found from all the particles in the swarm. *s*() is a uniform random function.

6.6 K-Mean Clustering

The K-mean algorithm uses yielding method to separate the in hand dataset to the required cluster form [12]. Each cluster is assigned with K-centroid. The K-centroid should be paced logically as it is the core part of the algorithm. The wrong placement can leads to different results. The basic theory is to place the K-centroid as far as possible from the neighbour one. Afterwards, they assign each dataset to the closest centroid. Once the points are nominated with K-centroid, preliminary part is completed. The next step is to recalculate K new centroids as barycenter of the clusters generating from the preceding one. A loop thus produced causes the change of K-centroids location [13]. This continues until no more changes are being done. To optimize the objective function J using K-mean algorithm shown in Eq. (3) until there is no further changes.

$$J = \sum_{i=1}^{m} \sum_{k=1}^{k} w_{ik} \left[x^i - \mu_k \right]^2 \quad (3)$$

7 Conclusion

In this paper, a review of various researches done in the areas of K-mean and PSO is presented, and it is concluded that when the document clustering takes place by combining K-mean with particle swarm optimization technique it produces coherent results in terms of efficiency and accuracy. From the study, it can be found out that as the proportion of the datasets is escalated, the novel hail churn-out remarkable refinement over the conventional clustering technique.

References

1. Lillesand T, Kiefer RW, Chipman J (2015) Remote sensing and image interpretation, 7th edn. Wiley, London
2. Otukei JR, Blaschke T (2010) Land cover change assessment using decision trees, support vector machines and maximum likelihood classification algorithms. Int J Appl Earth Observ Geo Inf 12:S27–S31
3. Wu K, Zhong Y, Wang X, Sun W (2017) A novel approach to sub pixel land-cover change detection based on a supervised back-propagation neural network for remotely sensed images with different resolutions. IEEE Geosci Remote Sens Lett
4. Jain AK, Murty MN, Flynn PJ (1999) Data clustering: a review. ACM Comput Surv 31(3):264–323
5. Shah N, Mahajan S (2012) Semantic based document clustering: a detailed review. Int J Comput Appl (0975–8887) 52(5)

6. Abualigah LM, Khader AT, Hanandeh ES (2018) A combination of objective functions and hybrid Krill herd algorithm for text document clustering analysis. Eng Appl Artif Intell 73:111–125
7. Pandian AP (2019) Review of machine learning techniques forvoluminous information management. J Soft Comput Paradigm (JSCP) 1(02):103–112
8. Sridevi UK, Nagaveni N (2011) Semantically enhanced document clustering based on PSO. Euro J Publ 57(3):485–493
9. Berkhin P (2002) Survey of clustering data mining techniques. Accrue Software Research Paper
10. Shaban KB (2009) A semantic approach for document clustering. J Softw 4(5)
11. Chuang L-Y, Lin Y-D, Yang C-H (2012) An improved particle swarm optimization for data clustering. In: Proceedings of the international multiconference of engineers and computer scientists, vol 1
12. Hartigan JA (1975) Clustering algorithms. Wiley, New York
13. Ye F, Chen C-Y (2005) Alternative KPSO Clustering algorithm. Tamkang J Sci Eng 8(2):165–174

Diagnosis of COVID-19 from Chest X-rays Using CNN and Determination of Its Severity by Text Analysis

Gopalam Pujitha, Potu Siva Parvathi, Lukka Venkata Siva Phaneendra, Nimmagadda Snehita, and Rizwan Patan

Abstract In India, the effect of COVID-19 has been worst because of various reasons like huge population, lack of necessary medical infrastructure, lack of awareness among people, inability to identify people with actual severe conditions and many more. Some people are waiting for more than a day to get the test results besides having rapid diagnosing kits. Due to a lack of awareness among people, patients with mild conditions are joining hospitals, leaving no place for severely infected patients. There is a need to automate the diagnosis of COVID-19 and identify the people with actual severe conditions so that those patients can be equipped with the required medical infrastructure and can potentially stop the process of spreading the disease and can even reduce the mortality rate. This need motivated us to propose a model which can diagnose COVID-19 and detect patients with severe conditions. Chest X-rays of individuals are efficient and can be used for rapid diagnosis of COVID-19 as X-ray centers are available even at rural areas. The proposed system automates the detection of COVID-19 and distinguishes the COVID-19 cases from other pneumonia and normal cases using a 11-layer Convolution Neural Network (CNN) model. We can use text analysis techniques on the patient's health condition which can be obtained by collecting details of the patient like age, body temperature, need for supplementary oxygen requirement, etc., we can identify the severity of the disease. The proposed CNN model achieved a 0.84 accuracy and on test data.

Keywords Classification · CNN · X-ray images · Text analysis · Severity

G. Pujitha (✉) · P. Siva Parvathi · L. V. S. Phaneendra · N. Snehita · R. Patan
Department of Computer Science and Engineering, V R Siddhartha Engineering College, Affiliated To JNTUK, Vijayawada, India

R. Patan
e-mail: rizwan@vrsiddhartha.ac.in

S. Shakya et al. (eds.), *Proceedings of Second International Conference on Sustainable Expert Systems*, Lecture Notes in Networks and Systems 351,
https://doi.org/10.1007/978-981-16-7657-4_44

1 Introduction

In India, COVID-19 started at the end of 2019. It continued showing its effect even in 2021. During the second wave, India remained the worst hit country in the world with an average number of 300,000 positive cases per day. The reasons behind this could be huge population in India, lack of proper awareness about the disease, poor medical facilities, huge number of people below the poverty line unable to have access to good medical facilities, etc. Even after the invention of rapid test kits and making many COVID test centers available to people for free, for a common man, it is taking more than a day to get the test results. This delay may spread the disease even more. Moreover, the rapid test results are not as accurate as expected. So, people are showing interest in waiting for long hours for accurate results rather than depending on the rapid tests. Many other alternatives like CT scan came into existence, but they are too costly to a common man. Hence, there is always a requirement for an accurate, budget-friendly and a rapid diagnosis procedure for COVID-19.

Manufacturing and transporting special testing kits are a costly and time-consuming process. So, there is a need to search for an efficient alternative to solve this problem. Here, chest X-ray images can come to our aid. X-ray centers are available even in rural areas all over India. As COVID-19 is an acute respiratory syndrome which cause lung infection, chest X-ray image can be used for automatic COVID-19 diagnosis [1–5]. Chest X-ray image of a person diagnosed with COVID-19 is as shown in Fig. 1 [6]. We can compare it with an X-ray image of a normal healthy person as shown in Fig. 2 [6]. Some of the features cannot be observed carefully by our naked eye. But an automatic system can be developed which can extract features

Fig. 1 Normal category

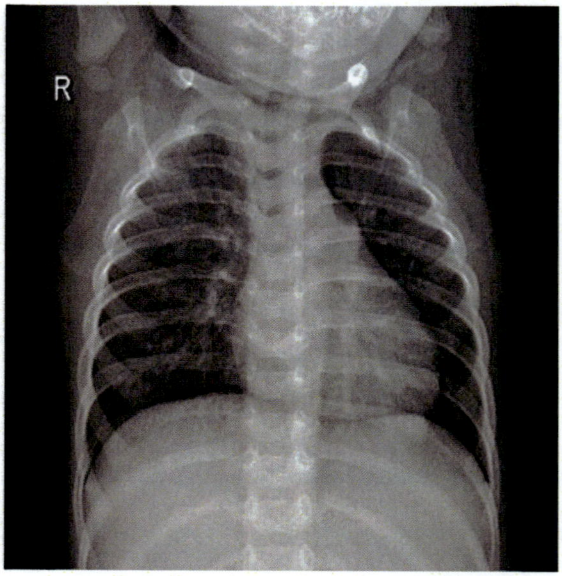

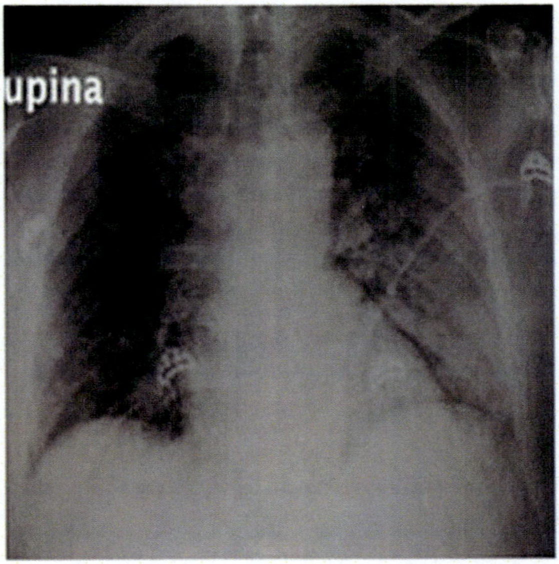

Fig. 2 COVID-19 category

from X-ray images. Convolution Neural Networks (CNN) can be used to extract the features the images automatically.

Hospitals in the country are filled with patients. But the actual thing is, without proper awareness and with the fear of the disease, people are joining hospitals immediately after testing positive. Due to this, people who are in actual need of hospitals are not able to avail the services. And as a result, mortality rate is increasing. As a solution to this problem, we can automate the process of determining the severity of the disease so that people come to know about their severity and join the hospital only if their condition is not mild. We used various text analysis techniques to analyze the health condition of a patient. And even hospitals can prioritize people based on severity.

2 Related Work

In the work presented in [1], transfer learning and model integration are used for the detection of COVID-19. They expanded the dataset 17 times and trained the model. They chose ResNet-101 and ResNet-152 based on accuracy and loss value. Pre-trained DenseNet, ResNet and Xception models have been implemented in [2]. In [3], transfer learning is used as there are only few images in the dataset taken. CNNs are used for feature extraction of X-ray images. The authors implemented various machine learning methods such as k-Nearest Neighbor, Bayes, Random Forest, Multilayer perceptron and Support vector machine on the CNN developed. SVM classifier gained the best performance out of all the classifiers implemented.

In the work published in [4], they demonstrated how CNN based transfer learning models can be selected to detect COVID-19 using multiple imaging modes like CT scan, X-ray and ultra sound. VGG 19 model is selected here. It performs in considerable levels with precision of up to 86% for X-ray, 100% for ultrasound and 84% for CT scan. CNN is used to detect COVID-19 in [5]. They used day-wise chest X-ray images to accomplish the task. They used only 224 chest X-ray images to train the model. Usually, large datasets are used to train CNNs for getting good results. Hence, a CNN model developed with small number of image dataset might not give good results. In [7], severity is predicted using various machine learning algorithms like SVM, LDA, NB, RF, LR, KNN and CART. Results are compared and analyzed. In [8], nearly 20 metrics like travel history, location, health records, etc., have been used to predict the likeliness of a person to get COVID-19 and to predict the severity of the case.

In [9], CNN is used for classifying traffic sign images. The work is entirely concentrated on a learning approach based on CNN, which is used here to extract the features of images and classify RGB-D images. They explored the "fine tuning technique" which reuses the layers trained on ImageNet and provided a four-class classification on traffic signs.

In [10], drinking habits of a cow on a daily basis are analyzed. CNN is used here to classify and detect the head of a cow above the trough from the video stream available. In [11], coronary artery disease (CAD) is predicted. Pooled area curve (PUC) is constructed in the machine learning method used to classify images even when weak pixels are surrounding the medical images. In [12], natural language processing (NLP) techniques based on deep learning were used in text to speech synthesis for communication recognition.

3 Methodology

The work flow of the project is shown in Fig. 3.

As shown above, a scanned X-ray image is taken as input form the user, and it is fed to the CNN model. CNN model classifies the image into one of the three categories available. Then, if the category obtained is COVID-19, we move on to find out the severity. Here, the required details are taken from the patient as input and using a function, and severity is determined.

It is important to breakdown a complex task into different modules for easier understanding. The proposed system consists of four phases as shown in Fig. 4.

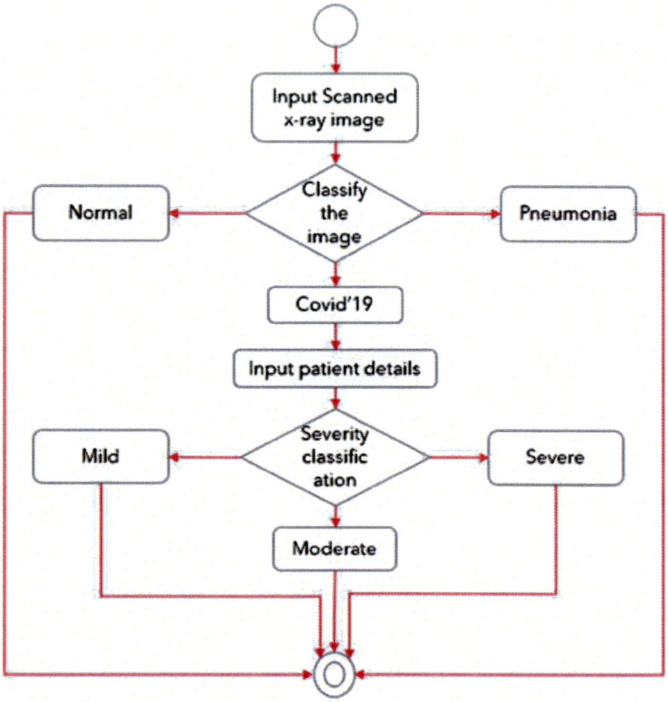

Fig. 3 Work flow of the system

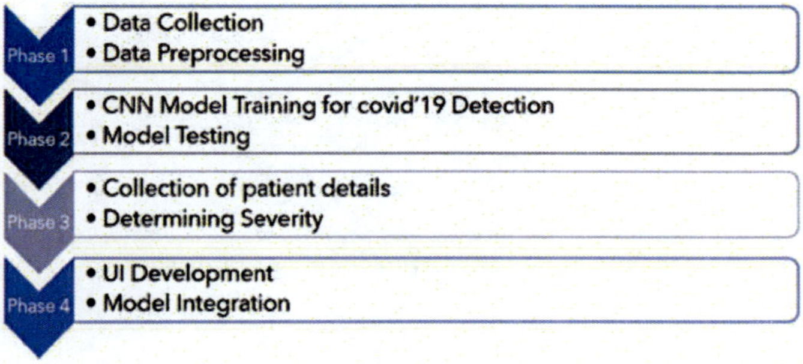

Fig. 4 Phases involved

Fig. 5 Phase 1

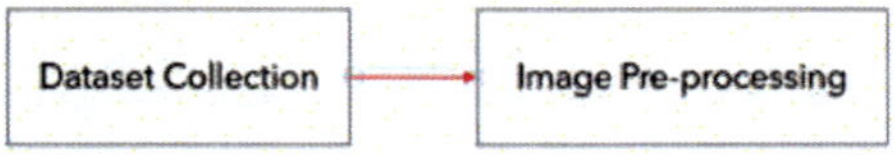

3.1 Phase 1

Phase 1 involves dataset collection and its cleaning techniques as shown in Fig. 5.

Dataset Collection

X-ray dataset is collected from Kaggle database [6]. It consists of a collection of X-ray images grouped into three categories namely, COVID-19, viral pneumonia and normal. The number of X-ray images in each category is shown in Table 1.

A sample X-ray image from the dataset collected that belongs to COVID-19 category is shown in Fig. 6.

For performing text analysis techniques, dataset is collected from Kaggle database. It consists of 371 rows and 28 columns out of which we used 12 columns. They are offset, sex, age, intubated, intubation_present, went_icu, in_icu,

Table 1 Composition of X-ray dataset

Category	Number of X-ray images
COVID-19	1200
Viral Pneumonia	1345
Normal	1341

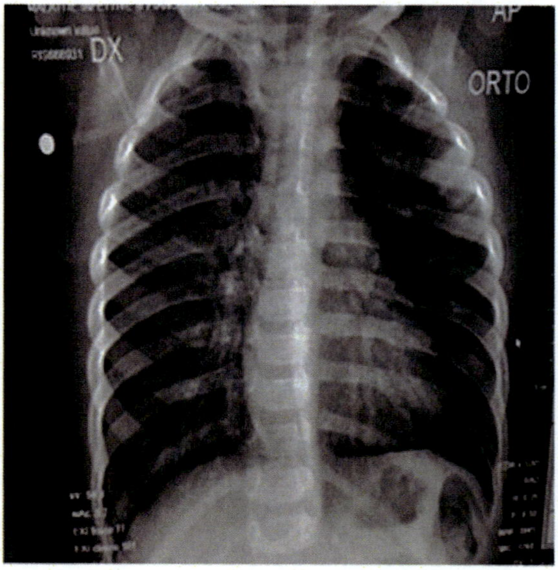

Fig. 6 Image categorized as COVID-19

needed_supplemental_O2, extubated, temperature, pO2_saturation, clinical_notes. We passed 12 columns of that database to our function which determines severity.

Image Pre-processing
This is an important step because the images uploaded by the user may be of different sizes and colors. An X-ray image can be best examined in gray scale when compared to that of others. And according to the readme .txt file obtained from the dataset source, the images in that dataset can be best used in 224 × 224 pixel resolution. So, two techniques are used here.

1. Image resizing (to 224 × 224)
2. Gray scale conversion

This pre-processed image is sent for further processing.

3.2 *Phase 2*

CNN Model Development
Phase 2 involves CNN model development. Process involved in phase 2 is shown in Fig. 7. The pre-processed images are used for used for training and testing purposes.

We implemented a 11-layer Convolution Neural Network 2D sequential model with 3 convolution layers, each followed by a pooling layer to improve the extraction capability of CNN network, 1 dropout layer, 1 flatten layer, 1 dense layer, 1 dropout layer, finally 1 fully connected layer dense layer. In every convolution layer, ReLu Activation function is applied. In the convolutional and pooling layers, X-ray features are extracted. Dropout layers are created with reduced density for better optimization. Flattening layer is used for reducing image to one-dimensional vector. Dense layers are deeply connected. In the last layer, softmax activation function is used. Softmax activation function gives out the probability of the image to belong to each and every class. The class with maximum probability is given as output. Dataset is split into train and test sets in 7:3 ratio.

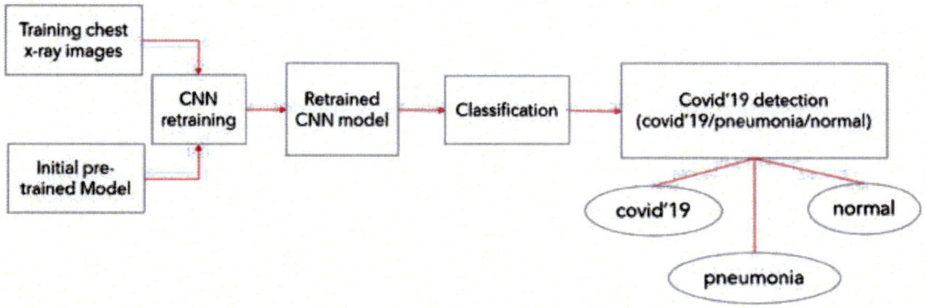

Fig. 7 Phase 2

- Number of training samples = 2720
- Number of testing samples = 1166

Setting the number of epochs value to 60 and batch size to 40 gave good results which will be discussed in the results section.

3.3 Phase 3

Determining Severity

This phase involves severity determination. Metrics used for evaluating patient condition include the following as shown in Fig. 8.

- Age
- Gender
- Offset (Number of days from which the patient is observing the symptoms)
- Body Temperature
- Intubation details
- Requirement for supplementary oxygen and its amount
- Health history

Using the clinical health management protocol released by the government of India [13] and some text analysis techniques, we developed a function to determine severity. The protocol clearly stated the conditions based on which a patient's condition is determined. It has general severe conditions based on patient's previous health history and age. It also contained the severe, moderate and mild conditions divided based on age group. Transforming all the conditions into a function that uses an if-else ladder can help us to determine the patient's condition. It compares the conditions stated in the protocol with those of the patient conditions and classifies him/her

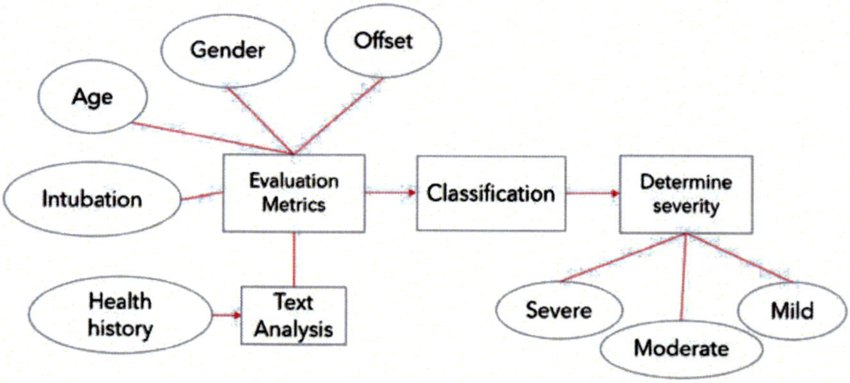

Fig. 8 Phase 3

into one of the three categories: severe, moderate and mild. Text analysis techniques used include the following:

1. Tokenization
2. Spell checking and correction
3. Removal of punctuation marks and stop words.
4. Case folding
5. Lemmatization

All the above steps must be applied to the text taken as input from the user and on the conditions mentioned in the health protocol wherever necessary. Then upon using simple if-else ladder on string comparison for condition checking, we can determine patient severity.

3.4 Phase 4

User Interface Development
A user interface is designed to enable the user to interact with the underlying model using HTML and CSS. UI contains an upload button to upload X-ray image as shown in Fig. 9.

This uploaded image is passed as input to the CNN model after image pre-processing. For taking patient details as input, a form is designed as shown in Fig. 10. All the details collected are passed to the function to determine patient's condition.

Model Integration
Model integration involves the process shown in Fig. 11. Firstly, CNN model structure is saved in a JSON file. JSON file format is used for storing data hierarchically. Model

Fig. 9 User Interface to upload X-ray image

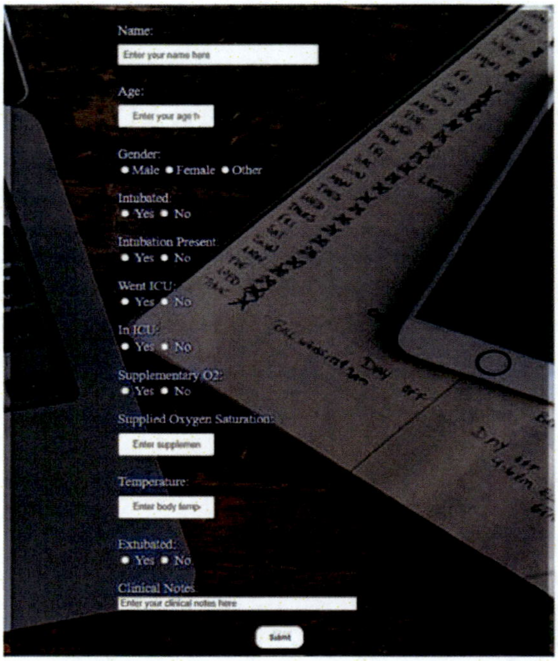

Fig. 10 User interface to upload X-ray image

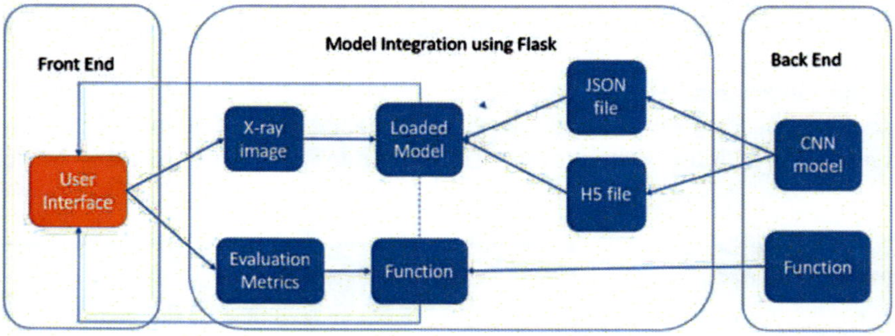

Fig. 11 Model integration

parameters are saved in a H5 file. A H5 file stores data in a hierarchical structure within a single file. Flask framework is used for model integration. The saved model is loaded using Keras library in Python. As shown in Fig. 11, input X-ray image is fetched from the user interface and sent to the loaded model after pre-processing.

Output of the model is shown on the user interface. Dotted line in Fig. 11 indicates that if the output of the model is COVID-19, then the function is called. Function prompts the user for input details like age, gender, body temperature, etc., known as evaluation metrics of the function as shown in Fig. 10. Then, the function determines the condition of the patient. The result is displayed on the user interface.

4 Results

Test split of the dataset is used for CNN model testing. The test results are as shown below. Confusion matrix on test data is as shown in Fig. 12. Based on the confusion matrix below, we calculated all the metrics like precision, recall, F1-score and support for all the classes using already existing formulae used for multi-class classification as illustrated below. Results are framed in Tables 2 and 3.

We can derive the following from it. For example, for COVID-19 class label, True Positive = 329, False Positive = 63, False Negative = 31, True Negative = 743.

- Precision = True Positive / (True Positive + False Positive)

Fig. 12 Confusion matrix

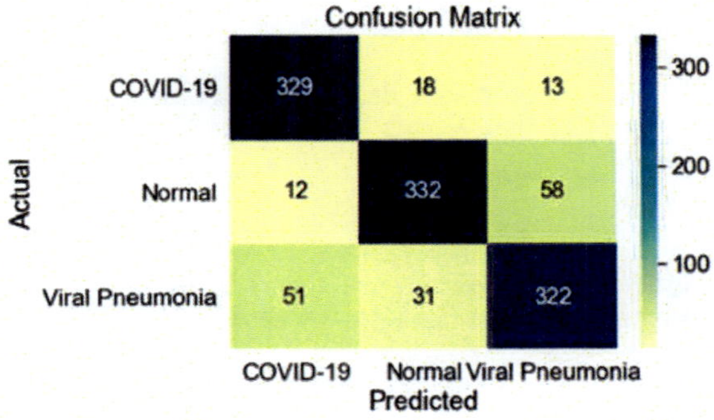

Table 2 Classification report on train data

	Precision	Recall	F1-score	Support
COVID-19	0.87731	0.96190	0.91766	840
Normal	0.97416	0.92332	0.94806	939
Pneumonia	0.90869	0.87779	0.89297	941
Accuracy	0.91949			2720
Macro avg.	0.92005	0.92101	0.91956	2720
Weighted avg.	0.92160	0.91949	0.91949	2720

Table 3 Classification report on test data

	Precision	Recall	F1-score	Support
COVID-19	0.83929	0.91389	0.87500	360
Normal	0.87139	0.82587	0.84802	402
Pneumonia	0.81934	0.79703	0.80803	404
Accuracy	0.84305			1166
Macro avg.	0.84344	0.84560	0.84368	1166
Weighted avg.	0.84344	0.84305	0.84269	1166

- Recall = True Positive / (True Positive + False Negative)
- F1-Score = 2 * (Precision * Recall) / (Precision + Recall)
- Support is the number of actual occurrences of a particular class in the given dataset.
- Macro average calculates the average of f1 score for each class label without considering the proportion of each label in the given dataset.
- Weighted average calculates the average of f1 score for each class label, considering the proportion of each label in the given dataset.

Classification report on the train data is shown in Table 2.

Classification report on test data is shown in Table 3.

Severity Analysis is done on the data [14]. Each row is passed to the function that determine severity. Bar plot on severity analysis is shown in Fig. 13. Bar plot is obtained by plotting the return values of the function that determine patient's condition.

Table 4 presents the details of the number of patients in each condition.

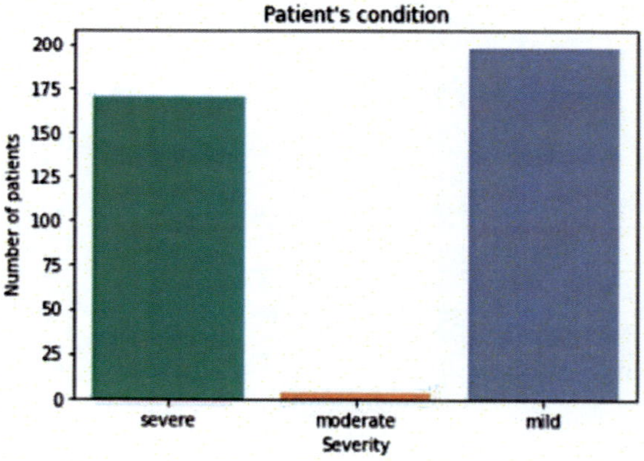

Fig. 13 Bar plot presenting severity analysis

Table 4 Severity analysis

Patient's condition	Number of patients
Severe	170
Moderate	4
Mild	198

5 Conclusion and Future Work

Using our proposed system, COVID-19 can be diagnosed with 0.84 accuracy using X-rays itself. As we used a large dataset to train our model, we can try using it in real-time. We can eliminate the delay in getting the test results and thus make the testing procedure available even at rural areas. The predictions can help public health experts to track COVID-19 patients. Using a few parameters and a description of patient's health condition, severity is analyzed. Even though proper test data is not available, we can reply on this approach of determining severity as it is designed based on the clinical health management protocol and people can monitor their health using this automated system.

We can extend this work to predict the patient's condition and recovery time based on periodic observation of clinical notes and chest X-rays. Artificial intelligence (AI) can be used to develop this prediction system.

References

1. Wang N, Liu H, Xu C (2020) Deep learning for the detection of COVID-19 using transfer learning and model integration. In: 2020 IEEE 10th international conference on electronics information and emergency communication (ICEIEC), Beijing, China, pp 281–284. https://doi.org/10.1109/ICEIEC49280.2020.9152329
2. Mporas I, Naronglerdrit P (2020) COVID-19 identification from chest X-rays. In: 2020 international conference on biomedical innovations and applications (BIA), Varna, Bulgaria, pp 69–72. https://doi.org/10.1109/BIA50171.2020.9244509
3. Ohata EF et al (2021) Automatic detection of COVID-19 infection using chest X-ray images through transfer learning. IEEE/CAA J Autom Sin 8(1):239–248. https://doi.org/10.1109/JAS.2020.1003393
4. Horry MJ et al (2020) COVID-19 detection through transfer learning using multimodal imaging data. IEEE Access 8:149808–149824. https://doi.org/10.1109/ACCESS.2020.3016780
5. Padma T, Kumari CU (2020) Deep learning based chest X-ray image as a diagnostic tool for COVID-19. In: 2020 international conference on smart electronics and communication (ICOSEC), Trichy, India, pp 589–592. https://doi.org/10.1109/ICOSEC49089.2020.9215257
6. Rehman T (2020) Covid-19 radio graphy database. Kaggle.com, September, 2020. (Online), Available:https://www.kaggle.com/tawsifurrahman/covid19-radiography-database. Accessed 27 Nov 2020
7. Gull H, Krishna G, Aldossary MI, Iqbal SZ (2020) Severity prediction of COVID-19 patients using machine learning classification algorithms: a case study of small city in Pakistan with minimal health facility. In: 2020 IEEE 6th international conference on computer and communications (ICCC), pp 1537–1541. https://doi.org/10.1109/ICCC51575.2020.9344984
8. Rohini M, Naveena KR, Jothipriya G, Kameshwaran S, Jagadeeswari M (2021) A comparative approach to predict corona virus using machine learning. In: 2021 international conference on artificial intelligence and smart systems (ICAIS), 2021, pp 331–337. https://doi.org/10.1109/ICAIS50930.2021.9395827
9. Jmour N, Zayen S, Abdelkrim A (2018) Convolutional neural networks for image classification. In: 2018 international conference on advanced systems and electric technologies (IC_ASET), pp 397–402. https://doi.org/10.1109/ASET.2018.8379889
10. Manoharan S (2020) Embedded imaging system based behavior analysis of dairy cow. J Electron 2(02):148–154

11. Chen JIZ, Hengjinda P (2021) Early prediction of coronary artery disease (CAD) by machine learning method—a comparative study. J Artif Intell 3(01):17–33
12. Adam EEB (2020) Deep learning based NLP techniques in text to speech synthesis for communication recognition. J Soft Comput Paradigm (JSCP) 2(04):209–215
13. Government of India (2020) Clinical management protocol for Covid-19 in Adults. mohfw.gov.in, 2020. (Online). Available: https://www.mohfw.gov.in/pdf/UpdatedDetailedClinicalManagementProtocolforCOVID19adultsdated24052021.pdf. Accessed 25 May 2021
14. Bachir (2020) Covid-19 Chest X-ray. Kaggle.com, May, 2020. (Online). Available: https://www.kaggle.com/bachrr/covid-chest-xray. Accessed 8 Feb 2020

Safe Encryption Algorithm for Secured Message Communication Using Dcombo: A New Sorting Technique

S. Muthusundari, A. Sonya, C. M. Nalayini, A. R. Sathyabama, and P. V. Rajasuganya

Abstract The DES and AES algorithms are often utilized to transfer messages. All those specifications can not appear to be sufficiently safe and rapid to deal with the communication issues. A modern technology for encrypting the sort message has already been postulated. This paper focuses on the safe platform used by encrypting and decrypting the text to address the security of data. In this document, we suggest a new sorting technique with an accurate encryption. The fundamental assumption of this model is to upgrade safety by changing the Bubble sorting method for encryption and decryption of a text message, which is much easier to perform and to enhance quality. DCombo sort is called the new sorting approach. In digital data communication it performs at least 1.5 times comparatively higher than previous technique.

Keywords Encryption · Decryption · Plain text · Cipher text · ASCII value · DCombo sorting technique · Higher performance

1 Introduction

That use of arithmetic's to encrypt and decrypt information; cryptography is used as a scientific skill. Encryption should be transformed into an unauthorized form in

S. Muthusundari (✉)
Department of CSE, R.M.D. Engineering College, Kavaraipettai, India
e-mail: sms.cse@rmd.ac.in

A. Sonya
Department of Information Technology, B. S. Abdur Rahman Crescent Institute of Science and Technology, Chennai, India

C. M. Nalayini · A. R. Sathyabama
Department of Information Technology, Velammal Engineering College, Chennai, India

P. V. Rajasuganya
Department of Artificial Intelligence and Data Science, Velammal Engineering College, Chennai, India

S. Shakya et al. (eds.), *Proceedings of Second International Conference on Sustainable Expert Systems*, Lecture Notes in Networks and Systems 351,
https://doi.org/10.1007/978-981-16-7657-4_45

Fig. 1 General cryptographic system

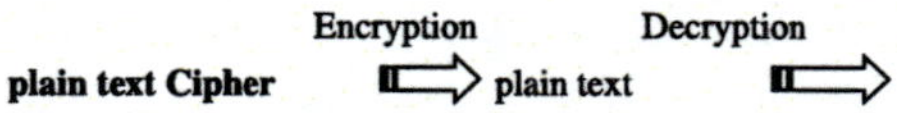

the simplified term. This protects the confidentiality of the transmitter to the recipient while delivering its data. The data can be decoded and returned to their appropriate method on the sender side [1–3]. This idea of this process needs additional data to be encrypted and decrypted. These are considered as key information.

Encryption also includes extensive protection for wireless communication. Encryption can be made using two fundamental elements known as substitution and transposition. Characters are modified by certain alphabets on substitution methods. The words are set up in another form in transposition ciphers (Fig. 1).

Information sent via the online platform on many databases, and/or devices and third gateway interception possibilities exist. It is not difficult to stop interference if not; documents should be made unrecognizable during transmission so that the authorized holder can receive propagation (decryption) [4].

Data encryption is a means of safeguarding data secrecy by encoding it into cipher text, which can only be decrypted using a unique decryption key generated at the moment of encryption or previously [5, 6]. Data encryption can be used during data storage or transfer, and it is usually combined with authentication services to ensure that keys are only given to those who need them.

Encryption helps preserve data integrity, preventing its contents from undocumented modification, and can be used to verify data's origin, reducing the danger of obtaining data from untrustworthy sources, in addition to the obvious benefit of greater privacy protection [7–9]. It is more accessible and desirable than it was earlier because of growing Internet use and the worth of data to hackers.

2 Existing Techniques

The two important principles are used in all cryptographic algorithms: substitution, where each component is mapped into plaintext and by transposition, the aspects of a plaintext are purely reorganized in reverse context with each other [10].

2.1 Algorithms for Encryption

Today, key size is commonly used to assess encryption strength. Regardless of how strong the method is, the encrypted data can be subjected to brute force assaults, which try all conceivable key combinations. The encryption will eventually be broken. The time it takes to crack most current ciphers with reasonable key lengths using brute force is estimated in millennia.

A. Transposition Cipher

These ciphers are frames that transform their location of specific alphabets or fragments on the code word. The sequence is splitted into symbols for encryption, and one key is used to identify how many possible permutations ($n!-1$). This is carried out by means of an inverse permutation that preserves the intensity values [5]. Transposition preserves the allocation of only one text frequency and destruct the flowcharts and allocations are at greater order. These technologies are also merged with other ciphers to make it safer [11]. The easiest ciphers are the key stream technology wherein the transmitted data is recorded as a vertical pattern and read as a sequential order. For instance: to encipher the statement—flee at once found—we published the latest intensity 3.

```
W . . . E . . . C . . . R . . . L . . . T . . . . E
. E . R . D . S . O . E . E . F . E . A . O . . C .
. . A . . . I . . . V . . . D . . . E . . . . N . .
```

The encrypted message: WECRLTEERDSOEEFEAOCAIVDEN

B. **Substitution Ciphers**

A replacement method is among the plaintext messages are substituted by other characters, icons or digits. When the plain text are the series of bytes and replace the data patterns with cipher text bits. Julius Caesar developed the well-recognized and most simple use of a substitution cipher. Every step in the sequence is replaced by the letter three places even farther beneath the letters [6]. For instance: plaintext: task finished

Ciphertext: wdvnfrpsohwhg. In the alphabet of this algorithm, *Y* and *Z* are wrapped to previous text is *Z* to *A*. By identifying all possibilities, we can describe the conversion

a b c d e f g h i j k l m n o p q r s t u v w x y z

Cipher:

D E F G H I J K L M N O P Q R S T U V W X Y Z A B C

3 Related Work

Nagarajan et al. [3] suggested a proper encryption system with 9's supplement. You have the 9 additions to configure the pad encrypted data on a one-time basis.

Kumar et al. [12] introduced the poly-substitution genetic technique to create the ASCII values of the given context and then convert it to cryptographic properties. Rather than one as with many easier devices, they recommended utilizing two or three simpler systems.

Kumar [13] described a secure and rapid data transfer pathway to address the issue using a circular queue theory.

The improvement of the proposed methods that 6 × 6 sequence substitution techniques were built by Babu et al. [14].

Kaur et al. [15] had proposed to fine-tune the hyperparameters of a five-dimensional chaotic map (TFCM). The input image is first split into subbands using a dual-tree complex wavelet transform (DTCWT) to implement TFCM. The secret key obtained from the optimized 5D chaotic map is then utilized to diffuse these subbands.

Wang [16] discussed with the local Shannon entropy, the suggested approach is faster and has a higher pass rate. The ant differential attack test results are closer to theoretical values and have less data fluctuation, and the cropping and noise attacks photos are crisper. As a result, the suggested algorithm provides increased security and resistance to a variety of threats.

Ma and Zhang [17] finally concluded that, to ensure the integrity of the encrypted private information, we employ a stream cipher and a dual-key method to perform a complete freedom nondestructive transition between plaintext and cipher text. Their experiments have shown that the suggested method takes less time to encrypt and decrypt data, produces better cipher text conversion output, and is less vulnerable to network attacks. The time it takes to encrypt and decrypt messages is faster than with other methods. In terms of calculating cost, the proposed solution saves about 14%. It also has a greater level of security and increases the security and integrity of the privacy data collection procedure.

Wang et al. [18] in their research, they offered a modified homomorphism encryption approach for retrieving multiple keywords. It is capable of efficiently resolving the issue of search keyword privacy leaking. In the multi-keyword retrieval method, retrieval performance is considerably increased. The novel approach is more efficient and accurate than other cipher text retrieval schemes, according to experimental data.

Zhu et al. [19] had discussed with the permutation-confusion-diffusion architecture, here they applied the two delayed systems to a novel picture encryption technique. To begin, the plain-image is scrambled using a multilayer of the nonlinear index to efficiently permute the pixel of the image. Second, the image is muddled by utilizing a chaotic matrix created by combining two chaotic sequences, and the cipher text is then turned into an integer. Finally, instead of employing a plaintext-related key stream, we use an unique diffusion method to improve plaintext sensitivity and make key management easier. The total of undiffused pixels and the cyclic bit-shift operation are included in the diffusion equation. The effectiveness of the suggested encryption algorithm is demonstrated by simulation results for the gray image.

Wang and Gao [20] suggested the matrix semi-tensor product (STP) is used to create a chaotic picture encryption technique with a compound secret key in this research. The first step is to create a new scrambling method. The pixels of the original plaintext image are divided into four blocks at random. Each block's pixels are then treated to a varied number of rounds of Arnold transformation, and the four blocks are combined to produce a jumbled image. After that, a compound secret key is created. To generate the true secret key, a series of pseudo secret keys is given and filtered by a synchronously updating Boolean network. To construct a chaotic sequence, this secret key is employed as the beginning value of the mixed linear-nonlinear coupled map lattice (MLNCML) system. Finally, the chaotic sequences and scrambled image are subjected to the STP procedure to produce an encrypted image. The approach suggested in this paper is safer and more effective than existing encryption algorithms, and it is also suited for color picture encryption.

Xu et al. [21] developed based on bit-plane matrix rotation and two hyper chaotic systems, they presented a new picture encryption algorithm. The algorithm decomposes the plain-image into eight bit planes and builds a three-dimensional (3D) matrix from there. The 3D bit-plane matrix's sub-matrix is then rotated in multiple directions using PRNS created by a hyper chaotic system. Finally, another key stream is used to change the pixel values of the intermediate image. Furthermore, the plain-MD5 image's hash function generates the starting values of diffusion and parameters relevant to generating chaotic sequences, which improves the correlation between the encryption process and the plain-image. In terms of key space, histogram, information entropy, key sensitivity, and neighboring pixels correlation index, simulation experiments are presented to examine the image encryption system. Theoretical analysis and actual findings show that the suggested algorithm performs well and provides adequate security.

Lu et al. [22] described the new picture encryption technique is based on the new S-Box and the chaotic key stream, and it consists of a round of permutation and two rounds of substitution. The plaintext image content is linked to the permutation and substitution key sequences, allowing the cryptosystem to withstand CPA. The proposed picture encryption scheme's efficiency was confirmed by simulation results and security research. The novel approach, in particular, exhibits clear efficiency benefits, indicating that it has greater application potential in real-time image encryption.

Patro et al. [23] suggested a new method bytes of the ciphered block are then further scrambled to make the scheme more diffused in the next phase. The block size changes as the key stream changes, which is influenced by the previous key stream as well as the plain-image. The size of the first block and the key stream are produced from a 64-byte secret key and the plain picture, respectively. Between the original and encrypted photos, the correlation and number of pixel change rate are 0.000479 and 99.620901, respectively. Both of the foregoing results, as well as other pertinent experimental data, significantly support the suggested scheme's resilience.

Naskar et al. [24] presented the utilisation of arbitrary sized multiple images to perform multiple image encryptions is the key feature of this approach. Another feature of this algorithm is that it solely uses PWLCM systems for permutation and diffusion operations, making it secure and efficient on both software and hardware platforms. The proposed cryptosystem's encryption results are good, according to the computer simulation. The proposed solution outperforms well-known security attacks, according to the security study.

Patro and Acharya [25] presented the novel approach, here instead of employing plaintext-related key stream, they used an unique diffusion method to improve plaintext sensitivity and make key management easier. The total of un-diffused pixels and the cyclic bit-shift operation are included in the diffusion equation. The effectiveness of the suggested encryption algorithm is demonstrated by simulation results for the gray image.

Algorithm
1. Print the n items
2. Split the items into two different subsets.
3. In case of intersection and compose as in a[1] and a[2] locations, the very first subgroup of a[1] is equated with a second set of [mid +1].
4. Return step 3 to all a [2], a [3] and a [mid].
5. Follow the first subsequence and do that for both step 3 and step 4 for the second group.
6. Repeat this procedure for the particular sub-sets again.
7. Use the Bubble sorting technique.

Fig. 2 DCombo sorting method

Kaur et al. [26] described in their presentation is focused on analyzing and evaluating valuable AES research publications. This article also goes through certain machine learning techniques, such as the artificial neural network (ANN) that can be used to protect images from side channel attacks. Aside from that, several other recent innovations have been addressed and researched in depth, such as the effective differential power analysis (DPA) technique and the hybrid encryption algorithm employing LZW. Finally, the study finishes with a thorough examination of several sorts of AES attacks, as well as possible future research directions. Finally, this paper provided a new attack detection model for AES.

4 Proposed Methodology

The device deals the safety use of a new encryption and decryption sorting methods.

4.1 DCombo Sorting Technique

(see Fig. 2)

4.2 Example

Get the *n* components.

$$N = 8$$

3 8 15 20 25 30 12 11

The process of this approach is given in Table 1.

Table 1 Process of the DCombo sorting algorithm

3	8	15	20	25	30	12	11	$N = 8$
3	8	15	20	25	30	12	11	Mid = 8/2 = 4,divide in to two subsets
3 3	8* 25	15 + 8	20■ 30	25 12	30* 15	12 + 11	11■ 20	Compare a [4] with a [mid + 1], and interchange if necessary and write it as combined
3 3	25* 8	8 25	30* 30	12 11	15* 12	11 15	20* 20	Divide the first subset into two subsets, and repeat step 3
3 3	8 11	25 8	30 12	11 15	12 25	15 20	20 30	Compare a [4] with a [mid + 1], and interchange if necessary and write it as combined
3 3	11 8	8 11	12 12	15 15	25 20	20 25	30 30	Apply Bubble sort method Pass 1 sorted data

Table 2 Bubble sort process

3	8	15	20	25	30	12	11	$N = 8$
3	8	15	20	25	12	11	30	Pass 1
3	8	15	20	12	11	25	30	Pass 2
3	8	15	12	11	20	25	30	Pass 3
3	8	12	11	15	20	25	30	Pass 4
3	8	11	12	15	20	25	30	Pass 5

4.3 *The New Sorting Technique Compares with Bubble Sort Method*

To get *n* objects, $N = 8$.

3 8 15 20 25 30 12 11

The procedure of Bubbles sort is given in Table 2.

4.4 *Performance Analysis on DCombo Sorting Technique*

As the above research paper relates the existing Bubble sort method where the DCombo sort process is required. The classifier of the new methodology enhances the framework five times more quickly than Bubble sort. This promotes productivity by 80 percent. We can therefore use our new trial scheme here for sorting methods.

5 Encryption Process

There are three stages for encoding of $C = E\ (K, P)$ with a suggested technique.

Steps

1. Allocate an integer Alphabetic ideals besides Fig. 2 syntaxes.
2. Schedule the numerical data by the new ordering software.
3. Replace the ordered set with the consonants.

5.1 Working Principle of Encryption

Step 1:

Each character is accessed and the ASCII value is initialized. For instance, one might configure the 'HELLO' sequence for records, and then allocate with the ASCII value (e.g., ASCII (A) = 65, ASCII (B), = 66, ASCII (C) = 67, etc.). Thus the numeral sequence is then passed through a channel and can then be decrypted by the transmitter to use the identical concept reciprocally (Fig. 3).

Step 2:

Replacement of symbols with ASCII digits is organized by the new DCombo sorting method in ascending order.
ASCII values by step1:
72 69 76 76 79
Set the sequence by new DCombo sort: 69 72 76 76 79

Step 3:

Substitute the sorted data with the ASCII values (Fig. 4).
Char(69) = E, Char(72) = H like so on.
Substitution of characters: E H L L O

ASCII	Hex	Symbol	ASCII	Hex	Symbol	ASCII	Hex	Symbol	ASCII	Hex	Symbol
64	40	@	**80**	**50**	**P**	96	60	`	112	70	p
65	41	A	**81**	**51**	**Q**	97	61	a	113	71	q
66	42	B	**82**	**52**	**R**	98	62	b	114	72	r
67	43	C	**83**	**53**	**S**	99	63	c	115	73	s
68	44	D	**84**	**54**	**T**	100	64	d	116	74	t
69	45	E	**85**	**55**	**U**	101	65	e	117	75	u
70	46	F	**86**	**56**	**V**	102	66	f	118	76	v
71	47	G	**87**	**57**	**W**	103	67	g	119	77	w
72	48	H	**88**	**58**	**X**	104	68	h	120	78	x
73	49	I	**89**	**59**	**Y**	105	69	i	121	79	y
74		J	**90**	**5A**	**Z**	106	6A	j	122	7A	z
75	4B	K	**91**	**5B**	**[**	107	6B	k	123	7B	{
76	4C	L	**92**	**5C**	****	108	6C	l	124	7C	|
77		M	**93**	**5D**	**]**	109	6D	m	125	7D	}
78		N	**94**	**5E**	**^**	110	6E	n	126	7E	~
79		O	**95**	**5F**	**_**	111	6F	o	127	7F	

Fig. 3 ASCII value for the characters

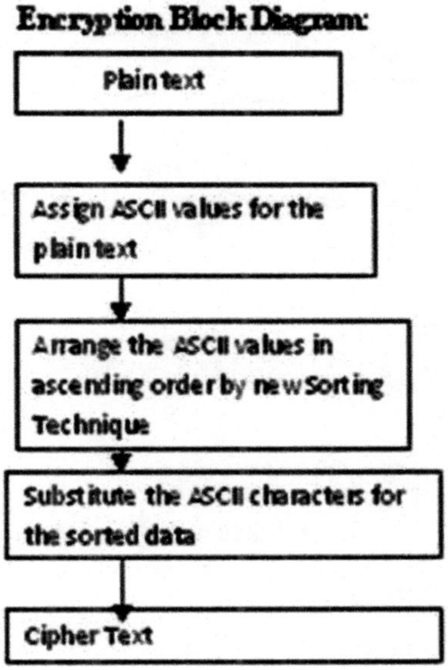

Fig. 4 Encryption block diagram

6 Decryption Process

The method for decryption process executes encryption functions that is $P = D(K, C)$. Made in 3 stages. The following process is:

1. Invert the text of the cipher. Then replace the unit of ASCII.
2. Assemble the previous step with new sort method to give ascending order. Then replace the secure record with the ASCII value.
3. When it fits the cipher data, the plain text will be shown to the recipient, or the defect report will be given to the sender if it aligns the cipher text. The consumer will not classify the text file.

6.1 Working Principle of Decryption

Step 1: Convert the encryption values to the cipher.

Cipher Text: E H L L O.
Return the text: O L L H E.
Replace the reverse cipher text ASCII principles: 79 76 76 72 69

Step 2: New DCombo sorting technology manipulates the characters to the backward text in increasing method.

Now the sequence order is: 69 72 76 76 79.
Substitute the ASCII values: E H L L O.

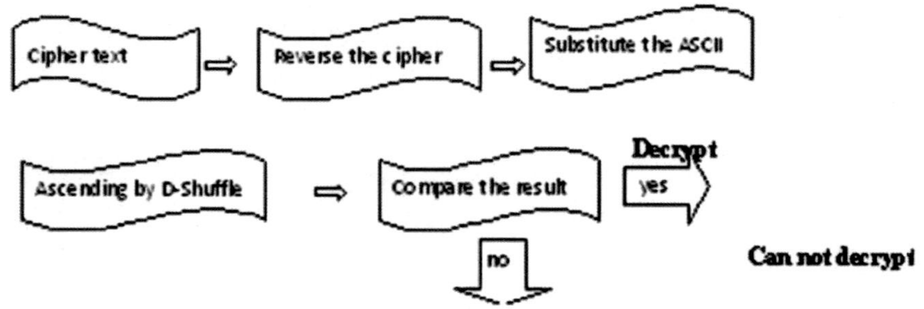

Fig. 5 Decryption block diagram

Step 3:

Evaluate phase 1 and 2 outcomes. If the two are equaled, the sensor detects the source code. It will not show text file alternatively. The recipient will not have the plain text as mentioned. Only the sender can decode once they know the functional area (Fig. 5).

Contrast the step 1 output = step 2 output

E H LL O = E H L L O

The similarity has now been paired. The person receiving can therefore authenticate and get text file.

Decrypt message: H E L L O

7 Case Study of Encryption Process and Decryption Process

Case 1: Encryption Process

Take the message as one term: S E C R E T

Step 1:

Set the ASCII code in plain text for each character set. Then we have,

ASCII (S) = 83 ASCII (E) = 69 ASCII (C) = 67 ASCII (R) = 82 ASCII (E) = 69 ASCII (T) = 84

PLAIN TEXT S E C R E T

ASCII VALUE 83 69 67 82 69 84

STEP 2: Reorder it by new sorting method

ASCII VALUE: 83 69 67 82 69 84

Ascending Order: 67 69 69 82 83 84

Step 3: Replace the upward order characters.

Ascending Order: 67 69 69 82 83 84
Cipher Text: C E E R S T
PLAIN TEXT : SECRET

CIPHER TEXT: CEERST

Case 1: Decryption Process

Step 1: Get the cipher text and alter it by encoded method.

CIPHER TEXT : C E E R S T
Reverse Cipher Text : T S R E E C

Step 2: Fix the opposite value in increasing order by D-Shuffle procedure.

Reverse Cipher Text : T S R E E C
Ascending order : C E E R S T

Step 3: Evaluate the content to step 2 outcome. If both match the recipient receives the message or the plain text is not received. The receiver does not decode the cipher text. Therefore, the algorithm ensures protection.

CIPHER TEXT: C EE R S Tis equal to
OUTPUT OF STEP 2: C EE R S T
The recipient will therefore receive the text file.
PLAIN TEXT: S E C R E T

8 Simulation and Experimental Results

To access our enhancements productivity, the Bubble sort [4] method calculates the encrypted data. For both methodologies we have incorporated C Language. The outcomes of the application display the achievement of our strengthening analysis tool better than the sort techniques (Figs. 6 and 7).

The samples from the execution of C are provided below.

A. Implementation Procedure

The value to be converted in a form of a message by use of a new sort technology, be scheduled in smallest to largest, the associated symbols are replacing the digit and thus generating the chip document.

The cipher text is provided in the sender's guidance and then decode the recipient by opposite text and again to the same invert encoding and replaces the ASCII code with a new sorting methodology. Then it corresponds it and then replaces the information, if several matches, the text is removed from the folder and displayed. The recipient cannot decrypt simple text on the other hand. Security is thus superior to other methods of encryption. Though the recipient knows how to decrypt and has to texting the key in the same way, the sender cannot achieve the plain text once any

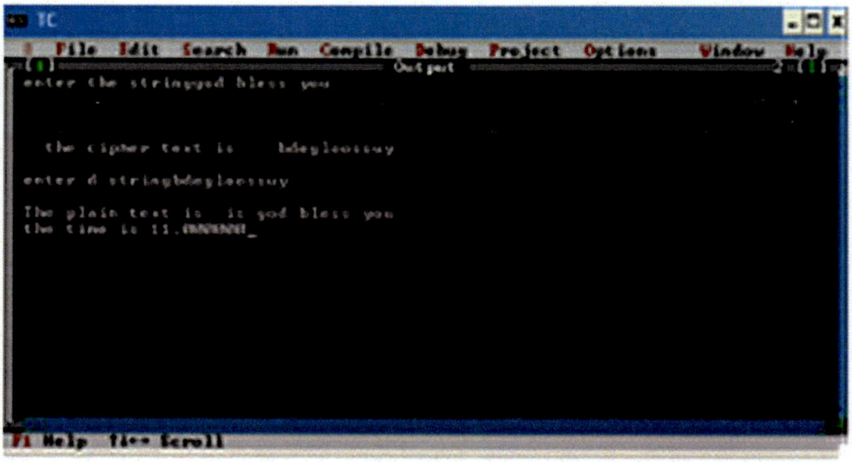

Fig. 6 Result of encryption using new sort

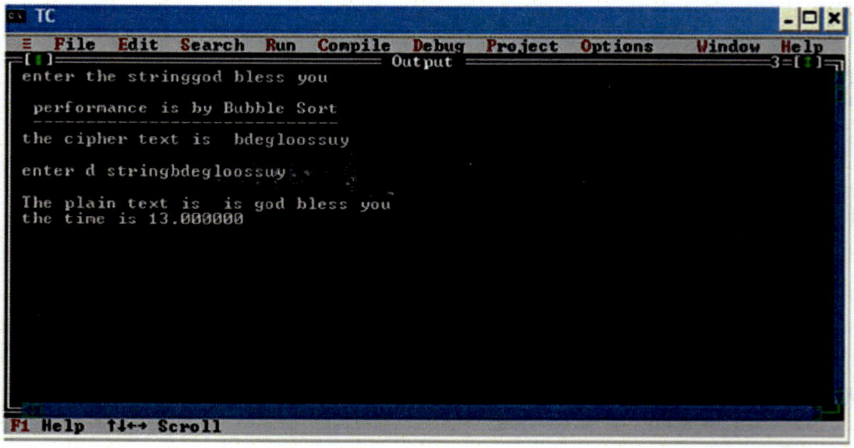

Fig. 7 Result of encryption using bubble sort

key error is entered. Nobody can reach the plain text, if the recipient cannot obtain. It ensures that 100% message security is passed efficiently.

The outcomes displayed here are carried out by new sorting. This is why the project is effectively executed.

B. Analysis the performance of new sort with Bubble sort for Encryption Process:

The C method has made with various data aspects as the insight and the efficiency of the outcomes has been evaluated. The results show our improvement in algorithm using new sort takes much less time with the various data experiments than the data encryption with Bubble sort. The findings with the run-time of these data are listed in

Table 3 Comparison of performance with data size

S. no.	Data size (Char)	New sorting technique (Time in sec)	Bubble sort (Time in sec)
1	10	11	16
2	25	28	38
3	35	26	48
4	50	34	67
5	75	48	95

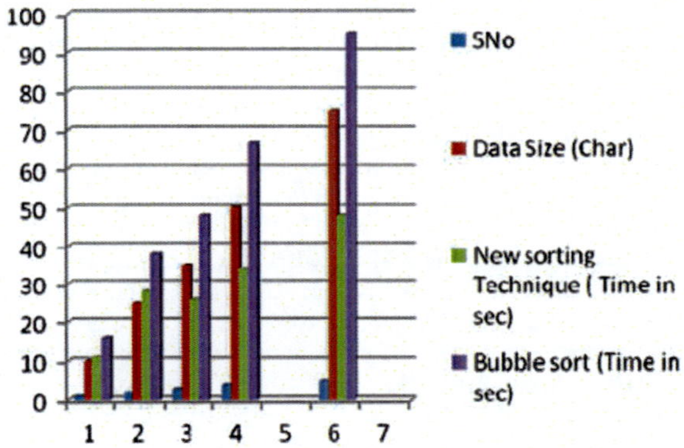

Fig. 8 Performance analysis of proposed Dcombo sorting method

Table 3. The level was calculated by using the operating system during the embedding process. Figure 3 presents the comparative analysis of the encryption algorithm with Bubble sorting efficiency (Fig. 8).

This process is used for DCombo is therefore uses Bubble sorting method at least 1.5 times faster than the encryption algorithm.

9 Conclusion

We developed an unique one in this document with DCombo sorting method to analyze the cryptional scheme. The primary aim of this automated system was to estimate its data size output. The findings seem to have an increase in ambiguity and safety of the framework.

The above-presented technique is a simple, straightforward, sorting algorithm. It is therefore quick and efficient for conditions requiring high speed encryption. This authentication method is inspired by ASCII value manuscripts. The mixture of algebraic replacement and filtering makes the process easy.

References

1. Yang K-H, Niu S-J (2009) Data safe transmission mechanism based on integrated encryption algorithm, Staffordshire University
2. Chen Tao XY (2005) Design and implementation of encryption algorithm based on dimension magic cube. J Inf 2:13–14
3. Nagaraj S, Kishore, Ramachandra (2010) Formal method of encryption using 9's complement. Int J Comput Appl 8(5):23–25
4. Muthusundari S, Santhosh Baboo S (2012) A new encryption method for secured message passing by sorting technique. In: Second international conference CCSIT, Springer, Bangalore
5. Lee MH (2007) Bounds on substitution ciphers. IEEE Inf Theor 6:2294–2296
6. Stinson DR (2003) Cryptogrphy theory and practice, 2nd edn.
7. Kaufman C et al (2003) Network security private communication in a public world. Prentice Hall of India Private Limited
8. Ravi Kumar B et al (2011) Data encryption and decryption process by using bit shifting and stuffing methodology. Int J Comput Sci Eng (IJCSE) 3(7):2816–2826
9. Shannon's C (1949) Communication theory of secrecy systems
10. Wikipedia "Encryption". http://en.wikipedia.org/wiki/Encryption. Modified on 13 December 2006
11. Chatterjee A, Aung KMM (2019) Sorting on encrypted data. In: Fully homomorphic encryption in real world applications. Computer architecture and design methodologies, Springer, Singapore. https://doi.org/10.1007/978-981-13-6393-1_3
12. Kumar J et al (2010) Poly substitution method for encryption and decryption. Int J Comput Sci Eng 02(05):1810–1812
13. Kumar K (2010) Efficient digital encryption algorithm based on matrix scrambling technique. Int J Netw Secur Appl (IJNSA) 2(4):30–41
14. Babu R et al (2011) An extension to traditional playfair cryptographic method. Int J Comput Appl (0975–8887) 17(5):34–36
15. Kaur M et al (2020) Color image encryption using non-dominated sorting genetic algorithm with local chaotic search based 5D chaotic map. Future Gener Comput Syst 107:333–350
16. Wang YXX (2021) Fractal sorting matrix and its application on chaotic image encryption. Inf Sci 547(8):1154–1169
17. Ma H, Zhang Z (2020) A new private information encryption method in internet of things under cloud computing environment. Wireless Commun Mobile Comput 2020:9, Article ID 8810987. https://doi.org/10.1155/2020/8810987
18. Wang X, Yin S, Li H, Teng L, Karim S (2020) A modified homomorphic encryption method for multiple keywords retrieval. Int J Netw Secur
19. Zhu S, Zhu C, Fu Y, Zhang W, Wu X (2000) A secure image encryption scheme with compression-confusion-diffusion structure. Multimedia Tools Appl 79(43–44):31957–31980
20. Wang X, Gao S (2020) Image encryption algorithm based on the matrix semi-tensor product with a compound secret key produced by a Boolean network. Inf Sci 539:195–214
21. Xu C, Sun J, Wang C (2020) A novel image encryption algorithm based on bit-plane matrix rotation and hyper chaotic systems. Multimedia Tools Appl 79(9–10):5573–5593
22. Lu Q, Zhu C, Deng X (2020) An efficient image encryption scheme based on the LSS chaotic map and single S-box. IEEE Access 8:25664–25678
23. Patro KAK, Soni A, Netam PK, Acharya B (2020) Multiple grayscale image encryption using cross-coupled chaotic maps. J Inf Secur Appl 52, Article ID 102470
24. Naskar P, Bhattacharyya S, Nandy D et al (2020) A robust image encryption scheme using chaotic tent map and cellular automata. Nonlinear Dyn 100:3
25. Patro KAK, Acharya B (2020) A novel multi-dimensional multiple image encryption technique. Multimedia Tools Appl 79(5)
26. Kaur J, Lamba S, Saini P (2021) Advanced encryption standard: attacks and current research trends. In: 2021 international conference on advance computing and innovative technologies in engineering (ICACITE), pp 112–116. https://doi.org/10.1109/ICACITE51222.2021.9404716

Domain Text Classification Using Machine Learning Models

Akula V. S. Siva Rama Rao, D. Ganga Bhavani, J. Gopi Krishna, B. Swapna, and K. Rama Sai Varma

Abstract Natural Language Processing (NLP) is used to classify text domains. In all applications where data is critical, such as News media, educational institutes, business organizations, research organizations, scientific, technology companies, and government organizations maintaining huge every day generated data is a serious challenge. Automated text classification has long been regarded as a critical way for managing and processing a large number of digital documents that are widely distributed and growing. In this paper to classify text domain four algorithms such as SVM, Naive Bayes, Decision Trees, and Random Forest evaluated and proved that SVM have achieved more accuracy than other algorithms.

Keywords Text classification · Machine learning · Stemming · Stopwords

1 Introduction

The Domain Text Classification project includes robotically assigning a record to 1 or greater club classes. In many cases, it is far an essential project. Since the inception of virtual files, computerized textual content type has been a key utility and studies

A. V. S. Siva Rama Rao (✉) · D. Ganga Bhavani · J. Gopi Krishna · B. Swapna · K. Rama Sai Varma
Department of Computer Science and Engineering, Sasi Institute of Technology and Engineering, (Affiliated To JNTU, Kakinada), Tadepalligudem, Andhra Pradesh 534101, India
e-mail: shiva.akula@sasi.ac.in

D. Ganga Bhavani
e-mail: bhavani534@sasi.ac.in

J. Gopi Krishna
e-mail: gopi548@sasi.ac.in

B. Swapna
e-mail: swapna516@sasi.ac.in

K. Rama Sai Varma
e-mail: ramasai505@sasi.ac.in

S. Shakya et al. (eds.), *Proceedings of Second International Conference on Sustainable Expert Systems*, Lecture Notes in Networks and Systems 351,
https://doi.org/10.1007/978-981-16-7657-4_46

area. Text type has end up essential on the grounds that we address this type of huge quantity of textual content files on a normal basis. Text type is primarily based totally on records analytics, and it is able to additionally function the engine for facts exploration. These findings may be implemented to emergent programs that assist decision-making.

Text classification can be divided into two types: topic-based text classification and text genre-based text classification. Topic-based text categorization organizes documents into categories based on their content [1]. Scientific publications, news stories, movie reviews, and advertising are just a few examples of text types. The majority of genre classification information is obtained from the Internet, newsgroups, message boards, and broadcast or printed news. Because they are multi-source, they come in a range of forms, vocabularies, and writing styles, even among works of the same genre. To put it another way, the data is heterogeneous.

To overcome these problems machine learning provides a solution to the automatic classification of text datasets. The categorization process is a very difficult task for search engines in identifying intended content, spam mail separation, text content management on the Web, in case eliminating unimport data.

The task of accumulating a capability from the preparation knowledge is referred to as supervised learning. The preparation data is made up of a variety of perceptions as well as the end outcome.

To predict categorical values, classification methods are used, and regression methods are used to predict numerical attributes [2]. Polynomial regression and linear regression are two well-known regression algorithms. K-nearest neighbors, Decision trees, and Logistic regression are three well-known classification algorithms.

Extracting the entities in the text is the most basic and useful technique in Natural Language Processing [3]. Unsupervised learning refers to a set of techniques that are used to make conclusions from data sets without using the results. Principal component analysis, *k*-means, and clustering are three prominent unsupervised learning algorithms. Reinforcement learning is a complex subject [4].

Machine learning play an important role in text classification process of every generated digital text corpora especially in process of identifying false news, spam mails, fake currency, and feedback applications [5].

A trained data corpora is necessary for all supervised machine learning processes, even when an item can be classified into multiple classes, only one class items is considered in this process. It also considered that other than the pure text that is hierarchical structure of text or date were not considered in order to offer strategies that make the most of the each item and to perform well.

Sebastiani's proposed text evaluating plan in well manner. He illustrated various steps for domain text classification as shown in Fig. 1. Apart from a brief description of the text classification, we refer to a few more current publications than those mentioned in Sebastiani's paper, as well as a few pieces that Sebastiani did not mention. The graphical representation of the text classification process is shown in Fig. 1.

Construction of classifier for document items is similar to machine learning projects. But the difficult task is representation of the document [6]. Unlike Numeric

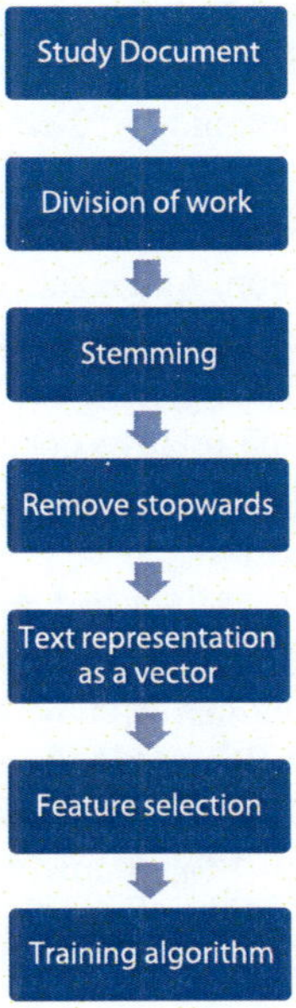

Fig. 1 Text classification process

classification the text classification different aspects has be considered such as words and phrases which involved in huge number, these words and phrases need to be converted in numbers during process, its emerged as challenging task.

This paper is organized into 7 sections, in Sect. 1 introduction of the paper is given. Section 2 discussed review of literature, Sect. 3 describes proposed models, in Sect. 4 results are presented, Sect. 5 explore the underlying meaning of your research, in Sect. 6 main finding were stated and in last Section list of references were given.

2 Review of Previous Work

Luo explained that text classification is a method for classifying any type of document into or out of a specified category. The support vector machines (SVM) model was used to classify English text and documents in this investigation. He used English papers in two analytical tests to test the selected classifiers. When the size of the feature set is small, the Rocchio classifier delivers the best performance results, but SVM beats the other classifiers, according to experimental results on a collection of 1033 text documents [5].

Sebastiani mentioned the principle gadget getting to know methods to textual content categorization. The advantages of this technique over the know how engineering technique (which includes guide definition of a classifier via way of means of area experts) encompass excessive effectiveness, substantial value financial savings in phrases of professional labor, and smooth portability to specific domains [7].

Kolluri et al. were used machine learning and deep learning models in their study procedure. Text classification can be done well utilizing machine learning and deep learning in their research [6].

Liu et al. discussed that how to remove noise in a huge number of text documents, such as irrelevant and redundant information, is a difficult task in text categorization. As a result, dimension reduction strategies such as feature extraction and feature selection have been developed to overcome this challenge. Finally, they compared the findings of various experiments conducted on six benchmark corpora, concluding that the relative document-term frequency difference (RDTFD) method based on independent feature space search is more resilient than other feature selection methods [8].

Aliwy and Abdul Ameer used the classification techniques in their study were Decision Tree, Support Vector Machine, K-Nearest Neighbors, Nave Bayes, and hidden Markov model. Many researchers revised all of these methods in order to achieve high precision. For text classification, the optimal algorithm was chosen based on previous analysis [9].

Mohammad et al. were said that there are three well-known strategies for classifying data. On an Arabic data set, these three well-known approaches are used [10].

3 Methodology

ML techniques can construct a model based on the standard training data. In other words, a machine learning computation fixes input instances of ordinary data and produces a model. This is then used to collect or categorize observed data along these lines as either common or unusual. In most cases, machine learning [4] models can deal with minor variations in the observed data.

On a regular basis, massive amounts of data are transmitted from one organization to another, resulting in unstructured data. Researchers typically employed machine learning techniques to categorize such large amounts of data. This also aids in the creation of well-organized and managed information in the form of classes or clusters.

It can also find interdependency criteria or deeper correlations between the metrics of the observed data. However, in order for this strategy to be effective, the typical training data must be representative of everyday behaviors; otherwise, there may be a lot of false positives when we use the model for prediction. Similarly, in the training stage, a considerable number of information is required.

In any event, once the model is produced, further inquiry is usually extremely fruitful [9]. The Naive Bayes family of algorithms, as well as support vector machines (SVM), are some of the most used text categorization algorithms.

3.1 Naive Bayes

The Naive Bayes family of statistical algorithms are among the most widely employed in text classification and analysis [11, 12]. The Naive Bayes algorithm is a method for computing classification probabilities. The Bayes' Theorem can be used to generate a collection of classification algorithms called Naive Bayes classifiers. These families of algorithms have a similar principle, for example, that each pair of features being classified independently from another.

Multinomial Naive Bayes (MNB) is one of those members, and it has the advantage of producing excellent results even when your dataset is small (a few thousand tagged samples) and computational resources are limited [13].

The Naive Bayes set of policies is based mostly on Bayes' Theorem, which al lows us to compute the conditional opportunities of occurrences depending on the opportunities of each individual event. So, for a given text, we calculate the threat of each tag and then output the tag with the very nice threat.

$$P(R/S) = (P(S/R) * P(R))/P(S) \tag{1}$$

If S is true, the probability of R equals the likelihood of S times the probability of R being true, divided by the probability of S being true.

Because of its accuracy and efficiency, Naive Bayes is commonly used in text categorization applications and experiments [10]. However, its performance is frequently hampered since it does not model text correctly. Schneider addressed the problems and showed how they can be handled with a few simple changes.

Klopotek and Woch said the findings of an empirical assessment of a Bayesian multinet classifier primarily based totally completely totally on a present day method for studying very big tree-like Bayesian networks with the beneficial useful resource of the usage of Klopotek and Woch [14]. According to the findings, tree-like Bayesian networks can control a textual content type undertaking with 100,000 variables with suitable tempo and accuracy.

3.2 Support Vector Machines

A Support Vector Machines (SVM) are an advanced textual content categorization gadget gaining knowledge of approach that, like Naive Bayes, calls for little schooling information to start generating accurate consequences [12]. SVM, on the alternative hand, calls for greater computing energy than Naive Bayes, however the consequences are quicker and greater accurate.

When implemented to textual content classification, support vector machines (SVM) offer accurate precision however low recall [15]. Adjusting the edge related to an SVM is one manner to personalize SVMs to growth recall. With advanced results,

Fig. 2 Decision trees

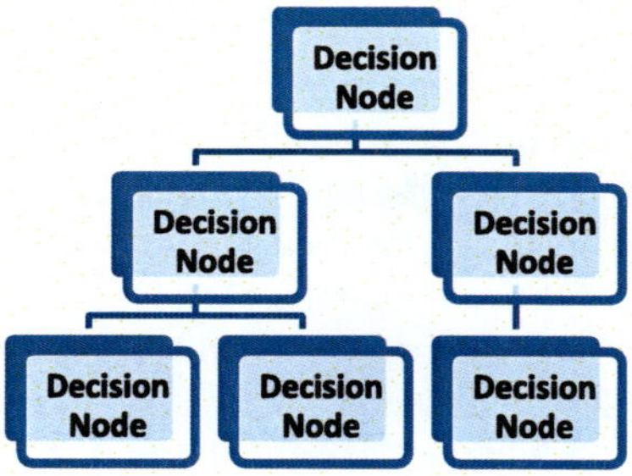

Shanahan and Roma defined an automated approach for enhancing the thresholds of time-honored SVM [15–17].

The vectors method is used for the classification of related documents items using the support vector machine (SVM) algorithm. Khamar et al. used this technique to form a two-class research topic based on the splitting of hyperplanes supplied by classes of data [18].

3.3 Decision Trees

It's a choice-making device that employs a hierarchical choice-making structure [3]. By incorporating simple alternative guidelines from the schooling data, those choice bushes try and create a version that predicts the corresponding classes. A choice tree is a device getting to know method that may be used to categorize or are expecting data. We will speak approximately it in phrases of classification, however it is able to additionally be implemented to regression. By segmenting the enter area into regions, a choice tree classifies inputs.

One of the most well-known ways for modeling data classification in classifiers is the decision tree classifier [8]. Machine learning algorithms for classification include a variety of approaches. The construction of decision trees is depicted in Fig. 2.

3.4 Random Forest Classifier

In the process of text classification, Random Forest classifiers are well suitable for coping with high-dimensional noisy input [19]. A random forest model is made up of a collection of decision trees, each of which is trained using a different collection of random features.

The random forest's class prediction for a given instance is created by a majority votes of all the trees in the forest predictions [20]. For different test examples, on the other hand, would have different values for the features used in the trees, and the trees need to contribute to the predictions in different ways. The general random forest classifiers do not considered the trees' various contributions.

4 Result

In this paper, we considered some datasets and use various machine learning methods to do text categorization on it. SVM, Naive Bayes, Decision Trees, and Random Forest Classifier are the four methods considered here. We used these datasets to test various algorithms to see which one is the best.

We begin by importing the necessary dataset and libraries. We must implement the source code for the algorithms we evaluate after importing the data set and relevant libraries. The goal of implementing the algorithm is to determine the accuracy of text classification performance.

First, we need to import dataset which contains some of the domains like sports, computers and vehicles which is shown in Table 1. Scikit-analyze is a loose gadget getting to know library for Python. It functions diverse algorithms like help vector gadget, random forests, and k-neighbors, and it additionally helps Python numerical and clinical libraries like NumPy and SciPy.

We did not have to install any packages to run the source code because scikit includes them all by default. We just use the import function to import the dataset. We simply validate the dataset's dimensions and output some of the information on it for confirmation purposes after it has been imported.

Text files are simply a collection of words (ordered). We must convert text files into numerical feature vectors in order to execute machine learning algorithms. A variety of algorithms can be used to classify text. We will begin with the most fundamental, "Naive Bayes" (NB).

Similarly, we must utilize SVM, Decision Trees, and the Random Forest Algorithm for the data provided on the given dataset. All of those algorithms can be readily implemented by importing their specific methods and functions. The entire implementation can be completed using simply the Python programming language. We will plot the performance graph of different algorithms to see which algorithm provides us better results in text classification. By comparing all of the graph's findings, we can easily decide which algorithm is the best. Comparing multiple models is often beneficial in achieving the best result.

Table 1 Analyzed datasets

List of datasets	Computers			Sports		Vehicles	
	Graphics	Windows	Mac	Base Ball	Hockey	Autos	Motor cycles
No of samples used	584	593	578	597	600	594	598

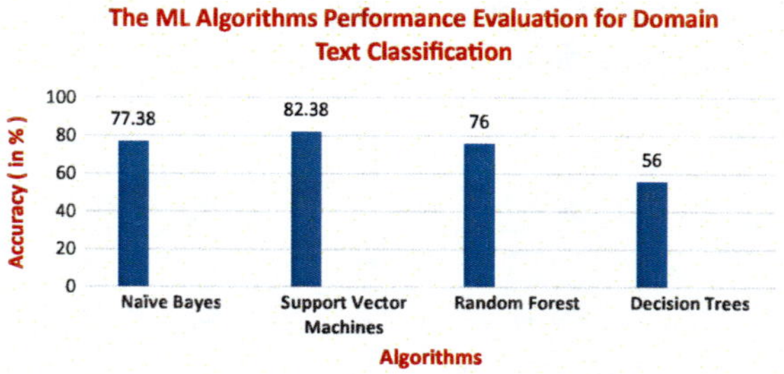

Fig. 3 Performance of text domain classification algorithms

We applied machine learning to find the most effective text classification technique in this study. For this study, we have to find the accuracy of algorithms. The outcomes are depicted in Fig. 3.

Figure 3 illustrates that the accuracy percentage of algorithms for performing text classification in the best possible approach. For performing text classification, we consider four types of algorithms and observed their corresponding accuracy levels. The techniques we explore for checking purposes are Naive Bayes, support vector machines, and random forest, decision trees. The accuracy of a measurement relates to how close it is to the true or acceptable value. The accuracy of the Naïve Bayes method is 77.38%, which is decent for a beginning and a Naive classifier. It is a clear and easy algorithm.

Support vector machines (SVM) has an accuracy of 82.38%. We obtain an accuracy score of 82.38%, which is a 5% improvement over Naive Bayes.

Random forest classifier has an accuracy of 76.0%, which is lower than support vector machines (SVM) and Naive Bayes. The accuracy of decision tress is 56.0%, which is lower than the other three algorithms.

5 Discussions

Text categorization algorithms are generally based on information exploration since it gives raw data an appropriate format. Pre-processing to remove tags, storage, access, feature extraction for removing irrelevant terms, over fitting, data imbalance, parameter estimate, stop words, and other issues are all addressed in text classification systems [21–23]. Based on the current research, a variety of classifiers have been identified. As a result, the few things that occur while creating a model include handling huge taxonomies in heterogeneous data, performance boosting for large datasets, and determining the ideal classifier.

6 Conclusion and Future Enhancement

Different "text classification" approaches have been identified based on this literature survey, along with their capabilities, advantages, and repercussions in extracting information from data. Understanding the problems with classification methods is critical at this stage, since it will make comparing different classifiers easier.

Our paper's main purpose is to present the best text classification algorithm. Figure 3 clearly shows which technique is appropriate for text classification. When the accuracy percentage of each method is considered, SVM outperforms the other three algorithms. Based on this data, we conclude that the support vector machines (SVM) technique is the most effective. In future the ML algorithms may be applied on biomedical terms datasets.

References

1. Yang Y (1999) An evaluation of statistical approaches to text categorization. J Inf Retrieval 1(1/2):67–88
2. Aseervatham S, Antoniadis A, Gaussier E, Burlet M, Denneulin Y (2011) A sparse version of the logistic regression for large-scale text categorization. Pattern Recogn Lett 32:101–106
3. Siva Rama Rao AVS, Ranjana P (2020) Empower good governance with public assessed schemes by improved sentiment analysis accuracy. Electron Government Int J 16(1/2), ISSN online: 1740–7508, ISSN print: 1740–7494
4. Allahyari M, Pouriyeh SA, Assefi M, Safaei S, Trippe ED, Gutierrez JB, Kochut KJ (2017) A brief survey of text mining: classification, clustering and extraction techniques
5. Luo X (2021) Efficient English text classification using selected machine learning techniques 60:3401–3409
6. Kolluri J, Razia S, Nayak SR (2021) Text classification using machine learning and deep learning model
7. Sebastiani F (2002) Machine learning in automated text categorization. ACM Comput Surv 34(1):1–47
8. Liu Y, Ju S, Wang J, Su C (2020) A new feature selection method for text classification based on independent feature space search 2020:1–14
9. Aliwy AH, Abdul Ameer EH (2017) Comparative study of five text classification algorithms with their improvements. Int J Appl Eng Res 12:4309–4319, ISSN 0973–4562
10. Mohammad AH, Alwada'n T, Al-Momani O (2016) Arabic text categorization using support vector machine, Naive Bayes and neural network. GSTF J Comput (JoC) 5(1):108
11. Ahmad S, Lavin A, Purdy S, Agha Z (2017) Unsupervised realtime anomaly detection for streaming data. Neurocomputing 262(1):134–147
12. Klopotek M, Woch M (2003) Very large Bayesian networks in text classification, ICCS 2003. LNCS 2657:397–406
13. Arar OF, Ayan K (2017) A feature dependent Naive Bayes approach and its application to the software defect prediction problem. Appl Soft Comput 59:197–209
14. Kim SB, Rim HC, Yook DS, Lim HS (2002) Effective methods for improving Naive Bayes text classifiers. LNAI 2417:414–423
15. Ghaddar B, Naoum-Sawaya J (2018) High dimensional data classification and feature selection using support vector machines. Eur J Oper Res 265(3):993–1004
16. Siva Rama Rao AVS, Ranjana P (2018) Machine learning based solution for homograph and auto-antonym ambiguities in social media corpora of sentiment analysis. J Adv Res Dyn Control Syst 1943–023x

17. Siva Rama Rao AVS, Ranjana P (2018) Machine learning based solution for homograph and auto-antonym ambiguities in social media corpora of sentiment analysis. J Adv Res Dyn Control Syst 10(3), SCOPUS INDEXED J ISSN 1943–023x
18. Shanahan J, Roma N (2003) Improving SVM text classification performance through threshold adjustment. LNAI 2837:361–372
19. Ke Y, Hagiwara M (2017) An English neural network that learns texts, finds hidden knowledge, and answers questions. J. Artif. Intell. Soft Comput. Res. 7(4):229–242
20. Xu BX, Li JJ, Wang Q, Chen XJ (2011) A tree selection model for ımproved random forest. In: Proceedings of the ınternational conference on knowledge discovery, pp 382–386
21. Breiman L (2001) Random forests. Mach Learn 45(1):5–32
22. Asadi S, Shahrabi J (2017) Complexity-based parallel rule induction for multiclass classification. Inf Sci 380:53–73
23. Cerchiello P, Giudici P (2012) Non parametric statistical models for on-line text classification. Adv Data Anal Classif 6:277–288

Named Entity Recognition Using Stanford Classes and NLTK

Akula V. S. Siva Rama Rao, P. V. V. Vamsi, N. Rashmika, K. Hemanth, and K. Aditya Kumar

Abstract The Information Extraction is a branch of natural language processing that focuses on extracting actual information from unstructured text. The goal of Information Extraction (IE) is to find a predetermined set of ideas, or a machine-understandable representation of the data. The traditional IE duties include the following: The challenge of identifying and classifying predefined categories of named entities is addressed by Named Entity Recognition (NER). The idea of Named Entity Recognition is to identify and categorized hard designators like Names, Locations, Time, medical terms, and temporal expressions from text. We conduct experiment using NLTK-NER, Stanford 3-Class NER, Stanford 4-Class NER and Stanford 7-Class NER model to extract Named Entities from text corpora and proved that Stanford class-3 model higher accuracy than other algorithms. Also proved that Stanford 7-Class more effective than others.

Keywords Named entity · Stanford NER · Stanford NER tagger · Natural language process

---

A. V. S. Siva Rama Rao (✉) · P. V. V. Vamsi · N. Rashmika · K. Hemanth · K. Aditya Kumar
Department of Computer Science and Engineering, SASI Institute of Technology and Engineering (Affiliated To JNTU, Kakinada), Tadepalligudem, Andhra Pradesh 534101, India
e-mail: shiva.akula@sasi.ac.in

P. V. V. Vamsi
e-mail: vamsi5a6@sasi.ac.in

N. Rashmika
e-mail: rashmika5a3@sasi.ac.in

K. Hemanth
e-mail: hemanth570@sasi.ac.in

K. Aditya Kumar
e-mail: yaswanth572@sasi.ac.in

S. Shakya et al. (eds.), *Proceedings of Second International Conference on Sustainable Expert Systems*, Lecture Notes in Networks and Systems 351,
https://doi.org/10.1007/978-981-16-7657-4_47

1 Introduction

NER is a part of information extraction that entails analysing structured and unstructured texts and finding phrases that relate to persons, locations, organizations, and businesses. Named Entity Recognition process is the subtask of Natural Language Processing. NLP is a branch of computing and AI that deals with computer–human (natural) language interactions. Natural Language Processing is subfield of Artificial Intelligence. Artificial Intelligence is technology that can be applied by machines to interpret and understand underlying meaning of the natural human languages [1]. This technology can be applied various applications such as information extraction, language translation, speech synthesis, data segmentation, Named Entity Recognition. It is used to apply machine learning algorithms to text and speech the first task of NER is to identify proper names in text, and the second task is to classify these names into a group of predefined categories of interest, like person names, organizations (companies, government organization, committees, etc.), locations (cities, countries, etc.), and date and time expressions. In the sixth Message Understanding Conference, the phrase "Named Entity" was first used (MUC-6). In reality, it was the MUC conferences that made a significant contribution to this field's study. It served as a baseline for named entity systems that were used to conduct a range of data extraction tasks.

Information Extraction is the process of detecting entities and extraction from textual corpora. The Named Entities and the relationships between them are stated in the text documents with the aim of organizing text corpora in a specific format that is more similar to database or flexible algorithms for mining [2]. Two primary approaches are utilized to handle the challenge of extracting information from texts: rule-based and machine learning.

The performance of any algorithm depends upon the purity of datasets. Some Named Entity Recognition easy to identify from text as they are proper nouns and the begin with capital letters, for machines it is difficult task to identify the entities, so that the has learn from that text.

Even the entities are identified, some of them may be shown irrespective of their meaning, i.e., if a person name is identified, it may determine it as some other entity. And in some cases it is difficult to categorize for a NER to determine whether the identified entity belongs to which category. This deals with the sense of the word. For example, if Stanford is the recognized word, it is difficult to say when Stanford is a university name and when it is a person's name.

The objective of named entity recognition and extraction is to extract and categorize names from text into specific categories based on their meaning.

2 Related Works

Many systems have been developed by the software community to extract named items from unstructured texts. Source codes, executables, and Web services are used to distribute systems, which vary in terms of licences, target domains, supported languages, processing methodology, recognized entity kinds, and input and output formats.

Nozza et al. demonstrated that embedding named items improves the semantic meaning of the feature space used to adapt from a source to a target domain [3]. The findings obtained on informal text benchmark datasets demonstrate that L2AWE not only outperforms various state-of-the-art models, but it can also handle the mistakes and uncertainties introduced by NER systems.

Patil et al. proposed a statistical named entity recognition system based on machine learning for the identification and classification of named entities present in Marathi language text [4]. In our system, named entities are identified and classified using conditional random fields (CRFs).

Chirila et al. used the Stanford NER and Scikit-learn tools to extract medical information and create a corpus to train the algorithms. From the testing, the best-performing tool proved to be Scikit-learn [5].

Won et al. proposed a model to extract LOCATION NAMES. They developed a system with 5 algorithm as ensemble algorithm by using voting system [6]. They got outstanding precision and recall performance with ensemble algorithm. Their results proved that the Named Entity Recognition system may not depend on pre-processing and modern English.

Jones and Robins said that using the precision, recall and $f1$ values to compare the NER tools the conclusion is Stanford NER gives the best performance results [7]. Stanford gives the precision value of 0.98 in comparison of NLTK's value of 0.94 and SpaCy's 0.57 value and NLTK's 0.48 recall value. It subsequently has the best $F1$ value of 0.85 with SpaCy in second with 0.65 and NLTK with the worst score of 0.61.

Costa—Named Entity extraction process examines the impact of each of the primary features available in the Stanford NER system on overall entity identification performance, allowing language model training to be fine-tuned to a specific corpus [8]. There were 91 tests run on the Stanford NER implementation's primary configurations/features, using the suggested configuration as a starting point and each test changing a numeric parameter or enabling/disabling a specific feature.

Jiang et al. compared Stanford NER, spaCy, Alias-LingPipe, and Natural Language Toolkit (NLTK) in a study. WikiGold, which has 145 categorised Wikipedia articles, is used as a text corpora. In terms of the entity types PERSON, LOCATION, and ORGANIZATION, the evaluation is susceptible to precise and partial matching. They used NLTK, LingPipe, and spaCy to test algorithms, and the overall average accuracy was 0.7609 $F1$-score [9].

Ekbal et al. analysed bibliographic texts and compared the tools such as OpenCalais, Alias-i LingPipe, Illinois NET, Stanford NER, NER WS etc. They analysed 247 Wikipedia articles and annotated to evaluate the tool's performance [10].

Landsbergen explained that NER algorithms can be used to identify the names and relationships of objects in voice and textual teaching [11].

Bird et al. proposed a novel model to extract named entities from the biographic text corpora. To analyse biographic text they used OpenCalais, LingPipe and Stanford Named Entity Recognizers and they observed low recall score because of not considering the uniform entity type and categories [12].

Mansouri et al.—Introduced natural language processing which allows a wide range of language methods, including predictive text and email filtering, as well as automatic summarization and translation [13].

Nadeau introduces a supervised learning approach for recognizing acronyms. After limiting the search space with a few hand-coded restrictions, the approach involves applying supervised learning to apply further constraints [14]. When the previously learned constraints no longer apply, the system may be simply retrained for a new corpus using this method. Although the hand-coded limitations lower the amount of acronym definition pair candidates that the supervised learning system must categorise, they are weak enough to be readily ported to a new corpus.

Minkov et al. investigated the performance of two machine learning-based methods for learning HMMs, conditional random fields and a perceptron-based method, on four novel annotated corpora. Experiments show that introducing email-specific learning elements improves baseline $F1$ performance significantly [15].

The CoNLL-2002 common task, language-independent named entity recognition, was described by Sang [16]. Data from two Western European languages, Spanish and Dutch, was analysed using twelve different systems. On both data sets, a boosted decision tree technique performed the best [2].

3 Proposed Methodolgy

What is a Named Entity?

A Named Entity is any word that symbolizes a person, organization location, or other entity. The process of identifying words that are named entities in a given raw text is known as Named Entity Recognition, which is a subtask of Information Extraction. It is also called as entity chunking or entity identification.

What is NLTK?

Natural Language Toolkit (NLTK) is a popular library in Python programming language. This library can be used for Natural Language Processing tasks. The hug data can analyse the data programmatically. Various text processing task can be performed by using NLTK.

3.1 Basic NLTK Algorithm

Word Segmentation or Tokenisation

Tokenisation is the process of breaking up a text's sequence of characters by determining where one word ends and another begins [17]. The words that are so recognised are commonly referred to as tokens in computational linguistics. Tokenisation is also known as word segmentation in written languages when no clear word boundaries are specified in the writing system, and this term is usually used interchangeably with tokenisation.

Word_tokenize module is used for basic word tokenization. Following example will split a sentence into words.

```
Ex: import nltk
    from nltk.tokenize import word_tokenize
```

Sentence Segmentation

The technique of finding the longer processing units made up of one or more words is known as sentence segmentation. Identifying sentence boundaries between words in different sentences is the goal of this assignment. Sentence segmentation is also known as sentence boundary detection. Since in written languages punctuation marks occurs at sentence borders. All of these phrases pertain to the same task: figuring out how to break a text into sentences for subsequent processing.

NLTK provides **sent_tokenize** module for this purpose.

```
Ex: import nltk
    from nltk.tokenize import sent_tokenize
```

3.2 Named Entity Recognition with Stanford NER Tagger and NLTK

The Stanford NER tagger is one of the CRF algorithm to extract Named Entities form textual corpora. This tagger able to extract more number of Named Entities from text corpora when compared with NLTK-NER it uses statistical complex model. Stanford Named Entity Recognizer is a named entity recognizer it was developed on linear chain Conditional Random Field (CRF) sequence models [15]. A big advantage of the Stanford NER tagger is that it is provided with a couple of different models for interrupts named entities.

Any of the following models can be used:

1. 3-class model recognizes locations, persons, and organizations
2. 4-class model recognizes locations, persons, organizations, and miscellaneous entities
3. 7-class model recognizes locations, persons, organizations, times, money, per cent and dates

Because the NER classifier is developed in Java, we will need to download the models and a jar file in order to proceed. The Stanford NER Tagger class is passed the following parameters:

1. Classification model path (3 class model used below)
2. Stanford tagger jar file path
3. Training data encoding (default of ASCII)

Two different NER classifiers are in our toolbox! However, should we go with NLTK's or Stanford's? Let us see what we can find out by conducting some tests.

CRF as a retrieval model:

$$p(y_t|x) = \frac{1}{Z}\exp\left(\sum_{k=1}^{K} w_k f_k(y_t, y_t - 1, x_t)\right) \tag{1}$$

We can not get rid of y dependence on $y_t - 1$ using pure mathematics, so we run to experiments to see if we can reduce this equation even more and quantify the accuracy loss.

$$p(Q|x) = \sum_{k=1}^{K} Q_k f_k(x) = \vec{Q}.\vec{f}(x) \tag{2}$$

We created a linear query model and calculated the difference between it and the more common CRF.

$$p(y|x) = \frac{1}{Z(x)}\prod_{t=1}^{T}\exp\left\{\sum_{k=1}^{K}\theta_k f_k(y_t, y_t - 1, x_t\right\}. \tag{3}$$

Above is the formula for CRF, where Y is the hidden state (ex, parts of speech) and X is the observed variable (entity).

It may be challenging, however NLTK provides a built-in feature for us [18]. You have two options with NLTK's named entity recognition either recognize all named entities or recognise named entities as their appropriate type, such as persons, places, and locations. When binary = True is selected, it signifies that something is either a named entity or not. There will be no further information.

It can observe a couple things right away if you set binary = False. When Binary is False, it picked up the same items, but it split up terms like Indian Institute into "Indian" and "institute" as if they were two separate entities, whereas when Binary = True, the named entity recognition was correct, and Indian Institute was part of the same named entity. It can use the binary choice as you want depending on objectives [16]. If you set binary to false, you can receive the following types of Named Entities:

Named Entity Type and Examples.

ORGANIZATION—Flipkart, Google
PERSON—Stark, Geetha
LOCATION—America, Asia
DATE—July, 2000-07-28
TIME—five thirty am, 2:45 pm
MONEY—130 Dollars, $455
PER CENT—thirty per cent, 75%
GPE—India, England

3.3 *Implementation and Setup*

How to check?
For Python
Open Command Prompt and type the command below.
python–version
For NLTK
Open python IDE or IDLE then write the code in it
import nltk
print nltk__version__
For Java
Enter the following command on Command Prompt.
java–version
Les get started.

1. Update Existing NLTK

   To upgrade your NLTK to the most recent version, type the following command in Command Prompt.

   Pip install-U nltk

2. Download the required files
   Download the following Stanford Parser packages.
   Stanford Parser,
   The Stanford Named Entity Recognizer (NER),
   Stanford Log-linear Part Of Speech Tagger

3. Extract each archive to separate folders

4. Open Environment Variables Window
   Go to search panel then write Environment Variables
   Open Environment Variable Window

5. Create two new variables
   CLASSPATH
   By clicking the New button, you may create a CLASSPATH variable with the following values.
   Drive:\path\to\stanford-ner-2020-11-17
   Drive:\path\to\stanford-parser-full-2020-11-17
   Drive:\path\to\stanford-postagger-2020-11-17
   STANFORD_MODELS
   Create another variable named STANFORD_MODELS
   Then add the following values to the variables
   Drive:\path\to\stanford-postagger-2020-11-17\models
   Drive:\path\to\stanford-ner-2010-11-17\classifiers

Initially the datasets are given as input to the algorithm which outputs the recognised entities. From that True Positives, True Negatives, False Positives and False Negatives are observed. The precision, recall and $F1$ values for the datasets can be calculated by applying formulas as shown in Fig. 1.

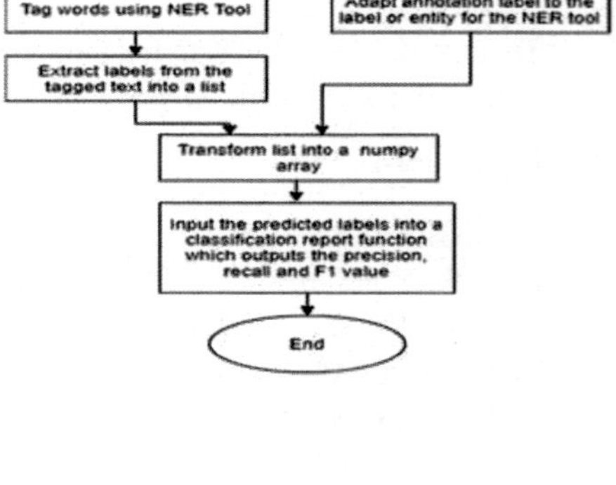

Fig. 1 Flow chart for precision, recall and $f1$ values for Stanford and NLTK

3.4 Algorithm

The 7-class model with Stanford NER Tagger method can be implemented like this:

```
1. import nltk
2. from nltk.tag import StanfordNERTagger
3. def get_Entities(text : TEXT):

        3.a) Generate 'Stanford Tagger' ensureClassification
             model path for 7-class and Stanford tagger jar file
        3.b) Split the given text
        3.c) Tag_Val_Types = Apply Stanford Tagger on splited
             words
        3.d) for Tag_Val_Type in Tag_Val_Types:
                tag_value = Tag_Val_Type[0]
                tag_type = Tag_Val_Type[1]
                if tag_type != 'O':
                        display(tag_value,tag_type)
def get_Entities(text) # calling
```

Similarly, it can be implemented to NLTK and other models also.

4 Results and Disussion

4.1 Accuracy Classification

Accuracy refers to how close something is to a known value. The number of successfully recognized entities divided by the total number of entities in the data is called accuracy.

$$\text{Accuracy} = (\text{TP} + \text{TN})/(\text{TP} + \text{TN} + \text{FP} + \text{FN}) \quad (4)$$

where TP denotes True Positive (correct positive case prediction), TN denotes True Negative (correct negative case prediction), FP denotes False Positive (incorrect positive case prediction), and FN is False Negative.

When compared to NLTK method, Stanford NER Tagger method is more accurate [19, 20]. The accuracy between Stanford NER Tagger and NLTK method can be plotted.

4.2 Confusion Matrix

Although the confusion matrix is not a performance indicator in and of itself, its components are critical for algorithm evaluation. It gives a matrix-like output with TP, TN, FP, and FN values, just like the accuracy metric.

a. Precision

Precision metric is computed by dividing the true positives by the sum of predicated positives, the accurate a classifier out of the predicated positives. Precision is known as the number of correct positive predictions (True Positive) divided by the number of correct positive predictions (True Positive) and wrong.

Positive predictions (False Positive). As a result, a high precision value represents that the algorithm produced a relevant result.

$$\text{Precision} = \text{TP}/(\text{FP} + \text{TP}) \tag{5}$$

b. Recall

Recall is the calculation of dividing the true positives (TP) by the total actual positives therefore, including the true positives (TP) and false negatives (FN). This means that when a model predicts a favourable outcome, the precision assures that the objects are classified as such. As a result, a high precision value indicates that the algorithm produced a meaningful result.

$$\text{Recall} = \text{TP}/(\text{TP} + \text{FN}) \tag{6}$$

c. $F1$ Score

$F1$-score values can be computed by using precision and recall values. The ratio of true positive and the total of true positive and false negative is finds by recall. It is computed by using the following formula.

$$F1\ \text{Score} = 2 * (\text{Precision} * \text{Recall})/(\text{Precision} + \text{Recall}) \tag{7}$$

From the above formulas and discussion we have calculated the precision, recall and $f1$ value and Accuracy of Stanford class-3, class-4, class-7 and NLTK models.

Class-3 Results:

The Stanford Class-3 model gives the following averages values of precision-0.978, recall-0.782, $F1$ score-0.866 and accuracy-0.958 which are the highest values when compared to values of remaining models. The chart representation and calculated values can be seen in Fig. 2 and Table 1.

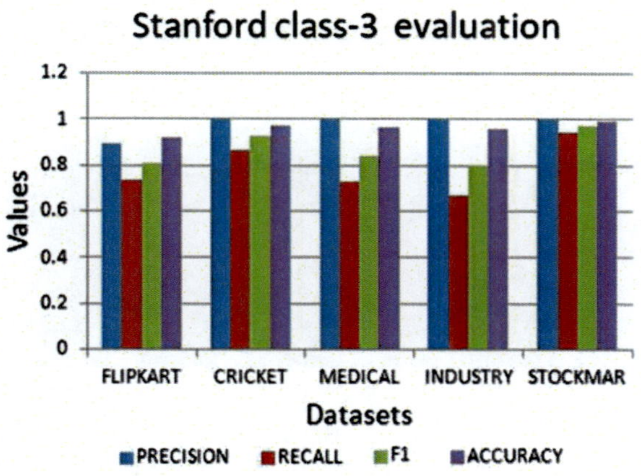

Fig. 2 Stanford class-3 evaluation

Table 1 Performance of class-3 model

	Precision	Recall	$F1$-score	Accuracy
FLIPKART	0.892857	0.73529	0.80645	0.918919
CRICKET	1	0.86486	0.92754	0.972678
MEDICAL	1	0.72727	0.84211	0.965714
INDUSTRY	1	0.66667	0.8	0.961111
STOCKMA	1	0.94737	0.97297	0.995968
AVERAGE	0.978	0.782	0.866	0.958

Class-4 Results:

From Table 2, we can observed that the Stanford class-4 model gives the following average values of precision-0.82, recall-0.81, $F1$ value-0.808 and accuracy-0.936.The chart representation for the respective values is shown in Fig. 3.

Table 2 Performance by Stanford class-4 model

	Precision	Recall	$F1$-score	Accuracy
FLIPKART	0.78379	0.93548	0.85294	0.932432
CRICKET	0.89744	0.625	0.73684	0.875
MEDICAL	0.91304	0.91304	0.91304	0.977143
INDUSTRY	0.52632	0.66667	0.58824	0.922222
STOCKMA	1	0.94737	0.97297	0.995968
AVERAGE	0.82	0.81	0.808	0.936

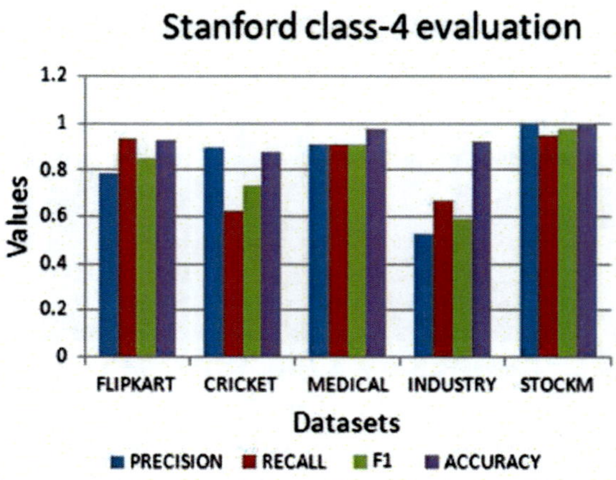

Fig. 3 Stanford class-4 evaluation

Table 3 Performance of Stanford class-7 model

	Precision	Recall	$F1$-score	Accuracy
FLIPKART	0.9	0.71429	0.79646	0.844595
CRICKET	0.904762	0.84444	0.87356	0.938889
MEDICAL	0.892857	0.86207	0.87719	0.96
INDUSTRY	0.757576	0.78125	0.76923	0.916667
STOCKMA	1	0.92683	0.96203	0.987903
AVERAGE	0.88	0.82	0.85	0.91

Class-7 Results:

The average values of precision, recall, $F1$-score and accuracy for the Stanford Class-7 model are 0.88, 0.82, 0.85 and 0.91, respectively, as shown in Table 3. The related chart representation for Stanford class-7 model is seen in Fig. 4.

NLTK Results:

From Table 4 and Fig. 5, we can observe the following averages of precision-0.56, Recall-0.88, $F1$-score-0.78 and accuracy-0.94 for NLTK model.

5 Conclusion

The evaluation metrics such as precision, recall and $F1$ Score of NLTK library, Stanford class NER shows that the Stanford class-3 is more accurate than other algorithms, i.e. Stanford class-3 model produces the best precision value 0.97 when compared with class-4's 0.82, class-7's 0.88 and NLTK's 0.56. But Stanford class-3

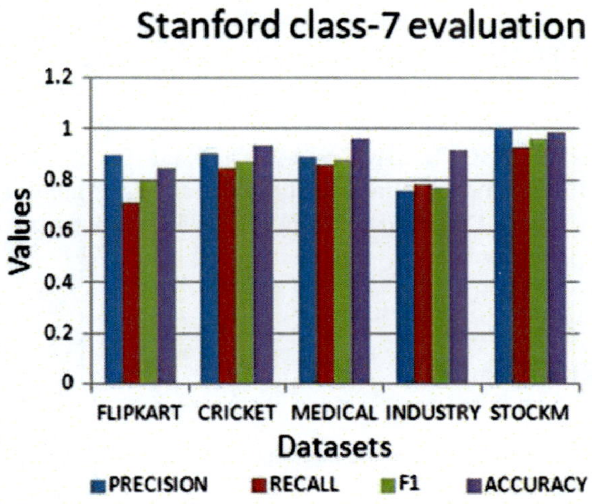

Fig. 4 Stanford class-7 evaluation

Table 4 Performance of NLTK

	Precision	Recall	$F1$-score	Accuracy
FLIPKART	0.44444	0.85714	0.58537	0.885135
CRICKET	0.82759	0.88889	0.85714	0.956284
MEDICAL	0.82353	0.93333	0.875	0.977143
INDUSTRY	0.8125	0.86667	0.83871	0.972222
STOCKMA	0.72727	0.88889	0.8	0.983871
AVERAGE	0.56	0.88	0.78	0.94

Fig. 5 NLTK evaluation

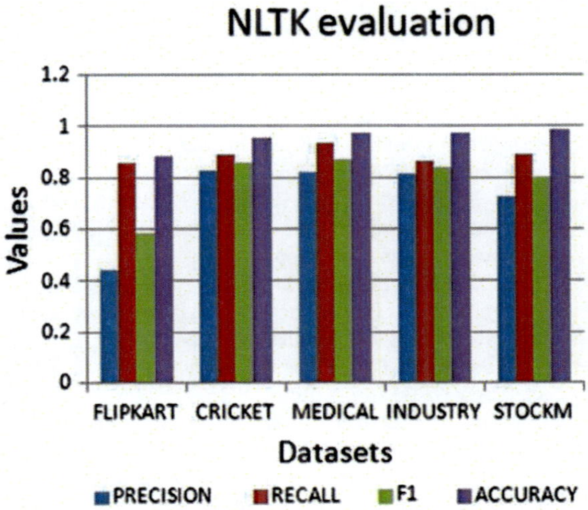

model disadvantage is it can extract a few entities such as LOCATIONS, PERSONS and ORGANIZATIONS, where class-4 extracts MISCELLANEOUS entities also, including class-3 entities, i.e. LOCATIONS, PERSONS, ORGANIZATIONS and MISCELLANEOUS entities whereas Stanford class-7 model extracts more entities like LOCATIONS, PERSONS, ORGANIZATIONS, TIMES, MONEY, PER CENTS and DATES and NLTK model extracts entities same as class-3 model such as LOCATIONS, PERSONS and ORGANIZATIONS but in a clear manner representing the whole name as one record.

6 Future Enhancements

Most of real world entities were not included in new version of Spacy NER, user defined entities can be added to NLP.

References

1. Suman C, Saichethan et al (2021) Why pay more? A simple and efficient named entity recognition system for tweets
2. NLTK project (2021) COLING-02: proceedings of the 6th conference on Natural language learning, August 2002, vol 20, pp 1–4. https://doi.org/10.3115/1118853.1118877
3. Nozza D, Manchanda P, Fersini E, Palmonari M, Messina E (2021) Learning to adapt with word embedding's: domain adaptation of named entity recognition systems
4. Patil N, Patil A, Pawar BV (2020) Named entity recognition using conditional random fields. Proc Comput Sci 167
5. Chirila OS, Chirila C-B, Stoicu-Tivadar L (2019) Named entity recognition for the contraindication and dosing sections of patient information leaflets with CRF classifier tools
6. Won M, Murrieta-Flores P, Martins B (2018) Ensemble named entity recognition (NER): evaluating NER tools in the identification of place names in historical corpora. pp 1–12
7. Jones C, Robins E (2018) Recognizing and disambiguating place names in text documents
8. Costa M (2017) Stanford NER is being evaluated for extracting assembly information from instruction manuals. In: IEEE international conference on autonomous robot systems and competitions (ICARSC), Coimbra, Portugal
9. Jiang R, Banchs RE, Li H (2016) Evaluating and combining named entity recognition systems. In: Proceedings of the sixth named entity workshop, joint with 54th association for computational linguistics, pp 21–27
10. Ekbal A, Saha S, Singh D (2012) Active machine learning technique for named entity recognition. In: Proceedings of the international conference on advances in computing, communications and informatics (ACM)
11. Landsbergen F (2011) Named entity work in IMPACT. Presented at the IMPACT final conference 2011
12. Bird S, Klein E, Loper E (2009) Natural language processing with python
13. Mansouri A, Affendey LS, Mamat A (2008) IJCSNS Int J Comput Sci Netw Secur 8(2)
14. Nadeau D (2007) Semi-supervised named entity recognition: learning to recognize 100 entity types with little supervision. Ottawa-Carleton Institute for Computer Science, School of Information Technology and Engineering, University of Ottawa

15. Minkov E, Wang RC, Cohen WW (2005) Extracting personal names from email: applying named entity recognition to informal text. In: HLT/EMNLP, pp 443–450
16. Sang EFTK (2002) Introduction to the CoNLL-2002 shared task: language-independent named entity recognition. In: 6th conference on natural language learning
17. Tanabe L, Xie N, Thom LH, Matten W, Wilbur WJ (2005) GENETAG: a tagged corpus for gene/protein named entity recognition. BMC Bioinf 6:S3
18. Tran VC, Nguyen NT, Fujita H, Hoang DT, Hwang D (2017) A combination of active learning and self-learning for named entity recognition on Twitter using conditional random fields
19. Zhang R, Gao Y, Yu R, Wang R, Lu W (2020) Medical named entity recognition based on overlapping neural networks 174
20. Zhou C, Li B, Sun X (2020) Improving software bug-specific named entity recognition with deep neural network

Enhancing the Quality of Service in MANET Based Optimization on Routing Protocol

Nedumaran Arappali, Daniel Getachew Tadesse, Abdulkerim Seid Endris, and P. Karthika

Abstract The QoS (quality of service) of a mobile adhoc network (MANET) is determined by a number of factors, and the most important of which is the quality of the links among the routes. The ratio of connection failure decreases throughput efficiency, which has a direct impact on QoS. Many industrial units provide a variety of services that can be retrieved through a MANET. This enables industrial things and users to access services through several mobile devices. Since the Manet nodes travel at different speeds and in different directions, the frequency of connection failure will always be higher. However, a variety of routing protocols has been addressed previously to support data transmission, but they struggle when it comes to achieving higher QoS efficiency. This paper presents an optimization-based route selection algorithm to increase the efficiency of route selection and maximize QoS. The method measures the likelihood of connection availability by using the position information, mobility speed, and direction of numerous nodes. In addition, the value of traffic at each node, the neighbors of each node, and the frequency of transmission in each hop are used to calculate the QoS Measure. These parameters are built by using the *k*-mean clustering algorithm, the final trust for each node will calculate, and nodes will classify based on the level of trust. The results show that the proposed service model has lower communication overhead, and thus, smaller energy consumption amount is achieved when compared with recently proposed models. The proposed approach decreases the energy consumed up to about 50% than the standard AODV protocol.

Keywords Mobile Adhoc network · Quality of service · *K*-means clustering · Optimization · Energy consumption

N. Arappali · D. G. Tadesse · A. S. Endris
School of Electrical and Computer Engineering, Wollo University, Kombolcha Institute of Technology, Kombolcha, Ethiopia

P. Karthika (✉)
Department of Computer Applications, Kalasalingam Academy of Research and Education, Krishnankoil, India

S. Shakya et al. (eds.), *Proceedings of Second International Conference on Sustainable Expert Systems*, Lecture Notes in Networks and Systems 351,
https://doi.org/10.1007/978-981-16-7657-4_48

1 Introduction

MANET is a collection of mobile nodes communicating without a static physical structure. MANETs have a variety of striking features such as topology, quick setup and wireless multihop communication. MANET is suited for several time-sensitive applications [1, 2] by all these features. Adhoc networking provides a promising connectivity facility that is difficult to construct physical infrastructure. Moreover, MANETs allow mobile nodes, with no physical structure or administrative activity, to exchange information. These networks are thus dynamic, automatically structured and automatically configured, enabling nodes to switch arbitrarily during communication. In Industry 4.0, too, MANETs play an important role. These networks can be expanded and combined with emerging skills such as cloud technology, IoT and engineering techniques in order to build intelligent industrial automation applications [3, 4]. Figure 1 shows the arrangement of the MANET. Implementing a stable routing protocol, ensure QoS, because of its complex topology, is a challenge in the MANET [5–7]. MANET offers an open communiquésubstructure which enables any multi-faceted node to join the network easily and contribute in data transmission [8]. This open communication infrastructure causes MANET safety violations [9].

IoT presents various Quality of Service (QoS) requirements which does not exist in conventional homogeneous wireless and wired networks [10]. A built-in QoS guarantees are required to provide a reliable end-to-end intelligent system that meets the requirement of a complete acquisition broadcast interpretation action loop. Therefore, different network mechanisms and protocols need to be developed for IoT. QoS is considered as one of the most critical networking issues that have obtained researchers substantial attention for both wired and wireless networks [11]. QoS is one of the leading research concerns in the field of IoT networks. However, the restriction of resources of IoT devices is considered as one of the main obstacles against developing a reliable QoS handling mechanism [12]. Interactions and data exchange between IoT devices can provide a reliable trust model where the QoS approach can be built upon [13].

Different approaches have been proposed to design the trust model most of these proposed mechanisms mainly provide security for IoT networks [14]. Based on state-of-the-art, there is a lack of study for build trust model by considering QoS trust and energy consumption. However, the trust model can be built to focus on

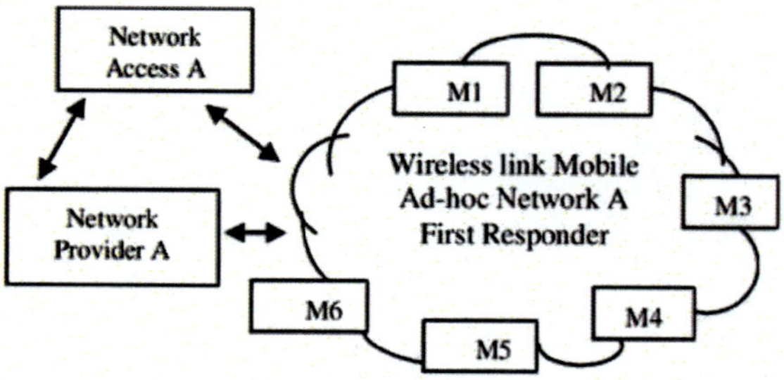

Fig. 1 MANET structure

QoS parameters, including throughput, delay, packet loss, and energy consumption. Scalability is also one of the main problems present in IoT networks that consist of a high number of devices [15]. The chief influence of this paper is the design of an energy aware QoS trust model for IoT healthcare devices that considers energy and QoS in building trust relation and reduce the energy consumption significantly, when updating the trust values between network members. Furthermore, this study could be effective for researchers who interested in developing IoT devices in the healthcare system.

2 Related Works

Chen et al. [16] the loT is decomposed into three layers: application layer, a core layer, and sensor layer, from sides of the network structure of loT. Trust management for a special purpose for each layer is controlled accordingly, and these purposes include self-organized, effective routing, and multi-service one by one. The proposed mechanism introduced a high computation overhead in different layers to build and support the trust model, where IoT limited node resources can be exhausted. In this case, QoS is also not considered where all data are handled similarly.

Chen et al. [17] extend this model by adding comprehensive simulation effectiveness, assess state-of-the-art related work. In light of this, a more advanced attacker model for analyzing the flexibility alongside these attacks arranges smart storage management planning for the ability to limit IoT devices. It is due to scalability with a comprehensive evaluation, addressing the better route to merge social parallel metrics to evaluate rates for application performance maximization. The proposed mechanism targets social networks and services. Also, data priority was not included in the trust management, where all data are handled similarly.

Asiri and Miri [18] employ distributed probabilistic neural networks (PNNs) to cluster trustworthy IoT devices from malicious nodes. In addition, it outfits the problem of the cold start in IoT networks by expecting ratings for newly joined nodes depending on their characteristics and information collected over time. The processing task is wholly distributed. Proposed model trying to conquer limitations of other trust models by training the eventuality neural network and adapt its weights count on "nodes" resemblance and the quality of data which is sent.

Saied et al. [19] the approach clarification aims to manage collaboration in a nonhomogeneous IoT structure encompass nodes with various resource abilities. To create an association, support and practicability of trusted elements in a group of cooperating services. A single point of failure approach has been used where trust manager failure can result in network trust model failure. Also, higher communication overhead and long computation are assumed in the bootstrapping period. QoS is also not considered in building the trust model.

Chen et al. [20] a generalized and unified mechanism was introduced to handle the issue of trust and reputation by utilizing the development of sensor nodes community in the Wireless Sensors Network (WSN) of IoT in cyber-physical systems. Furthermore, the proposed mechanism has been introduced for cyber-physical systems (CPS) an authentication mechanism is also included. However, QoS approach is not considered where the primary purpose is to deliver secure data delivery for all types of data in the same manner where high and low priority are handled in the same way.

Bao and Chen [21] proposed a scalable trust organization protocol for IoT. It emphasizes mainly on the social relationships between nodes. Multiple trust possessions are considered, with community-interest to account, cooperativeness, and honesty for social interaction. The proposed mechanism depends on social relationships where power consumption is not included in trust parameters. All traffic is handled in the same manner with low and high priority exchanged similarly.

3 Proposed System

Based on the state-of-the-art trust models and control approaches, most of these approaches are designed for a specific purpose environment. The emphasis is made more on security issues about applying a reliable trust model to prevent attacks. Other than that approaches introducing higher communication and processing overhead to building the trust model and exchange trust parameters reliant on a specific environment are discussed. However, there is an identifiable gap in the studies for QoS, and different priority traffic handling approaches that are not introduced as required metrics in building a trust model. Moreover, energy consumption needs to be considered in term of trust model designing for IoT devices as a fundamental factor due to its potentialities of effecting on the efficiency of these devices.

The goal of energy-saving routing protocols is to find a secure energy route from the Source Node to the BS for the appropriate period of the network lifetime and functionality. The network's main purpose is to align energy use (EC) using various techniques and approaches with the purpose of enlarging the functionality and NL. This paper introduces SFLA optimized *K*-means clustering and AODV routing. SFLA is used to optimize the *K*-means procedure with this hybrid-*K*-means SFLA-AODV method. The eight key steps in this hybrid-*K*-means-SFLA-AODV method include: (1) IoT Mobile Node Growth, (2) SFLA-based cluster selection, (3) CH weight-based router estimate, (4) AODV router estimate, (5) CH-received data, and (6) data receiving successfully. Figure 2 shows the flux map of the hybrid-K-means-SFLA-AODV method.

3.1 Clustering Based on SFLA with K-means

Initializing the particles, calculating fitness features and ultimately updating the speed and location in SFLA-based clustering algorithms. SFLA is an optimal iteration-based procedure, and the structure is prepared by modernizing the generations with

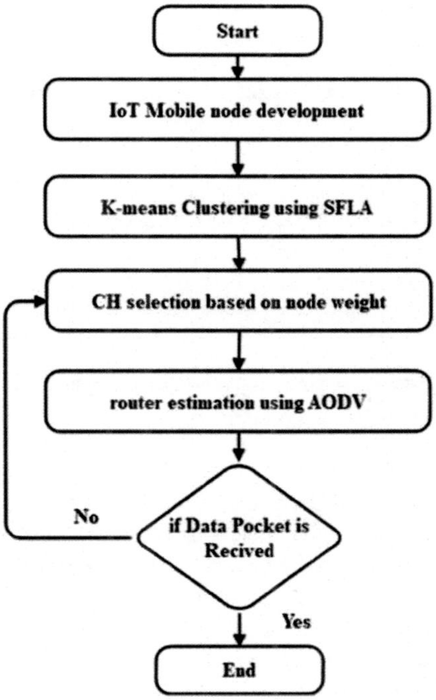

Fig. 2 Hybrid-*K*-means-SFLA-AODV procedure

a population of random decisions and optimal search. The particle changes its current position and discovers the best result through reiteration, combining local and global knowledge in the search phase of SFLA, based on its own historical and linked information on the neighboring particles. In each reiteration, the first finest rate is gotten by the qualification function, and the second most excellent value is obtained by the application of SFLA within the population. Elements are modified by each element. The best local value is determined using topological neighbors' particles. After the finest double standards are discovered, the particle modernizes its speed and position with following formulas.

$$V_{id} = WV_{id} + C_1\text{rand}(P_{id} - X_{id}) + C_2\text{Rand}(P_{gd} - X_{id}) \quad (1)$$

$$X_{id} = X_{id} + V_{id} \quad (2)$$

The particles position X_{id}, P_{id} and P_{gd} are the best position among altogether particles is where the particle velocity V_{id} is located. Two irregular [0, 1] spectrum features. And positive parameters are constantly called acceleration coefficients. The inertia weigh is a restriction which is defined by the user to monitor and influence the impact of the previous real fragment speed. Enormous inactivity gage for international research, while a smaller latency weight calibrates the quest area in flux and ebb (Fig. 3).

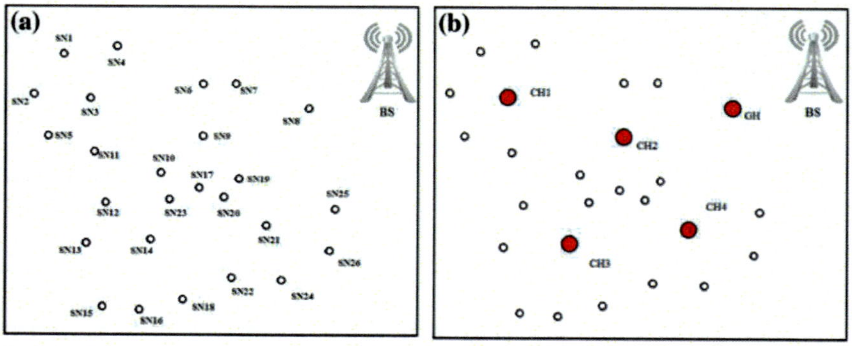

Fig. 3 **a** Unintended sensor employment, 2. **b** CH and GH detection, BS

Installation of the molecules: clusters are made using the K-implies group calculation for the use of centroids advanced by SFLA when the particles are randomly transmitted in the underlying advance. The isolation of the streamlined center with all the hubs in the area is the best aspect of CH. As for the base bunches are selected, and for several 20xNp molecules, the d best estimates are accumulated to gain the situation of the molecule. This process is revisited. The *d* gbst appreciation is achieved when this process is refreshed with specific cycles when the molecule's location is compared. The isolation of particles and the bunch center, if d est is as little as possible, is calculated as a wellness activity.

3.2 EE Clustering-Load Balancing Using SFLA

SFLA improves the grouping inspired by the characteristics of the ants of nature and the related area of SFLA, which identifies the problem in transportation systems for the quickest directions. SFLA is a calculation that streamlines the creation and operation of fluffy animals. A swarm molecule is a population of moving items, which can be pulled into better positions through the space of pursuit. Each winged creature refers to it as a "molecule," flying with a certain speed and moving to the best location worldwide. SFLA is a global search that has a strong ability to identify hopeful global results. The meeting ended with MANET's use of SFLA. Figure 4 provides basic SN clustering techniques. Burden adjustment aims to amplify the system throughput, limit the time of reaction and steer overhead. The survey network searched a wide range of sensor hubs to clarify the adaptability, vitality and lives of SNs.

In the picture 3, the SNs are organized into various cluster networks. The CH is chosen on the basis of the minimum grade value for each cluster network. The clusterhead is responsible for assembling data from its neighboring nodes and for transmitting it to another CH in various networks. The CHs transmit a confirmation

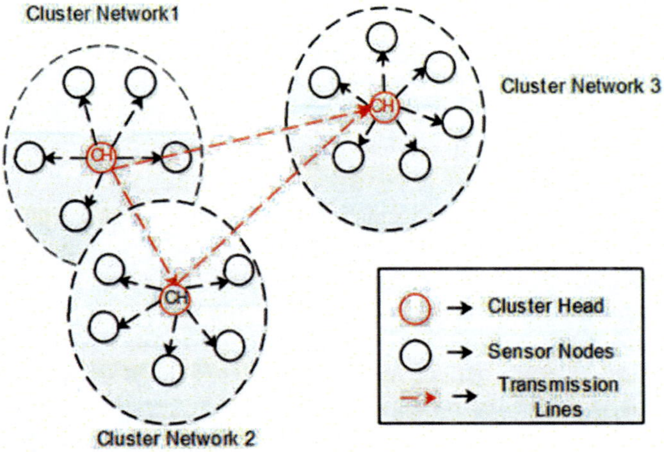

Fig. 4 Basic grouping approaches of SN

message which contains a time slot plan for their cluster members to be used during the stage of communication.

Given the adjustments, adjustments and topographic transmission of CH transmitters and recipients, groups are formed. Each hub knows when the ball can be transmitted, according to the availability schedule. The CHs collect messages from all of their groups, supply them and send the results to the BS.

Cluster Head—A head node among the same member of the cluster and other CHs is called CH.

Cluster Members—In the same cluster networks, cluster members share the same CH information.

3.3 *Shuffled Frog-Leaping Algorithm (SFLA)*

Another organically oriented meta heuristic approach was initially industrialized by Eusuff and Lansey and is based on the second clustering approach. Eusuff et al. [22] named SFLA. The SFLA principally developed as a population-based metaheuristic capable of performing a cognizant heuristic search through any mathematical function seeking solutions for combinatorial optimization problems. It can be employed to solve several complicated optimization problems, with the essence of the nonlinearity, non-differentiability, and multi-modality [23]. It associates the advantages of the genetic-based memetic algorithm with the PSO algorithm. In this algorithm, a virtual population of possible solutions is generated by a set of frogs (solutions), which segmented to dissimilar subgroups called memeplexes, by each a local search is performed. The separate frogs hold ideas inside any memeplex that may be influenced by other frogs' ideas. After a predefined number of paces of memetic evolution, ideas are transmitted through a process of memeplexes. The local quest and process of mixing continue until the specified convece requirements are met [23]. The SFLA can be briefly described thru the following steps:

1. **Initialization:** Select m and n, where m represents the amount of memeplexes and n denotes the amount of frogs in each memeplex. Thus, the total sample size, F would be $F = m \times n$.
2. **Population generation:** An initial population of F frogs is created random uniformly. For a S-dimensional problem, each frog i is represented by S variables, such as $Fi = (fi1, fi2, \ldots, fiS)$.
3. **Sorting and distribution:** The frogs are first classified according to their fitness values in descending order, after which the entire population is grouped into memeplex, each with n frogs. In this step, the first frog is shifted to the first memeplex, the second frog is shifted to the second, frogm shifts to memeplex.
4. **Memeplex evolution:** This phase is based on a local quest, in which the F_R and F_W are, respectively, found in each memeplex. The frog is identified separately as F_g with global best fitness. Equations (3) and (4) are then implemented to increase frog health to the worst possible exercise.

$$D_i = \text{rand}() \times \left(F_g - F_w\right) \tag{3}$$

$$F_w^{i+1} = F_w^i + D_i, \left(D_{\min} \le D_i \le D_{\max}\right) \tag{4}$$

   where rand () is random number among zero and 1, $D_{\min}$ and $D_{\max}$ are the minimum and the maximum displacement allowed in a frog's position, respectively. If this process generates a better solution, it substitutes the worst frog; else, if Eqs. (3) and (4) do not increase the worst solution, F_b of Eq. (3) is replaced with Fg adjusting to Eq. (5):

$$D_i = \text{rand}() \times \left(F_g - F_w\right) \tag{5}$$

5. **Shuffling:** The new population is classified according to their fitness in descending order after a predefining sum of steps to memeplex evolution all the memeplex frogs are collected.
6. **Stop condition:** The algorithm stops when a global solution or a fixed iteration sum has reached. Move to stage (2) otherwise and repeat.

3.4 AODV Routing Protocol

The routing protocol would be used in the ad hoc network by mobile nodes. The AODV is intended to reduce overhead and traffic dissemination. The AODV routing protocol covers two route discovery and route maintenance features. The route discovery role decides on the identification of the fresh route and route maintenance determines the detection of connection breakings and repair of an established route. The reactive protocol does not keep the route table permanently. AODV can easily analyze network topology changes.

4 Results and Discussion

By using the 2017 MATLAB software tool (for simulation purposes), the AODV-*K*-means-SFLA device has introduced i5 PC with 8 GB of RAM memory space. In that proposed method, SFLA used for energy efficient of load balancing and the routing is carried out by the AODV protocol. Table 1 demonstrations the simulation parameters which is used in the AODV-*K*-means-SFLA methodology. For knowing the better performance of this AODV-*K*-means-SFLA method, it is compared with the only AODV method.

Alive Nodes
Living nodes are called the node which has sufficient energy to transmit the information (chip text). A more efficient data transmission in MANET needs more live nodes.

Dead Nodes
The nodes with no energy for information transmission are called dead nos. It degrades the whole network's efficiency and should reduce dead nodes to prevent the packet loss.

Table 1 Simulation factors

Parameter	Value
Initial energy of sensor nodes	0.5 J
Area	250 * 250 m^2
Sensor nodes	200
Sum of simulation iterations	Length (msg)
Communication range of each node	50 nm
E_T	0.2 PJ/bit/m^2
E_R	0.1 PJ/bit/m^2
Packet size	4000 bits
Message size	200 bits

Energy Consumption

The total energy needed for each node to provide the message through the path extracted in the AODV routing (Cipher text).

$$E_c = E - (E_T - E_R) \tag{6}$$

where, MANET energy consumption is signified as E_c is defined as the total sum of energy, the transmitting and getting energy are signified as E_T and E_R, respectively.

Throughput

The performance is known as the amount of up-and-coming messages sent to the destination and should be high for an efficient data transmission.

$$T_H = N_T \times P_L \tag{7}$$

where, T_H. is signified as throughput of the MANET, total amount of rounds as signified as N_T and the packet length is signified as P_L.

Initially, Figs. 5, 6, 7, 8 and 9 show the graphical results of proposed optimization technique against AODV-*K*-means protocol in terms of energy consumption, total packet send, dead nodes, alive nodes, and throughput. The maximum rounds, i.e., 5000 rounds are considered in this experimental validation.

From the results itself, it is clearly stated that the proposed protocol along with SFLA optimization technique achieved better results than AODV protocol. The energy consumption is high, while implementing AODV-*K*-means protocol without optimization. In addition, the dead nodes are less and alive nodes are high in the proposed AODV-*K*-means protocol with optimization technique (SFLA). However, the AODV protocol without optimization has high dead nodes and low alive nodes, where the throughput is also less in existing AODV protocol.

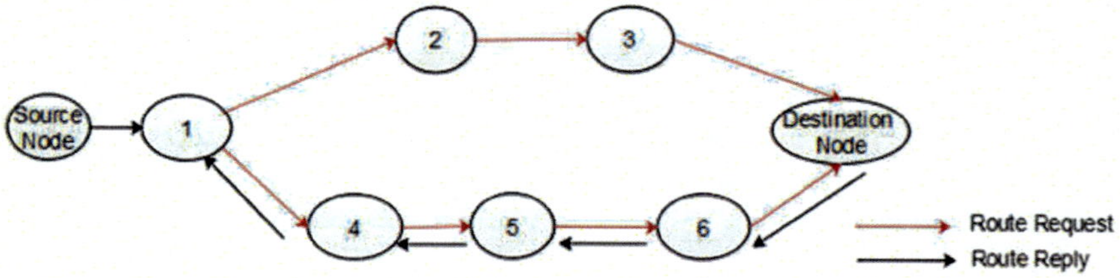

Fig. 5 AODV data transfer

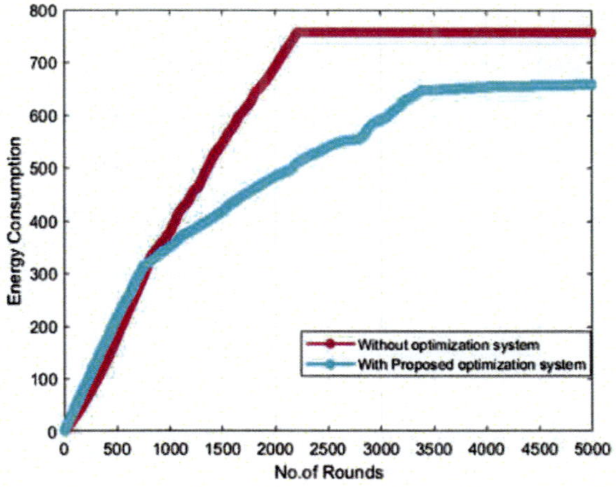

Fig. 6 Performance of proposed method in terms of energy consumption

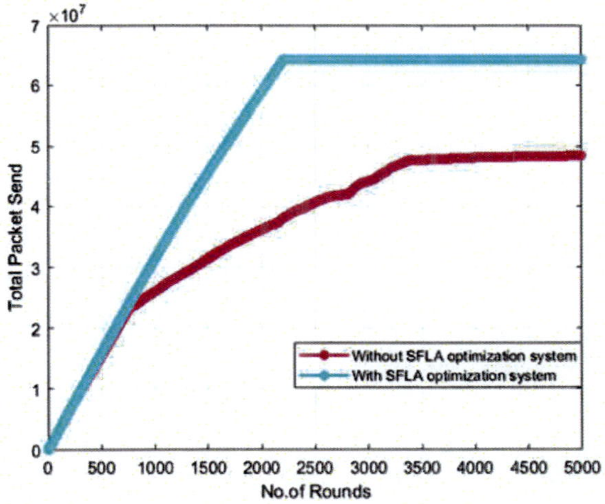

Fig. 7 Performance of proposed method in terms of dead nodes

5 Conclusion

MANET is a miscellaneous collection of versatile nodes communicating without a firm physical structure. Because of various remarkable features, such as dynamic topology, fast setup, and multihop data transmission, MANETs became more popular. These distinguishing features make MANETs appropriate for numerous applications such as ecological monitoring and disaster management. In order to attain the vision of Industry 4.0, MANETs can also be integrated with emerging skills, such as cloud computing and IoT. A secure and reliable data transmission that must satisfy a required QoS is required for all MANET sensitive realtime applications. MANET continuously studies minimizing energy consumption and protocol communication is identified as one of the main causes of increasing power consumption.

Fig. 8 Performance of proposed method by means of alive nodes

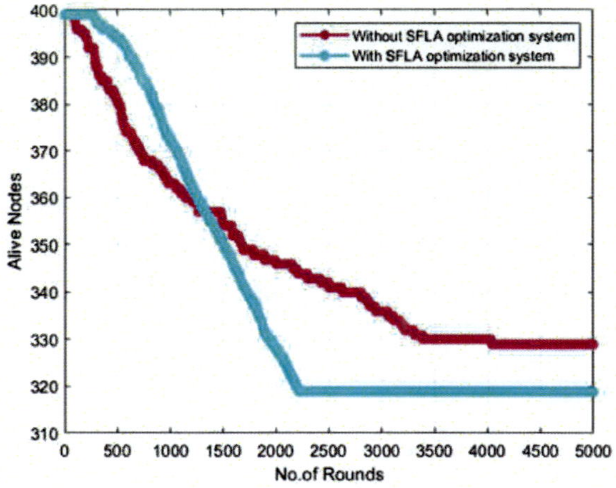

Fig. 9 Performance of proposed method in terms of throughput

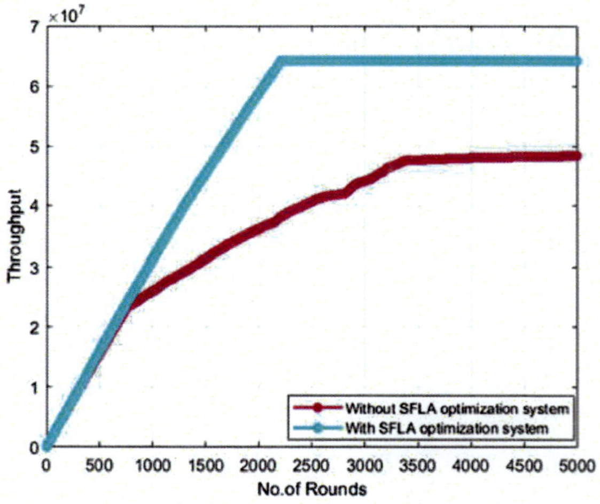

In order to achieve these difficult goals, a routing protocol needs to be developed to improve the MANET QoS. In this document, we have proposed a *K*-means clustering optimization-based routing protocol with the SFLA to improve the overall efficiency of the MANET. It takes into account a number of factors, such as congestion control, packet loss reduction, malicious node detection, and safe data transmission, in order to intensify the QoS of the MANET. Simulations in the MATLAB analyze the presentation of the proposed protocol. Our simulation results demonstrate the superior proposed routing protocol over the existing approach. In future, we will emphasize security algorithms through encryption, decryption, and blockchain solutions to ensure the MANET's high safety.

References

1. Sirajuddin MD, Rupa C, Prasad A (2016) Advanced congestion control techniques for MANET. Adv Intell Syst Comput 433:271–279
2. Femila L, Marsaline Beno M (2019) Optimizing transmission power and energy efficient routing protocol in MANETs. Wireless Pers Commun 106(3):1041–1056
3. Chen B, Wan J, Shu L, Peng L, Mukherjee M, Yin B (2018) Smart factory of ındustry 4.0: key technologies, application case, and challenges. IEEE Access 6
4. Kathiriya H, Pandya A, Dubay V, Bavarva A (2020) State of art: energy efficient protocols for self-powered wireless sensor network in IIoT to support industry 4.0. In: Proceedings of the 2020 8th ınternational conference on reliability, ınfocom technologies and optimization (Trends and future directions) (ICRITO), Noida, India, pp 1311–1314
5. Li T, Ma J, Sun C (2019) SRDPV: secure route discovery and privacy-preserving verification in MANETs. Wireless Netw 25(4):1731–1747
6. Singh T, Singh J, Sharma S (2016) Survey of secure routing protocols in MANET. Int J Mobile Netw Des Innov 6(3):142–155
7. Hossain S, Hussain MS, Ema RR, Dutta S, Sarkar S, Islam T (2019) Detecting black hole attack by selecting appropriate routes for authentic message passing using SHA-3 and diffie hellman algorithm in AODV and AOMDV routing protocols in MANET. In: Proceedings of the ICCCNT, Kanpur, India, pp 1–7
8. Sarbhukan VV, Ragha L (2020) Establishing secure routing path using trust to enhance security in MANET. Wireless Pers Commun 110(1):245–255
9. Moudni H, Er-rouidi M, Mouncif H, El Hadadi B (2016) Secure routing protocols for mobile adhoc networks. In: Proceedings of the IT4OD, Fez, Morocco, pp 1–7
10. Safari E, Babakhani M (2015) Web service and dynamic pricing competition. Int J Ind Eng Comput 6(1):99–116
11. El-Sayed H, Mellouk A, George L, Zeadally S (2008) Quality of service models for heterogeneous networks: overview and challenges. Ann Telecommun Ann des Télécommun 63(12):639–668
12. Awan I, Younas M, Naveed W (2014) Modelling QoS in IoT applications. In: Proceeding of the 17th ınternational conference on network based ınformation systems (NBIS), Salerno, Italy, pp 99–105
13. Guo J, Chen R, Tsai JJ (2017) A survey of trust computation models for service management in internet of things systems. Comput Commun 97(1):1–14
14. Al-Shammari BKJ, Al-Aboody N, Al-Raweshidy HS (2018) IoT traffic management and integration in the QoS supported network. IEEE Internet Things J 5(1):352–370
15. Alhasan A, Audah L, Alabbas A (2020) Energy overhead evaluation of security trust model for IoT application. J Theor Appl Inf Technol 98(1):69–77
16. Chen IR, Bao F, Guo J (2015) Trust-based service management for social internet of things systems. IEEE Trans Dependable Secure Comput 13(6):684–696
17. Chen IR, Guo J, Bao F (2014) Trust management for SOA-based IoT and its application to service composition. IEEE Trans Serv Comput 9(3):482–495
18. Asiri S, Miri A (2016) An IoT trust and reputation model based on recommender systems. In: Proceeding of the 14th annual conference on privacy, security and trust (PST), Auckland, New Zealand, pp 561–568
19. Saied YB, Olivereau A, Zeghlache D, Laurent M (2013) Trust management system design for the internet of things: a context-aware and multi-service approach. Comput Secur 39(11):351–365
20. Chen D, Chang G, Sun D, Li J, Jia J, Wang X (2011) TRM-IoT: a trust management model based on fuzzy reputation for internet of things. Comput Sci Inf Syst 8(4):1207–1228
21. Bao F, Chen IR (2012) Trust management for the internet of things and its application to service composition. In: Proceeding of the IEEE ınternational symposium on a world of wireless mobile and multimedia networks (WoWMoM). San Francisco, CA, USA, pp 1–6

22. Eusuff H, Lansey M, Pasha F (2006) Shuffled frog-leaping algorithm: a memetic meta-heuristic for discrete optimization. Eng Optim 38(2):129–154
23. Khorsandi A, Alimardani A, Vahidi B, Hosseinian SH (2011) Hybrid shuffled frog leapingalgorithm and nelder–mead simplex search for optimal reactive power dispatch. IET GenerTransm Distrib 5(2)

Movie Recommendation Using User Based, Item Based and LDA Analysis

Dharsan Vasant and N. Harini

Abstract A movie recommendation systems plays an important role in our social life due to its capability in providing various recommendations in an enhanced manner. This system can suggest various set of movies based on the user's preference or the popularity of the movies. This recommendation system works with the data using various kinds of algorithms, and they recommend movies in which the users might be likely to watch. There are different types of filtering in this movie recommender system which are demographic filtering, content-based filtering, collaborative filtering, and hybrid approach. This paper discusses the movie recommendation system and recommendation techniques such as item-based recommender and user-based recommender using the collaborative filtering in particular and also giving recommendation to the user based on online reviews. Recommendation engines are to recommend movies to the users and also to predict user ratings on a movie in social networks, thereby enabling recommendations. The idea behind this movie recommendation system is to help users to discover movies by predicting the user's rating for each item and displaying similar items that they might like, also if the user uses the online platform such as Twitter and Facebook, based on your likes or tweets the system should recommend movies for the users by using LDA analysis and TF-IDF methods.

Keywords Movie recommendation · User-based and item-based · Online reviews · Data mining · LDA analysis

---

D. Vasant · N. Harini (✉)
Department of Computer Science and Engineering, Amrita School of Engineering, Amrita Vishwa Vidyapeetham, Coimbatore, India
e-mail: n_harini@cb.amrita.edu

D. Vasant
e-mail: cb.en.p2aid19008@cb.students.amrita.edu

S. Shakya et al. (eds.), *Proceedings of Second International Conference on Sustainable Expert Systems*, Lecture Notes in Networks and Systems 351,
https://doi.org/10.1007/978-981-16-7657-4_49

1 Introduction

A recommendation system is a type of filtering system which attempts to predict the user's preferences and make suggestions based on the user's preferences. There are various kinds of applications for the recommendation systems. These systems have become popular day by day over the last few years, and they are used in many online platforms like Netflix, Prime Video, Hotstar which are used by the people for the video content. These recommendations provide the user the necessary recommendation, and they can all be from various social media handles. The various social media platforms have used these recommendation systems to provide the user with better recommendation.

For a large number of recommendations, it can make a recommendation based on the user likes by using the filtering process. Many people watch a particular movie in a thriller genre, and the system recommender will recommend a new user who selects thriller genre; due to some good advances in the recommendation systems, the users always expect the best recommendations. Every user has different tastes in movies so it is important to recommend the movies preferred on user likes. The problem that recommendation system possesses is that the recommendation system should work on different domains because to know more about the users preference. Different social media data provides different database using this recommendation filtering music, movies, or product whatever the user likes the system can recommend to the user. This paper discusses collaborative filtering techniques to recommend the movies for the user.

Collaborative filtering is the technique in which the recommendation process can be done in an effective manner. These collaborative filtering can make the prediction in an automatic manner in an effective way for the users. The collaborative filtering process has various types to recommend movies for the users, and they can be in similarity way or the user's preference, etc. Two types of CF filtering which user-based and item-based have been used in this paper, for example, if a user likes a movie of the particular genre, these filtering methods can make the recommendation of other movies in the same genre based on the user likes in this way the collaborative filtering execution carried out [1]. These content-based and collaborative-based filtering processes find similar movies to the users and predict user movie ratings as well [2].

In this collaborative filtering methods, there are different types of approaches that have been performed such as user-based collaborative filter and item-based collaborative filter. In user-based filtering process is used to predict the items that a user might like on the basis of ratings given to the item by the users who have similar taste with that of the target user. In the item-based approach based on the similarity between the items calculated using the rating users have given to the items, and also, it helps to solve issues that user-based collaborative filters suffers. In today's modern era, recommendation system plays a vital role for many companies success such as Amazon, Facebook, and Twitter.

The reason behind these companies success is that of providing the correct recommendation for the users based on their preferences. Most of the movie recommendations use content-based and collaborative-based filtering. In this project, the Twitter reviews [3] have been taken from the users and every time while see a new user it computes the vector of all his/her tweets and compare it to all users by computing their cosine similarity and also uses TF-IDF using all the tweets of each user as a document. Using these best matches and their liked movie genres, it can recommend a list of ordered movies to the new user.

The remainder of the paper is coordinated as follows: Sect. 2 covers the related literature review. Section 3 details the proposed system. Section 4 explains the experimental data utilized in this paper. Section 5 displays the outcomes along with the discussion. Section 6 concludes the work.

2 Dataset Description

The user- and item-based data were taken from the movie dataset, and it is used in this paper the Twitter dataset which consists of the users review about the movies through their which have been taken in it and those collected data have been taken and implemented some recommendations techniques so that the user can get the correct recommendation of the movies. The dataset consists of two phases: one is the user-based and item-based recommendation and other for the tweet analysis. The movie dataset consists of user id names and the movie names which have been included in the data and the other dataset consists of the tweets which were given by the users through online media like Facebook and Twitter. These data have been taken and used in the recommendation of movies through online reviews, and the user-based and item-based data carry the I'd names and movies they have liked through this the recommendation system works on the process.

3 Literature Survey

Harpreet Kaur et al. proposed a hybrid recommended system in that system uses both the recommendation system; i.e., both the filtering techniques in that the context of the movies have been identified and implemented in this paper also uses relation between the user and item relationship while recommending.

Li et al. [4] proposed a novel model which turns to the social networks and mine user preferences information expressed in microblogs for evaluating the similarity between online movies and TV episodes. This can be easily applied in online media streaming sites in order that intelligent recommendations of programs can be made to the customers through mining microblogs.

Utkarsh Gupta et al. proposed a user's specific recommendation and item-specific recommendation which is clubbed in a form of clusters, and it recommended movies to the users. This is based on the hierarchical clustering estimate the rate [5] if the item vote in which the function is used. This paper has a better clustering of similar items and has a good prediction of movies for the uses.

Luis M. Capos et al. proposed content-based and collaborative-based filtering which are traditional recommender systems as these filtering system have their own cons, respectively. Then the author implemented the combination of both Bayesian networks and the CF in which it gives better results [6].

4 Methodology

In this part, the overall proposed system has been shown in Fig. 1. The main objective of this project is to recommend movies using user-based and item-based filtering which is memory and model-based filtering and also recommend movies by online reviews. To recommend the movies used on the user's likes and preference. There are numerous recommendation systems like content-based, collaborative-based algorithms, and in this project, we used memory-based filtering techniques and also recommendation of the movies is based on the online reviews.

Figure 1 shows the process of the recommender of user- and item-based that have been implemented. In the user-based recommender by clicking the history user–item matrix determines the relation between the user's ratings and the item ratings and provides the suitable recommendation for the users accordingly. In the item-based recommender, it will create the item–item matrix relationship which determines the relationship between the users and the items, and it recommends the movies to the users. The process in which the user-based and item-based recommender possesses through various factors, and the relationship between the users is identified using the cosine similarity, and based on this, the similarities between the users were identified and given recommendations for the users [7, 8]. Also, the item-based recommender

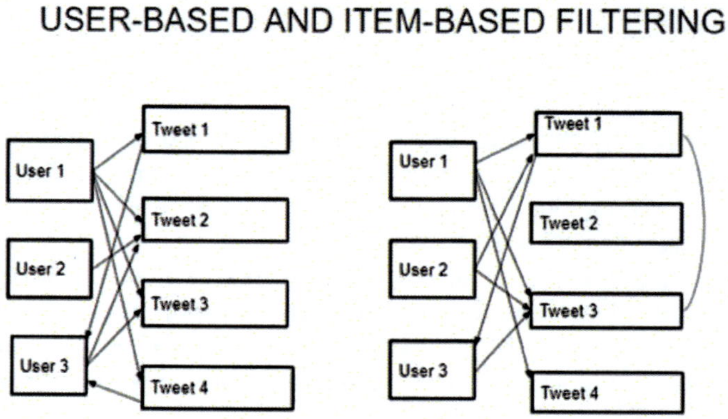

Fig. 1 Movie recommendation using user based and item based filtering

is more accurate than the user-based because of the correct prediction of movies to the users also it helps when user-based recommender has some issues.

In the social media reviews, the process the recommendation works through different processes in which the user can get the movie recommendation first process is to create the dataset for the users which is nothing but the users reviews about the movies then to create the movie rating matrix and then the user clustering by genre process takes place because of different user has different genre likes and preferences so that the clustering process has been taken place so that users and based on their liked genre have been specified through it. Then the TF-IDF method has been computed to data cleaning process such as removing numbers and punctuation lowering the numbers, symbols [9]. After the TF-IDF process, each document will be the contents of all the tweets for one user. Thus, we got a vector for each user that we could compare with the vectors of any other user in order to make good movie recommendations. An LDA analysis of the features is for each user. We want to find out if we can match the clusters found through LDA to the genres of movies. To check there is a link between what a user tweets and the genre of movies they watch. Then it recommends the movie to the users based on the similarities between the users' likes and preferences. Recommending movies for a given user and figure out the precision, recall, and success rate. We lock on the 50 closest users to our target and retrieve their movie ratings. Then the idea is to standardize by row in order to remove some of the bias by user (users who are rating always well versus always badly) and normalize the matrix to get a score between 0 and 1. Then we multiply each users' row by their cosine distance to the target. Finally, we simply sum up the scores for each movie and recommend the 20 best ones. Then the new user, if logged in about sharing his/her thoughts about the movies, then the system will recommend movies based on his/her liked movies.

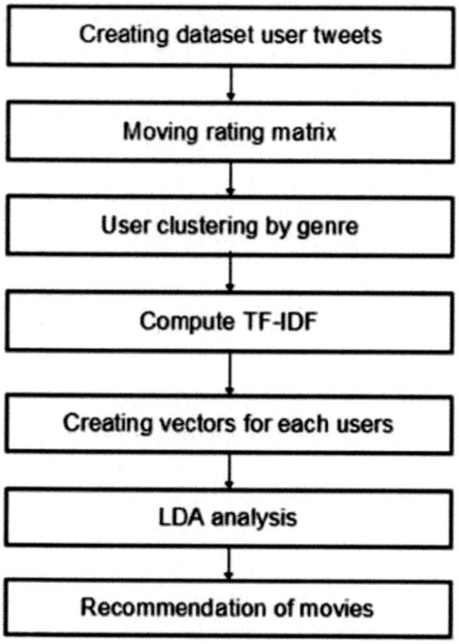

5 Result Analysis

The proposed system gives the recommendation for the users based on the user likes, and also for the new users, it recommends the movies based on their preference. The system also got the reviews from the online websites such as Twitter and Facebook, and it gives some recommendation once the system gets the users id and it is carried out. In the proposed work, the dataset which is used was movie dataset which consist of movie titles, genre, timestamp, etc. In the other dataset, it consists of users review about the movies whether it is positive or negatives all the reviews are taken into the dataset. It gives the recommendation for the users based on their likes.

```
User name: Favorite Movie: Iron Man 2 (2010)

Films you might enjooy based that you watched Iron Man 2 (2010)
```

	title	cosine_sim
1	X-Men: First Class (2011)	0.699524
2	Avengers, The (2012)	0.695324
3	Iron Man 3 (2013)	0.677566
4	Iron Man (2008)	0.641986
5	Guardians of the Galaxy (2014)	0.638427

```
*****************************************************************

Films you might enjooy with similar genre then Iron Man 2 (2010)
```

	title	cosine_sim	genre_similarity
1763	G.I. Joe: Retaliation (2013)	0.175867	1.00000
86	Matrix Revolutions, The (2003)	0.434889	1.00000
74	Matrix Reloaded, The (2003)	0.446645	1.00000
67	Spider-Man 3 (2007)	0.453103	1.00000
667	G.I. Joe: The Rise of Cobra (2009)	0.256834	0.89443

```
-----------------------------------------------------------------

Flims reccomended for you:
```

userId	Bernardo Bianchi	298	362	495	295	Mean_score
title						
Shawshank Redemption, The (1994)	0.0	3.5	4.5	5.0	5.0	4.500
Kill Bill: Vol. 1 (2003)	0.0	4.0	4.5	5.0	4.5	4.500
Kill Bill: Vol. 2 (2004)	0.0	4.0	4.5	5.0	4.5	4.500
Usual Suspects, The (1995)	0.0	3.5	3.5	4.0	5.0	4.000
Matrix, The (1999)	0.0	4.0	4.5	5.0	0.0	3.375

In the movie review dataset, the users reviews have been taken in the process of LDA analysis, TF-IDF and clustering based on the users genre, and it gives the output which it gives the recommendation of the movies to the user and also gives the recommendation to the new users if the user has given the review, based on that the system will give the recommendation.

```
Recall:  0.2525137476078656
Precision:  0.07321537522879727
Success rate:  0.7358145210494204
```

In this, recall and precision are standard metrics, but the success rate is defined as the top-20 recommendation metric. Meaning that we have a 74% chance that the user will pick at least of the recommended movies. The numbers can seem low; however, considering the sparsely of the data and the fact that we do not mix this model with another better recommender system, we still get a quite good top-20 metric, the precision being less important since we really are shooting for one movie, recall is more important in that sense.

This gives the recommendations based on the users review through online reviews. This gives a better recommendation for the user and also for the new user too.

```
predict_for_user('neiltyson')

['Life of Pi',
 'Man of Steel',
 'Skyfall',
 'Seven Psychopaths',
 'In the House',
 'The Lord of the Rings: The Return of the King',
 'Prometheus',
 'Elysium',
 'Django Unchained',
 'The Dark Knight Rises',
 'Avengers Assemble',
 'Cloud Atlas',
 'A Nightmare on Elm Street',
 'Ted',
 'The Hunger Games',
 'Warm Bodies',
 'The Hobbit: The Desolation of Smaug',
 'Flight',
 'End of Watch',
 'World War Z']
```

6 Conclusion

The movie recommender system gives the recommendation for the user likes and preferences, and the recommender gets the reviews from the user and gives the recommendation based upon their likes, and also, it recommends to the new user as well. In this process, the system identifies the user and their likes and the usefulness of this method is checked by a study of user experiments over a particular prototype of recommendation of the movies and the accuracy is satisfactory. This project can be used in the marketing field in social networking based on the profile of the user. In this paper, first it gives the user-based and item-based recommendation and also gives recommendation to the user. Also, the online reviews have been taken and some filtering techniques have been used such as movie rating matrix, clustering techniques, clustering based on the genre, TF-IDF method, and LDA analysis. Through this, the recommendation of the movies is given to the users and it is satisfactory. This project can be used in different social media platforms.

References

1. Chaudhary S, Anupama CG (2020) Recommendation system for Big Data software using popularity model and collaborative filtering. In: Advances in intelligent systems and computing, vol 1034, pp 315–324
2. Lavanya R, Bharathi B (2021) Systematic analysis of movie recommendation system through sentiment analysis. In: International conference on artificial intelligence and smart systems, ICAIS 9395854, pp 614–620
3. Chauhan A, Nagar D, Chaudhary P (2021) Movie recommendation system using sentiment analysis. In: Proceedings of international conference on innovative practices in technology and management, ICIPTM 9388340, pp 190–193
4. Li H, Cui J, Shen B, Ma J (2016) An intelligent movie recommendation system through group-level sentiment analysis in microblogs. Neurocomputing 210:164–173
5. Anwar T, Uma V (2021) Comparative study of recommender system approaches and movie recommendation using collaborative filtering. Int J Syst Assur Eng Manag
6. Singh VK, Mukherjee M, Mehta GK (2011) Combining collaborative filtering and sentiment classification for improved movie recommendations. In: Proceedings of the 5th international conference on multi-disciplinary trends in artificial intelligence, MIWAI'11, 2011. Springer, Berlin, Heidelberg, pp 38–50
7. Liu J, Choi W-H, Liu J (2021) Personalized movie recommendation method based on deep learning. Math Probl Eng 6694237
8. Kumar P, Kibriya SG, Ajay Y, Ilampiray (2021) Movie recommender system using machine learning algorithms. J Phys Conf Ser 1916(1):012052
9. Phelan O, McCarthy K, Smyth B (2009) Using twitter to recommend real-time topical news. In: Proceedings of ACM Recsys09, pp 385–388

Privacy Preservation Using Anonymity in Social Networks

Miryabelli Avinash and N. Harini

Abstract The COVID-19 pandemic has demanded voluminous data related to social network and health care be available for data analytics in the World Wide Web. Apart from security privacy preservation in data publishing has also become the need of the hour. The health information and personal information associated with an individual is supposed to be private data and not to be shared with others. Professional hackers even with very little background information would be able to identify individuals from the published data if data de-identification is not performed on the data set. Although data related to health condition is very important to perform accurate analysis on the spread of pandemic, it is equally important to ensure that the personal information of the user is not revealed. In this research, an initiative is taken to improve the usability of the data while still implementing data de-identification on health and social networking data sets. The proposed scheme in the paper prevents the adversary from learning about the target victim even when a small amount of background information about the victim is known. The work proposes to develop practical solutions to mitigate neighbourhood attack problems and to enhance privacy preservation using anonymity, diversity and closeness principles.

Keywords Social attacks · Neighborhood attacks · Healthcare · Interplanetary file system

1 Introduction

The pandemic has urged netizen's to be more participative in online social networking activities using Facebook, Instagram, LinkedIn, Twitter, etc. All users are informed prior about the security and privacy offered by these sites through the terms and

M. Avinash (✉) · N. Harini
Department of Computer Science and Engineering, Amrita School of Engineering, Amrita Vishwa Vidyapeetham, Coimbatore, India
e-mail: Cb.en.p2cse19013@cb.students.amrita.edu

N. Harini
e-mail: n_harini@cb.amrita.edu

S. Shakya et al. (eds.), *Proceedings of Second International Conference on Sustainable Expert Systems*, Lecture Notes in Networks and Systems 351,
https://doi.org/10.1007/978-981-16-7657-4_50

conditions message displayed while creating his/her personal account. These sites work on a centralized platform which makes the complete profile information about the account holders be available on the centralized server under the control of the site admin. Several instances of data leakage have been reported which makes personal information of individuals. This makes all individuals who are a part of these sites become more susceptible to be targeted by malicious attacks. The privacy concerns arise from not just the users posting their personal information but also the online social networks publishing the information for analysis. Most of the social network users are divulging endless details about their lives online. But there is a real need for sharing data among various applications. With most services being offered on the internet platform, it is very essential that there be a lot of data dependency among applications. For example, a pharmaceutical company has to depend on patient data available with hospitals to plan their business requirements but in reality the hospital data contains health information associated with individual person which is supposed to be private data and not to be shared with others. Even if the hospital management decides to remove the identity of the person and reveal only contents like age group, nature of illness, location, gender, etc. It is possible that these data can still be uniquely associated with an individual if combined with public data sets like voter database. So, there is a privacy breach as the identity of the person is revealed. And the breach becomes even worse if the attacker or hacker has sufficient background knowledge about the individual.

2 Related Work

Siddula et al. [1] present a survey on privacy-enhancement methods in the social networks. Privacy preservation is obtained by a simple approach through perturbation which is actually much simpler and less complex which in result larger networks can be dealt. And these methods are designed to work with big data specifically.

Backstrom et al. [2] deal with healthcare data that faces cyberattacks and to attain privacy by using blockchain as storage, it presents a healthcare data management system that is patient-centric. To protect patient's data, use cryptographic functions to attain pseudonymity.

Oliveira and Zaiane [3] propose a privacy preservation scheme for health data which is based completely on blockchain technology where the health data is encrypted to conduct good access control. The experimental results and security analysis show that the proposed health chain is applicable for a smart healthcare system.

Heatherly et al. [4] using blockchain technology and IPFS, an off-chain medical data storage is proposed. The framework provides easy access of healthcare data to the registered and authorized personnel while still preserving all the personal details of the patient. It also achieves consistency, integrity and availability.

Yin et al. [5] the problem of preserving the privacy of a participant in popular social media networks have been considered. Also, efficiency of all the algorithms considered for research have been evaluated using appropriate metrics. Also, privacy attack risk is raised considering the structure of social networks and it is termed as attribute couplet attack. Concept of *k*-couplet is proposed to deal with this attack.

Backstrom et al. [6] a clustering-based *K*-anonymity norm is proposed, and the outcome is optimized by incorporating parallel processing techniques. The experimental outcomes discussed in the work bring out the capability of the scheme in terms of outperformance metrics like managing information loss. The work also compares the outcome of the scheme with existing popular KACA and Incognito algorithms. Also, an initiative is taken towards privacy preservation in social media data. Specifically, identifying different types of privacy attacks: neighbourhood attacks. To protect the privacy against neighbourhood attacks, *k*-anonymity as well as *l*-diversity models extends to social media data from relational database. Backstrom et al. [6] propose a method to show that the problems of computing these neighbourhood attacks are NP-hard.

Raghavendra and Amit [7] propose a system which will provide sanitized outputs by taking user post as inputs where important keywords by using gibberish word substitution are sanitized. Also, a social network user is liable to attacks like screen scraping. The method adopts the principle of hiding the sensitive attributes by overwriting it with noisy data which appears to be real but actually represents false data of the same type is known as data sanitization.

Sahi et al. [8] based on the privacy requirements PHI/EHR patients are divided into components. A secure as well as privacy preservation e-Healthcare enterprise could be planned to implement heterogeneous protocols at various components of PHI/EHR that are capable of providing guarantee in terms of security and privacy requirements based on the sensitiveness of the attribute.

Tambe and Vora [9] evaluate the capacity of the scheme to resist attacks of different forms while still protecting the personal information of the individual from getting leaked in the hands of unauthorized users. Essentially de-identity of data is performed to prevent linkage of social network and healthcare data. The scheme claims to have incorporated the very fundamental principle of security that is privacy-by-design approach. This study focusses on insisting that security by design is more important for an application to be more successfully delivered in the internet medium which contains of volumes of data published about individuals, business, etc.

2.1 IPFS

The InterPlanetary File System (IPFS) is a p2p network for storage of the data. Content addressing is used by IPFS to identify each and every file by connecting all the computation devices. The distributed file system offers support to auto-convert the content and specifics of file into a form that is suitable for distribution in a networked environment. IPFS is by design a public network. Nodes participating in the network store the data with content addresses and advertise that they have those CID'S available for the nodes to use through publicly viewable distributed hashing tables. The distributed hash table (DHT) with key is used for record of data in the file system. A facility to download the stored hash file from any node is supported. It can be understood that only if the user is in possession of the cryptographic key necessary for the data access he/she would have a direct link to access and modify. These keys are normally monitored by the authorities who would share them only with trustable requestor's. This forms the backbone to support secure sharing of content in this file system. Also, the IPFS provides support for versioned File system and Merkle DAG, which enables secure permanent store of data. It is a fact that IPFS has the capacity to self-authenticate based on public-key cryptographic schemes for supporting secure exchange of data. The proposed work uses OpenPGP encryption standard for generating strong key. Kleopatra is used in the system to enable OpenPGP encryption for files/data. For secured storage and transfer of data, we use IPFS and encrypt the file so only the users who are registered can access the file.

The data shared is anonymized using ARX data anonymization tool to preserve privacy. *K*-anonymity uses transformation techniques such as generalization, global recording, suppression or being re-identified to anonymize the data. *L*-diversity anonymization preserves private data leak in the data sets by decreasing the granularity of the data. Also, the principal *l*-diversity model has been shown to overcome the weaknesses in *k*-anonymity model and offer better defence.

With data transformation, different generalization levels may be used for same attribute values in different records.

2.2 *Trust Evaluation*

A systematic procedure is used by the model to evaluate the trust relationship between the peers. The trust scores are computed using the results of supervised ML algorithm. Trust value is being calculated using the SVM scheme. The end result of the trust-recommender module is a quantified value that states the credibility of the peers which are connected to the user. The proposed model monitors all c participants in the connection graph, and this data is used to assess the level of trust one can place in the peer. This evaluated trust score in turn is used by the recommender engine to recommend the users with whom the information shall be shared and with whom it should not be shared. The engine uses SVM classifier, regression analysis and associated learning algorithms for monitoring the behaviour and quantifying belief.

2.3 *Blockchain*

Nodes in blockchain are connected peer-to-peer. You can connect a local node running in your laptop to the blockchain network seamlessly. In blockchain, an account with a key pair (public key and private key) is used a wallet for submitting transactions. There is no way of identifying a user or any other personal information by looking at a transaction. It is completely anonymous. As storage of huge volumes of data in blockchain is associated with more complexities, IPFS is used in addition to blockchain to offer a better platform for secure storage. The support for creating and working with smart contracts also makes this system more flexible for defining policies. Incorporating blockchain allows the users to hold their private data and exercise stringent control procedures to secure data the sensitive data in a better way. The decentralized and distributed principle of block chain facilitates immutability while storing the data. This also offers users with provision to own their data fully and exercise thorough control on how it can be shared with third parties and their peers. This overcomes the security flaws associated with the centralized systems. The blockchain is integrated with smart contracts.

The combination of the secure storage and distributed management using blockchain is performed to ensure quick secure and unmodifiable transfer of data among nodes. Hao et al. state that their further work was to extend the scheme to include support for secure storage, trust computation and defining security policies using smart contracts.

3 Proposed Architecture

Generally, to access an Internet service one has to register his/her details with the site. Depending on the nature of the service, one may have to provide personal information (e.g. social media). Most of the account creators fail to understand the importance of reading message that states the terms and conditions related to holding and sharing of information gathered by the site and thereby put themselves in trouble. This leads to award of full access rights to personal information (collected from profile) be made visible to other users who may/may not be hackers. Most of the hackers relate the published information with some other information like health data which is supposed to be private. So, there is a privacy breach. The work presented in this paper (Fig. 1) recommends a scheme to enhance privacy preservation considering the data linkage that can be made when relating data sets of health care and social media domain published in the web. Key manager component is a module responsible for generation, distribution and management of cryptographic keys used for securing the data. Only those individuals who possess the valid key will be able to access the information. The key will be distributed to the users in social network based on the suggestions from the trust manager which is an expert system that quantifies the trust score. The scheme in Fig. 1 initially classifies all available attributes of the user as public (insensitive) and private (highly confidential) and partly sensitive (quasi sensitive). The private information which requires high degree of security is piled in IPFS and given access rights are given only to those who have the key value to access this private data. Both the healthcare data and social network data of the user are collected, and attributes are categorized on their sensitivity and kept in a secure storage medium. All the transactions related to data sharing are recorded in a blockchain. The system continuously monitors the trust it can hold on the individuals

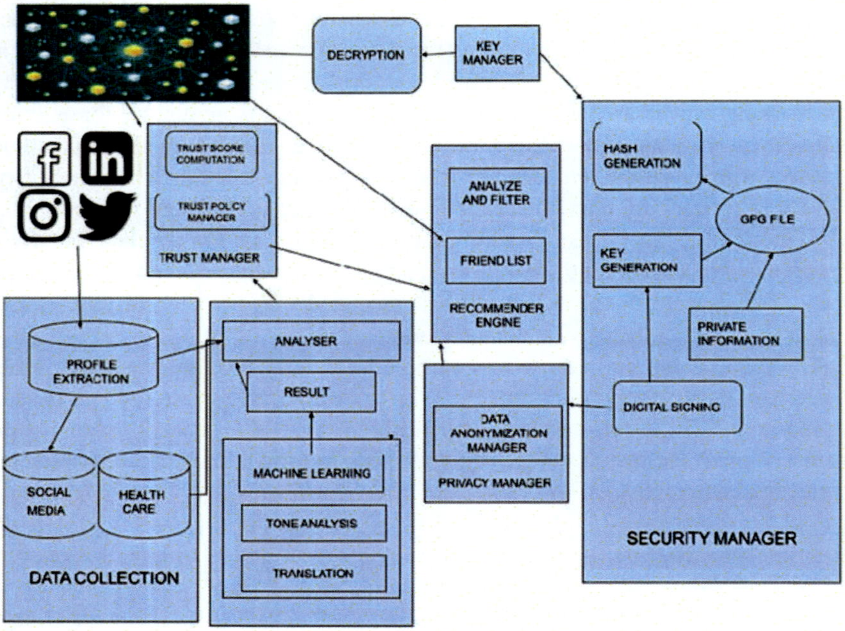

Fig. 1 Architecture diagram of the proposed model

who request for data and uses it as a basis to decide on what information can be shared and what should not be shared.

4 Results and Discussions

The most laborious process in this work was data collection from social network and health database. All the attributes were analysed individually for their security and privacy requirements, and their combinations were protected with keys and hash values.

The keys for access are generated, shared and managed by key manager component. The distribution of the key was decided by the trust manager key based on the trust score computed for individual connects of the user. One should understand that the possession of key is essential for gaining access to various attributes associated with the user. Highly trusted peers will be shared with a key that facilitates them to view secure data, and those not highly trusted would possess a key that shares only common data about the user. Periodically, the system monitors the behaviour of connects and updates the score which is then updated to the node in the blockchain.

The process of storage in IPFS is clearly shown in Figs. 2, 3 and 4. Owning the right key results in successful access to the attribute requested by the requestor.

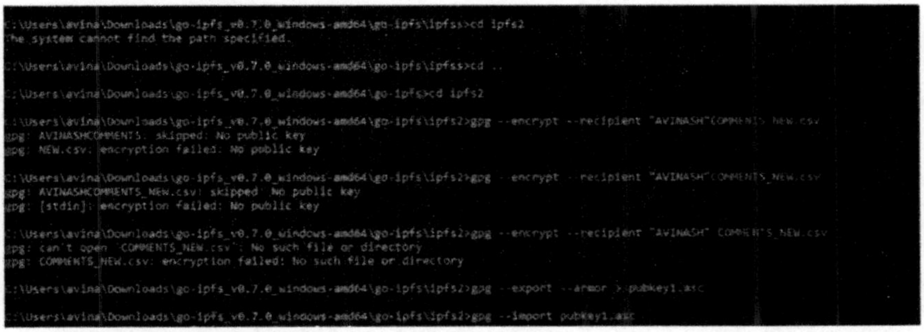

Fig. 2 Encryption in IPFS

Fig. 3 Addition of files to IPFS network

Although account holders are expected to perform an additional step of providing the key, it can be seen that the dangers associated with the leak of private information are prevented. A strict rule that user can access the information only if valid key is provided and information cannot be accessed if not which is enforced in the contract ensures privacy preservation.

IPFS is implemented even though there is blockchain because the data set used is huge. IPFS is linked with blockchain technology for storage in addition to smart contracts which can specify flow of info/data based on security rules is incorporated in the system. This will allow the user to own and control their data sharing with peers in a better way, and it will also allow the user to store data in nodes and possess them completely.

IPFS is linked with blockchain technology for storage and security rules and data/information flow directions specified in smart contracts are found to offer better degree of security and high rate of privacy preservation. This system permits the user to own and control their personal data by exercising better control on it. The

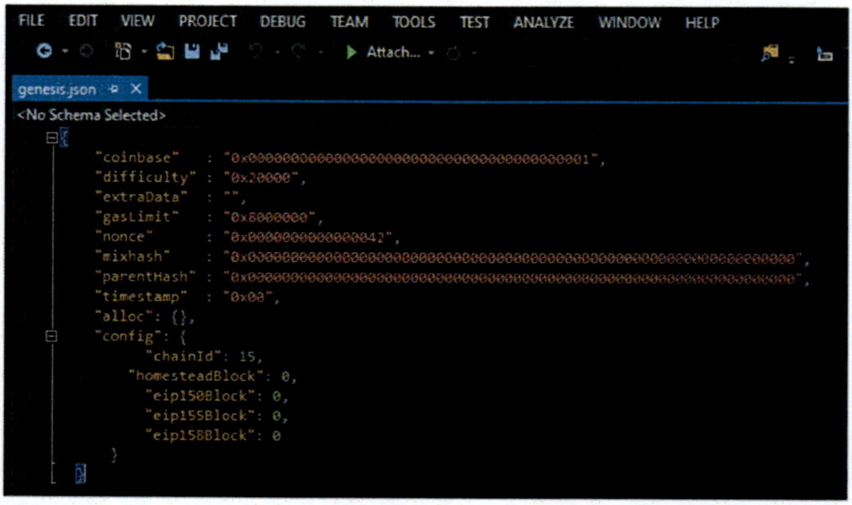

Fig. 4 Genesis block to start blockchain protocol

blockchain is used to store the trust value as transactional information and the policies related to data protection are defined in smart contracts solidity language. A clear observation of fast secure data transfer was also made, and this was due to the combination of blockchain with IPFS. The work has explored a method to preserve the privacy associated with an individual's data using data de-identification methods combined with distributed blockchain and secure storage.

References

1. Siddula M, Li L, Li Y (2018) An empirical study on the privacy preservation of online social networks. IEEE Access 6:19912–19922. https://doi.org/10.1109/ACCESS.2018.2822693
2. Backstrom L, Dwork C, Kleinberg J (2007) Wherefore art thou r3579x? anonymized social networks, hidden patterns, and structural steganography. In: WWW'07: Proceedings of the 16th international conference on World Wide Web, New York, NY, USA. ACM Press, pp 181–190
3. Oliveira SRM, Zaiane OR (2003) Protecting sensitive knowledge by data sanitization. In: Third IEEE international conference data mining, ICDM 2003
4. Heatherly R, Kantarcioglu M, Thuraisingham B (2013) Preventing private information inference attacks on social networks. IEEE Trans Knowl Data Eng 25(8)
5. Yin D, Shen Y, Liu C (2017) Attribute couplet attacks and privacy preservation in social networks. IEEE Access 5:25295–25305. https://doi.org/10.1109/ACCESS.2017.2769090
6. Backstrom L, Huttenlocher D, Kleinberg J, Lan X (2006) Group formation in large social networks: membership, growth, and evolution. In: Proceedings of the 12th ACM SIGKDD international conference on knowledge discovery and data mining (KDD), ACM Press, New York, pp 44–54
7. Raghavendra M, Amit KB (2015) A privacy preserving secure and efficient authentication scheme for telecare medical information systems. J Med Syst 39

8. Sahi MA et al (2018) Privacy preservation in e-healthcare environments: state of the art and future directions. IEEE Access 6:464–478. https://doi.org/10.1109/ACCESS.2017.2767561
9. Tambe P, Vora D (2016) Privacy preservation on social network using data sanitization. In: 2016 IEEE International conference on recent trends in electronics, information & communication technology (RTEICT), pp 751–753. https://doi.org/10.1109/RTEICT.2016.7807926

Comparison of Algorithms on Breast Cancer Detection in Developing Countries

Zahura Zaman, Md. Shakawat Al Sakib, Subarna Akter Liza, Nabanita Saha Joya, Afsana Taslim Farin, and Raka Moni

Abstract All over the world including Bangladesh Breast Cancer is one of the major cancer in women. According to a survey per minute, approximately 1 in 4 cancer patients in the world is who are diagnosed with breast cancer. Immoderate 41,000 women die because of breast cancer annually. Breast cancer is caused by several environmental factors and is caused by interactions between genetically sensitive hosts. Normally, other cells in the body divide and stop over and over again as needed. But in the cancer cell, this process occurs and does not stop. It can be one of the nothing but common cancers in both men and women, although women are in addition likely to get the disease. The risk of death can be greatly reduced by the early detection of cancer and with proper medical care. There are certain signs and symptoms of this cancer that can be used to predetermine if he/she is suffering from this disease. This paper aims to analyze the signs and symptoms with the help of machine learning. To diagnose the disease with better accuracy, the field of machine learning can help the medical professionals by using algorithms such as decision tree algorithm, random forest algorithm, linear regression algorithm, logistic regression algorithm, naive Bayes, and support vector machine. In this paper, we used a number of machine learning algorithms to detect the breast cancer patients. After using the algorithms, gradually and comparing these algorithms will find which algorithm gives principle accuracy. Using these algorithms for detecting patients can get accurate results.

Keywords Breast cancer · Survey · Diagnosis · Early detection · Machine learning · Dataset · Support vector machine · Decision tree · Random forest · Naive Bayes

---

Z. Zaman · Md. S. Al Sakib (✉) · S. A. Liza · N. S. Joya · A. T. Farin · R. Moni
Department of Computer Science and Engineering, Daffodil International University, Dhaka, Bangladesh

Z. Zaman
e-mail: zahura.taf@diu.edu.bd

S. Shakya et al. (eds.), *Proceedings of Second International Conference on Sustainable Expert Systems*, Lecture Notes in Networks and Systems 351,
https://doi.org/10.1007/978-981-16-7657-4_51

1 Introduction

People are affected by several kinds of cancer every day in the world. Breast cancer is the most impudent type of cancer after skin cancer. Inside human embryology, body tissues are made up of tiny cells. Sometimes the breast cells grow uncontrollably and turn into a lump, and these lumps are called tumors. During breast cancer, these tumor cells into a lump, and these lumps are called tumors. During breast cancer, these tumor cells begin to grow abnormally and turn into cancer. In human embryology, body tissues are made up of tiny cells. Sometimes the breast cells grow uncontrollably and turn into a lump, these lumps are called tumors. During breast cancer, these tumor cells begin to grow abnormally and turn into cancer. Both men and women can redeem breast cancer. However, women are more likely to be infected with this disease. The risk of breast cancer is increasing day by day. Generally, breast cancer increases with age, shorter period, delayed first childbirth, shorter time of breastfeeding, family history, if there is cancer or tumor in the breast before, unusually large breasts, hormonal treatment, previous radiotherapy in the breast, obesity, drinking vast amounts of alcohol. With early detection, we can reduce the death rate of breast cancer. There are some signs and symptoms by which we can predict. Some symptoms are given below:

It feels different than before since the breast lump or thickening tissues. Significant changes in the size, shape, and appearance of the breasts can be seen. Significant changes can be noticed over the skin of the breast such as lumpiness. Redness or pitting can be seen over the epidermis of the breast. If these signs, symptoms are noticed in personage, the victim should seek medical attention without any delay as soon as possible. Statistics show that eleven time's women are diagnosed with breast cancer but no symptoms are seen. As a result, cancer spreads and increases the risk of death. For this, regular breast cancer screening is important.

According to the recent data, the survival rate for the breast cancer diagnosed among women is:

91% at 5 years after diagnosis.
86% after 10 years.
80% after 15 years.

There are different stages and grades of breast cancer. Stages of breast cancer determine how substantial cancer has gotten or how far it has spread and how fast cancer has grown in the human body. If the cancer is detected at an early stage, the disease can be cured easily. But when cancer spreads, the risk of death will increase terribly. Using machine learning and its algorithm, we can detect cancer with more accuracy. In this paper, we used some algorithms for detecting, and these are support vector machine (SVM), decision tree algorithm (DT), random forest algorithm (RF), *K*-nearest neighbors algorithm (KNN). Above these seven algorithms, SVM algorithm gives the best accuracy.

2 Review Work

In this section, we will recap some review papers on breast cancer detection. Here are some short recap that given below:

In [1], Rivera-Franco and Leon-Rodriguez "Delays in carcinoma detection and treatment in developing countries" research 12 (2018): 1178223417752677. This paper aims to analyze the reasons for the delay in treatment and detection breast cancer. In [2], Marinovich et al. "Breast cancer screening using tom synthesis or mammography: a meta-analysis of cancer detection and recall" JNCI: 110.9 (2018): 942–949. This paper aims to analyze breast cancer screening using tomosynthesis or mammography to reduce recall. In [3], Charan et al. "Breast cancer detection in mammograms using convolutional neural networks" IEEE, 2018. This paper aims to analyze and implement the CNN on mammograms for cancer. In [4], Moore "Better breast cancer detection" IEEE. This paper aims to detect breast cancer by using technologies in digital X-ray, film reading by computers, and marks on chemical tumors. And analyze if these technologies can make it in the clinic. In [5], Nallamala et al. "Breast cancer detection using machine learning way" Int J Recent Technol Eng 8 (2019): 1402–1405.this paper aims to detect cancer.

3 Cancer Screening and Treatments

Mammography

"Mammography is the most challenging of the imaging technologies," said R. Janer. To detect any tumors or other tissue abnormalities, mammography can be used in the breast cancer screening setting [6]. A mammogram refers to an X-ray picture of the breast. To identify the early signs of breast cancer, the doctors use mammograms. Before patients experience the symptoms of a breast abnormality, it can be detected by screening mammography [7]. Low-dose X-rays are used in screening mammography. When the screening mammogram shows any suspiciousness, the technician will go for the X-rays to make sure whether it is cancer or not [8–10].

Digital Mammography

Also known as full-field digital mammography (FFDM). In this mammography system, the X-ray film is converted into an electric signal. By the digital version of this technology, radiologists can zoom in on particular areas of X-ray. They also can change the brightness or contrast for even greater visibility of mammograms. And the results can be read immediately after the screening. Digital mammograms allow for manipulation of subtler features in screening. It has more ability to differentiate between cancer cells and innocuous tissue [11–13]. This technology is often categorized into two classes:

- Indirect detection
- Direct detection.

Indirect detection can be called “scintillator X-ray conversion.” It will convert the X-rays into visible light which is picked up in a solid-state detector. Direct detection can be called electronic X-ray conversion. It will convert X-rays directly into electric signals.

Digital Breast Tomosynthesis (DBT)
The person from the global X-ray R&D Company explained about the IEEE Spectrum model. Because the upper sensitivity allows for lower doses of X-rays, those better-than-need-be characteristics may leave the detector’s use in breast tomosynthesis, a method of constructing three-dimensional images from scans taken at several angles [14–16]. Digital breast tomosynthesis (DBT) also referred to as 3D mammography used to convert two-dimensional images into a three-dimensional image of the breast. A traditional mammogram only captures a 2D image and provides a flat image. But tomosynthesis can look at multiple layers of the breast in a 3D image providing a different angle of a picture. “Tomosynthesis must take 11 images over 3 to five seconds,” said Schubert, but all 11 images must be taken employing a total dosage of radiation adequate to one standard film mammogram. By using tomosynthesis, we get better and more accurate results than mammograms.

Breast Ultrasound
Using the breast ultrasound process, we can test easily to create a picture in the inside of breast through sound waves. It can also be used to look at a suspicious area that was seen on a mammogram (Fig. 1).

Direct direction provides more advantages with better advanced technique with higher resolution. Approach X-rays are mostly pass through in indirect detection,

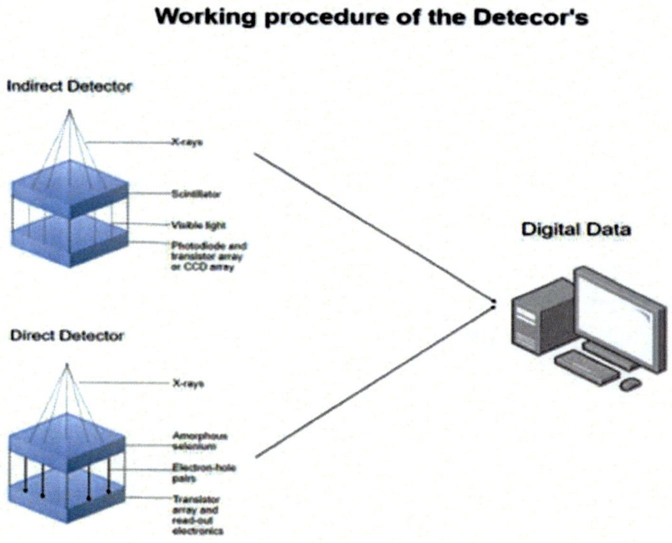

Fig. 1 Working procedure indirect detection direct detection

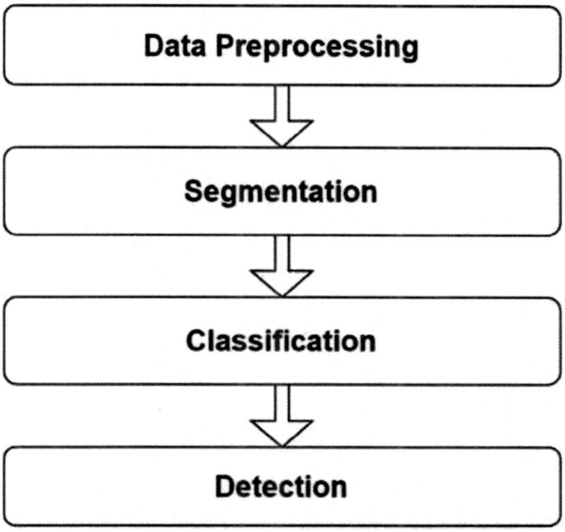

Fig. 2 Flowchart of computer-aided detection

and the X-rays will strike a scintillator. Scintillator is the part where the X-rays will absorb as well as ejaculate evident light. The scintillator throws the visible light to a photodetector array or connected to tiles of charge-coupled devices (CCDs) by tapered optical fibers. A thickness of scintillator is chosen for making more X-rays with the help of an important design consideration. Besides direct detection omits the metamorphosis of X-rays to visible light, thus removing the loss disperse from that light.

Computer-Aided Detection (CAD)

CAD is an artificial intelligence (AI) technique whose goal is to improve the accuracy of radiologists within a short time in the illustration of images. In the medical imaging field, CAD helps to make decisions rapidly. Medical imaging deals with information in image that helps the doctors to evaluate and analyze the abnormalities in breast within a shorter time period. Throwing out false positives using neural network, statistical learning, or other sorts of algorithms which will be optimized through training (Fig. 2).

4 Proposed System

In this paper, we collected data from online (kaggle.com) and found 357 of patients are benign and 212 patients are in malignant condition from all 570 patients. After taking the data for input variables, some influence factors and features are found out. The block diagram of the proposed work is shown in Fig. 3.

Fig. 3 Block diagram of the proposed work

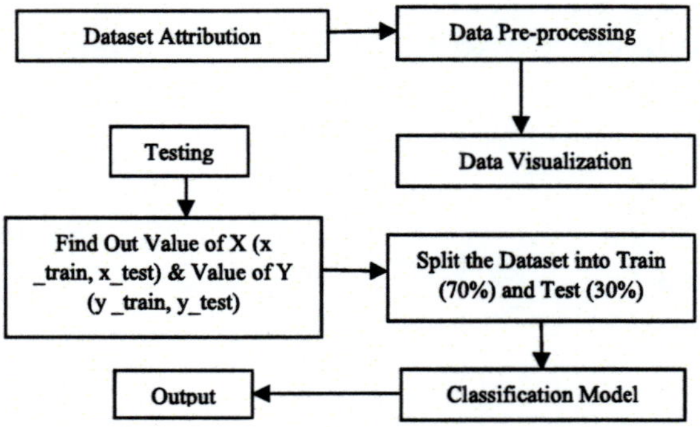

5 Requirements

A dataset is required to implement the proposal. For data pre-processing, around 570 data was taken. Almost 32 columns are added.

Here, "diagnosis" is target attribute in this dataset.

The required machine learning algorithm is presented below for classification:

- RF
- DT
- SVM
- KNN.

6 Implementations

We have collected our dataset from kaggle.com website. This dataset has total of 569 data and 32 columns. Of these 32 columns, diagnosis is the target attribute and the remaining 31 columns are the feature attribute. The target class of the dataset basically consists of two stages of breast cancer: the first is benign and the other is malignant, and this is what we know from data visualization.

First the dataset is collected then each column is numeric converted, and the diagnosis is converted by target class. Missing value is then handled and one last object and integer values are converted to float. Machine learning algorithms were applied in this section. Some results come out after implement (Figs. 4, 5, 6, and 7).

Here we can see four graphs, and these graphs refer to the machine learning algorithms that we used for implementation. There are SVM, decision tree algorithm, random forest algorithm, KNN algorithm. SVM gives 99% accuracy, random forest gives 96% accuracy, decision tree gives 95%, KNN gives 94% accuracy gradually. After analysis, we find that SVM is giving the best accuracy that is 99%. In this paper, it is found that it gives 99% accuracy (best accuracy) in breast cancer patients.

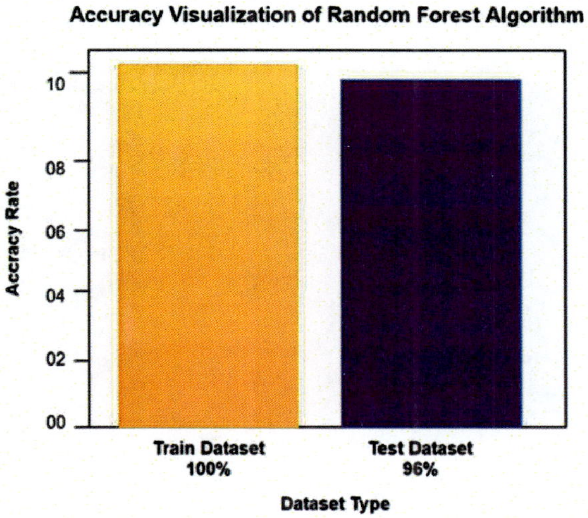

Fig. 4 Visualization of random forest algorithm

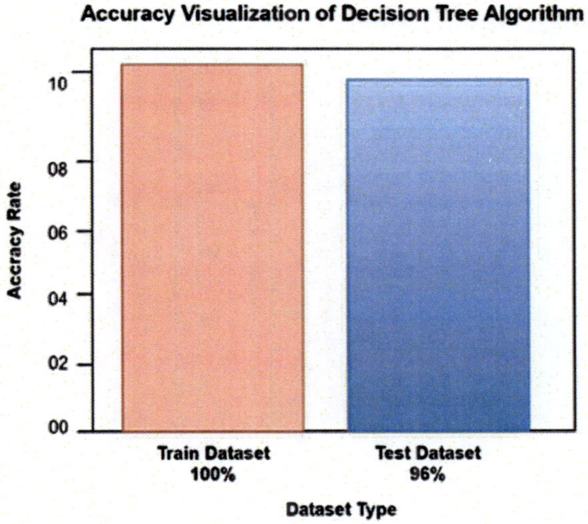

Fig. 5 Visualization of decision tree algorithm

7 Result

Accuracy: Accuracy is the calculation of how many accurate predictions can be made in a dataset.
Formula of Accuracy: $(TP + TN) \div (TP + TN + FP + FN)$.
Accuracy Result: 0.94.
Precision: Precision is a process that an algorithm actually declares breast cancer to be a proportion of the number of patients who have actually been diagnosed with breast cancer.

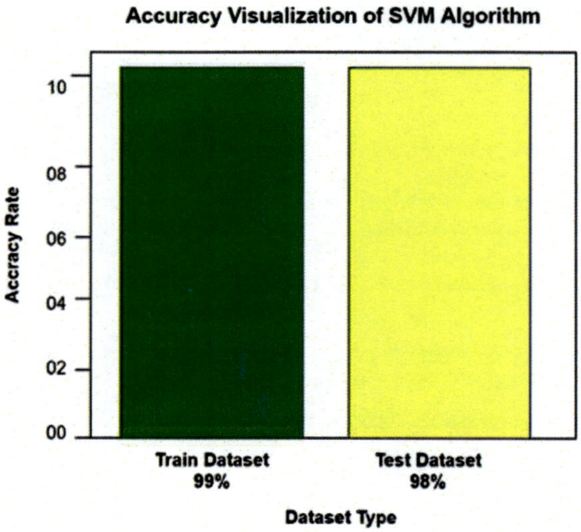

Fig. 6 Visualization of SVM algorithms

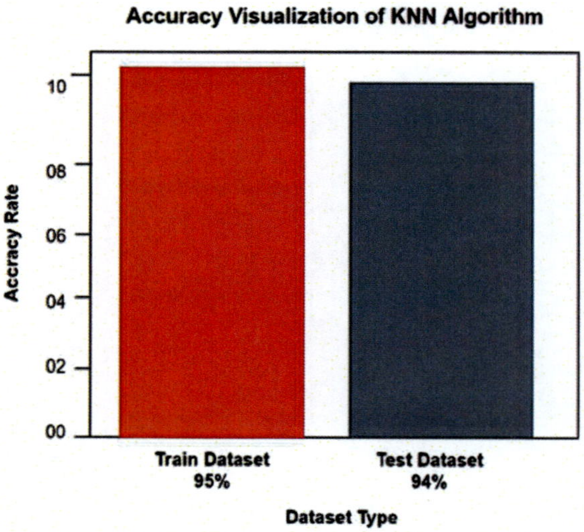

Fig. 7 Visualization of KNN algorithms

Formula of Precision: (TP) ÷ (TP + FP).

Precision Result: 1.0.

Recall: Recall is a process that an algorithm actually declares a number of patients to have breast cancer, and a proportion of those who do not actually have cancer.

Formula of Recall: (TP) ÷ (TP + FN).

Recall Result: 0.62.

F_1 Score: The 2 times multiplication of decision and recall is F_1 score.

Formula of F_1 Score: 2 * (Precision * Recall)/(Precision + Recall).

F_1 Score Result: 0.76 (Table 1).

Table 1 Accuracy of algorithms

Algorithms	Accuracy with train data (%)	Accuracy with test data (%)
SVM	99	98
Random forest	100	96
KNN	95	94
DT	100	96

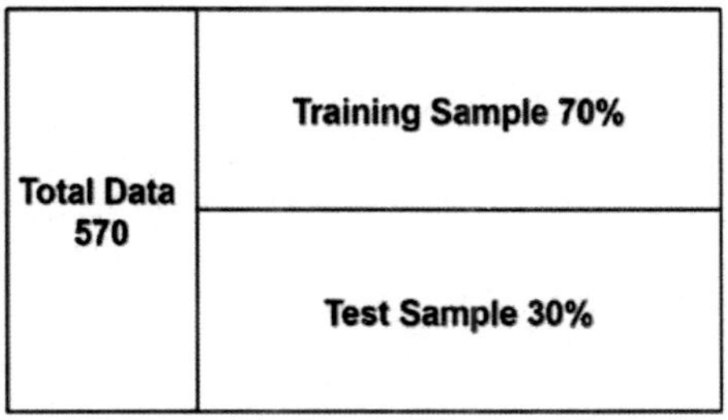

Fig. 8 Total output of train test sample

8 Input–Output Analysis

In the implementation part, we have used 31 columns id1, dia, rm, tm, pm, am, sm, cm, cnm, cpm, sym, fdm, rs, ts, ps, as, ss, cms, cs, cps, sys, fds, rw, tw, p_w, aw, sw, cmw, cnm, cpw, syw, fdw dataset as feature attribute. 1 column (diagnosis) is used as target attribute. Train data contains 70% while test data contains 30% and input the train value (x_train, y_train). The algorithm where feature scaling is required, input the train value as (x_train1, y_train1) and the machine gives the prediction output (Fig. 8).

9 Conclusion

From the study, it can be concluded that, some machine learning algorithms are proposed to predict and measure breast cancer detection. A dataset named breast cancer detection was collected from online (kaggle.com) and used here for doing this system organization and implementation. A proposed model was designed, developed, and implemented based on this dataset. It is included that among all the machine learning algorithms, and SVM performs best with 98% accuracy with test data and 99% accuracy with train data. But random forest, KNN, and DT also provide better accuracy with test data, respectively, 96%, 94%, 96% and with train data, respectively, 100%, 95%, 100% accuracy. Our dataset has 570 data which is not too much, but model can be able much better accuracy with many other algorithms on a large datasets, and model will be able more accurate result. Breast cancer treatment prediction can be done as future scope.

References

1. Manoharan S (2020) Early diagnosis of lung cancer with probability of malignancy calculation and automatic segmentation of lung CT scan images. J Innov Image Process (JIIP) 2(04):175–186
2. Chen J-Z (2021) Design of accurate classification of COVID-19 disease in X-ray images using deep learning approach. J ISMAC 3(02):132–148
3. Rivera-Franco MM, Leon-Rodriguez E (2018) Delays in breast cancer detection and treatment in developing countries. Breast Cancer Basic Clin Res 12:1178223417752677
4. Marinovich ML, Hunter KE, Macaskill P, Houssami N (2018) Breast cancer screening using tomosynthesis or mammography: a meta-analysis of cancer detection and recall. JNCI J Natl Cancer Inst 110(9):942–949
5. Charan S, Khan MJ, Khurshid K (2018) Breast cancer detection in mammograms using convolutional neural network. In: 2018 International conference on computing, mathematics and engineering technologies (iCoMET). IEEE, pp 1–5
6. Moore SK (2001) Better breast cancer detection. IEEE Spectr 38(5):50–54
7. Nallamala SH, Mishra P, Koneru SV (2019) Breast cancer detection using machine learning way. Int J Recent Technol Eng 8:1402–1405
8. Yaffe MJ, Mainprize JG (2004) Detectors for digital mammography. Technol Cancer Res Treat 3(4):309–324
9. Han X-L, Wu C-L (1996) Nuclear superdeformation data tables. At Data Nucl Data Tables 63(1):117–186
10. Misek DE, Kim EH (2011) Protein biomarkers for the early detection of breast cancer. Int J Proteomics
11. Durga KN, Krishnam Raju KV (2021) An improved prediction of breast cancer using deep neural networks (DNN)
12. Bassett LW, Bunnell DH, Jahanshahi R, Gold RH, Arndt RD, Linsman J (1987) Breast cancer detection: one versus two views. Radiology 165(1):95–97
13. Nioka S, Chance B (2005) NIR spectroscopic detection of breast cancer. Technol Cancer Res Treat 4(5):497–512
14. Kerhet A, Raffetto M, Boni A, Massa A (2006) A SVM-based approach to microwave breast cancer detection. Eng Appl Artif Intell 19(7):807–818
15. Chiarelli AM, Majpruz V, Brown P, Thériault M, Shumak R, Mai V (2009) The contribution of clinical breast examination to the accuracy of breast screening. JNCI J Natl Cancer Inst 101(18):1236–1243
16. Gilbey AM, Burnett D, Coleman RE, Holen I (2004) The detection of circulating breast cancer cells in blood. J Clin Pathol 57(9):903–911

An Approach for Building Content Recommendation System for Bilinguals

A. Bijay, Jahfar Ali, and C. C. Sobin

Abstract Language diversity in India is an extension of the vast cultural diversity that exists in Indian society. The globalization and presence of multinational companies made an inevitable need of English language, usage including in academics, though there are regional universities that use local languages as the medium of instruction. The emerging digital content market cannot elude the language diversity prevalent in the Indian population. The paper explores the possibility of finding common patterns among Hindi-speaking university-educated netizens regarding their first language preferences. These common patterns could be used to build a recommendation system for multilingual content-providing systems. The dataset that is used for building the system is collected as part of pilot work for creating a database in line with the Language Experience and Proficiency Questionnaire (LEAP-Q) for Indian languages. The questionnaire is formalized by the Language Technology Research Centre, Hyderabad, in collaboration with Pauranik Neuro Center, Indore. The questionnaire comprises questions related to language preferences and usage in different situations and reading habits. The questionnaire is devised to assess the Hindi language usage among bilingual Hindi native speakers. For this purpose, respondents are university-educated Native Hindi language users. The analysis shows that there is little relation to the language preferences between formal language usage and informal language usage. The reading habits of Hindi newspaper is observed as more of a personal choice, rather depends on any other language preferences. The analysis shows that three to four variables are enough to estimate preferences of the first language reading habit (Hindi). The proposed method could be used to recommend the right language content to bilingual readers at an average accuracy of around 50%.

Keywords Digital platforms · Bilinguals · Recommendation system

---

A. Bijay · C. C. Sobin (✉)
SRM University, Amaravati, Andhra Pradesh, India
e-mail: sobin.c@srmap.edu.in

J. Ali
LTRC, IIIT Hyderabad, Hyderabad, Telangana, India

S. Shakya et al. (eds.), *Proceedings of Second International Conference on Sustainable Expert Systems*, Lecture Notes in Networks and Systems 351,
https://doi.org/10.1007/978-981-16-7657-4_52

1 Introduction

The digital content market in India is going to spurt soon, particularly due to the digital inclusion policies of the Indian Government. The government policies made it possible to increase Internet connectivity and speed in rural areas of India. The penetration of smartphones made possible better content access to the majority of the Indian population. Digital technology platforms are going to be the major pillar in the socioeconomic development of emerging India. Indian government policies to create quality educational content in regional languages are to be read in this direction. Social equity in education is only possible by providing quality content in regional languages. India's linguistic diversity may be far more complex than any other country in the world. The census in 2011 lists 121 actively used languages among the Indian population [1]. Increasing movement of the people across different language states for business and occupation leads people to learn and use multiple languages. The census shows that bilingual language speakers are around 300 million in India. In India, almost 22 scheduled languages and around 500 dialects are already there. The studies showed that there are at least 50 languages in India, in which writing, and publishing has been done substantially. India has astounding language diversity and most of the Indians use multiple languages on different walks of life as the situation demands. Formal education is highly preferred in English medium, though the people have their own convenience and preferences on language contents. Seldom studies have been done in multilingualism of Indians, particularly in urban populations. The proficiency and competency of the languages being used by the people vary depending on their usages and situation demands.

In the Indian context, the content and language got an untainted relation due to the language diversity. The formal use of English is prevalent in territorial education across the Indian states as well as an official language in multinational companies. So the English language is considered to be important for better job and prosperity. In rural India, regional languages are more prevalent in usage and in their day-to-day activities.

UNESCO has conducted several studies related to learning outcomes against the language used for teaching. Based on those studies, UNESCO suggests home language for teaching at early years of teaching in school. It helps students to engage their ideas and thought with what they learned in school. It also advocates that multilingual education based on the mother tongue in schooling plays a vital role in respecting different cultures and learn from cultural diversity. Apart from this, UNESCO took immense efforts to foster mother tongue-based education for the development of indigenous people [2].

The census in 2011 lists 121 actively used languages among the Indian population [1]. Increasing movement of the people across different language states for business and occupation leads people to learn and use multiple languages. The census shows that bilingual language speakers are around 300 million in India. In India, almost 22 scheduled languages and around 500 dialects are already there. The studies showed that there are at least 50 languages in India, in which writing, and publishing has been done substantially. India has astounding language diversity and most of the Indians use multiple languages from different walks of life as the situation demands. Formal education is highly preferred in the English medium, though the people have their own convenience and preferences on language content. Seldom studies have been done in multilingualism of Indians, particularly in urban populations. The proficiency and competency of the languages being used by the people vary depending on their usages and situation demands.

The proposed study is based on responses from a language usage questionnaire. The questionnaire focuses on the language preferences of Hindi language speakers in different situations, such as formal, informal, professional life, and reading and other content consumption. The questionnaire is devised to understand Hindi language usage among the university-educated people in Hindi-speaking regions of India. The questionnaire helps to assess the proficiency, fluency of their first language (Hindi) in different usage contexts (Questionnaire is available in Appendix). Further analysis of the respondents' data gives the relationship between the different variables. It brings light to the people's preferences of their first language in reading and, watching the news and other content.

2 Related Work

Bilingualism has been focus of interest among multiple disciplines, such as psycholinguistics, linguistics, language development, and neurolinguistics.

There is little confusion about the concepts and observation among all these streams of disciplines. It is inevitable to bridge the disciplinary differences about the bilingual data interpretation. Conceptual and methodological issues are tried to meet, in order to share knowledge among different streams of discipline [3]. The language choice made by the multilingual people for their different functional and contextual interactions has been studied. It clearly shows that a single language cannot cater all the social needs of a multilingual society, particularly in Indian scenario [4]. A language usage questionnaire is devised to assess the language profile of multilingual. A rigorous statistical analysis has been done to validate the scoring mechanism against the self-rated fluency [5].

There is an inevitable need to identify the user preferences in Internet content or services, as the Internet covers more and more diverse demographics, and an enormous number of business models are operated over the digital platforms. There has been significant progress in the development of recommendation systems for predicting user preferences for Internet content and other service access interfaces [6]. There are lots of sites using various recommendation models for understanding their user preference and patterns. In this paper, we are looking to understand the users' language preference as most of the people are at least bilingual. There is not much paper on language preference but still being able to know some of the existing recommendation models helps us to know which technology/algorithm to use and would be beneficial. In this section, we describe some of the existing recommendation system-based techniques already applied in the language and user preference.

Mladenic [7] presented a vast survey on content-based recommendation approach. The author explains about building recommendation models or user profiles based on the preference and descriptions of objects or features of the items rated by that user. In this way, the profile portrays the user preference and interests, and the system can recommend new items based on the profile information. If a profile represents the user interest accurately, then it will ease the information access process. Using this, we can easily filter the search result based on user preference and also prevent users from displaying any search results that they do not prefer.

Similarly, Amin et al. [8] presented a recommender system to solve the problem by searching through a large proportion of strenuously generated data to provide users with customized content and services. For this study, the authors used the decision tree technique. The decision tree model is easy to interpret, visualize, and understand, is capable of handling both numerical and categorical data, and can be validated using statistical tests. The model helps to understand the dominant features of parameters in influencing the user interest and also their dependency in particular languages over others in multilingual cases.

Yoong et al. [9] developed a personality prediction model. The authors have taken 100 undergrad twitter profile information and utilizing each profile information as a list of capabilities. An option was provided to prepare the information utilizing an actualized choice tree. C4.5 algorithm was used for calculation to anticipate the understudies' character attributes. Based on this, every hub in the choice tree addresses the attributes of the Twitter profile and the character measurement of the extraversion.

Golbeck et al. [10] studied the relationship between user personality and preference to eventually build a recommendation system for movies and social media, respectively. They used the survey data of 73 Netflix users and with that data, they draw a correlation between personality and preference for specific movie genres. With the best features from the survey data, they were able to evaluate and improve the recommender system. Similarly, Alam et al. [11] proposed a paper on the user personality features recognition on social media to understand the nature, behavior, and preference of a person.

Ge et al. [12] proposed a new prediction approach that is based on an efficient linear regression model. The essential thought behind this methodology is that there exist patterns among various clients' preference. And we propose a linear regression model to interpret the inner connection between multiple user rating patterns.

Nwankwo [13] proposed a faceless AdvisorBot as a virtual support system model to improve the students support and academic advisor functions. The motive of the research is because of the lack of student support systems in tertiary institutions in Nigeria. This leads to replacing human advisor with virtual support by a bot who can support academic advising. The author has used unsupervised learning techniques to develop the real-time AdvisorBot along with data mining techniques.

Lombardo et al. [14] studied how to rank the words in a document based on their semantic relation to a given concept. The focus is on top-ranked words. So, the authors proposed a protocol to accurately evaluate the high ranks. The authors have also defined metrics for evaluating semantic model focusing on higher ranks. As part of pair-wise comparison, they have also defined a stochastic transitivity model. As part of future work, authors suggested to collect human annotations for testing the proposed protocol.

El-Ansari et al. [15] proposed a method for creating and modeling profiles of users using their browsing data. The authors have also worked on enhancing profile accuracy by combining different browsing data along with classifying the browsing pages, search queries, etc. Cella et al. [16] proposed a model for item-based recommendation based on machine learning techniques. The model uses item features along with past user experience for solving the cold-start problem prevailing in item-based recommendation system.

3 Dataset

Data is defined as an object collection (also known as record, item, point, sample, observation, or instance) and its attributes (also known as property, variable, field, characteristic, or feature). Amin et al. [8] collected data through observation and an online survey for building recommendation systems based on the preferences of backpackers when making decisions on the choice of the best accommodations.

The data used for the proposed system is the responses from a questionnaire devised by the Language Technology Research Centre, Hyderabad. The questionnaire is designed to collect responses from bilingual population languages preferences at their day-to-day life and formal learning of their first language as part of their education.

It is a part of initial steps being made on the lines of Language Experiences and Proficiency-Questionnaire (LEAP-Q) for Indian languages. The questionnaire consists of 15 questions, which collects information regarding three major aspects. Initially, the role of first language as part of respondent's formal education and professional life usages, second is preferred language in informal contexts with family and friends, and third is their first language preferences for reading newspaper, social media articles, and watching news and movies.

The dataset incorporates individuals' assessments who are from level 1 city, level 2 city, level 3 city, level 4 city, and town. School training is partitioned based on grade levels, i.e., lower essential, upper essential, auxiliary, senior optional, and tertiary. The dataset is categorized on the basis of level of study (intermediate, undergraduate, graduate, and Ph.D.) and field of study (science, technology, arts, business). It is also necessary to take note of their location. Location reflects the culture and understanding of the people. The location and orientation will also be important in determining the customer's preference in a more effective manner [17]. Each user is asked about their mother language, languages used in education, languages used to communicate at home, school, or colleges, with friends, etc. They were also asked about whether or not they read a newspaper, novel, or stories or an article in social media. These all questions were to know whether they have a language preference for reading or not and if yes then what are the factors or features influencing the results.

The dataset consists of mixed data types of samples with both categorical and numeric data. The samples were fuzzy and difficult to interpret. The categorical features were converted into dummy variables, and missing values were filled with the mean of their values. The final dataset consists of 21 columns and around 500 samples. The dataset includes the preference of languages of multilingual users at home, for education, at workplace, in social media, to read newspaper and other books, watching videos, and movies.

4 Proposed Method

The raw data generated from questionnaires had lots of discrepancies. After going through exploratory data analysis and feature engineering, a balanced and clean dataset was created and we built an ensemble-based machine learning model (Fig. 1) with two layers. The ensemble model helps to achieve better predictive performance. The first layer is built on Ward hierarchical clustering algorithms [18] to divide the dataset into clusters. Hierarchical clustering, which is a part of general agglomerative clustering, is an algorithm that groups identical objects into clusters called clusters. Agglomerative clustering is better to find small clusters from a dataset.

The second layer is regression analysis using a multivariate linear regression model. At first, we used Ward hierarchical clustering algorithms to divide the dataset

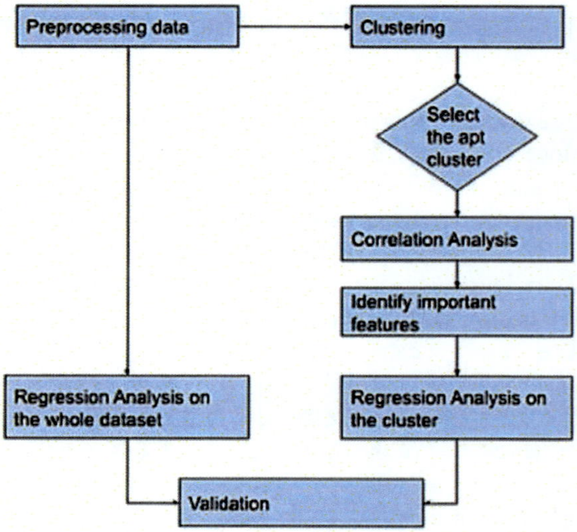

Fig. 1 Proposed model block diagram

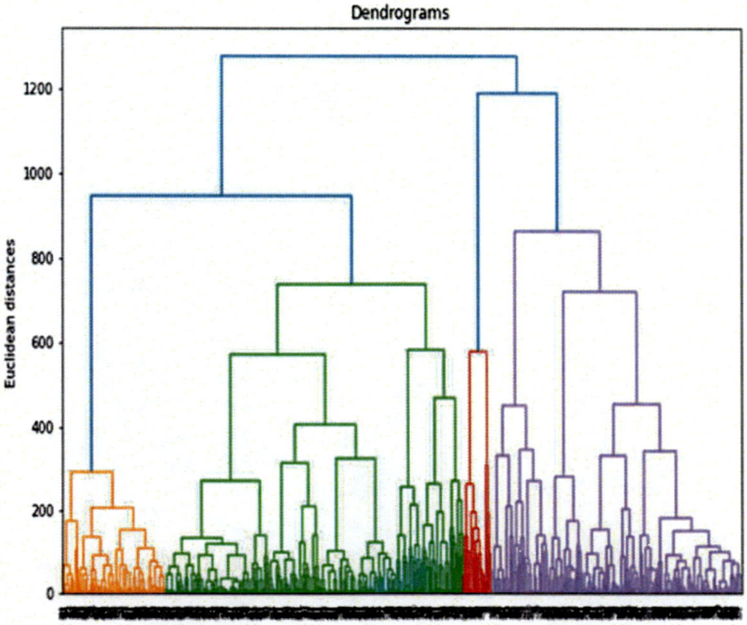

Fig. 2 Dendrogram

into clusters and separate the similar data and perform regression analysis on those data. The proposed model block diagram is shown in Fig. 1.

The interesting thing about the cluster is that it creates a subset of similar reader groups separating the other samples. Here, the cluster 4 (purple color-coded cluster) is a group of readers or who have reading habits. Since the purpose of the paper is to find the pattern or features that dominate among the similar readers. The clusters are made out of the dataset using a Dendrogram based on similar values as shown in Fig. 2.

After clustering the whole dataset, we get four different clusters and performed the correlation analysis in all the clusters but the results of the other three clusters were not good except the fourth cluster. There was one thing common between all the users in cluster 4 that was they have reading habit, they always or more often read newspapers or articles in social media in Hindi. So, we perform correlation analysis and extract out important features from it. The correlation coefficient matrix is shown in the form of heatmap in Fig. 3. The features are all possible independent sets identified from the correlation analysis. From the list of important features, only those features are grouped together that are not correlated with each other and by this mean, six cases of features are formed for regression analysis.

On the other hand, for comparison with the whole dataset, we performed the same feature extraction and grouping of the features in the whole dataset and did multivariate linear regression analysis on the whole dataset. From the regression

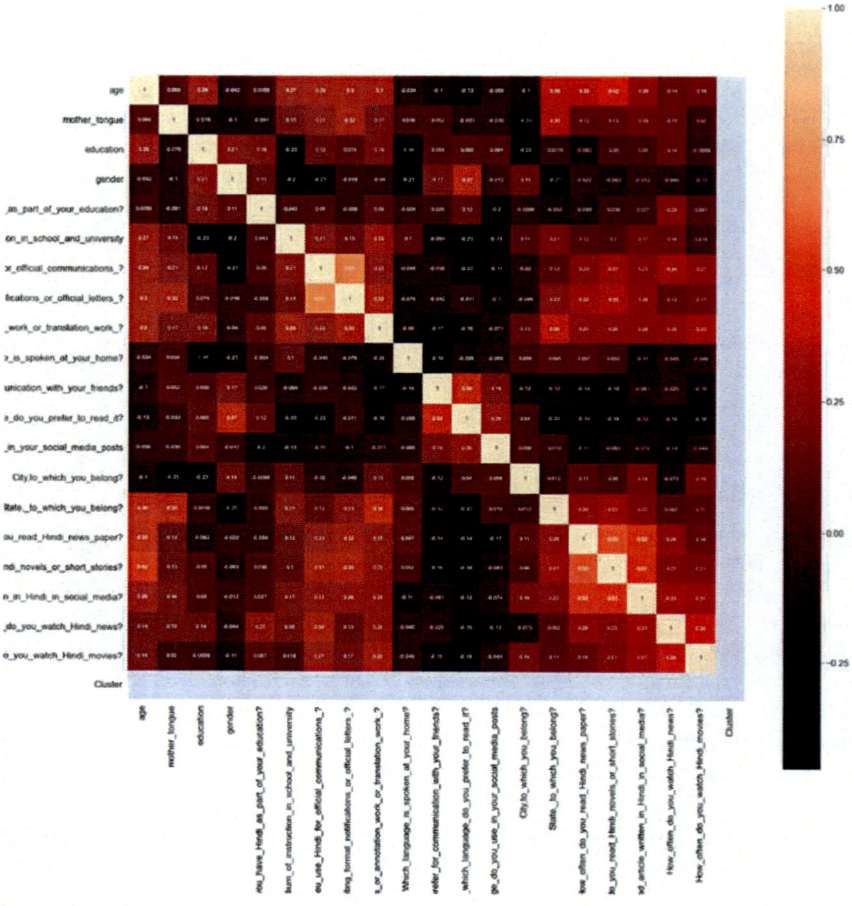

Fig. 3 Heatmap of correlation coefficient

analysis, we have calculated the R^2 value for all cases taking from the whole dataset and cluster 4. The five cases of sets of important features are given in Table 1.

Another important aspect was the 'city' feature. It was understandable that people from urban areas tend to read more content than villages, so we built a table and calculated all the R^2 values from all five cases with city variables in it and without city variables in it to compare and understand the difference in R^2 values.

5 Analysis and Evaluation

Coefficient of determination additionally called as R^2 score is utilized to assess the exhibition of a direct relapse model. It is the measure of the variety in the yield

Table 1 Cases described

Cases	Set of features
Case 1	'Which_language_do_you_use_in_your_social_media_posts', 'If_same_information_is_given_in_multiple_languages_which_language_do_you_prefer_to_read_it', 'Which_language_do_you_prefer_for_communication_with_your_friends', 'How_often_do_you_read_article_written_in_Hindi_in_social_media'
Case 2	'For_how_many_years_did_you_have_Hindi_as_part_of_your_education', 'How_often_do_you_read_article_written_in_Hindi_in_social_media'
Case 3	'Which_language_do_you_use_in_your_social_media_posts', 'Which_language_do_you_prefer_for_communication_with_your_friends', 'How_often_do_you_read_article_written_in_Hindi_in_social_media'
Case 4	'If_same_information_is_given_in_multiple_languages_which_language_do_you_prefer_to_read_it', 'How_often_do_you_read_article_written_in_Hindi_in_social_media'
Case 5	'For_how_many_years_did_you_have_Hindi_as_part_of_your_education', 'If_same_information_is_given_in_multiple_languages_which_language_do_you_prefer_to_read_it', 'How_often_do_you_read_article_written_in_Hindi_in_social_media'

Table 2 Evaluation metrics (R^2)

	With city variable		Without variable city	
	Whole dataset	Cluster 4	Whole dataset	Cluster 4
Case 1	0.477	0.626	0.480	0.646
Case 2	0.466	0.720	0.404	0.698
Case 3	0.513	0.655	0.500	0.683
Case 4	0.498	0.733	0.499	0.633
Case 5	0.510	0.628	0.463	0.627

subordinate property which is unsurprising from the information free variable(s). The whole dataset has mixed samples of readers and not readers of their first language, while the cluster 4 has only the samples of readers in their first language. Because of this, there is a significant improvement in the R^2 values among them.

The results in Table 2 show that the proposed method is good enough to predict the language preferences for reading their first language of educated netizens. These common patterns could be used to build a recommendation system for multilingual content-providing systems.

In the table mentioned above, we calculated the R^2 with and without city variable, and the model performs well with city variable in some of the cases like case 2 and case 4. The reason is it correlates with some other feature very well, i.e., 'how often you read article written in Hindi in social media'. It is difficult to visualize more than three features, but we tried with utmost two independent features. The multivariable linear regression plot in 3D with independent features 'city' and 'how often you read article written in Hindi in social media' and dependent features 'reading habit' is shown in Fig. 4.

The multivariable linear regression plot in 3D with independent features 'which language you use to communicate with friends' and 'how often you read article written in Hindi in social media' and dependent features 'reading habit' is shown in Fig. 5.

The multivariable linear regression plot in 3D with independent features 'If same information is given in multiple languages which language do you prefer to read it' and 'city' and dependent features 'reading habit' is shown in Fig. 6.

6 Conclusion

In the age of digitalization and the Internet, it is bliss to have a content recommendation system based on user preference. The content-based recommendation system helps in understanding in finding the similarity in language preference among different users. But language-based content recommendation has been left out due to the globalization of English language. Most people are bilingual or multilingual. They might prefer content like social media articles in other languages than English

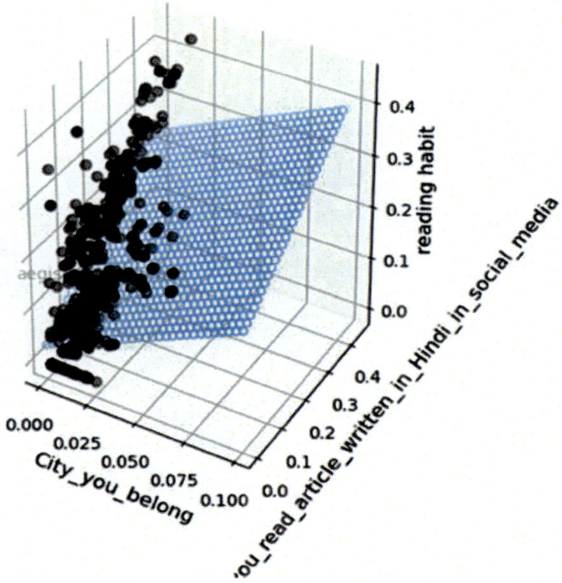

Fig. 4 Multivariate regression visualization

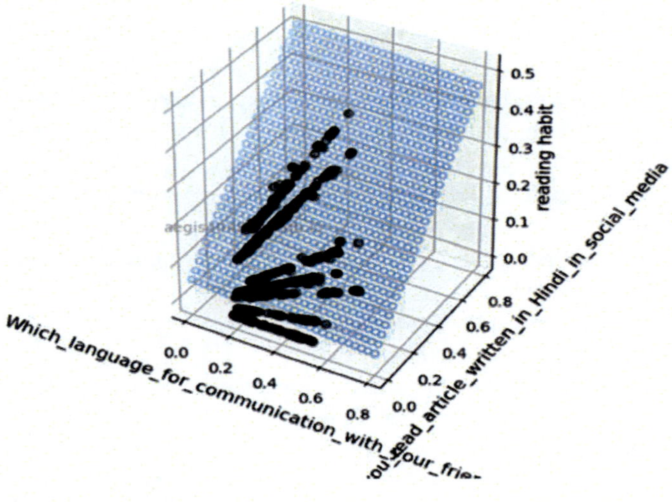

Fig. 5 Multivariate regression visualization

or they could prefer mixed languages like Hindi–English or Hindi–Bengali or any other. So, the insights drawn from the real user data about their language preference can help in analyzing the pattern in data and eventually building a preferred language-based content recommendation system.

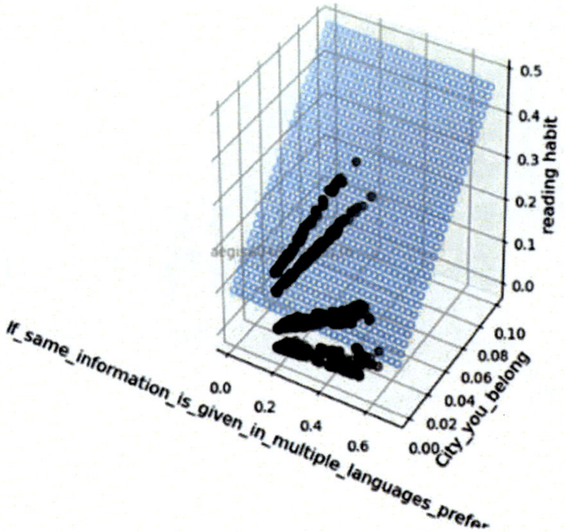

Fig. 6 Multivariate regression visualization

Appendix

1. For how many years did you have Hindi as part of your education?
2. Has Hindi been your medium of instruction in school and university? If yes, how many years (Physics, Chemistry, History are taught in Hindi language)
3. Do you use Hindi for official communication? If yes, how long you have been using it?
4. Do you use Hindi for writing formal notifications or official letters? If yes, for how long?
5. Do you use Hindi for language analysis or annotation or translation work? If yes, for how long?
6. Which language is spoken at your home?
7. At what age you started to learn Hindi?
8. Which language, do you prefer for communication with your friends?
9. If same information is given in multiple languages, which language do you prefer to read?
10. Which language do you use in your social media posts?
11. In which language do you affix your signature?
12. How often do you read Hindi newspapers?
13. How often do you read Hindi novels or short stories?
14. How often do you read article written in Hindi in social media?
15. How often do you watch Hindi Movies?
16. How often do you watch Hindi news?
17. In which language do you think or brood?
18. What is the common language in your dreams?

19. What is the approximate proportion you mix two languages simultaneously in your routine communications at home and at leisure?
20. In which language do you write your comment in visitor book?
21. In which language do you fill up official forms if options are available for Hindi along with English or other language?
22. In which language do you write letters to family and friends?
23. In which language do you write articles, blogs?
24. In which language do you write numerals beyond 20?
25. In which language do you do simple arithmetic for daily activities?
26. In which language do you read fiction literature?
27. In which language do you watch movies?
28. In which language do you listen songs?

References

1. https://censusindia.gov.in/2011-common/censusdata2011.html. Accessed on 21 Mar 2021
2. Ball J, Educational equity for children from diverse language backgrounds: mother tongue-based bilingual or multilingual education in the early years. http://hdl.handle.net/1828/2457
3. Grosjean F, Universiteâ de Neuchaâtel (1998) Bilingualism Lang Cogn 1(1998):131–149
4. Vasanta D, Suvarna A, Sireesha J, Bapi Raju S, Language choice and language use patterns among Telugu-Hindi/Urdu-English speakers in Hyderabad, India
5. Naik S, Bapi Raju S, Vasanta D, Alladi S, Rajan A, Gollahalli DR, Understanding multilingualism in language contact situations: a cluster analysis approach
6. Lops P, de Gemmis M, Semeraro G, Content-based recommender systems: state of the art and trends. https://doi.org/10.1007/978-0-387-85820-3_3
7. Mladenic D (1999) Text-learning and related intelligent agents: a survey. IEEE Intell Syst 14(4):44–54
8. Amin MM, Lan JYA, Makhtar M, Mamat AR, A decision tree based recommender system for backpackers accommodations. Int J Eng Technol 7(2.15):45–48
9. Yoong TLC, Ngatirin N, Zaino Z, Personality prediction based on social media using decision tree algorithm. Pertanika J Sci Technol 25:237–248
10. Golbeck J, Norris E (2013) Personality, movie preferences, and recommendations. In: Proceedings of the IEEE/ACM international conference on advances in social networks analysis and mining, Niagara, Ontario, Canada, pp 1414–1415
11. Alam F, Stepanov EA, Riccardi G (2013) Personality traits recognition on social network—Facebook. Computat Pers Recogn 6–9
12. Ge X, Liu J, Qi Q, Chen Z (2011) A new prediction approach based on linear regression for collaborative filtering. In: Eighth IEEE International conference on fuzzy systems and knowledge discovery (FSKD), vol 4, pp 2586–2590
13. Nwankwo W (2018) Interactive advising with bots: improving academic excellence in educational establishments. Am J Oper Manag Inf Syst 3(1):6–21
14. Lombardo P, Boiardi A, Colombo L, Schiavone A, Tamagnone N (2020) Top-rank-focused adaptive vote collection for the evaluation of domain-specific semantic models. arXiv preprint arXiv: 2010.04486
15. El-Ansari A, Beni-Hssane A, Saadi M (2020) An improved modeling method for profile-based personalized search. In: Proceedings of the 3rd international conference on networking, information systems & security, pp 1–6

16. Cella L, Cereda S, Quadrana M, Cremonesi P (2017) Deriving item features relevance from past user interactions. In: Proceedings of the 25th conference on user modeling, adaptation and personalization, pp 275–279
17. Joe CV, Raj JS (2021) Location-based orientation context dependent recommender system for users. J Trends Comput Sci Smart Technol (TCSST) 3(01):14–23
18. Murtagh F, Legendre P (2011) Ward's hierarchical clustering method: clustering criterion and agglomerative algorithm. arXiv preprint arXiv: 1111.6285

Upper Bounds of Zagreb Radio Indices

Anna Treesa Raj and Joseph Varghese Kureethara

Abstract Let $G = (V, E)$ be a simple connected graph with vertex set V and edge set E. This chapter consists of several bounds of the Zagreb radio indices of graphs such as the Primary Zagreb Radio Index, the First Zagreb Radio Index, the Second Zagreb Radio Index and the Third Zagreb Radio Index. The indices are defined for graphs after administering a radio labelling. In radio labelling, vertices are labelled with the positive integers such that the absolute difference of two vertex labels added to their distance should be at least one more than the diameter of the graph. In radio labelling, every vertex gets distinct labels. The least possible labels given to the vertices are used to create the radio indices.

Keywords Topological index · Chemical graph · Radio labelling · Radio number

1 Introduction

One of the most popular graph theoretical models is that of the communication networks. Radio/TV stations could be represented as vertices, and the physical adjacency within a certain distance between the stations could be represented as the edges between the vertices. Frequencies of the stations can be assigned based on the adjacencies between the stations avoiding the possible overlapping of the transmissions [4, 5]. Some of the most famous algorithms used are $L(h, k)$-labelling, Hamiltonian labelling and Radio labelling. Better algorithms and methods are coming up on a regular manner for efficient network management [7, 23].

A. T. Raj · J. V. Kureethara (✉)
Christ University, Bangalore 560029, India
e-mail: frjoseph@christuniversity.in

A. T. Raj
e-mail: annatreesa.raj@res.christuniversity.in

S. Shakya et al. (eds.), *Proceedings of Second International Conference on Sustainable Expert Systems*, Lecture Notes in Networks and Systems 351,
https://doi.org/10.1007/978-981-16-7657-4_53

Computer-based procedures have contributed a lot in the modern research in chemistry. Efficient computer programs are employed to calculate the structures and properties of molecules and solids. The topological index is a numerical quantity calculated from the molecular graph of a compound which defines various physiochemical and biological properties of the compound [6, 24]. Various topological indices have been defined concerning various parameters associated with the graph. Among them, topological indices based on the degree of the vertices and distance between the vertices are prominent [9, 13, 16, 17]. Radio labelling-based indices were studied in [14, 20, 22]. Some of the applications of the topological indices could be found in [1, 2, 12, 13, 18]. Inverse problems related to the topological indices were studied in [3, 15]. For fundamentals in graph theory, one may refer to [19, 25].

The chapter is presented in six sections other than the introduction. In Sect. 2, the basic definitions are presented. Four upper bounds for the Primary Zagreb Index are give in Sect. 3. Section 4 has the bounds related to the First Zagreb Radio Index. The bounds related to the Second and the Third Zagreb Radio Indices are given in Sects. 5 and 6, respectively. Finally, Conclusion is in the last section.

2 Basic Definitions

Gutman and Trinajstić had studied the relation between the total π-electron energy and the molecular structure. They had found out strong dependence of the former on the latter. In this connection, they had formulated the popular Zagreb indices in 1972 [10] and in 1975 [11]. For a graph G, the First and Second Zagreb indices are defined as follows,

$$M_1(G) = \sum_{u \in V(G)} d(v)^2$$

and

$$M_2(G) = \sum_{uv \in E(G)} d(u)d(v)$$

where $d(v)$ is the degree of vertex v in G [26]. The Third Zagreb index [8] is defined as

$$M_3(G) = \sum_{uv \in E(G)} |d(u) - d(v)|.$$

In the past couple of decades, there were introduced numerous variations of Zagreb indices. One such variation is Zagreb Radio Indices [14, 20–22]. The First Zagreb Radio Index was defined by changing the degree of the vertex in the First Zagreb Index to the radio labelling of the vertex. Since this definition neglects the adjacency of the graph, let us redefine the First Zagreb Radio Index by the following definition and rename the other one as the Primary Zagreb Radio Index.

A radio labelling c of a graph G is an assignment of positive integers to the vertices of G such that for every two distinct vertices u and v of G,

$$|c(u) - c(v)| \geq \text{diam}(G) + 1 - d(u, v)$$

where $d(u, v)$ is the distance between u and v. The span of the function c is the difference between the largest value of c and the smallest value of c. The radio labelling number of G is the minimum span among all radio labelings c of G. In all the following definitions, we use c as a radio labelling whose span is equal to the radio number of the graph.

Let G be a graph. The First Zagreb Radio Index of G, denoted by $M_{R1}(G)$, is the infimum of the sum $\sum_{xy\in E(G)}(c(x) + c(y))$ [20].

That is,

$$M_{R1}(G) = \inf\left\{\sum_{xy\in E(G)} (c(x) + c(y))\right\}.$$

We now define the primary Zagreb Radio Index of a graph. It is denoted as $M_{r1}(G)$. i.e.,

$$M_{r1}(G) = \inf\left\{\sum_{x\in V(G)} (c(x)^2)\right\}.$$

The Second Zagreb Radio Index is denoted as $M_{R2}(G)$. It is given as

$$M_{R2}(G) = \inf\left\{\sum_{xy\in E(G)} (c(x)c(y))\right\}.$$

The Third Zagreb Radio Index $M_{R3}(G)$ is

$$M_{R3}(G) = \inf\left\{\sum_{xy\in E(G)} |c(x) - c(y)|\right\}.$$

In this chapter, we establish certain bounds for these four indices.

3 Bounds of the Primary Zagreb Radio Index

Theorem 1 *Let n be the order of the graph G. Then,*

$$\frac{n(n+1)(2n+1)}{6} \leq M_{r1}(G). \tag{1}$$

The bound is sharp if and only if G admits consecutive radio labelling.

Proof Let us consider a graph G on n vertices. Let its radio number be r. Since $n \leq rn(G)$, $r \geq n$. Per radio labelling, each vertex of the graph G should get some distinct labels from 1 to r. Let 1, a_1, a_2, ..., a_{n-2} and r be the labels given to the vertices. Clearly, $1^2 + 2^2 + \cdots + n^2 \leq 1^2 + a_1^2 + a_2^2 + \cdots + a_{n-2}^2 + r^2$. Therefore,

$$\frac{n(n+1)(2n+1)}{6} \leq M_{r1}(G).$$

This completes the first part of the proof.

Now, suppose G admits consecutive radio labelling. Then by the definition of consecutive radio labelling, the labels are nothing but 1 to n.

Therefore,

$$M_{r1}(G) = \frac{n(n+1)(2n+1)}{6}.$$

Conversely, suppose $M_{r1}(G) = \frac{n(n+1)(2n+1)}{6}$, i.e., $M_{r1}(G) = 1^2 + 2^2 + \cdots + n^2$, i.e., the labels of the vertices are from 1 to n. Hence, we conclude that the graph G admits consecutive radio labelling.

Theorem 2 *Let n be the order of the graph G with diam(G) = 2. Let G has a pendant vertex and has a paw as an induced subgraph. Then,*

$$M_{r1}(G) \leq \frac{4n^3 - 12n^2 + 11n + 9}{3}. \tag{2}$$

The equality holds if and only if $G \cong K_{n-1} \bigoplus_u v$.

Proof Consider any graph G on n vertices with diameter 2 and having a paw as an induced subgraph. Consider the following labelling. Choose a vertex v such that v is a peripheral vertex of G and which allows to choose maximum number of peripheral vertices in a sequence, then allot it with the label 1. Select the second vertex which is at diametric distance from v and label it with 2. Since G is having paw as an induced subgraph, there will be at least one more vertex which is at a diametric distance from the second vertex, label the vertex with 3. Label the rest of the vertices such that the ith vertex selected will receive the label $2(i-1)-1$. Since the vertices having consecutive labels are at diametric distance and the distance between any other vertices is greater than or equal to 2, the diameter of the graph, the above labelling is a radio labelling. Thus,

$$\begin{aligned} M_{r1}(G) &\leq 1^2 + 2^2 + 3^2 + \cdots + (2(n-1)-1)^2 \\ &= \frac{4n^3 - 12n^2 + 11n + 9}{3}. \end{aligned}$$

Therefore,

$$M_{r1}(G) \leq \frac{4n^3 - 12n^2 + 11n + 9}{3}.$$

This completes the first part of the proof.

Consider the graph $K_{n-1} \bigoplus_u v$ which is obtained by adding a new vertex v and an edge between u and v, where u is a vertex in K_{n-1}. By calculation,

$$M_{r1}(K_{n-1}) = \frac{4n^3 - 12n^2 + 11n + 9}{3}.$$

Now, let us consider the converse part. Let n be the number of vertices of the graph G with diameter 2. Let and

$$M_{r1}(G) = \frac{4n^3 - 12n^2 + 11n + 9}{3}.$$

Then, the labels of the graph are $1, 2, 3, 5, 7, \ldots, 2(n-1) - 1$. Since three labels are consecutive, it has an induced paw, and the rest of the labels have difference equal to the diameter of the graph, which shows that it is the label for a clique. Therefore,

$$G \cong K_{n-1} \bigoplus_u v.$$

Theorem 3 *Let the graph G be with n vertices having radio number $rn(G) = r$. Then,*

$$M_{r1}(G) \leq \frac{r(r+1)(2r+1)}{6} \tag{3}$$

with equality if and only if $r = n$.

Proof Consider a graph G on n vertices with radio number r. Let the labels of the vertices are $1, a_1, a_2, \ldots, a_{n-2}, r$.

Clearly, $1^2 + a_1^2 + a_2^2 + \cdots + a_{n-2}^2 + r^2 \leq 1^2 + 2^2 + \cdots + r^2$.

Therefore,

$$M_{r1}(G) \leq \frac{r(r+1)(2r+1)}{6}.$$

To prove the sharpness, if $M_{R1}(G) = \frac{r(r+1)(2r+1)}{6}$, then the vertices of G have all the labels from 1 to r, which shows that $r = n$.

Conversely, if $r = n$, then by the definition G admits consecutive radio labelling and hence $M_{r1}(G) = \frac{r(r+1)(2r+1)}{6}$.

Theorem 4 *Let the graph G have n vertices. Let P be the set of all peripheral vertices of G. Let the diameter of G be d and have p peripheral vertices, such that if $u, v \in P$ then, $d(u, v) = d\ \forall u, v \in P$. Then,*

$$M_{r1}(G) \leq \sigma_1 p^3 + \sigma_2 p^2 + \sigma_3 p + \sigma_4 \tag{4}$$

where

$$\sigma_1 = d - \frac{d^2}{3} - \frac{2}{3},$$

$$\sigma_2 = d^2n - 2dn + n + \frac{d^2}{2} - 2d + \frac{5}{2},$$

$$\sigma_3 = dn^2 - d^2n^2 - d^2n + 5dn - 4n + \frac{5d^2}{6} + d + \frac{1}{6}$$

and

$$\sigma_4 = \frac{d^2n^3}{3} + \frac{d^2n^2}{2} - 2dn^2 + \frac{d^2n}{6} - 2dn + 4n - d^2 - 3$$

with equality if and only if $G \cong K_n \circ K_1$.

Proof Consider a graph G on n vertices with diameter d and p peripheral vertices satisfying the conditions in the theorem. Let us consider the following labelling. Choose any vertex other than the peripheral vertex and label it with 1. We now label the peripheral vertices from d to $d + p - 1$. Rest of the vertices will get the labels $2d + p - 2, 3d + p - 2, \ldots, (n - p)d + p - 2$. Clearly, the above labelling satisfies the condition for radio labelling.

Therefore, $M_{r1}(G) \leq 1^2 + d^2 + (d+1)^2 + \cdots + (d+p-1)^2 + (2d+p-2)^2 + (3d+p-2)^2 + \cdots + ((n-p)d+p-2)^2$.

Hence,

$$M_{r1}(G) \leq \sigma_1 p^3 + \sigma_2 p^2 + \sigma_3 p + \sigma_4$$

where $\sigma_1 = d - \frac{d^2}{3} - \frac{2}{3}$, $\sigma_2 = d^2n - 2dn + n + \frac{d^2}{2} - 2d + \frac{5}{2}$, $\sigma_3 = dn^2 - d^2n^2 - d^2n + 5dn - 4n + \frac{5d^2}{6} + d + \frac{1}{6}$, $\sigma_4 = \frac{d^2n^3}{3} + \frac{d^2n^2}{2} - 2dn^2 + \frac{d^2n}{6} - 2dn + 4n - d^2 - 3$.

From the calculation, we can show that

$$M_{r1}(K_n \circ K_1) = \sigma_1 p^3 + \sigma_2 p^2 + \sigma_3 p + \sigma_4$$

where $\sigma_1 = d - \frac{d^2}{3} - \frac{2}{3}$, $\sigma_2 = d^2n - 2dn + n + \frac{d^2}{2} - 2d + \frac{5}{2}$, $\sigma_3 = dn^2 - d^2n^2 - d^2n + 5dn - 4n + \frac{5d^2}{6} + d + \frac{1}{6}$, $\sigma_4 = \frac{d^2n^3}{3} + \frac{d^2n^2}{2} - 2dn^2 + \frac{d^2n}{6} - 2dn + 4n - d^2 - 3$.

4 Bounds of the First Zagreb Radio Index

Theorem 5 *Let G be a graph with radio number r. Then* $M_{R1}(G) \leq M_{R1}(K_r)$.

Since $n \leq r$ and $\Delta \leq n - 1$, the proof if immediate if we as seen in Theorem 2.

Theorem 6 *Let G be a graph with maximum degree* Δ *and radio number* r. *Then,*

$$M_{R1}(G) \leq \Delta\frac{r(r+1)}{2}. \tag{5}$$

The equality holds if and only if G is a regular graph admitting consecutive radio labelling.

Proof Consider a graph G. Let c assign the optimum radio labelling of G. Let $1, a_1, a_2, \ldots, a_{n-2}$ and r be the labels assigned by c.

Now,

$$\sum_{uv\in E(G)} c(u)+c(v) = d_1 1 + d_2 a_1 + \cdots + d_{n-1}a_{n-2} + d_n r$$
$$\leq \Delta\frac{r(r+1)}{2}.$$

If G is regular, then $d(v) = \Delta, \forall v \in V(G)$ and since G admits consecutive radio labelling $\sum_{uv\in E(G)} c(u)+c(v)$ becomes $\Delta\frac{r(r+1)}{2}$.

Conversely, suppose G is a graph such that $\sum_{uv\in E(G)} c(u)+c(v) = \Delta\frac{r(r+1)}{2}$. To prove that G is regular and admits consecutive radio labelling, assume the contrary. Since G is not regular $\delta \neq \Delta$. Also, at least one label from 1 to r is missing since it is not admitting consecutive radio labelling. Therefore,

$$\sum_{uv\in E(G)} c(u)+c(v) = \sum_{v\in V(G)} d(v)c(v) < \Delta\frac{r(r+1)}{2},$$

which is a contradiction. Hence, the proof follows.

Theorem 7 *Let G be a graph with minimum degree* δ *and G admits consecutive radio labelling. Then*

$$\delta\frac{n(n+1)}{2} \leq M_{R1}(G) \tag{6}$$

with equality if and only if G is regular.

The proof is similar to the above proof.

5 Bounds of the Second Zagreb Radio Index

Theorem 8 *Let the graph G of order n admit consecutive radio labelling. Then,*

$$M_{R2}(G) \leq \frac{n^4}{8} + \frac{n^3}{12} - \frac{n^2}{8} - \frac{n}{12} \tag{7}$$

with equality if and only if $G \cong K_n$.

Proof Obviously, the labels of the vertices of the graph G are from 1 to n. The Second Zagreb Index is the sum of the product of the labels of the end vertices of the edges. We obtain the maximum value when every vertex is adjacent to each other, which is the complete graph.

Therefore,

$$M_{R2}(G) \leq \frac{n^4}{8} + \frac{n^3}{12} - \frac{n^2}{8} - \frac{n}{12}$$

and the bound is sharp only when $G \cong K_n$.

Theorem 9 *Let G be any tree on* $n > 2$ *vertices. Then,*

$$\frac{n^2 + 3n - 4}{2} \leq M_{R2}(G) \tag{8}$$

with equality if and only if $G \cong S_n$.

Proof A graph on n vertices has the lowest radio number when it admits consecutive radio labelling. The only trees that admit consecutive radio labelling are K_1 and K_2. Obviously, K_1 satisfies the bound. Let G have order n.

Assume that a tree G has no consecutive radio labelling. Then, $rn(G) \geq n + 1$. Consider a graph G' having radio number $n + 1$.

Clearly, $M_{R2}(G') \geq (1 \times 3) + (1 \times 4) + \cdots + (1 \times (n + 1))$. i.e., $\frac{n^2+3n-4}{2} \leq M_{R2}(G')$.

For any tree G on n vertices with $rn(G) \geq n + 2$,

$$M_{R2}(G') \leq M_{R2}(G).$$

Therefore,

$$\frac{n^2 + 3n - 4}{2} \leq M_{R2}(G).$$

This completes the first part of the proof.

We have $M_{R2}(G) = \frac{n^2+3n-4}{2}$. i.e., $M_{R2}(G) = (1 \times 3) + (1 \times 4) + \cdots + (1 \times (n + 1))$.

So the labels of the vertices are $1, 3, 4, \ldots, n + 1$ and the vertex having label 1 is adjacent to all the other vertices and no other vertices are adjacent to each other, i.e., $G \cong S_n$.

Theorem 10 *Let n be the order of the graph G with* $diam(G) = 2$. *Let G has a pendant vertex and has a paw as an induced subgraph. Then,*

$$M_{R2}(G) \leq \frac{n^4}{2} - \frac{8n^3}{3} + 5n^2 - \frac{23n}{6} + 11 \tag{9}$$

with equality if and only if $G \cong K_{n-1} \bigoplus_u v$.

Proof Consider any graph on n vertices with diameter 2. The labelling is similar to that of Theorem 2.2. Since 1, 2 and 3 are consecutive labels and diameter of the graph is 2, the vertices having labels 1 and 3 cannot be adjacent to the vertex having label 2. Since our graph has a pendant vertex and to discover the extreme possibilities, let us allot the label 2 to the pendant vertex. Since Zagreb Radio Index is the infimum value, to minimise the sum the possible minimum value say 5 will be given to the vertex adjacent to the pendant vertex. Consider the case where all the vertices except the pendant vertex are adjacent to each other. Then by calculation

$$M_{R2}(G) \leq \frac{n^4}{2} - \frac{8n^3}{3} + 5n^2 - \frac{23n}{6} + 11.$$

From the first part of the proof, it is clear that $M_{R2}(G) = \frac{n^4}{2} - \frac{8n^3}{3} + 5n^2 - \frac{23n}{6} + 11$ if and only if $G \cong K_{n-1} \bigoplus_u v$.

6 Bounds of the Third Zagreb Radio Index

Theorem 11 *Let G be a graph that admits consecutive radio labelling. If G has n vertices, then*

$$M_{R3}(G) \leq \frac{n^3 - n}{6} \tag{10}$$

with equality if and only if $G \cong K_n$.

Proof As G has consecutive radio labelling, the labels radio labels are 1 to n. We get the upper bound when maximum number of edges contribute to the sum. The complete graph is the graph on n vertices with maximum edges. It admits consecutive radio labelling [21].

Therefore,

$$M_{R3}(G) \leq \frac{n^3 - n}{6}.$$

The bound is sharp only when $G \cong K_n$.

Theorem 12 *If the graph G admits consecutive radio labelling then,*

$$\frac{5n - 1}{2} \leq M_{R3}(G) \tag{11}$$

with equality if and only if $G = C_5$.

Proof Since $M_{R3}(K_n)$ gives the maximum value, let us consider $d \geq 2$. Therefore, no two adjacent vertices can have the label difference 1. Also, we cannot arrange n consecutive integers at a difference 2 alone when $d \geq 2$. So the smallest possibility is to arrange with differences 2 and 3 which happens only when $n = 5$.

Therefore,

$$\frac{5n-1}{2} \leq M_{R3}(G).$$

The cycle C_5 is the only graph on n vertices having $d \geq 2$ admitting consecutive radio labelling. Hence, the result follows.

Theorem 13 *For an even integer k, let G be a k-regular graph of order n having diameter d. Then,*

$$M_{R3}(G) \leq \frac{dk^2n}{4} + \frac{dkn}{2} - \frac{dk^3}{12} - \frac{dk^2}{4} - \frac{dk}{6}. \tag{12}$$

Proof Given that k is even. Let G be a graph order n with every vertex is of degree k (k-regular). Since k is even, let $k = 2m$. Arrange n vertices around a cycle and connect each vertex to m vertices on the right and m vertices on the left. Consider the following labelling. Choose any vertex and give the label 1. Then, allot the labels $1 + d, 1 + 2d, \ldots, 1 + (n-1)d$ to the rest of the vertices in the clockwise direction. The above labelling satisfies the condition for radio labelling and therefore $M_{R3}(G)$ is less than the value obtained from this labelling.

In this labelling, since 1 is the smallest label and every vertex is adjacent to exactly k, vertices 1 should get subtracted k times (using the Third Zagreb Radio Index). The label $1 + d$ will get added once and subtracted $(k-1)$ times since $1 < (1 + d)$. If $i \leq m$, then $1 + id$ will get added i times and subtracted $(k - i)$ times by the above argument. For $m \leq i \leq n - m - 1$, the label $1 + id$ will get added and subtracted m times since they have m labels less than that in the left side and m labels greater than them in the right side. For $1 \leq i \leq m$, the label $1 + (n - m + i - 1)d$ will get added $m + i$ times and subtracted $m - i$ times.

Therefore,

$$M_{R3}(G) \leq \frac{dk^2n}{4} + \frac{dkn}{2} - \frac{dk^3}{12} - \frac{dk^2}{4} - \frac{dk}{6}.$$

Theorem 14 *For an odd integer k, let G be a k-regular graph of order n having diameter d. Then,*

$$M_{R3}(G) \leq \frac{dn^2}{4} + dm^2n + dmn - \frac{2dm^3}{3} - dm^2 - \frac{dm}{3}. \tag{13}$$

Proof Given that k is odd. Let G be of diameter d having the order n. Since k is odd, let $k = 2m + 1$. Arrange n vertices around a cycle and connect each vertex to m vertices on the right and m vertices on the left and one vertex directly opposite. Consider the following labelling, choose any vertex and give the label 1. Then, allot the labels $1 + d, 1 + 2d, \ldots, 1 + (n-1)d$ to the rest of the vertices in the clockwise direction. The above labelling satisfies the condition for radio labelling, and therefore, $M_{R3}(G)$ is less than the value obtained from this labelling.

In this labelling, since 1 is the smallest label and every vertex is adjacent to other k vertices (The Third Zagreb radio index), 1 should get subtracted k times. The label $1 + d$ will get added once and subtracted $(k - 1)$ times since $1 < (1 + d)$. If $i \leq m$ then $1 + id$ will get added i times and subtracted $(k - i)$ times by the above argument. For $m \leq i \leq \frac{n-2}{2}$, the label $1 + id$ will get added m times and subtracted m+1 times since they have m labels less than that in the left side and $m + 1$ labels greater than them in the right side including the vertex directly opposite. For $\frac{n}{2} \leq i \leq n - m - 1$, the label $1 + id$ will get added m+1 times and subtracted m times since they have m labels less than that in the left side and m labels greater than them in the right side and a label less than that directly opposite. For $1 \leq i \leq m$, the label $1 + (n - m + i - 1)d$ will get added $m + i + 1$ times and subtracted $m - i$ times. Therefore,

$$M_{R3}(G) \leq \frac{dn^2}{4} + dm^2n + dmn - \frac{2dm^3}{3} - dm^2 - \frac{dm}{3}.$$

7 Conclusion

This chapter consists of certain bounds for the Zagreb Radio Indices of graphs. The Primary Zagreb Index is tested in the general case, for diameter 2 graphs, in relation with the radio number and for the number of peripheral vertices. The bounds of the First Zagreb Index are found in relation with the maximum degree, minimum degree and the radio number. The Second Zagreb Index and the Third Zagreb Index are worked out mainly for the graphs that admit consecutive radio labelling. Many of these bounds are found to be sharp.

References

1. Asok A, Kureethara JV (2018) The QSPR study of butane derivatives: a mathematical approach. Orient J Chem 34(4):1842–1846
2. Asok A, Kureethara JV (2019) QSPR analysis of certain degree and eccentricity based topological indices and butane derivatives. Rasayan J Chem 12(2):464–470
3. Asok A, Kureethara JV (2020) Characterization of the forgotten topological index and the hyper-Zagreb index for the unicyclic graphs. MATCH Commun Math Comput Chem 84(3):739–751
4. Bhalaji N (2020) A novel hybrid routing algorithm with two fish approach in wireless sensor networks. J Trends Comput Sci Smart Technol 2(3):134–140
5. Bindhu V (2020) Constraints mitigation in cognitive radio networks using cloud computing. J Trends Comput Sci Smart Technol 2(1):1–14
6. Devillers J, Balaban AT (2000) Topological indices and related descriptors in QSAR and QSPAR. CRC Press
7. Duraipandian M (2020) Long term evolution—self organizing network for minimization of sudden call termination in mobile radio access networks. J Trends Comput Sci Smart Technol 2(2):89–97

8. Fath-Tabar G (2011) Old and new Zagreb indices of graphs. MATCH Commun Math Comput Chem 65(1):79–84
9. Gutman I (2013) Degree-based topological indices. Croat Chem Acta 86(4):351–361
10. Gutman I, Trinajstić N (1972) Graph theory and molecular orbitals. Total π-electron energy of alternant hydrocarbons. Chem Phys Lett 17(4):535–538
11. Gutman I, Ruščić B, Trinajstić N, Wilcox CF (1975) Graph theory and molecular orbitals. XII. Acyclic polyenes. J Chem Phys 62:3399–3405
12. Haoer RS (2021) Wiener polarity and similar topological descriptors of some product graphs. Int J Quant Chem. https://doi.org/10.1002/qua.26796
13. Kok J, Sudev N, Mary U (2017) On chromatic Zagreb indices of certain graphs. Discrete Math Algorithms Appl 9(1):1750014
14. Kureethara JV, Raj AT (2018) Zagreb radio indices of graphs. Int J Pure Appl Math 118(10):343–352
15. Kureethara JV, Asok A, Cangul IN (2022) Inverse problem for the forgotten and the hyper Zagreb indices of trees. Commun Comb Optim
16. Li S, Wang L, Zhang H (2021) On ABC Estrada index of graphs. Discrete Math 344(11):112586
17. Li J, Zhang J (2019) On the second Zagreb eccentricity indices of graphs. Appl Math Comput 352:180–187
18. Naduvath S, Kok J (2020) On certain topological indices of signed graphs. J Math Comput Sci 10(2):248–261
19. Ore O (1962) Theory of graphs. Am Math Soc 38:206–212
20. Raj AT, Kureethara JV (2017) Zagreb radio indices of some graphs with diameter two. Int J Sci Res Math Stat Sci 4(6):7–11
21. Raj AT, Kureethara JV (2021) Consecutive radio labelling of graphs. In: Shukla S, Unal A, Kureethara JV, Mishra DK, Han DS (eds) Data science and security. Springer, Singapore, pp 317–323
22. Raj AT, Kureethara JV (2022) Zagreb radio indices and coindices. Lett Appl NanoBioSci 11(1):3006–3019
23. Suma V, Haoxiang WP (2020) Optimal key handover management for enhancing security in mobile network. J Trends Comput Sci Smart Technol 2(4):181–187
24. Thomas AS, Kalayathankal SJ, Kureethara JV (2017) An introductory note on the spectrum and energy of molecular graphs. Mapana J Sci 16(2):17–27
25. West DB (1999) Introduction to graph theory. Prentice Hall of India Pvt. Ltd
26. Xu K (2011) The Zagreb indices of graphs with a given clique number. Appl Math Lett 24(6):1026–1030

Interruption Detection for Detection and Quantification of QRS Fragmentation Based on Machine Learning and Deep Learning Technique

Afsana Banu and K. G. Manjunath

Abstract Apportioned QRS (fQRS) is an open biomarker and sign of myocardial scarring that can be perceived from the electrocardiogram (ECG) (Das in Heart Rhythm 4:1385–1392, 2007 [1]). ffQRS scoring is made dependent on seening premise, which burns-through time and prompts the abstract outcome. This strategy which has been proposes identifying and measuring ffQRS is a one single straight wau utilizing highlights which is separated from variational mode decomposition (VMD) and amended sign averaging. In the proposed framework, QRS structures in the ECG signals were first divided using VMD. By then, at that point, ten VMD- and PRSA-based features were figured and dealt with into prominent classifiers. Nowadays cyberattacks data collecting is often common and also imbalanced network traffic for large amounts of normal data. It gives a high robust solution to the user without any interaction, it makes difficult for data and network intrusion detection system (Zamani and Movahedi in Machine learning techniques for intrusion detection, 2013 [2]). This paper proposes a machine and deep learning technique to overcome imbalanced network traffic (Cieslak et al. in Combating imbalance in network intrusion datasets, 2006, pp. 732–737 [3]).

Keywords Traffic network · Deep learning · Machine learning · ECG SP · Segmentation · Variational mode decomposition

1 Introduction

Nowadays, technology is improving rapidly on wide applications like IoT, CC, 5G and other technologies. Network traffic is one which is becoming more complex and massive even attacks on the data also taking the diverse, and it has become

A. Banu (✉) · K. G. Manjunath
Department of Computer Science and Engineering, Siddaganga Institute of Technology, Tumkur, India

K. G. Manjunath
e-mail: kgmanjunath@sit.ac.in

S. Shakya et al. (eds.), *Proceedings of Second International Conference on Sustainable Expert Systems*, Lecture Notes in Networks and Systems 351,
https://doi.org/10.1007/978-981-16-7657-4_54

more challenging on cyberspace security. On different browsers NIIDS needs to be identified accurately the attacks and should give real-time monitoring dynamically and formulates the strategies.

ECG is diagnostic tool which is used to check the heart rate condition of a person, these signals might be single- or multi-lead, in medical domain 12 lead ECG's are commonly used in order to access the heart rate of the person [4]. ECG signals are used to find various conditions of the human body. So in this project, we are mainly monitoring the network traffic and attacks of the data based on machine and deep learning while transmitting ECG signals [5]. Also, ELM technique was created with sigmoidal function of biases in hidden nodes to conduct classification task [6].

Most of the data which transmitting will be normal data and will be in dominant position which creates a normal traffic while transmitting the data, but the data which contains malicious attacks, results in imbalance categories. In these kind of imbalance networks, intruder who has entered to hack the data will be facing tremendous pressure. This happens because attacks done on data can hide huge traffic on network while passing a data. So only machine learning technology is not enough to monitor since it can be misclassify easily, a deep learning framework in the study that used hardware methods and deep learning models were much superior in terms of detection precision, false alarms, and flexibility [7].

To monitor this imbalanced traffic network DSSTE algorithm has been proposed to tackle this problem. This method slightly reduces the problem and increase the sampling rate more effectively.

In our proposed approach, we are using machine and deep knowing technique to check the data, and the contributions are as follows:

1. We use NSL-KDD and CSE datasets, and based on this dataset, analysis will be conducted.
2. The work proposes DSSTE algorithm, which is mainly used for samples and to find augmenting in difficult set, to tackle the imbalance problem of the intrusion detection.
3. The algorithms used here are RF, SVM, XG Boost, and LSTM. Experiment has been divided into 30 epoch modules.

2 Database

As it has been discussed in Sect. 1, 12 lead ECG signals dataset has been used from 723 patients before implanting ICD in term to prevent cardiac death due to arrhythmia [8]. Signals which are read are digitized with sampling frequency 250 Hz within a duration of 10 s. Some signals reading from the patients are excluded because they have ECG morphologies.

3 Methods

The diagram drawn gives the proposed method of ffQRS in ECG signals which is named as Fig. 1. It has four fundamental parts: preprocessing, QRS division utilizing VMD, separating the highlights, and grouping. Every single step has been clarified exhaustively in beneath areas.

3.1 *Preprocessing*

This progression is the establishment of any machine or profound learning procedure execution, this cycle must be never really clamor from the ECG signals by going each lead through advanced fourth request butterworth high pass channel with low reach recurrence 0.5 Hz and max recurrence commotion segments of low reach recurrence 70 Hz. This cleaning is applied on both forward and reverse way to get non-stage contortion [9].

3.2 *QRS Segmentation*

The calculation called VMD which is talked about as of now, obviously this calculation is utilized to defined sign deterioration strategy [10]. This requires two parameters number nodes and bandwidth control. If the parameters is wrongly chosen which may lead to a problem like mode splitting or mode mixing. To overcome this problem, optimization technique has been used where it fixes the number of decomposition modes and balancing parameter [11] (Fig. 2).

Fig. 1 Proposed block diagram

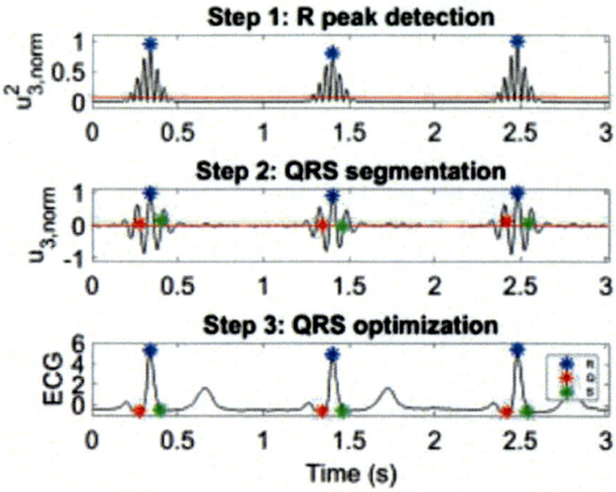

Fig. 2 Three main steps of segmentation

3.3 *Feature Extraction*

In previous section, how features were extracted using QRS segmentation has been described:

(1) PRSA-Based Features: Phase-outlined sign averaging plans to see and evaluate semi-eccentric upgrades covered by multi-part non-fixed signs. The method consists of three stages:

- Selection of anchor centers
- Window affirmation
- Averaging.

There are different steps to figure PRSA twist for common and not conventional beat as shown in Fig. 3. In this figure, constructive and gloomy centers are included in red, cyan, and green. Blue is used to include changes in the signs.

3.4 *Detecting Intrusion While Receiving Data and Extracting Feature*

Simply by using AI development, it fails to attract the thought due to nonappearance of limit and dealing with power. With mix of various developments like AI and others, the makers have applied ML to arrange security to get the data.

In the wake of defying the foolish convergence in network while passing data, we proposed DSSTE to pack the bigger part tests and augmentation. Diminishing number of data in planning set of ECG and intrusion recognizable proof can achieve more accuracy. So we are using different algorithms such as random forest, SVM, XGBoost, and LSTM classifiers.

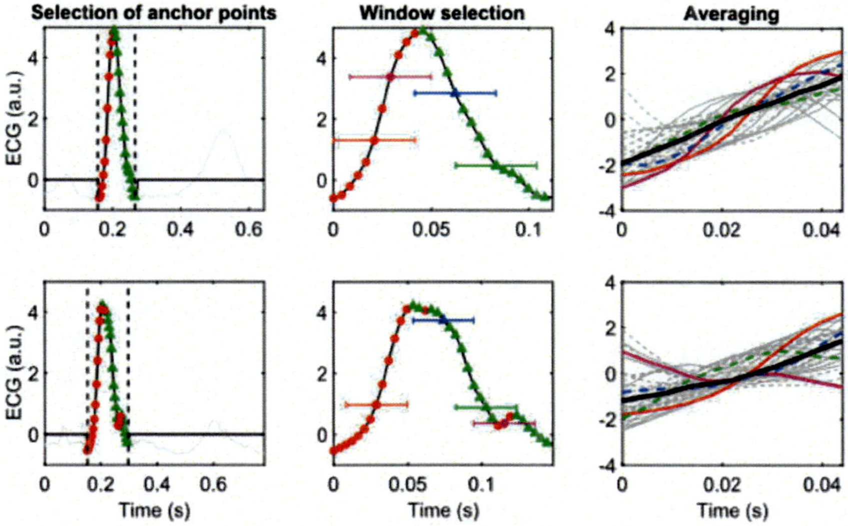

Fig. 3 Curves obtained for normal heart beat

We proposed the square exposure model showed in Fig. 4. Data pre-regulating from the start acted in our impedance revelation structure, including duplicate, exemption, and missing regard designing. By then, passing on the test set and the getting sorted out set, and the approach set managed for data changing using our proposed DSSTE estimation.

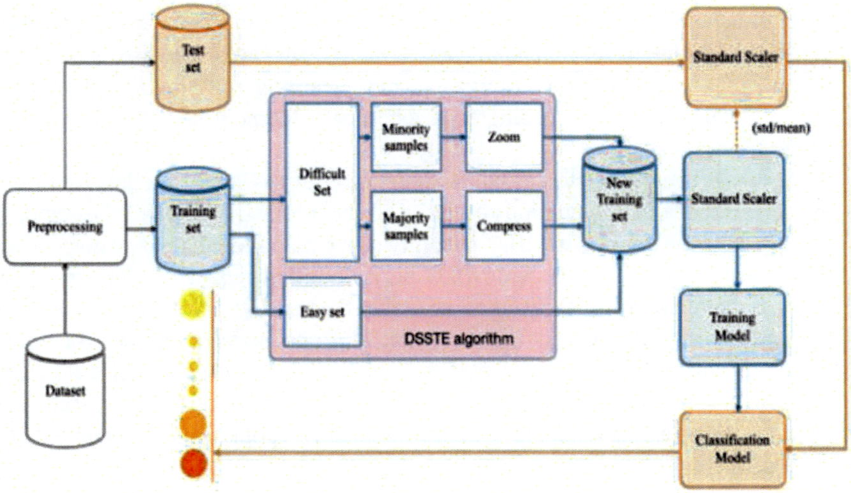

Fig. 4 The framework of network trespass finding mode

3.5 DSSTE Algorithm

In unseemly busy time gridlock, particular traffic has same kind of depiction, essentially a little attacks a be conceal over gigantic data sending, it is difficult to the classifier to get comfortable with the difference between planning measure by the machine.

Improper planning data to part into close neighbor and far-neighbor which has been proposed by ENN estimation. The data seems, by all accounts, to be essentially something similar, and it is difficult for the classifier to get to know the differentiation between groupings, so ENN has been referred, which made machine to conveniently perceive the little changes in the models.

The unseemly association traffic is basically isolated into ordinary and abnormal traffic by using decline, clustering, and course of action.

3.6 System Teach and Deep Teach Algorithms

Randomly forest, SVM, XGBoost, and memory of the long and short term are algorithms which are considered to train and test:

(1) RANDOM FOREST
This is one of the oversee algorithms which is used to train the model and used to do prediction in further process on certain test samples which has given in the dataset based on characteristic attributes. RF-based decision tree method to generate new sets.

(2) SUPPORT VECTOR MACHINE
This calculation shows various types of benefit in a little example, nonlinear, and high-dimensional pattern acknowledgment and likewise some all-encompassing capacities. Before profound learning come into an existence, SVM calculation was considered as a best calculation in AI KNN [12].

(3) XGBoost
Separated and the gradient boosting decision tree (GBDT) model, XGBoost defeats the bound evaluation speed and accu charming. XGBoost adds regularization to the important GBDT calamity capacity to keep the model away from overfitting [13]. The standard GBDT plays out a first-request Taylor expansion on the calculated difficulty cutoff and takes the negative point view as the extra worth of the current model. Then again, XGBoost plays out a second-request Taylor improvement to guarantee the accu shocking of the model. Likewise, XGBoost squares and sorts each fragment, making it conceivable to parallelize the estimation while searching for the best isolated point, which fundamentally accelerates the evaluation speed.

(4) LONG SHORT-TERM MEMORY
Like RNN, LSTM is universal algorithm in deep learning technique because it is suitable for weight matrix. This is used to monitor any element based on the network, LSTM is suitable for learning from experience.

3.7 Classification

The disposed of features were used as commitment to a classifier to perceive and assess QRS brokenness. It has been shown that help vector machines (SVMs) are in various types of issues better appeared differently in relation to three other prominent classifiers [14]. They are correspondingly known to have exceptional theory properties in supervised request issues. The key mulled over a SVM is to find a hyperplane what confines two classes by maximizing the edge between the classes. This ought to be conceivable in a speedy or non-straight way, by introducing piece works that change the part space. Right while stunning separation cannot be refined, misclassifications are allowed by introducing a tricky edge unsurprising that repulses zeros in that lie on some inadmissible side of the edge. The introduction of SVMs with a direct, polynomial, and winding clarification work parcel was stood separated from select the part which gets the best show.

4 Result and Discussion

Figure 5 shows ROC twists for AUC regards for all classifiers. Here, the test set includes the 20% excess signs with amazing understanding. The genuine class of these signs is viewed as certain so arrangement can be performed.

Figure 5 shows ROC turns for AUC regards for all classifiers. Here, the test set joins the 20% bounty consents to with surprising arrangement. The credible class of these signs is viewed as certain so approach can be performed.

Table 1 gives information about sensitivity, specificity, and accuracy results applied on proposed algorithm.

Table 1 are verified with same dataset. Parallely, when appraisal has been made for ECG data, besides outlined. We use the accuracy, prediction, recall, and F1-Score to consider the test model's show. These evaluation models reflect the showing of the impedance detection development's stream certificate precision rate, and joke alert rate.

The blend of the model figure results and the certifiable removing is taken into four sorts: false negative (FN), a positive model, which is wrongly settled as a negative model. Fake positive (FP), negativity models are overwhelmed as unequivocal models. Ensured negative (TN), far and away horrible, are totally settled as evil models.

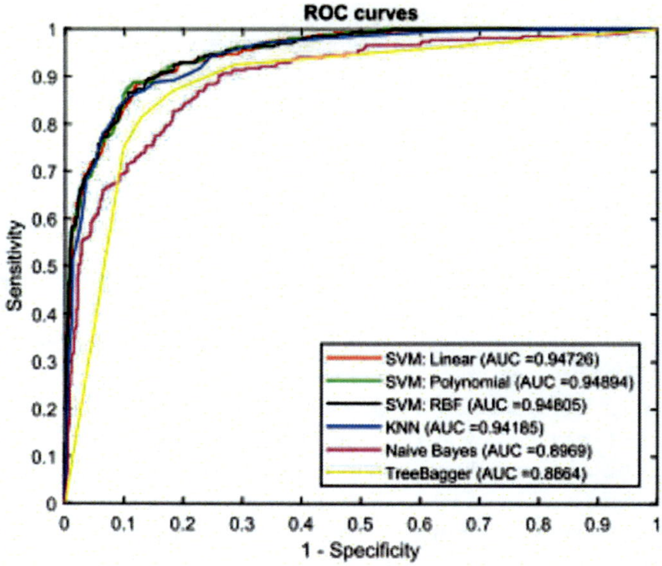

Fig. 5 Curves of the ROC for ffQRS with classifiers

Table 1 Results of algorithm applied

Method	Sens (%)	Spec (%)	Acc (%)
Proposed method	86	89	88
Jin et al. [5]	72	78	75
Maheshwari et al. [11]	84	87	85
Bono et al. [2]	75	78	77

5 Conclusion

AS break in the association extended, the strain on the spot is similarly extended. The issue caused in this is improper traffic in data conveying make it hard for interference disclosure systems to find poisonous attacks. In our proposed procedure, DSSTE computation is used which engages the request and strengthen the imbalanced association data learning.

The result shows the proposed fQRS in an incredible way with no noxious attacks and also partitioning of ECG signals. Comparison of existing methodology shows proposed procedure perform well.

References

1. Das MK et al (2007) Fragmented QRS on a 12-lead ECG: a predictor of mortality and cardiac events in patients with coronary artery disease. Heart Rhythm 4(11):1385–1392
2. Zamani M, Movahedi M (2013) Machine learning techniques for intrusion detection. arXiv: 1312.2177. [Online]. Available http://arxiv.org/abs/1312.2177
3. Cieslak DA, Chawla NV, Striegel A (2006) Combating imbalance in network intrusion datasets. In: Proceedings of IEEE international conference on granular computing, May 2006, pp 732–737
4. Pietrasik G, Zareba W (2012) QRS fragmentation: diagnostic and prognostic significance. Cardiol J 19(2):114–121
5. Shone N, Ngoc TN, Phai VD, Shi Q (2018) A deep learning approach to network intrusion detection. IEEE Trans Emerg Top Comput Intell 2(1):41–50
6. Mugunthan SR, Vijayakumar T (2021) Design of improved version of sigmoidal function with biases for classification task in ELM domain. J Soft Comput Paradigm (JSCP) 3(02):70–82
7. Shakya S, Smys S (2020) Anomalies detection in fog computing architectures using deep learning. J Trends Comput Sci Smart Technol 2(1):46–55
8. Maheshwari S et al (2013) An automated algorithm for online detection of fragmented QRS and identification of its various morphologies. J R Soc Interface 10(89):20130761
9. Flowers NC, Horan LG, Thomas JR, Tolleson WJ (1969) The anatomic basis for high-frequency components in the electrocardiogram. Circulation 39(4):531–539
10. Morita H et al (2008) Fragmented QRS as a marker of conduction abnormality and a predictor of prognosis of Brugada syndrome. Circulation 118(17):1697–1704
11. Jin F, Sugavaneswaran L, Krishnan S, Chauhan VS (2017) Quantification of fragmented QRS complex using DEEP. Biomed Signal Process Control 31:513–523
12. Shapoorifard H, Shamsinejad P (2017) Intrusion detection using a novel hybrid method incorporating an improved KNN. Int J Comput Appl 173(1):5–9
13. Bhattacharya S, Maddikunta PKR, Kaluri R, Singh S, Gadekallu TR, Alazab M, Tariq U (2020) A novel PCA-firefly based XGBoost classification model for intrusion detection in networks using GPU. Electronics 9(2):219
14. Pervez MS, Farid DM (2014) Feature selection and intrusion classification in NSL-KDD cup 99 dataset employing SVMs. In: Proceedings of the 8th international conference on software, knowledge, information management and applications (SKIMA), Dec 2014, pp 1–6

Negative Domination in Networks

Athira P. Ranjith and Joseph Varghese Kureethara

Abstract We introduce s-domination in signed graphs which is based on the number of negative edges between the dominating set and its complement. The s-domination in both the positive and negative homogeneous signed graph will be studied for each value of s. As a special case, the properties of s-domination in sum signed graphs will be analyzed. The maximum value of s for a graph for which the s-domination exists is identified.

Keywords Signed graph · Domination in networks · Sum signed graph

1 Introduction

Human society has several binary partitions. Male-Female, Literate-Illiterate, Have-Have Not, Rich-Poor, Prey-Predator, etc., are examples of some societal segregation. Given a network of people, is there a way to check a dominant group that is dominating the rest of the section. Accounting all the theoretical studies, graph theory has contributed a lot in this field. Network structures can be designed based on certain elements that are dominating the entire structure. Some of the dominating items include but not exclusively, servers, routers, supercomputers, workstations, etc. [8, 20, 21]. In this paper, we develop the concept negative domination in graphs and networks.

Domination is one of the major research areas in graph theory. The study of domination started in the year 1962 with Ore [16] and till date numerous research on various types of domination are being conducted. Even though the study of signed graphs was initiated in the year 1952 by Harary [10], it was not until 2013 that the concept of domination was introduced in signed graphs by Acharya [1]. In this paper,

A. P. Ranjith · J. V. Kureethara (✉)
Christ University, Bangalore 560029, India
e-mail: frjoseph@christuniversity.in

A. P. Ranjith
e-mail: athira.ranjith@res.christuniversity.in

S. Shakya et al. (eds.), *Proceedings of Second International Conference on Sustainable Expert Systems*, Lecture Notes in Networks and Systems 351,
https://doi.org/10.1007/978-981-16-7657-4_55

we define a domination on signed graphs known as s-domination which is based on the number of negative edges between the dominating set and its complement. For further details on domination in signed and unsigned graphs, one may refer to the articles [6, 7, 11–14, 22, 23].

Psychologists used square matrices with elements -1, 0 and 1 to represent disliking, indifference and liking among a group of individuals. Harary used this idea to introduce the concept of signed graphs. A signed graph is a graph with positive and negative edges which are usually represented using $+$ and $-$ signs or solid and dashed line segments, respectively [2–4]. A signed graph is said to be homogeneous if it is either all-positive or all-negative and heterogeneous otherwise. Some of the recent works in signed graphs are degree product induced signed signed graphs [5], signed distance [9], topological indices of signed graphs [15] and Laplacian matrices for skew gain graphs [19].

In a society, there will always exist a group of people who can have a huge impact on the society. These people can include politicians, community workers, etc. The impact of these people can be either of a positive or a negative nature. Considering the amount of positive and negative influence of the group on the people around them, the concept of s-domination is introduced.

Definition 1 Let $X \subseteq V(G)$ where G is a graph. Then, X is said to be a σ_s-dominating set with respect to a signed graph S of G if the following conditions are satisfied.

i. X is a dominating set of G.
ii. Among the signed edges between X and $V(G) - X$, there are exactly s negative edges.

The minimum cardinality of such a dominating set over all signed graphs of the graph G is called a s-domination number denoted by $\gamma_s(G)$.

From the definition of s-domination, it is clear that s-dominating set will not exist for every value of s. For example in the above illustration, the signed graph will not have a s-dominating set when $s > 5$ as the number of negative edges in the graph is 4.

Let $G = (V, E)$ always represent a simple connected graph unless mentioned otherwise. The degree of the vertex $v \in V$ denoted by $\deg(v)$ gives the number of vertices adjacent to v. The maximum and minimum degree in the graph is indicated by $\Delta(G)$ and $\delta(G)$, respectively. Let $S = (G, \sigma)$ be a signed graph of the underlying graph G where $\sigma : E(G) \longrightarrow \{+, -\}$. The set of negative and positive edges in the signed graph S is denoted by $E^-(S)$ and $E^+(S)$, respectively.

2 Negative Homogeneous Signed Graph

Theorem 1 *The s-dominating set of a negative homogeneous signed graph of a graph G will be $V(G)$ whenever $s = 0$.*

Proof Consider a negative homogeneous signed graph S of the graph $G = (V, E)$. Let $X \subseteq V(G)$ be a 0 dominating set. Then, there should exist 0 negative edges between X and $V(G)$. If possible assume that $X \neq V(G)$. Then, there will exist at least one negative edge between X and $V(G) - X$. Hence, $V(G)$ is the only 0-dominating set. If possible assume that $V(G)$ is a s-dominating set whenever $s \geq 1$. Then, there should exist s negative edges between $V(G)$ and its complement which is not the case.

Theorem 2 *The 1-dominating set of a negative homogeneous signed path P_n only dominates its pendant vertex.*

Proof Consider a negative homogeneous signed path S of P_n with n vertices. Then, there will exist $n - 1$ negative edges in S. Let $X \subseteq V(G)$ be a 1 dominating set of S. Then, there should exist only one negative edge between X and $V(P_n) - X$. If possible assume that $V(P_n) - X \neq \{v\}$ where v is the pendant vertex of P_n. Then, there will exist another vertex u of P_n or $V(P_n) - X$ does not contain any pendant vertices. In the first case, if $\{u, v\} \subseteq V(P_n) - X$ then, there will exist at least two negative edges between X and its complement since $\deg(v) = 1$ and $\deg(u) \geq 1$. In the second case, let w be a vertex other that pendant vertex and $\{w\} \subseteq V(P_n) - X$. Then, since $\deg(w) = 2$ there will exist at least 2 negative edges X and its complement. In both the situation X is not a 1-dominating set. Hence, $X = V(P_{n-1})$ (Fig. 1).

Remark 1 The 2, 3, 4, ..., $n - 1$ domination of a negative homogeneous signed path will contain a negative homogeneous path of length $n - 2, n - 3, n - 4, \ldots, 1$, respectively.

Theorem 3 *The negative homogeneous signed cycle C_n does not have a 1-dominating set.*

Proof Let C_n be a cycle with n vertices and S be its negative homogeneous signed graph. Clearly, $V(C_n)$ is not a 1-dominating set. Let X be a 1-dominating set such that its cardinality will be $n - 1$. Then, $|V(G) - X| = 1$ and there will exist 2 negative edges between X and $V(G) - X$ which is a contradiction. Similarly, the other cases can be explained when the cardinality of X changes.

This result leads to the following corollary.

Corollary 1 *The negative homogeneous graph without a pendant vertex does not have a 1-dominating set.*

Theorem 4 *A 1-dominating set for a negative homogeneous signed graph of a graph exist whenever there exist a pendant vertex for the graph.*

Proof If possible assume that there exists a graph G which does not have a pendant vertex whose negative homogeneous signed graph S has a 1-dominating set X. There exists only one negative edge between X and $V(G) - X$., i.e., there is a vertex v

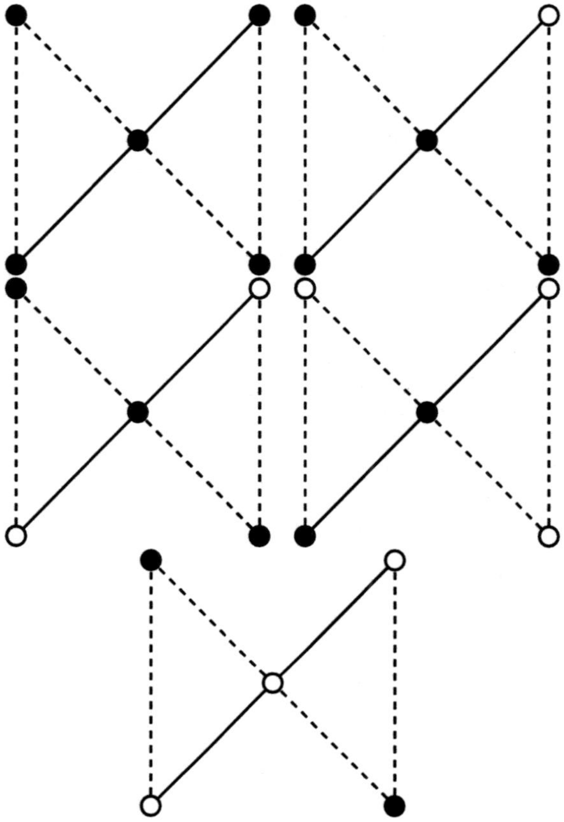

Fig. 1 The diagram clearly depicts 0, 1, 2, 3 and 4 domination. In the illustration, the colored vertices of the graph represent the dominating vertices

in X and u in $V(G) - X$ such that the edge uv is negative. Since G does not have a pendant vertex, $d(u) > 1$ and hence, u is adjacent to $N(u)$ vertices which has negative edges incident to them. Either $N(u)$ should be in X or $V(G) - X$. In the former case, there will be $1 + N(u)$ negative edges between X and $V(G) - X$. In the latter case, at least one of the $N(u)$ vertices will not be dominated since $\delta(G) \geq 2$ which is also a contradiction.

Theorem 5 *The minimum value of s other than 0 for which a s-dominating set exists for a negative homogeneous signed graph is* $s = \delta(G)$.

Proof Let G be a graph and u be the vertex with minimum degree $\delta(G)$. Then, $\deg(v) \geq \delta(G), \forall v \in G$. Let S be a negative homogeneous signed graph of G. If possible assume that S has a $\delta(G) - 1$-dominating set X such that $\delta(G) - 1 \neq 0$. Then, there should exist only $\delta(G) - 1$ negative edges between X and $V(G) - X$. Clearly, $V(G)$ is not a $\delta(G) - 1$-dominating set. Let $v \in X$ then, there exists $\deg(v)$ negative edges between X and $V(G) - X$ depending on the fact that X is a dominating set. i.e., $|X| \geq \deg(v) \geq \delta(G)$. Thus, there exists more that $\delta(G) - 1$ negative edges between X and $V(G) - X$.

3 Positive Homogeneous Signed Graph

Theorem 6 *For a positive homogeneous signed graph S of G, a s-dominating set exists whenever $s = 0$.*

Proof Let G be a graph and S be its positive homogeneous signed graph. If possible assume that there exists a dominating set X which is not a 0-dominating set. Then, there should be at least 1 negative edge between X and $V(G) - X$ which is not possible since S is positive homogeneous. Hence, any dominating set of G is a 0-dominating set of S. Whenever $s > 0$ there should exist s negative edges between the dominating set and its cardinality, which is not possible since the graph is positive homogeneous.

In the coming section, we will be discussing about $0, 1, 2, \ldots$ domination number of some of standard graphs. Since every s-dominating set does not exist for both negative and positive homogeneous signed graphs, we will discussing mostly the heterogeneous cases. For any graph, there exists a unique positive and negative homogeneous signed graph. But, there exists numerous heterogeneous signed graphs for the graph. To characterize the heterogeneous signed graph, we focus on a single type of signed graphs such as sum signed graphs introduced in [17, 18]. This reduces the number of heterogeneous signed graphs to be considered by an adequate amount.

4 The s-Domination in Sum Signed Graphs

The idea of labeling the vertices of signed graphs was introduced by Zaslavasky [23]. Thenceforth, different types of labeling in signed graph has originated. One such labeling in signed graphs is sum signed labeling.

Definition 2 ([17, 22]) A sum signed graph is a trio, $S = (G, f, \sigma)$ where G is a graph, $f : V(G) \longrightarrow \{1, 2, \ldots, \mid V(G) \mid\}$ is a bijective function and $\sigma : E(G) \longrightarrow \{+, -\}$ is a mapping such that $\sigma(uv) = +$, whenever $f(u) + f(v) \leq n$ and $\sigma(uv) = -$, whenever $f(u) + f(v) > n$.

Theorem 7 *For every non-trivial σ_0-dominating set X of sum signed graph of G, there exists at least one vertex u of G such that $N[u] \subseteq X$.*

Proof Let S be a sum signed graph of underlying graph G with n vertices and u be the vertex with the maximum label. Clearly, all the edges incident to u are negative. If possible assume that S contains a σ_0-dominating set such that $N[v] \nsubseteq X \; \forall v \in G$. If $u \in X$ then, the edges between u and $N[u]$ should be positive which is not the case. If $u \in V(G) - X$ then, there should exist at least one vertex w in X such that the edge uw is positive. Hence, u can not be placed in neither X nor $V(G) - X$ which makes our assumption wrong.

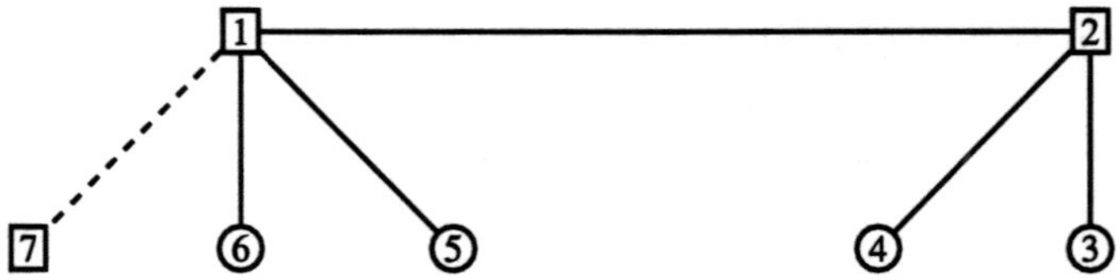

Fig. 2 A sum signed tree with 11 vertices with 0-dominating vertices. In the illustration, the squared vertices of the graph represent the dominating vertices

Corollary 2 *The vertex with the maximum label exists in every σ_0-dominating set of a sum signed graph S of the underlying graph G.*

Corollary 3 *If X is a minimal σ_0-dominating set of a sum signed graph of graph G, then, $V(G) - X$ is not a 0-dominating set.*

Corollary 4 *σ_0-dominating set of a sum signed graph is not an independent set.*

Theorem 8 *For any tree s satisfying sum signed labeling, $\gamma_0^-(T) = \gamma(T) + 1$.*

Proof Let s be a tree with n vertices and S be a sum signed graph of s such that the $|E^-(S)| = \sigma^-(T) = 1$. Since $|E^-(S)| = 1$, there exists only one negative edge associated with the maximum label n which will be incident to a pendant vertex u. If the maximum label n is not given a pendant vertex then, there will be at least 2 negative edges. Let, X be a dominating set of s having cardinality $\gamma(T)$. Since $|E^-(S)| = 1$, there exists only one negative edge associated with the maximum label n which will incident to a pendant vertex u. Let uv be the negative edge incident to the vertex u. If either $u \in X$ and $v \in V(T) - X$ or $v \in X$ and $u \in V(T) - X$ there will be a negative edge between X and $V(T) - X$ making X not a 0-dominating set. In the former case, X becomes a 0-dominating set when v is added to X, whereas in the later case, X becomes a 0-dominating set when u is added to X. In both scenarios the cardinality of 0-dominating set becomes $|X| + 1$, i.e., $\gamma(T) + 1$ (Fig. 2).

Theorem 9 *The domination number of a cycle with n vertices is* $\left\lceil \frac{n}{3} \right\rceil$.

The above statement is a known fact. Based on that, some observations are made which will be helpful in proving the next theorem.

Observation 10 *In the minimum dominating set of a cycle with $3m - 1$ or $3m - 2$ or $3m$ vertices, two of the vertices will either be joined by a path of length 1 or 2 or 3, respectively, where $m = 1, 2, 3, \ldots$*

Theorem 11 *For any cycle C_n,* $$\gamma_0^-(C_n) = \begin{cases} \gamma(C_n) + 2, & \text{if } n \text{ is a multiple of } 3, \\ \gamma(C_n) + 1, & \text{otherwise.} \end{cases}$$

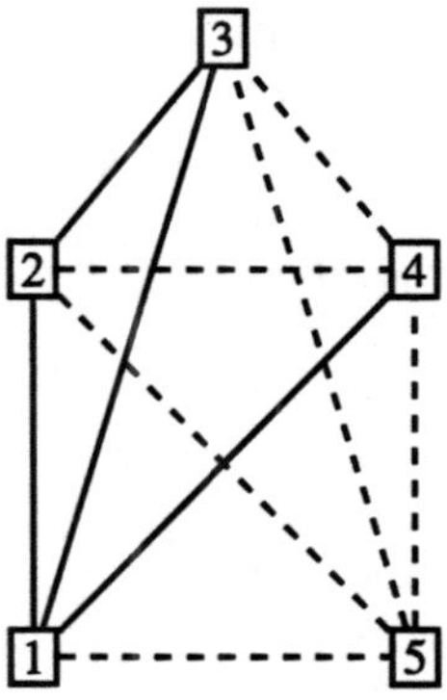

Fig. 3 In the illustration, the squared vertices of K_5 represent the 0-dominating vertices

Proof Consider a cycle C_n with n vertices and X be a dominating set of C_n with cardinality $\gamma(C_n)$. Let u and v be two vertices of X. By the above observation, when n is a multiple of 3 there exists two other vertices w and $z \in V(C_n) - X$ between u and v. Let S be a sum signed graph of C_n with minimum number of negative edges and let the maximum label n be assigned to vertex u. Then, the 2 edges incident to u will be negative, i.e., uw and ux where x is a vertex adjacent to u. The set X will become a minimal 0-dominating set when $\{w, x\}$ is added to X making the cardinality as $\gamma(C_n) + 2$.

In the other case when n is not a multiple of 3 then, either u and v are adjacent to each other or there exists a vertex $w \in V(C_n) - X$ between u and v. In the former case, let S be a sum signed with minimum number of negative edges and let the maximum label n be assigned to the vertex u. Then, the 2 edges uv and uz incident to u will be negative where z is a vertex adjacent to u. The set X will become a minimal 0-dominating set when $\{z\}$ is added to X making the cardinality as $\gamma(C_n) + 1$. In the later case, let S be a sum signed with minimum number of negative edges and let the maximum label n be assigned to the vertex w. Then, the edges uw and vw will be negative. Hence, the set X will become a minimal 0-dominating set when $\{w\}$ is added to X making the cardinality as $\gamma(C_n) + 1$.

Theorem 12 *For any complete graph K_n, $\gamma_0^-(K_n) = n$.*

Proof Let K_n be a complete graph with n vertices. Since complete graph can only satisfy one sum signed labeling, let S be a sum signed graph in which a vertex v_i receives the label i. Let X be a 0-dominating set. Then, clearly $|X| \leq n$. If possible assume that $|X| < n$, then there exists at least one vertex $u \in V(K_n) - X$. Since every vertex is adjacent to every other in K_n, at least one of the edges between X and $V(K_n) - X$ will be negative making X not a 0-dominating set. Hence, $|X| = n = \gamma_0^-(K_n)$ (Fig. 3).

Theorem 13 *For any connected graph G satisfying sum signed labeling, $\gamma(G) \leq \gamma_0^-(G) \leq \gamma(G) + \delta(G) + 1$ where $\delta(G)$ is the minimum degree of the graph G.*

Proof Let G be a connected graph with n vertices and u be the vertex with minimum degree $\delta(G)$. Let $\gamma(G)$ be the domination number of G. Let S be a sum signed graph of G with the minimum number of negative edges such that the $\gamma(G)$ vertices which dominates G also 0-dominates G. Then, u receives the maximum label n and all the edges incident to u will be negative. If possible assume that S contains a 0-dominating set X such that $N[v] \nsubseteq X \ \forall v \in G$. If $u \in X$ then, the edges between u and $N[u]$ should be positive which is not the case. If $u \in V(G) - X$ then, there should exist at least one vertex w in X such that the edge uw is positive. Hence, u can not be placed in neither X nor $V(G) - X$ which makes our assumption wrong. Hence, the $\gamma_0^-(G) \leq \gamma(G) + \delta(G) + 1$. The lower bound is clearly established by the definition.

Theorem 14 *For any star $K_{1,n}$, $\gamma_s^-(K_{1,n}) = \gamma(K_{1,n})$, $0 < s \leq n$.*

Proof Let $K_{1,n}$ be a tree with n pendant vertices and v be the central vertex. The domination number of $K_{1,n}$ is 1 and the minimum dominating set consists of u. The graph can admit labels from $\{1, 2, \ldots, n+1\}$ to satisfy sum signed labeling. Depending on the label of u the number of negative edges between the s-dominating set X and $V(K_{1,n} - X$ can be varied from 1 to n. Let u be labeled $i \in \{1, 2, \ldots, n+1\}$. Then, all the vertices incident to u having a label $> n + 2 - i$ will have a negative edge since, $i + n + 2 - i = n + 2 > n + 1$. Clearly, $\gamma_0^-(K_{1,n}) = \gamma(K_{1,n})$ by Theorem 8 (Fig. 4).

Theorem 15 *For any path P_n, $\gamma_s^-(P_n) = \gamma(P_n)$ whenever*

- $s = 1$ *and* $n = 2$
- $0 < s \leq 3$ *and* $n = 4$
- $0 < s \leq 2\left\lceil \frac{n}{3} \right\rceil$ *and* $n \neq 2, 4$.

Proof Let P_n be a path with n vertices where $n \neq 2, 4$. Let X be a minimum dominating set of P_n with cardinality $\gamma(P_n)$. Let the vertices of X be assigned labels from the set $\{1, 2, \ldots, n\}$ such that each sum signed labeling will have 1 negative edge, 2 negative edge and so on between X and $V(P_n) - X$. Since, the number of vertices in X is $\gamma(P_n)$ and each vertex dominates at most 2 vertices of $V(P_n) - X$. The maximum number of possible edge between X and $V(P_n) - X$ will be $2\gamma(P_n)$, i.e., there exists a maximum number $2\left\lceil \frac{n}{3} \right\rceil$ negative edges between X and $V(P_n) - X$. In the case of P_2 and P_4, the total number of edges is 1 and 3 which is different from the upper bound defined in other cases. Hence, $\gamma_s^-(P_n) = \gamma(P_n)$ whenever $s = 1$ and $s \leq 3$ for P_2 and P_4. In all cases, $\gamma_0^-(P_n) \neq \gamma(P_n)$ by Theorem 8.

Theorem 16 *For any cycle C_n, $\gamma_s^-(C_n) = \gamma(C_n)$ whenever $2 \leq s \leq 2\left\lceil \frac{n}{3} \right\rceil$.*

The proof is similar to as in the case of path.

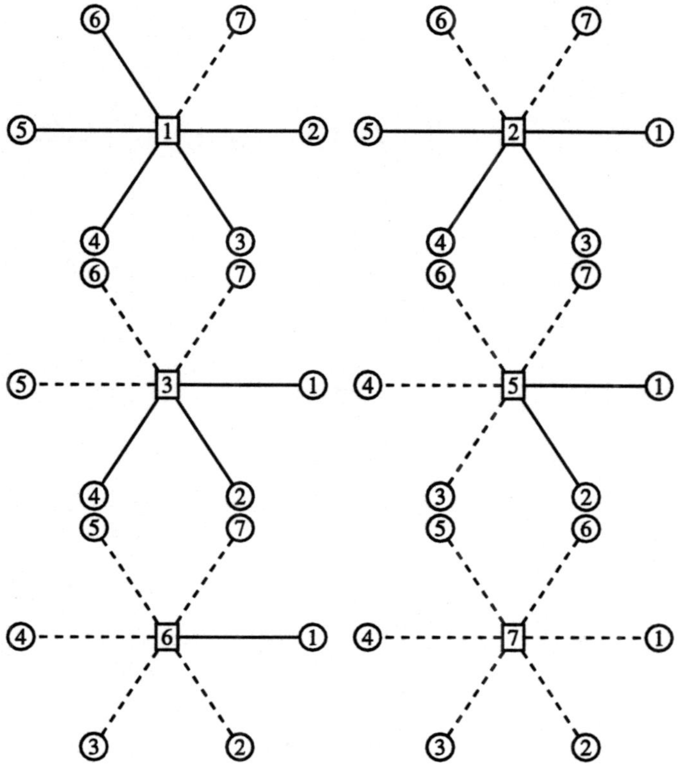

Fig. 4 The above illustration shows 0, 1, 2, 3, 4, 5 and 6 domination in the star graph, $K_{1,6}$. The squared vertices in the graph are the dominating vertices

Theorem 17 *For any complete graph* K_n, $\gamma_s^-(K_n) = \gamma(K_n)$, $1 \leq s \leq n-1$.

Proof Let K_n be a complete graph with n vertices. Since every vertex is adjacent to every other vertices in the graph it has a unique sum signed labeling. Let S be a sum signed graph of K_n whose vertices v_i is assigned the label i from the set $\{1, 2, \ldots, n\}$. Each vertex will have either i or $i-1$ negative edges incident to it. Then, the set $\{v_i\}$ will become an i or $i-1$ dominating set whenever $i < \left\lfloor \frac{n}{2} \right\rfloor$ and $i > \left\lfloor \frac{n}{2} \right\rfloor$, respectively. The vertex v_1 and v_n have 1 and $n-1$ negative edges incident to it, respectively, which is the minimum and maximum number of negative edges incident to a vertex of S. Hence, $\gamma_s^-(K_n) = \gamma(K_n)$ whenever $1 \leq s \leq n-1$.

Theorem 18 *The* 1 *and* 2*-dominating set of a complete graph is a singleton set.*

Proof Let K_n be a complete graph with n number of vertices. Since sum signed graph of K_n is unique, for simplicity let S be a sum signed labeling of K_n in which each vertex v_i receives i as its label where $i \leq n$. The vertex v_i is incident to i number

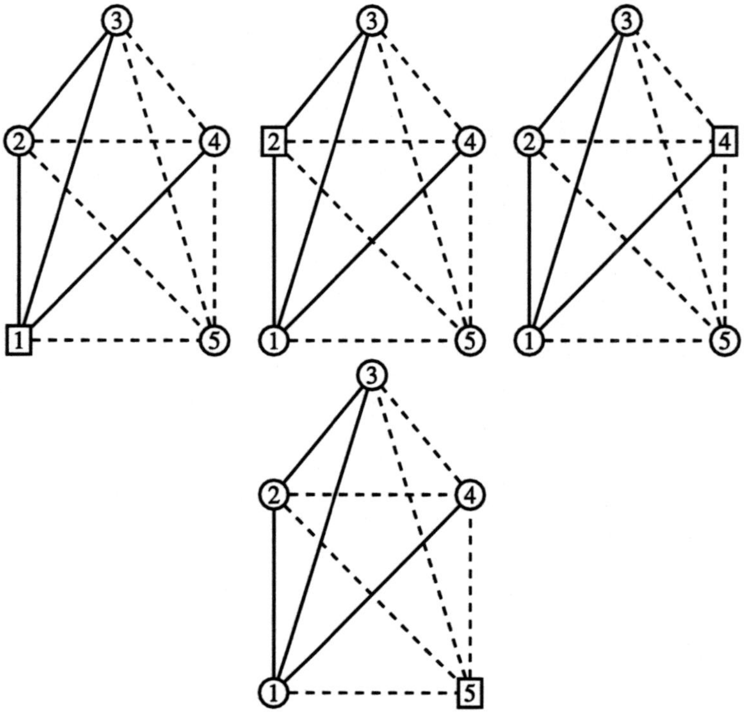

Fig. 5 The figure represents 1, 2, 3 and 4-domination in K_5. The squared vertices in the graph are the dominating vertices

of negative edges when $i < \left\lfloor \frac{n}{2} \right\rfloor$ and is incident to $i - 1$ number of negatives edges when $i > \left\lfloor \frac{n}{2} \right\rfloor$.

In every complete graph, there exists only one vertex v_1 which has exactly one negative edge incident to it. Hence, the *1*-dominating set of a complete graph is a singleton set.

Obviously, in every complete graph, the vertex v_2 forms a *2*-dominating set where $n > 3$. If possible assume that there exist another *2*-dominating set which consists of 2 vertices v_j and v_k. The vertex v_j and v_k is incident to j or $j - 1$ and k or $k - 1$ number of negative edges. When $j + k = 2$, $j = k = 1$ which is not possible. When $j - 1 + k - 1 = 2, j = k = 2, j = 3, k = 1$ or $j = 1, k = 3$ which are not possible. Similarly, the other two cases can be explained. Evidently, when $n = 3$, the vertex v_3 forms a *2*-dominating set. Hence, the *2*-dominating set for a complete graph is a singleton set (Fig. 5).

5 Bounds of *s*

Theorem 19 *For any graph G with n vertices, the value of s for* σ_t *domination can be at most* $\left\lfloor \frac{n}{2} \right\rfloor \left\lceil \frac{n}{2} \right\rceil$.

Proof To find the maximum value of s, a graph which has the maximum number of possible edges has to be taken into consideration. So, let us consider a complete graph K_n with n vertices. Since we are trying to maximize the number of negative edges between a dominating set and its complement, let S be the negative homogeneous signed graph of K_n. Let $v_1, v_2, \ldots, v_n$ be the n vertices of K_n. Each vertex of the graph is connected to every other hence, the set $\{v_1\}$ itself forms a dominating set X_1 and there exists $n-1$ negative edges between X_1 and $V(K_n) - X_1$ forming a $n-1$ dominating set. Let $X_2 = X_1 + \{v_2\}$ giving a total of $(n-2) + (n-2) = 2(n-2)$ negative edges between X_2 and $V(K_n) - X_2$. Proceeding like this, let $X_{\lfloor \frac{n}{2} \rfloor}$ be the dominating set with $\left\lfloor \frac{n}{2} \right\rfloor$ number of vertices then, there will exist $\left\lfloor \frac{n}{2} \right\rfloor \left(n - \left\lfloor \frac{n}{2} \right\rfloor\right)$ number of negative edges between $X_{\lfloor \frac{n}{2} \rfloor}$ and $V(K_n) - X_{\lfloor \frac{n}{2} \rfloor}$.

- *When n is odd:*

$$
\begin{aligned}
l\left\lfloor \frac{n}{2} \right\rfloor \left(n - \left\lfloor \frac{n}{2} \right\rfloor\right) &= \left\lfloor \frac{n}{2} \right\rfloor \left(n - \frac{n-1}{2}\right) \\
&= \left\lfloor \frac{n}{2} \right\rfloor \left(\frac{n+1}{2}\right) \\
&= \left\lfloor \frac{n}{2} \right\rfloor \left\lceil \frac{n}{2} \right\rceil
\end{aligned} \tag{1}
$$

- *When n is even:*

$$
\begin{aligned}
l\left\lfloor \frac{n}{2} \right\rfloor \left(n - \left\lfloor \frac{n}{2} \right\rfloor\right) &= \left\lfloor \frac{n}{2} \right\rfloor \left(n - \frac{n}{2}\right) \\
&= \left\lfloor \frac{n}{2} \right\rfloor \left(\frac{n}{2}\right) \\
&= \left\lfloor \frac{n}{2} \right\rfloor \left\lceil \frac{n}{2} \right\rceil
\end{aligned} \tag{2}
$$

If the same procedure is continued for $\lfloor \frac{n}{2} \rfloor + 1, \lfloor \frac{n}{2} \rfloor + 2, \ldots, n-1$, the number of negative edges between the dominating set and its cardinality will decrease or remain same as $\left\lfloor \frac{n}{2} \right\rfloor \left\lceil \frac{n}{2} \right\rceil$.

6 Conclusion

We have defined the concept of s-domination and have drawn some interesting characterization in cases of both positive and negative homogeneous signed graphs. To study s-domination in heterogeneous case, we have used the concept of sum signed graphs which gave some interesting results. Here, s-domination has been defined in consideration with negative edges between the dominating set and its complement. Some interesting results can be seen if we kept restrictions on the number of positive edges between the dominating set and its complement. This will be an interesting area worth exploring.

References

1. Acharya BD (2013) Domination and absorbance in signed graphs and digraphs I: foundations. J Comb Math Comb Comput 84:5–20
2. Acharya BD (1980) Spectral criterion for cycle balance in networks. J Graph Theory 4:1–11
3. Acharya M, Kureethara JV (2021) Parity labeling in signed graphs. J Prime Res Math 17(2):1–7
4. Acharya M, Kureethara JV, Zaslavsky T (2021) Characterizations of some parity signed graphs. Australas J Comb 81(1):89–100
5. Aniyan A, Naduvath S (2020) On degree product induced signed graphs of graphs. AIP Conf Proc 2261(1):030018
6. Ashraf PK, Germina KA (2016) Double domination in signed graphs. Cogent Math 3:1–9
7. Germina KA, Ashraf PK (2013) On open domination and domination in signed graphs. Int Math Forum 8:1863–1872
8. Chen JIZ, Lai K (2020) Machine learning based energy management at Internet of Things network nodes. J Trends Comput Sci Smart Technol 2020(3):127–133
9. Hameed SK, Shijin TV, Soorya P, Germina KA, Zaslavsky T (2021) Signed distance in signed graphs. Linear Algebra Appl 608:236–247
10. Harary F (1957) Structural duality. Behav Sci 2(4):255–265
11. Haynes TW, Hedetniemi ST, Slater PJ (1998) Fundamentals of domination in graphs. Marcel Dekker, New York
12. Haynes TW, Hedetniemi ST, Slater PJ (1998) Domination in graphs: advanced topics. Marcel Dekker, New York
13. Joseph M, Shreyas SR (2020) Characterization of signed paths and cycles admitting minus dominating function. Commun Comb Optim 5(1):61–68
14. Mathias A, Sangeetha V, Acharya M (2020) Restrained domination in signed graphs. Acta Univ Sapientiae Math 12(1):155–163
15. Naduvath S, Kok J (2020) On certain topological indices of signed graphs. J Math Comput Sci 10(2):248–261
16. Ore O (1962) Theory of graphs. Am Math Soc 38:206–212
17. Ranjith AP, Kureethara JV (2020) Sum signed graphs—I. AIP Conf Proc 2261:1–4
18. Ranjith, A. P., Kureethara, J.V.: Note on Sum Signed Graphs (Submitted). (2021)
19. Roy RT, Hameed SK, Germina KA (2021) On two Laplacian matrices for skew gain graphs. Electron J Graph Theory Appl 9(1):125–135
20. Shakya S, Pulchowk LN (2020) The robust routing protocol with authentication for wireless adhoc networks. J ISMAC 2(2):83–95
21. Suma V (2020) Automatic spotting of sceptical activity with visualization using elastic cluster for network traffic in educational campus. J Ubiquitous Comput Commun Technol 2:88–97
22. West DB (1999) Introduction to graph theory. Prentice Hall of India Pvt Ltd
23. Zaslavsky T (1982) Signed graphs. Discrete Appl Math 4(1):47–74

Analysis of IoT and Communication Technologies to Develop Waste Management Service Framework for Smart City

Mohd Anjum, M. Sarosh Umar, and Sana Shahab

Abstract The recent advancements in internet and the Internet of Things (IoT) technologies have engendered the usability of various electronic devices, namely radio frequency identification (RFID) tags, different sensors, and other intelligent devices. The ubiquitous internet connectivity and significant enhancement in data transfer speed have also increased the utilization of geographic information system (GIS) and global positioning system (GPS). Devices with significant computational capabilities and those have incorporated to make the objects intelligent/smart have been used to monitor and collect information about the environment of the city. All these devices have also been applied to build the service framework for smart city. Solid waste management (SWM) has identified as a criticial issue in prospect of enviromental pollution and smart city development. This study thoroughly analyses the literature and identifies various IoT and communication technologies to develop the novel framework for SWM services. The study also illustrates these technologies and suggest a novel SWM framework. The SWM framework comprises of three components, namely smart truck, smart bin and work server to perform the real-time monitoring of waste level in the bin and routes optimization for efficient collection of waste. The smart bin design incorporates various sensors such as ultrasonic, temperature, humidity, loadcell, proximity, and accelerometer along with radio frequency identification (RFID) tag and wireless communication module such as global system for mobile communication (GSM) and wireless fidelity (WiFi). The smart truck is

---

M. Anjum (✉) · M. Sarosh Umar
Department of Computer Engineering, Aligarh Muslim University, Aligarh, India
e-mail: mohdanjum@zhcet.ac.in

M. Sarosh Umar
e-mail: saroshumar@zhcet.ac.in

S. Shahab
College of Business Administration, Princess Nourah Bint Abdulrahman University, Riyadh, Saudi Arabia
e-mail: sshahab@pnu.edu.sa

S. Shakya et al. (eds.), *Proceedings of Second International Conference on Sustainable Expert Systems*, Lecture Notes in Networks and Systems 351,
https://doi.org/10.1007/978-981-16-7657-4_56

equipped with RFID reader, global positioning system (GPS), and wireless communication modules. The web server comprises web server, city cloud, database management system along route optimization and decision-making programs. The framework based on above devices will provide the significant advantage to the SWM services.

Keywords SWM · Smart city · IoT · RFID · GIS · GPS · GSM · WiFi · Sensors

1 Introduction

A smart city is described as: "A smart city is a city well performing in a forward-looking way in the following fundamental components (i.e., Smart Economy, Smart Mobility, Smart Environment, Smart People, Smart Living, and Smart Governance), built on the 'smart' combination of endowments and activities of self-decisive, independent and aware citizens" [1]. According to above definition, smart city is an idea that has been generally used to portray new patterns and objectives pursued to develop urban areas. It aims to highlight the role of IoT technologies and enable significant usage of spatial, acquisition, and identification technologies to prompt the urban development based on collective human resource and technological infrastructure by optimizing existing and new resources [2]. It integrates distinct technologies for proper management and sustainable infrastructure of urban development such as smart power grid, transportation, energy utilization, smart governance, environmental monitoring, resource utilization, health care services, etc. It also intervenes various aspects of technologies for efficient use of physical assets to attain new and communicable environment for the SWM system [3].

One of the building pillars of smart city is Smart Environment which is directly related to environmental pollution. The principal measure to limit the environmental pollution in a smart city is the IoT enabled waste monitoring, collection, and disposal system. Nowadays, SWM has emerged as one of the most attention seeking problems in smart city development due to tremendous growth in waste generation, population growth, migration to metropolitan, ineffective civic administration, deficiency in public cognizance and inadequate grant. Poor management practices lead to depreciation of resources and energy that can be extracted and recycled from a major segment of solid waste (SW). Several literatures have revealed that SW production rate in a smart city is immensely affected by socio-economic, demographic, and geographical factors. It has also been predicted that demographic factors, namely population, education, and employment ratio are more crucial than socio-economic factors [4]. According to the World Bank statistics 2012, worldwide SW production rate is approximately 1.3 billion tons per annum and estimated to be double in the next decade [5]. This huge amount triggers the system to displace from manual monitoring, collection, and transportation to advance technology-based efficient and effective solutions, landfill disposal to resource and energy retrieval solutions that also encourages recycling industry to participate in green environment initiatives. An

IoT based intelligent waste management and monitoring framework with service-oriented protocol seem to be the best alternative in the fast growing economies [6]. It is also designed to upgrade the functionalities of the prevailing system in favor of smart city vision. Moreover, IoT can be exploited to correlate the different operations and establish coordination among physical sensing devices, which have applications in many domains. Sensing devices with computing services and communication technologies have the power to accomplish the target of managing complex and dynamic system [6]. The well developed and appropriate infrastructure of IoT can facilitate parking problem, education system, disaster management, waste management services, etc., that serves as a backbone in smart city concept establishment to raise the living standard of the society.

IoT is a broad term and advances the conventional approach of standalone existence of the internet and physical objects by diffusion of sensing devices through the strong association of internet and communication links. Nowadays, IoT technologies have more pervasive and usable due to ubiquitous internet connectivity and significant enhancement in data transfer speed. Therefore, lot of researchers have exploited these advantages to build novel systems and applications for supporting smart city development [7]. In [8], pervasive computational model has been implemented to effectively manage health care wearable devices. The implemnted system has also emphasized on managing the mobility, apart from allocation of resources and distribution for proper data conveyance over these devices. The IoT technologies also handle the issues related to a wide diversification of internet infrastructure and web services by positioning of geographically distributed objects with inbuilt capabilities of identification, perception, acquisition and/or actuation to provide an advance set of web services and applications. With the growth of communicable physical devices, applications, and real-time services have emanated as a significant add-on in the field of networking. The vision of the IoT will be facilitated by creating inbuilt electronics into physical objects, building smart/intelligent objects which flawlessly coordinate with communication infrastructure [9]. This vision execution requires an architectural reference model that provides the ability to IoT resources to perform operations in coordination with every application domain based on open protocol and key enabling services. Consequently, the main focus of IoT technologies is to incorporate the worldwide interconnection of wireless sensor network from the existing Internet to IoT [3]. IoT technologies have also been merged with other computational technologies namely, cloud management, fog computing, big data, etc., to develop more effective system with large number of functionalities for smart city services [10, 11].

The main objective of this study is to thoroughly examine the research related to SWM and determine various IoT and communication technologies suitable for developing the novel SWM system. The study also draws the blueprint of a novel SWM system and explains distinct technologies in detail with their use in system development. The actual incorporation of identified devices at massive level will significantly decrease the operating cost of waste collection process through exchange of information in real time among different components of SWM system namely, smart bin, smart truck, and work server.

The paper is structured into five sections. Section 1 discusses the about SWM system and related technologies. Section 2 thoroughly examines various studies and tabulates their objective, prototype, IoT, and communication used. This section also identified potential devices to develop novel SWM framework. Based on examine literature and identified devices, a blueprint of novel SWM system is displayed in Sect. 3. Sections 4 and 5 explain about various technologies used in system design, and conclusion is drawn in Sect. 6.

2 Literature Survey

Smart city services rely upon data gathered from large set of sensors that highlight themselves on integration with communication technologies, and are visualized to communicate with server in synchronization with GPS, RFID reader, and GSM [3, 12]. It also establishes real-time monitoring of key elements of system such as waste collection truck tracking and routing, bin identification, positioning as well as level detection of waste amount [3, 12]. Furthermore, an upgraded system can intelligently visualize and automate the entire collection and transportation task by merging static parameters with current operational status to achieve intelligent services.

A real-time decision support system has been implemented to dynamically optimize the route of the waste collection truck in a smart city. This system has significantly reduced the energy consumption and optimized the operational efficiency [13]. The system is integrated with a data sharing model to exchange data in real time among truck drivers, and incorporates surveillance camera to capture the images of areas that have waste management related problems [13]. Above implemented system does not involve the bin status monitoring so truck drivers can visit empty or semi empty bins, which may cause the energy demand and reduce the system throughput. A multiagent-based waste management architecture has been simulated to determine the performance of the system in various scenarios. The system has three agents, namely bin incorporated with ultrasonic sensor and GSM module, waste collection trucks and citizens. The bin status is transferred to the central database and status-full, bin locations are used as nodes to determine the optimize path [14]. Misra et al. and Khan et al. have built a smart bin which comprises the ultrasonic sensor to automatically determine the garbage level inside the bin and gas sensor to sense the hazardous gases. The sensor data is transferred to the central web server along with cloud and utilized by the concern authority to take immediate action related to physical condition of the bin [15, 16].

Additionally, Sankeerth et al. have also developed similar type of smart bin without gas sensor [17]. A multi-tiered architecture for smart city development has been proposed and applied to the SWM services. This architecture has remarkably involved IoT and communication technologies to manage various services in smart city [18, 19]. Additionally, it has integrated SWM system with other services such surveillance, transportation, infrastructure and provided the design protocols of each layer [18, 19]. A novel approach, artificial intelligent of things, has been introduced for real-time monitoring of the trash bin. This approach implements the fuzzy expert system to determine the appropriate locations for trash bin installation and real time bin status data is utilized as an input in optimization algorithm to obtain optimal path for collecting vehicles [20]. Bin status data has also been used to train the machine learning algorithm which predicts the probabilities of different waste levels in bin [7, 21]. Additionally, graph theory and machine learning algorithm have been applied on same data to compute the shortest collection paths [7]. Ali et al. and one other have built hardware of smart bin using sensors, used GSM module to transfer data from bin to web server and GPS module to track the location of waste collecting vehicles [2, 22]. The location data is utilized to compute the total distance and sensors data is employed to estimate the forecasting model parameters [2, 23]. Baldo et al. have proposed the multi-layer infrastructure for smart city waste management using long range wide area network (LoRaWAN). RFID is incorporated to uniquely identify the bin along with various sensors. This system is exploited the data transferring capabilities of LoRaWAN and integrated with city surveillance which is used to detect fire and drop of garbage in proximity of bin [24]. Table 1 displays the various IoT, and communication technologies components employed in system development to accomplish the various tasks.

Based on the examined studies and research, Table 2 illustrates the spatial, IoT and communication devices that can be potentially incorporated in the SWM framework development for smart city. Applications of various devices involved in SWM framework are summarized in Table 3.

3 Novel SWM System Framework Design

The novel SWM system framework comprises three tier architectures namely, lower tier, middle tier, and upper tier. The main component of lower tier is sensing module (smart bin) which incorporates various sensors to measure waste level, lid condition, bin position, bin identity and environmental parameters temperature, humidity, hazardous gas. Middle tier is also called gateway which the GSM/GPRS network to transfer data from smart bin to cloud and work server. Upper tier is the central system of the SWM service framework which consists of two components namely, smart truck, and work server. Figure 1 displays the blueprint of novel SWM system.

3.1 Smart Bin

Smart bin has two tier architectures, microcontroller at the core and sensors, namely ultrasonic, proximity, load cell, accelerometer, temperature, and humidity, wireless communication module namely GSM module, GPS, RFID tag, powered with solar energy/battery at the periphery as shown in lower tier of Fig. 1. A programmable microcontroller process detective measures of the bin to periodically communicate

Table 1 Demonstration of key IoT and communication technologies devices involved in smart waste management

Refernce	Objective	Prototype	IoT and ICT components
[24]	Waste management infrastructure based on multi-layer LoRaWAN	Physical	Microcontroller, ultrasonic, temperature, tilt sensor, LoRaWAN module, RFID, and video surveillance unit Edge computing, image processing, web application
[23]	Smart bin monitoring and waste collection	Physical	Microcontroller, ultrasonic sensor, GSM module Web application, bin monitoring and collection algorithm
[2]	Smart bin building, waste collection, monitoring and prediction	Physical	Microcontroller, GSM module, ultrasonic, flame, temperature, humidity, and weight sensor MySQL, web application, distance calculation model, waste forecasting algorithm
[7]	Smart bin building, monitoring and collection	Physical	Microcontroller, ultrasonic sensor, LoRa module Machine learning algorithm for optimal route
[20]	Artificial intelligent of things based real time bin monitoring	Physical	Microcontroller, ultrasonic sensor, GPS, and GSM module Web application, cloud, fuzzy logic, route optimization algorithm
[18]	Waste collection and management infrastructure	Physical	Microcontroller, WiFi module, RFID tag and reader Web application, cloud database
[17]	Smart bin construction, and monitoring	Physical	Microcontroller, ultrasonic sensor, WiFi module Web application
[15]	Cloud based waste monitoring	Physical	Microcontroller, ultrasonic, gas sensor, WiFi module Web application, route optimization and cost analysis algorithm
[14]	Multiagent based smart waste collection and monitoring	Simulated	Microcontroller, GSM module, ultrasonic sensor Web interface, database, route optimization algorithm
[13]	Intelligent waste collection system	Simulated	Location: OpenStreetMap, routing: GraphHopper, web application

Table 2 Key devices identified to design SWM framework for smart city

Category	Identified technologies for SWM design
Spatial technolgies	Real time GIS and GPS
IoT and communication technologies	RFID, Barcode and QR Code, GSM, Bluetooth, WiFi, sesnsors: accelerometer, proximity, ultrasonic, load cell, bio sensor—gas sensor, environmental—temperature and humidity

Table 3 Key applications of identified IoT devices and technologies in SWM framework

Device name	Application in SWM framework
GIS	Bin allocation, collection truck route identification
GPS	Real-time location tracking of bin and truck
RFID, bar code, QR code	Bin identification
GSM, WiFi, bluetooth	Transfering the data from bin to web server
Ultrasonic, infrared sensor	Detect the waste fill level inside the bin
Loadcell sensor	Measure the weight of the waste in the bin
Accelerometer, proximity	Measure the bin lid state and orientation
Bio sensor, gas sensor	Detect the hazardous chemical and gas to prevent the scattering into the environment
Temperature	Detect the bin temperature to protect from fire
Humidity	Detect the humidity inside the bin to control the decomposition of biodegradable material

with central server enabling remote retrieval of data. An accelerometer is an electromechanical dynamic sensor capable of measuring force in one-, two- or three-dimensional space. Generally, it remains in idle mode and only responds with displacement of lid. Core functioning of the proximity sensor is to detect nearby objects and intimate the lid status to the microcontroller. Remaining sensors start transmitting status signals to microcontroller as soon as the lid is opened.

An ultrasonic sensor is a distance detector that record time lag between sending and bouncing back of an acoustic wave. Its basic function is to detect the level of waste in the bin and perceive the fullness of the bin. Another aspect to measure the bin overflow is achieved with a load cell sensor. It efficiently measures the weight by transforming the output into a voltage signal. The environmental sensor is necessary to provide ambient condition inside the bin for suitable functioning of other sensory activities. Along with the above features, bin encompasses GPS for describing its position and GSM module for interacting with an external network. Self-describing property of bin is implemented through RFID, and microcontroller is programmed such that bin responds intelligently on any activity of peripherals.

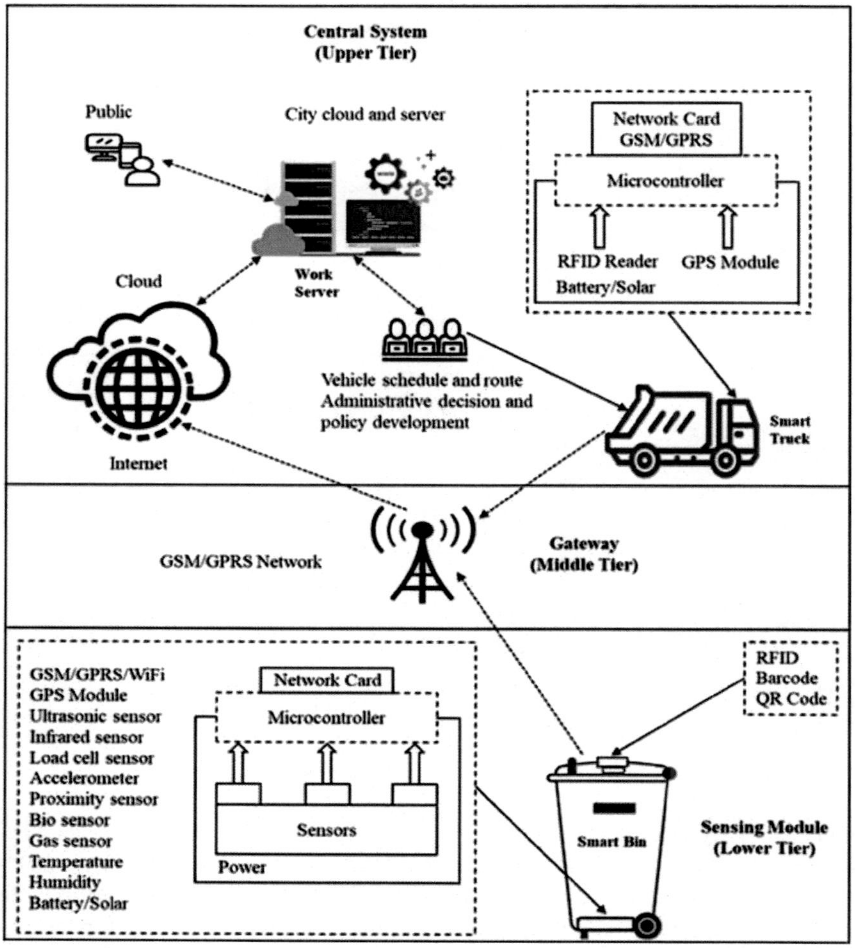

Fig. 1 Three tier architectures of novel SWM framework

3.2 *Smart Truck*

Nowadays, waste collecting trucks must be smarter with the incorporation of modern electronics and communication technologies for collecting and communicating essential data related to a daily schedule, driver personal information and bins visited for more productivity and transparency of fleet operations. Additionally, the everyday list of picking nodes is assigned to each truck related to its specific route. The truck with onboard microcontroller interfacing with GPS module, RFID reader, and GSM module to impart precise performance declaration and real-time communication as shown in upper tier of Fig. 1. RFID reader reads the bin unique id before unloading the waste and communicates bin unloading information along with bin id to work server. GPS module traces the path, followed by collecting vehicle and transmit over the network through GPS module.

3.3 Work Server

Work server hosts the waste management web application, route optimization program, and database management system. It contains continuously running multi-threaded programs that establish communication to handle the requests from the vehicles and bins. Real-time data is updated related to the bin's status, current position, and volume of waste collect by trucks. Database server program manages the database, which is used as input for the optimization program to generate daily schedules and routes for each truck. Web server program responses the requests from clients to access data related to the waste management system. It also maintains the dynamic web interface visualizing current bin status, route map, and citizen communication with system and complaint facility. Big data analytics tools are applied on dynamic sensing data to make decision related to waste management and administrative services. The solid waste optimization and collection problem considers static input and dynamic input, to evaluate and handle these inputs server framework is presented in upper tier Fig. 1.

4 Spatial Technologies: Real Time GIS and GPS for SWM

4.1 Geographic Information System

GIS is a recent computerized framework that can be coordinated with other platforms to get optimal solutions for proficient and adequate SWM planning. It is an approach for capturing or recovering, manipulating, analysing, and presenting geographically referenced data. It is a valuable tool in the field of engineering sciences due to its capability of including demographic, ecological, and economic information in its database.

Spatial data constitutes geometric and topological data. To represent a geographic object in GIS, the spatial data provides two different possibilities to represent a phenomenon, namely vector or raster form. In vector design, geographical components are communicated as feature vectors, represented as polygons, points or lines [25]. Raster format includes matrix format organized into rows and column with each cell containing information [26].

Applications incorporate waste management, spots for allocation of new stores, transportation tasks, crime mapping, map creation, road networking, power management, geological science as well as in risk analysis of natural hazards and emergency planning. A GIS model is intended to demonstrate its application in different areas related to SWM that reduce workload and amplify the profit engendered and furthermore enhances serviceability. The municipality involved in SWM will be clear about the functions, and their role with the assistance of information system. Appropriate logistics management and technology planning are done using GIS analysis for different layers of the SWM issues for managing the system efficiently. Additionally, the model is intended for efficient collection path for SW and optimal number of bin allocation. One of the most prominent issues in SWM is the route allocation for collection of waste. It is undoubtedly an inexorably imperative component as far as effectiveness and productivity for any municipality is concerned. Route network management application in view of GIS diminishes buffering and overlapping which consequently reduces the number of conveyances required for providing service. Furthermore guarantee, every other associated constraint is met in speedy and simple manner.

Using GIS, several programming software are available, however, standout among those is ArcGIS Desktop, which can be customized according to planners. It is an exceptionally flexible and straightforward tool for geographic display, query, and analysis. It gives boundless freedom to customize the interface according to business needs. The GIS is potentially used in site selection problems, bin allocation and route optimization of SWM system to obtain optimal results.

4.2 *Global Positioning System*

GPS initially came into existence in the 1970s in military administration with more elevated amount of encryption and later endorsed globally for civilian access. GPS service types include precise positioning service for military and standard positioning service for civilian access. It is made up of constellation of 24 satellites in 6 orbital planes of the United States that orbit the earth twice a day. Nowadays, 31 satellites are revolving in orbital plane.

GPS receiver occasionally receives radio signals through satellites. Received signal from at least three satellites at once broadcast latitude and longitude position using triangular strategy. Signal received from at least four satellites figure out latitude, longitude, and altitude position. GPS working principle includes computation of average speed and direction of signals to analysed location. Two-dimensional fix information is acquired from three satellites, while for three-dimensional fix, four satellites are required. Therefore, GPS is a key innovation for precise position fix.

GPS when integrated with navigation system serves as satellite-based radio navigation system that neglects the limitations of individual system. Commonly used navigation software are Doppler radar navigation and omega radio navigation systems. The GPS architecture is conceived of three segments. One of the consequential segments is space segment constituted of satellite constellation. The essential function is to receive and store information. Next is GPS control segment composed of 6 ground stations for determining the working condition of satellites and receiver for coordinate computation mounted on object as a part of user segment.

It serves as a precious asset to civilian with an accuracy of 11 m down to millimeters. Network is accessible everywhere universally under any climatic condition. To dropout position inaccuracy, differential GPS is used as an extension of traditional GPS. The differential GPS has two recipient stations designated as base station and rover station situated within a couple of hundreds of kilometers having known and unknown location, respectively. Therefore, if signal traveled same portion of atmosphere, then same errors are encountered virtually.

It has wide applications in numerous fields of engineering and management. GPS enabled road traffic congestion, fleet tracking and vehicle tracking serves as root nodes of SWM [27–29]. Moreover, it enables the SWM system to be more efficient with real-time monitoring of bin and vehicle. It is significantly used in mining, disaster management, agriculture planning, fishery industry, and astrometry calculation. Additionally, GPS monitor displacement of high-rise buildings under natural calamity.

5 IoT for SWM

IoT unifies physical objects/devices under established existing internet infrastructure, and it provides control over things along with real time information about state of an object. The concept of IoT is to transform objects into rational, smart, programmable, and more communicable with human and other objects. Generally, it alludes the network interconnection of heterogeneous sensing devices, often possess immanent intelligence. IoT technologies are becoming more dominating in the field of SWM due to contactless identification, object to object communication and real-time interaction of human with object. Nowadays, SWM problem is a complex problem in regard with collection, disposal, and transportation which is handled with robust and interactive IoT technologies. These are becoming significant due to inbuilt capabilities of identification, perception, acquisition, and actuation.

5.1 RFID

Identification technology is a mode of collecting information about an object without any physical contact with reader using radio frequency signals for transmitting data between tag and reader to authenticate object uniquely. The auto identification substantially decreases the labor time involved in manual data entry to enhance data acquisition accuracy. It is the fastest growing wireless and powerless automatic identification technology that enables reader to read the stored data on tags and transmit over network to work server. It does not require line of sight and range vary from 20 to 100 feet. Moreover, it has a high level of security in comparison with barcode.

An RFID system comprises of data storing microchip connected with an antenna called as transponder or tag and transceiver enable chip to read data and transfer information to central server for global connectivity and data processing. It is often classified according to the operating power of tags entitled as active and passive. Active tags have an internal battery that arouses when receiving radio signal from reader to conserve battery life and send back information to reader. Such property inherited tags are called transponders. Beacons fall under above category but powered on at pre-set intervals. Passive RFID tag does not have their own energy source instead becomes active when receive signal from reader to send information stored in tag. Modern technology has launched tags without silicon integrated circuit. The most recognized concept of achieving data is time domain reflectometry in which chip less tag receives pulses from reader and observes echoes sent by the tag that is used for encoding data.

RFID is not environmental friendly due to its non-recyclable and non-biodegradable properties but serves as green technology on integration with communication and sensing techniques in different areas such as waste control process, monitor different elements of environment for smart use of energy system, recycling process, monitor the impact of climate change and other decision making programs [30]. Moreover, its extension with sensors and spatial technology presents various advantages in resource allocation, tracking and identification in waste management architecture. In SWM, billing process and waste collections are non-automated which requires human intervention, and thus, the system is often subjected to errors. The probability of risk increases in system if the collection task is not properly supervised or undertaken by drivers individually. It plays a key role in system automation and abolishes manual practices prevailing from centuries.

RFID is a generic term which is available commercially and has countless applications in various multidisciplinary fields such as inventory tracing, bin identification in solid waste management system, contactless payments, electronic registration and authorization, consumer interactive marketing, livestock management and ticketing, medication flow, and disaster management [31].

5.2 Bar Code

RFID and barcode automate data acquisition that is both scan and interpret stored data for tracing assets but with few contradictions. Barcode technology consists of theoretical concepts of code generation, optoelectronics, information and communication technology, recognition algorithms and printing technologies. It is light weight, inexpensive, accurate, universally accessible, and easy to operate than RFID.

Barcode coding ability works as tracking technology for products. It consists of various patterns in the form of lines and then evolved into different polygonal shape (two dimensional) such as rectangles and hexagons having different width and reflectivity for unique representation. It replaces the manual data entry with automatic data transmission capability.

Linear bar code (one dimensional) commonly known as universal product code (UPC) represents information with thick or thin lines having encoding of 6–12 numerical digits. Due to linear structure, it does not contain information in the perpendicular direction and has low information density with relatively large size, which are sometimes unbearable. Data holding ability is maximum of 30 alpha numeric characters excluding Chinese and special character. Usually used in supermarket checkout system. Two-dimensional barcode diminishes the limitation of UPC with high information density and large data volume, referred as quick response (QR) code. Additionally, facilitates large coding domain which has potential to encode variety of digital information such as voice, image, thumb impression, character string and initials. Applications of QR code provide convenient facility to users such as anti-forgery traceability, inventory management, wireless advertising, trading and marketing for pre-sale strategy, product information tracking, phone scanning bar code for authentication, mobile security [32]. Nowadays, it is also used for identification of user application by scanning smart phone bar code along a line of sight. It exhibits environment monitoring applications to sense change in environmental parameters such as temperature, humidity, pressure, etc. As environmental parameters change, it shifts to other state. Hence, it is termed as dynamic and sensitive barcode. One major drawback is its messy black and white graphics that sometimes limits visual communication.

5.3 GSM

Cellular telecom on frequency division multiple access technology was commercially launched in 1970s identifiable as first generation. Technology upgraded with time and code division access for multiple user's environment deployed in Finland, based on standards of GSM which has been adopted worldwide. Primary benefits include digitally encrypted conversations and have significantly more efficient spectrum. Additionally, introduced data services for cellular communication and used digital modulation for improved audio quality. Evoluted 2.5 generation attained much faster data rate with superimposition of GPRS on GSM carriers. Furthermore, existing circuit switched network has been superseded by packet switched network which significantly improves data service. The next phase is called enhanced data rate for GSM evolution delivered much faster service using same GPRS infrastructure [33]. GSM module is integrated in bin to transfer the bin status data to web server in real time, and it is also mounted on truck to send the real-time location of truck and information of collected bin to web server.

5.4 WiFi

Wireless fidelity is a local area network communication technology designed using IEEE 802.11 family of standards to provide high-speed data transfer rate of 54 Mbps within range of 100 feet indoor and 1000 feet outdoor from the access point for broadband connectivity in homes, offices, and public places. 802.11a and 802.11b are several specifications of IEEE 802.11 standards that work on frequency-hopping spread spectrum concerned primarily over physical and data link layer. Standard 802.11b is suitable for 2.4 GHz band range with 11 Mbps data rate transmission, while 802.11g serves as an extension of predecessor which manages to provide speed of 54 Mbps within same band range using direct sequence spread spectrum and orthogonal frequency division multiplexing technology, respectively. Moreover, standard 802.11g is significantly different from 802.11a with specification of 5 GHz band width, orthogonal frequency division multiplexing technology and data transmission flow up to 54 Mbps and appears to be more flexible with less interference issue [34]. The application of WiFi is transfer the data from bin and truck to web server in real time.

5.5 Sensors

Conventional approach involves manual acquisition involving plenty of time, which has been replaced with automatic acquisition of data due to its high accuracy, low-running cost, efficient, and minimized human resource. So, to accelerate the acquisition of data from surrounding utilization of sensor is best alternative. It is an electronic device which perceives and quantizes any physical input and converts into electrical signal which can be stored, transmit, process, and regulate other devices. These are used to measure any physical and chemical property of object such as temperature, pressure, weight, concentration of gas, amount of bio-chemical material or any other environmental activity, and output is displayed in human understandable format. It is mainly made up of two components namely receptor and transducer element. The receptor actively reacts to a measured physical quantity while transducer element transforms into signal for further processing.

Sensors have wide applications due to their advance inherent characteristics such as high sensitivity, low-power consumption and linearity between input and output with high resolution and play a vital role in automation and control applications. They can be categorized based on signals namely analog and digital sensors. Among all accelerometer, ultrasonic, load cell, bio sensor and environmental sensor are commonly used in SWM applications for waste monitoring [35]. A sensory system has been designed for estimating bin content and route optimization based on multi-objective framework based on static and dynamic data [36]. Additionally, proposed two sets of parameters for routing model such as risk parameter and replenishment level parameter and analysed parameters dependency with waste production. Real-time management of logistics operation is displayed with synchronization of RFID and weight sensor [37]. Bin level detection is monitored using capacitive sensor [38], infrared sensor to take decision of bin visiting sequence for minimizing the overall distance traveled.

6 Conclusion

The study thoroughly examines numerous research articles related to SWM system development. Based on these studies, the study identifies and demonstrates various IoT devices and communication technologies which are appropriate to develop the SWM framework for smart city. Additionally, the blueprint of a novel SWM system is also demonstrated. These devices will provide the smart way for bin monitoring and waste collection. The GPS tracking of waste collection truck reduces the traffic congestion and noise pollution in the city. GIS is utilized to allocate the smart bin location in the city and determine the path for trucks. RFID is used to identify the bin identity and transmit to web server if waste is collected from the bin to change the bin status. Sensors transmit the real-time status of bin data to web server and locations of bins with full status are used to determine the optimal path for waste collection. This data is also used to forecast the future waste generation and taking various administrative decisions. The incorporation of IoT and communication technologies improve the operational efficiency, reduces human effort and system running cost. Finally, implementation of smart SWM at massive level will contribute to maintain the clean environment in the smart city. Additionally, the study can provide the deeper understanding of devices involved in SWM system to other researchers.

References

1. Giffinger R, Pichler-Milanović N, Smart cities: ranking of European medium-sized cities. http://www.smartcities.eu
2. Ali T, Irfan M, Alwadie AS, Glowacz A (2020) IoT-based smart waste bin monitoring and municipal solid waste management system for smart cities. Arab J Sci Eng 45:10185–10198. https://doi.org/10.1007/s13369-020-04637-w
3. Sanchez L, Munoz L, Galache JA, Sotres P, Santana JR, Gutierrez V, Ramdhany R, Gluhak A, Krco S, Theodoridis E, Pfisterer D (2014) Smart santander: IoT experimentation over a smart city testbed. Comput Networks 61:217–238. https://doi.org/10.1016/j.bjp.2013.12.020
4. Thanh NP, Matsui Y, Fujiwara T (2010) Household solid waste generation and characteristic in a Mekong Delta city, Vietnam. J Environ Manage 91:2307–2321. https://doi.org/10.1016/j.jenvman.2010.06.016
5. The World Bank (2012) What a waste: a global review of solid waste management. Urban develoment series—knowledge papers no. 15, World Bank, Washington DC
6. Anagnostopoulos T, Zaslavsky A, Kolomvatsos K, Medvedev A, Amirian P, Morley J, Hadjieftymiades S (2017) Challenges and opportunities of waste management in IoT-enabled

smart cities: a survey. IEEE Trans Sustain Comput 2:275–289. https://doi.org/10.1109/TSUSC.2017.2691049
7. Anh Khoa T, Phuc CH, Lam PD, Nhu LMB, Trong NM, Phuong NTH, Dung NV, Tan YN, Nguyen HN, Duc DNM (2020) Waste management system using IoT-based machine learning in university. Wirel Commun Mob Comput 1–13. https://doi.org/10.1155/2020/6138637
8. Shakya DS (2020) Computational enhancements of wearable healthcare devices on pervasive computing system. J Ubiquitous Comput Commun Technol 2:98–108. https://doi.org/10.36548/jucct.2020.2.005
9. Miorandi D, Sicari S, Pellegrini DF, Chlamtac I (2012) Internet of things: vision, applications and research challenges. Ad Hoc Netw 10:1497–1516. https://doi.org/10.1016/j.adhoc.2012.02.016
10. Zhang C (2020) Design and application of fog computing and Internet of Things service platform for smart city. Futur Gener Comput Syst 112:630–640. https://doi.org/10.1016/j.future.2020.06.016
11. Mugunthan SR (2020) Decision tree based ınterference recognition for fog enabled IOT architecture. J Trends Comput Sci Smart Technol 2:15–25. https://doi.org/10.36548/jtcsst.2020.1.002
12. Angelidou M (2014) Smart city policies: a spatial approach. Cities 41:S3–S11. https://doi.org/10.1016/j.cities.2014.06.007
13. Medvedev A, Fedchenkov P, Zaslavsky A, Anagnostopoulos T, Khoruzhnikov S (2015) Waste management as an IoT-enabled service in smart cities. In: Internet of Things, smart spaces, and next generation networks and systems. Springer, Cham, pp 104–115. https://doi.org/10.1007/978-3-319-23126-6
14. Likotiko ED, Nyambo D, Mwangoka J (2017) Multı-agent based IoT smart waste monitoring and collection architecture. Int J Comput Sci Eng Inf Technol 7:1–14
15. Misra D, Das G, Chakrabortty T, Das D (2018) An IoT-based waste management system monitored by cloud. J Mater Cycles Waste Manag 20:1574–1582. https://doi.org/10.1007/s10163-018-0720-y
16. Khan AA, Sajib AA, Shetu F, Bari S, Zishan MSR, Shikder K (2021) Smart waste management system for Bangladesh. In: ICREST 2021—2nd International conference on robotics, electrical and signal processing techniques. IEEE Inc., pp 659–663. https://doi.org/10.1109/ICREST51555.2021.9331159
17. Sankeerth VP, Markandeya VS, Ranga ES, Bhavana V (2019) Smart waste management system using IoT. In: International conference on ınventive computation technologies. Springer, Berlin, pp 661–668. https://doi.org/10.1007/978-3-030-33846-6_71
18. Marques P, Manfroi D, Deitos E, Cegoni J, Castilhos R, Rochol J, Pignaton E, Kunst R (2019) An IoT-based smart cities infrastructure architecture applied to a waste management scenario. Ad Hoc Netw 87:200–208. https://doi.org/10.1016/j.adhoc.2018.12.009
19. Sohag MU, Podder AK (2020) Smart garbage management system for a sustainable urban life: an IoT based application. Internet Things 11:100255. https://doi.org/10.1016/j.iot.2020.100255
20. Bano A, Ud Din I, Al-Huqail AA (2020) AIoT-based smart bin for real-time monitoring and management of solid waste. Sci Program 2020. https://doi.org/10.1155/2020/6613263
21. Varudandi S, Mehta R, Mahetalia J, Parmar H, Samdani K (2021) A smart waste management and segregation system that uses ınternet of things, machine learning and android application. In: 2021 6th International conference for convergence in technology, I2CT 2021. Institute of Electrical and Electronics Engineers Inc. https://doi.org/10.1109/I2CT51068.2021.9418125
22. Pardini K, Rodrigues JJPC, Diallo O, Das AK, de Albuquerque VHC, Kozlov SA (2020) A smart waste management solution geared towards citizens. Sensors (Switzerland) 20. https://doi.org/10.3390/s20082380
23. Roshan R, Rishi OP (2021) Effective and efficient smart waste management system for the smart cities using ınternet of things (IoT): an Indian perspective. Rising Threat Expert Appl Solut 473–479. https://doi.org/10.1007/978-981-15-6014-9_54

24. Baldo D, Mecocci A, Parrino S, Peruzzi G, Pozzebon A (2021) A multi-layer lorawan infrastructure for smart waste management. Sensors 21. https://doi.org/10.3390/s21082600
25. Rikalovic A, Cosic I, Lazarevic D (2014) GIS based multi-criteria analysis for industrial site selection. In: 24th DAAAM International symposium on ıntelligent manufacturing & automation 2013, vol 69, pp 1054–1063. https://doi.org/10.1016/j.proeng.2014.03.090
26. Bishop ID, Escobar FJ, Karuppannan S, Suwarnarat K, Williamson IP, Yates PM, Yaqub HW (2000) Spatial data infrastructures for cities in developing countries: lessons from the Bangkok experience. Cities 17:85–96
27. Lei S, Li Z, Wu B, Wang H (2016) Research on multi-objective bus route planning model based on taxi GPS data. In: 2016 Interrnational conference on cyber-enabled distributed computing and knowledge discovery, pp 249–255. https://doi.org/10.1109/CyberC.2016.56
28. Abbott H, Powell D (1999) Land-vehicle navigation using GPS. Proc IEEE 87:145–162
29. Dat PH, Drieberg M, Cuong NC (2013) Development of vehicle tracking system using GPS and GSM modem. In: 2013 IEEE Conference on open systems ICOS 2013, pp 89–94. https://doi.org/10.1109/ICOS.2013.6735054
30. Duroc Y, Kaddour D (2012) RFID potential impacts and future evolution for green projects. Energy Procedia 18:91–98. https://doi.org/10.1016/j.egypro.2012.05.021
31. Arebey M, Hannan MA, Basri H, Begum RA, Abdullah H (2010) Solid waste monitoring system integration based on RFID, GPS and camera. In: 2010 International conference on ıntelligent and advanced systems ICIAS 2010, pp 1–5. https://doi.org/10.1109/ICIAS.2010.5716183
32. Gao JZ, Prakash L, Jagatesan R (2007) Understanding 2D-bar code technology and applications in M-commerce—design and implementation of a 2D barcode processing solution. In: Proceedings—International computer software and applications conference, vol 2, pp 49–56. https://doi.org/10.1109/COMPSAC.2007.229
33. Bettstetter C, Vogel HJ, Eberspacher J (1999) GSM phase 2+ general packet radio service GPRS: architecture, protocols, and air interface. IEEE Commun Surv 2:2–14
34. Yinan G, Shuguo Z, Dawei X (2012) Overview of Wi-Fi technology. In: 2nd International conference on computer application and system modeling, pp 1293–1296. https://doi.org/10.2991/iccasm.2012.330
35. Mamun AA, Hannan MA, Hussain A, Basri H (2013) Wireless sensor network prototype for solid waste bin monitoring with energy efficient sensing algorithm. In: Proceedings—16th IEEE International conference on computational science and engineering CSE 2013, pp 382–387. https://doi.org/10.1109/CSE.2013.65
36. Faccio M, Persona A, Zanin G (2011) Waste collection multi objective model with real time traceability data. Waste Manag 31:2391–2405. https://doi.org/10.1016/j.wasman.2011.07.005
37. Prasanna KR, Hemalatha M (2012) RFID GPS and GSM based logistics vehicle load balancing and tracking mechanism. Procedia Eng 30:726–729. https://doi.org/10.1016/j.proeng.2012.01.920
38. Reverter F, Gasulla M, Areny RP (2003) Capacitive level sensing for solid-waste collection. In: Proceedings—IEEE Sensors, pp 7–11

Procedural Impact of Negotiations and Its Variations on Utilization of Resources in Cloud Environment

M. N. V. Kiranbabu, Yugandhar Garapati, Velagapudi Sreenivas, and V. Murali Mohan

Abstract Monitoring Cloud Computing enabled devices for services produces heterogeneous types of tools which will fit for several applications and outputs the performance characteristics for that specified software. Cloud Computing attributes like elasticity, authentication and cloning within the framework exhibit the back-up storage space, service grabbing, writing technologies and harmful attacks. We undertake the problem of negotiation in between the stack holders of the SAAS resource utilization and constitue an optimized algorithm.

Keywords Cloud auditor · QoS economics · Audit trail · Provenance data · Provider efficiency approach · Performance assessment

1 Introduction

Integrating issues towards Cloud Computing: The lucid use of data centres for acquiring the information retrieval adopts some tools for predicting architectures scalability. The pool of scripting webs that was embodied with different processing components which connects the other consumer with web solutions (languages) and outputs the info retrieval is call web of thing WoT. The notations of indexing, integrating and querying take an essential role in information retrieval. There are a few scoped bounds like information retrieval based on time factor, data abstraction level, processing of storage space.

The idea of mediating amongst Consumer and Provider by the Cloud Service Broker hires the network and resources for the Consumer needs. Many service

M. N. V. Kiranbabu (✉) · V. Murali Mohan
Koneru Lakshmaiah Education Foundation, Vaddeswaram, Guntur, Andhra Pradesh, India
e-mail: mnvkiranbabu@kluniversity.in

Y. Garapati
GITAM Deemed to be University, Hyderabad, India

V. Sreenivas
Dhanekula Institute of Engineering & Technology, Vijayawada, Andhra Pradesh, India

S. Shakya et al. (eds.), *Proceedings of Second International Conference on Sustainable Expert Systems*, Lecture Notes in Networks and Systems 351,
https://doi.org/10.1007/978-981-16-7657-4_57

oriented architectures SOA, which permit virtualization in several computing processing paradigms outputs the information for provisioning of SaaS represents the credibility of Cloud Computing at high level. Variable, scalable, integrating multitenant model features are used in this concern [1].

Intelligent nodes with variant interconnection in a network give up the idea of IOT. The triangle role of guidelines namely storage area resources, communication of resources and computation of resources provide scalability for the demanded facilities. Virtualization of data provides the opting of resources in a network for utilization and their instant access of communication is an optimal solution. The integration of the IOT towards Cloud Computing should meet up with the necessary criteria in mining of data, sensing, software of new process and energy consumption usage [2].

The demand for data mining, data removal gives deployment of new technology called big data with five characteristics. A data piece should have the worthiness, size and heterogeneity in it. This data piece must have veracity and high speed. There are various hurdles in big data they are storage space, monitoring and managing. The different types of files must have an integral value in mapping, and minimizing algorithms was initiated to get the required output [3].

The various analytical studies explore fog computing which states that VM migration and remittance of resource output with high temperature in Cloud Computing are strongly related. This provides clarification and attain best alternatives in the standards of energy use, data centres scalability, learning resource equipment, power usage and migration to other resources.

Within the context of web sites the Cloud Computing pinned the issues of web browser, data deletion, IP issues and some scripting problems. Inside the context of programme models, Cloud Computing displays hypervisor issues, dos problems, cookie breakdown and hacking, etc., are considered in this research [4].

Cloud Computing have good features like affordable multi tenanting. Virtualization is one which intercepts Cloud Computing security issues. The propagation of disorders at VM migrations, timing of source utilizations, and theft in billing data of resources will be the frequent addressing things. Embedded jobs in Cloud Computing have many interrogations on the virtualization service availableness, integrity factor between Customer and Cloud Provider. The loosely deployed segments in SLA's come up in raising the problems, accountability and privacy level issues [5].

2 Review of Literature

Cloud SIM is a tool made by Cloud labs to observe the actions of VMM's and various cloud. The brand new functionalities appended to the Cloud SIM are monitoring the patterns of inner network and data centre in various processing levels which is a best optimal strategy for real-time programme execution was gained. Federated Cloud SIM was very good in including of new strategic algorithms and SLA's. Active Cloud SIM tool displays the pool of resources performance during doubt situations with dynamic changes. Teach Cloud is principally oriented for student's research with custom-made network topologies and permits SLA's. Flexible SIM assists with monitoring resource provisioning algorithms and good GUI is launched. Cloud Analyst observes data centres at different physical locations and performance is assessed by means of charts tables. Mr. Cloud SIM for map reduces and big data can be materialized with good reports for doing simulation.

Cloud service agreement negotiations is a non stopping process in many scenario's of cloud computing. The part of auditor plays a vital role in assessment of QoS. A reputation system involves helping the Cloud Consumers based on the previous feedbacks and experiences. The grading of rank allotment was modelled with basic assessment of provider, user conditions, reputation assessment system, assessments of brokers, and assessment of auditors in error calculation which output the ranking [6].

A four step process of selecting the Consumer agent for the accommodation of resources, evaluates the utility functions of price, timeslot, grading the values for the specification of Consumer and recommendation is an optimal answer from article for agent based strategy and approach of cloud in commerce. Services are brokered with negotiations in most of the cloud computing scenarios. The architecture exhibits the functionalities of requirement requests, analyzer requests, and history of broker with time analyzer. A decision making is possible with game theory, and a pay off function is calculated [7].

The process of negotiation is carried with some functional parameters, procedures and protocols between parties or multiple parties which includes, trust ability, reputations, pricing schemes, security, decision-making procedures and principles. The negotiation model is appended to the broker who takes the delegate authority role, gives the services to the consumer and provider. In the context of negotiation, many research contributions was exhibited given an advantageous idea over the auction and bargaining procedures.

We involve our broker based negotiation task in an improved manner by three parts, one is selection of QoS attributes for negotiation strategy, second one is categorization of negotiation with giving possibility for the participation of multiparty, i.e. number of service providers and number of consumers of same kind. Awarding of weights on the consumers preference of QoS attributes and Providers Provisioning QoS attributes. The third part is structuring of the above inputs in the mathematical model to deliver the improvement when compared to the earlier contributions [8].

3 Problem Formulation

Accumulating the design of Improved Negotiation Algorithm: In the deploying of Improved Negotiation algorithm, theoretical basis was established into three aspects Collection aspect, Categorization aspect and Structuring aspect [9].

Collection aspect: As defined in the modelling of the problem definition, the broker undertakes the collecting of consumer preferences of QoS attributes requests and Provider provisioning offers of QoS attributes for negotiation. As our proposed framework states the SLA aware SaaS provisioning which holds the concept of negotiation to drive the optimal utilization of resources and services delivery. To our problem definition for negotiation the consumer preferences of QoS attributes are taken as Availability and Response Time. Only the normalized values are taken for computations, i.e. they should lie between 0 and 1.

Categorization aspect: In this concept, the broker differentiates consumer preferences of requests and provider provision of offers carefully and award weights to those both parties' preferences and offers individually by making benefit for the participants constructively in the negotiation process. Only the normalized values are taken for computations, i.e. they should lie between 0 and 1. The participation urge of parties was computed in the form of negotiation rate factor. This negotiation rate factor dynamically varies with the number of participation parties in consumer side and number of participation parties in provider side [10].

Structuring aspect: Broker awards the weights on consumer, provides QoS preferences and offerings. Awarding of weights is subjective to the condition delivered in the modelling of the problem definition, which states consumer preference of QoS attributes that are beneficiary based must have a high value compared to the beneficiary based QoS attributes of provider. The awarding of weights on QoS attributes of cost based to the consumer should possess less value compared to the QoS attributes of cost based to the provider. Negotiation rate factor computations accept the number of participants in negotiation process and their compelling urge of same kind. With prefixed time slots and rounds, the negotiation process is carried in between consumer and provider with proposals and counter proposals. Broker calculates the negotiation score for both parties by viewing the proposals and counter proposals and exhibit the value of negotiation at an instance of time and round [11–13].

4 Methodology

Formulation of Improved Negotiation Algorithm

Input

$Q = \{Q_1, Q_2, \ldots, Q_n\}$ //list of (QoS) Quality of Service attributes, where Q_1 may be BD or CD

C = {cb, cc} //set of Service Consumer Expectations, where cb represents Consumer preference for Beneficiary driven (BD) attributes and cc represents Consumer Preference for Cost driven (CD) attributes

P = {pb, pc} //set of Potential Service Provider Provisions, where pb represents Provider provision for Beneficiary driven (BD) attributes and pc represents Provider provision for Cost driven (CD) attributes

CW = {cbw, ccw} //SaaS Broker assigned weights for Service Consumer

PW = {pbw, pcw} //SaaS Broker assigned weights for Potential Service Provider

E.g., cbw = 0.8, ccw = 0.2

The constraint the sum of weights is 1 for both Consumer as well as provider.

Output

Negotiation done When Cscore > Sscore

Process

Step 1: Service Consumer will give his preference values for each parameter

$$\text{(BD and CD) i.e.,}\ \ C = \{\text{cb}, \text{cc}\}$$

Step 2: Potential Service Provider mentions his provisions of each parameter

$$\text{(BD and CD) i.e.,}\ \ P = \{\text{pb}, \text{pc}\}$$

Step 3: Let "nrf" be the Negotiation rate factor. For Service Consumer Side

$$\text{csnrfb} = (\text{No. of Potential Service Providers}/\text{No. of Service Consumers}) + \text{cbw}$$
$$\text{csnrfc} = (\text{No. of Potential Service Providers}/\text{No. of Service Consumers}) + \text{ccw}$$

For Potential Service Provider Side

$$\text{psnrfb} = (\text{No. of Potential Service Providers}/\text{No. of Service Consumers}) + \text{pbw}$$
$$\text{psnrfc} = (\text{No. of Potential Service Providers}/\text{No. of Service Consumers}) + \text{pcw}$$

Step 4: Service Consumer Negotiation is calculated as

$$\text{BD} = \text{cb} + (t/5)\text{csnrfb}$$
$$\text{CD} = \text{cc} - (t/5)\text{csnrfc}$$

Note: Here, "t" represents time slots [1, 2, 3, 4, 5] and 5 = "r" represents rounds. Like this negotiation process continues until the target condition is reached. The pair of Service Consumer Negotiation values are updated and used in further steps.

Step 5: Potential Service Provider Negotiation is calculated as

$$\text{BD} = \text{pb} - (t/5)\text{psnrfb}$$
$$\text{CD} = \text{pc} - (t/5)\text{psnrfc}$$

Note: Here, "t" represents time slots [1, 2, 3, 4, 5] and 5 = "r" represents rounds. Like this, negotiation process continues until the target condition is reached. The pair of Potential Service Provider Negotiation values are updated and used in further steps.

Step 6: Broker Calculates the Service Consumer Negotiation Score as Cscore

$$\text{Cscore} = ((\text{cb} - \text{pb Updated Value})/((\text{cb} + \text{pb})/2)$$

Table 1 Service consumer preference values for BD and CD

Consumer expectations		
	Availability	Response time
Camax	0.5	Ccmax
Camin	0.99997	Ccmin

Note In the process of negotiation beneficiary driven attribute Camax = consumer availability maximum value. Beneficiary driven attribute Camin = consumer availability minimum value. Cost driven attribute Ccmax = consumer response time maximum value. Cost driven attribute Ccmin = consumer response time minimum value. Where Camax and Ccmin are subjective to the conditions

$$+ \left((\text{pc Updated Value} - \text{cc}) / ((\text{cc} + \text{pc})/2) \right)$$

Step 7: Broker Calculates the Potential Service Provider Negotiation Score as Sscore

$$\begin{aligned}\text{Sscore} = {} & ((\text{pb} - \text{cb Updated Value})/((\text{cb} + \text{pb})/2) \\ & + ((\text{cc Updated Value} - \text{cc})/((\text{cc} + \text{pc})/2)\end{aligned}$$

Step 8: Finally, the Broker compares the Step 6 and Step 7 series of sequential values and marks the pivot value when Cscore exceeds Sscore

At that point the status is updated as negotiable.

Note: The above algorithm is a pseudo code which helps in selection of service provider precisely and where the programming code for this algorithm is not included here.

Working Illustration of Improved Negotiation Algorithm (Tables 1, 2 and 3).

Table 4 address the calculated values of Negotiation Rate Factor from Consumer side and Provider side.

Table 5 addresses Negotiations of Consumer and Provider with rounds and time the first column specify the rounds = 5, second column specify time slot [1 to 5],

Table 2 Potential service provider preference values for BD and CD

Service provider provisions		
	Availability	Response time
Samax	0.99999	Scmax
Samin	0.5	Scmin

Note In the process of negotiation, beneficiary driven attribute Samax = provider availability maximum value. Beneficiary driven attribute Samin = provider availability minimum value. Cost driven attribute Scmax = provider response time maximum value. Cost driven attribute Scmin = provider response time minimum value. Where Samin and Scmax are subjective to the conditions

Table 3 SaaS Broker assigned weights for consumer and provider

Broker weights	
Availability	RT
0.8	0.2
0.2	0.8

Note In the process of negotiation, broker award weights for consumer BD and CD as cbw and ccw. Broker award weights for provider BD and CD as pbw and pcw. Here, cbw is high compared to pbw and ccw is low compared to pcw subjective to the conditions

Table 4 Negotiation rate factor for consumer and provider side

Negotiation rate factor		
Consumer side (nci)		
No. of service providers	1	
No. of consumer	1	
Service provider side (nsi)		
No. of consumer	1	
No. of service providers	1	
	Availability	RT
Consumer side	1.8	1.2
Service provider side	1.2	1.8

Table 5 Negotiations of consumer and provider with rounds and time

		Consumer negotiations		Service provider negotiations	
Rounds = r	Time = t	Availability (min)>	RT (max)<	Availability (max)<	RT (min)>
		0.99997	0.5	0.99999	0.5
5	1	1.055159186	0.355044067	0.85503407	0.444810814
	2	1.247339096	0.022022784	0.52201278	0.252630904
	3	1.646062979	0.519705487	0.01971549	0.146092979
	4	2.315272293	0.245376513	0.74536651	0.523116334
	5	3.315272293	0.754623487	0.25463349	0.476883666

third and fourth column represents consumer negotiations values with time t. Fifth and sixth columns represent Providers negotiation values with time t.

Tables 6 and 7 address the Consumer and Provider negotiations, where the broker calculates the negotiation scores and compare them (Table 8).

Table 6 Broker calculates the consumer negotiation score as "C" score

Broker consumer negotiation score		
Availability	Cost	C Score
0.144938832	0.110378373	0.255317204
0.477966775	0.494738192	0.972704967
0.980274119	0.707814041	1.68808816
0.254608579	0.046232668	0.300841247
0.74535142	0.046232668	0.791584089

Table 7 Broker calculates the provider negotiation score as "S" score

Broker service provider negotiation score		
Availability	Cost	S Score
0.05517029	0.289911865	0.345082155
0.24735404	0.955954431	1.203308474
0.6460859	0.039410973	0.685496874
1.3153086	0.509246973	1.824555573
2.3153286	0.509246973	2.824575573

Table 8 Condition evaluation table for "C" score and "S" score with time *t* and rounds

Rounds = r	Time = t	Condition evaluation
5	1	Not negotiable
	2	Not negotiable
	3	Negotiable
	4	Not negotiable
	5	Not negotiable

5 Results

See Figs. 1, 2, 3 and 4.

The above depicted graphs on results exibit the beneficiary and cost driven based normalized values which are very sensitive data of the service provider. The representation of coefficient of mean deviation from medium values graphs is clear that the selection of SaaS Provider algorithm showcased with minimum variation rate compared to the other distinctive algorithms.

The exibition of our selection of SaaS Provider algorithm attains a peculiar characteristic which drives the maximum for BD and minimum CD of the top 10 service providers list order. The existing algorithm of negotiation doesn't specify the part of counter offers and proposals where, as our improved negotiaon algorithm depicted with five rounds with time series of consumer and provider.

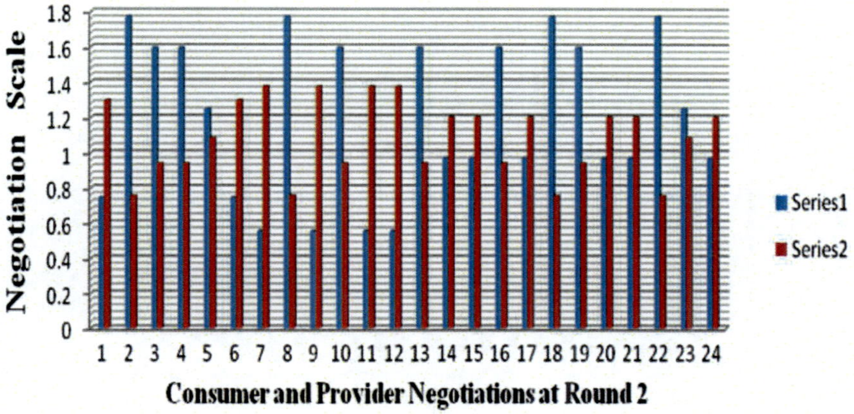

Fig. 1 Column graph represents consumer and provider negotiations at round 2. *X*-axis: consumer and provider negotiations *Y*-axis: negotiation scale series 1: consumer, series 2: provider

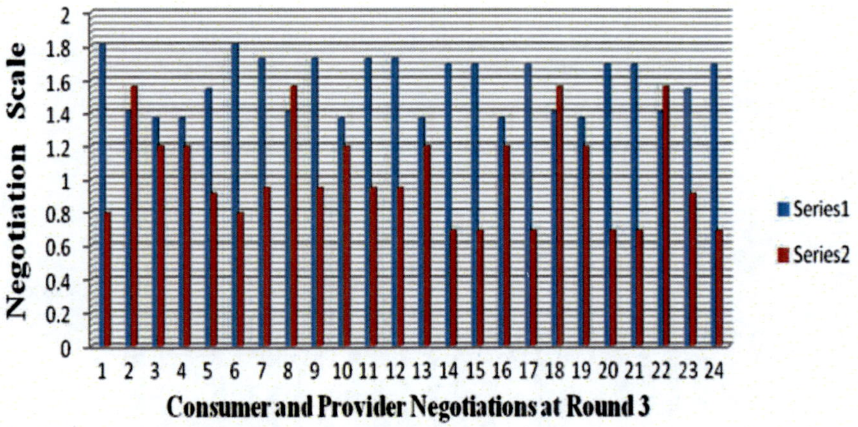

Fig. 2 Column graph represents consumer and provider negotiations at round 3. *X*-axis: consumer and provider negotiations, *Y*-axis: negotiation scale, series 1: consumer, series 2: provider

6 Conclusion

In this paper, we covered the concept of improved negotiation by taking the earlier research contributions as base step. As noticed in the modelling of the problem definition, negotiation process was carried with best outcomes. The process of negotiation was highlighted and compared to the other auction procedures. The application of mathematical model in delivering the negotiation process exhibit the flexibility to participants in the form of negotiation rate factor who are having the compelling

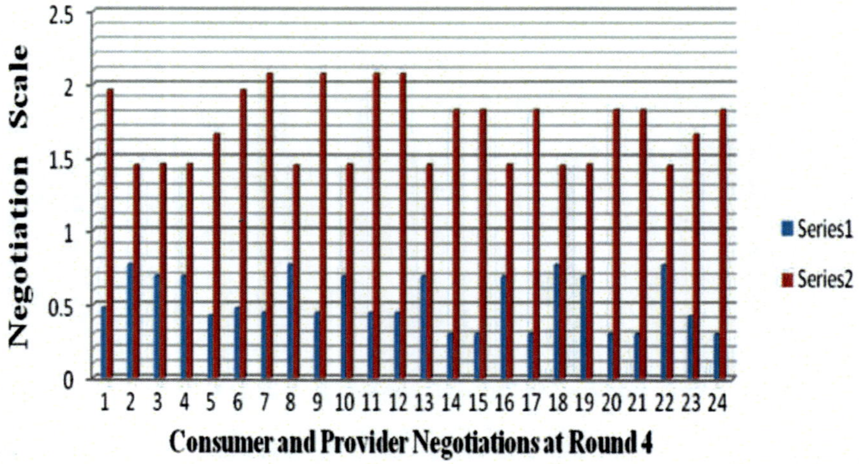

Fig. 3 Column graph represents consumer and provider negotiations at round 4. *X*-axis: consumer and provider negotiations, *Y*-axis: negotiation scale, series 1: consumer, series 2: provider

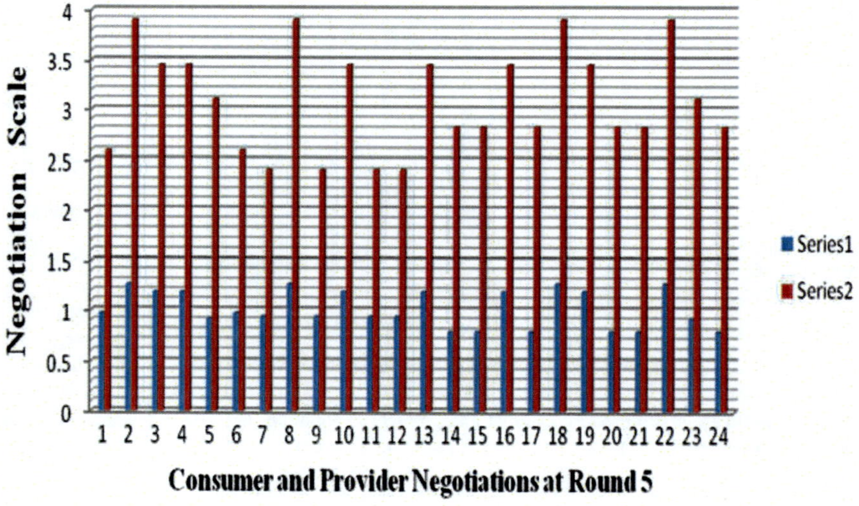

Fig. 4 Column graph represents consumer and provider negotiations at round 5. *X*-axis: consumer and provider negotiations, *Y*-axis: negotiation scale, series 1: consumer, series 2: provider

urge of same kind. An improved version of this negotiation process is that process of negotiation with time and rounds clearly stated.

References

1. Kiranbabu MNV, Satyanarayana KVV (2019) Inquisition the prospect of ranking cloud service provider using distinctive algorithms. Int J Innov Technol Exploring Eng 171–176
2. Kiranbabu MNV, Satyanarayana KVV (2017) A productive survey on resource usage and migration by client towards csp resource allotments in risk state of cloud computing based on SLA's. Int J Pure Appl Math 116(21):439–447
3. Bates A et al (2017) Transparent web service auditing via network provenance functions. In: Proceedings of the 26th international conference on world wide web
4. Meera G, Geethakumari G (2015) A provenance auditing framework for cloud computing systems. In: 2015 IEEE International conference on signal processing, informatics, communication and energy systems (SPICES). IEEE
5. Kiranbabu MNV, Satyanarayana KVV (2017) Appertaining "ABC analysis" on quality attributes for measly cost in utilization of CSP resources to clients accountability in cloud computing. Int J Pure Appl Math 115(8):389–394
6. Tsai W-T et al (2007) A new soa data-provenance framework. In: Eighth international symposium on autonomous decentralized systems (ISADS'07). IEEE
7. Haleema PK, Iyengar NChSN (2018) A mathematical model with flexible negotiation strategies for agent based negotiations in dynamic e-commerce environments. J King Saud Univ-Comput Inf Sci 30(4):547–560
8. Feng Y, Cai W (2007) Provenance provisioning in mobile agent-based distributed job workflow execution. In: International conference on computational science. Springer, Berlin, Heidelberg
9. Vadlamudi D, Thirupathi Rao K (2020) Provenance aware audit trail framework for saas providers efficiency assessment in cloud environment. J Crit Rev 7(12):1191–1196
10. Townend P, Groth P, Xu J (2005) A provenance-aware weighted fault tolerance scheme for service-based applications. In: Eighth IEEE international symposium on object-oriented real-time distributed computing (ISORC'05). IEEE
11. Lee I (2021) Pricing and profit management models for SaaS providers and IaaS providers. J Theor Appl Electron Commer Res 16(4):859–873
12. Kumar R, Hassan MF, Adnan MHM (2021) Intelligent negotiation agent architecture for SLA negotiation process in cloud computing. In: Proceedings of international conference on machine intelligence and data science applications. Springer, Singapore
13. Najjar A et al (2021) One-to-many negotiation QoE management mechanism for end-user satisfaction. IEEE Access 9:59231–59243

To Reduce the Water Wastage by Using Water Monitoring System in Cloud-Survey

R. Shobana and M. Nithya

Abstract Water plays an indispensable role in human lives as people cannot survive without it; moreover, water is very essential in agriculture for producing quality grains. Henceforth, it is highly required to develop a water monitoring system (WMS) to avoid water waste. The water level in the tank is shown by WMS using LED sensors. LED sensors are installed at three levels: LED 1 in the lower level of the water tank, LED 2 in the middle of the water tank, and LED 3 in the upper level of the water tank. Several devices, including mobile phones, laptop computers, and other devices save all sensor data in the cloud. This article mainly explores the efficient ways to decrease water waste in the cloud by utilizing WMA.

Keywords Cloud · LED · Sensor · Water · Water monitoring architecture · Water monitoring system

1 Introduction

The Earth's water level is decreasing day by day, and the scarcity of water is increasing at an unprecedented rate. People must use an efficient water monitoring system to minimize the wastage of water since water plays a highly vital in human lives. Figure 1 depicts the conceptual flow.

R. Shobana (✉)
Department of CSE, Vinayaka Missions Research Foundations, Salem, India

M. Nithya
Department of CSE, VMKV Engineering College, Vinayaka Missions Research Foundations, Salem, India

S. Shakya et al. (eds.), *Proceedings of Second International Conference on Sustainable Expert Systems*, Lecture Notes in Networks and Systems 351,
https://doi.org/10.1007/978-981-16-7657-4_58

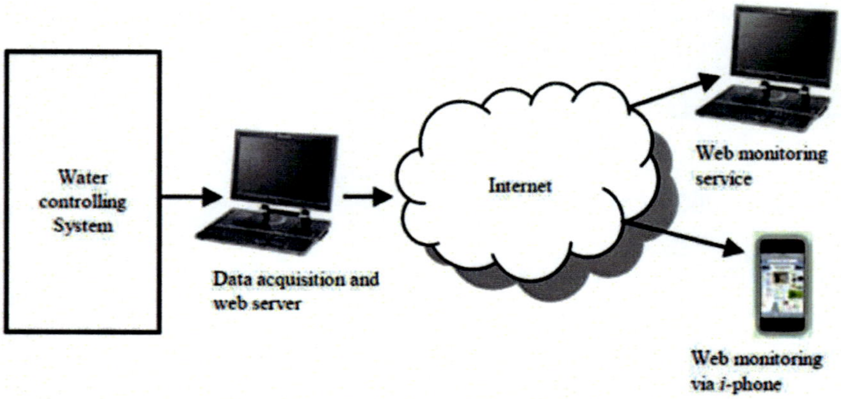

Fig. 1 Monitoring the water level

2 Related work

Ratan Mishra et al. addressed load balancing and cloud computing technologies. This balancing approach is used in cloud computing to reduce or optimize the performance of various factors such as memory capacity, CPU load, and delay load for different sizes [1]. Gulshan Soni et al. have briefly investigated load balancing in cloud computing. This load balancing approach was utilized to reduce resource usage, fail-over of the implementation, and eliminate bottlenecks in cloud computing [2]. Velagapudi Sreenivas et al. have discussed and compared about the load balancing in existing technique and proposed a novel cloud computing technique [3]. Jelena Mirkovic et al. have briefly studied the deployment of the cloud at 50 autonomous systems by utilizing the internet for reducing the spoofing effect [4]. Sharmin Rashid et al. discussed about the detection and prevention of IP spoofing. This method has used to detect and stop the attack and the communication is successfully secured [5]. Shiv Kumar explains about the awareness on the usage of the water and energy through smart meter [6]. Venkatesa Kumar et al. have studied about the security issues and attacks present in the cloud [7]. Deepiga et al. have studied about the water monitoring system in both residential and working places by implementing "wireless sensor network" digitally [8]. Yunlu Wang et al. have proposed the evolutionary game theory for performing load balancing and find the solution in short period [9]. Yingying Chen et al. have studied the effectiveness of the *K*-means detector and attacker of the spoofing [10]. Maisie Borg et al. had a main goal to initiate water supply in three areas, and increase the efficiency of water and conservation in indoor from those areas [11]. S. M. Khaled Reza et al. have proposed a water level monitoring through internet. This monitoring method has been used to avoid the water wastage [12]. Prabhjot Kaur et al. discussed about the prediction of the loads by using nodes [13]. Mayanka Katyal et al. have studied about the static and dynamic load balancing techniques [14].

Soumya Ray et al. have studied about the usage of cloud storage for load balancing [15]. Dharmesh Kashyap et al. compared the advantages and disadvantages of load balancing in cloud [16]. Further, Sidra Aslam et al. compared the different parameters of dynamic load balancing such as performance, response time, speed and throughput [17]. P. Das et al. mainly focused on security attacks present in cloud for lowest cost [18]. Bala et al. studied about the production and detection for IP spoofing [19], and it has been shown in Table 1 for different sensors. Various researchers are working with different applications for the wellbeing of people [20–29].

Table 1 Different sensors and working area

References	Context	Sensors	Display	Cloud server	Other technologies
S. Padmadevi et al.	To gauge the water level for sand	pH sensor, turbidity sensor	LDR	–	GSM
Pranita Vijaykumar Kulkarni et al.	Automatic control of motor on/off for water tank	Low sensor, middle sensor, high sensor	LCD	Adafruit	GSM
Thinagaran Peruma et al.	Alert message for water level	Ultrasonic sensor	LCD	Bespoke Cloud server	GSM
Sukriti et al.	To monitor the water level through mobile app	Ultrasonic sensor	LCD	Blynk server+	LED indicator
J. Infantial Rubalal et al.	To gauge the agriculture parameters for improve the production of the crop	Water level sensor	–	–	WSN
S. M. Khaled Reza et al.	To control the water level for using water pump	Water level sensor	–	–	LED
S. Ravichandran	Detection of the leakage	Load sensor	–	–	ZigBee
Bhad Vidya et al.	Monitoring the water level	Water level sensor	–	–	LED
Imothy Malche et al.	Monitoring the water level in smart village	Liquid level sensor	–	–	

3 Water Monitoring Architecture (WMA)

WMA has three level of indication. First, the water level in tank is indicated by using LED sensor. LED sensor has three ways to indicate: (1) indicates lower level of water in tank, LED; (2) indicates middle level of the water in tank, and LED; (3) indicates higher level of water present in the tank. When the water level reaches the minimum level, which is indicated by an LED sensor, the motor is turned ON automatically.

Similarly, when the water level reaches out the higher mark indicated by LED sensor, then automatically the motor gets switched OFF. Secondly, the water level quality is checked by using three parameters like temperature, pH, and turbidity sensor available inside the tank. Third, the detection of pressure sensor for pipe leakage of the water is also included. All sensors are connected in the wireless sensor network model. This network is used to communicate through mobile and laptop as shown in Fig. 2.

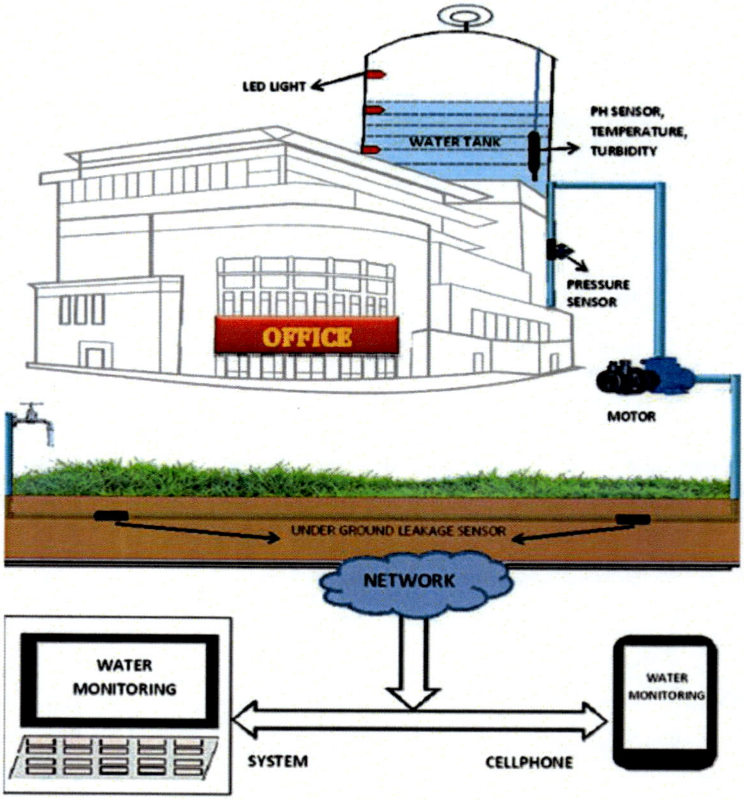

Fig. 2 Water monitoring architecture

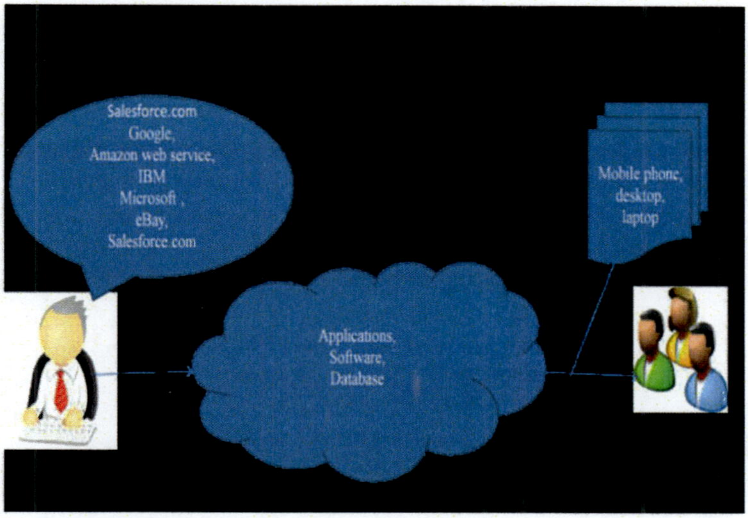

Fig. 3 Storage of cloud computing

4 Cloud Storage and Load Balancing

The cloud is highly essential in human lives since it allows for the storing of diverse data at any time and from any location. Figure 3 depicts how all sensor data is saved in the cloud by various devices such as mobile phones, laptop, computers, and other devices.

Load Balancing: Load balancing is divided into three levels: virtual server, resource pool, and node. There are two tiers of virtual servers: resource pool and node. Further, the resource pool has node process and each node has an IP address and a monitor. Figure 4 depicts the connections between all layers starting at the bottom.

5 Conclusion

The proposed research work has successfully developed a novel water monitoring and management system for assisting the production of agricultural grains. The LED sensor has successfully identified the water level of the intended tank. Totally, 3 LEDs are utilized to automatically sense the lowest, intermediate, and highest level in the water tank and switch OFF the motor. The cloud computing model has been successfully utilized to completely reduce the water wastage. Furthermore, all the sensor data has successfully saved in the cloud by using different devices such as smartphones, computers, and other smart devices.

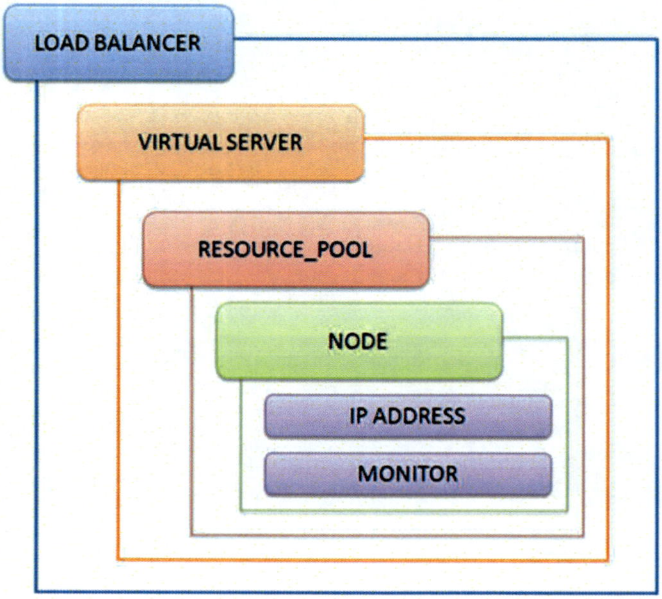

Fig. 4 Load balancing process

References

1. Mishra R, Jaiswal A (2012) Ant colony optimization: a solution of load balancing in cloud. Int J Web Semant Technol (IJWesT) 3(2):33–50
2. Soni G, Kalra M (2014) A novel approach for load balancing in cloud data center. In: IEEE International advance computing conference (IACC), pp 807–812
3. Sreenivas V, Prathap, Kemal M (2014) Load balancing techniques: major challenge in cloud computing—a systematic review. In: 2014 International conference on electronics and communication systems (ICECS), pp 1–6
4. Mirkovic J, Jevtic N, Reiher P (2006) A practical IP spoofing defense through route-based filtering
5. Rashid S, Paul SP (2013) Proposed methods of IP spoofing detection & prevention. Int J Sci Res (IJSR) 2(8):438–444
6. Kumar S (2015) Energy and water monitoring system for smart metering and consumer awareness. In: 2nd International electronic conference on sensors and applications, 15–30 Nov 2015
7. Venkatesa Kumar V, Nithya M (2014) Improving security issues and security attacks in cloud computing. Int J Adv Res Comput Commun Eng 3(10):8148–8151
8. Deepiga T, Sivasankari A (2015) Smart water monitoring system using wireless sensor network at home/office. Int Res J Eng Technol (IRJET) 02(04):1305–1314
9. Wang Y, Wu X, Haas H (2015) Distributed load balancing for internet of things by using Li-Fi and RF hybrid network. In: IEEE 26th International symposium on personal, indoor and mobile radio communications—(PIMRC): mobile and wireless networks, pp 1289–1294
10. Chen Y, Trappe W, Martin RP (2007) Detecting and localizing wireless spoofing attacks. In: 4th Annual IEEE communications society conference on sensor, mesh and ad hoc communications and networks, pp 193–202

11. Borg M, Edwards O, Kimpel S, A study of individual household water consumption. Water Sci Manag
12. Khaled Reza SM, Md. Tariq SA, Mohsin Reza SM (2010) Microcontroller based automated water level sensing and controlling: design and implementation issue. In: Proceedings of the World congress on engineering and computer science, 20–22 Oct 2010
13. Kaur P, Kaur PD (2015) Efficient and enhanced load balancing algorithms in cloud computing. Int J Grid Distrib Comput 8(2):9–14
14. Katyal M, Mishra A (2014) A comparative study of load balancing algorithms in cloud computing environment. Int J Distrib Cloud Comput
15. Ray S, De Sarkar A (2012) Execution analysis of load balancing algorithms in cloud computing environment. Int J Cloud Comput Serv Archit (IJCCSA) 2(5)
16. Kashyap D, Viradiya J (2014) A survey of various load balancing algorithms in cloud computing. Int J Sci Technol Res 3(11)
17. Aslam S, Shah MA (2015) Load balancing algorithms in cloud computing: a survey of modern techniques. In: National software engineering conference
18. Das P, Khilar PM (2013) VFT: a virtualization and Fault Tolerance approach for cloud computing. In: 2013 IEEE Conference on information and communication Technologies (ICT)
19. Bala A, Chana I (2015) Intelligent failure prediction models for scientific workflows. Expert Syst Appl 42:980–989
20. Banupriya S, Kottursamy K, Bashir AK (2021) Privacy-preserving hierarchical deterministic key generation based on a lattice of rings in public blockchain. Peer-to-Peer Netw Appl 1–13
21. Ponnusamy V, Kottursamy K, Karthick T, Mukeshkrishnan MB, Malathi D, Ahanger TA (2020) Primary user emulation attack mitigation using neural network. Comput Electr Eng 88:106849
22. Raja G, Kottursamy K, Theetharappan A, Cengiz K, Ganapathisubramaniyan A, Kharel R, Yu K (2020) Dynamic polygon generation for flexible pattern formation in large-scale UAV swarm networks. In: 2020 IEEE Globecom workshops (GC Wkshps). IEEE, pp 1–6
23. Mehra R, Pachpor K, Kottilingam K, Saranya A (2020) An initiative to prevent japanese encephalitis using genetic algorithm and artificial neural network. In: 2020 International conference on computational intelligence (ICCI). IEEE, pp 142–148
24. Saranya A, Kottilingam K (2021) A survey on bone fracture identification techniques using quantitative and learning based algorithms. In: 2021 International conference on artificial intelligence and smart systems (ICAIS). IEEE, pp 241–248
25. Nallarasan V, Kottilingam K (2021) Spectrum management analysis for cognitive radio IoT. In: 2021 International conference on computer communication and informatics (ICCCI). IEEE, pp 1–5
26. Prakash UM, Kottursamy K, Cengiz K, Kose U, Hung BT (2021) 4x-expert systems for early prediction of osteoporosis using multi-model algorithms. Measurement 180:109543
27. Cengiz K, Sharma R, Kottursamy K, Singh KK, Topac T, Ozyurt B (2021) Recent emerging technologies for intelligent learning and analytics in big data. In: Multimedia technologies in the internet of things environment. Springer, Singapore, pp 69–81
28. Raja G, Anbalagan S, Kottursamy K, Aparna GS, Kumaresan J, Ihsan M (2019) Authorized arming and safeguarded landing mechanism. IEEE Trans Commun 67(7):5024–5036
29. Rajagopal RD, Murugan S, Kottursamy K, Raju V (2019) Cluster based effective prediction approach for improving the curable rate of lymphatic filariasis affected patients. Clust Comput 22(1):197–205

KRB02 Method Used to Secure the Generalized Data

K. Karthik, S. Rajaprakash, B. Naga Yugandhar Reddy, V. Varun Kumar Reddy, and C. H. Maheswara Balaji

Abstract The current and future world is data world in a citizen's life; because data only decides one's existence in the world. Especially, data is more available in government, hospital, social media and, etc. This data does not have good security, due to which hackers can hack the data very easily. To address this problem, we are applying a new method namely KRB02. This method has four processes. The first process is applying for *T*-test technique and "0th" matrix value "start from end", followed by swap of the values in matrix; the second process is applying for odd operations; the third process is applying for even operations, and final process is applying for column operations. The KRB02 method gives high security, while compared to ChaCha method.

Keywords KRB02 · Encryption · Prime · *T*-test · ChaCha

1 Introduction

The huge data time has been bought with satisfactory opportunities for consistent improvements in clinical consideration, financial turn of events, improving the guidance system, and various sorts of delight. The assessment of enormous data needs to go through various stages to secure some critical characteristics, which joins a couple of stages like data compromise, data acquiring, information cleaning, information extraction, request planning, data illustrating, and interpretation. Each stage holds various challenges like heterogeneity, reasonableness, unpredictability, security, and insurance of individuals. One of the huge issues in colossal data is security and insurance due to its monstrous system like tremendous volume, speed, and assortment. The present and future world data is very important in world citizen's life; especially, data is more available in social media, hospital, populations, and, etc. This data does

K. Karthik (✉) · S. Rajaprakash · B. Naga Yugandhar Reddy · V. Varun Kumar Reddy · C. H. Maheswara Balaji
Department of CSE, Aarupadai Veedu Institute of Technology, Vinayaka Mission's Research Foundation, Salem, India
e-mail: karthik@avit.ac.in

S. Shakya et al. (eds.), *Proceedings of Second International Conference on Sustainable Expert Systems*, Lecture Notes in Networks and Systems 351,
https://doi.org/10.1007/978-981-16-7657-4_59

not have good security, so hackers can hack the data very easily. To this problem, we are applying the ChaCha method. This method effectively hacks the information from the computer programmers. The XOR for ChaCha is deficiency assault [1]. This creator utilized new hash idea for key speculating and ending condition [2]. Creator was presented through bricklayer assault for investigation of ChaCha [3]. They contribute the security for double A [4]. They made new plan for secure quick and adaptable calculation [5]. SRB18 strategy used to give security to information [6]. SRB21 Phase 1 and SRB21 Phase 2 strategy used to give security to information [7, 8]. CBB21, CBB22, CBB20, and RJB25 techniques are used to give security to information [9, 10, 12, 13]. Presented the novel strategy KRB (Karthik, Raja, Bagath) 02. Recent research domains utilize different parts of the algorithms for multiple purposes [11, 14–23].

2 Methods

In this part, we discuss about the encryption methods. The proposed KRB02 method has 8 phases. (1) To find the "secret" prime key S. (2) To find the X_1 and X_2 values from prime numbers. (3) To find the $\overline{X}_1$ and $\overline{X}_2$ values. (4) To find the standard deviation values with the help of Step 2 and Step 3. (5) To "swap" a and b "values" from left in a matrix and "0th matrix value start from end." (6) To "find" the T-test values and pair them that numbers from left to right. (7) To exchange the odd and even equation a and b in a given "matrix." (8) To execute the column operations and "swap the column values."

2.1 Encryption

- KRB02 method is Table 1 for encryption.
- Input of the matrix (IP).

$$\mathrm{IP} = \begin{bmatrix} 901/9 & 902/9 & 903/9 & 904/9 & 905/9 \\ 906/9 & 907/9 & 908/9 & 909/9 & 910/9 \\ 911/9 & 912/9 & 913/9 & 914/9 & 915/9 \\ 916/9 & 917/9 & 918/9 & 919/9 & 920/9 \\ 921/9 & 922/9 & 923/9 & 924/9 & 925/9 \end{bmatrix}$$

- Prime Numbers—1, 3, 5, 7, 11, 13, 17, 19, 23
- $X_1 = 3, 5, 7, 11; X_2 = 13, 17, 19, 23$

Using Eqs. (2) and (9) and Table 2

- $\overline{X}_1 = \sum X_1/N_1;$

Table 1 Encryption algorithm

Steps	Encryption procedure
I	To find a prime numbers in given matrix IP
II	Equally separate the two parts of prime numbers and apply those values in Eqs. (1), (2), and (3)
III	T-test formula $= (\overline{X}_1 - \overline{X}_2)/\sqrt{((S_1^2/N_1) + (S_2^2/N_2))}$ (1)
IV	To find the values for Eq. (1) with the help of Eqs. (2) and (3)
V	$\overline{X}_1 = \sum X_1/N_1$ $\overline{X}_2 = \sum X_2/N_2$ (2)
VI	$S_1 = \sqrt{\sum (X_1 - \overline{X}_1)^2/(N_1 - 1)}$ $S_2 = \sqrt{\sum (X_2 - \overline{X}_2)^2/(N_2 - 1)}$ (3)
VII	To pair the T-test value from left to right
VIII	To find the n value with the help of 'K' and K is secret key
IX	If n is odd number where $n = 2k + 1$ $a^n + b^n = (a + b)(a^{n-1} - a^{n-2}b + a^{n-3}b^2 \cdots + b^{n-2}a - b^{n-1})$ (4)
X	To apply the n value in Eq. (4)
XI	To merge the step x values from left to right and apply in the matrix
XII	If n is even number where $n = 2k$
XIII	$a^n + b^n = (a + b)(a^{n-1} - a^{n-2}b + a^{n-3}b^2 \cdots - b^{n-2}a + b^{n-1})$ (5)
XIV	To apply the n value in Eq. (5), then merge the values from left to right and apply in the matrix
XV	$\text{TCE} = C_i \leftrightarrow (C_{i+(n-m)})$ (6) where TCE is T—test column encrypted matrix, C is columns, i, n and m is column numbers

Table 2 X_1 and X_2 values

X_1	$(X_1 - \overline{X}_1)$	$(X_1 - \overline{X}_1)^2$	X_2	$(X_2 - \overline{X}_2)$	$(X_2 - \overline{X}_2)^2$
3	−3.5	12.25	13	−5	25
5	−1.5	2.25	17	−1	1
7	0.5	0.25	19	1	1
11	4.5	20.25	23	5	25
	$\sum (X_1 - \overline{X}_1)^2$	35		$\sum (X_2 - \overline{X}_2)^2$	52

- $\overline{X}_1 = (3 + 5 + 7 + 11)/4$; $\overline{X}_1 = 26/4$; $\overline{X}_1 = 6.5$
- $\overline{X}_2 = \sum X_2/N_2$;
- $\overline{X}_2 = (13 + 17 + 19 + 23)/4$ $\overline{X}_2 = 18$

Using Eq. (3)

- $S_1 = \sqrt{\sum (X_1 - \overline{X}_1)^2/(N_1 - 1)}$ $S_2 = \sqrt{\sum (X_2 - \overline{X}_2)^2/(N_2 - 1)}$
- $S_1 = \sqrt{(35/(4 - 1))}$ $S_2 = \sqrt{(52/(4 - 1))}$

- $S_1 = \sqrt{(35/3)}\ S_2 = \sqrt{(52/(3))}$
- $S_1 = 11.66\ S_2 = 17.33$

Using Eq. (1)

- T-test formula $= \left(\overline{X}_1 - \overline{X}_2\right)/\sqrt{\left(\left(S_1^2/N_1\right) + \left(S_2^2/N_2\right)\right)}$
- T-test formula $= (6.5 - 18)/\sqrt{\left(\left(11.66^2/4\right) + \left(17.33^2/4\right)\right)}$
- T-test formula $= 11.5/\sqrt{((135.95/4) + (300.32/4))}$
- T-test formula $= 11.5/\sqrt{((135.95 + 300.32)/4))}$
- T-test formula $= 11.5/\sqrt{(436.27/4)}$
- T-test formula $= 11.5/\sqrt{109.06}$
- "Pair the T-test value from left to right (1,1) (5,1) (0,9) (0,6) and swap it those numbers."

Step "1: (1,1)"

$$\text{TTE} = \begin{bmatrix} 901/9\ 902/9\ 903/9\ 904/9\ 905/9 \\ 906/9\ 907/9\ 908/9\ 909/9\ 910/9 \\ 911/9\ 912/9\ 913/9\ 914/9\ 915/9 \\ 916/9\ 917/9\ 918/9\ 919/9\ 920/9 \\ 921/9\ 922/9\ 923/9\ 924/9\ 925/9 \end{bmatrix}$$

where TTE is T-Test Encryption.

Step "2: (5,1)"

$$\text{TTE} = \begin{bmatrix} 901/9\ 902/9\ 903/9\ 904/9\ 905/9 \\ 906/9\ 907/9\ 908/9\ 909/9\ 910/9 \\ 911/9\ 912/9\ 913/9\ 914/9\ 915/9 \\ 916/9\ 917/9\ 918/9\ 919/9\ 924/9 \\ 921/9\ 922/9\ 923/9\ 920/9\ 925/9 \end{bmatrix}$$

Step "3: (0,9)"

$$\text{TTE} = \begin{bmatrix} 901/9\ 902/9\ 903/9\ 904/9\ 905/9 \\ 906/9\ 907/9\ 908/9\ 909/9\ 910/9 \\ 911/9\ 912/9\ 913/9\ 914/9\ 915/9 \\ 925/9\ 917/9\ 918/9\ 919/9\ 924/9 \\ 921/9\ 922/9\ 923/9\ 920/9\ 916/9 \end{bmatrix}$$

Step "4: (0,6)"

$$\mathrm{TTE} = \begin{bmatrix} 901/9 & 902/9 & 903/9 & 904/9 & 905/9 \\ 906/9 & 907/9 & 908/9 & 909/9 & 910/9 \\ 911/9 & 912/9 & 913/9 & 914/9 & 915/9 \\ 925/9 & 917/9 & 918/9 & 916/9 & 924/9 \\ 921/9 & 922/9 & 923/9 & 920/9 & 919/9 \end{bmatrix}$$

- To find the n value; $a = 2, b = 3, k = 2, n = 5$, n is odd number

Using Eq. (4)

- $a^2 + b^3 = (2 + 3)(2^{5-1} - 2^{5-2}3 + 2^{5-3}3^2 - 2^{5-4}3^3 - 2^{5-5}3^4 + 3^{5-4}2 - 3^{5-3}2 + 3^{5-2}2 - 3^{5-1})$
- $a^2 + b^3 = (5)\,(2^4 - 2^33 + 2^29 - 2^127 - 2^081 + 3^12 - 3^22 + 3^32 - 3^4)$
- $a^2 + b^3 = (5)(16 - 24 + 36 - 54 - 81 + 6 - 18 + 54 - 81)$
- $a^2 + b^3 = (5)(-146)$
- $a^2 + b^3 = (5,1), (4,6)$

Step "5: (5,1)"

$$\mathrm{TEB} = \begin{bmatrix} 901/9 & 906/9 & 903/9 & 904/9 & 905/9 \\ 902/9 & 907/9 & 908/9 & 909/9 & 910/9 \\ 911/9 & 912/9 & 913/9 & 914/9 & 915/9 \\ 925/9 & 917/9 & 918/9 & 916/9 & 924/9 \\ 921/9 & 922/9 & 923/9 & 920/9 & 919/9 \end{bmatrix}$$

Step "6: (4, 6)"

$$\mathrm{TEB} = \begin{bmatrix} 901/9 & 906/9 & 903/9 & 904/9 & 907/9 \\ 902/9 & 905/9 & 908/9 & 909/9 & 910/9 \\ 911/9 & 912/9 & 913/9 & 914/9 & 915/9 \\ 925/9 & 917/9 & 918/9 & 916/9 & 924/9 \\ 921/9 & 922/9 & 923/9 & 920/9 & 919/9 \end{bmatrix}$$

- To find the n value: $a = 2, b = 3, k = 2, n = 4$
- n is even number

Using Eq. (5)

- $a^2 + b^3 = (2 + 3)(2^{4-1} - 2^{4-2}3 + 2^{4-3}3^2 - 2^{4-4}3^3 + 3^{4-3}2 - 3^{4-2}2 + 3^{4-1})$
- $a^2 + b^3 = (5)\,(2^3 - 2^23 + 2^19 - 2^027 + 3^12 - 3^22 + 3^3)$
- $a^2 + b^3 = (5)(16 - 12 + 18 - 54 + 6 - 18 + 27)$
- $a^2 + b^3 = (5)(-17)$
- $a^2 + b^3 = (5,1), (7,0)$

Step "7: (5, 1)"

$$\text{TEB} = \begin{bmatrix} 901/9 & 902/9 & 903/9 & 904/9 & 907/9 \\ 906/9 & 905/9 & 908/9 & 909/9 & 910/9 \\ 911/9 & 912/9 & 913/9 & 914/9 & 915/9 \\ 925/9 & 917/9 & 918/9 & 916/9 & 924/9 \\ 921/9 & 922/9 & 923/9 & 920/9 & 919/9 \end{bmatrix}$$

Step "8: (7, 0)"

$$\text{TEB} = \begin{bmatrix} 908/9 & 902/9 & 903/9 & 904/9 & 907/9 \\ 906/9 & 905/9 & 901/9 & 909/9 & 910/9 \\ 911/9 & 912/9 & 913/9 & 914/9 & 915/9 \\ 925/9 & 917/9 & 918/9 & 916/9 & 924/9 \\ 921/9 & 922/9 & 923/9 & 920/9 & 919/9 \end{bmatrix}$$

Using Eq. (6)

$$\text{TCE} = \begin{bmatrix} 907/9 & 904/9 & 903/9 & 902/9 & 908/9 \\ 910/9 & 909/9 & 901/9 & 905/9 & 906/9 \\ 915/9 & 914/9 & 913/9 & 912/9 & 911/9 \\ 924/9 & 916/9 & 918/9 & 917/9 & 925/9 \\ 919/9 & 920/9 & 923/9 & 922/9 & 921/9 \end{bmatrix}$$

3 Result and Discussion

The proposed algorithm KRB02 encryption performance compared with ChaCha. ChaCha concept is all about moving diagonal values to the first column.

The three-by-three matrix has 24 bytes of file size; the six by six matrix has 76 bytes of file size; the ten by ten matrix has 312 bytes of file size; the fifteen by fifteen matrix has 812 bytes of file size; the twenty by twenty matrix has 1531 bytes of file size; and the forty by forty matrix has 6580 bytes of file size as shown in Table 3.

Table 3 KRB02 Encryption performance

File size	ChaCha	KRB02
24	1.69	2.2
76	1.29	2.6
312	2.73	2.9
822	2.64	3.1
1531	3.4	3.6
6580	2.27	3.9

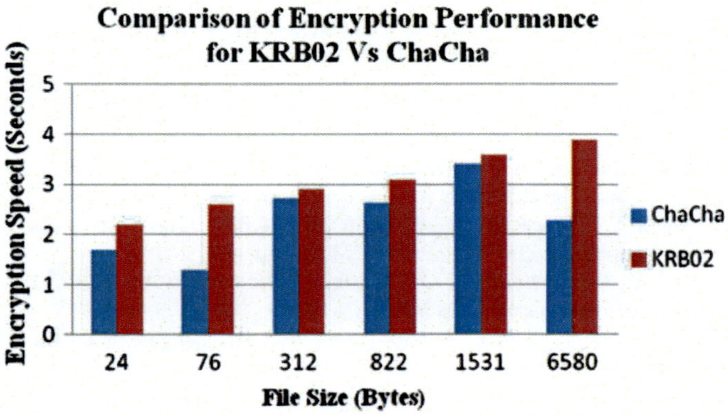

Fig. 1 Encryption performance

From Fig. 1, the KRB02 method and has compared the encryption speed in seconds. The encryption performance of the speed 2.2 (s), 2.6 (s), 2.9 (s), 3.1 (s), 3.6 (s) and 3.9 (s) for the KRB02 and 1.69 (s), 1.29 (s), 2.73 (s), 2.64 (s), 3.4 (s), and 2.27 (s) for the ChaCha. The KRB02 gives more protection of the data; when compared to existing techniques.

4 Conclusion

The current and future world is data in citizen's life; especially, data is more available in government, hospital, social media, and, etc. This data does not have good security, so hackers can hack the data very easily. To this problem, we are applying the new technique known as KRB02. The proposed KRB02 method has 8 phases. (1) To find the "secret" prime key S. (2) To find the X_1 and X_2 values from prime numbers. (3) To find the $\overline{X}_1$ and $\overline{X}_2$ values. (4) To find the standard deviation values with the help of Step 2 and Step 3. (5) To "swap" a and b "values" from left in a matrix and "0th matrix value start from end." (6) To "find" the T-test values and pair them that numbers from left to right. (7) To exchange the odd and even equation a and b in a given "matrix." (8) To execute the column operations and "swap the column values." The KRB02 method gives extraordinary security when compared to ChaCha method.

References

1. Dilip Kumar SV, Patranabis S, Breier J, Mukhopadhyay D, Bhasin S, Chattopadhyay A, Baksi A (2017) A practical fault attack on ARX-like cipher with a case study on ChaCha20. Workshop on fault diagnosis and tolerance in cryptography (FDTC), pp 33–40

2. Arun Babu P, Thomas JJ (2019) Freestyle, a randomized version of ChaCha for resisting offline brute-force and dictionary attacks. J Inf Secur Appl 102396
3. Adomnicai A, Jacques Fournier JA, Masson L (2017) Bricklayer attack: a side-channel analysis on the ChaCha quarter round. In: Progress in cryptology—INDOCRYPT 2017. Lecture notes in computer science, vol 10698. Springer, Cham, pp 65–84
4. Mazumdar B, Subidh Ali SK, Sinanoglu O (2015) Power analysis attacks on ARX: an application to Salsa20. In: IEEE 21st International on-line testing symposium, pp 40–43
5. Watt C, Renner J, Popescu N, Cauligi S, Stefan D (2019) CT-Wasm: type-driven secure cryptography for the web ecosystem. Proc ACM Prog Lang 77:1–29
6. Bagath Basha C, Rajapraksh S (2020) Enhancing the security using SRB18 method of embedding computing. Microprocess Microsyst
7. Bagath Basha C, Rajaprakash S (2019) Securing twitter data using phase I methodology. Int J Sci Technol Res 1952–1955
8. Bagath Basha C, Rajaprakash S (2020) Applying the SRB21 phase II methodology for securing twitter analyzed data. In: AIP Conference proceedings of the international conference on mechanical electronics and computer engineering
9. Bagath Basha C, Rajaprakash S (2020) Applying the CBB21 phase 2 method for securing twitter analyzed data. Adv Math Sci J 1085–1091
10. Bagath Basha C, Rajaprakash S, Harish VVA, Krishna MS, Prabhas K (2020) Securing twitter analysed data using CBB22 algorithm. Adv Math Sci J 1093–1100
11. Bagath Basha C, Somasundaram K (2019) A comparative study of twitter sentiment analysis using machine learning algorithms in big data. Int J Recent Technol Eng 591–599
12. Bagath Basha C, Rajaprakash S, Muthuselvan S, Saisatishsunder P, Alekhya Rani SVL (2020) Applying the CBB20 algorithm for twitter analyzed data. J Phys Conf Ser. (First international conference on advances in physical sciences and materials, Coimbatore, Tamil Nadu, India)
13. Rajaprakash S, Bagath Basha C, Muthuselvan S, Jaisankar N, Singh RP (2020) RBJ25 Cryptography algorithm for securing big data. J Phys Conf Ser. (First international conference on advances in physical sciences and materials, Coimbatore, Tamil Nadu, India)
14. Banupriya S, Kottursamy K, Bashir AK (2021) Privacy-preserving hierarchical deterministic key generation based on a lattice of rings in public blockchain. Peer-to-Peer Netw Appl 1–13
15. Ponnusamy V, Kottursamy K, Karthick T, Mukeshkrishnan MB, Malathi D, Ahanger TA (2020) Primary user emulation attack mitigation using neural network. Comput Electr Eng 88:106849
16. Raja G, Kottursamy K, Theetharappan A, Cengiz K, Ganapathisubramaniyan A, Kharel R, Yu K (2020) Dynamic polygon generation for flexible pattern formation in large-scale UAV swarm networks. In: 2020 IEEE Globecom workshops (GC Wkshps). IEEE, pp 1–6
17. Mehra R, Pachpor K, Kottilingam K, Saranya A (2020) An initiative to prevent Japanese encephalitis using genetic algorithm and artificial neural network. In: 2020 International conference on computational intelligence (ICCI). IEEE, pp 142–148
18. Saranya A, Kottilingam K (2021) A survey on bone fracture identification techniques using quantitative and learning based algorithms. In: 2021 International conference on artificial intelligence and smart systems (ICAIS). IEEE, pp 241–248
19. Nallarasan V, Kottilingam K (2021) Spectrum management analysis for cognitive radio IoT. In: 2021 International conference on computer communication and informatics (ICCCI). IEEE, pp 1–5
20. Prakash UM, Kottursamy K, Cengiz K, Kose U, Hung BT (2021) 4x-expert systems for early prediction of osteoporosis using multi-model algorithms. Measurement 180:109543
21. Cengiz K, Sharma R, Kottursamy K, Singh KK, Topac T, Ozyurt B (2021) Recent emerging technologies for intelligent learning and analytics in big data. In: Multimedia technologies in the internet of things environment. Springer, Singapore, pp 69–81
22. Raja G, Anbalagan S, Kottursamy K, Aparna GS, Kumaresan J, Ihsan M (2019) Authorized arming and safeguarded landing mechanism. IEEE Trans Commun 67(7):5024–5036
23. Rajagopal RD, Murugan S, Kottursamy K, Raju V (2019) Cluster based effective prediction approach for improving the curable rate of lymphatic filariasis affected patients. Clust Comput 22(1):197–205

Sentiment Analysis on Tweets Using Emojis to Help the Distressed

S. Maneesha, S. Srihari, and H. R. Mamatha

Abstract In the recent years, many people use social media such as twitter to express distressed emotions. Proofs establish that depressed people utilize Twitter to express their feelings. To provide help to the distressed, we believe sentiment analysis serves this purpose because it is a field of study that analyses sentiments and emotions from the text. However, there is a need for an additional factor to be considered while performing sentiment analysis to make the process more effective and accurate. Emojis can fulfil this purpose since it can express one's emotions effectively irrespective of their diversities. Sentiment analysis on tweets through emojis would classify the tweets into different category of emotions, thereby making the process of extraction of emojis indicating negative feelings trouble-free. From obtained results, we intend to help the affected through an in-platform reply to the tweet with the details of appropriate authorities who will do the needful.

Keywords Sentiment analysis · Twitter · Convolutional neural network · Emojis

1 Introduction

Sentiment analysis on tweets would order the tweets into various classes, for example, emoji's communicating bliss, misery and nonpartisan sentiments; it is simpler to scrape emojis showing dismal or negative emotions and subsequently use it in our site to offer help.

For the purpose of identifying mental health status which comprises of despair and hopelessness, Twitter is looked up as an efficient data provider.

S. Maneesha · S. Srihari · H. R. Mamatha (✉)
PES University, Bengaluru, India
e-mail: mamathahr@pes.edu

S. Shakya et al. (eds.), *Proceedings of Second International Conference on Sustainable Expert Systems*, Lecture Notes in Networks and Systems 351,
https://doi.org/10.1007/978-981-16-7657-4_60

1.1 Purpose

By virtue of sentiment analysis conducted on tweets indicating suicidal propensity, an in-platform reply approach designed to provide social assistance to individuals to alleviate negative feelings.

1.2 Objectives

Extract tweets consisting of genres such as depression, self-harm and convert them into emojis and analyse the emotions conveyed by them. Categorize them as a positive, negative or neutral term. Design an in-platform reply method that identifies suicidal tweets and offers support/assistance.

1.3 Benefits

To channelize the details of authorities such as NGOs and counsellors to the affected people in need of timely assistance.

1.4 Goal

Doing sentiment analysis from tweets on emojis and creating a platform that channels the details of the authorities to the concerned individual, which in turn offers a helping hand in building a stable and secure community.

1.5 Novelty

- In real world, sentiment analysis has been pre-dominantly used to capture, study and analyse the human reaction towards a product or service, and the main motive is to use the analysis for profits of organizations.
- Breaking the conventional and ongoing practice, our project utilizes the concept of sentiment analysis for a public cause targeting individuals who are in need of help and attention expressing suicidal tendency through twitter—one of the most widely used social media application.
- To sharpen the accuracy of our desired work, a new component—emoji—was used; though emojis had no prior vast usage in the field of sentiment analysis, we happened to view the potential embedding in it.

- By making use of the concept of emoji, we successfully carried out sentiment analysis on tweets and pulled out a meaningful pioneer attempt.

2 Literature Survey

The following are the key takeaways from the extensive literature survey carried out:

- It was found that the machine classifier replicated the accuracy of the human coders in being able to distinguish the level of concern among suicide-related tweets [1].
- Twitter is a rich source of data for opinion mining and sentiment analysis. The focus is on the problem of sentiment analysis for twitter by analysing emotion symbols—emoji ideograms, emotion tokens and emoticons. It is observed that emotion tokens are used very commonly as they directly express one's emotion irrespective of their linguistic differences. It is observed that the most fruitful approach to this process would be integration of NLP and symbol analysis [2].
- Emojis often represent emotions or sentiments that are not easily concisely expressed in words; therefore, emoji handling clearly increases one's ability to conduct effective Twitter sentiment analysis. Naive Bayes, maximum entropy, SVM, SNN and RNN were compared [3].
- In order to arrange opinion in tweets, we use Naive Bayes classifier on *n*-gram highlights. The pre-requisite step for acquiring *n*-gram highlights is to tokenize the content info [4].
- In spite of the number of studies on twitter have been increasing over the years, majority of them did not focus on emojis and emoticons. Upon keen observation, it could be concluded that the amount of research done in the domain of emojis is very negligible. Various approaches (keyword spotting, machine learning approach) that maybe applicable in the usage of emojis for sentiment analysis purposes are mentioned, by virtue of which a conceptual background is put up related to inspection of emojis for conduction of sentimental analysis in order to spot light on distinct prospects they offer [5].
- The sentiment analysis shows that majorly, the emojis would be placed at the end of the tweets, and most of the emojis that were annotated were positive. It is observed that there are no differences as such in the emoji ranking between the 13 languages and emoji ranking system [6].
- Self-destruction shirking is getting to know and making note of the ambiguous elements of danger and forewarning indications that may quicken the occurrence [7].
- Generally, while carrying out sentiment analysis, researchers have underused emotions such as fear and disgust as the major attention in performing training and supervision were given to emoji and hashtags [8].
- A bi-directional long short-term memory is used to capture the context of each word where a bidirectional long short-term memory has 1024 hidden units. The

DeepMoji model makes use of an embedding layer, where the model consists of 256 dimensions in order to project each word into a vector space. For the purpose of enforcing a constraint of each embedding dimension, a hyperbolic tangent function is used. At last, an attention layer (which helps in deciding the importance of a text) that take all of the present layers as an input using skip-connections is used. It is shown that millions of texts on social media with emojis can be utilized for pre training models, which would eventually facilitate them to learn representations of emotional content present in texts [9].

- Imitation of human activities by computer has been made possible by machine learning. One such primitive human activity is to understand what others say and give appropriate replies. Neural network is utilized to provide smart replies where the neural networks role is to make a decision whether or not to suggest responses. Long short-term memory (LSTM) neural network works in the following manner, the incoming message is processed first then the processed message is used to predict the most likely responses [10].
- Sentiment analysis is utilized in many corporations for review of products, comments from social media and from a small amount of it is utilized to check whether or not the text is positive, negative or neutral [12].

3 Procedure

Two datasets were used. One contained 11,000 tweets which was used for training purposes and another dataset contained 200 tweets (Fig. 1) collected in real time which is used in the working model for sentimental analysis.

	Name	Tweet	tweet_id	location
0	nareshkaliya	Sad to see go the #RaulVinci way in stutter...	1438191365750345729	Bengaluru, India
1	pranavcondur4	im sad that you've graduated but happy that i...	1438189064834859009	Chennai, India
2	EkanthEsha	Sad that a comment from bluemelon didn't bear...	1438186328747765762	Bengaluru, Karnataka - India
3	Befu_DA	oki ^^\nUff, ist der so sad? 😞	1438185962245349380	Panem
4	Akbarmanjyani	Sad to see him go. Such an iconic voice, he wi...	1438185031952842756	Hyderabad, India
...	...	...	...	...
195	1512Ram	Today, I had the unfortunate loss of a Tulsi p...	1438028211183517696	Chennai
196	MukeshWadhwa1	When you're happy, you enjoy the music. But wh...	1438026106485968898	Bangalore
197	GetlaPrasad	all the best brother. Don't feel sad this wil...	1438025836481761288	Hyderabad, India
198	praxisleaders	It has been 30 plus days.... We are not disap...	1438022040527650822	Karnataka, India
199	Anees_Offl	very sad....\n\nhelpline # 104 !!	1438021877977419786	Chennai, India

200 rows × 4 columns

Fig. 1 Real-time cleaned dataset

df_

	Tweets	Emojized
0	Sad to see go the #RaulVinci way in stutter...	😥
1	im sad that you've graduated but happy that i...	😥
2	Sad that a comment from bluemelon didn't bear...	😥
3	oki ^^\nUff, ist der so sad? 😥	❤️
4	Sad to see him go. Such an iconic voice, he wi...	😥
...	...	...
195	Today, I had the unfortunate loss of a Tulsi p...	😥
196	When you're happy, you enjoy the music. But wh...	🎼
197	all the best brother. Don't feel sad this wil...	👍
198	It has been 30 plus days.... We are not disap...	💔
199	very sad....\n\nhelpline # 104 !!	😥

200 rows × 2 columns

Fig. 2 Emojized form of tweets

3.1 Data Collection

- Using twitter API tweets were extracted containing the keywords 'depress', 'suicide' etc. using Cursor tweepy.
- Retweets, URLs and special characters were removed using regex for scraping the data.

3.2 Dataset Creation

For training the CNN model, 11,000 tweets were collected and cleaned.

3.3 Data Labelling

- VADER sentiment analysis algorithm to label the tweets automatically was applied on the dataset created for training purpose.
- Standardization of obtained compound value for distinct classification of tweets into three classes—positive, neutral and negative.

3.4 Emojize

- Tweets are tokenized and corresponding probability of each token is calculated.
- Emotion with the probability is identified and converted into emoji.
- A dataset containing tweet and its emojized format is returned (Fig. 2) .

3.5 CNN Accuracy Check on Images

1. CNN architecture:
   - Convolution neural network (CNN) is a type of deep neural network used in image recognition and image processing.
   - Constituting layers of CNN and their functionalities:
   - Convolutional layer: The convolutional layer identifies the features present in an image.
   - Same Padding: When an image is augmented and kernel is applied over it, the convolved matrix retains its original dimensions.
   - Rectified Linear Unit layer (ReLu): To increase the non-linearity of the image. For providing better feature extraction, an image is stripped of excessive fat.
   - Max Pool: Returns the maximum value from the portion of the image covered by the kernel, also performs de-noising.
   - Dropout: To avoid overfitting.
   - Flatten: The image is flattened into a column vector.
   - Softmax: Major purpose is to distinguish dominating and low-level features in images and classify them.
   - Adam optimiser: A replacement optimization algorithm which handles sparse gradient on noisy problems, its feature to get easily configured makes it feasible for usage.
   - Binary Cross Entropy: Computes the cross-entropy loss between true labels and predicted labels.
2. Training and testing set containing images of emoji were converted to RGB format and resized.
3. Resized array and the index corresponding to the class label were returned.
4. According to the following:

   batch_size * steps_per_epoch ≥ Number of samples in the training set (1)

   Number of epochs was chosen to be 35 as it provided accurate result, with a learning rate of 0.001.
5. Results were visualized.
6. $F1$ score, recall and precision were computed.

3.6 *Training: Sentiment Analysis on 11 k Tweets*

- 11 thousand tweets were passed into DeepMoji which converted the text into the most appropriate emoji. A data frame 'df_' containing the tweet and emoji representing it was returned.
- 'All' folder: This folder was created to enclose all the unique desired emojis.
- emoji_emocode excel file: This file was designed to contains 222 emojis, names of the emoji images and path connecting emoji to corresponding image present in All folder.
- 'emo' data frame: This data frame reads the emoji_emocode excel file and converts the provided path into a string format and pre-processes the emojis present in second column.
- Pre-processing [11]:

  (a) emoji is encoded into utf-8. e.g.: 😪is encoded as '\xf0\x9f\ × 98\xaa'.
  (b) from this, \x's were removed. i.e., 'f0 9f 98 aa' would remain.
  (c) each character from the above-obtained is converted to its ASCII equivalent. i.e., 240,159,152 called the 'emocode'.

- These emocodes are appended into a column in 'emo' data frame.
- Similar process is followed for the emojis present in df_. After pre-processing the emojis, the emocodes are stored in emocode column of df_ data frame.
- If the 'emocode' from the df_ and emo match, the path corresponding to this emocode connecting emoji to its respective image is fed into the CNN classifier [11].
- The classifier then converts the image into suitable form and predicts to which class the emoji would belong to.
- Each class is represented by the polarity:

  (a) −1 for negative
  (b) 0 for neutral
  (c) 1 for positive

- The obtained polarity is added into 'EmoPol' column of the df_ data frame.

3.7 *Real-Time: Sentiment Analysis on Real-Time Tweets*

- 200 tweets collected in real time by searching specific keywords were passed into DeepMoji which converted the text into the most appropriate emoji. A data frame df_ containing the tweet and emoji representing it was returned.
- Process of identifying corresponding images and passing into CNN model for classification is same as above.

3.8 Reply Within the Platform

- On completion of sentiment analysis using CNN on real-time tweets, the usernames and tweet IDs of the tweets classified as negative with 'EmoPol' value as -1 were identified.
- A tweepy functionality was used to send pre-structured reply to that particular tweet from Twitter platform itself, where the reply constituted of a statement with positive intent and provided the user with contacts of few professional helplines.

4 Results and Discussions

4.1 Collection of Tweets

A dataset was created with 200 tweets collected in real time based on specific key words and the respective user details were also created using Tweepy. Start date for the collection of tweets was specified for enhanced search results which connect to the recent times with the help of 'date_since' field. Retweets, handles, URL links and special characters were removed by a filter; hence, data was scraped.

4.2 Emojize

This picture depicts the 200 tweets dataset (df_) which was passed into DeepMoji for conversion from textual data to emojized form.

4.3 CNN Accuracy Check on Images

Figure 3 indicates the number of emojis in the three folders after passing the pictures of emojis in the **training set** into Open CV for restructuring the images into suitable form for easier analysis by CNN model. The three folders are namely—Positive, Negative and Neutral classified based on limited human-eye analysis.

Figure 4 indicates the number of emoji pictures in **testing data** in the three folders—Positive, Negative and Neutral as a result of following the similar procedure which was applied for the training data.

Figure 5 depicts the characteristics of the CNN model with information about the layers being used.

Figure 6 represents the data flow between different layers illustrating the CNN architecture.

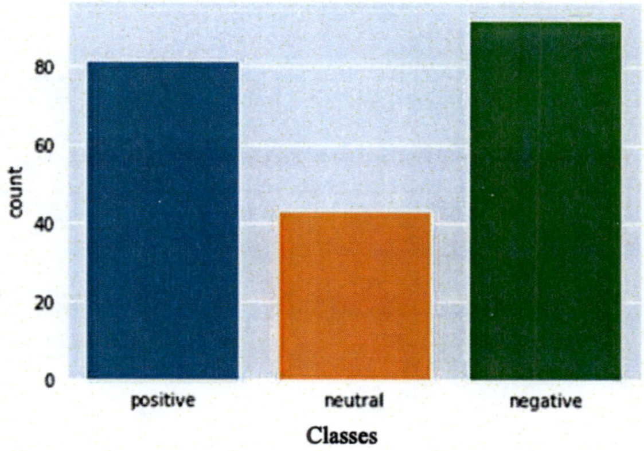

Fig. 3 Training data statistics

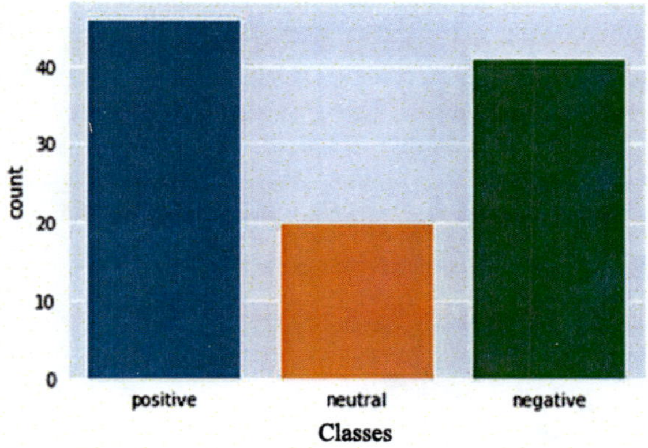

Fig. 4 Testing data statistics

Table 1 describes the accuracy and loss of the CNN model when computed for training and testing datasets.

The above results showcase the accuracy of the CNN model in classifying the images of emojis present in training and testing datasets when plotted against the number of epochs represented in Fig. 7.

4.4 *CNN on Emojized Tweets*

Figure 8 is of the dataset obtained after tweets are processed by DeepMoji and CNN model. It elucidates user details such as name, tweet_id and the location from which the tweet was sent. Also, it contains the tweet in textual form and 'Emojized' form along with emoji polarity computed by CNN.

```
Model: "sequential"
_________________________________________________________________
Layer (type)                 Output Shape              Param #
=================================================================
conv2d (Conv2D)              (None, 224, 224, 32)      896
_________________________________________________________________
max_pooling2d (MaxPooling2D) (None, 112, 112, 32)      0
_________________________________________________________________
conv2d_1 (Conv2D)            (None, 112, 112, 32)      9248
_________________________________________________________________
max_pooling2d_1 (MaxPooling2 (None, 56, 56, 32)        0
_________________________________________________________________
conv2d_2 (Conv2D)            (None, 56, 56, 64)        18496
_________________________________________________________________
max_pooling2d_2 (MaxPooling2 (None, 28, 28, 64)        0
_________________________________________________________________
dropout (Dropout)            (None, 28, 28, 64)        0
_________________________________________________________________
flatten (Flatten)            (None, 50176)             0
_________________________________________________________________
dense (Dense)                (None, 128)               6422656
_________________________________________________________________
dense_1 (Dense)              (None, 3)                 387
=================================================================
Total params: 6,451,683
Trainable params: 6,451,683
Non-trainable params: 0
_________________________________________________________________
```

Fig. 5 CNN model summary

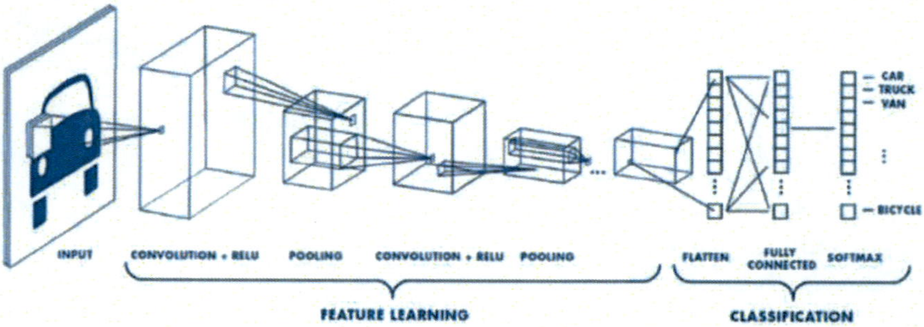

Fig. 6 CNN architecture

4.5 Top Emojis

This graph indicates the top five emotions that were conveyed after the tweet was converted to emoji form that were further analysed with CNN (Figs. 9 and 10).

Frequency of emojis obtained from the training dataset was analysed as above.

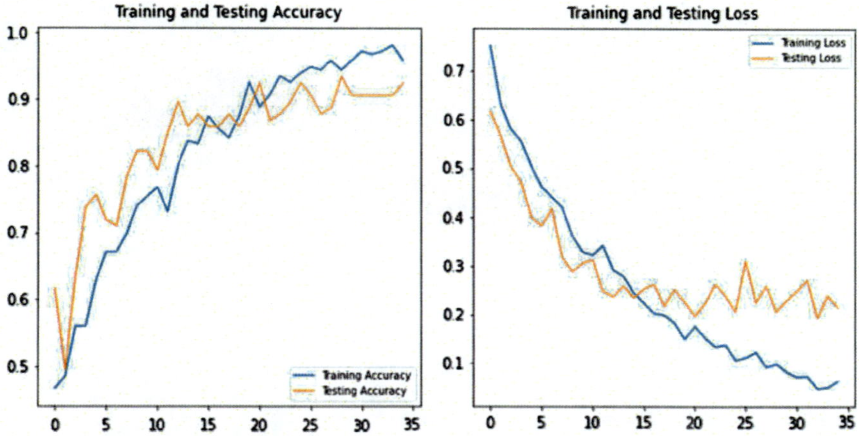

Fig. 7 Training and testing data accuracy and loss

	Tweets	Emojized	emocode	EmoPol
0	Sad to see go the #RaulVinci way in stutter...	😞	[102485710257569750]	-1.0
1	im sad that you've graduated but happy that i...	😞	[102485710257569750]	-1.0
2	Sad that a comment from bluemelon didn't bear...	😞	[102485710257569750]	-1.0
3	oki ^^\nUff, ist der so sad? 😞	❤️	[102485710257505751]	1.0
4	Sad to see him go. Such an iconic voice, he wi...	😞	[102485710257569750]	-1.0
...	...	...	...	...
195	Today, I had the unfortunate loss of a Tulsi p...	😞	[102485710257569750]	-1.0
196	When you're happy, you enjoy the music. But wh...	🎶	[102485710256101985 4]	0.0
197	all the best brother. Don't feel sad this wil...	👍	[102485710257495610 0]	1.0
198	It has been 30 plus days.... We are not disap...	❤️	[102485710257505752]	-1.0
199	very sad....\n\nhelpline # 104 !!	😞	[102485710257569750]	-1.0

200 rows × 4 columns

Fig. 8 CNN results with emoji polarities and user details

Table 1 CNN model accuracy

Model	Training set		Testing set	
	Accuracy	Loss	Accuracy	Loss
CNN	**0.98**	**0.06**	**0.92**	**0.26**

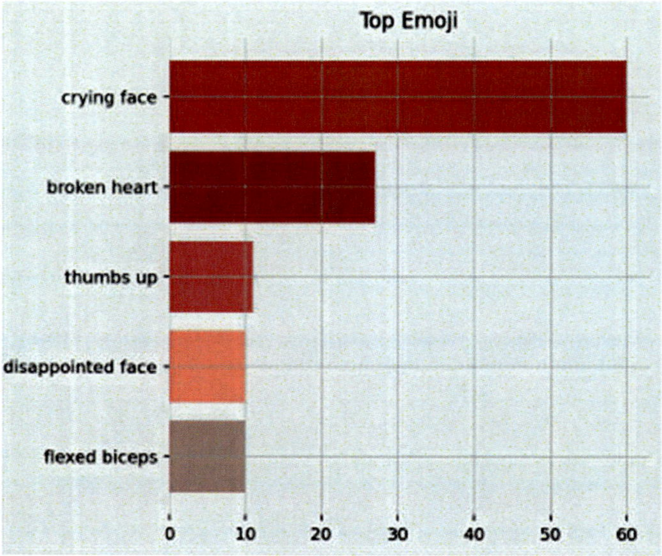

Fig. 9 Top five frequently occurred emojis

Fig. 10 Frequency of emojis

Fig. 11 List of emojis that were classified as negative by the model on real-time data

4.6 Number of Emojis Classified by CNN

Figure 11 represents all those emojis that were classified as negative after being processed by CNN obtained from real-time data.

Figure 12 represents all those emojis that were classified as positive after being processed by CNN obtained from real-time data.

Figure 13 represents all those emojis that were classified as neutral after being processed by CNN obtained from real-time data.

Insights:

- 50–52% of the negative tweets are correctly classified as negative by CNN.
- Nearly, 20% of the tweets from the dataset of 10,866 tweets were classified as positive.
- Around 30% of tweets were classified as neutral.

4.7 Word Cloud Representations

This word cloud represents words according to their weightage of occurrence from the tweets collected.

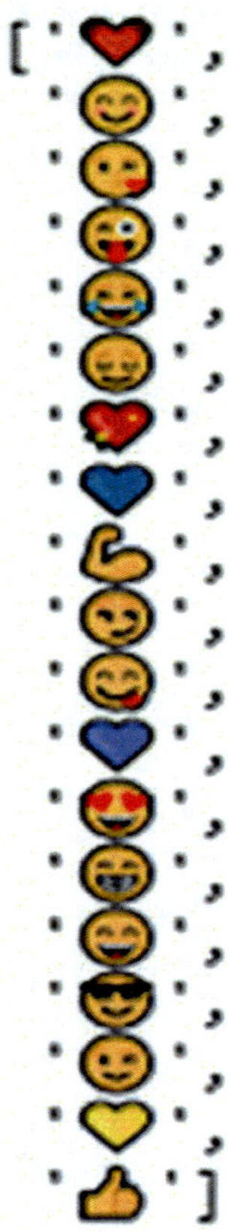

Fig. 12 List of emojis that were classified as positive by the model on real-time data

This word cloud indicates the words in the textual form of the tweet that are identified as negative by analysing the emoji polarity of the processed tweet (Figs. 14 and 15).

4.8 In-Platform Reply

Figure 16 illustrates the practical application of the project by providing an in-platform reply to the distressed from the account 'There's Always a Tomorrow' (Fig. 17).

Limitations:

Though all the 10,866 tweets were correctly identified to be negative by VADER sentiment analyser, the CNN model could classify only 50% tweets as negative because of:

- a cap defined on the amount of emojis considered.

Fig. 13 List of emojis that were classified as neutral by the model on real-time data

- DeepMoji's approach of considering certain keywords and their weightage while converting the text into just 1 emoji.
- errors in classification by CNN which are inexorable.

5 Conclusions and Future Works

5.1 Conclusions

- Extract tweets using Twitter API.

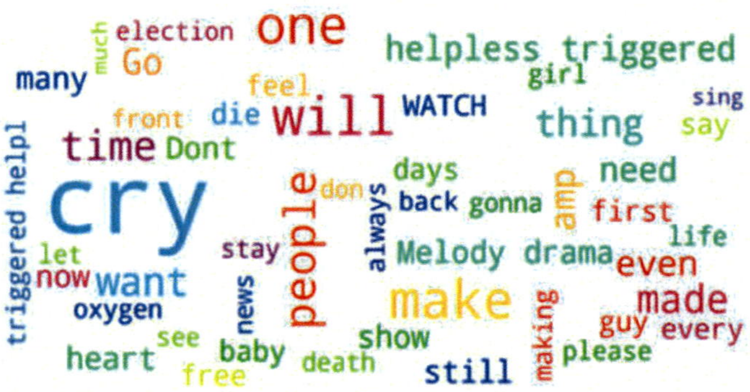

Fig. 14 Word cloud indicating frequently occurred words in tweets collected

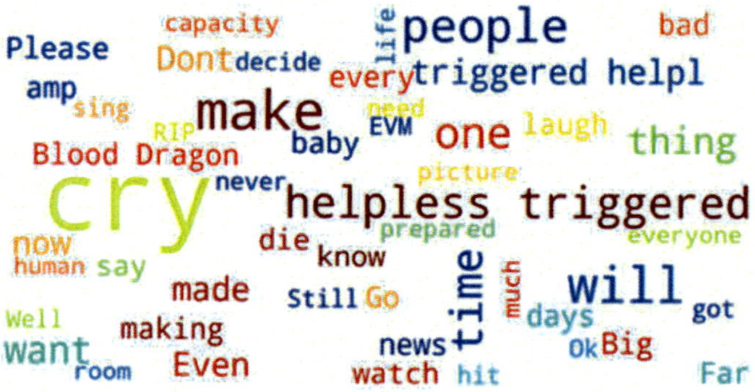

Fig. 15 Word cloud indicating frequently occurred words in tweets identified as negative

- Convert the tweets into emojis using DeepMoji.
- The whole project was an attempt of collaborating sentiment analysis and emoji to help the distressed people who put out their agony and pain in twitter.
- Twitter API was used for extraction of real-time tweets, and DeepMoji facilitated the process of conversion of tweets into emoji.
- The extracted tweets were classified into positive, negative and neutral with emoji polarities + 1, −1 or 0, respectively, with the help of images of the emoji using CNN model.
- Perform classification of tweets into positive, neutral or negative with emoji polarities of + 1, 0 or −1, respectively, with the help of images of emojis using CNN.
- Upon obtaining the details of tweets identified as negative such as the username, tweet ID and location, reply to those tweets with statements that offer support and hope along with the contact details of professional helplines were sent.

Fig. 16 Reply to the distressed

5.2 *Future Works*

- Detection and elimination of sarcasm in tweets.
- Negative tweets include emotions like anger and depression. Only the tweets expressing emotions bound around depression are considered for the project.

Fig. 17 Twitter account to provide in-platform reply

References

1. O'Dea B, Wan S, Batterham P, Calear A, Paris C, Christensen H (2015) Detecting suicidality on Twitter. Internet Intervent 2(2):183–188
2. Wiesław W (2016) Sentiment analysis of Twitter data using emoticons and emoji ideograms. Studia Ekonomiczne 296:163–171
3. Hankamer D, Liedtka D Twitter sentiment analysis with emojis
4. Daiphule L, Reddy B, Savith A, Apoorva TV (2019) Tracking suicidal tendency using twitter data and machine learning algorithms. Int J Eng Adv Technol 8:188–191
5. Guibon G, Ochs M, Bellot P (2016) From emojis to sentiment analysis. In: WACAI 2016, Brest
6. Kralj Novak P, Smailović J, Sluban B, Mozetič I (2015) Sentiment of emojis. PLOS ONE 10
7. Penchalaiah P, Nikhitha K, Devi R, Ramya S (2019) Detection of suicide related posts in Twitter data streams. Int J Res Eng IT Soc Sci 9:81–86
8. Billal B, Fatiha S, Mounir B, Hakim L (2020) Towards a multi-dataset for complex emotions learning based on deep neural networks. In: Language resources and evaluation conference, European Language Resources Association, Marseille, pp 50–58
9. Felbo B, Mislove A, Søgaard A, Rahwan I, Lehmann S (2017) Using millions of emoji occurrences to learn any-domain representations for detecting sentiment, emotion and sarcasm. In: Proceedings of the 2017 conference on empirical methods in natural language processing, pp 1615–1625
10. Mathew S (2020) On device deep neural networks for emoji and reply prediction. Int Res J Eng Technol (IRJET) 7:7818–7824
11. Karthik V, Nair D, Anuradha J (2018) Opinion mining on emojis using deep learning techniques. Proc Comput Sci 132:167–173
12. Mitra A (2020) Sentiment analysis using machine learning approaches (Lexicon based on movie review dataset). J Ubiquitous Comput Commun Technol 2(3):145–152. Available: https://www.irojournals.com/jucct/V2/I3/04.pdf. Accessed 15 September 2021

Study and Comparision of "*MIMO-OFDM*" *Under Acoustic Communication Systems*

S. Sankar Ganesh and S. Rajaprakash

Abstract High-velocity based underwater acoustic (UWA) correspondence is difficult due to the limited transfer speed, broadened multipath spread, significant Doppler effect, and massive transmission incompatibility. The use of different transducers at both the transmitter and the collector increases the spatial variation of the framework, and further orthogonal frequency-division multiplexing (OFDM) addresses the issue of limited transmission capacity due to its high unearthly proficiency. As a result, the combination of different information from multiple-input multiple-output (MIMO) and OFDM tends to raise the UWA correspondence to an unprecedented level, while increasing the planning difficulties for the turn of events of such frameworks. The signal contains intersymbol impedance, intercarrier obstruction, and co-channel impedance as a result of the long multipath and Doppler shift, as well as various types of commotions caused by the boisterous acoustic channel. As a result, numerous channel assessment and adjustment plans for the UWA MIMO-OFDM correspondence frameworks are developed. This paper discusses and compares the time complexity and performance for channel and equalization schemes.

Keywords Estimation of channel · Scheme for equalizers · Multiple-input Multiple-output (MIMO) · Orthogonal frequency-division multiplexing (OFDM) modulation · Underwater communication

1 Introduction

The sea assists on a shallow level by compelling winds, heat, and new water by the quiet further ocean, where this glow and new water is separated and then overall scales with longer time is liberated. The top layer is combined with the upper layer as a consequence of the climate, and a considerably depicted zone is mixed in the under layer, where vertical positions are hard. While the surface layer has been fully introduced to the environment in the end, water is highly crucial in clearly revealing

S. Sankar Ganesh (✉) · S. Rajaprakash
Department of CSE, Aarupadai Veedu Institute of Technology, Vinayaka Mission's Research Foundation, Salem, Tamil Nadu, India

S. Shakya et al. (eds.), *Proceedings of Second International Conference on Sustainable Expert Systems*, Lecture Notes in Networks and Systems 351,
https://doi.org/10.1007/978-981-16-7657-4_61

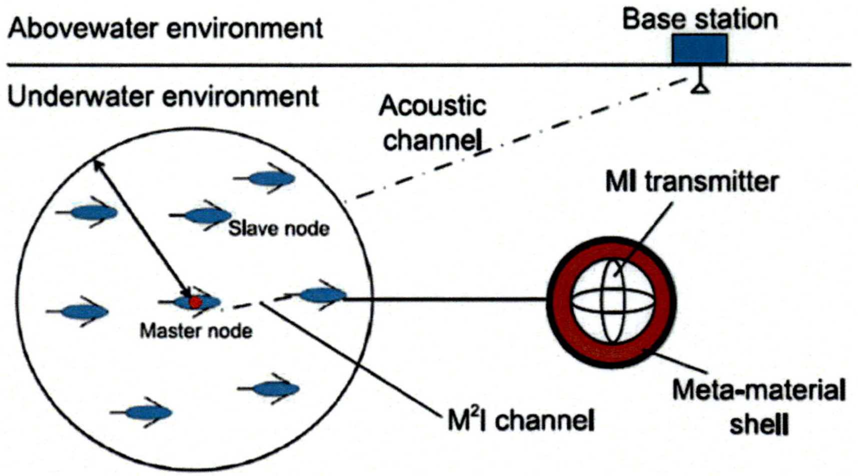

Fig. 1 Various MIMO systems

the mixed layer inside. The reflection of the layer is vertical of the surface development isn't changing instantly to obliging the climate anyway changes essentially by earlier forcing.

The driving actions that happened locally may be in the region or indirectly at various regions and moved via sea currents. This sea vertical plan describes the characteristics of the sea vertical development and examines how the stays are aware of them. They show various occasions of shapes changeability via air convincing of the sea plans is vertical. This vertical development taken on the assortment, and the one robot has multiple transmitters. Each robot hold the antenna for multiple MIMO is shown in Fig. 1.

The sea is vertical development is portrayed by pungency and temperature, and control of the structure of water thickness. The surface of the sea layers, several autonomous layers can be distinguished by various cycles with different timescales: Many layers are mixed with upper layer, intermittent pycnocline, and the amazingly strong pycnocline. Hence, the significance of the surface layer is portraying to change the water properties that are made comparative with the assessment of mass sea surface temperature (SST), and the primary idea of antenna signals in multipath is shown in Fig. 2.

2 Related Work

Zhangyu Li et al. proposed and implemented the testbed for software under the space–time coding and beamforming scheme [1]. Yanlong Li et al. analyzed the lens structure imaging of underwater imaging light visible "MIMO communication

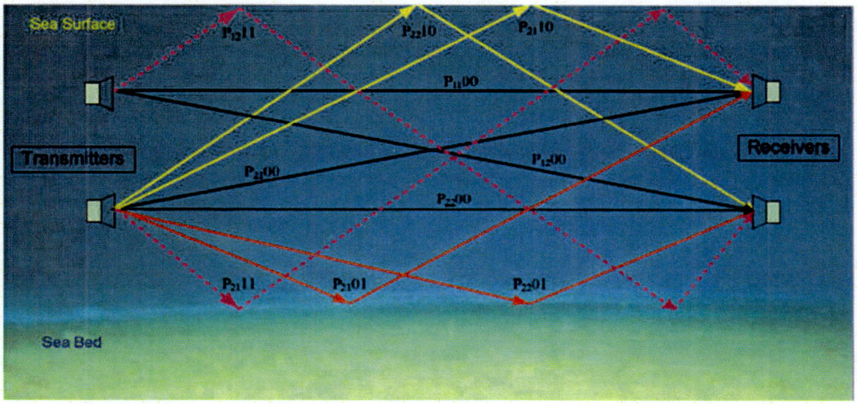

Fig. 2 Multiple effects

channel" for the cause of the spatial correlation [2]. Yuhang Song et al. proposed the MIMO-OFDM for "underwater wireless optical communication" and compared the performance of OFDM [3]. Gang Qiao et al. studied the four techniques such as equalization, channel estimation, detection, and coding. These techniques are used to compare the performance and complexity of the algorithms [4]. Yue-hai Zhou et al. proposed the continuous data blocks to improve the performance of the time-reversal receiver [5]. Aiping Huang et al. proposed the power allocation algorithm is to minimize the average for the BER of SMUOMIMO system [6]. Nelson Iruthayanathan et al. studied the high rate for spreading sequence in acoustic signal [7]. Prasad Naik Ramavath et al. proposed the RS-coded MIMO UWOC system is provide the efficiency is high in IoUT applications [8]. Alberto Alvarez Diaz et al. designed the MIMO controllers for the UUV dynamical model [9]. N. W. A. Sowerby et al. finding the two types of signals such as random noise and multi-level pseudorandom signals and compared them [10]. M. L. Corradini et al. have presented the underwater "Remotely Operated Vehicle (ROV)" for orientation control [11]. Xiang Pan et al. have proposed a model of sonar TR-MIMO for improving the detection of waveguide environment [12]. Gang Qiao et al. developed the equalization schemes and channel estimation for communication systems in the UWA MIMO-OFDM. These systems are used to compare the performance and complexity of the algorithms [13].

3 Discussions

This section discusses about the different schemes. This scheme has two methods such as channel estimation (CS) scheme and equalization (EQ) scheme. The CS scheme has identified the specific feature for different papers with time complexity

Table 1 Time complexity for channel estimation schemes

S. No.	Authors	Specialized feature	Time complexity
1	Li et al. [15]	"LS channel estimation"	CS
2	Huang et al. [20]	"CS based channel estimation"	CS
3	Berger et al. [22]	"OMP channel estimation"	CS
4	Sun et al. [16]	"STBC encoded LS estimation"	CS
5	Carrascosa and Stojanovic [14]	"Constant transfer function assumption"	CS
6	Stojanovic [17]	"Reduced Matrix size"	CS
7	Hua-Nan and Shu-Xu [23]	"CS based channel estimation"	CS
8	Kim [24]	"Angle domain, frequency selective and OMP"	CS
9	Bu et al. [25]	"Increased spectral efficiency"	CS
10	Zhu [18]	"Cost function based feedback"	CS
11	Zhang et al. [19]	"HRLS-DCD"	CS
12	Emre [27]	"Eliminated channel estimation"	CS
13	Eghbali et al. [26]	"Eliminated channel estimation"	CS

of the schemes as shown in Table 1. There are various science and applications, which use the modern-day research techniques [36–43].

The second scheme is an equalization scheme. This scheme also identified the specific feature for different papers with performance of the schemes are shown in Table 2.

4 Conclusion

This paper has provided a rundown of nearly every distributed works regarding the channel assessment and balance of UWA MIMO-OFDM frameworks. This research work has focused on various issues concerning channel assessment and equalization. Finally, this research work has discussed and compared the time complexity and performance of channel and equalization schemes.

Table 2 Performance for equalization schemes

S. No.	Authors	Specialized feature	Performance
1	Bu et al. [25]	"MSME with a priori information"	EQ
2	Lan et al. [28]	"SINR-OSIC-MMSE equalize"	EQ
3	Grythe and Hakegard [31]	"Extreme values for ICI and ISI"	EQ
4	Palou and Stojanovic [33]	"ZF-DFE"	EQ
5	Huang et al. [21]	"MMSE & MCMC"	EQ
6	Ma et al. [34]	"Blind Equalization"	EQ
7	Tu et al. [32]	"Multiuser MIMO"	EQ
8	Beheshti et al. [35]	"Used bank of time varying filters"	EQ
9	Tao [29]	"Channel estimation outside loop"	EQ
10	Zhang et al. [30]	"Space frequency block coding"	EQ
11	Zhang et al. [19]	"DA-TEQ"	EQ

References

1. Li Z, Desai S, Sudev VD, Wang P, Han J, Sun Z (2020) Underwater cooperative MIMO communications using hybrid acoustic and magnetic induction technique. Comput Netw 173
2. Li Y, Qiu H, Chen X, Fu J, Musa M, Li X (2019) Spatial correlation analysis of imaging MIMO for underwater visible light communication. Opt Commun 443:221–229
3. Song Y, Lu W, Sun B, Hong Y, Qu F, Han J, Zhang W, Xu J (2017) Experimental demonstration of MIMO-OFDM underwater wireless optical communication. Opt Commun 403:205–210
4. Qiao G, Babar Z, Ma L, Liu S, Wu J (2017) MIMO-OFDM underwater acoustic communication systems—a review. Phys Commun. https://doi.org/10.1016/j.phycom.2017.02.007
5. Zhou Y, Jiang W, Tong F, Zhang G (2017) Exploiting joint sparsity for underwater acoustic MIMO communications. Appl Acoust 116:357–363
6. Huang A, Tao L, Niu Y (2018) Underwater wireless optical MIMO system with spatial modulation and adaptive power allocation. Opt Commun 412:21–27
7. Iruthayanathan N, Vishvaksenan KS, Rajendran V, Mohankumar S (2015) Performance analysis of turbo-coded MIMO–OFDM system for underwater communication. Comput Electr Eng 43:1–8
8. Ramavath PN, Udupi SA, Krishnan P (2020) High-speed and reliable underwater wireless optical communication system using multiple-input multiple-output and channel coding techniques for IoUT applications. Opt Commun 461
9. Diaz AA, Cait A (2003) Evolutionary design of MIMO controllers for basic manoeuvring of unmanned underwater vehicles. Elsevier, Amsterdam
10. Sowerby NWA, Roberts GN (2003) A comparison of perturbation signals for mimo model-matching of underwater vehicles. Elsevier, Amsterdam
11. Corradini ML, Orlando G (1996) An adaptive mimo VSC algorithm, and its application to the control of an underwater vehicle. In: 13th Triennial world congress, San Francisco, USA
12. Pan X, Hui S, Wena X, Ross Chapman N (2014) TR-MIMO detection of a small target in a shallow water waveguide environment. Appl Acoust 79:16–22
13. Qiao G, Babar Z, Ma L, Ahmed N (2019) Channel estimation and equalization of underwater acoustic MIMO-OFDM systems: a review (Estimation Du Canal Et Legalisation Des Systems MEMS-MROF Acoustiques Sous-Marins: Une Revue). Can J Electr Comput Eng 42(4):199–208 (Fall 2019)
14. Carrascosa PC, Stojanovic M (2008) Adaptive MIMO detection of OFDM signals in an underwater acoustic channel. In: Proceedings of OCEANS, pp 1–7

15. Li B, Zhou S, Stojanovic M, Freitag L, Huang J, Willett P (2007) MIMO-OFDM Over an underwater acoustic channel. In: Proceedings of OCEANS, pp 1–6
16. Sun ZX et al (2012) Pilots updating channel compensation base on underwater MIMO-OFDM. Appl Mech Mater 198–199:1761–1767
17. Stojanovic M (2009) Adaptive channel estimation for underwater acoustic MIMO OFDM systems. In: Proceedings of the 5th IEEE signal processing education workshop (DSP/SPE), IEEE 13th digital signal processing workshop, Jan 2009, pp 132–137
18. Zhu A (2018) Feedback strategies for iterative channel estimation in OFDM underwater acoustic communications. High Technol Lett 24(4):396–401. https://doi.org/10.3772/j.issn.1006-6748.2018.04.008
19. Zhang Y, Zakharov YV, Li J (2018) Soft-decision-driven sparse channel estimation and turbo equalization for MIMO underwater acoustic communications. IEEE Access 6:4955–4973
20. Huang J, Huang J, Berger CR, Zhou S, Willett P (2009) Iterative sparse channel estimation and decoding for underwater MIMO-OFDM. In: Proceedings of the OCEANS, pp 1–8
21. Huang J, Berger CR, Zhou S, Huang J (2010) Comparison of basis pursuit algorithms for sparse channel estimation in underwater acoustic OFDM. In: Proceedings of the IEEE OCEANS, Sydney, NSW, Australia, May 2010, pp 1–6
22. Berger CR, Zhou S, Preisig JC, Willett P (2010) Sparse channel estimation for multicarrier underwater acoustic communication: from subspace methods to compressed sensing. IEEE Trans Signal Process 58(3):1708–1721
23. Hua-Nan Y, Shu-Xu G (2012) Channel estimation for MIMO-OFDM underwater acoustic communication based on compressed sensing. Syst Eng Electron 34(6):1252–1257
24. Kim S (2012) Angle-domain frequency-selective sparse channel estimation for underwater MIMO-OFDM systems. IEEE Commun Lett 16(5):685–687
25. Bu R, Shilian W, Xiaoyou Y (2016) Blind channel estimation and phase ambiguity elimination in MIMO-OFDM UWA communications. In: Proceedings of the IEEE international conference on signal processing, communications and computing (ICSPCC), Aug 2016, pp 1–6
26. Eghbali H, Stojanovic M, Muhaidat S (2014) Differential decoding for SFBC OFDM systems in underwater MIMO channels. In: Proceedings of the IEEE international conference on acoustics, speech and signal processing (ICASSP), May 2014, pp 8102–8105
27. Emre Y (2008) Multi-input multi-output OFDM for shallow-water UWA communications. In: Proceedings of the Acoustics, Paris, France, pp 5333–5338
28. Lan Z, Kang T, Song HC, Hodgkiss WS, Xu X (2013) MIMO OFDM acoustic communication in shallow water. In: Proceedings of the OCEANS, San Diego, CA, USA, pp 1–4
29. Tao J (2016) Turbo equalization for MIMO SC-FDMA underwater acoustic communications. In: Proceedings of the MTS/IEEE Monterey OCEANS, Sept 2016, pp 1–5
30. Zhang L, Han J, Huang J, Zhang Q (2016) Iterative channel estimation and equalization for underwater acoustic MIMO SFBC OFDM communication. In: Proceedings of the IEEE/OES China Ocean Acoustics (COA), Jan 2016, pp 1–6
31. Grythe K, Hakegard JE (2009) Non-perfect channel estimation in OFDM-MIMO-based underwater communication. In: Proceedings of the OCEANS Europe, pp 1–9
32. Tu K, Duman TM, Stojanovic M, Proakis JG (2013) Multiple resampling receiver design for OFDM over doppler-distorted underwater acoustic channels. IEEE J Ocean Eng 38(2):333–346
33. Palou G, Stojanovic M (2009) Underwater acoustic MIMO OFDM: an experimental analysis. In: Proceedings of the OCEANS, pp 1–8
34. Ma X, Zhao C-H, Qiao G (2009) The underwater acoustic MIMO OFDM system channel equalizer basing on independent component analysis. In: Proceedings of the WRI international conference on communications and mobile computing (CMC), pp 568–572
35. Beheshti M, Omidi MJ, Doost-Hoseini AM (2011) Joint ICI and IBI cancelation for underwater acoustic MIMO-OFDM systems. In: Proceedings of the 19th Iranian conference on electrical engineering, pp 1–5
36. Prakash UM, Kottursamy K, Cengiz K, Kose U, Hung BT (2021) 4x-expert systems for early prediction of osteoporosis using multi-model algorithms. Measurement 180:109543

37. Banupriya S, Kottilingam K (2021) An analysis of privacy issues and solutions in public blockchain (bitcoin). In: 2021 2nd International conference for emerging technology (INCET). IEEE, pp 1–7
38. Saranya A, Kottilingam K (2021) A survey on bone fracture identification techniques using quantitative and learning based algorithms. In: 2021 International conference on artificial intelligence and smart systems (ICAIS). IEEE, pp 241–248
39. Banupriya S, Kottursamy K, Bashir AK (2021) Privacy-preserving hierarchical deterministic key generation based on a lattice of rings in public blockchain. Peer-to-Peer Netw Appl 1–13
40. Nallarasan V, Kottilingam K (2021) Spectrum management analysis for cognitive radio IoT. In: 2021 International conference on computer communication and informatics (ICCCI). IEEE, pp 1–5
41. Indukuri CL, Kottursamy K (2021) Advanced accident avoiding, tracking and SOS alert system using GPS module and Raspberry Pi. In: Artificial intelligence techniques for advanced computing applications. Springer, Singapore, pp 167–178
42. Tharun KS, Kottilingam K (2018) Optimization load balancing over imbalance datacenter topology. In: International conference on computational vision and bio inspired computing. Springer, Cham, pp 397–407
43. Raja G, Kottursamy K, Theetharappan A, Cengiz K, Ganapathisubramaniyan A, Kharel R, Yu K (2020) Dynamic polygon generation for flexible pattern formation in large-scale UAV swarm networks. In: 2020 IEEE Globecom workshops (GC Wkshps). IEEE, pp 1–6

Comparative Study of Robust Feature Extraction Techniques for ASR for Limited Resource Hindi Language

Hetal Gaudani and Narendra M. Patel

Abstract Hindi is the most spoken language in India. It is desirable to have a communication system in a local language which permits ordinary people to communicate with machines via speech interface to retrieve information or to perform their daily activities. It is observed that conventional automatic speech recognition systems which use the Mel frequency cepstral coefficient or perceptual linear prediction as features do not perform well in noisy environments. Here, we have studied different feature extraction techniques in noisy environments, and experiment results show that gammatone frequency cepstral coefficients perform well in noisy environments. We have used DNN-HMM with a trigram language model to recognize Hindi speech. Compared to other approaches, our proposed algorithm outperforms. In comparison to MFCC and PLP characteristics, the basilar-membrane frequency-band cepstral coefficient likewise produces good results.

Keywords Automatic speech recognition · BFCC · PLP · GFCC · MFCC · HMM · GMM · Deep neural network

1 Introduction

Speech is a method by which humans communicate with one another. Nowadays, it is being used to communicate with machines also. In the discipline of signal processing, speech recognition is a challenging problem to solve. Speech recognition takes the audio signal as input and gives word sequence of an utterance as an output. Recognized words can be used as input to many voice automated applications. Hindi, India's official language, is the most frequently spoken language in the country. More than 425 million people speak Hindi as their primary language, with another 120 million

H. Gaudani (✉)
Gujarat Technological University, Ahmedabad, Gujarat, India
e-mail: hetalgaudani@gcet.ac.in

N. M. Patel
Birla Vishvakarma Mahavidyalaya, V. V. Nagar, Gujarat, India
e-mail: nmpatel@bvmengineering.ac.in

S. Shakya et al. (eds.), *Proceedings of Second International Conference on Sustainable Expert Systems*, Lecture Notes in Networks and Systems 351,
https://doi.org/10.1007/978-981-16-7657-4_62

using it as a secondary language. It is desirable to have such a communication system in a local language which permits ordinary people to communicate with machines to retrieve information. Although enormous research has been made during the last four decades in the domain of ASR, there is still a considerable gap between human and machine performance due to their lack of robustness in noisy environments [1, 2]. The presence of noise in speech inputs makes a standard ASR system challenging to deploy [3]. The speech signal's intelligence suffers as a result of the noise. In ASR system, the front-end is signal modelling which extracts acoustic properties from input speech signals using particular feature extraction methods. Noise resilient feature extraction strategies can help to construct robust ASR systems in this case [4–6]. Many techniques like MFCC [7], PLP [8], RASTA PLP [9–11] and LPCC [12] have been proposed to generate features from speech utterance, but MFCC [7] is most widely used and most accurate feature extraction method and in clean environment conditions, it functions admirably. In the presence of additive noise, however, it does not perform well. As a result of MFCC's lack of resilience to background noise, noise resilient features such as GFCC [13–16] and BFCC [17] have been used.

The goal of this research is to look at how noise affects a number of traditional and hybrid feature extraction methods in a controlled environment. We have used MFCC, BFCC [17], GFCC [13] and PLP feature extraction techniques for our study. MFCC and PLP algorithms have been widely used for feature extraction over many years. While GFCC is mostly used in speaker verification and speaker recognition. But experimental results show that it outperforms in speech recognition also. Here, distinct filter banks are utilised by different feature extraction strategies. The nonlinear Gammatone filterbank is utilised by GFCC, the Mel scale filter bank is used by MFCC, BFCC uses gammachirp filterbank and PLP uses barkscale filterbank. Generated features are matched with reference modals to generate the recognition result.

The following is how the paper is structured: Feature extraction techniques are briefly described in Sect. 2. Section 3 discusses the suggested architecture of ASR in detail, the experimental setup and findings are discussed in Sect. 4, and conclusion and future works are described in Sect. 5.

2 Feature Extraction Techniques

The purpose of feature extraction is to find a set of acoustically related properties in a voice signal. Feature extraction eliminates unnecessary and duplicated data. Various feature extraction approaches are described in this section.

2.1 Mel Frequency Cepstral Coefficient

The human vocal tract filters spoken sounds. The shape of vocal tract determines which sound is being produced. Specific shapes produce specific sounds of speech. The vocal tract's shape is embodied by the short-term power spectrum's envelope. MFCC accurately represents this envelope. The following steps are included in the MFCC feature extraction method, as shown in Fig. 1.

- The amount of energy in the high frequencies is increased via pre-emphasis. It has been noticed that lower frequencies have more energy than higher frequencies for voiced sounds like vowels. Boosting the high frequency will improve phone detection accuracy. Pre-emphasis is done by the first order high-pass filter as given below.

$$y(t) = x(t) - \alpha x(t-1) \quad \text{where } 0.95 < \alpha < 0.99 \tag{1}$$

- After pre-emphasis, the speech signal is segmented into short-time frames. A speech signal is a time-varying signal, yet the frequencies in a signal are stationary for a limited amount of time. For this reason, the signal is split into short frames. In speech processing, typical frame length ranges from 20 to 40 ms, with overlap between consecutive frames. If the signal is directly chopped off at the edges, then sudden fall in amplitude at the edges will produce noise in the high-frequency domain. So a Hamming window is applied in each frame to remove edge discontinuities.
- After windowing, the energy is separated into different frequency bands using the DFT. FFT is performed to each frame to extract frequency components of the

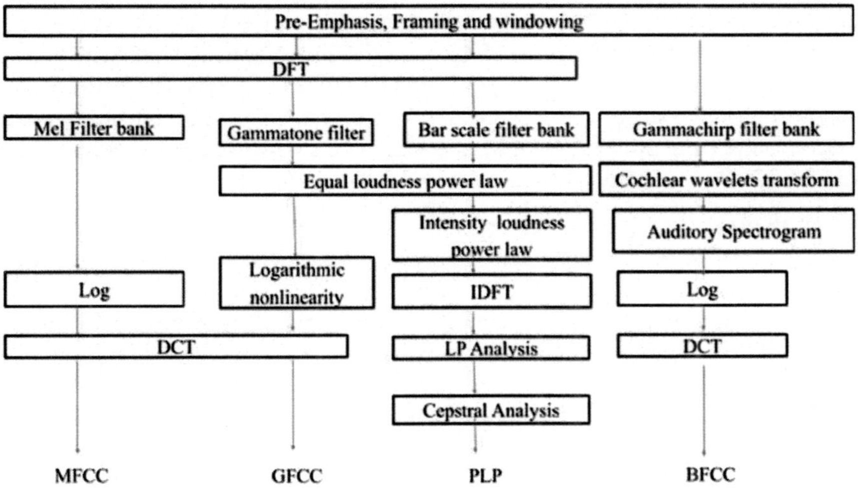

Fig. 1 Feature extraction methods

segmented voice stream. Next Mel spectrum of magnitude spectrum is computed by passing it through a Mel filter bank, which is a collection of band pass filters. Humans do not perceive pitch on a linear scale. Low-frequency pitch shifts are significantly easier for humans to detect than high-frequency pitch changes. The Mel scale compares the perceived frequency of a pre tone to its actual pitch measurement. When the Mel scale is applied to the power spectrum, the features are more closely related to what humans perceive. Frequency of speech signal can be converted to Mel scale using following formula.

$$\mathrm{Mel}(f) - 2595\log_{10}(1 + f/100) \tag{2}$$

- Resultant voice signals are converted into logarithmic forms to achieve the human perception of loudness. The energies of the filter banks are correlated. These energies are decorrelated by the DCT. The DCT is applied next to extract cepstral coefficients. The amplitudes of the resulting spectrum are the MFCCs. Only 12 coefficients are retained, and one more feature, the log energy of a frame, is added, resulting in a feature vector with 13 MFCC features. The power spectral envelope of a single frame is described by the MFCC feature vector, and delta (first-order derivative) and double delta are used to determine the MFCC coefficients' trajectories across time.

2.2 *Gammatone Frequency Cepstral Coefficients*

Most difficult task for any speech recognition system is to operate quite well in a noisy environment. Literature reveals that MFCC is the most common choice for building the ASR system. But it is observed that MFCC is sensitive to additive noise. Recent trends also indicate that due to better performance and robustness in noisy conditions, gammatone frequency cepstral coefficients (GFCC) replace the MFCC. GFCC is based on a collection of banks of gammatone filters and an equivalent rectangular bandwidth (ERB) scale. The gammatone filter bank's non-linear structure mimics the human hearing process. The filter bank is the main distinction between MFCC and GFCC. GFCC is calculated as a process shown in Fig. 1.

- Splitting the spoken signal into frames is the initial stage, followed by applying a hamming window. Using the discrete Fourier transform, convert the spoken signal to frequency domain.
- To obtain a time–frequency (T–F) representation, the Fourier transform output is passed through a 64-channel Gammatone filter bank, and each filtered output is properly corrected.
- The cubic root of the T–F representation is then extracted, and the cepstral coefficients are calculated using DCT. Following is the formula for the gammatone filter function [13]

$$g(f,t) = t^{n-1}\mathrm{e}^{-2\pi bt}\cos(2\pi ft), \tag{3}$$

where f is the frequency, b is the rectangle bandwidth, t is the time and $n = 4$ is the filter's order.

- The amplitudes of the resulting spectrum are the GFCCs. Only 13 coefficients are retained, and delta (first-order derivative) and double delta are used to determine the GFCC coefficients' trajectories across time.

2.3 Gammachirp Filterbank and BFCC

BFCC approach uses an auditory spectrogram based on a gammachirp wavelet transform to intimate the auditory response of the human inner ear to increase noise immunity, as opposed to Fourier spectrograms based on MFCC and GFCC methods. The gammachirp filter is a close approximation to the cochlea's frequency selective behaviour. It is an auditory filter that introduces asymmetry and level-dependent cochlear filter features, and it can be thought of as a generalisation and improvement of the gammatone filter. Figure 1 shows the basic flowchart for the proposed BFCC, in which the human inner ear can be approximated using a gammachirp filter [18] for high frequency selectivity.

- It does pre-processing and windowing, similar to MFCC and GFCC, to reduce discontinuities at edges. The following formula is used to normalise the speech signal.

$$s[n] = (X[n] - X_{\text{mean}})/X_{\text{max}} \tag{4}$$

where n is the sample index, $X[n]$ is the speech signal, X_{mean} is speech signal mean intensity, and X_{max} is the speech signal maximum intensity.

- To provide a more precise bandwidth for the auditory system, BFCC employs the Equivalent Rectangular Bandwidth (ERB) scale. ERB is determined using the following equation for any frequency value f (Hz):

$$\text{ERB}_f = (f/9.26) + 24.7 \tag{5}$$

- Gammachirp filter bank processes the normalised signal. The gammachirp filter's impulsional answer is given by the following function.

$$g(t) = t^{n-1}\lambda_n e^{-2\pi \text{ERB}(f_0 t)} e^{j(-2\pi f_0 t + c\ln(t) + \phi)} \tag{6}$$

- The Cochlear Wavelet Transformation (CWT) is used after the normalisation phase. As indicated in the following formula, CWT is calculated using normalised signal and the gammachirp function.

$$\text{CWT}(t) = \sum_{R}^{f=1} \frac{1}{\sqrt{n}} g(t) * s[n] \tag{7}$$

where n denotes the filter's order. The absolute value of CWT(t) is used to create an auditory spectrogram. Finally, to obtain BFCC characteristics, logarithmic and DCT operations are applied.

2.4 Perceptual Linear Prediction

PLP and MFCC are quite similar. Because PLP is inspired by auditory perception, it uses equal-loudness pre-emphasis and cube-root compression instead of log compression.

The fundamental flowchart for the PLP is shown in Fig. 1.

- The initial few steps are the same as GFCC, like splitting the spoken signal into frames, followed by applying a Hamming window. The discrete Fourier transforms are applied on segmented the spoken signals to convert into frequency domain.
- The Bark scale is set up so that each of the important bands of human hearing is one Bark wide. A closer correlation with spectral information processing in the ear is gained by representing spectral energy across the Bark scale. The bank of bark filters is applied to the power spectrum.
- Multiply the output of the filter banks by equal-loudness curve and compresses it to the cube-root to mimic the physiology of the human ear. Performed inverse DFT and used LP to get cepstral coefficient.

2.5 Comparison of Feature Extraction Techniques

All feature extraction algorithms, as seen in Fig. 1, perform pre-emphasis, framing and windowing, and then use different filters at a later stage. Two significant differences between MFCC and GFCC are frequency scale and non-linear rectification step before discrete cosine transform. In comparison to MFCC, GFCC uses an equivalent rectangular bandwidth scale, which provides finer precision at low frequencies and cubic-root compression which is more robust in noisy environments compared to logarithmic compression [19]. Unlike MFCC, PLP and GFCC, which employ the Fourier transform, BFCC uses a wavelet transform based on the gammachirp filter bank. In FFT, the frequency bands are linearly distributed; however, in the human cochlea, they are non-linearly distributed. In terms of frequency band distribution, the wavelet transformation approach used in BFCC chooses high frequency utilising a Gammachirp filter to represent the human ear, precisely replicating the human cochlea and resulting in better noise robustness compared to other mentioned techniques. Table 1 compares and contrasts the three techniques in terms of various parameters.

Table 1 Comparison of feature extraction techniques

Parameter	MFCC	PLP	GFCC	BFCC
Window function	Hamming	Hamming	Hamming	Hamming
Filterbank	Mel scale	Bark scale	Gammaton	Gammarchip
# of Frequency bands	23	23	64	64
Frequency scale	Mel	ERB	ERB	ERB
Cepstral liftering	Yes	No	No	No
Nonlinear rectification	Logarithmic	Cubic root	Cubic root	Logarithmic
Equal-loudness power law	No	Yes	Yes	No
Intensity-loudness power law	No	Yes	No	No
Technique to determine cepstral coefficient	DCT	LP model	DCT	DCT

3 Proposed Architecture

Speech recognition's main goal is to create a statistical model that can infer word sequences. If O is represents sequence of acoustic feature vectors and W denotes a word sequence, the most likely word sequence W^* for O is given by

$$W^* = \text{agr}\ \max(P(O/W)/P(W)) \tag{8}$$

The likelihood of feature O given the Word W is $P(W/O)$, while the word probability determined from language modal is $P(W)$.We find the text sequence with the highest likelihood by integrating the acoustic and linguistic models. The associated acoustic model for any given w is created by concatenating phone models to form words defined by a pronunciation modal. Words are mapped to phones using a pronunciation model. The acoustic model models a sequence of features to sequence of phones. The language model gives the likelihood of the word sequence. Given the preceding words, it predicts the following word. Here, we have used BFCC, GFCC PLP along with MFCC to extract features from utterances. To generate word sequence for given utterance, we have Gaussian Mixer-based HMM and DNN-based HMM model.

3.1 GMM-HMM Model

HMMs are the most popular models used in the area of continuous speech recognition due to its capability to model and match sequences that have inherent variability in length as well as acoustic characteristics. HMM represents a temporal pattern in the form of finite state networking [20]. In Hindi language, every word is expressed as a sequence of sub-word (or Phonemes) units. Every sub word unit is modelled by

HMM with several hidden states. Word HMMs are created by concatenating HMM of phonemes. Hindi is normally spoken using a combination of around 12 vowels and 36 consonants, so a total of 48 phonemes are used. So, a total of 48 HMM can be generated for a mono phone. But the acoustic phonetic context of speech changes depending on surrounding Phonemes. For example "/*n*" is not pronounced the same in "ten" and "tenth". Articulation of "/*n*" depends upon the phone previous to it and the phone next to it. To represent this characteristic, we have used context-dependent triphone HMM. It is often not possible to create a different model for each triphone. With an average of 48 phones, we'd need 48 × 48 × 48, or more than 110,000 models. We do not have enough data to view them all, so we use a decision tree to cluster them. Here, phonemes are represented by three state or five states HMM. HMM is characterized by following.

1. set of hidden states $S = s_1, s_2, \ldots s_n$
2. set of distinct observation symbol per state, $O = O_1, O_2, \ldots, O_M$
3. state transition probability, aij, which is the likelihood of moving from one state to another ($n \times n$ matrix)
4. Observation symbol probability B is the probability of on observation o_j generated form state s_i ($n \times m$ matrix)
5. The initial state distribution $\pi = \pi i$.

Given suitable values of S, O, A, B and π, HMM model is created for each phoneme of Hindi language. Acoustic modelling map feature vector to various HMM states, which is nothing but tri phoneme. The HMMs are trained to give the likelihood of phoneme at time t. Here, the observations likelihood are generated by Gaussian mixture models at each state. The audio features will be transcribed into a sequence of context-dependent phonemes by the GMM-HMM model.

The training of the HMM-GMM is done by Baum-Welch algorithm and expectation maximization (EM) to produce the best state sequence. After training, a WFST decoding graph is used to generate the best possible sequence of words. A combination of trained HMM models, pronunciation modals and language models is used to match patterns. Decoding graph takes the phoneme as input and gives lattices as output. WFST is described by

$$W = \min(\det(H\ C\ L\ G)), \tag{9}$$

where H represent the Hidden Markov Model, C represent phonetic context-dependency, L represents lexicon and G represent grammar, respectively, the $\circ$ represents the composition operation of WFST and determinization (det) is used to make deterministic searching easier and reduction (min) is used to decrease repetition in states and transitions. The input label on an arc in HCLG corresponds to a clustered context-dependent HMM state whereas the output label represents word. On both input and output labels, a special symbol ϵ may appear, indicating that no label is present. During the decoding, beam pruning is used to improve search efficiency.

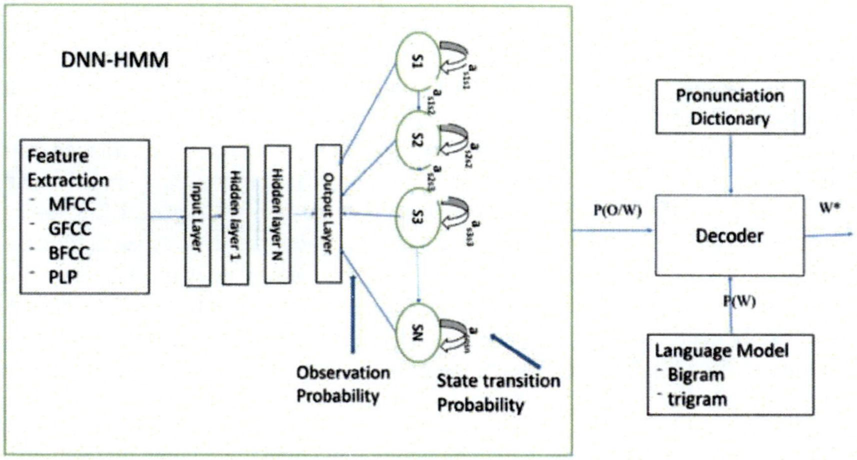

Fig. 2 DNN-HMM architecture

3.2 *Hybrid DNN-HMM*

The architecture of our suggested CD-DNN-HMMs is depicted in Fig. 2. The setup of this model is similar to the one proposed in [21]. Instead of using a Gaussian mixture model, we use a DNN to model the states of the hidden Markov model. The state transition probabilities are unchanged from the above triphone model but state emission probability will now be modelled by a deep neural network. Here, we have used two hidden layers and each hidden layer has 256 neurons in DNN. Softmax activation function is used in the output layer with an output dimension equal to the number of context-dependent states in the system. Each hidden layer in DNN uses sigmoid activation function and cross entropy loss function is used. Network is optimized through stochastic gradient descent optimization. The DNN is learned on LDA + MLLR features to produce emission probability for the HMM states. Input to the neural network is nine frames, four frames from the left side of the current frame and four frames from the right side of the current frame, with 40 dimension features for each frame.

4 Experimental Setup and Results

The availability of tagged speech data for training models under both clean and noisy environments determines the accuracy of any ASR system. A vast amount of speech corpus is available in languages like English for the creation of complicated ASR systems. Hindi is a language with a limited number of resources. It is lacking behind the availability of speech corpus for training. For our proposed system, well annotated and time aligned dataset prepared by [22] is used. This dataset contains data of seven speakers who speak out the 150 phonetically rich sentences in Hindi having maximum possible unique words and trying to capture distinct situations. Out of 150, twenty different mutually exclusive sentences from each speaker were used for testing. Total 140 utterances were used for testing and 910 utterances were used for training. Speech utterances are recorded with sampling frequency 16 kHz. We have utilized the MS-SNSD noise dataset constructed by Microsoft [23] which is openly available. This dataset contains a large collection of clean speech files and a variety of environmental noise files in wave format sampled at 16 kHz. Here, we have used babble noise and cafeteria noise for experiments which are non-stationary in nature. To generate noisy data, we have augmented clean speech data with babble and cafe noise [23]. The proposed system is evaluated using noise with different signal-to-noise ratio.

We have created a lexicon dictionary which maps words with phonemes. Bigram and trigram are used as language modals. Accuracy of ASR system is measures using word error rate which can be calculated as follow

$$\text{WER} = (S + D + 1)/N \tag{10}$$

where S is number of substitutions, D is number of deletion and I is number of insertion and N is number of words in the reference.

First, we have trained our GMM-HMM model on top of this features. Model transcribed the audio features into some sequence of context-dependent phonemes. Table 2 shows the result of an experiment that is evaluated on GMM-HMM acoustic model with bigram and trigram language model. Speech signals with a different signal-to-noise ratio are used for experiments. To extract features, the first speech signal is segmented into a 25 ms frame with 10 ms overlapping in subsequent frames. MFCC, GFCC, BFCC and PLP features are extracted for every frame. These features are evaluated on the GMM-HMM model. First experiment is carried out with the speech signal convoluted with babble noise at different signal-to-noise ratio while the second experiment is performed on cafeteria noise. Next, LDA is used to build the final features vector. The following are our baseline characteristics for LDA-based features: the 13-dimensional cepstral coefficient features are spliced in time with a context size of seven frames (i.e. three frame before and three frame after current frame), then de-correlation and dimensionality reduction using linear discriminant analysis (LDA) [24]. The maximum likelihood linear transform (MLLT) is used to further de-correlate the resultant features [25]. This 40-dimensional feature vector is used to train GMM-HMM and DNN-HMM models.

Tables 2 and 3 also show results of both acoustic models with babble and cafeteria noise, respectively. We have used two hidden layers of dimension 256 hidden nodes. Results in Tables 2 and 3 clearly show that GFCC-based systems out-perform the MFCC and PLP feature extraction method. Trigram language model also performs well compared to the bigram language model.

Table 2 Comparison of various feature extraction techniques on babble noise

Acoustic model	Feature extraction	BIGRAM			TRIGRAM		
		20 dB wer (%)	30 dB wer (%)	40 dB wer (%)	20 dB wer (%)	30 dB wer (%)	40 dB wer (%)
GMM-HMM	MFCC	14.87	8.22	6.94	10.79	6.65	5.36
	GCFF	9.8	6.82	5.42	7.76	5.66	4.08
	BFCC	14.87	8.1	6.52	10.79	6.12	5.23
	PLP	25.71	9.15	7.46	21.87	7.46	5.66
	MFCC + LDA	14.46	9.04	6.94	10.32	6.94	5.66
	GFCC + LDA	11.78	7.52	7	7.76	5.66	4.08
	BFCC + LDA	14.69	7.6	6.97	10.32	6.2	5.39
	PLP + LDA	21.4	8.45	8.63	17.49	6.41	6.24
DNN-HMM	MFCC + LDA	6.12	4.49	4.08	3.85	3.32	2.86
	GFCC + LDA	5.13	3	3.32	2.62	2.51	2.27
	BFCC + LDA	5.95	5.02	4.43	4.2	3.38	2.33
	PLP + LDA	12.71	4.66	3.97	3.92	3.21	2.97

5 Conclusion and Future Work

GFCC with deep neural networks improves the accuracy of speech recognition compared to MFCC-based traditional speech recognition model. Here, our proposed algorithm is trained on babble and cafeteria noise. In future, speech recognition can be made more robust by training it with other variety of noise and combined feature extraction methods. Various feature transformation, optimization and wavelet feature extraction techniques applied on the front-end and discriminatively training those features at the back-end of the ASR system can further extend this work.

Table 3 Comparison of feature extraction techniques on cafeteria noise

Acoustic model	Feature extraction	BIGRAM			TRIGRAM		
		20 dB wer (%)	30 dB wer (%)	40 dB wer (%)	20 DB wer (%)	30 dB wer (%)	40 dB wer (%)
GMM-HMM	MFCC	14.52	8.16	7.35	11.31	6.06	6.47
	GCFF	9.62	6.53	7.53	7.46	5.25	5.11
	BFCC	9.8	7.81	7.43	9.52	6.2	5.3
	PLP	18.43	8.22	6.59	15.92	7.11	5.42
	MFCC + LDA	12.13	8.92	6.59	9.1	7.11	5.66
	GFCC + LDA	9.5	7.23	6.59	7.41	5.89	5.66
	BFCC + LDA	10.52	7.56	6.47	9.98	5.95	5.72
	PLP + LDA	19.71	7.7	6.53	16.33	6.53	5.25
DNN-HMM	MFCC + LDA	4.72	4.49	4.49	2.86	3.15	2.92
	GFCC + LDA	4.62	3.73	4.66	2.74	2.51	2.74
	BFCC + LDA	4	4.08	3.32	2.87	3.38	3.21
	PLP + LDA	6.18	4.49	3.91	2.97	3.2	2.79

References

1. Acero A (1992) Acoustical and environmental robustness in automatic speech recognition, vol 201. Springer Science and Business Media
2. Acero A, Stern RM (1990) Environmental robustness in automatic speech recognition. In: International conference on acoustics, speech, and signal processing, IEEE
3. Rabiner L (1993) Fundamentals of speech recognition
4. Dua M, Sethi PS, Agrawal V, Chawla R (2021) Speaker recognition using noise robust features and LSTM-RNN. In: Panigrahi CR, Pati B, Pattanayak BK, Amic S, Li KC (eds) Progress in advanced computing and intelligent engineering. Advances in intelligent systems and computing, vol 1299. Springer, Singapore. https://doi.org/10.1007/978-981-33-4299-6_2
5. Kadyan V et al (2020) Developing in-vehicular noise robust children ASR system using Tandem-NN-based acoustic modelling. International J Veh Auton Syst 15(3–4):296–306
6. Naing HMS et al (2020) A front-end technique for automatic noisy speech recognition. In: 2020 23rd conference of the oriental COCOSDA international committee for the co-ordination and standardisation of speech databases and assessment techniques (O-COCOSDA), IEEE
7. Ittichaichareon C, Suksri S, Yingthawornsuk T (2012) Speech recognition using MFCC. In: International conference on computer graphics, simulation and modeling
8. Hermansky H (1990) Perceptual linear predictive (PLP) analysis of speech. J Acoust Soc Am 87(4):1738–1752
9. Hermansky H et al (1991) RASTA-PLP speech analysis. In: Proceedings IEEE international conference acoustics, speech and signal processing, vol 1

10. Gupta H, Gupta D (2016) LPC and LPCC method of feature extraction in speech recognition system. In: 2016 6th international conference—cloud system and big data engineering (confluence), IEEE
11. Këpuska V, Elharati H (2015) Robust speech recognition system using conventional and hybrid features of MFCC, LPCC, PLP, RASTA-PLP and hidden Markov model classifier in noisy conditions. J Comput Commun 3:1–9. https://doi.org/10.4236/jcc.2015.36001
12. Gulzar T, Singh A, Sharma S (2014) Comparative analysis of LPCC, MFCC and BFCC for the recognition of Hindi words using artificial neural networks. Int J Comput Appl 101(12):22–27
13. Dua M, Aggarwal RK, Biswas M (2020) Optimizing integrated features for Hindi automatic speech recognition system. J Intell Syst 29(1):959–976
14. Kadyan V, Shanawazuddin S, Singh A (2021) Developing children's speech recognition system for low resource Punjabi language. Appl Acoust 178:108002
15. Taherian H, Wang Z, Chang J, Wang D (2020) Robust speaker recognition based on single-channel and multi-channel speech enhancement. IEEE/ACM Trans Audio Speech Lang Process 28:1293–1302. https://doi.org/10.1109/TASLP.2020.2986896
16. Qu YQ, Long H, Duan Y (2021) An algorithm for calculating the contribution of acoustic features in speaker recognition. In: Yu Z, Patnaik S, Wang J, Dey N (eds) Advancements in mechatronics and intelligent robotics. Advances in intelligent systems and computing, vol 1220. Springer, Singapore. https://doi.org/10.1007/978-981-16-1843-7_33
17. Mishra AN et al (2011) Robust features for connected Hindi digits recognition. Int J Signal Process Image Process Pattern Recogn 4(2):79–90
18. Kuo HS, Sung PH, Lee SC, Kuan TW, Wang JF (2014) Auditory-based robust speech recognition system for ambient assisted living in smart home. In: 2014 IEEE international conference on orange technologies (ICOT), IEEE, Xian, China, pp 20–23
19. Jawarkar NP, Holambe RS, Basu TK (2015) Effect of nonlinear compression function on the performance of the speaker identification system under noisy conditions. In: Proceedings of the 2nd international conference on perception and machine intelligence
20. Samudravijaya K (2004) Automatic speech recognition. Tata Inst Fundam Res Arch
21. Veselý K et al (2013) Sequence-discriminative training of deep neural networks. In: Interspeech, vol 2013
22. https://github.com/KunalDhawan/ASR-System-for-Hindi-Language
23. Reddy CKA et al (2019) A scalable noisy speech dataset and online subjective test framework. arXiv preprint arXiv:1909.08050
24. Duda RO, Hart PE, Stork DG (2000) Pattern classification, Wiley
25. Gopinath R (1998) Maximum likelihood modeling with Gaussian distributions for classification. In: Proceedings of IEEE ICASSP, vol 2. pp 661–664
26. Sarma M, Sarma KK (2015) Speech recognition in Indian languages—a survey. In: Recent trends in intelligent and emerging systems, Springer, New Delhi, pp 173–187

Creating Sustainable Human Resource Management Systems for High-Performance Work Culture

Shweta Jha

Abstract The concept of sustainability revolves around concerted efforts by organisations to attain triple bottom line, i.e. people, profit and planet. In the last fifty years, sustainability has pervaded politics, governance, public policy, international relations and above all, the management of businesses and enterprises across the globe. Thus, the notion of sustainability has also transformed HR systems and practices over time. This study examines how sustainable HR systems have evolved and influenced organisational outcomes by creating high-performance work culture to meet and exceed the expectations of internal and external stakeholders without compromising profitability. Based on an integrative review of over 40 researches, the present study tries to find out the roadblocks in adopting sustainable HR systems, how sustainable HR can become part of mainstream management systems and what the leaders are doing about it. The study also presents a holistic view of the prospects of sustainable HR in the emerging scenario.

Keywords Human resource management · Sustainable HR systems · Sustainable HR practices · High-performance work culture

1 Introduction

The idea of sustainability revolves around 'meeting the needs of people today without compromising the ability of future generations to meet their own needs' [1]. In a typical business context, sustainability has been viewed as the capacity of an enterprise to accomplish its business goals and raise long-term shareholder value by integrating economic, environmental and social prospects into its corporate strategies. Several studies have indicated that business successes are closely linked to social and ecological outcomes [2–4]. In recent times, the investors have been particularly interested in the organisation's record vis-à-vis environment, health and governance. Even, the customers are more likely to patronise such companies, which come across

S. Jha (✉)
Apeejay School of Management, New Delhi, India
e-mail: shweta.jha@learn.apeejay.edu

S. Shakya et al. (eds.), *Proceedings of Second International Conference on Sustainable Expert Systems*, Lecture Notes in Networks and Systems 351,
https://doi.org/10.1007/978-981-16-7657-4_63

as environmentally friendly. Clamour for sustainable business practices has obligated several organisations to reorient their management models and develop sustainable human resource management systems (sustainable HR), sustainable sourcing, sustainable manufacturing processes, sustainable supply chain management systems, etc. In fact, the organisations have started putting principles of sustainability in operations, decision-making and stakeholder engagement. Sustainable HR has evolved over the years as a formidable champion of sustainability at the workplace.

Sustainable HR is a manifestation of one of the responses of enterprises to address the call of the world communities to reorient their business practices in line with the core concepts of sustainability. Sustainable HR entails the appropriation of people management strategies and techniques that ensure the accomplishment of the financial, social and ecological goals in the long run [5]. Promoting social and environmental objectives has been an integral part of sustainable HR functions [6]. Moreover, scholars have observed that sustainable HR policies are more likely to create high-performance work culture, leading to sustained competitive advantage through better organisational outcomes and success [7]. Sustainable HR is based on reorienting the traditional frameworks of employee participation, strategy formulation and implementation, competency and knowledge development to meet and exceed the expectations of all the internal as well as external stakeholders with an express objective of adding value in all the offerings of the organisation and enriching their experience. So, the mandate of sustainable HR is quite demanding and requires a significant paradigm shift vis-a-vis traditional HR systems and best practices. For example equity, justice, transparency and well-being were not part of the typical HR framework. Now, the organisations cannot do without these principles in running a formidable HR system.

Sustainable HR is, thus, geared towards enhancing the employee value propositions and organisational outcomes simultaneously. However, not many organisations in the world have adopted the sustainable HR system, which requires the willingness of the top management to modify traditional approaches to manage people. Sustainability-induced competitive advantage may be developed by hiring talented personnel, setting up motivational rewards mechanisms and building employees' competencies in sync with vision, mission and organisational strategy [8, 9]. Augmenting the value of human capital has tremendous potential to improve the operational efficacy of an organisation [10]. Even, the employees need to reorient themselves to leverage the nuances of sustainable HR. Employees' performance is crucial for sustained organisational effectiveness [11]. High-performance work culture has two building blocks: a sustainable HR system and motivated employees ready to go beyond their call of duty. Top management of organisations can create high-performance work culture to gain and maintain a competitive advantage by adopting sustainable HR and inspiring the employees to get along with the alternative way of people management to meet and exceed the expectations of all the major stakeholders.

2 Review of Related Literature

Sustainable HR is an alternative way of organising the people management functions. Many published resources are available on sustainable HR revolving around its essential features, typology, advantages, weaknesses and best practices. It is crucial to understand the nuances of sustainable HR to develop a holistic perspective on the subject. Scholars have drawn a comprehensive list of characteristic features of sustainable HR with matching insinuations for practice, as reflected in Table 1 [12].

Apart from the well-articulated characteristic feature of sustainable HR, which enables a better understanding of the concept, the scholars have also provided a typology based on best practices adopted by organisations having diverse purposes, key concerns, processes and outcomes, as reflected in Table 2 [13]. Sustainable HR

Table 1 Characteristic features of sustainable HR

Characteristic features of sustainable HR	Defining aspects
Long-term orientation	Identification of the availability of human resources in future; identification of the needs of the future employees; elimination of the 'hire and fire' approach
Care of employees	Health and safety management; work-life balance
Care of environment	Evaluating the employee performance according to environment-related criteria; fostering 'eco-career'; employee rewarding according to environment-related criteria
Profitability	Share programmes
Employee participation and social dialogue	Different types and forms of participation
Employee development	Job rotation; different training formats and methods; the transfer of experience; focus on future skills and employability
External partnership	Cooperation with education system; partnership with all external stakeholders
Flexibility	Flexible working arrangements; job rotation
Compliance beyond labour regulations	Involvement of employee representatives in decision-making process; financial and non-financial support
Employee cooperation	Team spirit, congenial relations between managers and other team members
Fairness and equality	Fostering diversity; respectful mutual relationships; fairness in compensation and benefits as well as career growth opportunities within the organisation

Source Stankevičiūtė and Savanevičienė [12]

Table 2 Sustainable HR typology

Socially responsible HRM	Green HRM
Purpose: inside-out, i.e. economic and social purpose	Purpose: inside-out, i.e. economic and environmental
Key concerns: generating an awareness about the impact of business on people beyond organisational boundaries and present time frames	Key concerns: implementing green workplace practices
Process: socially responsible HR behaviour, practices and culture, e.g. health and well-being initiatives	Process: green HR behaviour, practices and culture, e.g. green recruitment
Outcomes: enhanced reputation, employer branding	Outcomes: growing employee engagement with reduced carbon footprints
Triple bottom line HRM	*Common good HRM*
Purpose: inside-out, i.e. economic, environmental and social purpose	Purpose: outside-in
Key concerns: using HR competencies to create win–win scenarios	Key concerns: using HR competencies for common good
Process: behaviour, practice and culture geared for attaining triple bottom line	Process: HR practices and behaviour enhancing common good
Outcomes: economic, environmental and social well-being	Outcomes: decent working conditions, workplace democracy

Source Dyllick and Muff [13]

typology provides a broad range of alternative approaches that the HR managers can choose as per the typical needs and philosophies of their respective organisations.

There is a general consensus among the HR leaders that sustainable HR is current imperative as well as the way forward. Sooner or later, all the organisations across the world are likely to adopt some form of sustainable HR system. Already, a lot of innovations are taking place to transform the HR practices by leveraging artificial intelligence and machine learning [14]. Other functions of HR such as training and development, performance management, compensation and benefits management, career development and exit management are also becoming sustainable by means of emerging technologies duly supported by artificial intelligence, blockchain and machine learning [15–19]. Besides the run of the mill HR functions which are now being assigned to technologies driven by artificial intelligence, companies are now using chatbots to gauge the mood of the employee or keep them engaged. Such developments are clear indications about the future of HR. Interestingly, artificial intelligence-driven technologies are helping the HR reduce their carbon footprints and emerge as champions of sustainability without compromising on employee experience and quality of HR services and offerings.

An organisation can build all-encompassing sustainable HR systems by investing in innovative HR practices, forward-looking HR practices, integrated HR systems and autonomy-based HR systems to ensure desirable business results aligned with

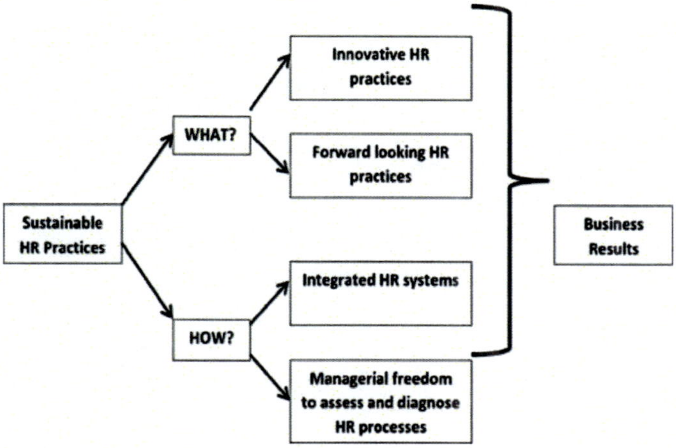

Fig. 1 Conceptual model of sustainable HR practices. *Source* Mishra and Sarkar [20]

vision, mission and strategic objectives. Innovative, forward-looking and integrated HR practices have tremendous potential to improve employee experience as well as the employee value proposition. Examples of such out of the box HR practices may include gamification of performance appraisal systems, use of artificial intelligence and machine learning in talent acquisition/training needs assessments, paid time off for volunteering in community-based projects, employee wellness programmes. Besides, the organisations also need to provide job autonomy to leverage the benefits of implementing sustainable HR systems. An all-inclusive conceptual model of sustainable HR is presented in Fig. 1 [20].

A close look at the extant literature on sustainable HR reveals that there is equal emphasis on concerns other than the ones related to protecting the environment, reducing carbon footprints and addressing climate change issues. For example equity, justice, transparency can be part of sustainable HR systems. Following observation of a practising manager is illustrative of the broader dimensions of sustainable HR:

> Sustainable HRM includes all issues related to the employment relationship and further linked to the collective agreements and laws. In the same vein, the proper management of employment relationships is a signal of the responsibility the company has to its employees. [21]

Transparent HR practices are reflected in impartial recruitment and selection, career development opportunities, human resource allocation, rewards, flexibility and participation in decision-making [21]. Likewise, justice and equality in the context of sustainable HR systems include reasonable compliance to law and regulations, management of diversity and inclusion and ethicality as a basis of all HR practices [21]. In the context of sustainable HR systems, profitability can be long-term view, proactiveness in action and integration of HR with corporate strategy [21]. Furthermore, it has been observed that ‘organisations that look beyond short-term financial goals to engage in long-term commitments to the betterment of society are more efficient and are better able to market products and less susceptible to expensive public relations blunders' [22].

3 Findings and Discussion

The concept of sustainable HR has matured over the years. The following definition of sustainable HR illustrates its broad approach and coverage:

> ...the adaption of HRM strategies and practices that enables the achievement of financial, social and ecological goals, with an impact inside and outside of the organisation and over a long-term time horizon while controlling for unintended side effects and negative feedback. [23]

It is interesting to note that the organisations have started setting social and environmental goals in addition to the traditional financial profits due to increased focus on social responsibility and sustainability performance [24]. Based on a survey of 2800 global companies, it was found that 70% of organisations included sustainability as a primary issue in their strategic plans and agendas [25]. Furthermore, sustainable HR systems create appropriate mechanisms for achieving social and ecological goals and ensure an equitable, just and transparent workplace for the employees. Thus, the interests and concerns of both internal and external stakeholders are treated with equanimity.

Based on the principles of equity, justice and transparency, sustainable HR has emerged as the mainstay of high-performance work culture. Sustainable HRM systems have integrated several value-added HRM practices to attain a picture-perfect fit with the needs of the organisations as well as that of the employees. It has also been observed that the employees respond to the value-added sustainable HR systems by stretching themselves towards attaining organisational goals and living up to the stakeholders' expectations. The success of sustainable HR depends on sustainability-oriented values, support of corporate leaders at all levels, systems and metrics aligned with the core concept of sustainability and stakeholder engagement [7].

Sustainable HR has tremendous potential to energise the employees by promoting personal sustainability programmes to boost in-company environmentally friendly initiatives. For example Wal-Mart's 'Personal Sustainability Programme' has enabled more than 500,000 workers to fulfil their commitments towards protecting the planet by using less water or biking to the workplace [26]. Other researchers have also found that green HR practices result in an improved in-role and extra-role environmentally friendly behaviour of the employees [27, 28].

True, sustainable HR has a great appeal among all the internal and external stakeholders. However, implementing sustainable HR systems is plagued by several roadblocks. The scholars have listed some of the barriers in the successful implementation of sustainable HR under three categories, viz. external factors (linking external partnerships, funding and support of organisations), organisational factors (internal partnerships, organisation's mission and leadership) and programme specific factors (alignment of HR programmes and corporate mission, administrative support and developmental evaluation) [29]. Also, limited financial means, cost of implementing and maintaining sustainable HR systems are the most significant impediments lack of a comprehensive plan for implementing sustainable HR systems, ambiguity about benefits of sustainable HR, lack of incentives to engage in environmentally friendly activities, low competencies of the management concerning sustainable HRM and a culture based on economic values have been viewed as roadblocks in implementing sustainable HR systems in organisations [30–33].

In spite of inherent constraints in implementing sustainable HR systems, processes and practices, a good number of organisations have come forward to reorient their approaches and world views with regard to people management. Several researchers have found evidences regarding interconnectedness between implementation of sustainable HR systems and organisational outcomes [34–38]. Sustainable HR systems help the organisations reduce their operational cost by way of improved productivity and thereby increasing competitiveness [39, 40]. Sustainable HR systems also help the organisations create an ecosystem conducive for innovations and creative problem solving [41, 42].

4 Conclusion

The existing literature indicates a significant linkage between sustainable business practices and increased profitability as well as greater stakeholder satisfaction. However, sustainable HR systems have not been adopted in a big way by organisations worldwide, although they can also deliver the same value as other sustainable business practices. Hence, the HR leaders across the sectors must be sensitised about their role in creating a high-performance work culture and transforming the organisations to gain and maintain competitive advantages while pursuing social, economic and environmental goals. Further, the HR leaders need to be proactive in persuading the top management to usher in sustainable HR systems for the benefit of all the internal as well as external stakeholders. In the years to come, organisations will have to come to terms with the demands of environmentally sensitive customers, investors and communities in the wake of rising degradation of natural resources and concomitant climatic changes threatening the very survival of the human race.

Already, people face immense difficulties due to the severe scarcity of potable water, especially in developing countries, caused due to heavy industrialisation and indiscriminate exploitation of waterbodies. Pollution due to industries and the increased number of vehicles have rendered the quality of air unbreathable in many parts of the developing world. Moreover, fast fashion has put tremendous pressure on the ecologies of the world. But, things can be stopped from getting worse if the companies adopt the *mantra* of sustainability and start pursuing profits responsibly. Sustainable HR can play a role of catalyst in converting the run of the mill organisations into sustainable companies. HR leaders can also play a crucial role in building sustainability-oriented culture, collaborative framework, competencies, reward system, and facilitating robotics, artificial intelligence and machine learning to reduce carbon footprints. As a result of the enabling ecosystem created by HR leaders, the companies can showcase their best practices to inspire others in the industry, write their own success storeys and continue to remain competitive in the long run and deliver incredible shareholder value. The onus now lies on the HR leaders.

References

1. World Business Council for Sustainable Development (2005) Driving success: human resources and sustainable development. University of Cambridge Programme for Industry
2. Semerad D (2017) How to leverage sustainability to increase business profitability. Forbes: https://www.forbes.com/sites/forbesagencycouncil/2017/10/31/how-to-leverage-sustainability-to-increase-business-profitability/?sh=17ff267a3440
3. World Economic Forum (2020) Sustainability and profitability can co-exist. Here's how. https://www.weforum.org/agenda/2020/01/sustainability-profitability-co-exist/
4. Bonini S, Swartz S (2014) Profits with purpose: how organizing for sustainability can benefit the bottom line. McKinsey: https://www.mckinsey.com/~/media/McKinsey/Business%20Functions/Sustainability/Our%20Insights/Profits%20with%20purpose/Profits%20with%20Purpose.ashx
5. Ehnert I, Parsa S, Roper I, Wagner M, Muller-Camen M (2016) Reporting on sustainability and HRM: a comparative study of sustainability reporting practices by the world's largest companies. Int J Hum Resour Manag 27:88–108
6. Kramar R (2014) Beyond strategic human resource management: is sustainable human resource management the next approach? Int J Hum Resour Manag 25:1069–1089
7. Wirtenberg J, Harmon J, Russell WG, Fairfield KD (2007) HR's role in building a sustainable enterprise: insights from some of the world's best companies. People Strat 30(1):10
8. Paré G, Tremblay M (2007) The influence of high-involvement human resources practices, procedural justice, organizational commitment, and citizenship behaviors on information technology professionals' turnover intentions. Group Org Manag 32(3):326–357
9. Edvardsson IR (2008) HRM and knowledge management. Empl Relat 30(5):553–561
10. Scheel T, Rigotti T, Mohr G (2014) Training and performance of a diverse workforce. Hum Resour Manage 53(5):749–772
11. Melián-González S (2016) An extended model of the interaction between work-related attitudes and job performance. Int J Product Perform Manag 65(1):42–57
12. Stankevičiūtė Ž, Savanevičienė A (2018) Designing sustainable HRM: the core characteristics of emerging field. Sustainability 10(12):4798
13. Dyllick T, Muff K (2016) Clarifying the meaning of sustainable business: introducing a typology from business-as-usual to true business sustainability. Org Environ 29(2):156–74
14. Jha SK, Jha S, Gupta MK (2020) Leveraging artificial intelligence for effective recruitment and selection processes. In: International conference on communication, computing and electronics systems, pp 287–293. Springer, Singapore
15. Gulliford F, Dixon AP (2019) AI: the HR revolution. Strat HR Rev 18(2):52–55
16. Altemeyer B (2019) Making the business case for AI in HR: two case studies. Strat HR Rev 18(2):66–70
17. DiClaudio M (2019) People analytics and the rise of HR: how data, analytics and emerging technology can transform human resources (HR) into a profit center. Strat HR Rev 18(2):42–46
18. Wheeler AR, Buckley MR (2021) HR without people?: industrial evolution in the age of automation, AI, and machine learning. Emerald Group Publishing
19. Fernandez J (2019) The ball of wax we call HR analytics. Strat HR Rev 18(1):21–25

20. Mishra RK, Sarkar S (2020) Sustainable HRM practices: a drive towards sustainability (The case of NLCIL). In: Vanka S, Rao MB, Singh S, Pulaparthi MR (eds) Sustainable human resource management. Springer, Singapore. https://doi.org/10.1007/978-981-15-5656-211
21. Järlström M, Saru S, Vanhala S (2020) Sustainable human resource management with salience of stakeholders: a top management perspective. J Bus Ethics 152:703–724
22. Bulmash J (2011) HR plays critical role in sustainable organisation. Canadian HR Reporter, 17 Jan 2011
23. Ehnert I (2009) Sustainable human resource management: a conceptual and exploratory analysis from a paradox perspective. Physica-Verlag, Heidelberg
24. Elkington J (1997) Cannibals with forks: the triple bottom line of the 21st century. Capstone, Oxford
25. Kiron D, Kruschwitz N, Haanaes K, von Streng Velken I (2012) Sustainability nears a tipping point. MIT Sloan Manag Rev 53(2):69–74
26. Winston A (2009) Green recovery. Harvard Business Press
27. Aboramadan M (2020) The effect of green HRM on employee green behaviors in higher education: the mediating mechanism of green work engagement. Int J Org Anal
28. Amjad F, Abbas W, Zia-UR-Rehman M, Baig SA, Hashim M, Khan A, Rehman H (2021) Effect of green human resource management practices on organizational sustainability: the mediating role of environmental and employee performance. Environ Sci Pollut Res 28:28191–28206. https://doi.org/10.1007/s11356-020-11307-9
29. Russ-Eft DF (2014) Human resource development, evaluation, and sustainability: what are the relationships? Hum Res Dev Int 17(5):545–559
30. Jafri S (2012) Green HR practices: an empirical study of certain automobile organisations of India. Hum Resour Manage 42:6193–6198
31. Fayyazi M, Shahbazmoradi S, Afshar Z, Shahbazmoradi M (2015) Investigating the barriers of the green human resource management implementation in oil industry. Manag Sci Lett 5(1):101–108
32. Guerci M, Carollo L (2016) A paradox view on green human resource management: insights from the Italian context. Int J Hum Resour Manag 27(2):212–238
33. Bombiak E (2020) Barierrs to implementing the concept of green human resource management: the case of Poland. Eur Res Stud J 23(4):68–81
34. Cherian J, Jacoband J, Farouk S (2019) Impact of green HR practces on envıronmental sustaınabılıty of a UAE based organization. Ecol Environ Conserv Pap 25(2):475–484
35. Rawashdeh AM (2018) The impact of green human resource management on organizational environmental performance in Jordanian health service organizations. Manag Sci Lett 8:1049–1058
36. Ren S, Jiang K, Tang G (2021) Leveraging green HRM for firm performance: the joint effects of CEO environmental belief and external pollution severity and the mediating role of employee environmental commitment. Hum Resour Manag. https://doi.org/10.1002/hrm.22079
37. Benevene P, Buonomo I (2020) Green human resource management: an evidence-based systematic literature review. Sustainability 12(15):5974
38. Andjarwati T, Budiarti E, Audah AK, Khouri S, Rębilas R (2019) The impact of green human resource management to gaın enterprıse sustaınabılıty. Polish J Manag Stud 20(2):93–103
39. Deshwal P (2015) Green HRM: an organizational strategy of greening people. Int J Appl Res 1(13):176–181
40. Aravamudhan NR (2012) Green HR getting into the business of green. HRM Rev 12(1)
41. Ellefsen J (1999) Managing green teams: environmental change in organizations and networks. In: Moxen J, Peter A (eds) Strachan Greenleaf Publishing, 256 pages, $65.00 Sheffield, England, 1998. Corp Environ Strat 4(6):427–429
42. Beard C, Rees S (2000) Green teams and the management of environmental change in a UK county council. Environ Manag Health 11(1):27–38

Algorithms: Supervised Machine Learning Types and Their Application Domains

N. Divyashree and K. S. Nandini Prasad

Abstract In today's world, millions and trillions of data are available from anywhere and everywhere. However, these data have no use if not processed right away for obtaining unseen and meaningful information from it. Machine learning technique, a subfield under artificial intelligence, aims to train the computer systems to learn and apply knowledge from the given set of training data. Machine learning algorithms are used to extract unseen trends and patterns from the data for deriving meaningful insights and foresights to make future decisions in business, manufacturing, government and many more. This survey provides a complete view on supervised machine learning algorithms, their pros and cons along with their applications in specific areas under each machine learning class.

Keywords Machine Learning · Supervised Learning · Artificial Intelligence · Regression · Classification · Decision Trees · Deep Learning · Artificial Neural Networks

1 Introduction

Machine learning (ML), one of the subfields of artificial intelligence (AI), is used to understand the different patterns and structures of the given data and convert those understanding into models and use it to generalize over the future unseen data of the same domain. ML algorithms learn from the data without any explicit programme to perform specific task compared to traditional algorithms that are explicitly programmed. ML algorithms have the ability to learn, adapt, interpret

N. Divyashree (✉) · K. S. Nandini Prasad
Department of Information Science and Engineering, Dr. Ambedkar Institute of Technology, Bengaluru, India
e-mail: 1da19pis02.is@drait.edu.in

K. S. Nandini Prasad
e-mail: nandini.is@drait.edu.in

S. Shakya et al. (eds.), *Proceedings of Second International Conference on Sustainable Expert Systems*, Lecture Notes in Networks and Systems 351,
https://doi.org/10.1007/978-981-16-7657-4_64

and improve progressively to predict/perform from the experience (historical data). ML algorithms are classified into three groups, namely supervised, unsupervised and reinforcement learning.

1.1 Parametric and Non-parametric ML Algorithms

ML algorithms can be further analysed under supervised learning as parametric and non-parametric.

(i) **Parametric ML Algorithms**
A learning algorithm that allows to represent predictive models with a fixed number of parameters to summarize the given data is called parametric. The number of parameters is directly proportional to the number of input features being fed as input to the learning algorithm, and the amount of data (number of training examples) have no effect on the number of parameters being used to obtain a model.

(ii) **Non-parametric ML Algorithms**
A learning algorithm allows to represent a model with any number of parameters, and these parameter sizes grow with the increase in data (number of training examples). Supervised algorithms that come under parametric and non-parametric are shown in Fig. 1.

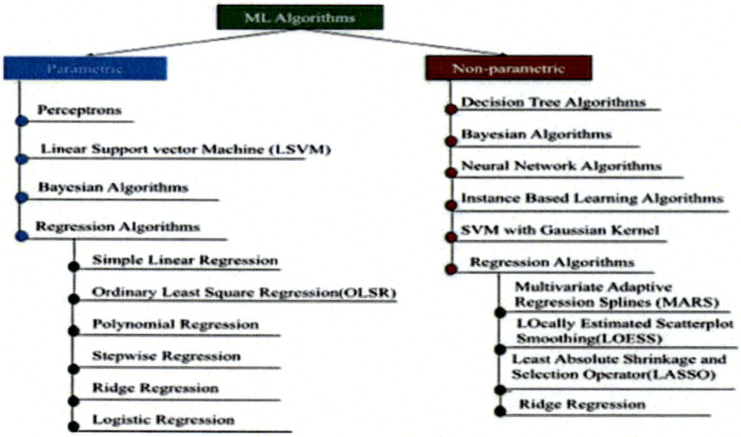

Fig. 1 Parametric and non-Parametric ML algorithms

1.2 Common Machine Learning Terminologies

The most common ML terminologies used are:

1. *Training dataset*: it is a collection of $(X^{(i)}, Y^{(i)})$ input/output pairs.
   where
   $X = \{X^{(1)}, X^{(2)}, X^{(3)} \ldots X^{(m)}\}$ is the set of input datasets.
   Each input data $X^{(i)} = \{x_1^{(i)}, x_2^{(i)}, x_3^{(i)} \ldots x_n^{(i)}\}$ for $i = 1$ to n is the #attributes of each input data.
   $y = \{y^{(1)}, y^{(2)}, y^{(3)}, \ldots, y^{(m)}\}$ for $i = 1$ to m output/target values w.r.t m input datasets.
2. *Attributes*: they are the features which define each input instance.
3. *Instance*: each $(X^{(i)}, Y^{(i)})$ training example is called an instance.
4. *Validation set*: collection of $(X^{(i)}, Y^{(i)})$ is the unseen datasets which are used to evaluate the fitness of a learned model.
5. *Model/hypothesis*: the learning outcome (generalization) over the given training instances.
6. *Test data*: A new, unknown instance whose outcome is to be predicted by the learned model.
7. *Label*: the solution part $(Y^{(i)})$ in the training dataset.
8. *Learning rate*: the updation stride taken by any optimization techniques like gradient descent.

In this survey, Sect. 1 is an introduction about ML and its class, followed by parametric and non-parametric algorithms under supervised learning and the terminologies of ML. Literature survey is carried out in Sect. 2, Sect. 3 describes the supervised machine learning algorithms, their pros and cons and applications in detail, and finally, the paper includes conclusion.

2 Literature Survey

The word "machine learning" was coined in 1959 by Arthur Samuel, a discoverer in artificial intelligence (AI) and computer gaming field. Samuel's computer checkers game was the first successful and ground-breaking self-learning programme. The term ML was defined by Mitchell [1] as "a computer programme is said to learn from experience E with respect to some class of tasks T and performance measure *P* if its performance at tasks in *T*, as measured by *P*, improves with experience *E*". Turing [2] quotes a question saying "Can machines think?" should be replaced with "Can machines do (as thinking entities) what we can do?" and proposes a theoretical concept that any digital computer can simulate any other digital machines behaviour. The main objective of any ML algorithms is to generalize (accurate performance on new unseen tasks/examples) from its experience [3]. ML algorithms under the supervised learning [4] category build a mathematical model by learning through the given large number of training data (contains input and desired output data). The model is then used for performing on new unseen tasks. Depending on the type of task to be executed, supervised ML algorithms can be divided as regression, classification [5] and active learning [6] algorithms. Madhuri et al. [7] compared in different regression techniques on a regression task of house price prediction, and it is concluded that linear regression combined along with gradient boosting technique has high predicting accuracy compared to other regression techniques such as ridge regression, multiple linear regression elastic net regression and LASSO regression. Mohammad et al. [8] have tabulated few drawbacks of different techniques of classification algorithms used for customer churn prediction in a telecom industry and concluded that logistic regression predicts better compared to other techniques such as random forest, AdaBoost and naive Bayes. Classification and regression trees (CART's) an algorithm under decision trees category is used as a common algorithm for performing both machine learning tasks of classification and regression [9]. The McCulloch-Pitts neuron model introduced in 1943 became the prototype of artificial neural network models [10]. Pouyanfar et al. surveyed the state-of-the-art algorithms in deep learning [11]. Eck et al. [12] consider using the recurrent neural networks (RNN's) for predicting whether a user will leave or complete the web survey based on the user behaviour. The most frequent and exhaustively used ML algorithms under supervised learning are explained, and their flavours are listed in the next section.

3 Supervised Machine Learning Algorithms

In supervised learning, the ML algorithms will be fed with the preprocessed and labelled (input features and the output variable) training datasets. Every input feature (attribute) in the training dataset is called as an instance, and the output variable is the target variable to be learned based on the problem context by the respective ML algorithm. ML algorithms try to learn from the dataset (every instance) being fed to it and come out with a learned model (or sometimes called hypothesis) which can be used for forecasting/predicting on the future unseen dataset (without target variable). Majority of the ML algorithms come under supervised learning type, and most of them are discussed from Sects. 3.1–3.6 with their introduction, pros and cons and few applications of them.

3.1 Regression Algorithms

Regression algorithms try to find hidden patterns in the data by extracting the relationship among the input features and output/target variable. It is used in statistical modelling for just finding the relationship among input and output variables and is extended as an ML model for identifying the hidden patterns and to obtain useful insights. Regression algorithms can be further extended under regression tasks and classification tasks. One such simple yet strong algorithm under regression task is the linear regression as described here.

(i) Simple Linear Regression

It is a simple ML algorithm used to find a linear relationship among the input and the output/target variables and thereby extending its learned model to predict the outcome on the unseen instances. It implements cost function and gradient descent technique to finalize interpretable parameters for the model [13].

Algorithm for Linear Regression

Input: $\left(X^{(i)}, Y^{(i)}\right) = \left\{\left(X^1, Y^1\right), \left(X^2, Y^2\right), \ldots.(X^m, Y^m)\right\}$ *is m training instances.*
Output: a linear regression model

$$h_\theta(x) = \theta_0 + \theta_1 x_1 + \theta_2 x_2 + \theta_3 x_3 + \cdots \theta_n x_n.$$

where
$h_\theta(x)$ *is the predicted model.*
$< \theta_0, \theta_1, \theta_2, \theta_3, \ldots \theta_n >$ *is the n interpretable parameters.*
$X^{(i)} =< x_1, x_2, x_3, \ldots x_n >$ *is the n input features of an instance.*

Algorithm

Initialize: The interpretable parameters $\theta = < \theta_0, \theta_1, \theta_2, \theta_3, \ldots\ldots\ldots \theta_n >$ *with small values*
Repeat
{
Compute:
$\hat{y} = h_\theta(x^{(i)}) = \theta o + \theta_1 x_1 + \theta_2 x_2 + \theta_3 x_3 + \ldots\ldots\ldots \theta_n x_n$
Compute Cost:
$J(\theta) = \sum_{i=1}^{m}(h_\theta(x^{(i)}) - Y^{(i)})^2$
Where,
$J(\theta)$, *is the Cost function and* 'α', *is the learning rate.*
Compute: Gradient Descent Update for automatic tuning of parameters (θ)
{
$$\theta_j = \theta_j - \alpha \frac{\delta}{\delta_j} J(\theta)$$
Where,
θ_j, *is the partial derivative w.r.t the cost function,*
$$\frac{\delta}{\delta_j} J(\theta) = \sum_{i=1}^{m} (h_\theta(x^{(i)}) - Y^{(i)}) * x_j^{(i)}$$
}
} *until Convergence*

Output the model: $h_\theta(x) = \theta_0 + \theta_1 x_1 + \theta_2 x_2 + \theta_3 x_3 + \cdots \theta_n x_n$, with final interpretable parameters.

(ii) **Other flavours of regression task algorithms**

1. Ordinary least square regression (OLSR)
2. Multivariate adaptive regression splines (MARS)
3. Stepwise regression, locally estimated, scatterplot smoothing (LOESS)
4. Least absolute shrinkage and selection operator (LASSO) [14] and
5. Ridge regression.

The only algorithm for classification task is logistic regression.

(iii) **Logistic Regression**
Logistic regression is a class of ML algorithms used for classification tasks. It is a simple binary classification (only two classes) algorithm which tries to learn the relationship among input features and the target variable from the training instances to identify the class of any new instance. The logistic regression applies sigmoid function over the obtained linear regression model which squashes the obtained result to lie between the range 0 and 1, thereby predicting the class of an instance.

Algorithm for Logistic Regression

Input: $\left(X^{(i)}, Y^{(i)}\right) = \left\{\left(X^1, Y^1\right), \left(X^2, Y^2\right), \ldots (X^m, Y^m)\right\}$ *is m training instances.*

Output: $\begin{cases} P(y == 1|X), & \text{for } P(y) \geq \text{threshold} \\ P(y == 0|X), & \text{for } P(y) < \text{threshold} \end{cases}$

where threshold is chosen by the user as per the problem requirement.

Algorithm

Initialize: The interpretable parameters $< \theta_0, \theta_1, \theta_2, \theta_3, \ldots\ldots.. \theta_n >$ *with small values*

Repeat

{

Compute: $P(y) = P\left(h_\theta(x) = {}^1/_{1 + e^{-z}}\right)$

Where,

$$Z = \theta o + \theta_1 x_1 + \theta_2 x_2 + \theta_3 x_3 + \ldots\ldots.. \theta_n x_n$$

$< \theta_0, \theta_1, \theta_2, \theta_3, \ldots\ldots.. \theta_n >$, *are the n interpretable parameters*

$X^{(i)} =< x_1, x_2, x_3, \ldots\ldots.. x_n >$, *are the n input features of an instance.*

Compute Cost:

$$J(\theta) = \sum_{i=1}^{m} y^{(i)} \log(h_\theta(x^{(i)}) - (1 - y^{(i)}) \log(1 - h_\theta x^{(i)})$$

Where,
$J(\theta)$ *Is the Cost function and* $'\alpha'$, *is the learning rate.*

Log function is used to penalize the model for wrong prediction
Compute: Gradient Descent Update for automatic tuning of parameters (θ)
{

$$\theta_j = \theta_j - \alpha \frac{\delta}{\delta_j} J(\theta) \qquad \textit{Where,}$$

θ_j, *is the partial derivative w.r.t the cost function*
} *Until Convergence*

3.1.1 Pros and Cons of Regression Algorithms

Pros:

1. Provides interpretability of a model.
2. Takes less time for model training.
3. Suitable for small training datasets.

Cons:

1. Not suitable for large number of attributes and huge amount of datasets.
2. Susceptible to outliers.
3. Needs prior assumptions and Not suitable for nonlinear attribute dependencies.

3.1.2 Applications

Some of the applications of regression algorithms are: weather prediction, house price prediction, diabetes prediction, cancer disease prediction, business market prediction, population growth prediction, estimating life expectancy, advertisement popularity prediction and many more.

3.2 Decision Tree Algorithms

Decision trees are very powerful algorithms used in ML, data mining and statistics as a predictive model that are used for solving regression and classification tasks for prediction and pattern extraction. Decision tree is depicted as an inverted tree structure, with internal nodes representing the decision (split) attributes and leaf nodes representing the target concept. Traversal of the decision tree starting from root node to leaf node represents the interpretability of the decision model.

3.2.1 Decision Tree Terminologies

Along with the terminologies mentioned under ML terminologies, other terminologies explicit to decision trees are as listed.

1. *Root Node*: it represents the starting node of a decision tree which is further divided into two or more homogeneous sets.
2. *Child and Parent Nodes*: nodes that are divided into sub-nodes are parent nodes, and sub-nodes which came from the parent nodes are child nodes.
3. *Decision Node(s)*: any parent node that splits into sub-nodes is decision nodes.
4. *Terminal/Leaf Nodes*: nodes with no split are terminal or leaf nodes.
5. *Sub-tree/Branch*: subsection/part of a complete tree is sub-tree or branch.
6. *Pruning*: removal of one or more sub-nodes of a decision node is pruning, and the opposite of pruning is called splitting.

All decision tree algorithms are non-parametric, and one such algorithm of decision trees is iterative dichotomiser 3 (ID3).

(i) **Iterative Dichotomiser 3 (ID3)**

The ID3 algorithm works on the principle of Occam's razor states that "simple hypothesis is preferred over complex one" and builds possibly a smallest decision tree which is used for predicting/making future decisions on unseen events.

Algorithm for ID3

Input: $\left(X^{(i)}, Y^{(i)}\right) = \left\{\left(X^1, Y^1\right), \left(X^2, Y^2\right), \cdots (X^m, Y^m)\right\}$ *is m training instances.*

Output: leaf nodes containing the learned target concepts.

Algorithm

Create a Start (Root) node for the tree.
If training instances are univariate (only single attribute)
Then,

Return that single attribute tree Root with leaf nodes depicting the classification classes based on the common attribute value for splitting the decision class.

Else

REPEAT for choosing every decision attribute A

{

If all the training instances w.r.t the attribute A belongs to the same class

Then,

Return the tree with leaf nodes indicating the class label.

Else,

Calculate the total entropy E(S) w.r.t the number of training instances available for the decision attribute A:

$$Entropy(S) = -\left(\frac{pi}{pi} + ni\right) * log2\left(\frac{pi}{pi} + ni\right) - \left(\frac{ni}{pi} + ni\right) * log2\left(\frac{ni}{pi} + ni\right),$$

pi: positive instance, ni: negative instance

Calculate the individual entropy of all other attributes (v):

$$E(Sv) = -\left(\frac{pi_v}{pi_v} + ni_v\right) * log2\left(\frac{pi_v}{pi_v} + ni_v\right) - \left(\frac{ni_v}{pi_v} + ni_v\right) * log2\left(\frac{ni_v}{pi_v} + ni_v\right)$$

Compute Gain of all the attributes:

$$Gain(S, A) = E(S) - E(Sv)$$

Choose that decision attribute A which has the highest Gain value.

} UNTIL all leaf nodes are obtained

(ii) **Other flavours of decision tree algorithms**

1. C4.5
2. C5.0
3. Classification and regression tree (CART)
4. Chi-squared automatic interaction detection (CHAID)
5. Decision stump [15]
6. Conditional decision trees and
7. M5.

3.2.2 Pros and Cons of Decision Trees

Pros:

1. Used for both classification and regression tasks and are non-parametric
2. It handles both categorical and numerical data and nonlinear relationships
3. Easy to visualize, understand and interpret
4. Can be used for identifying significant features
5. Robust to outliers and less prone to data skews.

Cons:

1. Can lead to overfitting model
2. Does not always converge to global optimal
3. Does not give effective prediction for continuous variables
4. Extrapolation is not possible
5. Susceptible to high variance (can be reduced through bagging and boosting techniques).

3.2.3 Applications

Some of the applications of decision tree algorithms are: loan fraud detection, asset failure prediction, sports win prediction, business decisions and many more.

3.3 Bayesian Learning

Bayesian analysis is a process of obtaining probability prediction on the occurrence of an event with some prior assumptions of conditions under certainty/uncertainty related to the event. Bayesian learning algorithms help in encoding certain prior beliefs on how models should look irrespective of what the data conveys. Bayesian algorithms are chosen in cases when there are not much data to train a model confidently. These algorithms work well if any prior knowledge on part of a model is available. All Bayesian algorithms are derived on the basis of Bayes theorem which is a principal component to derive conditional probability. The most commonly used Bayesian learning algorithm is the naive Bayes.

Bayes theorem:

$$P(A|B) = P(A) * P(B|A)P(B) \tag{1}$$

3.3.1 Bayesian Learning Terminologies

The common terminologies under Bayesian learning are as listed.

1. Posterior probability or posteriori: *P*(*A*|*B*)
2. Likelihood or conditioned probability: *P*(*B*|*A*)
3. Prior probability or a priori: *P*(*A*)
4. Evidence or marginal: *P*(*B*)
5. Joint probability: probability of two or more simultaneous occurrence of events, denoted as *P*(*A* and *B*) or *P*(*A*, *B*).

(i) Naive Bayes Algorithm

A simple probabilistic classifier also known as "simple Bayes" and "independence Bayes" used for classification tasks which applies "Bayes theorem" and "naive" a prior assumption on all features is to be conditionally independent of each other. Naive Bayes uses Bayes theorem to predict the class(c) of the given training example. Naive Bayes assumption: given the class (to be classified), all attributes/features are conditionally independent of each other.

Algorithm for Naive Bayes

Input:

$\left(X^{(i)}, Y^{(i)}\right) = D = \left\{\left(X^1, Y^1\right), \left(X^2, Y^2\right), \ldots (X^m, Y^m)\right\}$ *is m training instances.*
t = *new instance whose class is to be predicted,*
C = *set of classes used for instance labelling, where* $c \in C$
Output: Class $C_t \in L$*, the class of "t"*

Algorithm

Step1: Training Phase
For each attribute value xi of X from training set D
{
For each class C in Y
{
Compute: Prior probability of each class

$$P(Y_C) = \frac{sum(Y==c)}{n(D)}$$

Compute: likelihood

$$P(x_1, x_2, x_3, \dots \dots .. x_n | Y_C) \prod_{i=1}^{n} P(x_i | Y_C)$$

Compute: Evidence (will be same for all classes, hence it's optional)

$$P(x_i) = \frac{sum(x_i=v)}{n(D)}$$

}
}
Training phase Output: prior probabilities and likelihood
Step2: Testing Phase
For each class C in Y
{

$$P(Y = c) = P(Y_C) * \prod_{i=1}^{n} P(t_1 | Y == c) * P(t_2 | Y == c) \dots . P(t_3 | Y == c)$$

}
Output: $Class\ (Y_C) = \arg max_{c=C}(P(Y_C) * P(t\ | Y_C)$

(ii) **Flavours of naive Bayes algorithms**

1. Gaussian naive Bayes
2. Bernoulli naive Bayes
3. Multinomial naive Bayes.

(iii) **Other flavours of Bayesian learning algorithms**

1. Bayesian network
2. Bayesian belief network (BBN) and
3. Averaged one-dependence estimators (AODE).

3.3.2 Pros and Cons of Bayesian Learning

Pros:

1. Requires less training and easy to implement
2. Highly scalable
3. Robust to outliers and irrelevant features.

Cons:

1. Strong assumption on attribute independence
2. Susceptible to "zero frequency" problem but can be solved using Laplace estimation
3. Sensitive to the input data.

3.3.3 Applications

Some of the applications of Bayesian Learning are: image/medical image classification, stocks and business models, weather forecasting, environmental changes prediction, text classification and spam filtering.

3.4 Artificial Neural Networks

Artificial neural networks also called just "neural networks" are intelligent computing agents inspired from biological neural networks of the human/animal brain. Each basic processing unit is called a neuron, and the collection of neurons stacked in layers with unique connectivity between the neurons of every other layer forms a network; hence, the name is neural network. Each connection is identified by certain weights which specifies the strength/importance of input features in the connectivity. Any neural network consists of three major parts, namely the input layer (I), hidden layer (H) and the output layer (O). There can be more than one hidden layer in the neural network, and more hidden layers mean more complex patterns a neural network can learn. Each input unit in the input layer defines the features of the input training samples.

3.4.1 Neural Network Terminologies

Along with the terminologies stated under common machine learning terminologies, other terminologies explicit to neural networks are as listed.

1. *Neuron*: a single unit in an input/hidden/output layer of the deep neural network.
2. *Input layer*: consists of input units (the neurons) which accepts input (training examples) to be fed to the deep neural network. There can be more than one input unit in the input layer.
3. *Hidden layer*: consists of several neurons per each layer. It is also called a computational layer which learns and understands about the patterns and insights present in the data (training data). Hidden layers lie in between the input and output layers.

4. *Output layer*: produces the output which is the outcome of the learned model of a deep neural network. There can be more than one output unit in the output layer.
5. *Weights*: describes the strength/importance of an input attribute of an instance (training example) for its contribution to obtain a supervised output.
6. *Bias*: specifies importance to some of the input features for better generalization over the training examples.
7. *Connectivity*: weights associated between two neurons of different layers.
8. *Activation function*: they are the mathematical equations added to every neuron in the neural network in order to make them learn complex and nonlinear patterns present in the input data. Linear, binary step, bipolar step, sigmoid, binary sigmoid, bipolar sigmoid, tanh, ReLU, ramp and softmax are the seven activation functions available.
9. *Learning rate*: denoted as "α" is used for adjusting weights at each training step.
10. *Epochs*: single iteration over the entire training set is called one epoch.
11. *Batches*: the entire training set is divided in equal size into several chunks and fed to the training algorithms in order to obtain a better generalized model that best fits the training examples.

All artificial neural networks are non-parametric algorithms that are used to perform regression, classification and pattern recognition tasks. One such algorithm is the simple feedforward neural network.

(i) Single-layer Perceptron

A single-layer perceptron neural network having single hidden layer (can contain more than one neural units) is the most simple form of neural networks where the information propagates only in one direction, i.e. from input layer to hidden layer and finally to output layer without any loops or cycles in the network. Each layer (input, hidden and the output) uses different activation functions for supervised learning.

Algorithm of single-layer perceptron with single output class.

Input: $\left(X^{(i)}, Y^{(i)}\right) = \left\{\left(X^1, Y^1\right), \left(X^2, Y^2\right), \ldots..(X^m, Y^m)\right\}$ *is the m datasets.*

Output: linear/nonlinear model based on the type of task.

Algorithm

Algorithms: Supervised Machine Learning Types ...

Initialize: weights (w), bias (b) and learning rate α
Repeat {

For every training input $\left(X^{(i)}, Y^{(i)}\right)$
{do

Compute net input:

$$Y_{in} := b + j = \sum_{j=1}^{n} X_j * w_j$$

Compute Predicted Output:

$$\hat{Y} := \begin{cases} f(Y_{in}) = 1 \; if \; Y_{in} > \theta \\ f(Y_{in}) = 0 \; if \; Y_{in} < \theta \end{cases}$$

Here, binary activation function is used with some threshold value θ

If $(Y \neq \hat{y})$
{

$$w_j^{(new)} := w_j^{(old)} + \alpha * Y * X_j$$
$$b^{(new)} := b^{(old)} + \alpha * Y$$

}
Else
{

$$w_j^{(new)} := w_j^{(old)}$$
$$b^{(new)} := b^{(old)}$$

}

}

} Until Convergence

(ii) Flavours of neural network algorithms under supervised learning

1. Feedforward neural networks (perceptron)
   (a) Single-layer perceptron
   (b) Multilayer perceptron
2. McCulloch-Pitts neuron model [16]
3. Adaptive linear neuron (ADALINE) neural network
4. Multiple adaptive linear neuron (MADALINE) neural network,
5. Radial basis function neural network (RBFNN)
6. Feedforward neural network with backpropagations algorithm
7. Recurrent neural network (RNN)
   (a) Fully recurrent neural network
   (b) Hopfield network [17]

    (c) Independently RNN
    (d) Echo state RNN
    (e) Long short-term memory (LSTM)
8. Modular neural network.

3.4.2 Pros and Cons of Artificial Neural Network

Pros:

1. Parallel processing capability and fault tolerant
2. Can work with vague and incomplete data and are non-parametric
3. Self-adaptive capability with huge storage capability
4. Does not start with any assumption over the model structure.

Cons:

1. Computation time is more compared to other algorithms
2. Does not provide model interpretability
3. Requires huge amounts of data to train the neural network to obtain a good model.

3.4.3 Applications

Some of the applications of artificial neural networks are: social media marketing, machine translation, pattern recognition, novelty detection, named entity recognition, cyber security, robotics, recognition of handwritten digits and many more.

3.5 Deep Learning

Deep learning is viewed as a subset of ML algorithms which uses several multiple layers (hence, the name "deep") of artificial neural units to progressively extract and represent features automatically [18] from raw and mostly unstructured data (i.e. text, audio, image, video, webpage, meta-data). Deep learning algorithms can be applied to any class of learning algorithms (i.e. supervised, semi-supervised and unsupervised).

i. **Deep Neural Networks**

Deep neural networks are nothing but ANN's with a large number of hidden layers to progressively extract input features from raw input.

Algorithm for deep neural network (forward propagation with single output unit)

Notations:

1. $(X^{(i)}, Y^{(i)}) = \{(X^1, Y^1), (X^2, Y^2), \ldots (X^m, Y^m)\}$ *is m datasets*
2. $X = \langle x_1, x_2, x_3, \ldots x_n \rangle$ is *n* features/attributes of an input instance
3. X_0: *bias at the input layer*
4. $a_i^{(j)}$: neural unit "*i*" in layer "*j*"
5. $\Theta(j)$: weight matrix between each layer j to next layer $(j-1)$
6. $g(Z^{(j)})$: *activation function of layer j*
7. $Z^{(j)}$: *linear computation of input value of layer j and weight matrix* $\Theta(j)$
8. $h_\Theta(X)$: *final output at the single output layer*

Input: $(X^{(i)}, Y^{(i)})$
Output: $h_\Theta(X)$

Algorithm

For each $(X^{(i)}, Y^{(i)})$ *training instance*
{
Set $a^{(1)} := \{x_0, x_1, x_2, x_3, \ldots\ldots\ldots x_n\}$
For each hidden layer, l=2 to L-1
{
Compute $Z_p^{(j)} := \sum_{p=1}^{h} \theta^{(j)} a_p^{(j)}$
Compute $a_p^{(j)} := \sum_{p=1}^{h} g(Z_p^{(j)})$
}
Compute: output $h_\theta(X) \coloneqq \sum_{j=2,p=1}^{L-1,h} g(a_p^{(j)}$
}

3.5.1 Flavours of Deep Learning Algorithms

1. Deep neural networks (DNN) [19]
2. Recurrent neural network (RNN)
3. Deep belief network (DBN) [12]
4. Convolutional neural network (CNN) [6, 20]
5. Deep Boltzmann machine (DBM) [21]
6. Stacked autoencoders.

3.5.2 Pros and Cons of Deep Learning

Pros:

1. Automatic feature extraction and handles high-dimensional data
2. Applicable across different domains like speech, computer vision
3. Ability to represent complex models and robust to outliers
4. Flexibility in adapting to new problems.

Cons:

1. Difficult to understand how to network learns
2. Computationally expensive
3. Needs lot of data to be trained properly.

3.5.3 Applications

Some of the applications of deep learning are: paraphrase identification, mobile advertising, automatic question and answering system, social network analysis (SNA), customer relationship management (CRM), image separation and enhancement, information retrieval, emotions recognition in speech, sentiment analysis, medical image classification, bioinformatics, speech recognition, autonomous driving, drug design, board games programme and many more.

3.6 *Instance-Based Learning*

Instance-based learning algorithms, also called "memory-based learning", are class of ML algorithms that performs classification and regression tasks by comparing it with the stored training instances based on the similarity of the feature/attribute values between the new/test instance and the training instances and outputs a class prediction/label. One such algorithm is the KNN algorithm.

(i) **KNN Algorithm**

KNN algorithm, also called as "lazy learning algorithm", is used for both supervised and unsupervised learning tasks. It is a robust and versatile classifier that does not produce any model rather and just stores the training set in the learning phase, and whenever it gets new instance for execution, it uses one of the distance measures (Euclidian, Manhattan, Minkowski) to evaluate how far the new instance is from the existing training instances and employs majority voting mechanism depending on the chosen k value to obtain final output class prediction/label. It is used as a benchmark very often for complex classifiers like ANN and support vector machine (SVM).

Algorithm

Input: D = set of training instances, k = count of nearest neighbours to be considered, x = new instance whose class of prediction/label to be determined, L = set of classes used for instance labelling

Output: Class$C_x \in L$, the class of x

For (each training instance i in D)
{do
Compute distance(x, i) and store in set S
}
Arrange values of set Sin ascending order
Select k neighbours from the set S (i.e., the first k instances that are near to x as per the computed distances)
Output $C_x = argmax \sum_{i \in S}^{k} i_v = C_v, where\ v \in L$

(ii) **Other flavours of Instance-based learning algorithms**

1. Locally weighted learning (LWL)
2. Self-organizing map (SOM)
3. Learning vector quantization (LVQ).

3.6.1 Pros and Cons of Instance-based Learning

Pros:

1. Quick learning, easy to implement and understand
2. It is non-parametric
3. Learns both continuous- and discrete-valued concept.

Cons:

1. Not all distance measures lead to the same results
2. Needs large storage, and its execution time is more
3. Existence of noise degrades its accuracy
4. Susceptible to curse of dimensionality
5. Does not have any technique to adjust decision boundaries after storing the training data.

3.6.2 Applications

Some of the applications of instance-based learning are: economic forecasting, concept searching and text mining, genetics, agriculture, credit rating, stock price prediction, currency exchange rates, predicting financial distress, bank bankruptcies, money laundering analysis, financial risk prediction, trading futures, data compression and many more.

4 Conclusion

Extensive survey has been done on frequently used supervised learning algorithms of linear regression, logistic regression, ID3, naive Bayes, single-layer perceptron's, deep neural network and KNN, and the remaining algorithms are listed. Pros and cons along with applications of all supervised ML algorithms are discussed. Further broad surveys can be done on unsupervised learning ML algorithms.

Acknowledgements This work was financially supported by the "KSTEPS, Department of Science and Technology (DST), GOVT. OF KARNATAKA". The funders had no role in study design, data collection and analysis, decision to publish or preparation of the manuscript. I also thank Dr. Ambedkar Institute of Technology for their guidance and support with all the facility that was required.

References

1. Druck G, Settles B, McCallum A (2009) Active learning by labeling features. In: Proceedings of the 2009 conference on empirical methods in natural language processing, vol 1—EMNLP '09
2. Turing AM (2021) Computing machinery and intelligence (1950). In: Ideas that created the future. The MIT Press, pp 147–164
3. Russell SJ, Norvig P (2010) Artificial intelligence: a modern approach. Prentice Hall
4. Pouyanfar S et al (2019) A survey on deep learning: algorithms, techniques, and applications. ACM Comput Surv 51(5):1–36
5. Mohammad NI, Ismail SA, Kama MN, Yusop OM, Azmi A (2019) Customer churn prediction in telecommunication industry using machine learning classifiers. In: Proceedings of the 3rd international conference on vision, image and signal processing
6. Krizhevsky A, Sutskever I, Hinton GE (2017) ImageNet classification with deep convolutional neural networks. Commun ACM 60(6):84–90
7. Alpaydin E (2020) Introduction to machine learning, 4th edn. MIT Press, London
8. Christopher M (2011) Cram101 textbook reviews. Study guide for pattern recognition and machine learning by bishop. ISBN 9780387310732. Cram101
9. Buskirk TD (2018) Surveying the forests and sampling the trees: an overview of classification and regression trees and random forests with applications in survey research. Surv Pract 11(1):1–13
10. Mitchell TM, Carbonell JG, Michalski RS (1986) Machine learning: a guide to current research, 1986th edn. Springer, New York
11. Divyashree N, Nandini PKS (2021) Cognitive computing technologies, products, and applications. In: Evolutionary computing and mobile sustainable networks. Springer, Singapore, pp 693–701
12. Hinton GE, Osindero S, Teh Y-W (2006) A fast learning algorithm for deep belief nets. Neural Comput 18(7):1527–1554

13. Madhuri CHR, Anuradha G, Pujitha MV (2019) House price prediction using regression techniques: a comparative study. In: 2019 International conference on smart structures and systems (ICSSS)
14. Signorino CS, Kirchner A (2018) Using LASSO to model interactions and nonlinearities in survey data. Surv Pract 11(1):1–10
15. Venkata Reddy PS, Nandini Prasad KS, Puttamadappa C (2020) SASI: smart agriculture system based on IoT. Eur J Mol Clin Med 07(08):3113–3120
16. McCulloch W, Pitts W (2021) A logical calculus of the ideas immanent in nervous activity (1943). In: Ideas that created the future. The MIT Press, pp 79–88
17. Horne B, Jamshidi M, Vadiee N (1990) Neural networks in robotics: a survey. J Intell Robot Syst 3(1):51–66
18. Smys S, Raj JS (2021) Analysis of deep learning techniques for early detection of depression on social media network: a comparative study. J Trends Comput Sci Smart Technol (TCSST) 3(01):24–39
19. Ioannidou A, Chatzilari E, Nikolopoulos S, Kompatsiaris I (2017) Deep learning advances in computer vision with 3D data: a survey. ACM Comput Surv 50(2):1–38
20. Karpathy A, Toderici G, Shetty S, Leung T, Sukthankar R, Fei-Fei L (2014) Large-scale video classification with convolutional neural networks. In: 2014 IEEE conference on computer vision and pattern recognition
21. Eck A, Soh LK (2017) Sequential prediction of respondent behaviors leading to error in web-based surveys

A Review on Database and Transaction Models in Different Cloud Application Architectures

N. Priya and E. Punithavathy

Abstract Cloud computing applications are completely focused on scalable applications and optimized resource utilization. Due to the growing demands in the cloud, the application architecture gets evolved with many features, which are suitable to work with distributed systems. When comparing the monolithic, SOA and microservice architecture, the microservice is found to be the favoring architecture of the cloud. Similar to application architecture, the database model has also taken a shift from RDBMS to NoSQL and NewSQL models. This paper reviews the application architectures analyze their characteristics based on their performance when working with the suitable database models and their associated transactional models. It also examines some of the successful cloud applications, deployed under the distributed systems, with suitable architecture that favors the demands of the cloud.

Keywords Cloud application architectures · Database models · Transactions · Distributed systems · Microservice

1 Introduction

Cloud computing is a model that delivers computing services, like databases, servers, software, networks, etc., based on the demands. The host manages, modifies and improves cloud computing and its services continuously [1]. The applications developed in cloud require higher rate of scalability, reliability, deployment and minimum downtime [2]. At present, Cloud providers offer Container Service Platforms (CSPs), which support the flexible orchestration of containerized applications [3].

The cloud application architecture is focused up on, easier constructive of applications, interoperability among hardware and the software components, and also

N. Priya (✉)
Research Department of Computer Science, Shrimathi Devkunvar Nanalal Bhatt Vaishnav College for Women, Chennai, India

E. Punithavathy
Department of Computer Applications, Madras Christian College, Chennai, India
e-mail: punithavathy@mcc.edu.in

S. Shakya et al. (eds.), *Proceedings of Second International Conference on Sustainable Expert Systems*, Lecture Notes in Networks and Systems 351,
https://doi.org/10.1007/978-981-16-7657-4_65

promotes ease of deployment, which increases their efficiency [4]. Similarly, application architecture refers to the design of the application that is to be built and used as per the expectations of cloud service providers.

Databases play a major role in handling structured or unstructured data depending on the demand of the IT provider. Varied types of databases have evolved by the increasing demand in the cloud. Moreover, many transaction techniques have evolved in order to deal with the data stored in databases. The transactional method is determined based on the database. ACID method is implemented, while working with traditional database.

With the arrival of containers, the architectural design of applications under cloud computing has shifted its phase from monolithic to microservice architecture.

This paper reviews the frequently used cloud application architectures for the development of cloud applications, which has to be deployed under the demands from the providers. They also list the recent cloud applications, which are deployed in distributed systems with its favorable architecture.

2 Cloud Application Architecture

The cloud application architecture is the key principle used by all service providers in IT, for more efficiency. The application architecture acts as a protocol for designing applications. It also describes the pattern that is preferred in deploying applications, based on the dependencies required on the existing demands from the IT sector. The application architecture plays a vital role in developing the cloud applications, how they exhibit their roles and adapt to the cloud environment. The application architectures can be categorized into different types, but the most preferred architectures are: Monolithic, Service oriented and Microservice.

2.1 Monolithic Architecture

Monolithic applications are described as a single-tiered software application in which different components are combined into a single program from a single platform. It follows the principle of Open-Close, where, the written code can be extended later, but not modifiable [5]. It is a traditional style of application architecture which does not consider modularity as a design principle [6]. The Monolithic application could be difficult to understand and modify. The whole team must coordinate all development and redeployment efforts [7]. They suffer from a technology stack, in which all the modules are linked with same technology. Since all the modules are from the same programming language they have high coupling [8]. Horizontal scaling is done to enhance the performance which results in creating multiple copies behind the load balancer [9]. The entire application incurs downtime if any error occurs in a module. In case of frequent updates, deployment has to be done for the entire application at regular intervals [10]. The development phase and error tracing phase are relatively easy. Communication among the components is easier to control. It depends on the single server. During the scaling phase, all its instances have to be updated. Due to centralized concept, it doesn't have the flexibility to implement RDBMS and NoSQL databases in a single application [11]. The application developed using this architecture has greater security when compared with microservice and SOA. Moreover, due to the revolution of digital technology, it becomes complex to get shifted toward the new technologies.

2.2 *Service Oriented Architecture*

SOA (Service Oriented Architecture) follows the principle of "share as much as possible" ensuring reusability among the services [12]. It has a great flexibility in integration, and also promotes reusability among the services due to its coarse—grained modular structure [13].

SOA principles strictly follow DevOps skills, since many technologies are involved [14]. They normally use smart routing mechanisms for communicating among the services [15]. They are not focused on self-deployable units that are uniting the data and the related database as a single unit. The service implementation can be done in single or multiple servers. There are provisions to work with RDBMS and NoSQL databases in the same application. Hence, it requires more skills in the development and has complex management. It is comparatively costlier than monolithic architecture. The Enterprise Service Bus, which acts as a communication tool, might reflect in slow communication and less performance. So, the probability of failure of services is greater in this architecture [16].

Maintenance of SOA applications is difficult due to the following factors:

- Heterogeneity of applications, which requires knowledge from different languages.
- Distributed ownership of services, due to the unavailability of source code.
- Not well coordinated due to different driving forces and multiple versioned services [17].

2.3 *Microservice*

The Microservice follows the principle of "share as little as possible," which ensures that resources are not shared among modules [12]. This architecture is evolved from Service Oriented Architecture [7]. The most common characteristics of cloud computing properties like isolation state, distributed nature, elasticity, and automated nature have been satisfied by this architecture. Isolation is achieved in microservice using the concept of containerization, in which application has its own resource and no interaction is allowed among the resources in the environment [18]. It has its own storage in the form of a private database.

Microservice focuses on simple routing mechanisms such as REST. They allow scaling of agile processes by dividing the large team into small teams [19]. It encourages server-less-based technologies through the use of Faas (function as a service) [20]. The major problem of "dependency hell" is solved in this architecture [15]. This architecture has the capacity to eliminate the need for servers [21].

The CI (Continuous Integration) and CD (Continuous Deployment) which is enabled by microservices ensure the fastest update of an application. They follow eventual consistency in order to substitute the ACID transaction properties [22]. It can work with NoSQL databases and NewSQL databases, which work with modified RDBMS patterns and implements ACID models. The major drawback of this architecture is that it is comparatively costlier than the other two architectures. Due to an increase in the number of services, the complexity of the application is also increased. Finally, the security of the application is the major disadvantage of this architecture. When there is an increase of data exchange, there is less possibility of strong security [16]. They are also prone to many types of network attacks such as Denial of Service, Man in the Middle. Distributed systems, might expose much information at the endpoints, which might result in exposing the attack surface [23].

Figure 1 shows monolithic architecture has a tightly coupled module, with the common database, whereas in SOA the modules are more loosely coupled than monolithic with some basic interactions. In the case of microservice, the modules are very loosely coupled to SOA without any interactions. Similarly, Table 1 shows the basic difference among the cloud features when the applications are deployed.

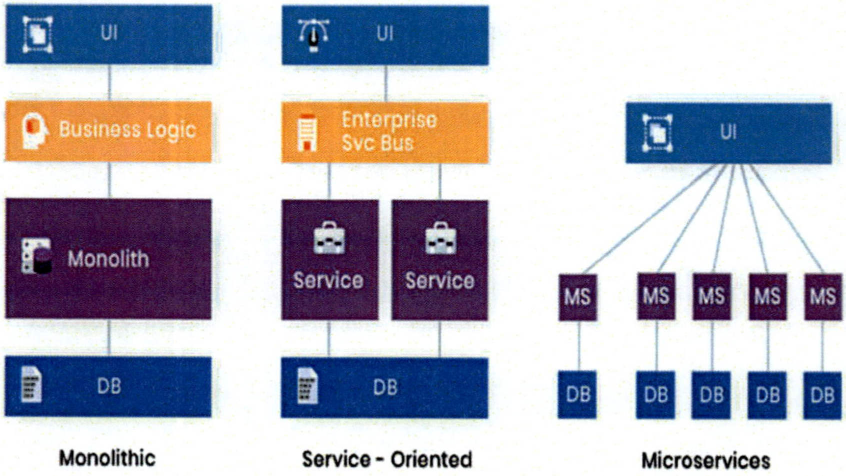

Fig. 1 Monolithic, SOA, microservice architectures [24]

Table 1 Comparison of monolithic, SOA and microservice based on cloud features

Features	Monolithic	SOA	Microservice
Maintenance	Difficult to maintain	Slow to maintain	Easier to update
Scalability	Performed in one dimension	Depends upon the services	Performed in two dimensions
Dependency	Full of dependencies	Minimizes dependencies	No dependencies
Deployment	It is difficult	It is too risky	It is done easily
Technology stack	Devoted to single technology	Supports different languages	Supports multiple technologies
Coupling	High coupling	Low coupling	Low coupling
Reliability	Less reliable	Less reliable	Highly reliable
Testing	Complex for the large code	Time consuming task	Easy to test
Isolation	No isolation	It is partial	Complete isolation
Decentralized	Centralized pattern	Centralized pattern	Decentralized pattern
Change management	Entire application needs to be updated	Frequent updates are not easy	Updation is easier
Communication	Internal communications	SOAP with enterprise service bus	REST and HTTP protocols
Transaction model	ACID	ACID	Eventual consistency

3 Database Models

Storage space has become a need of the hour due to the rapid growth of digital technologies. This leads to the development of cloud databases. The cloud database needs to satisfy the features such as scalability, high availability, optimized resource allocation and multi-tenancy. Considering the database of clouds, they primarily follow; two primary DBMS architectures. They are Shared-Nothing and Shared-Disk architecture. The Shared-Nothing architecture is distributed by nature, which is self-sufficient and independent of its memory and disk storage. The Shared-Disk architecture has its own memory but shares the disk storage space [25]. A cloud database can be of any type either structural or nonstructural. Traditional DBMS deals with structured data.

Due to the increasing demands, cloud databases also need to work with unstructured, semi structured or structured data. They should support the changing requirements of applications that involve unstructured data, such as documents and photos [26]. The different database models available in the cloud are (a) RDBMS (b) NoSQL Database and (c) NewSQL.

3.1 RDBMS

The Idea of a traditional database or Relational Database Management System (RDBMS) is found to be the substitute for a flat file database system. The RDBMS uses tables for storing data. The fields are represented in columns and records are represented in the form of rows. The main purpose of using RDBMS is to maintain a relation among the data stored.

SQL (Structured Query Language) is the programming language, which is used as a querying language in RDBMS. It is the most preferred technique used by Database Management Systems. Even though it is easy to work with, it has inability toward scalability, performance and availability problems [26]. It follows shared-disk architecture. The RDBMS follow the ACID transaction properties. These databases are not well suited for working with cloud. In traditional databases, strong consistency is supported by using conflict resolution at write time (using read and write locks) which leads to scalability problems [27].

The RDBMS ensures that data is always available, semi-handled and unstructured data and rapidly adapt to the changing requirements with frequent updates and new features [28]. In a cloud environment, it is necessary to maintain the consistency of data. It also has great strength in controlling the structure of the data. The ACID model ensures that transactions are completed in a single step, if not it is rolled back to its previous state. The characteristics of RDBMS are:

- Structured—It follows structured data, and all the data is stored in table
- ACID—It ensures whether they follow ACID transaction model
- Shared disk—It shares the database
- Consistency—It ensures whether data is consistent in all the steps.

PostgreSQL, a popular open-source implementation of RDBMS for online applications.

3.2 NoSQL

In order to overcome the problems faced by RDBMS, NoSQL is implemented. NoSQL can be defined as dealing with both structured and unstructured data. It follows Shared-Nothing architecture. It avoids many complexities that are followed in traditional ways to ensure scalability. They are more flexible to work with distributed data [26].

The most important feature of the NoSQL database is that it does not support, update, delete and joints [29]. They do not follow ACID properties which is the heart of RDBMS, instead they follow BASE principles [27]. The characteristics of NoSQL are:

- schema less—they do not necessarily form a structure.
- shared Nothing—It maintains its own disk and doesn't share with others.

- elasticity—It represents dynamic expandability.
- Sharing—practice of separating one table row into multiple tables.
- Asynchronous replication—it's a store with forward approach to backup and protect data.
- BASE instead of ACID.

Various categories of NoSQL databases are:

1. Key-value stores,
2. Document databases,
3. Wide-Column stores,
4. Graph databases.

Key-value store

The key-value represents the hash table model, for pointing a particular data using a unique key. Those values are usually stored as uninterrupted byte arrays [28]. Each data item is converted into a key using some unique formula which is stored in the lookup table or directory. When the data is needed, the key is converted into the location of the data and accordingly data is retrieved [30]. Fast accessibility and highly scalable to retrieve the values from the data store is the benefit of key-value [27].

Document Databases

In document stores the data are stored as a collection of key-value pairs, but now they are compressed as a document, i.e., entities that provide some structure and encoding of the managed data. Document stores do not have any schema restrictions and support big sized data collections in forms such as text documents, email messages and XML documents [27, 28].

Wide-Column Stores

It is a combination of NoSQL and relational databases. Unlike relational databases, they give preference to column structure. Column stores can be used to build high-performance applications. This data model is widely used in special functions such as batch processing, analytics prediction and for distributed data processing. It is widely implemented in the field of data storage, sorting, manipulation and compression [25, 30].

Graph Database

The graph databases store relational information in the tuples [30]. These databases are efficient in managing highly linked data and heavily connected data [28]. The best example that uses Graph database is Facebook, and it is suitable for maintaining highly interconnected data and traversing through entities. It is suitable for (a) Location-based services (b) Recommendation Engines, and (c) Complex network-based applications such as Social, Information, Technological and Biological networks.

Table 2 Comparison of NoSQL databases [29]

Data model	Ability	Functionality	Scaling	Elasticity	Intricacy
Key value	Strong	Varying	Strong	Strong	Weak
Column	Strong	Minimal	Strong	Moderate	Weak
Document	Strong	Varying	Varying	Strong	Weak
Graph	Varying	Graph theory	Varying	Strong	Strong

Table 2 shows the different characteristic features that exist in between the different NoSQL databases and their unique features. When considering the complexity of the database, the graph database is stronger than the other relevant databases.

3.3 NewSQL

The NewSQL solution aims to bring the combination of relational data model and the ideology of NoSQL. The NewSQL is next generation scalable relational database management system designed especially for transactions of processing systems. They follow ACID properties along with BASE properties. They break the conventional RDBMS restrictions by following the features of NoSQL such as Column oriented data storage and distributed architectures.

The characteristics of NewSQL:

- The fundamental method implemented in this model is SQL.
- It also relies on a transactional model named, ACID.
- It controls a non-locking concurrency control mechanism.
- The architecture that is supported in this model (dbshards) is more reliable than the traditional.
- They have the capacity to extend their performance without any bottle-neck during their execution.
- Those systems are approximately 50 times faster than the traditional OLTP RDBMS [31].

Table 3 shows the RDBMS type of data, which has a limited scalability when compared with the databases such as NoSQL and NewSQL. And it also has many limitations subject to working with distributed systems in the cloud.

4 Transaction Models in Cloud

These transaction models ensure a successful transaction across the database. They guarantee the safe state, availability, consistency and isolation state of the data. In

Table 3 Comparison of database models

Features	RDBMS	NoSQL	NewSQL
ACID/BASE	ACID	BASE through CAP	ACID, BASE
Architecture	Shared-disk	Shared-nothing	Shared-nothing
Scalability	Vertical	Horizontal	Horizontal
Consistency	Strong	Weak	Strong
Availability	Available	Available	Available
Query complexity	High	Low	Very high
DBMS type	Relational	Non-relational	Relational
Schema	Table	No specific structure	Supports both
OLTP	Partial	Successful	Successful
SQL	Supported	Not supported	Supported
Machine dependency	Single	Distributed	Distributed
Security	Very high	Low	Low
Cloud support	Partial	Supported	Fully supported

cloud database, all the database models should make use of the properties for their OLTP (Online Transaction Processing System). Based on the cloud requirements, there are different types of transaction models such as ACID, CAP and BASE.

4.1 ACID

The ACID model is essentially followed to achieve the reliability in databases. It determines the success of a transaction through various principles [32]. ACID refers to a set of guarantees for each transaction to be processed by the TP system: Atomicity, Consistency, Isolation and Durability.

- Atomicity—A transaction consisting of multiple operations is executed as a whole or not at all ("all or nothing").
- Consistent—Each transaction transforms the database from one consistent, valid state to another, adhering to all validation rules and database integrity constraints.
- Isolation—Concurrent transactions are executed by maintaining isolation, that is, by executing them as if they were sequential.
- Durability—Once a transaction has been committed, the results become permanent [33].

ACID follows the rule of one database with many users. The transactions are made in such a way that only one transaction will be performed at a time. It adopts pessimistic locking for transactions of higher level. The ACID ensures that the integrity of data is maintained at all times. If any transaction has to take place, it must strictly adhere to the rules of ACID. If any transaction avoids the rule then that particular transaction is made to roll back to the previous state [32]. The traditional database or RDBMS usually follows this architecture since they follow shared-disk architecture. Even though ACID is closely related to RDBMS, it cannot handle unstructured data and when transaction processing has to be shifted to the distributed nature, and these properties cannot be preserved.

4.2 CAP Theorem

The CAP theorem is framed to overcome the problems on Distributed Computing. It states three important components namely: Consistency, Availability and Partition Tolerance System. These components are very essential for a successful development of applications under distributed computing.

- Consistency—It ensures that all the users get the similar view of the database.
- Availability—It guarantees that the database versions are available for the users at any time.
- Partition Tolerance—It states that the database can be partitioned to many parts, so that failures will not affect other systems.

In order to follow CAP in distributed systems, either it can be CA (Consistent, Available, and No Partitions), or CP (Consistent, Partitions, not Available) and AP (Available, Partition, No consistent). There are many difficulties in following these systems. In CA, the system is consistent and available but it lacks network partitioning. Whereas CP maintains consistency and acquires network partitions but lacks availability. Occasional availability issues due to network partitions are a major threat in this. The NoSQL database usually follows this model in distributed systems [32]. Using this theorem, it is impossible to guarantee all the three properties. The unavailability of the system leads to financial loss [34].

4.3 BASE

BASE databases use strategies to have consistency, atomicity and partition tolerance "eventually." BASE does not violate CAP theorem, but works around it. BASE is usually defined as—Basically Available, Soft state and Eventual consistent. It is a model entirely proposed by cloud systems. Its properties are defined as

- Basically Available—A system is basically available when supporting partial failures without total system failure.
- Soft state—The state of the system is "soft" in that it can change over time even if no further updates are made.
- Eventual Consistent—The system will eventually become consistent, if no new updates are made to the system [33].

Table 4 Different cloud based applications and algorithm in distributed system

Application	Technique	Algorithm/framework	Dataset	Results
Banking [35]	Monolithic to microservice		Real-time data	scalability is increased
Banking [36]	Monolithic to microservice		Real-time data	Response time is decreased and reusability is increased
Book application [37]	Microservice	M3 (monitoring framework)	–	Monitors in multi-cloud environment
–	Microservice	Agnostic controller auto scaling algorithm [38]	Real-time data	Auto scaling of applications deployed in Kubernetes engine
Complex social network [39]	Microservice	Pishahang	–	Orchestrates the services using network function tools in the cloud
Urban risk management [40]	Microservice	Genetic architecture design based on microservice	–	Easy implementation of application and protection of sensitive data in the cloud
Legacy system [41]	Monolithic to microservice	–	Real-time data	Removes the complexity of the system
–	Microservice	DeathStarBench [42]	–	Open-source bench mark suite for monitoring in cloud
–	Monolithic to Microservice	Service Fabric [43]	–	Distributed environment to build microservice applications
–	Microservice	Micado (framework) [44]	–	Automatic management of applications in the cloud
Smart surveillance system [45]	Microservice and block chain	–	–	Scalable, secure and maintainable surveillance system

Availability refers to both performance and fault tolerance. Though availability fails, the application has to run and deliver the responses. In BASE transaction model, consistency should be governed by the developer, and it should not be associated with the database. In BASE model, the operation is divided into many sub tasks, which run parallel. The failure of one subtask will not affect other tasks. In eventual consistency, all read operations may return different data from the last completed written operation, but as time goes on, "In a steady state," the system will eventually return to the last written value. Clients may therefore face an inconsistent state of data as updates are in progress [29].

Table 4 lists out the recent cloud applications that are hosted in distributed systems, with various combinations of database models. They also list the different algorithms and frameworks that are suited to work with this architecture. From this study, it is known that the existing monolithic architecture based applications are converted to microservice based applications based on the requirements in the distributed system. It also noted that there is a decline in the development of applications in monolithic and SOA.

5 Conclusion

It can be concluded application architecture monolithic and database model RDBMS are not suitable for developing applications and deploying it in distributed systems in the cloud. The application architectures such as microservice, along with the database model such as NoSQL, NewSQL database are the most preferable to work with unstructured data like photo and documents. It can also deploy the applications in distributed nature. Even though the microservice is implemented in a distributed environment the consistency of data is difficult to maintain at all stages of transaction. Further, to develop a suitable distributed application under the microservice; major issues such as security and resilient patterns have to be analyzed for a successful deployment of applications.

References

1. Pandian AP, Smys S (2020) Effective fragmentation minimization by cloud enabled back up storage. J Ubiquitous Computi Commun Technol (UCCT) 2(01):1–9
2. Slimani S, Hamrouni T, Ben Charrada F (2021) Service-oriented replication strategies for improving quality-of-service in cloud computing: a survey. Cluster Comput 24:361–392. https://doi.org/10.1007/s10586-020-03108-z
3. Tomarchio O, Calcaterra D, Modica GD (2020) Cloud resource orchestration in the multi-cloud landscape: a systematic review of existing frameworks. J Cloud Comput 9(1). https://doi.org/10.1186/s13677-020-00194-7
4. Eldein S, Elhaj AI (2019) A container-based architecture for the design of portable cloud applications. Dissertation of Sudan University of Science & Technology
5. Hatami-Alamdari E, Etzioni Z (2019) Monolithic architecture vs. multi-layered cloud-based architecture in the CRM application domain
6. Laigner R et al (2020) From a monolithic big data system to a microservices event-driven architecture. In: 2020 46th Euromicro conference on software engineering and advanced applications (SEAA). IEEE
7. Faradj R (2018) The run-time impact of business functionality when decomposing and adopting the microservice architecture

8. Lloyd W, Ramesh S, Chinthalapati S, Ly L, Pallickara S (2018) Serverless computing: an investigation of factors influencing microservice performance. In: 2018 IEEE international conference on cloud engineering (IC2E)
9. Martinfowler [Online]. https://martinfowler.com/articles/microservices.html
10. Chris Richardson [online]. https://microservices.io/patterns/monolithic.html
11. .Chawla H, Kathuria H. Database Design for Microservices. Building Microservices Applications on Microsoft Azure, 225–244. https://doi.org/10.1007/978-1-4842-4828-7_8@@
12. Richards M (2016) Microservices vs. service-oriented architecture. O'Reilly Media, Inc.
13. Niknejad N, Ismail W, Ghani I, Nazari B, Bahari M, Hussin ARBC (2020) Understanding service-oriented architecture (SOA): a systematic literature review and directions for further investigation. Inf Syst 101491. https://doi.org/10.1016/j.is.2020.101491
14. Pulparambil S, Baghdadi Y, Al-Hamdani A, Al-Badawi M (2017) Exploring the main building blocks of SOA method: SOA maturity model perspective. Serv Orient Comput Appl 11:217–232
15. Cerny, Tom &Donahoo, Michael &Trnka, Michal. "Contextual understanding of microservice architecture: current and future directions" , ACM SIGAPP Applied Computing Review,vol.17,pp. 29–45,Jan (2018)
16. Ianculescu M, Alexandru A (2020) Microservices——a catalyzer for better managing healthcare data empowerment. Stud Inf Control 29(2):231–242
17. Haoxiang W, Smys S (2021) Big data analysis and perturbation using data mining algorithm. J Soft Comput Paradigm (JSCP) 3(01):19–28
18. Dragoni N, Giallorenzo S, Lafuente AL, Mazzara M, Montesi F, Mustafin R, Safina L (2017) Microservices: yesterday, today and tomorrow. In: Present and ulterior software engineering. Springer, Cham
19. Wolff E (2016) Microservices: flexible software architecture. Addison-Wesley-Professional
20. Jamshidi P, Pahl C, Mendonça N, Lewis J, Tilkov S (2018) Microservices: the journey so far and challenges ahead. IEEE Softw 35:24–35
21. Schneider T (2016) Achieving cloud scalability with microservices and DevOps in the connected car domain. Softw Eng
22. Kalske M (2018) Transforming monolithic architecture towards microservice architecture. CS thesis, University of Helsinki, Faculty of Science. Department of Computer Science
23. Yu D et al (2019) A survey on security issues in services communication of microservices-enabled fog applications. Concurr Comput Pract Exp 31(22):e4436
24. Antasia D (2019) Best software architecture for MVP, Apr 2019, pp 32–23. Available: https://rubygarage.org/blog/monolith-soa-microservices-serverless
25. Alam M, Shakil KA (2013) Cloud database management system architecture. UACEE Int J Comput Sci Its Appl 3(1). ISSN 2250-3765
26. Ali W et al (2019) Comparison between SQL and NoSQL databases and their relationship with big data analytics. Asian J Res Comput Sci 1–10
27. Moniruzzaman ABM, Hossain S (2013) NoSQL database: new era of databases for big data analytics—classification, characteristics and comparison. Int J Database Theory Appl 6
28. Makrisa A, Tserpesa K, Andronikoub V, Anagnostopoulosa D (2016) A classification of NoSQL data stores based on key design characteristics. In: Distributed to complete computing, CF2016, pp 18–20, Madrid, Spain, Oct 2016
29. Lotfy AE, Saleh AI, El-Ghareeb HA, Ali HA (2016) A middle layer solution to support ACID properties for NoSQL databases. J King Saud Univ Comput Inf Sci 28(1):133–145. ISSN 1319-1578. https://doi.org/10.1016/j.jksuci.2015.05.003
30. Ganesh Chandra D (2015) BASE analysis of NoSQL database. Fut Gener Comput Syst 52(C):13–21
31. Murazzo M, Gómez P, Rodríguez N, Medel D (2019) Database NewSQL performance evaluation for big data in the public cloud. In: Cloud computing and Big Data, JCC&BD 2019, vol 1050. Springer, Cham
32. Mapanga I, Kadebu P (2013) Database management systems: a NoSQL analysis. Int J Modern Commun Technol Res (IJMCTR) 1:12–18

33. Tai S, Eberhardt J, Klems M (2017) Not ACID, not BASE, but SALT: a transaction processing perspective on Blockchains. In: 7th International conference on cloud computing and services science, Apr 2017
34. Pokorný J (2011) NoSQL databases: a step to database scalability in web environment. Int J Web Inf Syst 9:278–283
35. Bucchiarone A, Dragoni N, Dustdar S, Larsen ST, Mazzara M (2018) From monolithic to microservices: an experience report from the banking domain. IEEE Softw 35(3):50–55. https://doi.org/10.1109/MS.2018.2141026
36. Megargel A, Shankararaman V, Walker DK (2020) Migrating from monoliths to cloud-based microservices: a banking industry example. In: Software engineering in the era of cloud computing. Springer, Cham, pp 85–108
37. Noor A et al (2019) A framework for monitoring microservice-oriented cloud applications in heterogeneous virtualization environments. In: 2019 IEEE 12th international conference on cloud computing (CLOUD). IEEE
38. Khaleq AA, Ra I (2019) Agnostic approach for microservices autoscaling in cloud applications. In: 2019 international conference on computational science and computational intelligence (CSCI). IEEE
39. Kouchaksaraei R, Hadi TD, Karl H (2018) Pishahang: joint orchestration of network function chains and distributed cloud applications
40. Kramer M, Frese S, Kuijper A (2019) Implementing secure applications in smart city clouds using microservices. Fut Gener Comput Syst 99
41. Knoche H, Hasselbring W (2018) Using microservices for legacy software modernization. IEEE Softw 35(3)
42. Gan Y et al (2019) An open-source benchmark suite for microservices and their hardware-software implications for cloud and edge systems. In: Proceedings of the twenty-fourth international conference on architectural support for programming languages and operating systems
43. Kakivaya G et al (2018) Service fabric: a distributed platform for building microservices in the cloud. In: Proceedings of the thirteenth EuroSys conference
44. Kiss T et al (2019) Micado—microservice-based cloud application-level dynamic orchestrator. Fut Gener Comput Syst 94:937–946
45. Marie-Magdelaine N, Ahmed T, Astruc-Amato G (2019) Demonstration of an observability framework for cloud native microservices. In: 2019 IFIP/IEEE symposium on integrated network and service management (IM). IEEE

Effective Design Approach of Yard Trimmer

R. Seetharaman, M. Tharun, S. S. Sreeja Mole, and K. Anandan

Abstract A yard trimmer is a machine that utilizes at least one edge to slice grass to even tallness. The edges might be fueled either by hand pushing the cutter forward to work the mechanical sharp edges or may have an electrical engine or an interior burning motor to turn the edges. An electrical grass trimmer is more appropriate and simpler to use than the yard cutter with a motor, so we are making an electrical grass trimmer with an electrical engine which is giving fast revolution to the edges. Cordless electrical cutters are fueled by factor numbers (commonly 1–2) of 12 V battery-powered batteries. Commonly more batteries mean more run time and force. Batteries can be either positioned inside the machine or outside. In the event that outwardly, drained batteries can be immediately traded with the re-energized batteries. The tallness of the cut grass is flexible and can be changed by a switch, explicitly single expert level or nut and fastener. Numerous lawnmowers incorporate different capacities, for example, mulching the cut grass or gathering their grass clippings in a removable sack or container.

Keywords Yard trimmer · Batteries · Sensors · Cordless electrical cutter · Movable blade

1 Introduction

In current occasions, each interaction is being robotized decreasing human obstruction as the work has been progressively increasing thus universally, we change to the advanced innovation. The computerized grass shaper is working proficiently and

R. Seetharaman (✉) · M. Tharun
Department of ECE, CEG Campus, Anna University, Chennai 600025, India
e-mail: rseetharamanece@annauniv.edu

S. S. Sreeja Mole
Department of ECE, Christu Jyothi Institute of Technology & Science, Jangaon 506167, India

K. Anandan
PS Division of E & E, TUV SUD South Asia Pvt Ltd., Chennai 600041, India

S. Shakya et al. (eds.), *Proceedings of Second International Conference on Sustainable Expert Systems*, Lecture Notes in Networks and Systems 351,
https://doi.org/10.1007/978-981-16-7657-4_66

time utilization is dealt with viably. Computerized grass shaper module is a savvy item as it eliminates man powers, where work efficiently. The reel is joined to the trimmer's wheels with the help of pinion wheels, the reel turns a few times quicker than the elastic sort wheels turn when the cutter is pushed forward and these kinds of reel trimmers offer the advantage of zero contamination being created [1]. This is not recommended particularly for enormous yards where the energy is devoured from the clients and this is the hardest undertaking. These trimmers rely upon the arrangement of the reel frequently can't ready to slice grass exceptionally near yard obstructions, similar to trees, carports, edging, and so on, and furthermore require ordinary level grass surface to work appropriately without reaching as far down as possible the shaper bar [2].

The grasscutter contains line wire which assists with characterizing the region to be cut and surprisingly the robot utilizes it in other applications, for example, to find the region to be managed and to find a re-energizing deck [3]. Mechanical trimmers are fit for keeping up with up to 5 sections of large area of grass [4]. Automated yard trimmers are progressively modern, are generally self-docking, and contain downpour sensors, almost taking out human cooperation for taking care of lawn and Multiple mechanical cutters can be utilized to cut a significantly bigger region [5].

The most prime thing is to ensure that the engine is focused when the engine is mounted to the trimmer deck. The engine ensured it was focused by estimating an equivalent separation from the sides of the engine to the external distance across the trimmer and protected the engine to the deck with four 3/8″ bolts with washers [6]. Adding washers between the engine and the deck permitted us to raise various sides of the engine to even out. It expected to put two washers on the front two bolts to bring the engine up to even out [7]. Robots with enormous wheels are regularly more attractive than robots with little wheels. The right wheel size is vital and cares we have taken to pick the right one. The wheels were mounted to the engines; the wheels utilized are appended with the engines and didn't need any exceptional changes for the establishment. They were connected just mounted to the shaft of the engines. We attempted with two sorts of wheels, one made of plastic and the other made of wood. Each enjoys its own benefits and inconveniences.

2 Proposed Design

The expense of fuel is getting expanded, the impact of gases being transmitted from the consumed fuel is debilitating the ozone layer and fossil fuel product has likewise become a major test in this day and age [8]. In this manner, by utilizing sunlight-based force, we can decrease carbon impressions. Yard cutters are becoming famous these days. Grass cutting being dreary work for individuals can be streamlined by utilizing a programmed grass shaper, which can be performed without an actual individual is available. The primary source of energy is the solar panel. It receives the power from the sun and transfers it to the battery of 12 V and 7.5 Ah. The battery transfers the power to the microcontroller and to other equipment which requires power. The

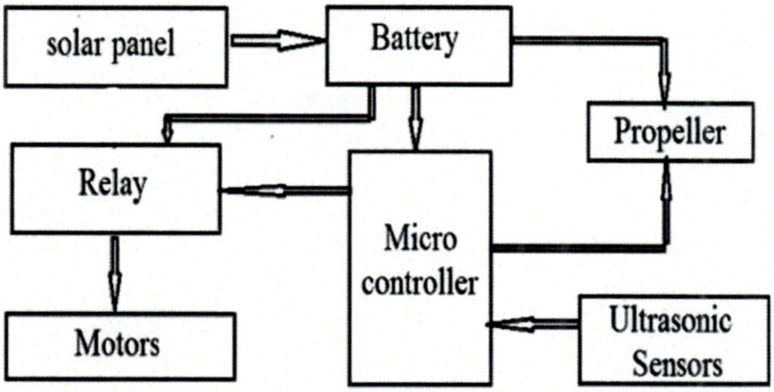

Fig. 1 Block diagram of the module

relay is used to control the direction of the wheels and the relay is controlled by the Arduino. Because of this turn, a simpler, quicker and more dependable way for controlling a lawnmower is accessible for everybody and has added to the general ecological tidiness of the general public [9]. The Arduino microcontroller gives control signals to relay and the propeller which is used to cut the grass. The ultrasonic sensors are used to detect boundaries and to send the distance of the obstacle to the microcontroller again to take decisions [10]. Youngsters, too as old people can do the errand of trimming a grass, absent a lot of exertion. It is an eco-accommodating trimmer that utilizes sustainable wellsprings of energy and this framework discovers its application where grass slicing is needed to be done for an enormous scope which is tedious and exorbitant in nature [11]. The proposed design is shown in Fig. 1.

3 Motor Characteristics

3.1 DC Motor

A DC engine is a precisely commutated electric engine controlled from direct flow (DC). The stator is fixed in space and the current in the rotor is exchanged by the observer to likewise be fixed in space. This is the way the relative point between the stator and rotor attractive transition is kept up with almost a quarter of 360°, which creates the most extreme force.

3.2 Torque

Force is the inclination of a power to turn an item about a hub, support, or turn. Similarly, as a power is a push or a force, thus the torque can be considered as a bend or rotate an item. Mathematically,

$$\boldsymbol{\tau} = \mathbf{r} * \mathbf{F}$$
$$\tau = ||\boldsymbol{r}||||\boldsymbol{F}||\mathbf{sin}\theta$$

$\boldsymbol{\tau}$ is the torque vector.
$\mathbf{r}$ is the position vector.
$\mathbf{F}$ is the force vector.
$\boldsymbol{\theta}$ is the angle between the force vector and the level arm vector.

$$\text{Power} = (\text{Torque}) * (\text{Angular speed})$$

Initially, we tried with low torque motors but as per the design, our robot is with the solar panel so in order to support the entire body we replaced it with the high torque motors. Each of 10 kg torque to drive the wheels. Torque can be perceived to be the work behind the work done and rakish speed determines the rate at which it occurs. So, more the force, i.e., torque, less is the rakish speed needed for a similar pace of work utilization. In the event that the information power is consistent, force and rakish speed become conversely corresponding. Really at that time does the precise speed diminishes as the force increments. It is a sort of a tradeoff just when steady torque input is free.

3.3 RPM

It is a proportion of the recurrence of pivot, explicitly the quantity of turns around a proper hub in one moment. It is utilized as a proportion of rotational speed of a mechanical speed. We experimented with many motors of varying RPM. We finally fixed 200 RPM motors and we thought of decreasing it as it moves faster in open field.

3.4 Shaft

A drive shaft, driving shaft, propeller shaft (prop shaft), or Cardan shaft is a mechanical segment for communicating force and pivot, typically used to interface different segments of a drive train that cannot be associated directly on account of distance or the need to take into consideration relative development between them. There are two types of shafts design in the motor.

(1) Single shaft and
(2) Side shaft.

4 Proposed Work

This section manages the stream diagram of the task, ideal arrangement of sensors and its working, and the working of transfer circuit. There are numerous strategies and calculations that exist to cover the yard for cutting. We study the issue of discovering briefest visits/ways for "grass cutting" and "processing" issues: Given an area in the plane, and given the state of a "shaper" (regularly, a circle or a square), track down a most limited visit/way for the shaper to such an extent that each point inside the locale is covered by the shaper at some situation along the visit/way [12]. In the processing adaptation of the issue, the shaper is obliged to remain inside the district and the processing issue emerges normally in the space of programmed instrument way age for NC pocket machining [13]. The grass cutting issue emerges in optical review, splash painting, and ideal pursuit arranging.

Every one of the strategies enjoys their own benefits and hindrances. The principal factor to consider is distance, and this one is handily precluded by the primary law of trimming a grass: Don't go over a similar fix twice [14]. This given takes out numerous well-meaning, yet plainly wasteful courses on an unobstructed fix of grass. This compact grasscutter can be utilized to keep up with and trim grass in gardens, homes, schools, or yards [15].

5 Results and Discussion

We performed many different approaches to conclude the efficient way. Figure 2 shows a unidirectional approach and few turns. Figure 3 is bidirectional and quite similar to Fig. 2. Figures 4, 6 and 7 are similar to each other with large number of turns. Figures 5 and 8 are less distance and highest number of turns. Figure 9 is the simplest approach with equal number of left turns and right turns which is the efficient algorithm.

Fig. 2 Left turns—0, right turns—30 total turns = 30

Fig. 3 Left turns—16, right turns—15 total turns = 31

Fig. 4 Left turns—32, right turns—30 total turns = 62

Fig. 5 Left turns—120, right turns—120 total turns = 240

Fig. 6 Left turns—20, Right turns—18 Total turns = 38

Fig. 7 Left turns—33, Right turns—38 Total turns = 68

Fig. 8 Left turns—101, Right turns—102 Total turns = 203

Fig. 9 Left turns—14, Right turns—16 Total turns = 30

Mower Test 1. In the beginning, a simple wired bot is tested, expected result is that bot will be free to run and take left and right. The test result is a success.

Mower Test 2. Tried to cut the grass below the bot body, but resulted in failure because the mower blade is touching the ground shown in Fig. 10.

Battery Test 1. Battery being in sufficient power, we expect the mower to move and cut the grass. The result is success.

Sensor Test 1. Whenever a person comes in the front, the bot immediately stops moving forward. Hence the test result is success.

Sensor Test 2. When an obstacle is below the sensors placing level, the bot will not detect it. Thus, the test result is failure shown in Fig. 11.

The major goals we have set in the objective have been met in the final design of our project. This project eliminated the physical power required in pushing the lawn

Fig. 10 Mower test 2

Fig. 11 Sensor test 2

mower without sacrificing safety of the humans. The lawn mower makes the decision on its own by using the sensor values thereby autonomous movement is achieved. All three sensors are working as per the design. Our big challenge is to cover the lawn area effectively and we practically examined the system, the mower traces the lawn as exactly as our algorithm we have used. In addition to all this provision for various trimming levels of grass is also included in our project for user friendly. Due to the power demand, we choose renewable energy (solar) which is also eco-friendly. So design needs like safety, reliability, cost efficiency, user friendliness, effectiveness, durability, and eco-friendly all are met.

6 Future Work

We can attach any other mechanism to collect the grass which is already cut in the lawn. We can arrange a place to dump the grass which is already cut and can make the bot go that place and dump it. We can increase the torque of the motor to pull more load. We can use any suitable material to grip the wheels to move smoother on the lawn. We can use higher watt solar panels to quickly charge the battery. We can use quick response sensors to work according to the algorithm well. We can use any other sensors to detect water in case of rain to protect the bot and to move to a safer place automatically.

References

1. Palve SV, Panchal K, Chipkar R, Patil A, Sonawane GL (2016) Solar powered automated grass cutter machine. Sci Technol Eng IRJET 05(04):576–580
2. Nagarajan N, Sivakumar NS, Saravanan R (2017) Design and fabrication of lawn mower. Asian J Appl Sci Technol AJAST 01(04):50–54
3. Baloch TM, Kae TTC (2008) Design and modelling a prototype of a robotic lawn mower. Int Symp Inf Technol IEEE 04:1–5
4. Mallikarjun M, Vishwa T, Srujan K, Praveen K (2018) Automatic solar grass cutter. Int J Res Appl Sci Eng Technol 06(04):1148–1151
5. Hameed IA (2017) Coverage path planning software for autonomous robotic lawn mower using Dubins' curve. In: 2017 IEEE international conference on real-time computing and robotics (RCAR). IEEE, pp 517–522
6. Okafor B (2013) Simple design of self-powered lawn mower. Int J Eng Technol 03(10):933–938
7. Chon W, Amano RS (2005) Investigation of flow behavior around corotating blades in a double-spindle lawn mower deck. Int J Rotating Machin 01:77–89
8. Kishore PSV, Kumar PS, Dash S, Ramesh K (2018) Solar powered obstacle avoiding lawn mower. J Electr Eng 12(02):1–5
9. Ocholi F, Agom I, Koldim R, Edeoja AO (2019) Design of a solar-assisted lawnmower with an arduino-based mobile application control system. TAJET 01(02):61–67
10. Carullo A, Parvis M (2001) An ultrasonic sensor for distance measurement in automotive applications. IEEE Sensors J 01(02):143
11. Sathya S, Arunmozhi P, Pavithra E, Anusuya N, Gowthaman KS (2020) Renewable energy based robotic grass cutter using IOT. Renewab Energy 07(03)
12. Arkin EM, Fekete SP, Mitchell JS (2000) Approximation algorithms for lawn mowing and milling. Comput Geometry 17(1–2):25–50
13. De Assis IR, De Souza CC (2011) Experimental evaluation of algorithms for the orthogonal milling problem with turn costs. In: International symposium on experimental algorithms. Springer, Berlin, Heidelberg, pp 304–311
14. Lerman SB, Contosta AR, Milam J, Bang C (2018) To mow or to mow less: lawn mowing frequency affects bee abundance and diversity in suburban yards. Biolog Conserv 221:160–174
15. Sivagurunathan R, Sivagurunathan L, Hao JCJ (2017) Design and fabrication of low cost portable lawn mower. Schol J Eng Technol 05(10):584–591

A Comprehensive Assessment of Privacy Preserving Data Mining Techniques

Kundeti Naga Prasanthi and M. V. P. Chandra Sekhara Rao

Abstract The process of knowledge extraction from large datasets is the primary task in Data Mining. Data warehouse is one of the central repositories of data. When data is available only at a single location, a number of serious security problems may crop up. So, various data mining techniques are applied to the data located in data warehouse to extract useful patterns from the data and protect privacy of sensitive data. In recent times, attacks on people have increasingly become common. So, people are reluctant to share their data for loss of privacy. In such scenarios, privacy preserving data mining techniques play a major role in ensuring security of sensitive data. Sometimes, non-sensitive data may also convey sensitive information to attackers. This paper reviews various privacy preserving data mining techniques in literature specifying their details. A table providing a comparison of such techniques is also presented here.

Keywords Sensitive data · Privacy preserving techniques · Data mining · Data perturbation · Blocking based technique · Cryptographic technique · Micro-data publishing

1 Introduction

Accumulation of large volumes of data from different data sources has become a common phenomenon in the last few years. Such voluminous data can be used for knowledge extraction like data mining. Data mining refers to "the method of extracting useful knowledge from large amounts of data". Using data mining, knowledge can be retrieved from bulky volumes of data present in various repositories like databases, data warehouses etc. High-level information, unique patterns and

K. N. Prasanthi (✉)
Department of Computer Science and Engineering, Lakireddy BaliReddy College of Engineering, Mylavaram, India

M. V. P. Chandra Sekhara Rao
Department of Computer Science and Engineering, R. V. R. & J. C. College of Engineering, Chowdavaram, India

S. Shakya et al. (eds.), *Proceedings of Second International Conference on Sustainable Expert Systems*, Lecture Notes in Networks and Systems 351,
https://doi.org/10.1007/978-981-16-7657-4_67

novel data can be extracted or viewed in different parlance by using data mining. The discovered knowledge can be used for decision making by higher authorities. Organizations release data so that researchers can utilize the data and extract useful knowledge from it, which is known as micro-data publishing. Data mining discovers valuable and subtle information from large databases. But the data may be misused by attackers and sensitive information may be stolen. So, there is a conflict between data publishing and privacy. Privacy can be defined as the quality or condition of being secluded from the presence or view of others [1]. Privacy means hiding personal data about an individual or a group of individuals from other unauthorized person(s). Privacy is said to be violated if a person's personal information is used negatively by others. Revelation of personal information is the main concern in data publishing. If personal information is revealed, it can be misused by attackers. For example, personal details such as date of birth, father's name, sex etc. are revealed, then there is little threat. But if other details like SSN, voter ID are disclosed, then, it results in identity stealing. Privacy preserving data mining techniques provide privacy of data in data publishing. Privacy of data is ensured when data is hidden by using different methods, while releasing anonymous data. The data released can still be used for extracting useful patterns. In fact, Data Mining results can be similar in both cases, i.e. using original data from organizations or the transformed data. This paper reviews different approaches and techniques available in Privacy Preserving Data Mining (PPDM). The paper is organized as follows: Sect. 1 discusses the concept of data mining and privacy, while Sect. 2 discusses various privacy threats. Section 3 discusses classification schemes and evaluation criteria for privacy preserving data mining techniques. Section 4 describes privacy preserving data mining techniques and their limitations. In Sect. 5, a tabular form of various PPDM techniques is presented, while, Sect. 6 deals with the conclusion of the paper.

2 Privacy Threats

Micro-data publishing may cause different privacy threats like identity revelation, membership revelation and attribute revelation to appear over personal data sharing space. Data mining results released to public may also cause privacy threats. Privacy risks result in more disclosure attacks. PPDM techniques can protect the data from different disclosure risks and provide better utility.

(i) Identity Disclosure

Generally, an individual can be linked to a published table by the attacker when he accesses his/her background knowledge. If the identity of an individual is known, then any sensitive information about his/her may be revealed. For example, An attacker can link voter data with hospital data, release and identify the medical details of a person.

(ii) Attribute Disclosure

When information about individual records is revealed, then attribute disclosure becomes possible. Individual's attributes should be made confidential while releasing data. According to Vani et al. [2], matching multiple buckets is crucial for protecting attribute exposure.

(iii) Membership Disclosure

The membership information in a released table can be identified by using different linkage attacks. Moreover, membership can be revealed if sensitive attribute value is used in selection criteria.

3 Privacy Preserving Data Mining Techniques Categorization Methods and Assessment Criteria

There are several PPDM (privacy preserving data mining) techniques available in literature. Some techniques are suitable to be applied in certain scenarios, while useful in some specified objectives. A classification scheme and evaluation criteria for these techniques is presented in [3].

A. *Classification Scheme*

Privacy preserving techniques can be categorized based on following features:

- Data Mining Scenario
- Data Mining Activities
- Data Distribution
- Types of Data
- Definition of Privacy
- Protection Method.

We illustrate these classification features as follows:

(i) *Data Mining Scenario*: There are two fundamental data mining scenarios. In the first scenario, data is released by the organization for data mining, without any restriction for accessing data. In this scenario, privacy is provided by data modification. In the second scenario, data sets are not released by the organization, though data mining tasks are allowed. Data is applied with cryptographic techniques to provide privacy here.

(ii) *Data Mining Task*: Data mining task involves extracting useful patterns from huge amounts of data by using techniques like clustering, association, classification etc. Privacy preserving techniques can be applied to data mining activities and statistical investigations to present privacy. However, only a few data mining activities support privacy preserving techniques. Based on this support, privacy preserving techniques are divided into several categories.

(iii) *Data Distribution*: Data Mining can be applied to centralized as well as distributed data sets. In case of centralized data set, a single party owner deals with the data, which is generally available at the computational site or that may be sent to the computational site. In case of distributed data, data set is divided between two or more parties. Here, the parties may not trust each other and may wish to carry out data mining on combined data. The distributed dataset may be either analogous or diversified. Horizontally partitioned data contains same set of attributes that can be termed as homogeneous or analogous data sets. The vertically partitioned data which contains a different set of attributes at each site is termed as heterogeneous or diversified data set.

(iv) *Data Types*: The data set may contain several types of attributes that may include numerical or categorical types. Boolean data takes two probable values of '0' and '1'. As categorical values do not follow orders, the privacy preserving techniques can be applied based on the difference between categorical and numerical values.

(v) *Privacy Definition*: Based on the context, privacy may have different definitions. Privacy may be provided to either individual's data or data mining results like association or classification rules. Various privacy preserving techniques can be applied based on different definitions of privacy.

(vi) *Protection Methods*: There are various ways of providing privacy in data mining like data alteration and Secure Multiparty Computation. Privacy preserving techniques may vary depending on the protection method applied.

B. *Evaluation Criteria*

There are various evaluation criteria for selection of privacy preserving techniques. Some of them are:

(i) *Versatility*: Versatility means the potential applicability of a technique to various data mining activities while catering to the various levels of privacy and different types of data sets. Versatility and usefulness are proportional to each other. Versatility features are shown below:

- Privacy: Data versu Knowledge Patterns
- Data Sets: Centralized or Distributed (Vertically or horizontally)
- Attributes: Categorical or Numerical
- Data Mining activities.

(ii) *Disclosure Risks*: Disclosure risk means the likelihood of disclosure of sensitive data by a malicious user. Disclosure risk is inversely proportional to the level of privacy provided by the technique. Several factors influence disclosure risks like nature of technique, supplementary knowledge of an intruder etc. The main aim of privacy preserving technique is minimization of the disclosure risk. So, it may be noted that disclosure risk evaluation is indispensable.

(iii) *Information Loss*: When privacy preserving techniques are applied, there is a chance of information loss. Information loss is determined by taking into account the noise added and the level of security provided. Data quality decreases in case of more information loss. Privacy preserving data mining techniques aim to secure high quality data. Hence, providing more security to data at the cost of data quality is unnecessary.

(iv) *Cost*: Computational cost and communication cost between cooperating parties together constitute the total cost of the technique. Computational cost is the summation of price for preprocessing (e.g. perturbed values) and execution cost (e.g. processing fixed cost). Communication cost plays major role in case of distributed data sets.

4 PPDM Techniques

Various Privacy Preserving Data Mining (PPDM) techniques are present in literature. These techniques include data randomization, data perturbation, cryptography-based techniques and blocking based techniques.

A. Data Perturbation

In this method, a random process is applied to data to modify it [4]. Data perturbation is also termed Data distortion or Data noise. This technique is applicable to different types of attributes like int, boolean, char and type of classification. When X denotes a set of data records, it is represented as $X = \{x_1, x_2, \ldots x_n\}$.

When X refers to a set of data records, it is denoted by $X = \{x_1, x_2, \ldots x_n\}$. For record x_i, a noise component is added from probability distribution $f_y(Y)$. These noise components are denoted by $y_1, y_2 \ldots y_n$. Now, the distorted noise records become $x_1 + y_1, x_2 + y_2, \ldots x_n + y_n$. However, the actual data records can't be recovered from the modified records. Data distribution can only be obtained. As Randomization technique is simple and does not require distribution of data records, it can be executed during data compilation.

Data can be distorted by adding noise or adding unknown values or by using data transpose matrix. A new distribution based data mining algorithm can be generated for DM (data mining) tasks like classification, clustering or association rule mining. Singular value decomposition (SVD) [5] and sparsified singular value decomposition (SSVD) are two different methods of data perturbation. Original data set and distorted data set are compared by using different matrices. More Data utility occurs when SSVD technique is applied. One of the drawbacks of the perturbation technique is that it modifies each attribute independently, though each data dimension allocation is independently recreated. Inter attribute correlation plays a major role in data mining algorithms.

B. Blocking Based Technique (BBT)

BBT (Blocking based technique) [6] implements a methodology for hiding sensitive data. This technique comprises two methods viz., (1) Identifying transactions of sensitive rule and (2) Replacing known values with unknown values (?). The transactions containing sensitive rules are identified by scanning the database. Confidential data in every transaction is substituted with anonymous value. Kargupta et al. [7] have substituted '1' by '0' or '0' by '1' or with any unidentified (?) values in transactions.

III. Cryptographic Technique

In this technique, data is hidden by applying cryptographic algorithm. The data can be read only by the proper recipient. Two parties can communicate securely by using this technique. A variety of cryptographic algorithms that provide security of data are available. Many cryptography-based algorithms are available in literature. Some of these techniques can be applied to horizontally partitioned data as well. An efficient cryptographic technique is presented in [8]. This technique is very popular as it grants safety of confidential attributes. There are some disadvantages of cryptography algorithms. But, some of the algorithms [8] are inefficient on large databases as they consume a lot of computation time.

IV. Condensation Approach

Condensation was first introduced by Aggarwal and Philip [9]. In this technique, Data is condensed into multiple groups of pre-defined size. Pseudo data is generated from condensed clusters using statistics of cluster. The data set containing these synthetic groups of data is similar to the original data. Groups of pre-defined size are formed by applying contraction or condensation. Each group retains certain statistical variation. Each group of data has a specific size 'k'. This k refers to the level of privacy preserving technique. More privacy can be provided by using higher values of k but leading to less utility of data. This approach provides better privacy protection when compared with other PPDM techniques. However, the main disadvantage of this technique is loss of information.

E. Hybrid Technique

In recent times, privacy preserving data mining has become a prolific research area, wherein various hybrid algorithms are projected to provide privacy of data. A Hybrid technique is evolved by combining one or more privacy preserving techniques, while in most cases, these techniques have shown better accuracy. Lohiya and Ragha [10] proposed a hybrid technique. This Hybrid technique involves application of two different techniques namely randomization and generalization. There are other hybrid techniques also which provide more privacy as given in [11–13].

5 Comparison Between Different Techniques

A vast collection of techniques are proposed in the PPDM (Privacy Preserving Data Mining) area. Each technique has its own advantages and disadvantages. The techniques can be categorized based on different criteria like implementation, usefulness of data, execution cost, difficulty of implementation, forbearance against data mining techniques etc. Table 1 shows prominent works done by eminent authors in the field

Table 1 Comparison of PPDM techniques

S. No.	Authors	Year of publication	Technique used for PPDM	Approach	Result
1	Lindell and Pinkas [8]	2000	Cryptographic technique	Technique involves encrypting sensitive data	This technique is costly for large databases and databases involving multiple parties
2	Sweeney [1]	2002	K-Anonymity	A record is indistinguishable from $k - 1$ records	This technique provides privacy
3	Vaidya and Clifton [14]	2002	Association rule	This technique involves distributing data into vertical segments	This technique provides privacy
4	Kargupta et al. [7]	2003	Data perturbation	Random noise is added to data to provide privacy. The randomized data still gives accurate data mining results	Random matrices are generated using randomization-based technique
5	Aggarwal and Yu [9]	2004	Condensation	This technique utilizes pseudo data instead of original data	The data mining algorithms do not require redesign even with pseudo data
6	Machanavajjhala et al. [4]	2006	L-diversity	Each class contains l-well represented values for sensitive attribute	This technique provides more privacy than K-Anonymity
7	Kisilevich et al. [15]	2010	Anonymization	Generalization and Suppression techniques are used as part of Anonymization	This technique suffers from background knowledge and homogeneity attacks
8	Deivanai et al. [16]	2011	Hybrid technique	This is a combination of different techniques	Anonymization and suppression are used to provide privacy
9	Mathew and Obradovic [17]	2011	Decision tree	This technique is graph based	This technique protects sensitive information

(continued)

Table 1 (continued)

S. No.	Authors	Year of publication	Technique used for PPDM	Approach	Result
10	Parmar et al. [6]	2011	Blocking based technique	This technique replaces sensitive data with unknown values	Data reconstruction is difficult
11	Mumtaz et al. [18]	2011	Distortion based perturbation technique	This technique distorts all cells in OLAP data cube	This technique provides utmost privacy for range sum queries
12	Huang and Fang [19]	2011	Histogram based technique	This technique is based on concept of reversibility	This technique is mainly used for X-rays or medical images
13	Liu et al. [20]	2011	Rating based privacy	This technique shuns curse of dimensionality	Provides more privacy than K-anonymity and L-diversity
14	Alotaibi et al. [21]	2012	Multi-dimensional scaling	Non-linear dimensionality reduction technique	This technique uses non metric MDS transformation
15	Komishani and Abadi [22]	2012	Trajectory data	This technique involves generalization of trajectories and sensitive attributes w.r.t. moving objects	This technique is useful in trajectory data publishing
16	Jahan et al. [5]	2012	Data perturbation using SSVD	This technique uses specified singular value decomposition	Applying SSVD is more successful
17	Karthikeswarant et al. [23]	2012	Association rule	This technique uses sliding window algorithm to provide privacy	This technique modifies database to hide sensitive data
18	Kumbhar and Kharat [24]	2012	Association rule by horizontal and vertical distribution	This technique implements different approaches in Association rules	Models are analyzed in terms of privacy
19	Lohiya and Ragha [10]	2012	Hybrid technique	This technique is combination of K-Anonymization and Randomization	This technique provides better accuracy

(continued)

Table 1 (continued)

S. No.	Authors	Year of publication	Technique used for PPDM	Approach	Result
20	Beck and Marhöfer [25]	2012	Anonymizing demonstrator	This technique performs anonymization	Utility can be enhanced by applying swapping and recording

of PPDM in chronological order. Comparison among different algorithms is done based on parameters like algorithm used for PPDM, its approach, consequences and precision are taken into consideration.

6 Conclusion

In recent times, protection of sensitive data has become a major challenge. People are reluctant to share their data due to the fear of several types of threats making their data vulnerable. Data perturbation and Cryptography methods achieve better performance than other methods. In the realm of data encryption, cryptography is widely considered the best technique. In particular, sensitive data encryption can be carried out effectively by using Cryptography. Application of data perturbation results in better handling of Sensitivity of data. Moreover, privacy protection is efficiently handled by application of Utility of data. However, Cryptographic techniques provide less data utility than other techniques. After reviewing the existing literature in the field of privacy preservation, our study proves that no single technique is efficient in all aspects of performance. Finally, this article observes that hybrid techniques achieve better performance than individual techniques.

References

1. Sweeney L (2002) K-anonymity: a model for protecting privacy. Proc Int J Uncertain Fuzziness Knowl-Based Syst
2. Vani B, Jayanthi D (2013) Efficient approach for privacy preserving microdata publishing using slicing. IJRCTT
3. Verykios VS, Bertino E, Fovino IN, Provonza LP, Saygin Y, Theodoridis Y (2004) State-of-the-art in privacy preserving data mining. SIGMOD Rec 33(1):50–57
4. Machanavajjhala A, Gehrke J, Kifer D, Venkitasubramaniam M (2006) l-diversity: privacy beyond k-anonymity. Proc Int Con Data Eng (ICDE):24
5. Jahan T, Narsimha G, Guru Rao CV (2012) Data perturbation and features selection in preserving privacy. In: Proceedings of 978-1-4673-1989-8/12. IEEE
6. Parmar A, Rao UP, Patel DR (2011) Blocking based approach for classification rule hiding to preserve the privacy in database. In: Proceedings of international symposium on computer science and society. IEEE

7. Kargupta H, Datta S, Wang Q, Sivakumar K (2003) On the privacy preserving properties of random data perturbation techniques. In: Proceedings of the third IEEE international conference on data mining. IEEE
8. Lindell Y, Pinkas B (2000) Privacy preserving data mining. Proc J Cryptol 5(3)
9. Aggarwal C, Yu PS (2004) A condensation approach to privacy preserving data mining. In: Proceedings of international conference on extending database technology (EDBT), pp 183–199
10. Lohiya S, Ragha L (2012) Privacy preserving in data mining using hybrid approach. In: Proceedings of 2012 fourth international conference on computational intelligence and communication networks. IEEE
11. Prasanthi KN, Chandra Sekhara Rao MVP (2018) Accuracy and utility balanced privacy preserving classification mining by improving K-anonymization. Int J Simul Syst Sci Technol 19(6):51.1–51.7
12. Prasanthi KN, Chandra Sekhara Rao MVP (2019) A novel method of privacy preserving classification mining balancing utility and accuracy. J Adv Res Dyn Control Syst 11(5):577–587
13. Prasanthi KN, Chandra Sekhara Rao MVP (2019) Combined approach for privacy preserving classification mining. J Eng Appl Sci 14(1):188–194
14. Vaidya J, Clifton C (2002) Privacy preserving association rule mining in vertically partitioned data. In: The eighth ACMSIGKDD international conference on knowledge discovery and data mining, Edmonton, Alberta, CA, IEEE, July 2002
15. Kisilevich S, Rokach L, Elovici Y, Shapira B (2010) Efficient multi-dimensional suppression for k-anonymity. Proc IEEE Trans Knowl Data Eng 22(3):334–347
16. Deivanai P, Jesu Vedha Nayahi J, Kavitha V (2011) A hybrid data anonymization integrated with suppression for preserving privacy in mining multi party data. In: Proceedings of international conference on recent trends in information technology. IEEE
17. Mathew G, Obradovic Z (2011) A privacy-preserving framework for distributed clinical decision support. In: Proceedings of 978-1-61284-852-5/11/$26.00. IEEE ©2011
18. Mumtaz S, Rauf A, Khusro S (2011) A distortion based technique for preserving privacy in OLAP data cube. In: Proceedings of 978-1-61284-941-6/11/$26.00. IEEE
19. Huang HC, Fang WC (2011) Integrity preservation and privacy protection for medical images with histogram-based reversible data hiding. In: Proceedings of 978-1-4577-0422-2/11. IEEE
20. Liu J, Luo J, Huang JZ (2011) Rating: privacy preservation for multiple attributes with different sensitivity requirements. In: Proceedings of 11th IEEE international conference on data mining workshops. IEEE
21. Alotaibi K, Rayward-Smith VJ, Wang W, de Laiglesia B (2012) Non-linear dimensionality reduction for privacy-preserving data classification. In: Proceedings of ASE/IEEE international conference on social computing and international conference on privacy, security, risk and trust. IEEE
22. Komishani EG, Abadi M (2012) A generalization-based approach for personalized privacy preservation in trajectory data publishing. In: Proceedings of 6'th international symposium on telecommunications (IST'2012). IEEE
23. Karthikeswarant D, Sudha VM, Suresh VM, Sultan AJ (2012) A pattern based framework for privacy preservation through association rule mining. In: Proceedings of international conference on advances in engineering, science and management (ICAESM-2012). IEEE
24. Kumbhar MN, Kharat R (2012) Privacy preserving mining of association rules on horizontally and vertically partitioned data: a review paper. In: Proceedings of 978-1-4673-5116-4/12/$31.00_c. IEEE
25. Beck M, Marhöfer M (2012) Privacy-preserving data mining demonstrator. In: Proceedings of 16th international conference on intelligence in next generation networks. IEEE
26. The free dictionary. Homepage on privacy [Online]. Available http://www.thefreedictionary.com/privacy

Mathematical Model for Handling Jamming Attacks in Wireless Sensor Networks

E. Jayabalan and R. Pugazendi

Abstract The nodes In a wireless sensor network communicate and send information. This can be used for a variety of applications like agriculture, health care. In setting up a sensor network the major challenges include energy and integrity of data. The nodes are prone to active and passive attacks that are hard to handle. This paper proposes a solution for handling Jammer attacks with a probability density function for the same. Nodes are largely varying as master or gateway nodes due to their functional characteristics. The nodes in the source select a path that has unjammed nodes with the help of the packets received from the sink node. Here a master-slave approach is followed to make sure a high network lifetime and less transmission power energy consumption in WSNs in the presence of jammers. With the proposed model, each node identifies the optimal transmission power level E_n^*. The proposed model ensures more than 60% reduction in energy consumption, and the lifetime of the network increases to 70%.

Keywords Energy consumption · Transmission power · Probability function

1 Introduction

Jammer is one of the attackers of WSN. The frequency of the radio signal of the jammer matches the sensor node's, so the jammer affects the node's communications and reduces the lifetime of the network by making multiple interrupt transmissions between nodes. The lifetime of the network can be improved by either avoiding the jammed node in the neighbor set of sensor nodes or stopping the jammer's transmissions [1]. Either part is discussed mathematically in our study [2].

E. Jayabalan (✉) · R. Pugazendi
Department of Computer Science, Government Arts College (A), Salem, Tamil Nadu 636007, India

S. Shakya et al. (eds.), *Proceedings of Second International Conference on Sustainable Expert Systems*, Lecture Notes in Networks and Systems 351,
https://doi.org/10.1007/978-981-16-7657-4_68

A mathematical control model is developed on the basis of transmission power level of the sensor nodes that are non-jammed. The selection of an optimal transmission power level of sensor nodes increases the lifetime of the network. The sensor nodes will decide the optimal path, ignoring the jammed nodes while knowing the list of affected nodes [3].

Consider a model which has N number of sensor nodes form a WSN. And let the id's of each sensor notes are x_i, $i = 1, 2 \ldots n$ and can be deployed in an area of $R \times R$. The maximum range of any node can be defined as R_n, beyond which it cannot connect to any node, and consider the sink node lie within a maximum range of R_n and also it is not affected by jamming. Here it is assumed that all the nodes have their neighbor nodes N_n and is well aware of the destination node, and J is the collection of jammers in the network [4]. Each node informs the sink node their neighbors' list periodically or in response to the sink's request, and each node can vary the transmission power level to make sure that it reduces the energy consumption of the nodes, and increase the network lifetime. The sensor nodes are continuously broadcasting and receiving the data, they tend to lose their energy, and their energy level is updated to the sink node periodically. If the energy of the sink becomes lower than a node, then the node is treated as a jammed one [5]. The jammed nodes lose their energy by receiving unrelated data. So there is a control measure is needed to optimize the transmission power level. The range of the power level of each sensor node is considered as (E^-, E^+), where E^- is the minimum power level of transmission and E^+ is the maximum power level of transmission, so the optimum power level of transmission E_n^* of each sensor node n needs to satisfy the condition $E^- \le E_n^* \le E^+$.

Lemma 1 *Let E_n be the power level of transmission of a sensor noden, then the function $T: R_n \rightarrow E_n$ is a surjective* [6].

Proof For each value of E_n, transmission power level, there is a sensor node with the communication range R_n. Thus by definition of surjective $f : r_n \rightarrow p_n$is a surjective function.

Lemma 2 *Let E_n be the transmission power level of a sensor node n, then the function $T: R_n \rightarrow E_n$ is a injective* [7].

Proof It is clear that for each rang ($\le R_n$) there is a distinct E_n value, that is, distinct notes within the range having distinct transmission power levels. Hence by the definition of an injective function $T: R_n \rightarrow E_n$ is an injective function.

Theorem 1 *The function $T: R_n \rightarrow E_n$ is a bijective function.*

Proof From Lemmas 1 and 2, $T: R_n \rightarrow E_n$ is both injective and surjective, and hence the function T is a bijective function.

Therefore, for k number of transmission power levels there is k number communication range R,$(R \le R_n)$.

So, we can represent the WSN as a graph G with its vertices sets V is the collection of the sensor nodes and the edge set $E = \{e_{ij} \,/\, i, j \in N\}$, where.

$$e_{ij} = \begin{cases} 1, \text{ if } d(i, j) \le R_n \\ 0 \text{ otherwise} \end{cases},$$

$d(i, j)$ is the distance between nodes $i, j \in N$.

2 Series Network

A network can be described as a set of interconnected sensor nodes, where each node represents a service facility of communication. Let C_i be the number of packets at each node $i \in N$. In general, the packet can arrive from outside the network region to any node and may leave the network region from any node. Therefore, a packet may enter the network region at some node, can transverse from node to node in the network region, and finally can leave the network region at some other node. It is noted that "not all packets necessarily enter and leave at the same node or take the same path once after entering the network region". Also, packets can return to nodes already visited, skip some nodes (jammed) entirely, or even choose to remain in the network region forever.

Theorem 2 *The inter-departure times of packets in nodes are exponential with parameter* λ.

Proof Let $L(t)$ be the number of pockets in the network region at a time "t" after the last processed packet. In the case of steady state, we have.

$$\text{Prop}\{L(t) = n\} = p_n \tag{1}$$

Let ω be the time between successive processing.
Let

$$D_n(t) = \text{Prop}\{L(t) = n \text{ and } \omega > 1\} \tag{2}$$

Therefore, $D_n(t)$ is the probability that there are n packets in the network region at a time t after the last process and that $t < \omega$ (another process has not yet occurred) [8]. The cumulative distribution of the random variable ω, denoted by $F(t)$ is given by.

$$\begin{aligned} F(t) &= P(\omega \le t) = 1 - P(\omega > t) \\ &= 1 - \sum_{n=0}^{\infty} D_n(t) \quad \left[\because \sum_{n=0}^{\infty} D_n(t) = P(\omega > t)\right] \end{aligned} \tag{3}$$

For $C > n$, (C = number of sensor nodes)

$$D_n(t+\Delta t) = (1-\lambda\Delta t)(1-C\mu\Delta t)D_n(t) + \lambda\Delta t(1-C\mu\Delta t)D_{n-1}(t) + 0(\Delta t) \quad (4)$$

For $1 \le n \le C$,

$$D_0(t+\Delta t) = (1-\lambda\Delta t)(1-n\mu\Delta t)D_n(t) + \lambda\Delta t(1-n\mu\Delta t)D_{n-1}(t) + 0(\Delta t) \quad (5)$$

and

$$D_0(t+\Delta t) = (1-\lambda\Delta t)D_0(t) + 0(\Delta t) \quad (6)$$

Equations (4), (5), (6) can be written as

$$\frac{D_n(t+\Delta t) - D_n(t)}{\Delta t} = -(\lambda + C\mu)D_n(t) + \lambda D_{n-1}(t) + \frac{0(\Delta t)}{\Delta t} \quad (C \le n)$$

$$\frac{D_n(t+\Delta t) - D_n(t)}{\Delta t} = -(\lambda + n\mu)D_n(t) + \lambda D_{n-1}(t) + \frac{0(\Delta t)}{\Delta t} \quad (1 \le n \le C)$$

Also

$$\frac{D_0(t+\Delta t) - D_0(t)}{\Delta t} = -\lambda D_0(t) + \frac{0(\Delta t)}{\Delta t}.$$

As $\Delta t \to 0$, we get

$$D_n'(t) = -(\lambda + C\mu)D_n(t) + \lambda D_{n-1}(t) \quad (C \le n) \quad (7)$$

$$D_n'(t) = -(\lambda + n\mu)D_n(t) + \lambda D_{n-1}(t) \quad (1 \le n \le C) \quad (8)$$

$$D_0'(t) = -\lambda D_0(t) \quad (9)$$

From the boundary condition

$$D_n(0) = \text{Prop}\{L(0) = n \text{ and } \omega > 0\} = \text{Prop}\{L(0) = n\} = p_n \quad \text{(by (1))}$$

By using the same method, we can derive the solution of the Eqs. (7), (8), (9) as follows,

$$D_n(t) = p_n e^{-\lambda t}$$

Now,

$$\begin{aligned}
F(t) &= 1 - \sum_{n=0}^{\infty} D_n(t) \\
&= 1 - \sum_{n=0}^{\infty} p_n e^{-\lambda t} \\
&= 1 - e^{-\lambda t} \sum_{n=0}^{\infty} p_n \\
F(t) &= 1 - e^{-\lambda t} \qquad [\Sigma p_i = 1]
\end{aligned}$$

The probability density function is given by

$$\begin{aligned}
f(t) &= F'(t) \\
&= \frac{\mathrm{d}}{\mathrm{d}t}\left(1 - e^{-\lambda t}\right) = \lambda e^{-\lambda t}.
\end{aligned}$$

3 Mathematical Model

In wireless sensor networks, the nodes are transmission power energy constrained. The source sensor nodes select paths to the sink, in a distributed manner. Thus, if any intermediate nodes are in a jamming affected area, and the source nodes do not have any information about the jamming affected nodes, the nodes might deplete energy by sending multiple duplicate packets. In such a scenario, the nodes can choose a path consisting of unjammed nodes, while having information of the jammed nodes. In this model, the sink node helps the normal nodes, while identifying the presence of jammer. We propose a game model for increasing network lifetime in the presence of jammers, through the sink acts as the master, and the deployed sensor nodes act as the slaves, attempts to identify the optimum value of power energy and minimize the jammers activation [9].

Now, consider a wireless sensor network with multiple jammers. The sink node acts as the master, and the sensor nodes act as the slaves (followers). Also consider that the jammers are nonslaves' nodes, which disrupt the network performance while deciding strategies locally. The sink indicates to the slaves about the jammed nodes

[10]. Then the slaves choose a path to the sink without jammed node and find an optimal transmission power through the 1-hop neighbor list [11].

Consider the game as follows:

$$\mathbb{G} = \{N, G, E_n, R_n, N_n, \delta_n(.,.), S(\cdot), J(\cdot)\} \quad (10)$$

where

1. $\mathbb{N}$ is the set of deployed sensor nodes in WSN.
2. $G = (V, E)$ is a graph with $V = N$ and $E = e_{ij}$.
3. E_n is the transmission power level of sensor node n.
4. R_n is the communication range of the node n corresponding to the E_n.
5. N_n is the set of neighbor nodes of sensor node n.
6. $\delta_n(.,.)$ is a joint distribution function of E_n andN_n, for each $n \epsilon N$.
7. $S(.)$ is a revenue function of N_n and find by the sink node.
8. $J(.)$ is a function of N_j, which is defined by the jammers, and having information about the loss of energy of the jammed nodes.

Each sensor node $n \in N$, needs to decide the optimum transmission power level E_n^* in order to the function $\delta_n(E_n, N_n)$ maximize the payoff of the game. While considering that it must have optimum number of neighbor nodes. If there is any neighbor node is jammed by a jammer $j \in J$, then whose transmission power level reduces to minimum $E_i = E_-$.

To find the optimum power level E_n^* of the node $n \in N$, the function $\delta_n(E_n, N_n)$ should satisfies the following conditions,

$$\frac{\partial}{\partial E_n}(\delta_n(E_n, N_n)) = 0; \frac{\partial}{\partial E_n}(\delta_n(E_n, N_n)) \leq 0.$$

Each node i has more possibility to choosing next hop of the neighbor list with an increase in $|N_i|$. Therefore we consider that the $|N_i|$ varies proportionally with the transmission power level E_i,

$$\text{(i.e.)} \quad N_i = \alpha.E_i, \alpha > 0, i \in N \quad (11)$$

and if a node i have two transmission power level E_i and $E_i^{'}$ with neighbor set N_i and $N_i^{'}$ respectively, and $E_i < E_i^{'}$, and we claim that $\mathbb{N}_i \subseteq \mathbb{N}_i^{'}$, then

$$E_i^* = \begin{cases} E_i \text{ if } \mathbb{N}_i \equiv \mathbb{N}_i^{'} \\ E_i^{'} \text{ otherwise} \end{cases}$$

Let E_n and $E_n^{'}$ are two successive transmission power levels of a chosen by sensor node n, and let $|N_n^{'}|$, $|N_n|$ and $(|N_n|)^+$ are the cardinality of neighbor nodes list with transmission power levels $E_n^{'}$, E_n and E^+ respectively [12].

Thus, each node n tries to maximize its payoff, at the same time satisfying the following constraints: $E_{-} \leqslant E_n^{'}, E_n \leqslant E^{+}$ and .

Now, define $\delta_n(E_n, N_n)$ as follows

$$\delta_n(E_n, N_n) = \frac{E_n}{\mathbb{S}(\mathbb{N}_n).E^{+}}\left[\frac{|\mathbb{N}_n'| - |\mathbb{N}_n|}{(|\mathbb{N}_n|)^{+}}\right] + \tan^{-1}\left(e^{\left(1-\frac{E_n}{E_n'}\right)}\right) \quad (12)$$

The first derivative of $\delta_n(E_n, N_n)$ with respect to E_n:

$\frac{\partial}{\partial E_n}(\delta_n(E_n, N_n)) = \frac{\Delta E_n - E_n}{(E^{+})^2} - \frac{1}{E_n'}\left[\frac{e^{\left(\frac{\nabla E_n}{E_n'}\right)}}{1+e^{\left(\frac{2\nabla E_n}{E_n'}\right)}}\right]$; where $\nabla E_n = E_n' - E_n$

$$= \frac{\Delta E_n - E_n}{(E^{+})^2} - \frac{1}{E_n'}\left[\frac{1}{e^{\left(\frac{\nabla E_n}{E_n'}\right)} + e^{-\left(\frac{\nabla E_n}{E_n'}\right)}}\right]$$

$= \frac{\Delta E_n - E_n}{(E^{+})^2} - \frac{1}{E_n'}\left[\frac{1}{2+\left(\frac{\nabla E_n}{E_n'}\right)^2}\right]$; Since $e^x + e^{-x} \cong 2 + x^2$

$$\frac{\partial}{\partial E_n}(\delta_n(E_n, N_n)) = \frac{E_n' - 2E_n}{(E^{+})^2} - \frac{1}{3E_n' - 2E_n + \left(\frac{E_n^2}{E_n'}\right)} \quad (13)$$

The second derivative of $\delta_n(E_n, N_n)$ with respect to E_n:

$$\frac{\partial^2}{\partial E_n^2}(\delta_n(E_n, N_n)) = -2\left[\frac{1}{(E^{+})^2} + \frac{\left(1+\frac{E_n}{E_n'}\right)}{\left(3E_n' - 2E_n + \left(\frac{E_n^2}{E_n'}\right)\right)^2}\right]$$

$$= -2\left[\frac{1}{(E^{+})^2} + \frac{\left(1+\frac{E_n}{E_n'}\right)}{E_n'\left(2+\left(\frac{\nabla E_n}{E_n'}\right)^2\right)}\right]$$

Clearly

$$\frac{\partial^2}{\partial E_n^2}(\delta_n(E_n, N_n)) < 0 \quad (14)$$

From the Eqs. (13) and (14) we can say that the function $\delta_n(E_n, N_n)$ has a maximum value. And to find the maximum value we have to solve the equation $\frac{\partial}{\partial E_n}(\delta_n(E_n, N_n)) = 0$. The maximum value is the optimum transmission power, E_n^{*}, chosen by each node n [13].

From Eq. (13) we get

$$\frac{E_n' - 2E_n}{(E^+)^2} - \frac{1}{3E_n' - 2E_n + \left(\frac{E_n^2}{E_n'}\right)} = 0$$

$$(E_n' - 2E_n)\left(3E_n' - 2E_n + \left(\frac{E_n^2}{E_n'}\right)\right) - (E^+)^2 = 0$$

$$\left(\frac{4}{E_n'}\right)E_n^3 - 6E_n^2 + 8E_n'E_n = 3E_n^2 - (E^+)^2$$

$$\left(\frac{4}{E_n'}\right)E_n^3 + (-6)E_n^2 + (8E_n')E_n + \left((E^+)^2 - 3E_n^2\right) = 0$$

$$(i.e.)\quad C_1E_n^3 + C_2E_n^2 + C_3E_n + C_4 = 0 \tag{15}$$

where $C_1 = \left(\frac{4}{E_n'}\right)$, $C_2 = -6$, $C_3 = 8E_n'$, and $C_4 = \left((E^+)^2 - 3E_n^2\right)$.

To solve the Eq. (15), put $E_n = y - \frac{C_2}{3C_1}$

Now, the Eq. (15) becomes

$$C_1y^3 + \left(C_3 - \frac{C_2^2}{3C_1}\right)y + \left(C_4 + \frac{2C_2^3}{27C_1^3} - \frac{C_2C_3}{3C_1}\right) = 0$$

$$y^3 + \left(\frac{3C_1C_3 - C_2^2}{3C_1^2}\right)y + \frac{2C_2^3 - 9C_1C_2C_3 + 27C_1^2C_4}{27C_1^3} = 0 \tag{16}$$

The above Eq. (16) is rewritten as follows:

$$y^3 + 3My - 2N = 0 \tag{17}$$

where $y = A + B$, $AB = -M$. Therefore, we get the following:

$$(A + B)^3 + 3Q(A + B) - 2N = 0$$
$$\Rightarrow \left[A^3 + B^3 + 3AB(A + B)\right] - 3AB(A + B) - 2N = 0,$$

as $M = -AB$.

$$\Rightarrow A^3 + B^3 = 2N \tag{18}$$

$\Rightarrow A^3 - \frac{M^3}{A^3} - 2N = 0$, as $B = -\frac{M}{A}$.

$$\Rightarrow A^6 - 2NA^3 - M^3 = 0 \tag{19}$$

From (18), we have:

$$A^3 = \frac{2N \pm \sqrt{4N^2 + 4M^3}}{2}$$
$$\Rightarrow A^3 = N \pm \sqrt{N^2 + M^3}$$
$$\Rightarrow A = \left(N + \left(N^2 + M^3\right)^{\frac{1}{2}}\right)^{\frac{1}{3}} \tag{20}$$

And $A^3 + B^3 = 2N$.

$$\Rightarrow B^3 = 2N - A^3$$
$$\Rightarrow B^3 = 2N - \left[N + \sqrt{N^2 + M^3}\right]$$
$$\Rightarrow B^3 = N - \left[\sqrt{N^2 + M^3}\right]$$
$$\Rightarrow B = \left(N - \left(N^2 + M^3\right)^{\frac{1}{2}}\right)^{\frac{1}{3}} \tag{21}$$

By the Eqs. (20) and (21), we have:

$$y = \left(N + \left(N^2 + M^3\right)^{\frac{1}{2}}\right)^{\frac{1}{3}} + \left(N - \left(N^2 + M^3\right)^{\frac{1}{2}}\right)^{\frac{1}{3}} \tag{22}$$

$$E_n = \left(N + \left(N^2 + M^3\right)^{\frac{1}{2}}\right)^{\frac{1}{3}} + \left(N - \left(N^2 + M^3\right)^{\frac{1}{2}}\right)^{\frac{1}{3}} - \frac{C_2}{3C_1} \tag{23}$$

Hence, we get that the optimum transmission power level, p_n^* is as follows:

$$E_n^* = \left(N + \left(N^2 + M^3\right)^{\frac{1}{2}}\right)^{\frac{1}{3}} + \left(N - \left(N^2 + M^3\right)^{\frac{1}{2}}\right)^{\frac{1}{3}} - \frac{C_2}{3C_1} \tag{24}$$

where $E_{-} \leqslant E_n^* \leqslant E^{+}$, $|N_n| = f\left(E_n^*\right)$ and f is a uniform distribution function.

Now, the utility function $S(N_n)$ is defined by.

$$S(N_n) = \frac{\left|\bigcup_{n=1}^{N} N_n\right|}{|N|} \text{ such that } \left|\bigcup_{n=1}^{\mathrm{N}} \mathrm{N}_n\right| \leq |\mathrm{N}| \tag{25}$$

The 1st and 2nd derivatives of $S(N_n)$ gives

$$\frac{\partial}{\partial |N_n|}(S(N_n)) \geq 0 \text{ and}$$
$$\frac{\partial^2}{\partial |N_n|^2}(S(N_n)) < 0$$

Thus the sinks utility function $S(N_n)$ also having maximum point [14].

Now the utility function of each jammer $J(N_J)$ is defined by

$$J(N_J) = |N_J| \quad (26)$$

Clearly $J(N_J)$ is also a maximum function. Each jammer always maximizes their utility function $J(N_J)$ through jamming more number of nodes. At the same time, the sink tries to minimize the jammer utility function $J(N_J)$ by informing the presence of the jammed node in N_i of sensor nodes.

4 Assumptions

1. Let n, be the initial node that begins communication, and is elected as the sink (Master) node, which does not communicate with Jammer.
2. Nodes that receive the sinks id become its members, and if the nodes wish, they can either leave the network by checking its distance with the sink.
3. Each node $n \in N$ is capable of varying its transmission power level E_n, and communication range R_n.
4. The transmission power level E_n of all sensor nodes are periodically checked by the sink and the node with a low transmission power level is identified as a jammed node.
5. The sink node transmits the presence of the jammer nodes in the neighbor list N_n to all the nodes within the range.
6. Each sensor node $n \in N$ chooses a path by omitting the jammed node to the sink node. And calculate the packet delivery ratio $\mathbb{P}_{\mathbf{dr}} = \frac{M_d}{M_s}$, where M_d is the number of data delivered to the destination, and M_s is the number of data sent by the sender.
7. If $\mathbb{P}_{\mathbf{dr}}$ value increased then the network lifetime also increases.

5 Algorithm

Step 1	Take a G = WSN and construct the graph of the network
Step 2	Let N be set of deployed nodes
Step 3	Let J be set of Jammers in the network
Step 4	Let N_n be set of neighbor nodes of $n \in N$
Step 5	Let N_n^* be the optimum set of neighbor nodes $n \in N$
Step 6	Let E be set of transmission power levels
Step 7	Let E^- and E^+ be the minimum and maximum transmission power level
Step 8	Let E_n^* be the optimum transmission power level of $n \in N$
Step 9	Let N_j be set of jammed nodes, $j \in J$
Step 10	Evaluate $\bigcup N_n$
Step 11	Evaluate $S(N_n)$

Step 12 Evaluate $N_j^+ = N \bigcap \left(\bigcup N_n\right)^c$
Step 13 Find N_m for the node whose transmission power level is E^+
Step 14 $E_n \leftarrow E^+$
Step 15 $E_n' \leftarrow E_n$
Step 16 Find N_n'
Step 17 Select $E_n \in E$ such that $E_n \neq E_n'$
Step 18 Find N_n
Step 19 If $\delta_n(E_n, N_n) < \delta_n\left(E_n', N_n'\right)$ goto step 15
Step 20 $E_n^* \leftarrow E_n$
Step 21 $N_n^* \leftarrow N_n$
Step 22 Print (i) The optimum transmission power level is E_n^*
(ii) The optimum neighbor node set is N_n^*.
(iii) The set of Jamming affected nodes N_j^+.
Step 23 Stop

6 Applications

In recent years, wireless sensor networks play a vital role in various fields, like medical, defense, environment, transportation, structural, etc. [15].

In the medical application, the WSNs are doing the following services:

- Investigation and maintain the patient history
- Diagnosing
- Drug administration
- Integrated patient monitoring and management,
- Tele-monitoring of human physiological information,
- Tracking and monitoring medical practitioners.

Hospital management, etc.

Wireless body sensor networks (WBSN) helps in monitoring in-patients, patients' healthcare, body medical checkups, etc. Wireless sensor networks ensure the privacy and security of medical data. And also maintain data availability, data efficiency, and data suitability in various medical-sub fields.

In defense, WSNs are involved in the surveillance, probe, and communication of secrete data.

The Environmental Sensor Networks (ESN) is also a WSN in environmental study. The sensor nodes of ESNs sense and control the environmental changes like temperature, air pressure, etc., and communicate to the base station to maintain the safety measures. ESNs sense seas, oceans, celestial spheres, volcanoes, etc. The air and water contamination, forest fires detection, and landslide detection, etc., are discovered by the biosensors of the latest WSNs.

The GPS is one of the WSN, which updates the traffic statistics and alerts the users about the possible congestion and traffic difficulties instantaneously in the transportation application.

The WSNs influence structural works, like the construction projects of flyovers, bridges, tunnels, etc., and this would reduce the expenditure and completion time incurred from physical activates.

7 Justification of Proposed Model

The sensor nodes of the WSNs have several challenges. In our proposed model, we discussed the challenge of the presents of jammers in the sensor region. To overcome the challenge, the source nodes choose a path consisting of unjammed nodes with the help of the packets received from the sink node. Here we follow a master-slave approach to ensure less transmission power energy consumption and high network lifetime in the presence of jammers in WSNs [16]. By the proposed model, each node fids the optimal transmission power level E_n^*. The proposed model ensures more than 60% reduction in energy consumption, and the network lifetime increases to 70% [17].

8 Conclusion

In this work, we proposed a mathematical model to find the optimum transmission power level and optimum neighbor set of the sensor nodes by choosing an optimum transmission path consisting of no affected nodes from a source node to the destination node in a WSN [18]. Each node sends the packets through the optimal transmission path with less transmission power level. The proposed mathematical model improves the lifetime of the network. Also, we can identify the presence of the jammer in the given network.

References

1. He J, Li H, Tan J (2007) Real time daily activity classification with wireless sensor networks using hidden markov model. In: Proceedings of 29th annual international conference on IEEE engineering medical biology society, pp 3192–3195
2. Murugesan S, Senthil Kumar T, Priyanka US, Abinaya K (2013) Towards an approach for ımproved security in wireless networks. Int J Comput Appl 1:9–13
3. Gao T, Pesto C, Selavo L, Chen Y, Ko J, Lim J (2008) Wireless medical sensor netowrks in emergency response:ımplementations and pilot results. In: IEEE international conference on technology homeland security, pp 187–198
4. Baalamurugan KM, Vijay Bhanu S (2019) An efficient clustering scheme for cloud computing problems using metaheuristic algorithms. Cluster Comput 22(5):12917–12927
5. Kiani F, Rad A, Sis MK, Kut A, Alpkocak A (2013) EEAR: an energy effective-accuracy routing algorithm for wireless sensor networks. J Life Sci J 10(2):39–45

6. Lorincz K, Malan D, Fulford-Jones J, Nawoj A, Clavel A, Shnayder V, Mainland G, Mouton S, Welsh M (2004) Sensor networks for emergency response: challenge and opportunities. J IEEE Pervasive Comput 3(4):16–23
7. Li XY, Wan PJ (2001) Constructing minimum energy mobile wireless networks. J ACM J Mobile Comput Commun Rev 5(4):55–67
8. Baalamurugan KM, Vinotha D, Premkumar S (2020) An efficient ant colony-based self-adaptive routing strategy for IoT network model. In: Bindhu V, Tavares JMRS, Boulogeorgos AAA, Vuppalapati C (eds) International conference on communication, computing and electronics systems, lecture notes in electrical engineering, vol 733. Springer, pp 753–764
9. Latre B, Braem B, Moeman L (2011) A survey on wireless body area networks 17(1):1–18
10. Daniel A, Baalamurugan KM, Vijay R, Arjun KP (2021) Energy aware clustering with multi-hop routing algorithm for wireless sensor networks. Intell Autom Soft Comput 29(1):233–246
11. Singh M, Prasanna VK (2004) A hierarchical model for distributed collaborative computation in wireless sensor networks. J Int Found Comput Sci 15(3):485–506
12. Marks IIR, Das A, El-Sharkawi M, Arabshani P, Gray A (2002) Minimum power broadcast trees for wireless networks: optimizing using the viability lemma. In: Proceedings of IEEE international symposium on circuits and systems, pp 245–248
13. Virone G, Wood A, Selavo L, Cao Q, Fang L, Doan T (2006) An advanced wireless sensor network for health monitoring. In: Proceedings of transdisciplinary conference on distributed diagnosis home healthcare, pp 95–100
14. Wieselthier J, Nguyen G, Ephremides A (2000) On the construction of energy efficient broadcast and multicast trees in wireless networks. In: Proceedings of IEEE Infocom, pp 585–594
15. Benaloh J, Chase E, Horvitz E, Lauter K (2009) Patient controlled encryption: patent privacy in electronic medical records. In: Proceedings of ACM computation security workshop, pp 103–114
16. Chu T, Nikolaidis I (2002) Energy efficient broadcast in mobile ad hoc networks. In: Proceedings of ad-hoc networks and wireless, pp 177–190
17. Younis M, Youssef M, Arisha K (2002) Energy-aware routing in cluster based sensor networks. In: Tenth ACM international symposium on modeling analysis and simulation of computer and telecommunication systems
18. Egecioglu O, Gonzales T (2001) Minimum energy broadcast in simple graphs with limited node power. In: Proceedings of IASTED ınternational conference on parallel and distributed computing and systems, pp 334–338

E. Jayabalan is a Research Scholar and working as Assistant professor in the Department of Computer Science, Government Arts College, Salem, 636007, Tamil Nadu, India. He has more than 24 years of teaching experience. His research interests include Wireless Networking with Deep Learning concepts.

Dr. R. Pugazendi is a Research Supervisor and working as Assistant professor in the Department of Computer Science, Government Arts College, Salem, 636007, Tamil Nadu, India. He has more than 25 years of teaching experience. His research interests include Computer Networking and Internet of Things. He is the author of several scholarly research papers. He has published papers in several international journals and conferences.

Heart Disease Prediction Using Ensemble Model

Bikal Adhikari and Subarna Shakya

Abstract Heart is a vital organ of the human body. The mortality rate due to cardiovascular or heart disease is in the rising trend. From the review of existing literature, timely diagnosis of the potential heart disease is the key to preventing the risk of early death. But due to the lack of proper health infrastructure in different parts of the country (especially in the rural area), timely diagnosis or the early diagnosis of heart disease has been a problem in Nepal. In this thesis, ensemble learning is used to predict the potential heart risk in patients. First different supervised learning models such as Logistic Regression, SVM, Decision Tree, KNN, and Gaussian Naive Bayes were built out of UCI heart disease dataset. These models gave accuracy of 82.46%, 87.34%, 97.67%, 89.94% and 78.57% respectively. An ensemble model was built by combining aforementioned five supervised ML models. The voting and averaging based ensemble model gave an accuracy of 96.10% and 96.43% respectively. Upon comparison, the ensemble model gave overall better accuracy. A module capable of predicting heart disease for a patient was implemented.

Keywords Heart disease · Machine learning · Ensemble learning · Prediction

1 Introduction

Heart is one of the vital organs of the human body. The main function of the heart is to pump the oxygenated blood to all parts of the body. Healthy heart is the mandatory requirement for living a healthy lifestyle.

B. Adhikari (✉)
Department of Information System Engineering, HIST College, Kathmandu, Nepal
e-mail: bikaladhikari@kathford.edu.np

Department of Computer and Electronics and Communication Engineering, Kathford Int'l College of Engineering and Management, Lalitpur, Nepal

S. Shakya (✉)
Department of Electronics and Computer Engineering, Pulchowk Campus, Lalitpur, Nepal
e-mail: drss@ioe.edu.np

S. Shakya et al. (eds.), *Proceedings of Second International Conference on Sustainable Expert Systems*, Lecture Notes in Networks and Systems 351,
https://doi.org/10.1007/978-981-16-7657-4_69

Heart disease or cardiovascular disease is a major cause of death worldwide. As per WHO, in 2016, 17.9 million people died due to cardiovascular diseases. It represents in total 31% of all global deaths. Among several types of cardiovascular diseases, heart attack is the major cause of death. In case of the US, someone dies of heart attack every 40 s. In case of Nepal, the mortality rate due to cardiovascular disease has increased from 124.1 per thousand (1990 AD) to 164.7 per thousand (2017) over the span of 27 years [1–4].

This shows that the mortality rate due to cardiovascular or heart disease is in the rising trend. From the review of existing literature, timely diagnosis of the potential heart disease is the key to preventing the risk of premature death. But due to the lack of proper health infrastructure in different parts of the country (especially in the rural area), timely diagnosis or the early diagnosis of heart disease has been a problem in Nepal.

In this context, we can make a hypothesis that a Machine Learning model trained with the dataset of the heart patients can be used to make the prediction of the potential heart risk in the patients from the area where there is lack of proper health infrastructure.

In this research, five different supervised machine learning models (Logistic Regression, Support Vector, Decision Tree, K Nearest Neighbor, and Gaussian Naive Bayes) were built using UCI heart disease dataset. The models are then used to build the ensemble model using voting and averaging method. The ensemble model thus built can be used to predict the chance of heart disease for a patient.

2 Review of Relevant Works

In a research by [5], the authors have used the ensemble model for the prediction of Parkinson's disease. Different performance measures of machine learning such as accuracy, log loss, $F1$ score, etc. have been used to evaluate the models. The authors found that the ensemble model based on stacking and voting gave better performance.

In a research by [6], the authors used the concept of machine learning and deep learning for the prediction of heart disease. Their model was based on UCI heart disease dataset with 14 attributes. The authors reported that they achieved 94.2% accuracy using the approach of deep learning.

In a research by [7], the authors implemented Naïve Bayes, Decision Tree, KNN and Random Forest algorithm based on the dataset provided by UCI heart disease. The authors reported that the highest accuracy was achieved using KNN algorithm.

In a research by [8], the authors implemented several supervised machine algorithms such as Decision Tree, KNN, SVM, and Random Forest. They reported the accuracy of 79%, 87%, 83%, and 84% respectively. Similarly, the authors reported AUC values of 71.6%, 88.5%, 90.4 and 90.8% respectively.

In a research by [9], the authors found that different classification algorithms gave high accuracy for the prediction of heart disease. The authors also mentioned that for the prediction of heart disease, we can use limited number of attributes instead of considering all the attributes in the dataset.

3 Supervised Machine Learning Models

Machine Learning helps to create different models based on the dataset and helps to solve different problems such as classification problem, clustering problem, prediction problem, etc. Machine learning helps to analyze the data without writing codes explicitly. In general, machine learning models can be classified as supervised models and unsupervised models. Supervised models are the ones that use labeled data while the unsupervised models are the ones that use unlabelled data [10]. In this research, five different supervised machine learning models namely Logistic Regression, Support Vector Machine, Decision Tree, KNN, and Gaussian Naïve Bayes have been implemented using Python based on UCI heart disease dataset. These models are then combined to implement the ensemble model [11–13].

4 Ensemble Model

In an ensemble model, diverse predictors/classifiers are used for classification. Different types of ensemble models are voting, averaging, weighted averaging, stacking, blending, bagging, boosting, etc. In this research, the voting and averaging ensemble models have been used. Figure 1 shows the concept of ensemble model using voting approach.

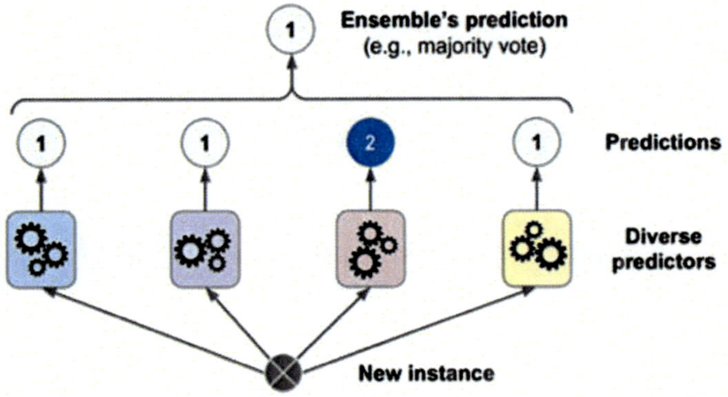

Fig. 1 Ensemble method of prediction using voting technique [14]

5 Methodology

Figure 2 shows the flow diagram of the methodology adopted. First task was problem formulation. It identified the research problems. To justify the problem formulation and to validate the research purpose, detailed study of available literature on the topic was carried out. In literature review, relevant materials were studied.

Following literature review, data sets were collected and an ensemble learning model was developed on the basis of training data set. The machine learning modeling process is shown in Fig. 3 [15, 16].

After the implementation, the model was tested with the test data set and accuracy of the model was found.

In this research, first existing classification models such as Logistic Regression, Support Vector Machine, KNN, Decision Tree, etc. are used. Then ensemble learning models are built upon these models to improvise the accuracy of the model. After the model was ready, it was able to predict the result for individual patients. This is illustrated in Fig. 4.

Fig. 2 Flow diagram of methodology adopted

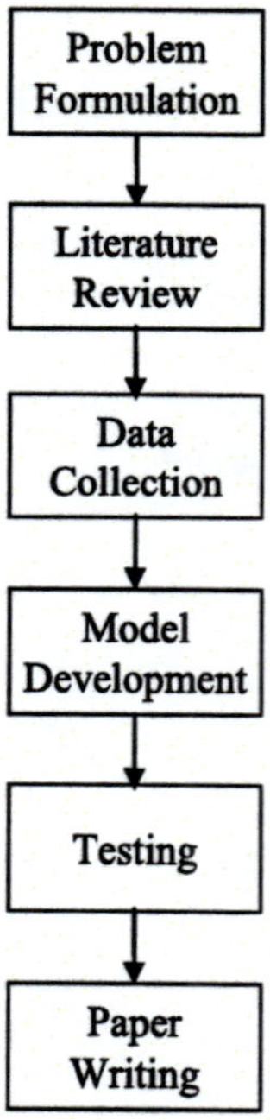

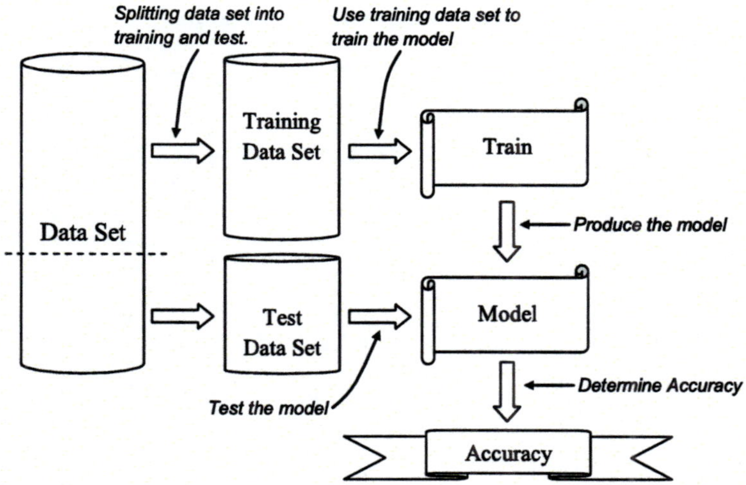

Fig. 3 Machine learning process

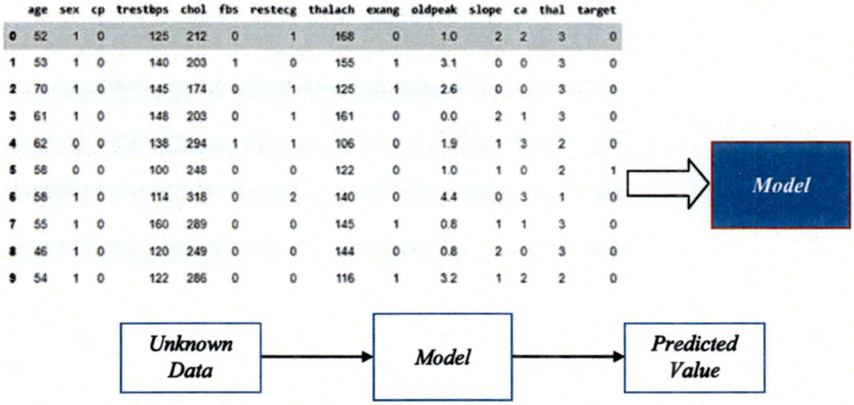

	age	sex	cp	trestbps	chol	fbs	restecg	thalach	exang	oldpeak	slope	ca	thal	target
0	52	1	0	125	212	0	1	168	0	1.0	2	2	3	0
1	53	1	0	140	203	1	0	155	1	3.1	0	0	3	0
2	70	1	0	145	174	0	1	125	1	2.6	0	0	3	0
3	61	1	0	148	203	0	1	161	0	0.0	2	1	3	0
4	62	0	0	138	294	1	1	106	0	1.9	1	3	2	0
5	58	0	0	100	248	0	0	122	0	1.0	1	0	2	1
6	58	1	0	114	318	0	2	140	0	4.4	0	3	1	0
7	55	1	0	160	289	0	0	145	1	0.8	1	1	3	0
8	46	1	0	120	249	0	0	144	0	0.8	2	0	3	0
9	54	1	0	122	286	0	0	116	1	3.2	1	2	2	0

Fig. 4 Use of ML model to predict for unknown data

6 Results, Analysis and Comparison

Following sections presents the results obtained from the application of different classification model to the UCI heart disease dataset.

6.1 Result from Logistic Regression

Following is the result obtained from the Logistic Regression.

	precision	recall	f1-score	support
0	0.85	0.77	0.80	145
1	0.81	0.88	0.84	163
accuracy			0.82	308
macro avg	0.83	0.82	0.82	308
weighted avg	0.83	0.82	0.82	308

Fig. 5 Results obtained from logistic regression

	precision	recall	f1-score	support
0	0.86	0.87	0.87	145
1	0.88	0.88	0.88	163
accuracy			0.87	308
macro avg	0.87	0.87	0.87	308
weighted avg	0.87	0.87	0.87	308

Fig. 6 Results obtained from SVM

Figure 5 shows the result obtained from the application of logistic regression classification model. For target 0 precision is 0.85, recall rate is 0.77, $f1$-score is 0.80 and support is 145. Similarly, for target 1, precision is 0.81, recall rate is 0.88, $f1$ score is 0.84, and support is 163. The accuracy of the model was determined to be 82.46%.

6.2 *Result from Support Vector Machine*

Following is the result obtained from the Support Vector Machine (SVM).

Figure 6 shows the result obtained from the application of SVM classification model. For target 0 precision is 0.86, recall rate is 0.87, $f1$-score is 0.87 and support is 145. Similarly, for target 1, precision is 0.87, recall rate is 0.88, $f1$ score is 0.88, and support is 163. The accuracy of the model was determined to be 87.34%.

6.3 *Result from Decision Tree*

Following is the result obtained from the Decision Tree.

Figure 7 shows the result obtained from the application of Decision Tree based classification model. For target 0 precision is 0.96, recall rate is 0.98, $f1$-score is 0.97 and support is 145. Similarly, for target 1, precision is 0.98, recall rate is 0.96,

	precision	recall	f1-score	support
0	0.96	0.98	0.97	145
1	0.98	0.96	0.97	163
accuracy			0.97	308
macro avg	0.97	0.97	0.97	308
weighted avg	0.97	0.97	0.97	308

Fig. 7 Results obtained from decision tree

	precision	recall	f1-score	support
0	0.85	0.96	0.90	145
1	0.96	0.85	0.90	163
accuracy			0.90	308
macro avg	0.90	0.90	0.90	308
weighted avg	0.91	0.90	0.90	308

Fig. 8 Results obtained from KNN classifier

$f1$ score is 0.97, and support is 163. The accuracy of the model was determined to be 97.07%.

6.4 *Result from KNN*

Following is the result obtained from the KNN classifier.

Figure 8 shows the result obtained from the application of KNN classifier. For target 0 precision is 0.85, recall rate is 0.96, $f1$-score is 0.90 and support is 145. Similarly, for target 1, precision is 0.96, recall rate is 0.85, $f1$ score is 0.90, and support is 163. The accuracy of the model was determined to be 89.94%.

6.5 *Result from Gaussian Naive Bayes*

Following is the result obtained from the Naive Bayes classifier:

Figure 9 shows the result obtained from the application of Gaussian Naive Bayes classification model. For target 0 precision is 0.78, recall rate is 0.76, $f1$-score is 0.77 and support is 145. Similarly, for target 1, precision is 0.79, recall rate is 0.81, $f1$ score is 0.80, and support is 163. The accuracy of the model was determined to be 78.57%.

	precision	recall	f1-score	support
0	0.78	0.76	0.77	145
1	0.79	0.81	0.80	163
accuracy			0.79	308
macro avg	0.79	0.78	0.78	308
weighted avg	0.79	0.79	0.79	308

Fig. 9 Results obtained from Naive Bayes classifier

6.6 Result from Ensemble Model

Following is the result obtained from the voting based ensemble model.

Figure 10 shows the result obtained from the application of voting classifier based ensemble model. For target 0 precision is 0.94, recall rate is 0.98, $f1$-score is 0.96 and support is 145. Similarly, for target 1, precision is 0.98, recall rate is 0.94, $f1$ score is 0.96, and support is 163. The accuracy of the model was determined to be 96.10%.

Similarly, the result obtained from the average based ensemble model is.

Figure 11 shows the result obtained from the application of average classifier based ensemble model. For target 0 precision is 0.97, recall rate is 0.96, $f1$-score is 0.96 and support is 145. Similarly, for target 1, precision is 0.96, recall rate is 0.97,

	precision	recall	f1-score	support
0	0.94	0.98	0.96	145
1	0.98	0.94	0.96	163
accuracy			0.96	308
macro avg	0.96	0.96	0.96	308
weighted avg	0.96	0.96	0.96	308

Fig. 10 Results obtained from ensemble model (voting based)

	precision	recall	f1-score	support
0	0.97	0.96	0.96	145
1	0.96	0.97	0.97	163
accuracy			0.96	308
macro avg	0.96	0.96	0.96	308
weighted avg	0.96	0.96	0.96	308

Fig. 11 Results obtained from ensemble model (average)

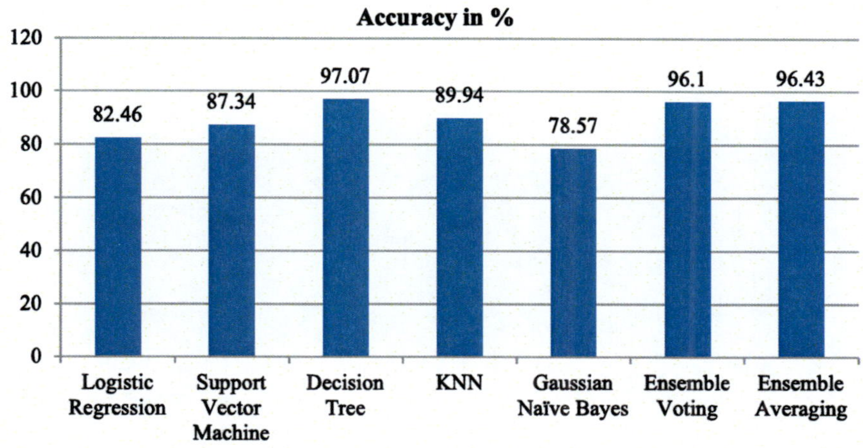

Fig. 12 Comparison between different models used for prediction

$f1$ score is 0.97, and support is 163. The accuracy of the model was determined to be 96.43%.

6.7 *Comparison of Different Models*

The comparison between different models used for prediction is given in Fig. 12.

Figure 12 shows the comparison between the different prediction models used for the heart disease prediction. The Decision Tree gave the highest accuracy for the given dataset. However, in totality, the ensemble model gave overall better accuracy. The ensemble models give better performance in prediction.

6.8 *Making Prediction for Single Patient*

In this research, a module capable of predicting chances of heart disease using a feature set for single patient has been implemented. Figures 13 and 14 shows the module making predictions for a patient who is fine or not likely to have heart disease.

Similarly, Figs. 15 and 16 shows the module making predictions for a patient who is likely to have heart disease.

```
Please Provide the input variable respectively :
52
1
0
125
212
0
1
168
0
1
2
2
3
```

Fig. 13 Taking input the feature set for single patient who is fine

```
The patient is fine!
```

Fig. 14 Output showing that patient is fine

```
Please Provide the input variable respectively :
108
0
0
60
148
1
1
100
0
1
1
0
5
```

Fig. 15 Taking input the feature set for single patient likely to have heart disease

```
The patient is likely to have Heart Disease!
```

Fig. 16 Output showing that patient is likely to have heart disease

7 Conclusions

The machine learning model based on existing classifiers such as logistic regression, support vector machine, decision tree, KNN, Gaussian Naive Bayes were built using UCI heart disease dataset. Ensemble model using voting classifier and average classifier were built by using the aforementioned models. The accuracy of the models was determined. Logistic regression yielded an accuracy of 82.46%, SVM yielded an accuracy of 87.34%, Decision Tree yielded an accuracy of 97.07%, KNN yielded an accuracy of 89.94%, Gaussian Naive Bayes yielded an accuracy of 78.57%. The ensemble learning model based on voting classifier and probability was developed. The voting and averaging based ensemble model gave an accuracy of 96.10% and 96.43% respectively. Upon the comparison of the models, the ensemble models gave overall better accuracy. A module capable of predicting heart disease using the feature set for single patient was also implemented.

8 Recommendations

Since the ensemble model can help in making the prediction of the potential heart disease, this modeling approach can also be used to predict the chances or likelihood of other diseases as well. These types of models are useful for the early detection of disease and can be used in the rural sector where there is lack of adequate number of qualified doctors and medical equipment.

References

1. WHO (2017) Country cooperation strategy at a glance. World Health Organization
2. Murphy SL (2018) Mortality in the United States, 2017. Centers for Disease Control and Prevention
3. Bhattarai S (2017) Cardiovascular disease trend in Nepal. IJC Heart and Vasculature
4. Anil OM (2019) Prevalence of cardiovascular risk factors in apparently healthy urban adult population of Kathmandu. J Nepal Health Res Counc 16(41):438–445
5. Sipai S, Mali D, Shakya S, Mali R (2021) Parkinson's disease data analysis and prediction using ensemble machine learning techniques. In: Mobile computing and sustainable informatics. Lecture notes on data engineering and communications technologies, vol 68
6. Shabaz M, Dhiman G, PandeS, Singh P, Bharti R, Khamparia A (2021) Prediction of heart disease using a combination of machine learning and deep learning. In: Computational intelligence and neuroscience, vol 2021
7. Bharti SK, Shah D, Patel S (2020) Heart disease prediction using machine learning techniques. SN Computer Science
8. Choubey DK (2020) Heart disease prediction using machine learning and data mining. Int J Recent Technol Eng 9:212–219
9. Almustafa KM (2020) Prediction of heart disease and classifier's sensitivity analysis. BMC Bioinform
10. Theobald O (2017) Machine learning for absolute beginners: a plain English introduction. Scatterplot Press
11. Borges DM (2021, July) Kdnuggets [Online]. https://www.kdnuggets.com/2019/01/ensemble-learning-5-main-approaches.html
12. Santarcangelo J (2020) Machine learning with python: a practical introduction. edX course
13. Chowdary DH (2020) Towards AI [Online]. https://towardsai.net/p/programming/decision-trees-explained-with-a-practical-example-fe47872d3b53
14. OGIQ (2021) OpengenusIQ [Online]. https://iq.opengenus.org/gaussian-naive-bayes/

15. UCI (2021) UCI machine learning repository [Online]. https://archive.ics.uci.edu/ml/datasets/heart+disease
16. Sayed S (2021) An introduction to data science [Online]. https://www.saedsayad.com/decision_tree.htm

IoT Technologies and Machine Learning Algorithms on Diabetes Prediction-Survey

G. Thilagavathi, N. K. Karthikeyan, G. Priyadharshini, and G. Lavanya

Abstract Diabetes disease is characterized by gradual increase of blood glucose in body. Early prediction of diabetes is needed which can help to shorten the illness and lessen its severity. Serious health problems may occur due to Diabetes. Diabetes has no cure but the steps taken earlier can lower the chance of developing diabetes-related health problems. Predicting the accuracy of diabetes is considered the most challenging factor. In this survey, the various types of machine learning algorithms and IoT techniques were analyzed for predicting diabetes. The attributes which are chosen as important to predict diabetes were considered and different performance metrics are selected to classify diabetes to achieve a higher level of accuracy of prediction.

Keywords Glucose · Metabolic · Machine learning · Diabetes prediction

1 Introduction

Machine learning is considered as part of Artificial Intelligence that builds a model with minimal human intervention. The main objective is to classify the input data and make the prediction of future outcomes based on the models. The model is mainly focused on sample data for making decisions or predictions. Machine learning techniques can be applied to numerous scenarios in which outcome mainly focused on thousands of factors that are not possible for human to monitor. A machine learning algorithm learns from historical data to predict the outcome and uncover the pattern not easily spotted by humans.

Machine learning algorithms [1] are classified as supervised, unsupervised, and Reinforcement learning based on the feedback available to the system. Supervised

G. Thilagavathi (✉) · N. K. Karthikeyan · G. Priyadharshini
Coimbatore Institute of Technology, Coimbatore 641014, India
e-mail: thilagavathi.g@cit.edu.in

G. Lavanya
Sri Ramakrishna Engineering College, Coimbatore, India

S. Shakya et al. (eds.), *Proceedings of Second International Conference on Sustainable Expert Systems*, Lecture Notes in Networks and Systems 351,
https://doi.org/10.1007/978-981-16-7657-4_70

learning builds a model for data with input and desired outputs. The rules generated by machine learning algorithm will map the input to output. Unsupervised learning will have known inputs and it forms the pattern for grouping similar data. Reinforcement learning focuses on maximizing the reward by taking suitable action in a particular situation. The main applications of machine learning algorithms include Agriculture, Health care, Banking, Financial Market, Internet fraud detection, etc.

Internet of things [2] refers to system of devices that are connected that can collect and exchange the data among them without involving any human intervention. Internet of things Devices acts on the data collected by sensors through machine learning. Internet of things is an essential technology in business and industry as it provides real time glimpse into the working of company systems. The basic architecture and working of diabetes prediction is given in Fig. 1.

Initially, sensors are used to get the input from human body in real time. The following sensors such as temperature, pressure, Blood glucose, heart rate, etc. used to record the data. Then the input is passed on to preprocessing phase. Real world data can be missing or noisy or inconsistent. In preprocessing phase, data should be refined properly to get accurate results. After that, it moves on to feature extraction phase in which it builds the data to be informative and non-redundant. When the large input data need to be processed, in case if it is redundant that can be removed

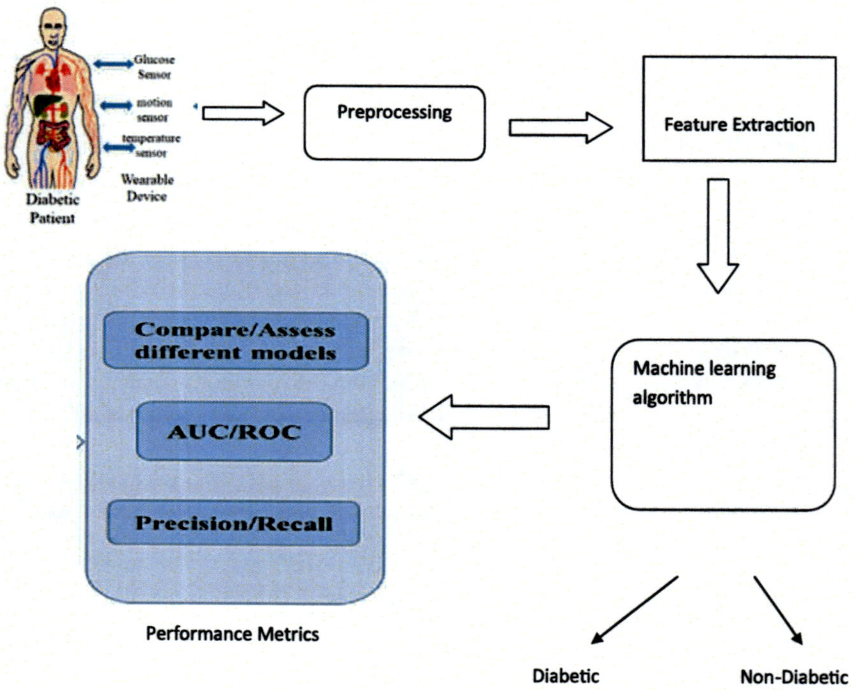

Fig. 1 Architecture of diabetes prediction

by Feature Extraction. Many dimensionality reduction techniques such as Principal component analysis, Nonlinear dimensionality reduction, Multifactor dimensionality reduction, Multilinear Principal Component Analysis, Independent component analysis can be used for different applications. The features which are selected are then passed on to Machine Learning Algorithms in which it predicts diabetes based on the classification of input. Various performance measures can be used to predict the accuracy of classification.

2 IOT Technologies for Diabetic Prediction

IoT Technologies connect different types of sensors, alarms, lights, and cameras throughout the day in which can be controlled by smart phone. IoT Technology has been divided into four basic layers which makes IoT work. Device Hardware, Software, Communication, and Platform will form these layers. Hardware acts as interface between real and digital world. It mainly depends on the sizes, shapes, and levels of technological complexity of hardware to coordinate with IoT devices. Software is responsible for making real time analysis of data collected and implementing those data to communicate with cloud. Communication layer involves communication between hardware and software. IoT platform is a place where data can be gathered, processed, analyzed, and presented in a user friendly way.

2.1 Blood Glucose Sensor

It is a device used to monitor glucose levels in blood. Person monitors or manages diabetes by pricking their finger using glucometer for testing daily. It is similar to that of continuous glucose monitoring system (CGM) and can be placed under the skin or external to the body.

2.2 Pedometer

Physical activity of diabetic patients can be monitored by Pedometer [3]. Physical activity will make us use energy in our body by doing walking or playing games. Physical activity is considered as most important for diabetic patients to manage their glucose level and to improve their health.

2.3 Temperature Sensor

DS18B20 Sensor is the most widely used temperature sensor. The temperature can be air, liquid, or solid. This sensor can work in parallel but more than sensors are not allowed. In case of medical applications, Infrared (IR) temperature sensors provide accurate result by measuring skin temperature.

2.4 Blood Pressure Sensor

This sensor is mainly to monitor human blood pressure. It measures the systolic, mean arterial, and pulse rate. photoplethysmograph (PPG) is the most popular pressure sensor.

2.5 Wearable Band

It consists of acceleration sensor and 3-axis sensor for identifying user wakeup and sleep times. It is also used for calculating pre-and post-exercise time.

2.6 Smart Food Tray

Smart food tray [4] will share the patient's mealtime information with a glucometer. When the patient starts eating a meal using food tray, they should press start and stop to monitor the beginning and end time of the meal. Weight difference of each food which was taken as intake can be monitored by food tray.

3 Machine Learning Algorithms for Diabetes Prediction

3.1 Ensemble Algorithm

In order to improve the prediction of diabetes, combining various machine learning algorithms to form a predictive model called Ensemble algorithm is used. Ensemble algorithm combines the decision of various models to improve the performance. By considering different parameter metrics such as the sensitivity, false omission rate, and AUC. The following machine learning algorithms as (Naive Bayes, Random Forest, AdaBoost, k-nearest Neighbor, Decision Trees, and XGBoost were used and Multilayer Perceptron (MLP) [5] also employed. From this, they predicted

Ensembler classifier performs well and acts as best classifier. The prediction of diabetes can be improved by making use of ensembling or combining different ML Models. Ensemble Perceptron Algorithm always works well with the combination of pocket algorithm with Boosting algorithm to predict diabetes by making use of weight vector as a factor.

3.2 Random Forest

It is a classification algorithm based on decision trees. The more accurate results will be produced by merging the trees. It will handle different types of data including numerical, nominal, and binary. Here overlapping of random forest trees will be taken place and it can be analyzed easily. Based on the probability machine learning model is constructed. By fixing random thresholds more accurate results will be identified on large dataset. Overfitting issue can be solved using Random Forest algorithm [6]. The measures such as sensitivity and specificity were used to predict diabetes and achieved an accuracy of 74.7% of prediction.

3.3 Support Vector Machines

It is a supervised learning algorithm. In this, each data point will be plotted as point in n-dimensional space, and n is taken as a feature. Each feature will have the value of particular coordinate. It works well by finding the best hyperplane which differentiates the classes properly. Diabetes is predicted by setting BMI and blood glucose as indicators and that is considered as input to SVM [7]. Diagnosis is based on predicting the variable. Radial-based kernel is used for training and prediction.

3.4 K-Nearest Neighbors

KNN is a supervised learning algorithm that is mainly used for regression problems. Based on the similarity between the data points, available and new data will be classified. KNN algorithm consists of training and classification phase. In training phase, features and corresponding class labels will be stored. In classification phase, class will be defined based on feature similarity. Classification process can be finalized by majority of voting. K-nearest neighbor algorithm [8] based diabetes prediction has considered the optimal number of neighbors which can be used to build the disease risk prediction model utilizing data.

3.5 *Naïve Bayes*

In Naïve Bayes classifier algorithm, features are selected for classification which is independent of each other. In this, no pair of features should be dependent on another. Secondly, each feature must have same weight. It works well for missing values and unbalancing problems. It mainly employs bayes theorem. Naïve Bayes [9] has an accuracy of 76.3 when compared to that of decision tree and SVM.

3.6 *Logistic Regression*

Logistic Regression is a classification method derived from Linear Regression in which target variable will be considered as binary class. For predicting Diabetes, classification of data is done based on the model which uses Logistic Regression function. It identifies the relationship between the categorical response variable and covariates.

3.7 *K-Means*

K-Means [10] is an unsupervised Learning algorithm that identifies the k number of centroids. By keeping smallest distance between the centroids, the data point can be allocated to nearest cluster. If K value is small and numbers of variables are also large, then k-means will perform faster.

3.8 *Learning Vector Quantization (LVQ)*

LVQ [11] is one of artificial neural network [12] which make use of combination of supervised and neighbor pattern classifiers [13]. It is necessary to get the near possible value of distribution of input vector for minimizing the error during classification. For achieving high performance it is necessary to calculate the distance which can be Euclidean dist

Algorithm used	Attributes	Parameters	Accuracy achieved
Ensembler algorithm (Naive Bayes, random forest, AdaBoost, k-nearest neighbor, decision trees, and XGBoost)	BMI, glucose level, blood pressure, age, pedigree function, insulin, triceps skin fold thickness	Area Under ROC Curve (AUC)	95% of prediction
Artificial neural network, random forest, and K-means clustering algorithm	BMI, glucose level, blood pressure, and age	Accuracy and the AUROC curve	74.7% of prediction
Support vector machine	Age, BMI, blood glucose concentration	Accuracy	65.6% of prediction
Naïve Baye's classifier	BMI, glucose level, blood pressure, age, pedigree function, insulin, triceps skin fold thickness	Accuracy, *F*-measure, recall, precision, and ROC (receiver operating curve) measure	76.3 of prediction
J48 algorithm	Glucose level, temperature, motion	precision, receiver-operating characteristics (ROC), and accuracy	99.17% of prediction
Learning vector quantization	BMI, glucose level, blood pressure, age, pedigree function, insulin, triceps skin fold thickness	Accuracy	75% of prediction

ance between the input and weight vector. LVQ has very good performance and the model produced can work well and updation on the model is also possible.

4 Conclusion

In this paper, various IoT Technologies and different machine learning algorithms and their importance has been discussed. Different Machine learning algorithms have been compared based on their accuracy level. Level of accuracy has been achieved using the attributes such as BMI, Glucose level, Blood Pressure, Age, Pedigree Function, Insulin, Triceps skin Fold thickness. In Future, the accuracy level can be increased by considering other important parameters such as glycosylated hemoglobin that will diagnose and produce the correct result. It will further improve patient health by controlling their illness and health status.

References

1. https://medium.com/lets-talk-ml/model-training-components-in-azure-machine-learning-an-overview-38592e767e0b
2. https://researchspace.csir.co.za/dspace/handle/10204/8128
3. Rghioui A, Lloret J, Harane M, Oumnad A (2020) A smart glucose monitoring system for diabetic patient. Electronics 9:678. https://doi.org/10.3390/electronics9040678
4. Park J-C, Kim S, Lee JH (2006) Self-care IoT platform for diabetic mellitus. Appl Sci 11:2006. https://doi.org/10.3390/app.11052006
5. Kamrul Hasan M, Ashraful Alam M, Das D, Hossain E, Hasan M (2020) Diabetes prediction using ensembling of different machine learning classifiers. In: IEEE access special section on deep learning algorithms for internet of medical things, 7 May 2020
6. Alam TM, Ibal MA, Wahab A, Ijaz S (2019) A model for early prediction of diabetes. Inform Med Unlocked 16:100204
7. Viloriaa A, Herazo-Beltranb Y, Cabrerac D, Pinedad OB (2020) Diabetes diagnostic prediction using vector support machines. In: The 11th international conference on ambient systems, networks and technologies (ANT), Warsaw, Poland, 6–9 Apr 2020
8. Sarker IH, Faisal Faruque M, Alqahtani H, Kalim A (2020) K-nearest neighbor learning based diabetes mellitus prediction and analysis for eHealth services. EAI Endorsed Trans Scal Inf Syst 7(26)
9. Sisodia D, Sisodia DS (2018) Prediction of diabetes using classification algorithms. Int Conf Comput Intell Data Sci (ICCIDS 2018)
10. Raihan M, Tanvir Islam M, Farzana F, Golam Morshed Raju M, Mondal HS (2019) An empirical study to predict diabetes mellitus using K-means and hierarchical clustering techniques. In: 10th ICCCNT 2019, IIT—Kanpur, 6–8 July 2019
11. Sonawane JS, Patil DR (2014) Prediction of heart disease using learning vector quantization algorithm. In: IEEE conference on IT in business, industry, and government (CSIBIG)
12. Putri NK, Rustam Z, Sarwinda D (2019) Learning vector quantization for diabetes data classification with chi-square feature selection. In: 9th annual basic science international conference 2019, materials science and engineering, vol 546, pp 052059. https://doi.org/10.1088/1757-899X/546/5/052059
13. Badbe V, Londhe V, Shirole G (2016) Analysis of heart disease by LVQ in neural network. Int J Recent Innov Trends Comput Commun 4:603–607

FIGSI—Facial Image Generation for Suspect Identification

Prem Bhajaj, Pratik Panchal, Neil Patel, Rucha Shinde, and Aruna Gawade

Abstract Currently, there is no such application used by the Forensic departments which generates faces automatically and quickly. We propose a system to generate a 2-D image of a face by using a novel technology called Conditional Generative Adversarial Networks (C-GANs). The system considers a few basic inputs like gender, hair color, face shape, nose size, eye size, and other detailed attributes. The user can enter this information through a form-like user interface and a face preview is displayed in real time. The face preview corresponds to the attributes selected by the user. Moreover, a face matching feature with the criminal database is also integrated so as to check whether the suspect already exists. The model has been trained on over 100 k images of the CelebaFaces Attributes dataset after preprocessing it. The model performance was evaluated by using Histogram Euclidean Distance and Frechet Inception Distance. Use of C-GANs for generating facial images will help in creating realistic images.

Keywords Generative adversarial networks (GANs) · C-GAN · Face generation · Suspect ıdentification · Applications of GAN

1 Introduction

A photo of any suspect helps the cops to search him/her effectively. Crime departments hugely rely on suspect photos, which could either be real images, hand drawn sketches, or software generated faces. Real images can only be obtained through surveillance camera footage or if the cops know the identity of the suspect and are in contact with his/her known ones. A facial sketch can be generated with the help of a sketch artist. This approach is sluggish and completely depends on the interpretation between the forensic artist and the witness. It becomes a strenuous and recurrent task for artists to pencil the facial characteristics of the suspect, as there are cases where the witness cannot pinpoint the exact description of the feature. Consequently,

P. Bhajaj (✉) · P. Panchal · N. Patel · R. Shinde · A. Gawade
Computer Engineering, Dwarkadas J. Sanghvi College of Engineering, Mumbai, India

S. Shakya et al. (eds.), *Proceedings of Second International Conference on Sustainable Expert Systems*, Lecture Notes in Networks and Systems 351,
https://doi.org/10.1007/978-981-16-7657-4_71

the gap is filled according to the best interpretation the artist feels. Another way to create the face involves the use of computer software like SketchCop® FACETTE [1], Ultimate flash face [2] or Adobe Photoshop, but it requires a trained operator. Presently, there is a lack of an application, which is easy to use and can be used by Forensic departments to generate faces automatically and quickly.

To overcome this problem, an evolutionary model is proposed that generates faces of the suspects, based on the attributes selected. These attributes are presented through the user interface to the witness, eliminating the 3rd party needed for the interpretation. The model focuses on the user-friendly platform enabling a smooth transition. The model provides a range of attributes for each feature, for example, eyes: baggy eyes, rounded eyes, sleepy eyes. After the selection for each feature, the face is created in real time with the help of the GAN model [3]. Since it is real time, if the user does not find a match between the generated face and the desired face, he/she can tweak the attributes accordingly. Through an iterative cycle of inputs and preview, the creation of the face is completed. The project focuses on the Forensic Department for criminal identification enhancement. Additionally, the suspect image will not only be generated but will also get matched with an existing criminal database. Due to this, the investigation process can be taken one step ahead. This will lead to finding the suspect's personal information such as his name, address, marital status, and age.

The proposed system uses Generative Adversarial Networks (GANs) as a technology, which empowers the computer to think. GANs is a class of neural network that is a part of unsupervised learning. It is a combination of the Generator and Discriminator network. Starting with the generator network, it defines how data is generated w.r.t. a probabilistic model. The generator is trained in such a way that it tries to maximize the probability of the Discriminator is making a mistake. Its output is connected to discriminator's input. This Discriminator acts as a loss function to the generator. The output of the discriminator is used to adjust the weights of the Generator as well as Discriminator in an alternating manner. It is based on a model that estimates the probability that the sample that it got, is actually received from the training data and not from the Generator. Hence, these networks can be tweaked in such a way that it generates a human face based on facial attributes.

The input to the generator is an N-dimensional uniform random variable, also called noise, which it uses to generate fake faces. On the other hand, the discriminator captures a probability of the outcome that the generated face is real. If the output of the generator is $P(A)$, then the discriminator calculates the conditional probability $P(B/A)$. Here, A is the input to the generator and B is the label (real/fake) that the discriminator gives as output.

The alternating training of the competing networks can be technically achieved as follows. The trainable parameter of the discriminator is set to *false*. So, while training the discriminator explicitly, the weights of the discriminator are changed. But, when the complete GAN network is trained with the discriminator acting as a loss function, its weights will not change.

2 Related Work

The USA police department uses SketchCop® FACETTE software. It contains defined presets of each attribute. The trained officer along with the witness sits together, the witness is asked from the collection of presets the one which is applicable. Through trial and error, the face is generated; moreover, the final touch up is done in the lighter version of photoshop that 2 comes along with the software for the result. Even though the process is substantially shortened, it is time-consuming and done by the expert [1].

Klare BF. and others discuss the face attributes of the suspect. The proposed approach defines 46 facial attributes to extract every aspect of the face attributes. The paper also shows a comparison between the automated sketch and manual sketch and concludes where it is shown that the accuracy achieved through machine generated is almost twice as compared to 5 the manual human sketch. By surveying and analyzing this paper, it was proven that machine generated sketch has a higher accuracy as compared to human sketch. As a result, the need for an automated tool for developing sketches was proven by analyzing the aforementioned research paper [4].

Jha M. and others propose a model that will help the police department in investigating crimes by drawing sketches. The model can generate images based on the textual description provided as input. The model was trained on 400 images belonging to Labeled Face in Wild Dataset, and the description of the images was taken from Face2Text. The system can work as an intermediary for sketch artists, who can work on the output. The limitation found in the research was that the number of images generated for a description was 16. In order to get more images, the description needs to be modified. This served as the motivation for us to provide a slider to the user to manipulate the noise passed as input to the generator. This helps the user to precisely adjust the face without changing the attributes. The idea of implementing *face matching with the criminal database* feature was obtained through this paper [5].

Mingming Hu and Jingtao Guo propose a model that translates sketches to images. By analyzing this paper, it was found that the C-GAN model delivers a similar output (face as per user-provided attributes) for a sketch as input. The research proposes a network architecture that consists of a facial attributes classifier apart from the generator and discriminator. The research was a motivation to use C-GAN and it helped in finding the dataset for the proposed system developed by us [6].

Muhammad Zeeshan Khan and others put forth a model that generates a face from a textual description. The model uses 2 discriminators, $D0$ and $D1$. The need for the second discriminator is to focus on the nose region, eyes region, and lips region. However, it costs more time, effort, and resources to train the additional discriminator, making the architecture complex at the same time. Hence, we decided to implement the system with just one discriminator, while trying to maintain the quality of the training by adding more layers and neurons to the discriminator. This research also used the CelebA dataset as one of the datasets for training, which made us firm on choosing this dataset [7].

Xiang Chen and others propose a model, a fully-trained GAN(FTGAN) for text-to-image synthesis, that trains the text-encoder and image-decoder simultaneously. The paper contributes in three distinct areas (1) Image synthesis from text description with the help of a Fully-trained Generative Adversarial Network FTGAN (2) Building of text2face dataset for synthesis task 3. The image quality of the synthesized faces is evaluated by calculating the similarity between generated faces and ground truth. The results show us that FTGAN could achieve higher-quality images and more similar faces to the ground truth. The paper presents face generation through a detailed description using FTGAN. The text-to-image synthesis framework used for the image generation is not stable in their training process [8].

Yaohui Wang and others propose a model to generate faces based on attribute labels. To achieve this, they suggest using a deep conditional convolutional generative adversarial network (DCGAN). The model is trained on 2D GAN with the dataset CelebA with the attribute selection of high descriptiveness. In addition, they expand the 2D GAN model into a 3D GAN model by training it on the UvA-NEMO dataset. The 3D GAN model generates facial smiling video restricted for the attributes—gender and age. The paper presents us the 2D and 3D GAN models through the use of DCGAN. The result of facial image generation in the 2D GAN model against the attribute estimator and benchmark quality scores suggests the models' ability to generate realistic faces from attribute labels can be instrumental in the suspect's visualization. The limitation of this model is that it considers few attributes while generating faces, hence the user shall not be able to precisely input the appearance [9].

Ting Zhang and others have done a comparative study on their Deep Convolution and Deep Transpose Convolution Neural Network with Deep Convolution Generative Adversarial Networks (DCGAN) and Wasserstein Generative Adversarial Network (WGAN). The dataset which was used is again the same CelebA dataset, which even further motivated us to use this dataset. They have also introduced a hyper-parameter that balances the generator and discriminator to properly distinguish between generated samples and the real ones. The use of Face_Recognition project on Github to test the model performance encouraged us to use opencv based face_recognition to search the criminal database with the generated face [10].

Youngjoo Jo and Jongyoul Park have proposed an image editing system with the help of GAN. The sole idea of this system was to complete the missing patches on a facial image. The patches are the editing that is provided by the user in a simple free-form sketch or color. The proposed network uses an SN-PatchGAN discriminator. Additionally, the system is also trained on style loss which helps the model to edit the parts of the face. The dataset used for the training was CelebA-HQ. This model can be incorporated with the system that we propose, to modify certain attributes such as cap, earrings, etc. that are currently not in the scope of the model [11].

3 Accomplishment of GAN

The accomplishment of GAN can be understood by looking at the following applications:

- Generating cartoon characters for animated movies: The proposed application converts images of real-world scenes to animated images. The system uses cartoon images and unpaired photos for training purpose. Here, the generator's architecture is a modified version of SRResNet which contains 16 Res-Blocks. Also, 3 sub-pixel CNN is used to improve the feature map [12].
- Face aging: This projects the future looks of any person by keeping the vital facial features intact. It uses a version of GAN called Age Conditional Generative Adversarial Network (Age-cGAN). The highlight of the model to retain the original person's identity in the reconstruction is its "Identity-Preserving" latent vector optimization approach [12].
- Creating three-dimensional models based on two-dimensional images: By using SliceGAN it is possible to produce an accurate 3D dataset using 2D images. This approach is more promising as it solves the availability issues of 3-D training data. Also, 2-D imaging methods are quicker, offer better resolution, and are readily available [13].
- Changing expressions of a person in an image: This application uses SliderGAN, which changes facial attributes based on the values of a statistical blendshape model of facial motion. The GAN edits the facial image provided according to the speech and expression blendshapes [14].
- Reconstructing distorted parts of an image. The SC-FEGAN used in the image editing system is able to restore and modify regions of the image which are of significant size. The system can edit face image features such as mouth, shape of face, eyes, and hairstyle [11].
- Generating the look of pizza: An impressive application of this technology was shown by a group of researchers who used it to add or remove ingredients from a pizza and predict how it would look after each modification. The motto of this system is to teach a machine to make a pizza with the help of GAN. In this way, a visual appearance of the pizza can be displayed by adding or removing the ingredients. This system is able to segment the pizza toppings in supervised fashion or remove them, revealing the area underneath [15].
- Another noteworthy application is Drug Discovery where the GAN model tries to generate a remedy for a disease incurable to date. The generator is trained in such a way that it accurately samples a drug from a Drug database. Post training the generator is able to generate a drug for a disease that has not been cured to date. Additionally, the Discriminator is used to determine whether the generated drug actually cures the disease [12].

4 Method

The below architecture of the system can be described as follows (Fig. 1).

Face generation stage:

1. Initially, the user opens the web application and chooses various attributes describing the face to be generated.
2. The application processes these inputs and sends it to the generator on the server.
3. The generator generates an image based on the attributes passed and sends it back to the application.

Training stage:

4. The generator has been trained to generate such images using the dataset containing facial attributes of several images. These attributes in the dataset have been preprocessed according to the needs of the generator.
5. During the training phase, the generator's output is passed to the discriminator. The discriminator also receives the attributes from the dataset.
6. The discriminator then classifies the image as real or fake, based on the attributes. A real image is an image from the dataset and a fake image is a generated image.
7. This classification is passed to the loss functions of the discriminator, which changes the weights of the neurons of the discriminator in order to train it.
8. The classification output is also passed to the loss function of the generator to adjust the weights of the generator's neurons.
9. The model's objective function V is:

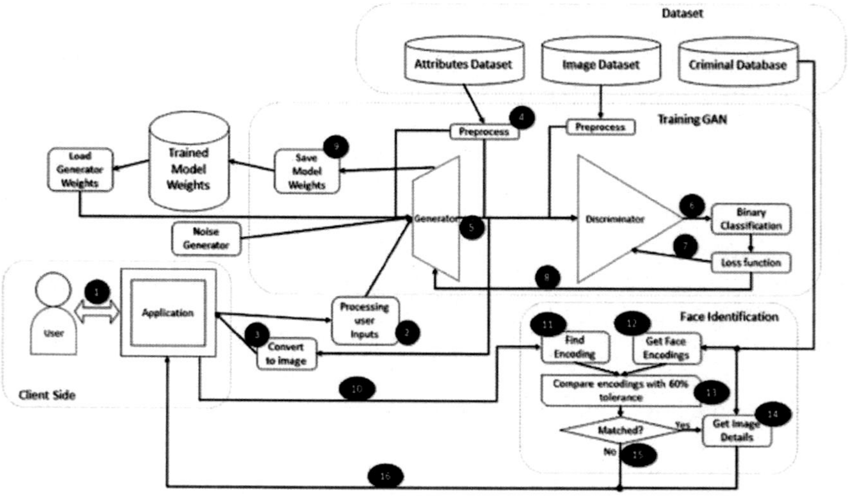

Fig. 1 Architecture of proposed system

$$\min_{G}\max_{D} V(D, G) = E_{x,y}\big[\log D(x, y)\big] + E_{z,y}\big[\log(1 - D(G(z, y), y))\big]$$

- $D(x, y)$ is the discriminator's estimate of the probability that real data instance x, y is real.
- $G(z, y)$ is the generator's output when given noise z and labels y.
- $D(G(z, y), y))$ is the discriminator's estimate of the probability that a fake data instance $G(z, y), y$ is real.
- $E_{x,y}$ is the expected value over real data instances.
- $E_{z,y}$ is the expected value over random inputs to the generator.

10. The generator periodically saves its weights in a file which can be loaded again later.

Face matching stage:

11. Once the user is satisfied with the face generated, he/she can choose to get the identity of the face from the criminal records.
12. When such a request is received on the server, the server finds the encoding of the generated image.
13. It also gets the encodings of the faces stored in the criminal database.
14. The encoding of the generated image is compared with those of the criminal database to find a match with 60% tolerance.
15. If a match is found, the details associated with the image are retrieved from the criminal records.
16. In case there is no match, the server responds with "No Match Found".

5 User Interface

The representation in Fig. 2 is the page visible to the user for capturing the facial characteristics, such as eyes, nose, ears, etc. The facial description selected by the user through the GUI is used to generate faces in real time with close resemblance to the feature description. The face generated is displayed in the Preview section of the interface. The user can also tweak the attributes selected and observe the difference in the Preview section.

Each time an attribute value is changed, the Generator generates a new face with the updated set of attributes, and this face gets displayed on the UI (Fig. 2).

This is the last section of the UI (or the process) where the user is provided with 7 sliders. These sliders are used to further fine tune the face without changing the attributes selected. Technically, the values selected for the sliders are passed to the generator as a random vector of size 7. Each slider has a range from − 3 to + 3, with a step size of 1×10^{-8}. So, the user can precisely adjust the slider to get the perfect face.

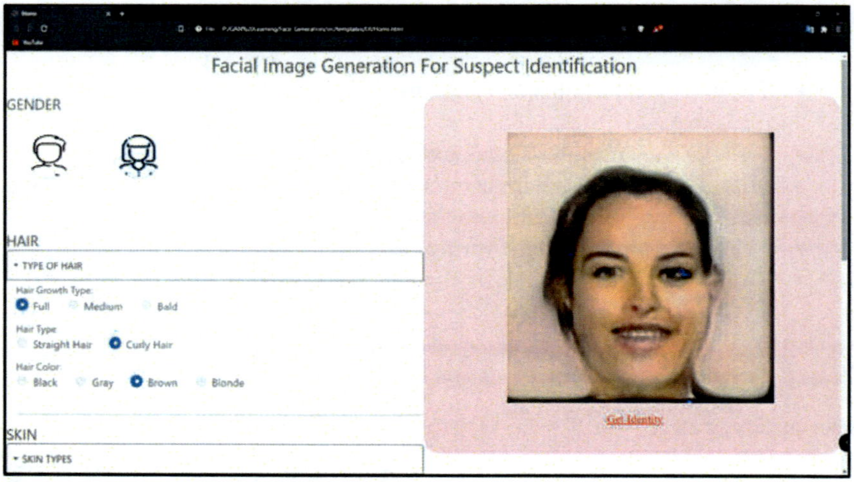

Fig. 2 User Interface for face generation

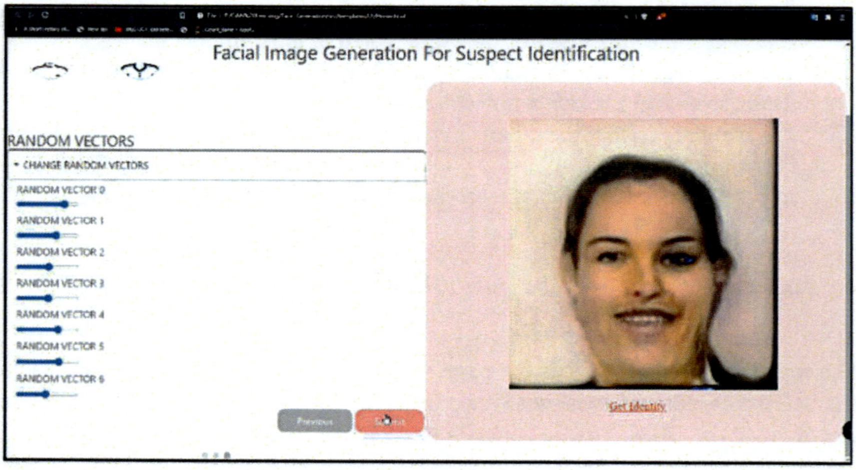

Fig. 3 User interface section for changing look of the face without changing the attributes

6 Experimentation

6.1 *Data Preprocessing*

The raw CelebA dataset [16] had many inconsistencies that could have impacted the model performance in a negative way. Hence, it was mandatory to process the data and rectify these inconsistencies before passing it to the GAN model. Also, the model accepts data in specific format therefore transforming the dataset into required state was necessary. Following are the steps that were performed under data preprocessing.

- Image Resizing: The raw images from the data set had resolution of 178 × 218 which were resized to 128 × 128 uniformly. This step was performed to remove unwanted parts of the images and ensure that the faces are focused upon.
- Normalizing Attributes: Scaling the values down to a range from 0 to 1.
- Analyzing the CSVs: The dataset used for the project contained 3 CSV files, out of which only one file had the desired list of attributes. Hence, the other files were ignored.
- Deleting Improper images: As the dataset had many blurred images the performance of the model could have been impacted badly. Thus, it was necessary to delete those records before giving it to the model. It became a lot easier as those images are tagged as blurred in one of the corresponding csv. The dataset also contained some improper records such as some female faces having their mustache or beard attributes as yes in the CSV file, therefore such records were deleted.
- Merge Attributes: Certain attributes which were split across multiple columns such as gray hair, blond hair, brown hair, and black hair, were merged as a single attribute, hair color.
- Attribute Selection: The raw dataset consisted of 40 facial attributes wherein few attributes were not so essential. Feeding these attributes created a lot of ambiguity in the model during the training process. Therefore, each attribute was closely analyzed, and grouping of similar attributes was done. Additionally, the attributes which do not seem to be useful were dropped out.
- Deleting images where the face is unclear: The dataset had lots of images where the face was not clearly visible. Such records were discarded by using a face detecting algorithm in OpenCV.
- Balancing Male to Female ratio: After performing the above-mentioned steps, it was found that the ratio of males: females was unbalanced. This could have led the model's output to be gender biased. So, the dataset was balanced by deleting extra images at random.

6.2 Implementing GAN Architecture

6.2.1 Generator

The Generator model begins by taking 2 inputs: Random Noise Vector and Attributes Vector. The random vector is a linear vector with size 7, and the attributes vector is also linear with size 25, indicating the values of the 25 processed facial attributes. Both the inputs are passed to separate dense layers and then reshaped to (16, 16, x), where x indicates a variable value that can be altered as per the model performance. Then, the outputs of the Reshape layers are merged and provided as input to the next layer. Then, a series of LeakyReLU and Convolutional layers are used to get a vector of size 128 × 128 × 3, representing an RGB image.

6.2.2 Discriminator

The Discriminator model also takes two inputs: Attributes Vector and an image. The image is either produced by the generator or taken from the dataset. The Attributes vector is linear with size 25, and the image is of size 128 × 128 × 3. The attributes vector is passed to a Dense layer and then reshaped to match the shape of the Image. The reshaped output is merged with the output of input_4 layer and passed as input to the next layer. Next, a series of convolutional, Leaky rectified linear activation functions and dropout layers are used to get an output of 5 × 5 × 256. This output is then flattened in the next layer and then passed to the final layer with one neuron. The final layer decides whether the image is real or fake. Its output is between 0 and 1, where 0 indicates the image is fake, and 1 indicates the image is real.

6.2.3 Generator Loading Module

Once the model is trained sufficiently, the discriminator is discarded as only the generator is needed to generate faces. Therefore, to load the generator model, a loading module was created, in which the generator version is specified. Later, the same module is used to generate faces by passing facial features and the noise vector.

6.3 Face Matching Module

This module is responsible for searching the face from the Criminal Database. This module makes use of the face_recognition library which is based on OpenCV to match the face. While matching the face it also generates the encoding of the target face and compares it with already encoded faces with some tolerance. Here, the tolerance is set to 60%. It then returns the id of the image whose match has been found.

7 Results

7.1 Comparison of Real and Generated Face

Figures 4 and 5 show a sample output of the trained Generator model. In this system, 75% of the data has been used for training the model and the remaining 25% is used for testing the model. In the above figure shown, two images from the testing set have been considered and the attributes of those images as shown in Table 1 has been passed to the generator model to generate the image. In Figs. 4 and 5 the image on

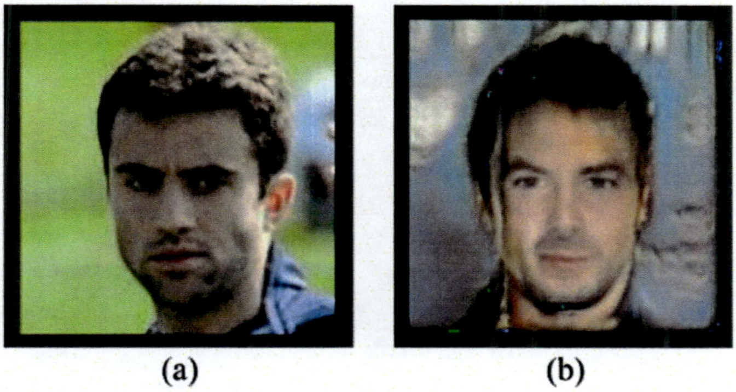

Fig. 4 Generated male face **a** original image, **b** generated image

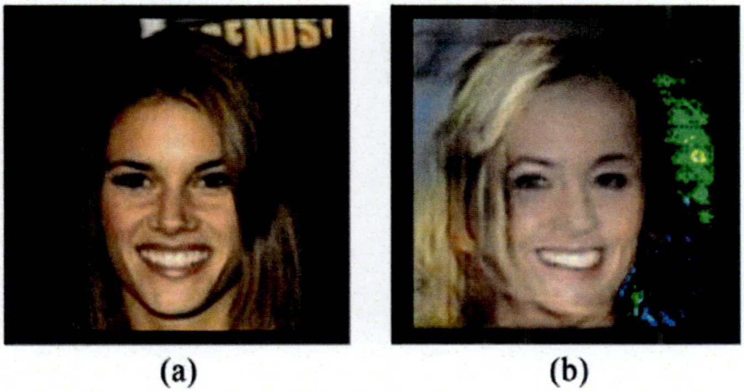

Fig. 5 Generated female face **a** original image, **b** generated image

the left (a) is the original image present in the dataset and the image on the right (b) is the image generated by the model.

Our model has been trained on Nvdia GTX 1070 Ti GPU with 8 GB RAM and 2432 cuda cores. The process of training took approximately 30 h to sufficiently train the generator to produce the images with 128 * 128 resolution as shown above.

7.2 *Performance Evaluation*

After training the GAN network the most challenging part is to evaluate the performance of the model. To do so, we have implemented two different techniques FID (Frechet Inception Distance) [17] and HED (Histogram Euclidean Distance). Both these techniques work on the same principle, they calculate the score (also known as similarity index) between real image and generated image (from each epoch) but

Table 1 Attributes considered while testing trained model

Attribute	Figure 4	Figure 5
5_o_Clock_Shadow	Present	Not present
Bags_Under_Eyes	Not present	Not present
Big_Lips	Not present	Not present
Big_Nose	Not present	Not present
Chubby	Not present	Not present
Double_Chin	Not present	Not present
Goatee	Present	Not present
Heavy_Makeup	Not present	Present
High_Cheekbones	Not present	Present
Male	Present	Not present
Mustache	Not present	Not present
Narrow_Eyes	Not present	Not present
No_Beard	Not present	Present
Oval_Face	Present	Present
Pale_Skin	Not present	Not present
Pointy_Nose	Not present	Present
Rosy_Cheeks	Not present	Not present
Sideburns	Not present	Not present
Smiling	Not present	Present

using two different techniques. Later this score is analyzed as the training progress and the more the score is close to 0 the better the model performance is.

- FID: This technique works on covariance and mean. The formula for FID is as follows.

$$\text{FID} = \|\text{mu}(r) - \text{mu}(f)\|^2 + \text{Trace}\Big(\text{cov}(r) = \text{cov}(f) - 2\sqrt{rf}\Big)$$

  - r is the grayscale matrix of real image
  - f is the grayscale matrix of fake image
  - mu is the mean of image matrix
  - cov is the covariance matrix of image matrix
  - Trace is the sum of main diagonal of square matrix

- HED: This technique first finds the histogram equalization and then a Euclidean distance is calculated. The formula is as follows.

$$\text{HED} = \sqrt{\sum_{n=0}^{L-1} (\text{Pr}_n - \text{Pf}_n)^2}$$

- L is the maximum pixel intensity, i.e., 255 here
- Pr and Pf is the normalized histogram of real and fake images respectively, it is calculated as follows

$$Pn = \frac{\text{number of pixels with intensity } n}{\text{total number of pixels}}$$

- n represents the intensity of pixels

7.3 *Identifying Generated Face*

The details of the criminal are displayed if there is a match. The details include but are not limited to demographic information such as date of birth, marital status, and address with state and country. In case there is no match in the criminal database the page displays a message "No Match Found" with the demographic fields displayed empty (Fig. 6).

The matching of the generated face is handled by the face_recognition library in python which is based on OpenCV. The first step of face matching is to locate the face in the image. Once the coordinates of the face are fetched, it creates a face encoding which is a vector of size 256. To match the generated face with the available ones

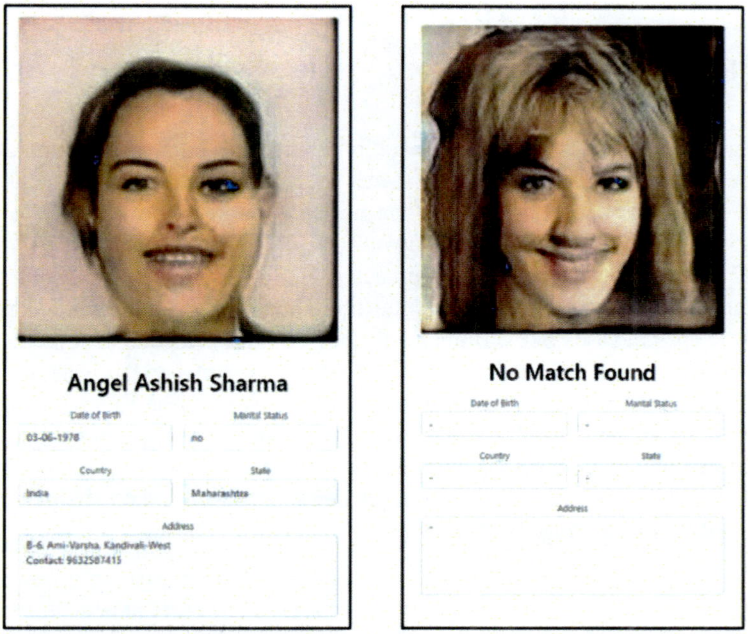

Fig. 6 Identification of generated page

in the database, these face encodings are compared with some tolerance. Here, the tolerance was 0.6, i.e., 60%. Once the match is found it returns a face ID, this ID is further used to fetch the aforementioned details from the database.

8 Conclusion

Thus, in the proposed project, we perform face generation with the help of an evolutionary generative neural network. The characteristics of the face are described by the user as per the facial features such as gender, skin tone, eyes, etc. These facial features, which are captured through the graphical user interface, are provided to a conditional GAN model to generate a face. Once the face is generated, the system offers the user the choice to finalize it or make modifications to it, to get a face that is almost identical to the actual one. Suspect face generation using a description, based on a neural network (GAN), is an unconventional concept brought into actuality. This GAN based technique works on the similarity of a generated face with an existing one and evolves through several generations for a perfect match. In each generation, the face appears to be closer to the actual one. The approach not only saves the prolonged time required by other methods but also provides more realistic results. We have also implemented face matching using face_recognition library. This helps in identifying criminals and getting their details from criminal records.

9 Discussion

Currently, the system is limited to generation of the faces as per the user inputs and searching it in a criminal dataset for a match with the generated face. Other external attributes such as hats, earrings, etc. are not considered in the system. However, Youngjoo Jo and Jongyoul Park's study on SC-FEGAN proved that it is possible to control these attributes using their approach [11]. So, their model can be integrated with the proposed system to overcome the limitation. Further, more study is needed to handle special cases like birthmarks or face deformations due to some external factor. The work can also be extended by adding new features to the system, such as Linking of Aadhaar details, wherein the generated face will be compared with the Aadhar faces to find a match and retrieve details of the suspect seamlessly. For this, explicit permission from the government will be required along with access to their Aadhaar records. An emerging study on reading brain signals could also play a vital role to get the exact image from the witness' mind [18, 19].

References

1. Streed MW (1941) SketchCop facial composite system. FinchTech [Online]. Available https://sketchcop.com/
2. Brigert A. Ultimate flash face [Online]. Available http://flashface.ctapt.de/
3. Goodfellow IJ, Abadie JP, Mirza M, Xu B, Farley DW, Ozair S, Courville A, Bengio Y (2014) Generative adversarial nets. Adv Neural Inf Process Syst
4. Klare BF, Klum S, Klontz JC, Taborsky E, Akgul T, Anil JK (2014) Suspect identification based on descriptive facial attributes. In: IEEE international joint conference on biometrics, Clearwater, FL, USA, 2014
5. Jha M, Jobanputra S, Kamath D, Hole V (2020) Creation of face for aiding in forensic investigation based on textual description. In: 2020 fourth international conference on inventive systems and control (ICISC), Coimbatore, India, 2020
6. Hu M, Guo J (2020) Facial attribute-controlled sketch-to-image translation with generative adversarial networks. EURASIP J Image Video Process 2020
7. Khan MZ, Jabeen S, Khan MUG, Saba T, Rehmat A, Rehman A, Tariq U (2020) A realistic image generation of face from text description using the fully trained generative adversarial networks. IEEE Access
8. Chen X, Qing L, He X, Luo X, Xu Y (2019) FTGAN: a fully-trained generative adversarial networks for text to face generation
9. Wang Y, Dantcheva A, Bremond F (2019) From attribute-labels to faces: face generation using a conditional generative adversarial network. In: Computer vision—ECCV 2018 workshops, Cham, Springer International Publishing, 2019, pp 692–698
10. Zhang T, Tian W-H, Zheng T-Y, Li Z-N, Du X-M, Li F (2019) Realistic face image generation based on generative adversarial network. In: 2019 16th international computer conference on wavelet active media technology and information processing, Chengdu, China
11. Jo Y, Park J (2019) SC-FEGAN: face editing generative adversarial network with user's sketch and color. In: 2019 IEEE/CVF international conference on computer vision (ICCV), Seoul, Korea (South)
12. Hamed A, Manolya K-T, Gulshan K (2019) Applications of generative adversarial networks (GANs): an updated review. Arch Comput Methods Eng 2019
13. Kench S, Cooper SJ (2021) Generating three-dimensional structures from a two-dimensional slice with generative adversarial network-based dimensionality expansion. Nat Mach Intell 2021
14. Ververas E, Zafeiriou S (2020) SliderGAN: synthesizing expressive face images by sliding 3D Blendshape parameters. Int J Comput Vis 2020
15. Papadopoulos PD, Youssef T, Ferda O, Ingmar W, Antonio T (2019) How to make a pizza: learning a compositional layer-based GAN model. In: 2019 IEEE/CVF conference on computer vision and pattern recognition (CVPR), Long Beach, CA, USA, 2019
16. Li J (2018) CelebFaces attributes (CelebA) dataset. Kaggle, 1 June 2018 [Online]. Available https://www.kaggle.com/jessicali9530/celeba-dataset
17. Heusel M, Ramsauer H, Unterthiner T, Nessler B, Hochreiter S (2017) GANs trained by a two time-scale update rule converge to a local Nash equilibrium. Adv Neural Inf Process Syst 30
18. Isaak K, Simone P, Concetto S, Giordano D, Mubarak S (2017) Brain2Image: converting brain signals into images. In: ACM
19. Isaac AC, Yufei H (2018) Deep EEG super-resolution: upsampling EEG spatial resolution with generative adversarial networks. In: IEEE EMBS international conference on biomedical and health informatics (BHI), Las Vegas, NV, USA, 2018

Review on Log-Based Anomaly Detection Techniques

Pooja Raut, Akanksha Mishra, Shreya Rao, Saloni Kawoor, Sushila Shelke, Mahendra Deore, and Vivek Kumar

Abstract Right from the operating system to the actual application, every part of a system produces logs on an everyday basis. Instead of just purging these files, smart algorithms can be deployed to extract meaningful information from these logs to identify anomalies in the system. Considering the volume and frequency of log data produced, manual analysis can be a tedious task, especially when the file runs into gigabytes. This paper aims to comprehensively compare and contrast the various methods that have been proposed for anomaly detection (AD) using logs. The comparative analysis in this paper reviews previously applied approaches by studying their datasets, techniques, and defined benchmarking outcomes. This paper introduces machine learning algorithms including supervised methods like the k-nearest neighbor (k-NN), decision tree (DT), naive Bayes (NB), support vector machine (SVM), and unsupervised methods like clustering algorithms (K-means), isolation forest, local outlier factor (LOF). The paper then describes novel deep learning approaches like basic multilayer perceptrons (MLPs), recurrent neural net-

P. Raut · A. Mishra (✉) · S. Rao · S. Kawoor · S. Shelke · M. Deore
Department of Computer Engineering, MKSSS's Cummins College of Engineering for Women, Karve Nagar, Pune, India
e-mail: akanksha.mishra@cumminscollege.in

P. Raut
e-mail: pooja.a.raut@cumminscollege.in

S. Rao
e-mail: shreya.rao@cumminscollege.in

S. Kawoor
e-mail: saloni.kawoor@cumminscollege.in

S. Shelke
e-mail: sushila.shelke@cumminscollege.in

M. Deore
e-mail: mahendra.deore@cumminscollege.in

V. Kumar
Veritas Technologies LLC, RMZ Icon, Baner-Pashan Link Road, Pune, India
e-mail: Vivek.Kumar8@veritas.com

S. Shakya et al. (eds.), *Proceedings of Second International Conference on Sustainable Expert Systems*, Lecture Notes in Networks and Systems 351,
https://doi.org/10.1007/978-981-16-7657-4_72

work (RNN) as long short-term memory (LSTM), and convolutional neural network (CNN). No single technique is optimal for all anomaly detection use cases as each method lends itself differently to each use case. This paper aims to go beyond the purpose of a generic review paper by additionally serving as a comprehensive guide for developers who wish to adopt any of the anomaly detection methods presented and benefit from this paper's findings.

Keywords Anomaly detection · Log analysis · Deep learning · Machine learning

1 Introduction

Logs are automatically produced files that store the record of all events and keep track of what's happening in the background ensuring that, if an error occurs within a complex system, the user can access a detailed timeline of events that happened before the malfunction. Log files usually have the extension .log or .txt. Systems today are generating logs at speeds nearing 50 GB (approximately 120–200 million lines) per hour [1]. The following are commonly found log types:

- Event logs: Event logs offer an audit trail of execution of a software that can help understand the activity and figure out the problem and provide overview of the working of complex systems, e.g., server applications which do not have a lot of user interaction.
- Transaction logs: Transaction logs record any changes that take place in the stored data of the database and allow it to recover in case of a system crash or errors in data. It is not human readable but helps maintain the consistency in a database.
- Message logs: It is used to check operations and monitor errors. They are outputs of the EADs server or commands.
- Thread dump: It is used to show information about any threads that are currently running within Java processes [2].

An anomaly can be many things—a software bug, an outlier, noise, an exception, or even a deviation from the norm. But at its core, it is a fault or error in a technical program or system/server which results in unexpected, incorrect, or unintended consequences. There are multiple techniques that are used for anomaly detection, and their performance depends on the dataset and the parameters considered to be analyzed. This paper provides a detailed comparative analysis of a few popular techniques mapped to diverse datasets [3].

The flow the of paper is as follows. Section 2 lists the challenges faced in log-based anomaly detection. Section 3 gives background of the generic methodology used for anomaly detection. Section 4 reviews previously applied approaches by studying various papers with different ML and DL anomaly detection techniques. Section 5 summarizes various publicly available datasets, and Sect. 6 explains the commonly used evaluation metrics. Section 7 is a study of various papers with their corresponding datasets, models used, and their findings as a comparative analysis table. Finally, Sect. 8 concludes the paper.

2 Challenges in Log-Based Anomaly Detection

In light of today's complex systems and the nature of logs produced, the following are a few key challenges faced:

2.1 Handling the Nature of Modern Systems

As we move toward big data, we also move toward the complexity associated with these systems too. Each developer cannot have a 360 °C understanding of how the system functions, thus making it a challenge to identify issues from huge logs.

2.2 Burst in Volume of Logs

As cloud-based, IoT-based, and big data platforms increase, the volume of log data increases too; but by many folds. Earlier, traditional methods such as manual searches or keyword matching techniques like grep were used for log analysis. But today, if we even begin to consider the frequency and volume at which logs are generated; manual analysis seems like an inconceivably arduous task.

2.3 Deriving Relationships from Keywords

Every log line consists of keywords. If a developer was to go through these lines manually, the human brain would make logical correlations between the output keywords, thus making it easier to identify the source and reason of the anomaly or error. However, in case of automated anomaly detection systems, models need to derive these deep relationships artificially in order to make the system more efficient and helpful to users.

2.4 Unstructured Data

Right from the operating system to the front end of the application—each part of the system generates logs in varying structures and formats, thereby making centralized collection of this data and processing another big challenge [4].

2.5 *Availability of Dataset*

Anomalies are rare data instances; therefore, it is difficult to collect a large amount of labeled anomalies (abnormal instances) [5]. Most system logs consist of crucial information, and their contents are of great value to third parties. Hence, most owners would not like to make their datasets public. The scarcity in the public availability of system logs hinders the associated research work.

2.6 *Identifying Unprecedented Anomalies*

Anomalies are the realm of unknowns, e.g., instances with abrupt behaviors, distributions, and data structures—they remain concealed until they actually occur. Hence, anomaly detection algorithms should be developed in such a way that heterogeneous classes and previously unknown; i.e., new kinds of anomalies should be identified to prevent any system damage.

2.7 *Requirement of Domain Knowledge*

Hypothetically even if we assure that there exists a perfect anomaly detection system which can identify all kinds with 100% accuracy—the output as anomaly plots are open to interpretation; Only a person will full system knowledge can correctly interpret the plots and take the necessary consequential steps to fix the anomaly/error identified.

3 Methodology for Log-Based Anomaly Detection

The cumbersome process of log-based anomaly detection can be broken down into the following four basic steps. Each step is equally crucial to ensure high accuracy in anomaly detection

1. **Log Collection**: Most new-age systems are programmed to routinely generate logs that record all the events occurring within. Each log line consists of a timestamp and its corresponding log message. This data can be managed [6] and used to extract valuable insights that could be utilized for multiple objectives (e.g., usage patterns, detection of anomalies).
2. **Log Parsing**: Logs from different systems have different formats. Log parsing [7, 8] extracts a group of templates making the data more structured. There are two methods of parsing: Clustering-based—the distance between logs is calculated to form clusters, and further, event templates are generated for each group. Heuristic-based—the frequency of words is counted and most occurring words are selected and labeled as event candidates.

3. **Feature Extraction**: Feature extraction is used to extract the most valuable features of the log dataset that can be further fed to the anomaly detection models. The parsed logs are first grouped into sets of log sequences by using various windowing techniques [9] (e.g., sliding windows, fixed windows, and session windows). Then, for each sequence a feature vector is generated that represents the occurrence count of every event. All these vectors together form a feature matrix.
4. **Anomaly Detection**: The feature matrix obtained above is given to machine learning models for training, and an anomaly detection system [10] is generated. This model can be trained using supervised (with labeled training data) or unsupervised (with unlabeled training data) machine learning methods. This model can be used to determine if the incoming log sequence is an anomalous or not.

4 Literature Review

This section explains generic concepts and explains different techniques with corresponding papers. The techniques have been classified as ML and DL approaches.

4.1 Machine Learning

Machine learning is a component of artificial intelligence that learns hidden trends in the data and uses these patterns to classify or predict events. Multiple machine learning techniques have been tried and tested for log-based anomaly detection and are known to give good accuracy with minimal complexity. They can further be categorized as unsupervised and supervised ML methods.

4.1.1 Supervised Methods

Supervised anomaly detection [11] techniques use a dataset fully labeled that marks instances as labeled as "normal" and "abnormal" and also involves training a classifier.

Support Vector Machines In support vector machines [12] (SVM), a hyperplane is constructed in a high-dimensional space to classify different data points. If a new data point is located above the hyperplane, it is considered to be an anomaly He et al. [13] use supervised methods (logistic regression, SVM, decision tree) for anomaly detection. Dataset: BGL + HDFS. Experimental setup: Used Linux server with Intel Xeon E5-2670v2 CPU, 128 GB DDR3 1600 RAM, with 64-bit Ubuntu 14.04.2 with Linux kernel 3.16.0. The final result reported was the cumulative average of five experiment runs. Outcome: Achieved high training accuracy (over 0.95). The accuracy on HDFS data is higher than BGL data because the HDFS system records relatively simple operations with lesser event types.

Decision Tree Decision Tree (DT) [14] is a useful supervised learning method which uses its branches to represent the state of a particular instance. It is constructed using a top-down approach where the root node of the DT consists of all the data. A tree node is built using the current "best" attribute, which is chosen to maximize the attribute's value gain. This splitting continues till the node contains only one class or no further split is possible. After constructing the whole tree, branches with low informational gain are pruned to avoid overfitting. Cao et al. [15] implemented machine learning algorithms like (SVM, HMM, and DT) on Scikit using IT log data as their dataset. Dataset: Experiments with 4,690,000 log messages from industry situations. Experimental Setup: Used a server with 8 GB memory, 8 cores at 3.30 GHz and 1 TB hard drives and also used Scikit. Outcome: After manual adjustment of the parameters, it was found that the accuracy when only using the DT algorithm is only 74.19%, whereas this can jump to 90.32% when using a HMM based on the previous result.

Bayes Classifier Simply put, the Bayes theorem says next = old × new. Bayes classifiers [16, 17] are called naive because of the assumption that the frequencies of different attributes are independent. Tharshini et al. [18] analyze static and dynamic access logs using naive Bayes multinomial text algorithm (since logs are basically strung input data with special characters). Dataset: The web sever-based dataset is 2.09 MB with 18,277 log lines. Experimental Setup: The dataset is loaded, and the experiment is run on Weka. Outcome: According to the paper, naive Bayes multinomial text algorithm gives great classification accuracy with the error rate of less than 10%.

K-Nearest Neighbor (KNN) The algorithm [20] assumes that similar events exist in close vicinity. The algorithm selects a value for "K" which represents the number of neighbors then it searches for observations which are closest to the target and uses the mode value from its K-nearest neighbors as the intermediate result. For parameter tuning the value of K is initialized to 1 then KNN searches for the nearest observation until the best value of K is found and it increments it by one. Wang et al. [21] propose an improved K-nearest neighbor (KNN) algorithm that applies the existing mean-shift clustering algorithm. Dataset: Liberty data, Thunderbird, and Blue Gene/P. Outcome: The results indicate that the improved KNN (after applying the mean-shift clustering)-based approach is suitable and can increase the efficiency and accuracy even if the samples in the dataset are irregular.

4.1.2 Unsupervised Methods

Unsupervised methods are given an unlabeled test dataset (i.e., only input data, no available output observation). The anomalies are detected with the assumption that the majority of the instances follow a certain pattern and are close to the normal.

K-means Clustering This unsupervised algorithm is applied to group different objects into clusters [22]. It is a centroid-based algorithm that iteratively tries to separate all the data points into predefined clusters K. It is also referred to as a distance-based algorithm in which we calculate the distances between the clusters using Euclidean distance or Manhattan distance to assign the data point to a cluster. Dani et al. [23] implement the K-means clustering algorithm and compare it to other clustering algorithms like dbscan and hclust. Dataset: High-performance computing (HPC) facility's system logs. Outcome: The clustering accuracy of K-means is greater than the other approaches and that of DBSCAN being the lowest with heavy processing. Results also showed that K-means took less time (few minutes) for execution, whereas dbscan took several hours.

Isolation Forest The algorithm Isolation forest [24] isolates the outliers by randomly picking a feature from the given feature set, then a random value is selected between the minimum and maximum values of the chosen feature as a split value. Using this split value, random partitioning of features takes place and smaller paths in the isolation trees are produced. This is done recursively and all points with a shorter path in the tree are considered anomalous. Farzad and Gulliver [25] propose an unsupervised model using isolation forest and two deep autoencoder networks for feature extraction and anomaly detection. Dataset: BGL, OpenStack, ThunderBird. Experimental Setup: Used Compute Canada Cedar cluster with 24 CPU cores, four P100 GPUs, and 125 GB of memory. All experiments were performed 10 times. Outcome: Precision for positive logs are reliable with an accuracy of 90–99%; i.e., the number of negative logs predicted to be positive is low using isolation forest and one autoencoder.

Local Outlier Factor This algorithm is a density-based outlier detection method that works by computing the local deviation of every data point, making it ideal uneven distributions too. The density between each data instance and its neighboring instances is used to determine whether it is an outlier. The lower the point's density, the more likely it is to be marked as an outlier [26]. Wang et al. [27] propose a deep neural network which implements models such as OneClassSVM, Isolation Forest, and LocalOutlierFactor for detecting abnormal behavior. Dataset: NetEngine40E router's real log data over a period of one month. Experimental Setup: Intel(R) Xeon(R) E5-2620 was used as the CPU with 32GB memory and GPU GTX 1080Ti 11GB running on Ubuntu OS. Outcome: The proposed model can detect the abnormal behavior and sudden increase in log number and gives the root cause of the anomaly too.

4.2 Deep Learning

As modern systems get more sophisticated by the day, developers have started to leverage novel deep Learning techniques for identifying deeper semantic correlations between loglines and unveiling concealed relationships in extracted features.

Long Short-Term Memory Long short-term memory (LSTM) [28–30] is a special kind of recurrent neural network that is used to overcome the problems of exploding and vanishing gradient during training using long sequence data. There are three stages [31] within LSTM: forget stage, selective memory stage, and output stage. The forget gate decides what information is to be passed to the current input from the previous state. The selective memory input gate defines how much information is added to the cell state from the current input. The output gate determines the output depending on the updated cell state. Since log data has time characteristics and is collected in time sequence, LSTM is among the most commonly used neural network models for anomaly detection use cases. Wang et al. [32] selected Naïve Bayes, LSTM, and GBDT algorithms. Dataset: ThunderBird. Experimental Setup: Uses Word2vec skip-gram model to train the word vectors (training vector dimension 100). LSTM algorithm uses with a batch size 32 and one embedding layer and dropout with the probability of 0.5 for regularization. Outcome: Post feature extraction using Word2vec, LSTM achieves a greater performance with 0.99 accuracy. Meng et al. [33] propose an end-to-end framework—LogAnomaly which uses a LSTM network. Dataset: BGL, HDFS. Experimental Setup: Conducted on Linux server with 64 GB memory, Intel Xeon 2.40 GHz CPU. Implemented DeepLog and LogAnomaly with Keras 2.1 and Python 3.6. Outcome: Implementation with BGL dataset [33] achieves an accuracy of 0.97, and the HDFS dataset [34] achieves 0.96 accuracy.

Convolutional Neural Network A convolutional neural network [35] (ConvNet/CNN) is a type of deep learning algorithm which normally takes 2-D data as input and allocates importance (biases and learnable weights) to various aspects of the input. It is a network of convolutional layers which perform the "convolution" operation which performs multiplication of input array data and array of weights generates a filter or a kernel. This kernel is then applied repeatedly to give output values called a "feature map" which represent input filtering.

Multilyer Perceptron Multilayer perceptrons (MLPs) [36] are the most classical type of feed-forward artificial neural network (ANN). They are composed of three layers: the input layer, one or more hidden layers (providing levels of abstraction) and output layer/visible layer which gives the results. Each hidden layer is associated with a nonlinear function which activates it. The real power of MLP comes from its ability to learn the representation in the dataset and how it can relate it to the output variable. Lu et al. [37] use CNN which automatically learns relationships between events in system logs and detect anomalies. The neural network is made up of a dropout layer, logkey2vec embeddings, max-pooling, and three 1D convolutional layers. Dataset: HDFS data. Experimental Setup: TensorFlow has been used for implementing CNN. Outcome: MLP model converges slowly and is time-consuming when compared to other models like CNN or LSTM but the relative ease of development and training of the model makes MLP a good baseline for anomaly detection.

Table 1 Dataset summary table

References	Dataset	Environment	Total lines	Log size
[13, 33, 37]	HDFS	Hadoop File System	11,175,629	1.55 GB
[13, 21, 25, 33]	BGL	Blue Gene/L supercomputer	4,747,963	708 MB
[25]	OpenStack	OpenStack System	207,820	58.61 MB
[21, 25, 32]	ThunderBird	Thunderbird Super Computer	211,212,192	29.60 GB
[23]	HPC	High Performance Cluster	433,489	32.0 MB
[27]	NetEngine40E	NetEngine40E Router	18,727,517	1.09 GB

5 Datasets

The type of log dataset selected acutely affects the method and results of log anomaly detection. Log data is collected from individual servers, software, supercomputer logs, or distributed system logs. Table 1 summarizes a collection of real-world datasets which are publicly available with real anomalies [38, 39] and cover a broad range of popular domains from various data sources.

6 Evaluation Metrics

Evaluation metrics [40] help determine how well the model is performing, and different ratios give insights about what aspect of the model needs improvement. The following are the most commonly used evaluation metrics used:

True positive (TP) count of real anomalies that are accurately detected by the model.

True negative (TN) count of normal cases that are accurately identified as non-anomalous.

False positive (FP) count of normal cases that are incorrectly classified as anomalous.

False negative (FN) count of anomalous log lines that are identified as normal.

Using these four metrics, three insightful measures are calculated.

Precision (P) percentage of correctly classified anomalies in total reported anomalies.

Recall the ratio of correctly classified true anomalies to all actual anomalies.

F_1-measure harmonic average of precision and recall.

$$P = \frac{\mathrm{TP}}{\mathrm{TP} + \mathrm{FP}} \tag{1}$$

$$\mathrm{Recall} = \frac{\mathrm{TP}}{(\mathrm{TP} + \mathrm{FN})} \tag{2}$$

$$F_1\text{-measure} = \frac{2P.\mathrm{Recall}}{(P + \mathrm{Recall})} \tag{3}$$

7 Comparative Analysis

This section surveys existing log-based anomaly detection methods through different papers (Table2).

Machine learning methods are suitable for use cases where the major requirement is prompt response with an acceptable accuracy. Supervised approaches are a good option when the main target is to achieve a higher accuracy rate based on a fixed feature set. Although supervised models seem appealing because of extremely high accuracy rates, they come with an overhead of labeling enormous datasets. Hence, unsupervised approaches also become an attractive option as they also possess the ability to identify previously unknown anomalies with low latency. Deep learning approaches are best suited for use cases where handling huge data loads while parallely understanding hidden semantic relationships are the primary focus.

8 Conclusion

In today's advanced distributed systems, logs are extensively used to determine anomalies. However, manual inspection of logs isn't feasible considering the increased volume and scale of logs. This paper is an addition to the recent research as a consolidation of the work done in the domain. The massive developments in ML have highlighted the importance of preprocessing data and how it impacts the output. However, when log-based anomaly detection is the use case selection of the right anomaly detection model becomes as crucial as preprocessing the data well. In this paper, we survey different ML and DL algorithms that have been implemented for log analysis. Supervised techniques provide the option of specific feature extraction; however, unsupervised training models eliminate the heavy task of manually labeling the datasets. Traditional supervised machine learning techniques are easy to implement and cost-effective; log analysis using dep learning is a rapidly advancing practice for knowledge extraction. DL models when optimized well can give us much

Table 2 Comparative analysis table

Citation	Challenges	Model	Dataset	Outcome
Farzad and Gulliver [24]	Not feasible to do anomaly detection manually for large frameworks	Isolation Forest (IF) and Autoencoders	BlueGene Openstack Thunderbird	Autoencoder enhances Isolation Forest results
Wang et al. [20]	Handles unbalanced characteristics of log data. Selection of training dataset to counter uneven distribution of log data	KNN algorithm improved with mean-shift clustering	5 log sets from supercomputer systems Thunderbird, BGL, BGP, spirit, and liberty size: 89.4 GB (315, 512, 424 faults)	The improved KNN algorithm achieves higher than that of the traditional KNN (spirit: 92.63%, BGL: 91.8%, liberty: 91.03%)
Wang et al. [26]	Causes for increasing volume of Log files and extracted time-period AD	LSTM Isolation Forest, OneClass SVM & LocalOutlierFactor	Router log of Netengine 40E	The LSTM model can detect a sharp increase in data and utilizes features like status value for training unsupervised algorithms
Meng et al. [32]	Worked to keep the semantic relation of logs. To detect sequential and quantitative anomalies simultaneously	LSTM	HDFS an BGL data	High precision (0.8039), recall (0.9319) and F_1 score (0.8632)
Wang et al. [31]	Use of NLP techniques like TFIDF and word2vec for feature extraction for better anomaly detection in log files	LSTM	ThunderBird logs	The model achieves accuracy 99% and overall F_1 score 0.99
Lu et al. [37]	Managing massive log loads. Handling the covered complex relationship in logs	CNN and MLP and LSTM	HDFS log generated in Amazon EC2 over a period of 200 d (1.55 GB and contains 11,197,954 log entries)	CNN can learn hidden relationships better due to multiple filters with 97 % accuracy MLP (no context info) with 98% accuracy LSTM (minimal context info) with accuracy 95% accuracy

(continued)

Table 2 (continued)

Citation	Challenges	Model	Dataset	Outcome
Cao et al. [15]	Traditional ML models cannot recognize new types of attacks leading to many false positives	Decision Tree and HMM	The data belongs to an IT security company's log audit project	The accuracy when only adopting DT is 74.19%, but is 90.32% using the HMM
Dani et al. [22]	Classical approaches do not provide the nature or cause of the issue and identifying faulty components	K-means, DBSCAN, Hclust	System logs of HPC.300 loglines of heterogeneous logs and 1 million used for anomaly detection	Clustering accuracy: K-means = 80%, DBSCAN = 20% Hclust = 64% execution time: DBSCAN more than K-means
Tharshini et al. [18]	Handling textual and categorical data from logs. Used dynamic implementation to identify DoS or DDoS attacks	Rules classifier and Naive Bayes and K-means	Access log files in arff format (2.09 MB consisting of 18,277 log lines)	Accuracy: rules classifier (64.3%) Naive Bayes (94.8%) K-means with k-means++ for initialization (65%)
He et al. [13]	With the growing volume of logs, it is infeasible to detect anomalies in logs manually	supervised (SVM, DT) and unsupervised (invariants mining)	HDFS and BGL data	Found that supervised methods (SVM, DT) achieve high training accuracy (over 0.95) compared to unsupervised methods

better accuracy at the price of heavy training costs. Since DL is fairly new to the industry and hasn't been explored to its fullest potential, there is a lot of scope for creating supremely efficient models with high accuracy by optimizing hyperparameters.

References

1. Mi H, Wang H, Zhou Y, Lyu MR-T, Cai H (2013) Toward fine-grained, unsupervised, scalable performance diagnosis for production cloud computing systems. IEEE Trans Distrib Syst. https://doi.org/10.1109/TPDS.2013.21
2. https://www.sumologic.com/glossary/log-file/
3. https://en.wikipedia.org/wiki/Anomaly_detection
4. Yadav RB, Kumar PS, Dhavale SV (2020) A survey on log anomaly detection using deep learning. In: 2020 8th International conference on reliability, Infocom technologies and optimization (trends and future directions) (ICRITO). https://doi.org/10.1109/icrito48877.2020.9197818

5. Pang G, University of Adelaide, Shen C, University of Adelaide, Cao L, University of Technology Sydney, Van Den Hengel, University of Adelaide (2020) Deep learning for anomaly detection: a review. arXiv:2007.02500v3
6. Madani A, Rezayi S, Gharaee H (2011) Log management comprehensive architecture in security operation center (SOC). In: 2011 International conference on computational aspects of social networks (CASoN). https://doi.org/10.1109/cason.2011.6085959
7. He P, Zhu J, He S, Li J, Lyu MR (2016) An evaluation study on log parsing and its use in log mining. In: 2016 46th Annual IEEE/IFIP international cconference on dependable systems and networks (DSN). https://doi.org/10.1109/dsn.2016.66
8. He P, Zhu J, He S, Li J, Lyu MR (2017) Towards automated log parsing for large-scale log data analysis. In: IEEE Transactions on dependable and secure computing, vol 1. https://doi.org/10.1109/tdsc.2017.2762673
9. Akidau T, Bradshaw R, Chambers C, Chernyak S, Fernández Moctezuma RJ, Lax R, McVeety S, Mills D, Perry F, Schmidt E, Whittle S (2015) The dataflow model: a practical approach to balancing correctness, latency, and cost in massive-scale, unbounded, out-of order data processing. In: PVLDB'15: Proceedings of the VLDB endowment, vol 8. VLDB Endowment, pp 1792–1803
10. Du M, Li F, Zheng G, Srikumar V (2017) Deeplog: anomaly detection and diagnosis from system logs through deep learning. In: Proceedings of the 2017 ACM SIGSAC conference on computer and communications security, pp 1285–1298
11. Nasteski V (2017) An overview of the supervised machine learning methods. Horizons B 4:51–62. https://doi.org/10.20544/HORIZONS.B.04.1.17.P05
12. Tao P, Sun Z, Sun Z (2018) An improved intrusion detection algorithm based on GA and SVM. IEEE Access 6:13624–13631. https://doi.org/10.1109/ACCESS.2018.2810198
13. He S, Zhu J, He P, Lyu MR (2016) Experience report: system log analysis for anomaly detection. In: 2016 IEEE 27th International symposium on software reliability engineering (ISSRE). https://doi.org/10.1109/issre.2016.21
14. Chen M, Zheng AX, Lloyd J, Jordan MI, Brewer E (n.d.) Failure diagnosis using decision trees. In: International conference on autonomic computing, 2004. Proceedings. https://doi.org/10.1109/icac.2004.1301345
15. Cao Q, Qiao Y, Lyu Z (2017) Machine learning to detect anomalies in web log analysis. In: 2017 3rd IEEE International conference on computer and communications (ICCC). https://doi.org/10.1109/compcomm.2017.8322600
16. Rish I, An empirical study of the Naive Bayes classifier. T.J. Watson Research Center
17. Zhanga W, Gaoa F (2011) An improvement to Naive Bayes for text classification. Xi'an Jiaotong University, China
18. Tharshini M, Ragavinodini M, Senthilkumar R (2017) Access log anomaly detection. In: IEEE 2017 Ninth international conference on advanced computing (ICoAC), Chennai, India, 14 Dec 2017–16 Dec 2017. https://doi.org/10.1109/ICoAC.2017.8441194
19. Wang B, Ying S, Yang Z (2020) A log-based anomaly detection method with efficient neighbor searching and automatic K neighbor selection. Sci Program 2020:17, Article ID 4365356
20. Wang B, Ying S, Cheng G, Wang R, Yang Z, Dong B (2020) Log-based anomaly detection with the improved K-nearest neighbor. Int J Softw Eng Knowl Eng. https://doi.org/10.1142/s0218194020500114
21. Kapil S, Chawla M (2016) Performance evaluation of K-means clustering algorithm with various distance metrics. In: IEEE 2016 IEEE 1st International conference on power electronics, intelligent control and energy systems (ICPEICES), Delhi, India, 4 July 2016–6 July 2016, pp 1–4. https://doi.org/10.1109/ICPEICES.2016.7853264
22. Dani MC, Doreau H, Alt S (2017) K-means application for anomaly detection and log classification in HPC. In: Benferhat S, Tabia K, Ali M (eds) Advances in artificial intelligence: from theory to practice, IEA/AIE 2017. Lecture notes in computer science, vol 10351. Springer, Cham
23. Liu FT, Ting KM, Zhou Z-H (2008) Isolation forest. In: 2008 Eighth IEEE international conference on data mining. https://doi.org/10.1109/icdm.2008.17

24. Farzad A, Gulliver TA (2020) Unsupervised log message anomaly detection. ICT Express. https://doi.org/10.1016/j.icte.2020.06.003
25. Cheng Z, Zou C, Dong J (2019) Outlier detection using isolation forest and local outlier factor. In: Proceedings of the conference on research in adaptive and convergent systems, RACS'19. ACM Press the Conference, Chongqing, China, 24 Sept 2019–27 Sept 2019, pp 161–168. https://doi.org/10.1145/3338840.3355641
26. Wang X, Wang D, Zhang Y, Jin L, Song M (2019) Unsupervised learning for log data analysis based on behavior and attribute features. In: Proceedings of the 2019 international conference on artificial intelligence and computer science, AICS 2019. https://doi.org/10.1145/3349341.3349460
27. Hochreiter S, Schmidhuber J (1997) Long short-term memory. Neural Comput 9(8):1735–1780. https://doi.org/10.1162/neco.1997.9.8.1735
28. Vinayakumar R, Soman KP, Poornachandran P (2017) Long short-term memory based operation log anomaly detection. In: 2017 International conference on advances in computing, communications and informatics (ICACCI). https://doi.org/10.1109/icacci.2017.8125846
29. Pal S, Ghosh S, Nag A (2018) Sentiment analysis in the light of LSTM recurrent neural networks. Int J Synth Emot 9(1):33–39. https://doi.org/10.4018/ijse.2018010103
30. Karim F, Majumdar S, Darabi H, Chen S (2018) LSTM Fully convolutional networks for time series classification. IEEE Access 6:1662–1669. https://doi.org/10.1109/access.2017.2779939
31. Wang M, Xu L, Guo L (2018) Anomaly detection of system logs based on natural language processing and deep learning. In: 2018 4th International conference on frontiers of signal processing (ICFSP). https://doi.org/10.1109/icfsp.2018.8552075
32. Meng W, Liu Y, Zhu Y, Zhang S, Pei D, Liu Y, Chen Y, Zhang R, Tao S, Sun P et al (2019) Loganomaly: unsupervised detection of sequential and quantitative anomalies in unstructured logs. In: Proceedings of the twenty-eighth international joint conference on artificial intelligence, IJCAI-19. International joint conferences on artificial intelligence organization, vol 7, pp 4739–4745
33. Oliner AJ, Stearley J (2007) What supercomputers say: a study of five system logs. In: IEEE International conference on dependable systems and networks (DSN'07), pp 575–584
34. Xu W, Huang L, Fox A, Patterson D, Jordan M (2009) Largescale system problem detection by mining console logs. In: Proceedings of SOSP'09
35. Albawi S, Mohammed TA, Al-Zawi S (2017) Understanding of a convolutional neural network. In: 2017 International conference on engineering and technology (ICET). https://doi.org/10.1109/icengtechnol.2017.8308186
36. Wilson E, Tufts DW (n.d.) Multilayer perceptron design algorithm. In: Proceedings of IEEE workshop on neural networks for signal processing. https://doi.org/10.1109/nnsp.1994.366063
37. Lu S, Wei X, Li Y, Wang L (2018) Detecting anomaly in big data system logs using convolutional neural network. In: IEEE 2018 IEEE 16th International conference on dependable, autonomic and secure computing. https://doi.org/10.1109/DASC/PiCom/DataCom/CyberSciTec.2018.00037
38. https://www.usenix.org/cfdr/
39. https://github.com/logpai/loghub
40. Junker M, Hoch R, Dengel A (1999) On the evaluation of document analysis components by recall, precision, and accuracy. In: Proceedings of the fifth international conference on document analysis and recognition, ICDAR'99 (Cat. No. PR00318). https://doi.org/10.1109/icdar.1999.791887

Author Index

A
Adhikari, Bikal, 857
Aditya Kumar, K., 583
Aji, S., 89
Ali, Jahfar, 643
Anandan, K., 823
Anbarasan, A., 273
Anjum, Mohd, 691
Anudeep, Ch., 509
Arappali, Nedumaran, 599
Arun, A. K., 167
Ashish Narayan, T., 509
Augustine, Jacob, 153
Avinash, Miryabelli, 623

B
Banu, Afsana, 669
Bhajaj, Prem, 877
Bhandari, Kaushal, 485
Bhasin, Anuj, 1
Bijay, A., 643
Bodavarapu, Pavan Nageswar Reddy, 509

C
Chandra Sekhara Rao, M. V. P., 833
Chandrasekhar, U., 389
Chithra, S. M., 211

D
Das, Bhaskarjyoti, 373
Deepak Bharadwaj, M. S., 43
Deore, Mahendra, 893
Deshmukh, Priyanka V., 499
Deshmukh, V. M., 459
Deshpande, S. A., 485
Devaraj, G., 273
Divyashree, N., 787

E
Endris, Abdulkerim Seid, 599

F
Faizi Al, Fawaz, 167
Farin, Afsana Taslim, 633

G
Ganga Bhavani, D., 573
Garapati, Yugandhar, 709
Gaudani, Hetal, 763
Gawade, Aruna, 877
Gopala Reddy, K., 403
Gopi Krishna, J., 573
Greeshma, Burla, 21
Guchhait, Subhadeep, 153

H
Hadzhikoleva, S., 11
Hadzhikolev, E., 11
Hans, Arnold Sachith A, 153
Harini, N., 613, 623
Harish, Vishnu Narayanan, 167
Harshan, Vijay, 261
Hemanth, K., 583

S. Shakya et al. (eds.), *Proceedings of Second International Conference on Sustainable Expert Systems*, Lecture Notes in Networks and Systems 351,
https://doi.org/10.1007/978-981-16-7657-4

J
Jaichandran, R., 245, 253, 261, 273
Jayabalan, E., 843
Jayalakshmi, S., 143
Jayapandian, N., 329
Jayaram, Rekha, 57
Jeevitha, S., 199
Jha, Shweta, 777
Jini, K. M., 447
Jogendra Kumar, M., 475
John, Tony K., 245
Joya, Nabanita Saha, 633

K
Kalaivani, M. S., 143
Kale, Akshay, 485
Kalpana Devi, P., 525
Kanavalli, Anita, 315
Kapse, A. S., 499
Karan, K. P., 245
Karthika, P., 599
Karthikeyan, N. K., 869
Karthik, K., 729
Kawoor, Saloni, 893
Kiranbabu, M. N. V., 709
Kotturi, Sai Hemanth Kumar, 229
Krishnan, Aravind, 245
Krishnan, R., 57
Kulkarni, Chaitanya, 43
Kulkarni, Chetana, 373
Kumar, Chandan, 31
Kumar, Krish, 31
Kumar, Vivek, 893
Kureethara, Joseph Varghese, 657, 679

L
Lanka, Surekha, 1
Lavanya, G., 869
Leelavathy, S., 261
Lengerov, A., 11
Liza, Subarna Akter, 633

M
Magesh Sunder, R., 261
Mahendra Chozhan, R., 343
Maheswara Balaji, C. H., 729
Mahima, V., 373
Mallegowda, M., 315
Mamatha, H. R., 737
Maneesha, S., 737
Manjula, R., 115
Manjunath, K. G., 669
Manoj, P., 297
Mantripragada, Krishna Harika, 329
Mary Anita, E. A., 427
Meghana, M., 115
Mishra, Akanksha, 893
Mittal, Ankush, 31
Mohana Geetha, D., 427
Mohan, D., 115
Mohanraj, V., 447
Monika, P., 43
Moni, Raka, 633
Murali Mohan, V., 709
Murugan, N., 71
Murugesan, R., 343
Muthamil Selvan, S., 253
Muthusundari, S., 559

N
Naganjaneyulu, Satuluri, 297
Naga Yugandhar Reddy, B., 729
Nalayini, C. M., 559
Nalini, N., 79
Nambiar, Adarsh R., 167
Nandini Prasad, K. S., 787
Nanditha, J., 403
Natu, Prachi, 127
Nijhawan, Rahul, 31, 413
Nikitha, M. A., 329
Nithya, M., 721

P
Panchal, Pratik, 877
Pardakhe, Nilima V., 459
Patan, Rizwan, 545
Patel, Narendra M., 763
Patel, Neil, 877
Pavan Kalyan, Bathula, 281
Phaneendra, Lukka Venkata Siva, 545
Pooja, H. K., 403
Poosarla, Anvitha, 373
Prasanthi, Kundeti Naga, 833
Preetham, Bhupalam, 281
Priyadharshini, G., 869
Priya, N., 185, 199, 809
Priya, S. Shanmuga, 185
Pugazendi, R., 843
Pujitha, Gopalam, 545
Punithavathy, E., 809

R
Rachovski, T., 11
Raghavendra Sai, N., 475
Raj, Anna Treesa, 657
Rajaprakash, S., 729, 755
Rajasuganya, P. V., 559
Rama Devi, B., 297
Rama Sai Varma, K., 573
Ramya, V. V. S. S., 389
Rani Roopha Devi, K. G., 343
Ranjith, Athira P., 679
Rao, Shreya, 893
Rashmika, N., 583
Raut, Pooja, 893
Ravikumar, T., 71
Reddy, Gunupati Lakshmi Sai Vikas Kumar, 253
Rehna, R. S., 535
Renjith, G., 89
Renukadevi, M., 427
Reshma, Shaik Afifa, 389
Roy, Vikas Kumar, 413

S
Sai, Medagoni Rohan, 253
Sai Swetha, B. S., 329
Sakib Al, Md. Shakawat, 633
Samreen, Afrah, 389
Samuel Manoharan, J., 305
Sandeep Kumar, S., 475
Sandhu, Harshit, 127
Sankar Ganesh, S., 755
Sarashetti, Pooja, 315
Sarosh Umar, M., 691
Sathyabama, A. R., 559
Seetharaman, R., 823
Senthilkumar, J., 447
Shabeer, H. Abdul, 245
Shahab, Sana, 691
Shah, Chinmay, 127
Shah, Jaineel, 127
Shaikh, Maruf, 127
Shakya, Subarna, 857
Shantha Shalini, K., 281
Shashank, M. P., 281
Shelke, Sushila, 893
Shinde, Rucha, 877
Shobana, R., 253, 273, 281, 721
Shobha, 79
Shruthi, S., 43
Shunmuganathan, K. L., 261
Shwetha, M. N. V. R., 403
Sireesha, Moturi, 21
Sivakumar, V., 273
Siva Parvathi, Potu, 545
Siva Rama Rao, Akula V. S., 573, 583
Snehita, Nimmagadda, 545
Sobin, C. C., 643
Sonya, A., 559
Sree, Ch. Likhitha, 115
Sreeja Mole, S. S., 823
Sreenivasu, S. V. N., 229
Sreenivas, Velagapudi, 709
Sridevi, S., 211
Srihari, S., 737
Srinivasa Rao, B., 101
Srinivas, P. V. V. S., 509
Sriram, M., 71
Sriram, Y., 297
Subbulakshmi, S., 167
Suhashini, J., 71
Suresh, Y., 447
Swapna, B., 573

T
Tadesse, Daniel Getachew, 599
Tarun, G., 297
Thakare, V. M., 499
Thakur, Vasu, 413
Tharun, M., 823
Thejaswini, S., 359
Thilagavathi, G., 869
Thirumala Rao, S. N., 21

U
Uday, D., 43

V
Vamsi, P. V. V., 583
Varun Kumar Reddy, V., 729
Vasant, Dharsan, 613
Vinushree, S., 359

W
Walke, Prathviraj, 485
Wanave, Omkar, 485

Y
Yuvaraj, T., 525

Z
Zaman, Zahura, 633
Zulfa, Saima, 403